UNDERSTANDING PHYSICS

Series Editors
John P. Ertel
Robert C. Hilborn
David Peak
Thomas Rossing
Cindy Schwarz

Springer
New York
Berlin
Heidelberg
Hong Kong
London
Milan
Paris
Tokyo

UNDERGRADUATE TEXTS IN CONTEMPORARY PHYSICS

Cassidy, Holton, and Rutherford, *Understanding Physics*

Enns and McGuire, *Computer Algebra Recipes: A Gourmet's Guide to the Mathematical Models of Science*

Hassani, *Mathematical Methods: For Students of Physics and Related Fields*

Holbrow, Lloyd, and Amato, *Modern Introductory Physics*

Roe, *Probability and Statistics in Experimental Physics, Second Edition*

Rossing and Chiaverina, *Light Science: Physics and the Visual Arts*

UNDERSTANDING PHYSICS

David Cassidy
Gerald Holton
James Rutherford

With 571 Illustrations, with 6 in Full Color

Springer

David Cassidy
Professor of Natural Science
Natural Science Program
Hofstra University
Hempstead, NY 11549
USA
chmdcc@Hofstra.edu

Gerald Holton
Mallinckrodt Professor of Physics and History of Science, Emeritus
358 Jefferson Physical Laboratory
Harvard University
Cambridge, MA 02138
USA

James Rutherford
Education Advisor
American Association for Advancement of Science
Washington, DC 20005
USA

Series Editors

John P. Ertel
Department of Physics
United States Naval Academy
572 Holloway Road
Annapolis, MD 21402-5026
USA
jpe@nadn.navy.mil

Robert C. Hilborn
Department of Physics
Amherst College
Amherst, MA 01002
USA
rchilborn@amherst.edu

Cindy Schwarz
Department of Physics and Astronomy
Vassar College
Poughkeepsie, NY 12601
USA
schwarz@vaxsar.vassar.edu

David Peak
Department of Physics
Utah State University
Logan, UT 84322
USA
peakd@cc.usu.edu

Thomas Rossing
Department of Physics
Northern Illinois University
De Kalb, IL 60115
USA
rossing@physics.niu.edu

Library of Congress Cataloging-in-Publication Data

Cassidy, David C., 1945–
Understanding physics / David Cassidy, Gerald Holton, F. James Rutherford.
p. cm. — (Undergraduate texts in contemporary physics)
Rev. ed. of: The project physics course. 1971.
Includes bibliographical references and index.
ISBN 0-387-98756-8 (pbk. : acid-free paper)
1. Physics. I. Holton, Gerald James. II. Rutherford, F. James (Floyd James), 1924–
III. Harvard Project Physics. Project physics course. IV. Title. V. Series.

QC23.2 .C37 2002
530—dc21 2002020937

ISBN 0-387-98756-8 Printed on acid-free paper.

Printed in the United States of America.

9 8 7 6 5 4 3 2 1 SPIN 15110102

Springer-Verlag New York Berlin Heidelberg
A member of BertelsmannSpringer Science+Business Media GmbH

To the students and instructors whose advice, while using
the draft versions of the text and guides, helped to improve our work greatly;
and to Joan Laws, whose quiet expertise all these years
helped to bring the project to success.

Science is an adventure of the whole human race to learn to live in and perhaps to love the universe in which they are. To be a part of it is to understand, to understand oneself, to begin to feel that there is a capacity within man far beyond what he felt he had, of an infinite extension of human possibilities. . . .

I propose that science be taught at whatever level, from the lowest to the highest, in the humanistic way. It should be taught with a certain historical understanding, with a certain philosophical understanding, with a social understanding and a human understanding in the sense of the biography, the nature of the people who made this construction, the triumphs, the trials, the tribulations.

—I.I. Rabi, Nobel Laureate in Physics

Preface

Understanding Physics is a completely revised, updated, and expanded edition of the *Project Physics Course*. It is an integrated introductory physics course, developed with funding from the Carnegie Corporation and the Sloan Foundation and with the close cooperation of Springer-Verlag New York.

In approach and content, *Understanding Physics* follows the trail blazed by the earlier versions, but it includes more recent developments in physics and a stronger emphasis on the relationships among physics, technology, and society. We have sought especially to incorporate the salient lessons of recent physics education research and practical experience gained in the classroom.

The Audience

Understanding Physics is written primarily for undergraduate college students not intending (at least initially) to enter careers in science or engineering. These may include liberal-arts students, business majors, prelegal, and prospective architecture students. We have found that when the course is taken with laboratory work, it has been deemed suitable by medical schools for premedical students.

An important group that this course is intended to serve are persons who plan to teach, or are already teaching, in K–12 classrooms. As has been widely discussed, there is a special need for improvement in the science education of current and future teachers as an important step toward achieving greater scientific literacy in general. Many states have recently incorporated the contextual approach used in *Understanding Physics* into state science education criteria. It is in part to meet the challenges of teacher education that this course was developed as a resource, along with the usual pedagogical training.

Since college students in introductory science courses usually represent a wide spectrum of expertise in science and mathematics, this book assumes no prerequisites in science or mathematics beyond high-school algebra, geometry, and general science. In this text we have taken great care to derive all necessary equations very patiently, but whenever possible we have used narrative text instead of equations to convey the meanings of laws and concepts. Even if students have taken physics in high school, they often still lack proficiency in even the most basic concepts and techniques. One of the aims of this course is to enable all students to gain experience and confidence with physical-science concepts and quantitative methods, and with an understanding of the nature of science itself. Of course, for classes in which the students are sufficiently prepared, instructors may decide to place more emphasis on quantitative or other aspects of physics as appropriate. The course is designed with such flexibility in mind.

The Approach

A unique feature of this text, like its predecessor, is that it places the fundamental concepts of physics within the broader humanistic and historical contexts in which they arose, but without handicapping students in tests that compare their performance with students who have taken a less broadly conceived, conventional physics course. Research has shown that students exposed to our approach gain a much deeper understanding of both the content and the processes of scientific research, as well as an appreciation not only of what we know, but also of how and why we think we know it. This approach has been endorsed by several national organizations, including the National Science Foundation, the Research Council of the National Academy of Sciences, and Project 2061 of the American Association for the Advancement of Science. The National Research Council stated in its *National Science Education Standards*:*

> In learning science, students need to understand that science reflects its history and is an ongoing, changing enterprise. The standards for the history and nature of science recommend the use of history in school science programs to clarify different aspects of scientific inquiry, the human aspects of science, and the role that science has played in the development of various cultures.

Thus, *Understanding Physics* operates on two levels, providing both the fundamental concepts of physics and the humanistic and intellectual contexts in which the concepts developed. In addition to the necessary concepts and equations, intentionally developed patiently, and using easy-to-visualize examples, it aims to convey a real sense of the nature of scientific

---

* Washington, DC: National Academy Press, 1996; p. 107.

thinking, the way intuitions about science had to be, often painfully, acquired by scientists, and what our current concepts really mean. However, this text is not intended to be used by itself, but rather as part of a program as integrated as possible with hands-on activities, small-group discussions either in or out of class, and other encouragements to active learning that enable the subject matter to come alive. Some of these and other possible activities are suggested in the accompanying *Student Guide.*

Understanding Physics is divided into two parts. Each part is self-contained, with enough material for a course lasting at least one semester. Each part encompasses topics in classical physics along with one of the two contemporary nonclassical physics: relativity theory and quantum mechanics. Both parts taken together may serve for a full one-year course. We have sought from the beginning to provide instructors with maximum flexibility in adapting the course to students of different backgrounds, to different educational settings, to different semester time frames, and to different preferences for course topics. Some suggestions for different scenarios are provided in the accompanying *Instructor Guide.* All of the course's printed materials, as well as links to many related Web sites for both instructors and students, may also be accessed on the World Wide Web via the publisher's site at http://www.springer-ny.com/up.

Acknowledgments

We are grateful to the Carnegie Corporation of New York and to the Sloan Foundation for their generous and timely support. Thomas von Foerster and the staff of Springer-Verlag New York provided much-appreciated encouragement, support, and helpful advice during the years of preparation and testing of these materials in draft form. We also thank David Couzens, our developmental editor, whose outstanding work on the illustrations and his suggestions regarding the content contributed greatly to this work, and Edwin F. Taylor for his very helpful comments regarding the chapter on relativity theory. We are indebted to our colleagues at Hofstra University and to the other instructors who thoroughly tested the draft of the text and guides in their classes, and offered many helpful comments and suggestions. We also thank our colleagues and the individuals who carefully reviewed the materials and provided insightful suggestions. Last but not least, we thank the students at all testing sites for their valuable suggestions and encouragement. Without the contributions of all of these individuals and institutions, this work would not have been possible.

David Cassidy
Gerald Holton
James Rutherford

Brief Contents

Contents

CHAPTER 2. MOVING THE EARTH 57

CHAPTER 3. UNDERSTANDING MOTION 117

CHAPTER 13. PROBING THE ATOM 585

CHAPTER 14. THE QUANTUM MODEL OF THE ATOM 621

CHAPTER 15. QUANTUM MECHANICS 661

PART ONE

MATTER AND MOTION

Prologue to Part One

1. LIVING IDEAS

The purpose of this course is to explore the development and content of the major ideas that have led to our understanding of the physical universe. As in any science course you will learn about many of the important concepts, theories, and laws that make up the content of the science, physics in this case. But this course goes beyond that; it presents science as *experience*, as an integrated and exciting intellectual adventure, as the product of humankind's continual drive to know and to understand our world and our relationship to it.

Not only will you learn about the many ideas and concepts that make up our understanding of the physical world today but, equally important, these ideas will come alive as we look back at how they arose, who the people were who arrived at these ideas in their struggle to understand nature, and how this struggle continues today. Our story has two sides to it: the ideas of physics *and* the people and atmosphere of the times in which these ideas emerged. As you watch the rise and fall of physical theories, you will gain an appreciation of the nature of science, where our current theories came from, the reasons why we accept them today, and the impact of these theories and ideas on the culture in which they arose.

Finally, you will see how physics came to be thought of as it is today: *as an organized body of experimentally tested ideas about the physical world.* Information about this world is accumulating ever more rapidly as we reach out into space, into the interior of matter, and into the subatomic domain. The great achievement of physics has been to find a fairly small number of basic principles which help us to organize and to make sense of key parts of this flood of information.

2. OUR PLACE IN TIME AND SPACE

Since the aim of this course is to understand the physical world in which we live, and the processes that led to that understanding, it will help to begin with some perspective on where we are in the vast ocean of time and space that is our Universe. In fact, the Universe is so vast that we need a new yardstick, the *light year*, to measure the distances involved. Light in empty space moves at the fastest speed possible, about 186,000 miles every second (about 300,000 kilometers every second). A light year is not a measure of time but of distance. A light year is defined as the distance light travels in one year, which is about five trillion miles. The tables that follow provide an overview of our place on this planet in both space and time.

Current Estimates of Our Place in Time and Space

Time	*Years since start*
Age of the Universe	about 15 billion years
Age of our Sun and Earth	5 billion
Beginning of life on Earth	3.5 billion
Extinction of dinosaurs (Jurassic Age)	65 million
First humanoids	5 million
First modern humans	100,000
Rise of civilization	30,000
End of the last Ice Age	12,000
Height of Hellenic Greece	2500
Rise of modern science	400

Distance (from the center of the Earth)	
Edge of the Universe	about 15 billion light years
Nearest spiral galaxy (Andromeda)	2.2 million light years
Radius of our galaxy (Milky Way)	100,000 light years
Nearest star (Alpha Centauri)	4.3 light years, or 25 trillion miles
Distance to the Sun	93 million miles (150 million kilometers)
Distance to the Moon	239,000 miles (384,000 kilometers)
Radius of the Earth	3963 miles (6,370 kilometers) (about 1.5 times the distance between New York and Los Angeles)

You may be amazed to see from these tables that, within this vast ocean of the Universe measuring billions of light years across, a frail species evolved on a ball of mud only about 4000 miles in radius, orbiting an average star, our Sun, in an average corner of an average galaxy—a species that is nevertheless able, or believes it is able, to understand the most fundamental properties of the universe in which it lives. Even more astonishing: this frail species, which first appeared in contemporary form only about 100,000 years ago, invented an enormously successful procedure for focusing its mind and its emotions on the study of nature, and that procedure, modern science, is now only a mere 400 years old! Yet within that brief span of just four centuries science has enabled that species—us—to make gigantic strides toward comprehending nature. For instance, we are now approaching a fairly good understanding of the origins of matter, the structure of space and time, the genetic code of life, the dynamic character of the Earth, and the origins and fate of stars and galaxies and the entire Universe itself. And within that same period we have utilized the knowledge we have gained to provide many members of our species with unparalleled comforts and with a higher standard of living than ever previously achieved.

Take a moment to look around at everything in the room, wherever you are right now. What do you see? Perhaps a table, a chair, lamp, computer, telephone, this book, painted walls, your clothes, a carpet, a half-eaten sandwich . . . . Now think about the technologies that went into making each of these things: the electricity that makes the light work; the chemical processes that generated the synthetic fabrics, dyes, paints, plastics, processed food, and even the paper, ink, and glue of this book; the microtransistors that make a computer work; the solid-state electronics in a television set, radio, phone, CD player; the high-speed networking and software that allows you to read a Web page from the other side of the Earth.

All of these are based upon scientific principles obtained only within the past few centuries, and all of these are based upon technologies invented within just the past 100 years or so. This gives you an idea of how much our lives are influenced by the knowledge we have gained through science. One hardly dares to imagine what life will be like in another century, or even within a mere 50, or 25, or 10 years!

Some Discoveries and Inventions of the Past 100 Years

airplane	structure of DNA
automobile	microchip
expansion of the Universe	organ transplants
penicillin	first human landing on the Moon
motion picture with sound	laser
elementary particles	MRI and CT scan
plate tectonics	personal computers
nuclear weapons	Internet
polio vaccine	planets around stars other than our Sun
first artificial satellite (Sputnik)	human genome

Let's look at some of the fundamental ideas of modern physics that made many of these inventions and discoveries possible.

3. FIRST THINGS FIRST

The basic assumptions about nature, the procedures employed in research today, and even some of our theories have at bottom not changed much since the rise of modern physics. Some of these assumptions originated even earlier, deriving from the ancient world, especially the work of such Greek thinkers as Plato, Aristotle, and Democritus.

What set the Greeks apart from other ancients was their effort to seek nonanimistic, natural explanations for the natural events they observed and to subject these explanations to rational criticism and debate. They were

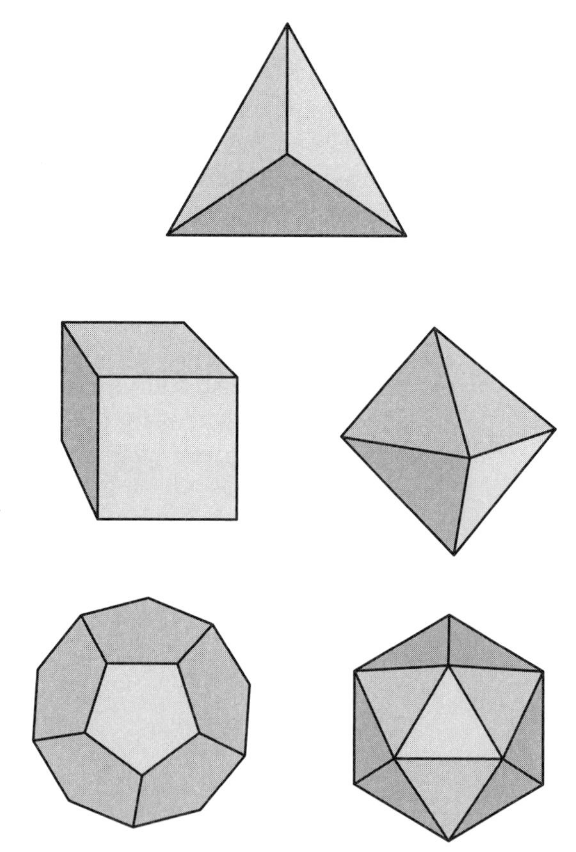

The five "regular solids" (also called "Pythagorean figures" or "Platonic solids") that appear in Kepler's *Harmonices Mundi (Harmony of the World)*. The *cube* is a regular solid with six square faces. The *dodecahedron* has 12 five-sided faces. The other three regular solids have faces that are equilateral triangles. The *tetrahedron* has four triangular faces, the *octahedron* has eight triangular faces, and the *icosahedron* has 20 triangular faces.

also the first to look for rational, universal first principles behind the events and phenomena they perceived in nature. On the other hand, the use of experimental investigation, now a fundamental tool of modern science, was invoked by only a few of the Greek thinkers, instead of being built in as an indispensable part of their research.

In seeking the first principles, Greek thinkers utilized the notion that all things are made up of four basic "elements," which they called earth, water, air, and fire. In many ways they viewed these elements the way we might view the three states of matter: solid, liquid, and gas, with heat (fire) serving as the source of change. (Some added a fifth element, called "quintessence," constituting the celestial objects.) The Greek philosopher Plato (427?–347 B.C.), regarded mathematical relationships as constituting the permanent first principles behind the constantly changing world that we observe around us. As such, Plato associated the five elements with the five Platonic solids in solid geometry. (Refer to pg. 6.) Although we no longer hold this view, scientists today often do express physical events, laws, and theories in terms of mathematical relationships. For instance, the physicist Albert Einstein wrote in 1933:

> I am convinced that we can discover by means of purely mathematical constructions the concepts and the laws connecting them with each other, which furnish the key to the understanding of natural phenomena. . . . Experience remains, of course, the sole criterion of the physical utility of a mathematical construction. But the creative principle resides in mathematics. In a certain sense, therefore, I hold it true that pure thought can grasp reality, as the ancients dreamed.*

The Greek thinker Democritus (fl. c. 420 B.C.) and his followers offered a quite different account of the permanent first principles constituting the elements that give rise to observed phenomena. For them, the elements are not made up of abstract geometrical figures but of individual particles of matter that they called "atomos," Greek for "indivisible." Democritus is said to have thought of the idea of atoms when smelling the aroma of freshly baked bread. He surmised that, in order to detect the smell, something had to travel from the bread to his nose. He concluded that the "something" must be tiny, invisible particles that leave the bread carrying the smell of the bread to his nose—an explanation that is quite similar to the one we have today! For the "atomists" down through the centuries, all of reality

* A. Einstein, *Ideas and Opinions* (New York: Crown, 1982), p. 274.

Albert Einstein (1879–1955).

and everything that can be perceived with their senses could be explained in terms of an infinite number of eternally existing indivisible atoms, moving about and clumping together in infinite empty space to form stars, planets, and people.

Like Plato's notions, the views of the ancient atomists bore some striking similarities to our current views. We too have a relatively small number of "elements" (92 naturally occurring elements) which we associate with different types of atoms, as you can see from the periodic table. And we too attribute the properties of everyday matter to the combinations and interactions of the atoms that constitute the matter. However, our atoms have been shown to be divisible, and they, along with the elements, behave quite differently from Greek atoms and elements. Moreover, our atomic idea is no longer just a speculation but an accepted theory based firmly upon experimental evidence. Since the days of Plato and Democritus, we have learned how to bring reason and experiment together into the much more powerful tool of research for exploring and comprehending atomic properties underlying the phenomena we observe in nature.

Unfortunately, both Plato and Aristotle rejected the atomic hypothesis of Democritus and his followers. Aristotle, Plato's pupil, also rejected Plato's

PLATO'S PROBLEM

Like many ancient thinkers, Plato believed that the celestial bodies must be perfect and divine, since they and their motions are eternal and unchanging, while the components of the earthly, terrestrial world are constantly changing. Thus, for him, analysis of the motions of the heavenly bodies according to mathematical principles became a quest for divine truth and goodness. This was the beginning of modern mathematical astronomy—although of course we no longer seek divine truth and goodness in celestial motions. But his idea was also the beginning of a split in the physical world between the Earth on the one hand and the rest of the Universe on the other, a split that was healed only with the rise of modern science.

It is said that Plato defined an astronomical problem for his students, a problem that lasted for centuries until the time of Johannes Kepler and Galileo Galilei, over 350 years ago. Because of their supposed perfection, Plato believed that the celestial objects move around the Earth (which he regarded as the center of the Universe) at a perfectly uniform, unchanging speed in what he regarded as the most "perfect" of all geometrical figures, the circle. He chose the circle because it is unending yet bounded, and encompasses the largest area inside a given perimeter. The problem Plato set for his followers was to reduce the complicated motions of the Sun, Moon, planets, and stars to simple circular motions, and to show how the complexity of their observed motions can arise from the interaction of mathematically simple perfect circles rotating with constant speeds.

Plato's problem, applied to the observed motions of the planets, as well as to the other celestial objects, was a problem that occupied most of the best mathematical astronomers for centuries. During the Renaissance, people found that Plato's assumption of perfectly circular motions at constant speed was no longer useful and did not agree with more precise observations.

theory. Instead, he offered a much more appealing and more fully worked-out system as an alternative to both Plato and the atomists. As a result, Aristotle's views dominated scientific thought for centuries, and Plato's penchant for mathematics and Democritus's atomic hypothesis were set aside for centuries.

4. ARISTOTLE'S UNIVERSE

The Greek philosopher Aristotle (384–322 B.C.) argued that we should rely on sense perceptions and the qualitative properties of bodies, which seem far more real and plausible than abstract atoms or mathematical formulas.

After all, we can see and touch a glob of earth, and feel the wetness of water or the heat of fire, but we can't see or touch an atom or a triangle. The result was an amazingly plausible, coherent, and common-sense system that naturally appealed to people for centuries.

As did Plato, Aristotle divided the Universe into two separate spheres: the celestial sphere, the heavens above where unchanging perfection resides; and the terrestrial sphere here below, where all change and imperfection and corruption and death are found. The upper boundary of the terrestrial sphere is the Moon, which is obviously imperfect, since one can see dark blotches on it. All change, such as comets, novae (exploding stars), and meteors, must occur below the Moon, which is also the limit of the reign of the four basic elements. Above the Moon are the perfect celestial bodies. These, to the naked eye, display no markings at all. So Aristotle attributed to them Plato's fifth element, quintessence, which fills all of space above the Moon. One of the assumed properties of quintessence was that it moves by itself in a circle. (In one of Aristotle's other writings he further argued that since every motion requires a mover, there must be a divine being—an "unmoved mover"—outside the whole system, who keeps it spinning.)

Aristotle argued that the spinning motion of the heavens around the Earth at the center caused a spinning motion of the terrestrial sphere—like an object in a giant washing machine—which in turn caused the four elements to separate out according to their weight (or density). In this system the "heaviest" element, Earth, coalesced in the center. On top of that came the next heaviest element, water, which covers much of the Earth in the form of oceans, lakes, and rivers. Then comes air, and finally fire, the lightest element. The terrestrial sphere is completely filled with these four elements, while the celestial sphere from the Moon outward is completely filled with quintessence. There is no empty space, or vacuum, anywhere.

Aristotle's system seemed quite plausible. A natural vacuum is extremely rare in daily experience, while in the whirling motion of a system of tiny objects of different densities (representing different elements) the objects actually do separate as he indicated. Einstein later explained that the pressure in a fluid mixture during rotation of materials of various densities forces the most dense material to the center, followed by the next dense material, and so on—resulting in layers of materials according to density, just as Aristotle had argued!

Aristotle applied his arrangement of the elements to explanations of practically everything. According to Aristotle, as a result of the whirling motion of the cosmos, each of the four elements ended up in a special place where it "belongs" according to its "weight" (really density): earth at the center, followed by water, then air, then fire, just as we see around us. However, because of imperfections in the system below the celestial objects, the

The four ancient "elements," shown superimposed on the Earth at the center of the whole Ptolemaic Universe.

separation of the four terrestrial elements was not quite complete, trapping some of the elements in the "wrong" place. If they are freed, they will head straight "home," meaning to the place where they belong—straight being in a vertical direction, either straight up or straight down. Such motions require no explanation; they are simply natural. (This is discussed further in Section 3.1.)

Mixing the elements and their natural motions helped to explain some of the changes and events one can see all around us. For instance, a stone lifted from the earth and released will drop straight down through air and water to reach the earth where it "belongs" at the bottom of a pond. A flame lit in air will move straight upward, as does a bubble of air trapped under water. Water trapped in the Earth will emerge onto the surface as springs or geysers; air emerges from the Earth by causing earthquakes; fire trapped in the Earth breaks forth in volcanoes. Oil, he believed, contains air in addition to earth and water, so it floats on water. Clouds, according to Aristotle, are condensed air mixed with water. They are densest at the top, Aristotle claimed, because they are closest to the source of heat, the Sun. Wind and fire squeezed out of the cloud produce thunder and lightning—a far cry from an angry Zeus hurling thunderbolts!

As you can see, Aristotle's explanations are all "commonsensical"—plausible, and reasonable, if you don't ask too many questions. Everything fit together into a single, rational cosmic scheme that could explain almost everything—from the behavior of the cosmos to the appearance of springs of water. Although the wide acceptance of Aristotle's system discouraged the consideration of more fruitful alternatives, such as those of Plato and the atomists, the dominance of his views for centuries encouraged the domination of the search for rational explanations of natural events in plausible, human terms that is one of the hallmarks of modern science. Aristotle was considered such an authority on the rational workings of nature that he was called for centuries simply "the Philosopher."

But This Is Not What We Today Would Call Science

Seen from today's perspective, the problem is not chiefly with the content but with the approach. For Aristotle, a theory was acceptable if it was logically sound, if all of the ideas were consistent with each other, and if the result was plausible. That is fine as far as it goes, and it is found in all theories today. But he did not take a necessary step further. He could not provide precise, perhaps even *quantitative*, explanations of the observed events that could be tested and confirmed, for example, in a laboratory. He offered only *qualitative* descriptions. For instance, things are not just hot or cold, but they have a precise temperature, say −16°C or +71°C. Nor did Aristotle think of explanations of events, no matter how logically sound, as being tentative hypotheses that must be tested, debated, and compared with the experimental evidence. Also, he rejected the approach of Plato and the atomists in which explanations of phenomena should involve the motions and interactions of invisible individual elements. Without resting on experimental research or more general underlying principles, Aristotle's philosophy lacked the capability of modern science, in which experiment, mathematics, and the atomic hypothesis are brought together into a powerful instrument for the study of nature.

WHAT IS SCIENCE?

The American Physical Society, the leading society of professional physicists, has issued the following statement in answer to the question "What is Science?":

> Science extends and enriches our lives, expands our imagination, and liberates us from the bonds of ignorance and superstition. The American Physical Society affirms the precepts of modern science that are responsible for its success.
>
> Science is the systematic enterprise of gathering knowledge about the Universe and organizing and condensing that knowledge into testable laws and theories.
>
> The success and credibility of science are anchored in the willingness of scientists to:
>
> 1. Expose their ideas and results to independent testing and replication by other scientists. This requires the complete and open exchange of data, procedures, and materials.
> 2. Abandon or modify accepted conclusions when confronted with more complete or reliable experimental evidence.
>
> Adherence to these principles provides a mechanism for self-correction that is the foundation of the credibility of science.

And when these elements were brought together, especially in the study of motion, modern physics emerged.

SOME NEW IDEAS AND CONCEPTS

animism
atoms
elements
first principles
terrestrial sphere

FURTHER READING

G. Holton and S.G. Brush, *Physics, The Human Adventure, From Copernicus to Einstein and Beyond* (Piscataway, NJ: Rutgers University Press, 2001), Chapters 1 and 3.

D.C. Lindberg, *The Beginnings of Western Science* (Chicago: University of Chicago Press, 1992).

K. Ferguson, *Measuring the Universe: Our Historic Quest to Chart the Horizons of Space and Time* (New York: Walker, 1999).

STUDY GUIDE QUESTIONS

1. Living Ideas

1. What are the "living ideas"? What makes them alive?
2. What is the twofold purpose of this course?
3. Why did the authors of this book choose this approach, instead of the standard emphasis on laws, formulas, and theories that you may have encountered in other science courses?
4. What is your reaction to this approach?

2. Our Place in Time and Space

1. How would you summarize our place in time and space?
2. In what ways is technology different from science? In what ways is it the same?

3. First Things First

1. Why, in this chapter, did we look back at the Ancient Greeks before introducing contemporary physics?
2. What was so special about the Ancient Greeks, as far as physics is concerned?
3. What types of answers were they seeking?
4. What did the word "elements" mean to the Greeks?
5. What are the two proposed solutions to the problem of change and diversity examined in this section?

4. Aristotle's Universe

1. What did Aristotle think was the best way to find the first principles?
2. What types of principles did he expect to find?
3. Describe Aristotle's cosmology.
4. Why is Aristotle's system not yet what we call science? What are the characteristics of science as presently understood?
5. Describe how Aristotle explained one of the everyday observations.
6. How would you evaluate Aristotle's physics in comparison with physics today?
7. A researcher claims to have reasoned that under certain circumstances heavy objects should actually rise upward, rather than fall downward on the surface of the Earth. As a good scientist, what would be your reaction?

CHAPTER 1

Motion Matters

1.1 MOTION

One of the most important properties of the objects that make up our physical world is the fact that they can move. Motion is all around us, from falling leaves and tumbling rocks, to moving people and speeding cars, to jet planes, orbiting space satellites, and planets. Understanding what motion is, how it can be described, and why it occurs, or doesn't occur, are therefore essential to understanding the nature of the physical world. You saw in the Prologue that Plato and others argued that mathematics can be used as a tool for comprehending the basic principles of nature. You also saw that we can use this tool to great advantage when we apply it to precise observations and experiments. This chapter shows how these two features of modern physics—mathematics and experiment—work together in helping us to understand the thing we call motion.

Motion might appear easy to understand, but initially it's not. For all of the sophistication and insights of all of the advanced cultures of the past, a really fundamental understanding of motion first arose in the scientific "backwater" of Europe in the seventeenth century. Yet that backwater was experiencing what we now know as the Scientific Revolution, the "revolving" to a new science, the science of today. But it wasn't easy. At that time it took some of the most brilliant scientists entire lifetimes to comprehend motion. One of those scientists was Galileo Galilei, the one whose insights helped incorporate motion in modern physics.

1.2 GALILEO

Galileo Galilei was born in Pisa in 1564, the year of Michelangelo's death and Shakespeare's birth. Galileo (usually called by his first name) was the son of a noble family from Florence, and he acquired his father's active interest in poetry, music, and the classics. His scientific inventiveness also began to show itself early. For example, as a medical student, he constructed a simple pendulum-type timing device for the accurate measurement of pulse rates. He died in 1642 under house arrest, in the same year as Newton's birth. The confinement was the sentence he received after being convicted of heresy by the high court of the Vatican for advocating the view that the Earth is not stationary at the center of the Universe, but instead rotates on its axis and orbits the Sun. We'll discuss this topic and the results later in Chapter 2, Section 12.

FIGURE 1.1 Galileo Galilei (1564–1642).

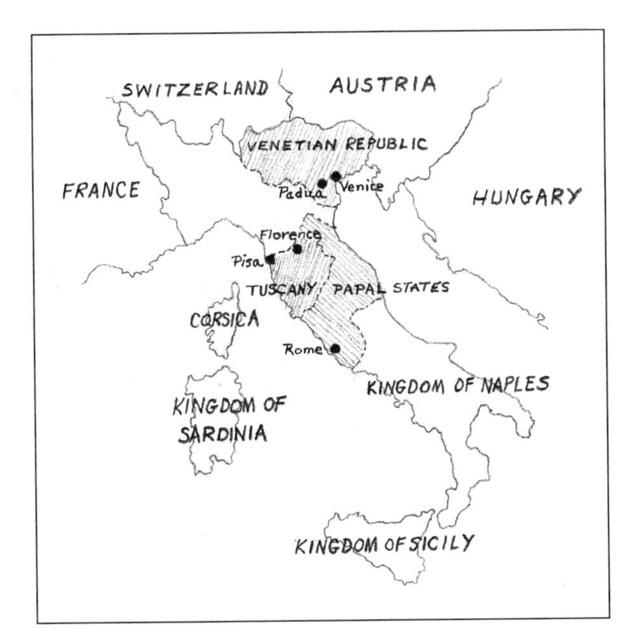

FIGURE 1.2 Italy, ca. 1600 (shaded portion).

After reading the classical Greek philosopher–scientists Euclid, Plato, and Archimedes, Galileo changed his interest from medicine to physical science. He quickly became known for his unusual scientific ability. At the age of 26 he was appointed Professor of Mathematics at Pisa. There he showed an independence of spirit, as well as a lack of tact and patience. Soon after his appointment he began to challenge the opinions of the older professors, many of whom became his enemies and helped convict him later of heresy. He left Pisa before completing his term as professor, apparently forced out by financial difficulties and his enraged opponents. Later, at Padua in the Republic of Venice, Galileo began his work on astronomy, which resulted in his strong support of our current view that the Earth rotates on its axis while orbiting around the Sun.

A generous offer of the Grand Duke of Tuscany, who had made a fortune in the newly thriving commerce of the early Renaissance, drew Galileo back to his native Tuscany, to the city of Florence, in 1610. He became Court Mathematician and Philosopher to the Grand Duke, whose generous patronage of the arts and sciences made Florence a leading cultural center of the Italian Renaissance, and one of the world's premier locations of Renaissance art to this day. From 1610 until his death at the age of 78, Galileo continued his research, teaching, and writing, despite illnesses, family troubles, and official condemnation.

Galileo's early writings were concerned with *mechanics*, the study of the nature and causes of the motion of matter. His writings followed the stan-

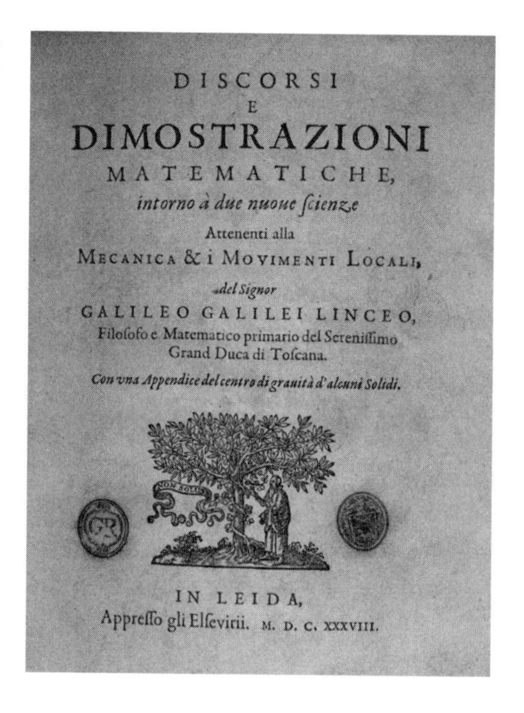
DISCORSI
E
DIMOSTRAZIONI
MATEMATICHE,
intorno à due nuove ſcienze
Attenenti alla
MECANICA & i MOVIMENTI LOCALI,
del Signor
GALILEO GALILEI LINCEO,
Filoſofo e Matematico primario del Sereniſſimo
Grand Duca di Toſcana.
Con una Appendice del centro di grauità d'alcuni Solidi.

IN LEIDA,
Appreſſo gli Elſevirii. M. D. C. XXXVIII.

FIGURE 1.3 Title page from Galileo's *Discourses and Mathematical Demonstrations Concerning Two New Sciences Pertaining to Mechanics and Local Motion* (1638).

dard theories of his day, but they also showed his awareness of the shortcomings of those theories. During his mature years his chief interest was in astronomy. However, forbidden to teach astronomy after his conviction for heresy, Galileo decided to concentrate instead on the sciences of mechanics and hydrodynamics. This work led to his book *Discourses and Mathematical Demonstrations Concerning Two New Sciences* (1638), usually referred to either as the *Discorsi* or as *Two New Sciences*. Despite Galileo's avoidance of astronomy, this book signaled the beginning of the end of Aristotle's cosmology and the birth of modern physics. We owe to Galileo many of the first insights into the topics in the following sections.

1.3 A MOVING OBJECT

Of all of the swirling, whirling, rolling, vibrating objects in this world of ours, let's look carefully at just one simple moving object and try to describe its motion. It's not easy to find an object that moves in a simple way, since most objects go through a complex set of motions and are subject to various pushes and pulls that complicate the motion even further.

Let's watch a dry-ice disk or a hockey puck moving on a horizontal, flat surface, as smooth and frictionless as possible. We chose this arrangement so that friction at least is nearly eliminated. Friction is a force that will impede or alter the motion. By eliminating it as much as possible so that we can generally ignore its effects in our observations, we can eliminate one complicating factor in our observation of the motion of the puck. Your instructor may demonstrate nearly frictionless motion in class, using a disk or a cart or some other uniformly moving object. You may also have an opportunity to try this in the laboratory.

If we give the frictionless disk a push, of course it moves forward for a while until someone stops it, or it reaches the end of the surface. Looking just at the motion before any remaining friction or anything else has a noticeable effect, we photographed the motion of a moving disk using a camera with the shutter left open. The result is shown in Figure 1.4. As nearly as you can judge by placing a straight edge on the photograph, the disk moved in a straight line. This is a very useful result. But can you tell anything further? Did the disk move steadily during this phase of the motion, or did it slow down? You really can't tell from the continuous blur. In order to answer, we have to improve the observation by controlling it more. In other words, we have to experiment.

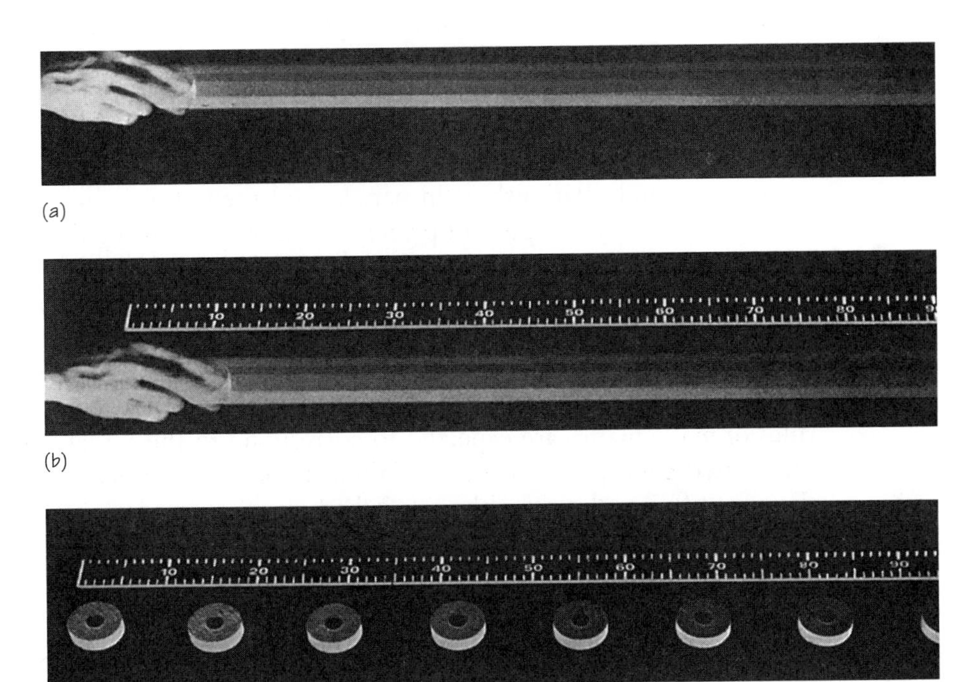

FIGURE 1.4 Time exposures of disk set in motion.

It would be helpful to know where the disk is at various times. In the next photograph, Figure 1.4b, we put a meter stick on the table parallel to the expected path of the disk, and then repeated the experiment with the camera shutter held open.

This photograph tells us again that the disk moved in a straight line, but it doesn't tell us much more. Again we have to improve the experiment. In this experiment the camera shutter will be left open and everything else will be the same as last time, except that the only source of light in the darkened room will come from a stroboscopic lamp. This lamp produces bright flashes of light at short time intervals which we can set as we please. We set each flash of light to occur every tenth of a second. (Each flash is so fast, one-millionth of a second, that its duration is negligible compared to one-tenth of a second.) The result is shown in Figure 1.4c.

This time the moving disk is seen in a series of separate, sharp exposures, or "snapshots," rather than as a continuous blur. Now we can actually see some of the positions of the front edge of the disk against the scale of the meter stick. We can also determine the moment when the disk was at each position from the number of strobe flashes corresponding to each position, each flash representing one-tenth of a second. This provides us with some very important information: *we can see that for every position reading of the disk recorded on the film there is a specific time, and for every time there is a specific position reading.*

Now that we know the position readings that correspond to each time (and vice versa), we can attempt to see if there is some relationship between them. This is what scientists often try to do: study events in an attempt to see patterns and relationships in nature, and then attempt to account for them using basic concepts and principles. In order to make the discussion a little easier, scientists usually substitute symbols at this point for different measurements as a type of shorthand. This shorthand is also very useful, since the symbols here and many times later will be found to follow the "language" of mathematics. In other words, just as Plato had argued centuries earlier, our manipulations of these basic symbols according to the rules of mathematics are expected to correspond to the actual behavior of the related concepts in real life. This was one of the great discoveries of the scientific revolution, although it had its roots in ideas going back to Plato and the Pythagoreans. You will see throughout this course how helpful mathematics can be in understanding actual observations.

In the following we will use the symbol d for the *position reading* of the front edge of the disk, measured from the starting point of the ruler, and the symbol t for the amount of *elapsed time* from the start of the experiment that goes with each position reading. We will also use the standard abbreviations cm for centimeters and s for seconds. You can obtain the values of some pairs of position readings d and the corresponding time readings t directly from the photograph. Here are some of the results:

Position	d *(in centimeters)*	t *(in seconds)*
1	6.0	0.1
2	19.0	0.2
3	32.0	0.3
4	45.0	0.4
5	58.0	0.5
6	71.0	0.6
7	84.0	0.7

From this table you can see that in each case the elapsed time increased by one-tenth of a second from one position to the next—which is of course what we expect, since the light flash was set to occur every one-tenth of a second. We call the duration between each pair of measurements the "time interval." In this case the time intervals are all the same, 0.1 s. The distance the disk traveled during each time interval we call simply the "distance traveled" during the time interval.

The time intervals and the corresponding distances traveled also have special symbols, which are again a type of shorthand for the concepts they represent. The time interval between any two time measurements is given the symbol Δt. The distance traveled between any two position readings is given the symbol Δd. These measurements do not have to be next to each other, or successive. They can extend over several flashes or over the entire motion, if you wish. The symbol Δ here is the fourth letter in the Greek alphabet and is called "delta." Whenever Δ precedes another symbol, it means "the change in" that measurement. Thus, Δd does *not* mean "Δ multiplied by d." Rather, it means "the change in d" or "the distance traveled." Likewise, Δt stands for "the change in t" or "the time interval." Since the value for Δt or Δd involves a change, we can obtain a value for the amount of change by subtracting the value of d or t at the start of the interval from the value of d or t at the end—in other words, how much the value is at the end minus the value at the start. In symbols:

$$\Delta d = d_{\text{final}} - d_{\text{initial}},$$

$$\Delta t = t_{\text{final}} - t_{\text{initial}}.$$

The result of each subtraction gives you the difference or the change in the reading. That is why the result of subtraction is often called the "difference."

Now let's go back and look more closely at our values in the table for the position and time readings for the moving disk. Look at the first time interval, from 0.1 s to 0.2 s. What is the value for Δt? Following the above definition, it is $t_{\text{final}} - t_{\text{initial}}$, or in this case 0.2 s − 0.1 s, which is 0.1 s. What is the corresponding change in the position readings, Δd? In that time interval the disk's position changed from 6.0 cm to 19.0 cm. Hence, the value for Δd is 19.0 cm − 6.0 cm, which is 13.0 cm.

What would you expect to find for Δd if the disk had been moving a little faster? Would Δd be larger or smaller? . . . If you answered larger, you're right, since it would cover more ground in the same amount of time if it's moving faster. What would happen if it was moving slower? . . . This time Δd would be smaller, since the disk would cover less ground in the given amount of time. So it seems that one way of describing how fast or how slow the disk is moving is to look at how far it travels in a given time interval, which is called the "rate" that the distance changes.

Of course, we could also describe how fast it goes by how much time it takes to cover a certain fixed amount of distance. Scientists in the seventeenth century made the decision not to use this definition, but to use the first definition involving the distance traveled per time interval (rather than the reverse). This gives us the "rate" of motion, which we call the *speed*. (The idea of *rate* can apply to the growth or change in anything over time, not just distance; for example, the rate at which a baby gains weight, or the rate of growth of a tomato plant.)

We can express the rate of motion—the speed—as a *ratio*. A ratio compares one quantity to another. In this case, we are comparing the amount of distance traveled, which is represented by Δd, to the size of the time interval, which is represented by Δt. Another way of saying this is the amount of Δd per Δt. If one quantity is compared, or "per," another amount, this can be written as a fraction

$$\text{speed} = \frac{\Delta d}{\Delta t}.$$

In words, this says that the speed of an object during the time interval Δt is the ratio of the distance traveled, Δd, to the time interval, Δt.

This definition of the speed of an object also tells us more about the meaning of a ratio. A ratio is simply a fraction, and speed is a ratio with the distance in the numerator and time in the denominator. As you know, a fraction always means division: in this case, the rate of motion or the speed given by distance traveled, Δd, *divided by* the time interval, Δt.

There is still one small complication: we don't know exactly what the disk is doing when we don't see it between flashes of the light. Probably it is not doing anything much different than when we do see it. But due to

■ NOW YOU TRY IT

The Tour de France is a grueling test of endurance over the varied terrain of France. The total distance of the bicycle race is 3664 km (2290 mi). Lance Armstrong, cancer survivor and winner of the 1999 Tour de France, set a new record in covering this distance in 91.1 hr of actual pedaling. The race included breaks each night along the way. From the data given, what was Armstrong's average speed for the entire race while he was riding? (This speed broke the old record for the course of 39.9 km/hr.) He repeated his win in 2000.

FIGURE 1.5 Lance Armstrong.

friction it may have slowed down just a bit between flashes. It could also have speeded up a bit after being hit by a sudden blast of air; or perhaps nothing changed at all, and it kept right on moving at the exact same rate. Since we don't know for certain, the ratio of Δd to Δt gives us only an "average," because it assumes that the rate of increase of d has not changed at all during the time interval Δt. This is another way of getting an average of similar numbers, rather than adding up a string of values and dividing by the number of values. We give this ratio of Δd to Δt a special name. We call it the *average speed* of the disk in the time interval Δt. This also has a special symbol, v_{av}:

$$\frac{\Delta d}{\Delta t} = v_{av}.$$

These symbols say in words: *The measured change in the position of an object divided by the measured time interval over which the change occurred is called the average speed.*

This *definition* of the term *average speed* is useful throughout all sciences, from physics to astronomy, geology, and biology.

To see how all this works, suppose you live 20 mi from school and it takes you one-half hour to travel from home to school. What is your average speed?

Answer: The distance traveled Δd is 20 mi. The time interval Δt is 0.5 hr. So the average speed is

$$v_{\text{av}} = \frac{\Delta d}{\Delta t} = \frac{20 \text{ mi}}{0.5 \text{ hr}} = 40 \frac{\text{mi}}{\text{hr}}.$$

Did you actually travel at a steady speed of 40 mi/hr for the entire half-hour? Probably not. There were probably stop lights, slow traffic, corners to turn, and stretches of open road. In other words, you were constantly speeding up and slowing down, even stopping, but you averaged 40 mi/hr for the trip. Average speed is a handy concept—even though it is possible that very rarely during your travel you went at exactly 40 mi/hr for any length of the road.

Back to the Moving Disk

Let's go back to the disk to apply these definitions. (See the table on page 21.) What is the average speed of the disk in the first time interval? Substituting the numbers into the formula that defines average speed

$$v_{\text{av}} = \frac{\Delta d}{\Delta t} = \frac{13.0 \text{ cm}}{0.1 \text{ s}} = 130 \frac{\text{cm}}{\text{s}}.$$

What about the next interval, 0.2 s to 0.3 s? Again, Δt is 0.1 s and Δd is 32.0 cm − 19.0 cm = 13.0 cm. So,

$$v_{\text{av}} = \frac{\Delta d}{\Delta t} = \frac{13.0 \text{ cm}}{0.1 \text{ s}} = 130 \text{ cm/s}.$$

The average speed is the same. Notice again that in finding the change in d or in t, we always subtract the beginning value *from* the ending value.

We don't have to consider only successive time readings. Let's try a larger time interval, say from 0.2 s to 0.7 s. The time interval is now Δt = 0.7 s − 0.2 s = 0.5 s. The corresponding distance traveled is Δd = 84.0 cm − 19.0 cm = 65.0 cm. So the average speed is

$$v_{\text{av}} = \frac{\Delta d}{\Delta t} = \frac{65.0 \text{ cm}}{0.5 \text{ s}} = 130 \text{ cm/s}.$$

What can you conclude from all of these average speeds? Our data indicate that the disk maintained the same average speed throughout the entire experiment as recorded on the photograph (we don't know what it did before or after the photograph was made). We say that anything that moves at a constant speed over an interval of time has a *uniform speed.*

1.4 PICTURING MOTION

Most sports involve motion of some sort. Some sports, such as swimming, jogging, bicycling, ice skating, and roller blading, involve maintaining speed over a given course. If it's a race, the winner is of course the person who can cover the course distance in the shortest time, which means the fastest average speed. Here the word "average" is obviously important, since no swimmer or biker or runner moves at a precisely uniform speed.

Let's look at an example. Jennifer is training for a running match. Recently, she made a trial run. The course was carefully measured to be 5000 m (5 km, or about 3.1 mi) over a flat road. She ran the entire course in 22 min and 20 s, which in decimal notation is 22.33 min. *What was her average speed in kilometers per minute during this run?*

FIGURE 1.6

The distance traveled Δd was 5 km; the time interval Δt was 22.33 min. So her average speed was

$$v_{av} = \frac{\Delta d}{\Delta t} = \frac{5.00 \text{ km}}{22.33 \text{ min}} = 0.224 \text{ km/min}$$

Did Jennifer actually run the 5 km at the constant speed of 0.224 km/min? Very likely not. To find out more about her average speed on different parts of the road, let's compare her overall average speed with her average speed during different segments of the run, just as we did earlier with the disk—only this time the distance covered (Δd) will be fixed, rather than having a fixed time interval Δt. To do this, we stationed five observers at 1-km intervals along the course. Δd will be the distance between neighboring observers. Each had a stopwatch that started at the start of her run. As Jennifer passed each position, the observer there read on the stopwatch the elapsed time from the start and recorded the results. The results, in decimals, are shown below. Take a moment to study this table and try to understand what each column refers to and how these numbers were obtained.

d *(in km)*	t *(in min)*	Δd *(km)*	Δt *(min)*	v_{av} *(km/min)*
0.00	0.00	—	—	—
1.00	4.40	1.00	4.40	0.227
2.00	8.83	1.00	4.43	0.226
3.00	13.40	1.00	4.57	0.219
4.00	18.05	1.00	4.65	0.215
5.00	22.33	1.00	4.28	0.234

Once you understand these data you can, for instance, determine Jennifer's average speed for the first kilometer and for the last kilometer separately:

average speed for the first kilometer is

$$\frac{\Delta d}{\Delta t} = \frac{1.0 \text{ km}}{4.40 \text{ min} - 0.00 \text{ min}}$$
$$= 0.227 \text{ km/min},$$

average speed for the last kilometer is

$$\frac{\Delta d}{\Delta t} = \frac{1.0 \text{ km}}{22.33 \text{ min} - 18.05 \text{ min}}$$
$$= 0.234 \text{ km/min}.$$

The results for v_{av} for each 1-km distance are given in the table on page 26, rounded off to the third decimal place. As you can see, none of these average speeds for each kilometer distance turned out to be exactly the overall average, 0.224 km/min. You can also see that Jennifer varied her speed during the run. In fact, on the average she slowed down steadily from the first to the fourth kilometer, then she speeded up dramatically as she approached the end. That last kilometer was the fastest of the five. In fact, it was faster than the overall average, while the intermediate length segments were covered more slowly.

Another way of observing Jennifer's run—a way favored by athletic trainers—is not just to look at the position and time readings in a table, but to look at the pattern these pairs of numbers form in a picture or "graphical representation" of the motion, called a "graph." By placing the position readings along one axis of a sheet of graph paper and the corresponding time readings along the other axis, each pair of numbers has a unique place on the graph, and together all the pairs of numbers form an overall pattern that gives us a picture of what happened during the overall motion. Usually the time is placed on the horizontal axis or x-axis, because as time increases to the right we think of the pattern as "progressing" over time. (Some common examples might be daily temperature data for a period of time, or sales of products by quarters for the past years.)

Two graphs of Jennifer's run are shown in Figures 1.7 and 1.8. The first shows the labeled axes and the data points. The second shows line segments connecting each pair of dots and the "origin," which is the point $d = 0$, $t = 0$ in the lower left corner, corresponding to the start of the run.

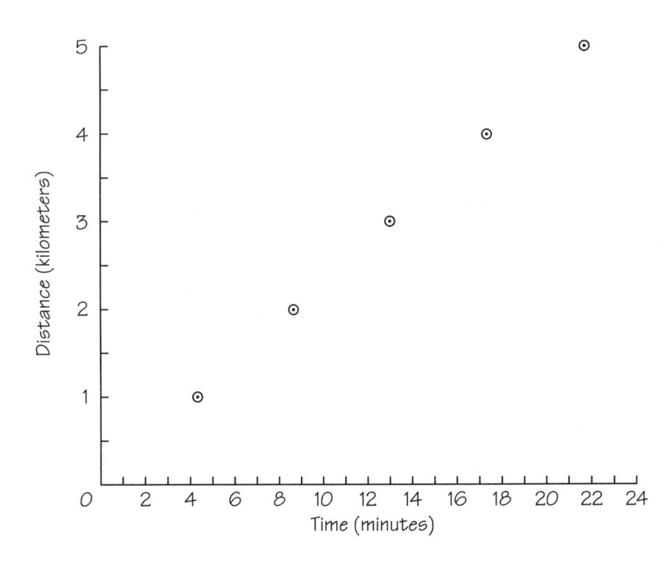

FIGURE 1.7 Distance vs. Time graph.

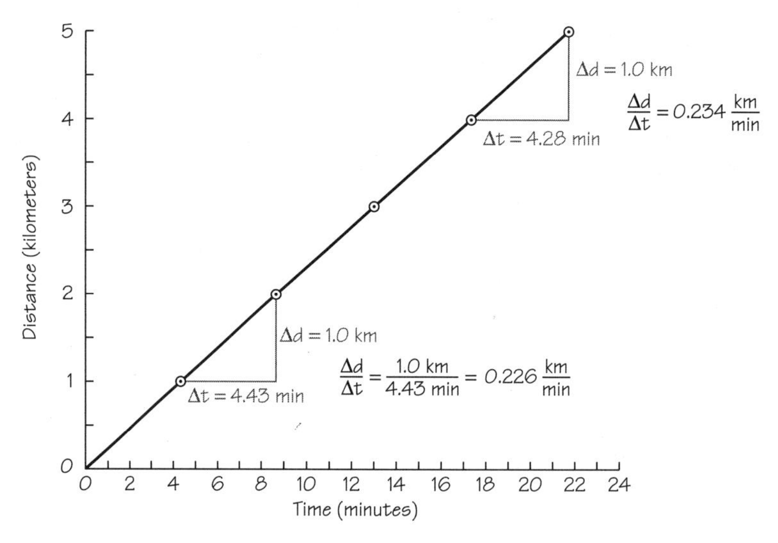

FIGURE 1.8 Distance vs. Time graph with slopes.

What can you observe, or "read," from the second graph? First of all, you can see (by putting a straight edge to it) that the first line segment is fairly steep, but that the last one is the steepest of all. There was a comparatively short time interval for the first leg of 1 km, and an even shorter one for the last leg of 1 km in length. In other words, Jennifer got off to a relatively fast start during the first kilometer distance; but she went even faster during the last one. In between, the steepness of each line declines, indicating that it took her longer and longer to cover the same distance of 1 km. She was slowing down. You can also obtain this result from the fourth column in the table (Δt). Notice how, for each kilometer covered, the time intervals Δt increase in the middle, but are less for the first and last kilometers. The distance covered stays the same, but the time to do so varies. This is seen as a change in average speed for each kilometer, as you can see from the last column of the table, and this agrees with the changing steepness of each line segment. We can conclude from this: *The steepness of each line segment on the graph is an indication of the average speed that Jennifer was moving in that interval.* The faster she ran, the steeper the line segment. The slower she ran, the less steep the line segment.

Looked at in this way, a graph of distance readings plotted on the y-axis against the corresponding time readings on the x-axis provides us with a visual representation of the motion, including the *qualitative* variations in the speed. But this kind of representation does not tell us directly what the *quantitative* speed was at any particular moment, what we can think of as the actual momentary speed in kilometers per minute, similar to the information on the speedometer of a moving car. We will come back to this notion later, but for now we will look only at the average speed in each time interval. Since the steepness of the graph line is an indication of the average speed during that segment of the graph, we need a way of measuring the steepness of a line on a graph. Here we must turn again to mathematics for help, as we often will.

The steepness of a graph line at any point is related to the change in vertical direction (Δy) during the corresponding change in horizontal direction (Δx). By definition, the ratio of these two changes is called the "slope":

$$\frac{\Delta y}{\Delta x} = \text{slope of a line.}$$

Slope can be used to indicate the steepness of a line in any graph. In a distance–time graph, like the one for Jennifer's run, the distance from the start is plotted on the vertical axis (d in place of y) and the corresponding times are plotted on the horizontal axis (t in place of x). Together, d and t make a series of points forming a line in the plane of the graph. In such a graph, the slope of a straight line representing the motion of a person or an object is defined as follows:

$$\text{slope} = \frac{\Delta d}{\Delta t}.$$

Does this ratio look familiar? It reminds us of the definition of average speed, $v_{av} = \Delta d/\Delta t$. The fact that these two definitions are identical means

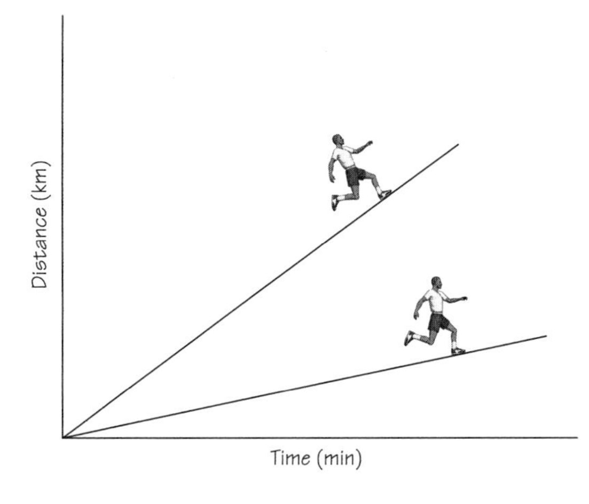

FIGURE 1.9 Distance vs. Time graphs for runners at two different speeds. Which one is faster?

that v_{av} is numerically equal to the slope on distance–time graphs! In other words, *the slope of any straight-line part of a graph of distance versus time gives a measure of the average speed of the object during that time interval.*

In short, we can use simple geometry to "capture" an observed motion.

1.5 SPEED AND VELOCITY

You may wonder why we used the letter v instead of s for speed. The reason is that the concept of speed is closely related to the concept of *velocity*, from which the symbol v arises. However, these two are *not* the same when used in precise technical terms. The term "velocity" is used to indicate *speed in a specified direction*, such as 50 km/hr to the north or 130 cm/s to the right. Speed indicates how fast something is moving regardless of whether or how it may change in direction. But velocity indicates how fast it is moving *and* the direction it is moving. In physics, anything that has both a size or magnitude and a direction is called a *vector.* Since a vector points in a definite direction, it is usually presented by an arrow in diagrams, such as the one in Figure 1.10. The direction of the arrow indicates the direction of the velocity. The length of the arrow indicates its magnitude, the speed.

Here is an example: one car is traveling west on a road at 40 mi/hr. Another identical car is traveling east on the same road at 40 mi/hr. Is there any difference between the motions of these two cars? Both are going at 40 mi/hr, so there is no difference in their speeds. The only difference is that one car is going west and the other is going east. That difference is obviously important if the two cars happen to be traveling on the same side of the road and meet! Obviously, the direction of the motion of each car, as well as its speed, is important in describing what happens in situations such as this. That is why it is necessary to use the velocity vector, which incorporates both speed and direction, in situations where both of these factors are important. When the direction of the motion is not important, the scalar speed is usually sufficient.

The velocity vector also has a special symbol. Following standard practice in textbooks, we will represent the velocity, or any vector, in bold font.

FIGURE 1.10 Cart traveling at velocity **v**.

So the symbol for the velocity vector is **v**. You will see other examples of vectors in this and later chapters.

When a direction is not specified, and only the size, or magnitude, is of interest, we'll remove the bold font and just use the italic letter v, which is the speed. This symbol represents only the *magnitude* of the velocity—how fast it is regardless of the direction. Speed is an example of what is known in technical terms as a *scalar*. It has no direction, just a reading on some scale. Some other examples of scalars are mass, temperature, and time.

1.6 CHANGING THE SPEED

In the previous sections we looked at two examples of motion. One was a disk moving, as far as we could tell, at constant or uniform speed; the other was a runner whose speed clearly varied as she went over a given distance. There are many examples in nature of moving objects that undergo changes of speed and/or direction. As you walk to class, you may realize you are late and pick up your pace. An airplane landing at an airport must decrease its altitude and slow its speed as it lands and comes to a halt on the runway. Cars going around a curve on a freeway usually maintain their speed but change the direction of their motion. A growing tomato plant may spurt up over a number of days, then slow its growth in height as it starts to bud. Many other such examples come to mind.

Motions that involve changes in speed and/or direction over a period of time are obviously an important part of the motions that occur in nature. A change in the velocity of a moving object, during an interval of time, is known as the *acceleration*. Since an interval of time is the amount of time that elapses from one instant to another, we need to know the velocity of the object at each instant of time, in order to find the acceleration. How can we examine changes of speed from one instant to another? So far, we have talked only about *average* speeds over a time interval. One way to study changes of speed from one instant to another is to take advantage of a modern device that measures speed at any given time for us by converting the speed of, say, the wheels of a car into a magnetic force that turns a needle on a dial. This is the basic principle behind the speedometer. Although this refers only to the motion of a car, the general principles will be the same for other moving objects.

The speedometer in Figure 1.11 reads in kilometers per hour. This car was driven on an open highway in a straight line, so no changes of direction occurred. As the car was traveling on the road, snapshots were made of the speedometer reading every second for 10 s after an arbitrary start, where we

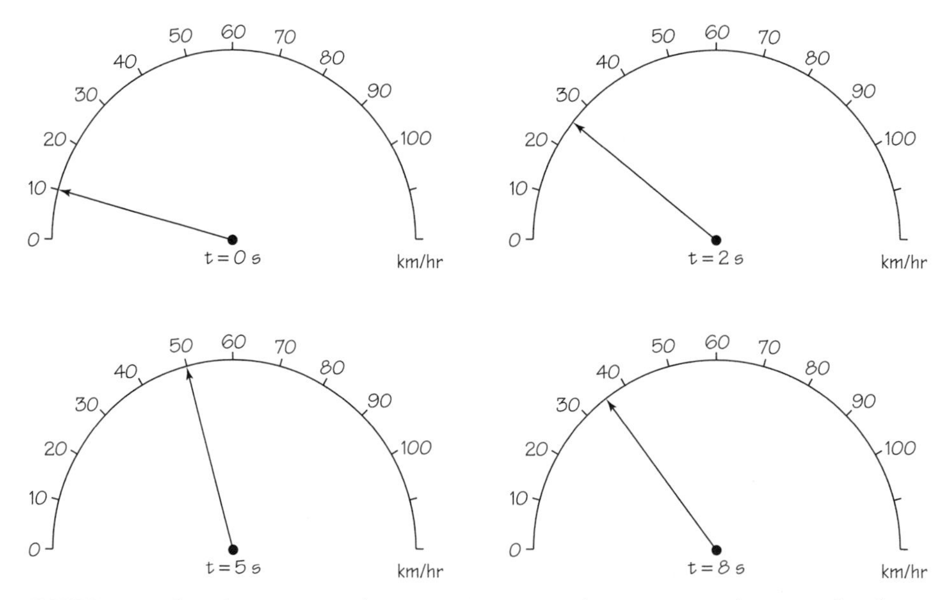

FIGURE 1.11 Speedometer snapshots. t = 0 s at an arbitrary point when car already in motion.

set $t = 0$. Some of the snapshots are shown below. Similarly to the earlier distance and time readings for the disk and runner, this experiment provides us with ten pairs of speed and time readings for the motion of this car. We have no data or snapshots for the motion of the car before or after this experiment. The results are given in the table below:

v (km/hr)	*t (s)*
10	0
18	1
26	2
34	3
42	4
50	5
50	6
50	7
35	8
20	9
5	10

We have to determine first what these numbers really mean before we can analyze what the car was actually doing. This is a common procedure in scientific research: determine first the precise meaning of the data be-

The instantaneous speed may be defined as the limit, as the time interval approaches zero, of the ratio of distance traveled per time interval, or:

$$v = \lim_{\Delta t \to 0} \frac{\Delta d}{\Delta t}.$$

fore attempting to analyze them. As you know from the previous sections, the average speed is defined as the ratio of the distance traveled to the time interval. In this case, the speedometer mechanism used very small increments of time to obtain these speeds. The increments of time and the corresponding distances the car traveled are so small that, within the limits of precision of this experiment, we can assume that for all practical purposes they are zero.

So we can say that, within the limits of accuracy of this experiment, the speeds shown in the table on page 32 indicate the "instantaneous speed" of the car, and the speed at the instant of time shown in the second column. Of course, with faster photographic equipment or a faster speedometer mechanism, we might see a slightly different instantaneous speed, but that would be beyond the limits of accuracy of *this* experiment.

Now we are ready to interpret what this car was actually doing, according to the data that we have. Similarly to the case of distance and time for the disk or runner, we can look at the change in speed in each time interval, which will help us to see how the motion is changing. Let's add three more columns to the table. The first will be the change Δv in the instantaneous speeds measured by the speedometer; the second will be the time interval, Δt, as before; and the third will be their ratio, $\Delta v/\Delta t$.

v (km/hr)	t (s)	Δv (km/hr)	Δt (s)	$\Delta v/\Delta t$ (km/hr/s)
10	0	—	—	—
18	1	8.0	1.0	8.0
26	2	8.0	1.0	8.0
34	3	8.0	1.0	8.0
42	4	8.0	1.0	8.0
50	5	8.0	1.0	8.0
50	6	0.0	1.0	0.0
50	7	0.0	1.0	0.0
35	8	−15.0	1.0	−15.0
20	9	−15.0	1.0	−15.0
5	10	−15.0	1.0	−15.0

At first glance this looks really puzzling! What was this car doing? To find out, let's "analyze" the motion, that is, let's take a closer look at the motion piece by piece. Starting at the top (when the experiment began) at time 0, the car was already going at 10 km/hr. From 0 to 5 s, the speed increased by an additional 8.0 km/hr in each time interval of 1.0 s (see the third column). As with distance and time, the ratio of Δv to Δt gives us a quantitative value for the average rate of change of the speed in each time interval. As you can see from the last column, during the first 5 s the speed of the car increased at the average rate of 8.0 km/hr/s, that is, 8.0 km/hr in each second.

Since it is so important in describing changing speeds, the ratio $\Delta v/\Delta t$ also has been given a special name. It is called the *average acceleration* in each time interval. This has the symbol a_{av}:

$$\frac{\Delta v}{\Delta t} = a_{av}.$$

These symbols say in words: *The change in the instantaneous speed of an object divided by the time interval over which the change occurs is defined as the average acceleration of the object.* In this case, the car maintained a constant average acceleration for the first 5 s. This behavior is called "uniform acceleration."

We have to add just one more idea. Since the speeds above were all in the same direction, we didn't need to refer to the velocity. But sometimes the direction of motion does change, as when a car goes around a corner, even though the speed stays the same. We can expand our definition of acceleration to include changes of direction as well as changes of speed by simply replacing the speed v in the above formula by the velocity vector, $\mathbf{v}$. Since velocity is a vector, the acceleration is also a vector, so we have the following definition:

$$\frac{\Delta \mathbf{v}}{\Delta t} = \mathbf{a}_{av}.$$

These symbols say in words: *The change in the velocity of an object divided by the time interval over which the change occurs is defined as the average acceleration vector.* In the example so far, we can say that the car velocity increased uniformly in the forward direction, so it maintained a uniform average acceleration vector of 8 km/hr/s in the forward direction for the first 5 s.

Now let's go on to the sixth and seventh seconds. What is the car doing? The change in the speed is zero, so the ratio of $\Delta v/\Delta t$ gives zero, while the direction remains unchanged. Does that mean the car stopped? No, what it means is that the *acceleration* (not the speed) stopped; in other words, the car stopped changing its speed for 2 s, so the average speed (and average velocity) remained constant. The car cruised for 2 s at 50 km/hr.

Now what happened during the last 3 s? You can figure this out yourself . . . but, here's the answer: the car is moving with *negative* accel-

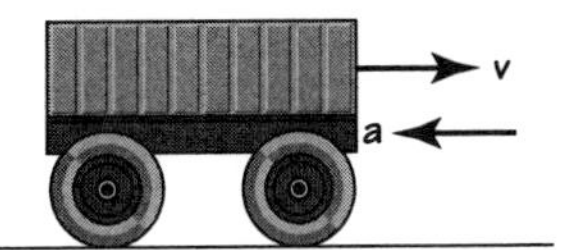

FIGURE 1.12 Cart traveling at v with acceleration opposite.

eration. (This is sometimes called the "deceleration.") Probably the driver put on the brakes, slowing the car down gradually to 5 km/hr over 3 s. How do we know this? Since Δ always means the ending value minus the beginning value, a negative result for Δv means that the final speed was *less than* the initial speed, so the car slowed down. This leads to a negative value for $\Delta v/\Delta t$. Since the speed decreased at the same rate for the last 3 s, this was uniform negative acceleration, i.e., deceleration. If we look at the vectors involved, the negative value of the acceleration means that the acceleration vector is now opposing the velocity vector, not helping it. So the two vectors point in the opposite directions, as in Figure 1.12.

Can you summarize what the car was doing during this entire experiment? . . . Here is one way of putting it: When the experiment began, the car was already going at 10 km/hr, and accelerated uniformly to 50 km/hr in 5 s at an average rate of 8.0 km/hr/s (8.0 km/hr in each second). The acceleration and velocity vectors pointed in the same direction. It then cruised at that speed for 2 s, so the acceleration was zero. Then it braked uniformly at the average rate of -15 km/hr/s for 3 s, ending with a speed of 5 km/hr. The acceleration and velocity vectors were pointing in the opposite directions during the slowing of the car.

Just as with distance and time, we can obtain a "picture" of the speed during this overall motion by drawing a graph of the motion, with the instantaneous speeds (neglect direction for now) on the vertical axis and the corresponding times on the horizontal axis. The result for this example is shown in Figure 1.13. We have also connected the data points together with straight lines. Again, the steepness of the lines has an important meaning. We can see that the graph starts out at 10 km/hr on the v-axis and climbs steadily upward to 50 km/hr at 5 s. During this time it had positive acceleration. Then the graph becomes "flat." As time increases, the speed does not change. There is no rise or fall of the line, so the acceleration is zero. Then the graph starts to fall. Speed is changing, but the change in values is *downward*, so the motion involves slowing, so the acceleration is negative. The steepness of the line and its upward or downward slopes appear to indicate the amount of positive or negative acceleration. When it is horizontal, the acceleration is zero.

As you learned earlier, we can obtain a quantitative measure of the "steepness" of any line on a graph by obtaining the "slope." Remember that the

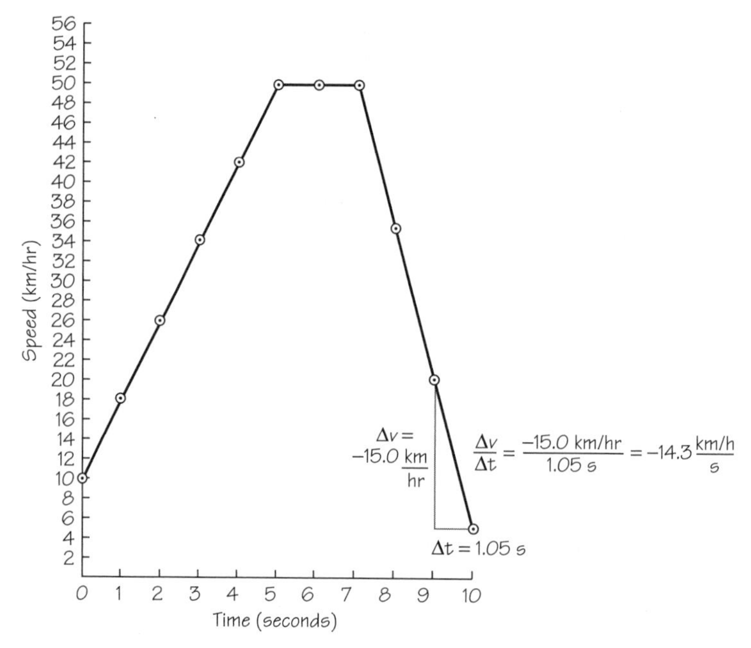

FIGURE 1.13 Speed vs. time graph.

slope is defined as the change in the y coordinate divided by the change in the x coordinate, slope $\equiv \Delta y/\Delta x$. In this case v is along the y-axis and t is along the x-axis, so the slopes of the line segments, $\Delta v/\Delta t$, are equal to the average acceleration, a_{av}, as we discussed before. In other words, *the slope of any straight-line part of a graph of instantaneous speed versus time gives a measure of the average acceleration of the object during that time interval.*

You can confirm the results in the last column of the table on page 33 by calculating the slopes of the lines in the graph.

1.7 FALLING FREELY

One of the most common occurrences of accelerated motion—yet one of the most difficult to study—is free fall, an object falling freely to the ground. An example is a ball dropped to the floor from a certain height. Just watching it fall, it is extremely difficult to see what is going on without modern equipment, since the ball drops so quickly. Happily today we do have equipment that helps us analyze the motion. The picture in Figure 1.14 is taken with the stroboscopic flash set at intervals of just 0.035 s and a camera with

We can obtain some simple equations *for the case of uniform acceleration*. If the average acceleration a_{av} is uniform, we can treat it as a mathematical constant and give it the symbol a. Thus,

$$\frac{\Delta v}{\Delta t} = a_{av} = a \quad \text{if } a_{av} \text{ is uniform.}$$

So,

$$\Delta v = a\,\Delta t$$

or

$$v = v_0 + a\,\Delta t.$$

In words, this result says that—for uniform acceleration—the instantaneous velocity (v) after a time interval of Δt is equal to the initial instantaneous velocity (v_0) plus the uniform acceleration (a) times the time interval (Δt).

Now, if the initial velocity was zero and if the clock was started at a reading of 0 s, this can be reduced even further to the simple formula

$$v = at.$$

But, as before, in order to use this simple formula the situation must satisfy all of the "ifs" in the discussed text (such as time starting at 0 s). Otherwise, when we don't know if the situation satisfies all of these "ifs," or when it satisfies only some of them, we must use the most general formula to define acceleration, which is

$$a_{av} = \frac{\Delta v}{\Delta t}.$$

an open shutter. This shows the position of the ball every 0.035 s against a metric ruler in the background.

How do you know that this ball is accelerating? Go back to the definition of acceleration in the previous section: a constantly increasing speed during equal time intervals. You can tell that the average speed is increasing because the distance traveled in each time interval is getting larger and larger. Free fall is an example of acceleration. In fact, as you will see, it is ideally an example of uniform acceleration. But to understand all of this, and without access to modern-day cameras and stroboscopic flashes, it took a person of the stature and genius of Galileo Galilei, who lived in Italy during the late sixteenth and early seventeenth centuries. What made Galileo different from many of his predecessors, especially those who followed Aristotle, and enabled his breakthrough, is his discovery that experimentation is the proper way to investigate nature, and that mathematics is the proper language for understanding and describing the laws of physics. "The book of nature is written in mathematical symbols," Galileo once said. Rather than qualitative arguments, Galileo relied upon the quantitative investiga-

FIGURE 1.14 Stroboscopic photograph of a ball falling next to a vertical meter stick.

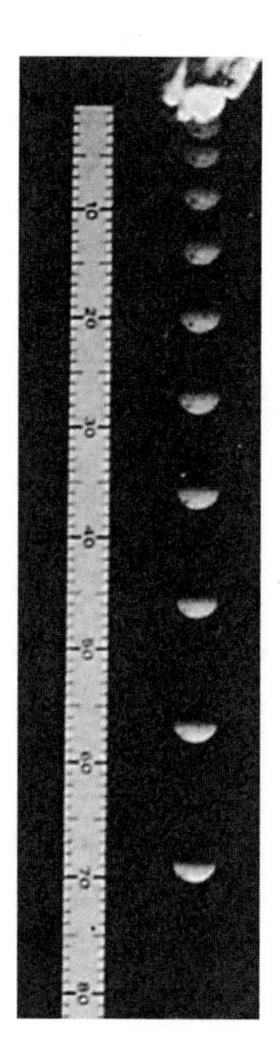

tion of physical events, just as physicists do today. We will follow his reasoning and discoveries, because he was the one who laid the foundations for the modern science of motion. In so doing, his view of nature, his way of thinking, his use of mathematics, and his reliance upon experimental tests set the style for modern physics in general. These aspects of his work are as important for understanding today's physics as are the actual results of his investigation.

1.8 *TWO NEW SCIENCES*

Galileo was old, sick, and nearly blind at the time he wrote *Two New Sciences*, which presented the new understanding of acceleration and free fall, and many other topics regarding motion. Yet, as in all his writings, his style is lively and delightful. He was also one of the very few authors of that time to write and publish in the vernacular, that is in Italian, rather than in the scholarly Latin. This indicated that he was writing as much for the educated Italian public as he was for a circle of academic specialists.

Probably influenced by his reading of Plato's dialogues, Galileo presented his ideas in *Two New Sciences* in the form of a dialogue, or conversation, among three fictional speakers. One of the speakers, named Simplicio, represented Aristotle's views. The proximity of his name to "simplicissimo," "the most simple one" in Italian, was surely no accident, although he was not made out to be a fool but a sophisticated Aristotelian philosopher. The other two fictional characters were Salviati, who represented Galileo himself, and Sagredo, a man of good will and open mind, eager to learn. Eventually, of course, Salviati leads Sagredo to Galileo's views and away from Simplicio's Aristotelian ideas.

The three friends first tackle the difficult problem of free fall. Aristotle's views on this subject still dominated at that time. According to Aristotle, each of the four elements has a natural place where it "belongs" and to which it will return on a straight line if removed from its natural place. Thus, a stone raised up into the air will, when released, drop straight down to the Earth. The heavier it is, the faster it will drop because it has more "earth" element in it, although air resistance will slow it down a little. Thus, Simplicio argued in Galileo's book, when a cannonball and a bird shot are dropped simultaneously from the same height, the cannonball will hit the ground much sooner than the bird shot.

This does sound very reasonable, and in fact different bodies falling from the same height may not reach the ground at exactly the same time. But the difference is not the huge difference predicted by Aristotle, but a minor difference which Galileo correctly attributed to the effect of air resistance on bodies of different size and weight. It is a further characteristic of Galileo's genius that he was able to recognize that the effects of air resistance and friction, though present in most real experiments, should be neglected so that the important feature of free fall—that in the absence of air resistance all objects fall with the same acceleration—is not overlooked. (We, too, neglected any friction and air resistance in the earlier disk experiment in order to observe the essential features of the motion.)

Aristotle regarded air resistance as such an important component of free fall that for him it had a major impact on the motion. He was right when we compare, for instance, a falling sheet of paper with a falling book. The book *does* reach the ground long before the paper does! But here again, it takes a special insight to realize that, while air resistance is a major factor for the falling paper, it is not so for the falling book. Hence the book and the paper are falling under two different circumstances, and the result is that one hits the ground a lot sooner than the other. Make the air resistance on the paper equivalent to that on

Aristotle: Rate of fall is proportional to weight divided by resistance.

FIGURE 1.15 A falling leaf.

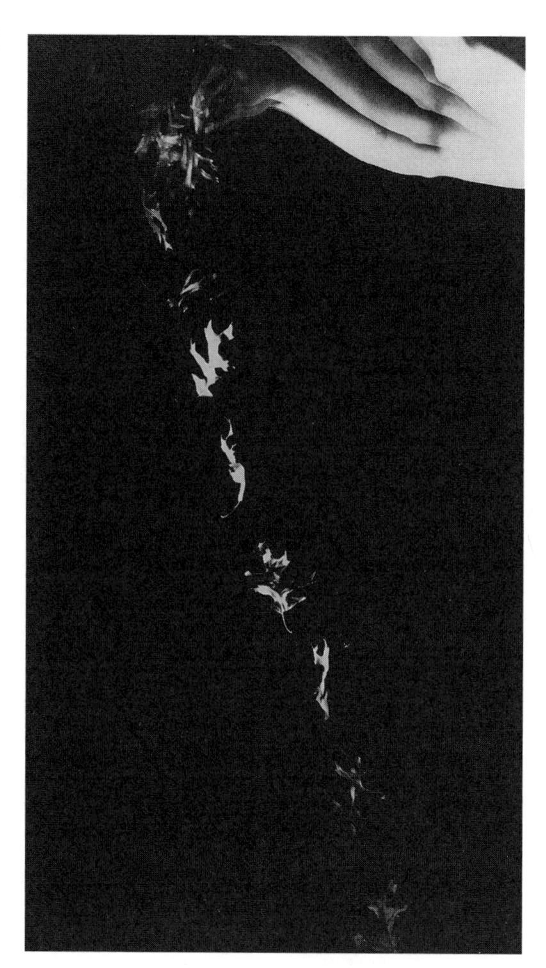

the book by crumpling the paper into a tight ball and then try the experiment again. You will see a big difference from the previous case!

Galileo's conclusion that all falling objects fall with the same acceleration if air resistance is neglected depended on his being able to imagine how two objects would fall if there were no air resistance. His result seems simple today, when we know about vacuum pumps and the near vacuum of outer space, where there is no air resistance. But in Galileo's day a vacuum could not be achieved, and his conclusion was at first very difficult to accept.

A few years after Galileo's death, the invention of the vacuum pump allowed others to show that Galileo was indeed right! In one experiment, a feather and a heavy gold coin were dropped from the same height at the same time inside a container pumped almost empty of air. With the effect of air resistance eliminated, the different bodies fell at exactly the same rate and struck the bottom of the container at the same instant. Centuries later, when the Apollo astronauts landed on the Moon, they performed a famous experiment in which they dropped a feather and a hammer (a gold coin was not available . . . or deemed too expensive) simultaneously from the same height in the vacuum of space on the surface of the Moon. Just as Galileo would have predicted, the feather and the hammer hit the Moon's dusty surface at the exact same time!

1.9 FALLING OBJECTS

After establishing that under ideal conditions falling objects hit the ground at the same time when dropped from the same height, Galileo then had his alter ego, Salviati, hypothesize that a freely falling cannonball or any heavy object falls at *uniform* acceleration. But how exactly should one define uniform acceleration (after all, Galileo was the first one to put these ideas on paper)? Should one base the definition on the distance the object travels, or on the time it takes to travel? It was up to Galileo to decide, and he chose the definition that has been accepted ever since. Galileo declared in the book, *Two New Sciences*:

> A motion is said to be uniformly accelerated when, starting from rest, it acquires equal increments of speed Δv during equal time intervals Δt.

In other words, for uniform acceleration the ratio $\Delta v/\Delta t$ would be constant during any portion of the accelerated motion.

Galileo then set out to show that this definition held in the case of falling objects. But he knew that there was a practical problem. Suppose you drop a heavy ball from a given height to test whether the ratio $\Delta v/\Delta t$ for its motion really is constant for the entire path of the falling ball. To obtain the value of the ratio, you would have to measure the instantaneous speed of the ball and the corresponding elapsed time at different points along the path of fall. You would then obtain the values of Δv and Δt for pairs of points along the path and divide the results to see if they yield the same constant value.

In reality this experiment was impossible to perform. Even with modern instruments, such as those that enabled us to create the earlier photographs, it is difficult to measure the speed of a falling object. Furthermore, the time intervals involved are so short that Galileo could not have measured them accurately with the timing devices available to him. Even a ball dropped off a 10-story building takes less than 6 s to cover the entire distance to the ground. So a direct test of the ratio $\Delta v/\Delta t$ was impossible then, and it is not easy even today.

This did not stop Galileo. He turned to mathematics, the "language of nature," in order to obtain from his hypothesis that uniform acceleration governs the fall of objects some other relationship that could be checked by measurements with the equipment available to him. Using a little bit of geometry then—we would use algebra today (see the discussion in the *Student Guide*)—Galileo eliminated the change of speed Δv from the defining formula for acceleration. He did this by expressing the uniform acceleration a for an object starting from rest in a relationship involving the distance traveled Δd and the elapsed time Δt, both of which are quantities that are easier to measure than Δv:

$$\Delta d = \tfrac{1}{2}\, a(\Delta t)^2$$

or, regrouping terms,

$$a = \frac{2\Delta d}{\Delta t^2}.$$

This is the kind of relation Galileo was seeking. It relates the total distance traveled Δd and the total elapsed time Δt to the acceleration a, without involving any speed term.

Before finishing, though, we can simplify the symbols in either equation to make the equations easier to use. If you measure distance and time from the position and the instant that the motions starts, then $d_{\text{initial}} = 0$ and $t_{\text{initial}} = 0$. Thus, the intervals Δd and Δt have the values given by d_{final} and t_{final}. We can then write the first equation above more simply as

$$d_{\text{final}} = \tfrac{1}{2}\, a t_{\text{final}}^2.$$

Or, if we simply write d_{final} as d and t_{final} as t, we have

$$d = \tfrac{1}{2}\, at^2.$$

This is the most well-known form of Galileo's famous result, but remember that it is a very specialized equation. It gives the total distance fallen as a function of the total time of free fall, but it does so only if the motion starts from rest ($v_{\text{initial}} = 0$), if the acceleration is uniform (a = constant), and if time and distance are measured from the start of the fall ($t_{\text{initial}} = 0$ and $d_{\text{initial}} = 0$).

The Meaning of Galileo's Result

Let's take a moment to look at what the equation $d = \frac{1}{2}at^2$ really means. It says in words that, for constant acceleration (in free fall or any other motion), the total distance traveled is equal to one-half the acceleration times the square of the total elapsed time. Since a is a constant, the equation can also be expressed as a proportion, using the proportionality sign $\propto$:

$$d \propto t^2.$$

In this case, $\frac{1}{2}a$ is the constant of proportionality. For example, if a uniformly accelerating car moves 10 m in the first second, it will move 40 m in the second second, 90 m in the third second, and so on. All objects accelerating with an acceleration equal to that of this car will have the same results. Or, expressed differently, the ratio of d to t^2 will have the same value

$$\frac{d}{t^2} = \text{constant} = \tfrac{1}{2}a.$$

Conversely, any motion for which this ratio of d and t^2 is constant for different distances and their corresponding times is a case of *uniform acceleration*. Galileo used this last statement to test if the acceleration of free fall is indeed uniform.

Galileo Turns to an Indirect Test

Galileo encountered another practical problem in examining the acceleration of free fall: how to measure the very short time intervals as an object fell the different distances? (See the insert "Dropping a ball from the Leaning Tower of Pisa.") A direct test of the hypothesis was still not possible. Ingeniously, Galileo turned to a clever indirect test. He decided to test an object that was falling under the influence of gravity but not falling freely. He proposed a new hypothesis:

> If a freely falling body has uniform acceleration, then a perfectly round ball rolling down a perfectly smooth incline will also have a constant, though smaller, acceleration.

Galileo's Salviati described just such an experimental test in *Two New Sciences*. Others who repeated this experiment have obtained results very similar to those he described. (You might perform this experiment yourself, or a similar one, in the laboratory.)

DROPPING A BALL FROM THE LEANING TOWER OF PISA

Although there is no certain evidence that Galileo actually dropped a ball from the Leaning Tower of Pisa, we can use his equation to find out how long it would take for a ball to reach the ground.

The Leaning Tower is about 58.4 m in length, but the actual distance to the ground is less, because of the lean. Today the amount of lean of the top with respect to the base is about 5.20 m (this may change soon, when efforts are made to stabilize the tower). Using the Pythagorean theorem, today the distance straight down from the top can be calculated to be 58.2 m. The acceleration of the ball is the acceleration of gravity, which has the symbol g. The acceleration of gravity is $g = 9.8\ \text{m/s}^2$.

Using Galileo's equation for a ball dropped from rest, $d = \frac{1}{2}at^2$, the time for the ball to fall this distance, neglecting air resistance, may be found as follows:

$$d = \tfrac{1}{2}at^2,$$

$$t = \sqrt{\frac{2d}{g}} = \sqrt{\frac{2(58.2\ \text{m})}{9.8\ \text{m/s}^2}}$$

$$= \sqrt{11.9\ \text{s}^2}$$

$$= 3.5\ \text{s}.$$

How fast would the ball be moving the instant before it hits the ground (again, neglecting air resistance)?

$$v = at = gt = (9.8\ \text{m/s}^2)(3.5\ \text{s})$$

$$= 34.3\ \text{m/s}.$$

This is very fast, indeed!

First, keeping a constant angle on an inclined plane, Galileo allowed a smooth round ball to roll different distances down the inclined plane from rest. He measured the elapsed time in each case using a water clock, a device that measures elapsed time by the amount of water that flows at a steady rate from a vessel. If d_1, d_2, and d_3 represent distances reached by the ball, measured from the starting point on the incline, and t_1, t_2, and t_3 represent the corresponding times taken to roll down these distances, then he could see whether the accelerations had the same value in each of the cases by dividing each d by the corresponding t^2. If these ratios all had the same value, then, as just discussed, this would be a case of uniform accel-

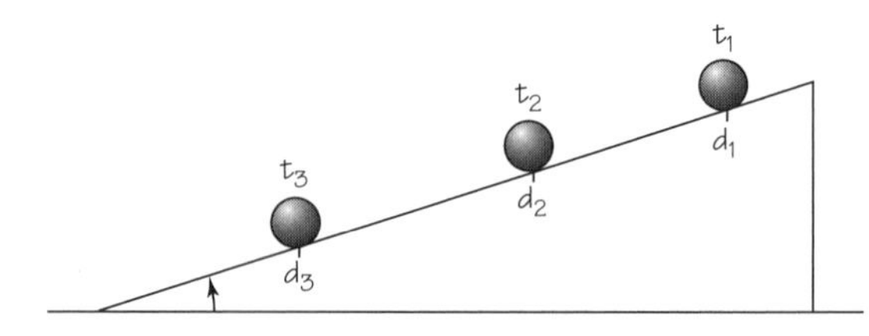

FIGURE 1.16 Ball rolling down inclined plane.

FIGURE 1.17 This 1841 painting by Guiseppi Bezzuoli attempts to reconstruct an experiment Galileo is alleged to have made during his time as lecturer at Pisa. Off to the left and right are men of ill will: the blasé Prince Giovanni de Medici (Galileo had shown a dredging machine invented by the prince to be unusable) along with Galileo's scientific opponents. These were leading men of the universities; they are shown here bending over a book of Aristotle, in which it is written in black and white that bodies of unequal weight fall with different speeds. Galileo, the tallest figure left of center in the picture, is surrounded by a group of students and followers.

eration. (In another account, Galileo kept time by singing a song and beating time with his fingers while rolling cannonballs on the inclined plane.

Whichever method he used for measuring time, Galileo obtained precisely the predicted result for the relationship between the distance traveled and the elapsed time when the angle of the incline is unchanged. He concluded from this that since the value of d/t^2 was constant for a given angle of incline, the acceleration for each incline is indeed uniform.

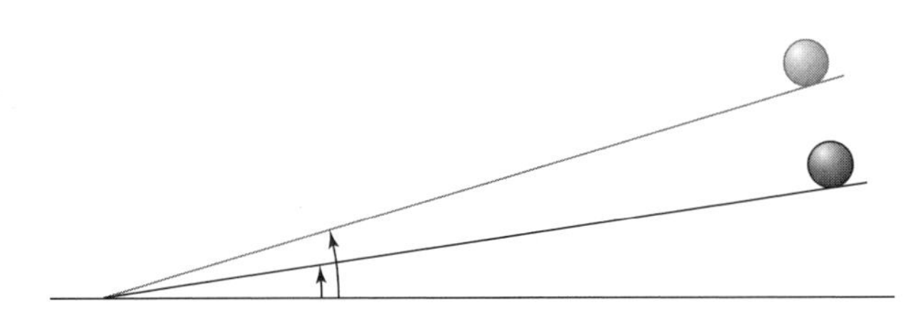

FIGURE 1.18 Ball rolling down inclined planes.

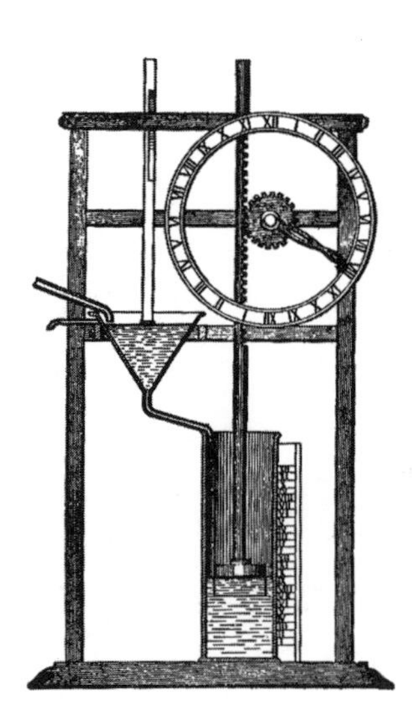

FIGURE 1.19 An early water clock.

In a second series of experiments, Galileo examined what happens when the angle of inclination of the plane is changed. As the angle of incline increased, Galileo reported that the ratio d/t^2 also increased.

After obtaining his results for small inclines, Galileo then tried a "thought experiment," that is, an experiment that he performed only logically in his mind, since it could not be carried out in reality. In this thought experiment, Galileo extrapolated his results for small angles of incline to steeper and steeper angles where the ball moves too fast for accurate measurements of t, and finally to the angle of inclination of 90°, when the ball would be moving straight down as a freely falling object. Since this was just the extreme case of motion on the incline (an "incline" that is actually vertical), he reasoned that d/t^2 would still be constant even in that extreme case.

In short, by the experiment on the inclined plane Galileo had found that a constant value of d/t^2 would be characteristic of uniform acceleration

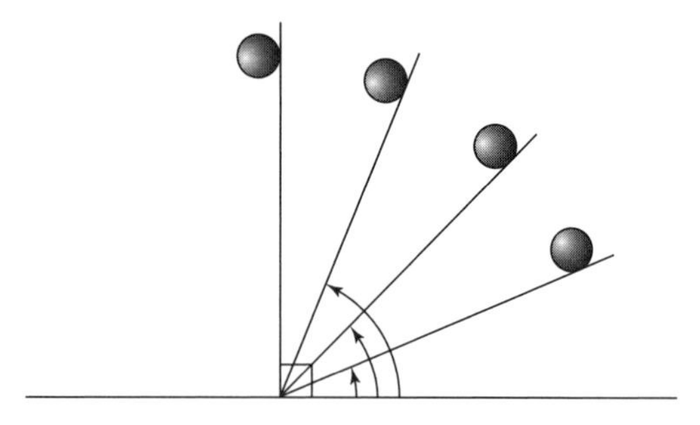

FIGURE 1.20 Spheres rolling down planes of increasingly steep inclination. For each angle, the acceleration has its own constant value. At 90 degrees, the inclined plane situation looks almost like free fall. Galileo assumed that the difference between free fall and "rolling fall" is not important (in most real situations, the ball would slide, not roll, down the steep inclines).

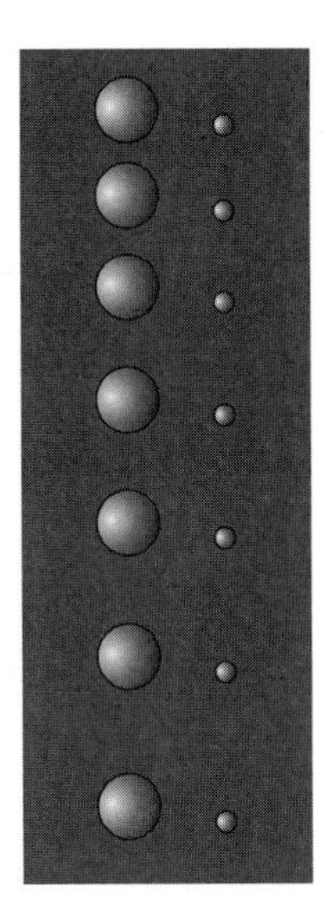

FIGURE 1.21 Two falling balls (unequal weights) in vacuum.

starting from rest. By extrapolation that experimental situation which he could measure, he concluded at last that free fall, which he could not examine directly, was an example of uniformly accelerated motion. Therefore, neglecting air resistance, Galileo concluded:

> all freely falling objects, whether heavy or light, will cover the same distance in the same amount of time, moving with the same constant acceleration. They will hit the ground together if they are dropped together from the same height.

This conclusion has been tested many times with modern apparatus (e.g., Figure 1.21), and it is sometimes called "Galileo's law (or rule) of free fall."

1.10 THE CONSEQUENCES

The results of Galileo's work on speed, acceleration, and falling bodies were most important to the development of physics, and they are now a fundamental part of today's physics. But they could scarcely have brought about a revolution in science by themselves. No sensible scholar in the seventeenth century would have given up a belief in Aristotle's cosmology only because some of its predictions had been disproved. Moreover, Galileo did not explain *why* objects fall and move as they do; he was providing only a *description* of these motions. In technical terms, Galileo provided what is known as the *kinematics* (description) of motion, not the *dynamics* (causes) of motion. Together, kinematics and dynamics form the field of *mechanics* in physics.

Still, even without going further, Galileo's understanding of free-fall motion helped to prepare the way for a new kind of physics, and indeed for a completely new cosmology, by planting the seeds of doubt about the basic assumptions of Aristotle's science and providing an alternative description.

The most disputed scientific problem during Galileo's lifetime was not in mechanics but in astronomy. A central question in astronomy was whether the Earth or the Sun was at the center of the Universe. Galileo supported the view that the Earth and other planets revolved around the Sun, a view entirely contrary to Aristotle's cosmology in which the Earth was at the center and the Sun and planets revolved around it. But the new astronomy required a new physical theory of why and how the Earth itself moved. Galileo's work on free fall and other motions turned out to be just what was needed to begin constructing such a theory. His work did not have its full effect, however, until it had been combined with studies of the causes of motion (forces) by the English scientist Isaac Newton. As Galileo had done earlier, Newton combined reason, mathematics, and experiment into the extremely capable tool of research in physics that we have today. Galileo himself foresaw that this was going to happen. In his book *Two New Sciences*, after summarizing the new method, Galileo had his alter ego Salviati, proclaim:

> We may say the door is now opened, for the first time, to a new method fraught with numerous and wonderful results which in future years will command the attention of other minds.

SOME NEW IDEAS AND CONCEPTS

acceleration
average speed
dynamics
Galileo's law of free fall
instantaneous speed
kinematics
mechanics
scalar
speed
thought experiment
uniform
vector
velocity

SOME FORMULAS FOR MOTION AND THEIR MEANING

This list is provided only as a reference. It is not intended that you merely memorize these formulas; but you should know what they mean.

Formula	Meaning
$\Delta d = d_{\text{final}} - d_{\text{initial}}$	The distance traveled is the final position minus the initial position.
$\Delta t = \text{t}_{\text{final}} - \text{t}_{\text{initial}}$	The time interval is the finishing time minus the starting time.
$v_{\text{av}} = \dfrac{\Delta d}{\Delta t}$	The average speed is the ratio of the distance traveled per time interval.
$\text{slope} = \dfrac{\Delta y}{\Delta x}$	The slope of a line on a graph between two points is defined as the ratio of the change in the y coordinate to the change in the x coordinate between the two points.
$v = \lim_{\Delta t \to 0} \dfrac{\Delta d}{\Delta t}$	The instantaneous speed is the limit, as the time interval approaches zero, of the ratio of the distance traveled per time interval.
$a_{\text{av}} = \dfrac{\Delta v}{\Delta t}$	The average acceleration is the ratio of the change in speed and the time interval.
$\mathbf{a}_{\text{av}} = \dfrac{\Delta \mathbf{v}}{\Delta t}$	The average vector acceleration is the ratio of the change in velocity per time interval.
$v = at$	If an object is accelerating uniformly from zero initial speed starting at time zero, the speed at time t is the acceleration times the time.
$\Delta d = \frac{1}{2} a(\Delta t)^2$	For uniform acceleration only and an object starting from rest, the distance traveled is one-half the acceleration times the square of the time interval.
$d = \frac{1}{2} at^2$	For objects accelerating uniformly from rest and from an initial distance and time of zero, the distance traveled during time t is one-half the acceleration times the square of the elapsed time.

FURTHER READING

G. Holton and S.G. Brush, *Physics, The Human Adventure* (Piscataway, NJ: Rutgers University Press, 2001), Chapters 6 and 7.

Web site

See the course Web site at: http://www.springer-ny.com/up.

STUDY GUIDE QUESTIONS*

1. Summarize in writing the main points of each section.
2. Compare the sections on the basic concepts of motion in this chapter with your laboratory work.
3. Write down any concepts or formulas or paragraphs that you do not understand and ask your instructor for help.
4. Look at each of the tables in this chapter. Explain each of these tables to a friend, including the meaning of each row and column and how the numbers were obtained.

Sections 1.1–1.7

1. What is meant by the Scientific Revolution?
2. Who was Galileo and what did he do?
3. Why didn't Galileo continue to write about astronomy during his lifetime?
4. Define in your own words the following terms: speed, velocity, average speed, average acceleration, uniform acceleration, and the symbol Δd.
5. Give an example of each of the concepts in Question 4.
6. What is free fall? What is "free" about free fall? Why is it so difficult to determine if free fall is an example of uniform acceleration?
7. How did Galileo's approach to the problem of motion differ from that of Aristotle and of the Aristotelians in his own day?
8. Which of the following properties do you believe might affect the observed rate of fall of an object: weight, size, mass, color, shape, or density.
9. A ball is rolling on the floor with a constant speed of 130 cm/s. It starts rolling at time 0.0 s and at distance 0.0 cm.
   (a) Calculate where it would be at every one-tenth of a second up to 1.0 s. Construct a table of distance and time to show your results.
   (b) Add additional columns to your table showing the distance covered in each time interval, and the average speed in each time interval.

---

* These questions are intended as an aid to study. Your instructor may ask you to work on these in groups, or individually, or as a class.

(c) Draw a graph of total distance traveled against the elapsed time.
(d) From the table, predict what the slope of the line should be.
(e) Find the slope of the line and compare the result with your prediction. Explain your result.

10. A car is increasing its speed uniformly at the rate of 8.0 km/hr each second. It starts out at 0 km/hr at time 0 s.
    (a) Calculate what its instantaneous speed would be at every second up to 10 s. Construct a table of speed and time to show your results.
    (b) Add additional columns to your table showing the increase of speed in each time interval, and the average rate of increase in each time interval.
    (c) Draw a graph of instantaneous speed against the elapsed time.
    (d) From the table, predict what the slope of the line should be.
    (e) Find the slope of the line and compare with your prediction. Explain your result.

Sections 1.8–1.10

1. What was Aristotle's theory about falling objects?
2. What did Galileo claim would happen if he dropped two different heavy objects simultaneously from the same height?
3. Why was Galileo slightly wrong? Why didn't this force him to change his mind?
4. If you drop a sheet of paper and a book simultaneously from the same height, they don't reach the ground together. Why not? Could one change the experiment so that they do reach the ground together?
5. What is the definition that Galileo chose for uniform acceleration?
6. What was Galileo's hypothesis about free fall?
7. Why couldn't Galileo test his hypothesis directly?
8. What procedure did he use to test it indirectly?
9. What is Galileo's law of free fall and how did he arrive at this conclusion?
10. What are some of the consequences of Galileo's work?
11. What are some important elements that go into scientific research? How are they represented in Galileo's work?

DISCOVERY QUESTIONS*

†1. Why do you think motion is so important to understanding nature?

†2. During the course of one day notice the different types of motion that you encounter and write down your observations in a notebook. Try to classify these motions according to the type of motion involved: uniform speed, uniform velocity, accelerated motion, or combinations of these.

---

* Some of these questions go beyond the text itself.

† These may be performed before or during the reading of this chapter.

3. Which of the two graphs below, for two different objects, has the greater slope.

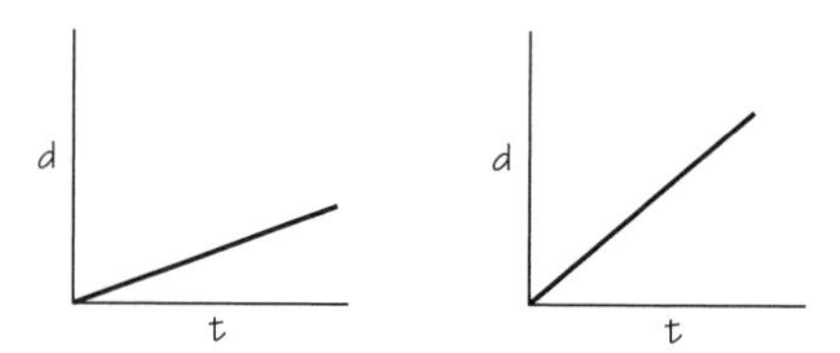

4. Explain the difference between average speed and instantaneous speed.
5. Shown below are graphs representing different motions of a disk on a tabletop. Explain what is happening in each case, and how you arrived at your answer. The axes and the slope of the lines will help to answer this question.

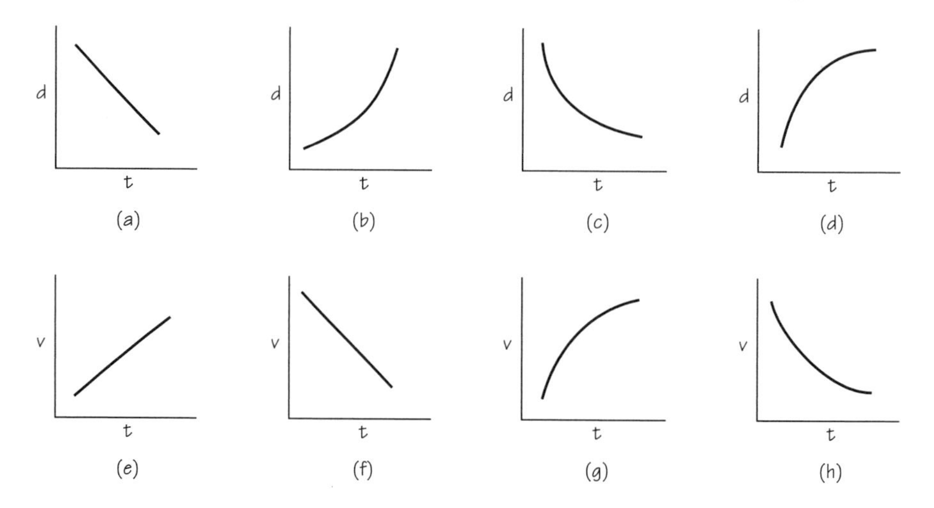

6. Make up a story to go along with the motion that is represented by the distance–time graph below.

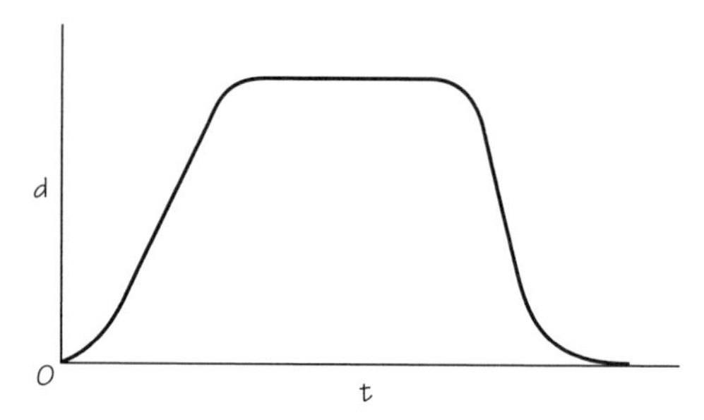

7. Galileo wrote: "A motion is said to be uniformly accelerated when, starting from rest, it acquires during equal time intervals, equal increments of speed." Using modern symbols, this says: uniform acceleration involves equal values

of Δv in equal time intervals Δt. Which of the following are other ways of expressing the same idea?
(a) Δv is proportional to Δt;
(b) $\Delta v/\Delta t$ = constant;
(c) the speed–time graph is a straight line;
(d) v is proportional to t.

8. Suppose Galileo performed his famous inclined plane experiment on the Moon, where the acceleration owing to gravity is less. Would he have obtained the same law of free fall?

Quantitative

1. Time yourself in going from one class to another, then the next time pace off the distance to obtain a rough estimate of the total distance. Find your average speed from these results.
2. A certain person who walks for exercise usually walks 2 mi (3.23 km) in about 28 min. What is his/her average speed in kilometers per second and miles per hour?
3. Amtrak runs a high-speed train, the Acela Express, between New York and Washington. The train can cover the distance of 194 mi in 2 hr, 43 min. What is its average speed?

Amtrak Acela Express.

4. (a) A world record for the mile, set in 1999, was 3 min, 43 s. What was the average speed of the runner in miles per hour?
   (b) The cheetah is the fastest animal on Earth. It can run at a sustained average speed of 80 mi/hr. What would be its time for 1 mi?

5. The winner of the last stage of the 1999 Tour de France, a French cyclist, covered the distance of 143.5 km from Arpajon to Paris in 3 hr 37 min 39 s. What was his average speed in kilometers per hour?

6. What is the average acceleration of an airplane that goes from 0 to 100 km/hr in 5 s during take off? What is the average acceleration during landing, going from 200 km/hr in 12 s?
7. (a) A rollerblader starts out at the top of a hill with a 200 m slope. She rolls, without skating, to the bottom of the hill in 15 s. What was her acceleration?

(b) What was her speed at the bottom of the hill?
(c) The rollerblader skates halfway back up the hill, then rolls down to the bottom. What was her acceleration on the way down? What was her speed at the bottom?

8. (a) Light waves travel through a vacuum in a straight line at a speed of about 3×10^8 m/s. How far is a "light year"?
   (b) The nearest star, Alpha Centauri, is about 4×10^{16} m distant from Earth. This star does not have any planets, but if it did and intelligent beings lived on them, how soon, at the earliest, could we expect to receive a reply after sending them a light signal strong enough to be received there?
9. The speed of sound in air is about 330 m/s (0.33 km/s). Suppose lightning strikes 1 km away. How long does it take before you see the flash? (The speed of light is given in 8a.) How long before you hear the thunder? On the basis of your results, can you set up a rule for determining how far away a lightning flash occurred if you count the seconds between the sight of the flash and the arrival of the sound of thunder?
10. Ask a friend who has access to a car to drive with you on a quiet stretch of straight road with little or no traffic or pedestrians. Take data on the car's speed. After attaining a speed of about 30 mi/hr, ask that the car slow down and record the speedometer readings in 5-s intervals.
    (a) After returning, create a table of speed and corresponding time increments.
    (b) Add the following columns to your table: the change in speed Δv, the time interval Δt, and the average acceleration in each time interval $\Delta v/\Delta t$.
    (c) Create a graph of speed and time for the motion of the car. Find the average acceleration of the car from the graph. How does it compare with the results from your table?

CHAPTER 2

Moving the Earth

2.1 ASTRONOMY, MOTION, AND MECHANICS

Galileo's discoveries about motion formed a major part of a much larger development across all of the sciences, a development now known as the *Scientific Revolution*. In the study of the physical world, the science of motion, or mechanics, joined with the science of astronomy to form the basic approach to modern physics. Paralleling the revolution in mechanics, the revolution in astronomy involved an extremely difficult transition for most people from the common-sense view of the Universe in which the Earth is stationary at the center of the Universe to our current, more abstract, view that the Earth is actually spinning on its axis as it orbits around a star, our Sun, as the third planet. Since the Earth was now seen as a moving object, the revolution in mechanics helped to encourage the revolution in astronomy, and vice versa. This chapter looks at the parallel developments in astronomy, before turning to the *causes* of motion in the next chapter.

Once again, Galileo played a key role in terms of both the new science and the new issues that it raised.

2.2 THE SCIENTIFIC REVOLUTION

Think back to what you may have learned in a social studies or history course about the period from A.D. 1550 to about 1700. This was the period of the Renaissance—the word for "rebirth" in French—that spread out from Italy across the Western world. The Renaissance movement brought new forms of art, music, and courageous new ideas about the Universe and humanity's place in it. Curiosity and a questioning attitude became acceptable, even prized.

The art of Botticelli, Rembrandt, and other great masters showed an enthusiasm for exploring the natural world that paralleled a similar enthusiasm in science and the actual explorations and discoveries of the seafaring explorers. New instruments, such as the telescope and the microscope,

FIGURE 2.1 Rembrandt's *The Three Trees* (1643).

opened up new worlds that had never been seen before. New devices, such as the barometer, thermometer, vacuum pump, and mechanical clock, enabled more sophisticated experimental research, and the invention of the printing press enabled the rapid dissemination of a researcher's works to an ever-growing audience. Among these works were also the newly discovered writings of such important Greek thinkers as Plato and Archimedes. Many of these were first encountered by Europeans in Arabic translation during and after the Crusades. In awe of what they found, European scholars eagerly translated and studied these ancient works, both as curiosities and as alternatives to Aristotle. Within a few generations there arose a new ideal of humankind, the confident individual full of curiosity and the joy of living.

Along with the new enthusiasm for learning about the natural world came a new freedom of thought encouraged by the Protestant Reformation and a new freedom from economic and social constraints encouraged by the rise of a new commercial middle class. The growth of cities as commercial centers in Europe and the breakdown of the hierarchical feudal order in society enabled the rise of a middle class that could afford to send its sons to the universities to learn about the heavens, instead of sending them to the fields to work from dawn to dusk. The growing numbers of these young men (there were only a few women in science at that time), their growing science, and their growing economic and cultural impact led to the founding of state-sponsored societies and academies of science. Here, amidst debate and critical peer review, the new scientists established the methods and the content of today's physics and many other contemporary sciences. From Europe the new sciences and the new scientific approach quickly spread throughout the entire world, constantly growing and progressing ever since through the contributions of many different cultures and peoples around the globe.

Let's look at one of the beginnings of the new physical sciences: astronomy.

2.3 COPERNICUS

Within the emerging Renaissance culture lived a Polish church official and mathematical astronomer named Nikolaj Koppernigk (1473–1543), better known as Nicolas Copernicus. Copernicus became famous for presenting the first viable, quantitative argument for the so-called *heliocentric theory* of the Sun and planets, which we accept today. In the heliocentric theory (from the Greek word "helios," for sun), the Earth is not the center of the

FIGURE 2.2 Nicolaus Copernicus (1473–1543).

Universe, but instead it and all the other planets orbit the Sun. The Earth orbits the Sun in one year, while at the same time the Earth rotates on its axis once a day. Copernicus's ideas were so revolutionary at the time that his work is often known as the Copernican Revolution.

Copernicus was a student in Poland when Columbus discovered America in 1492. An outstanding astronomer and mathematician, he was also a talented and respected churchman, jurist, diplomat, physician, and economist. During his further studies in Italy he had read the newly discovered writings of Plato and other early philosophers and astronomers. Plato provided a welcome alternative to Aristotle for those seeking new answers. What better way to challenge one old master (Aristotle) than with another (Plato)!

As a canon (priest) at the Cathedral of Frauenberg, Copernicus was busy with church and civic affairs, but he found time to work on astronomy and calendar reform. It is said that on the day of his death in 1543 he saw the first copy of the great book on which he had worked most of his life. It was this book, *On the Revolutions of the Heavenly Spheres*, that opened a whole new vision of the Universe.

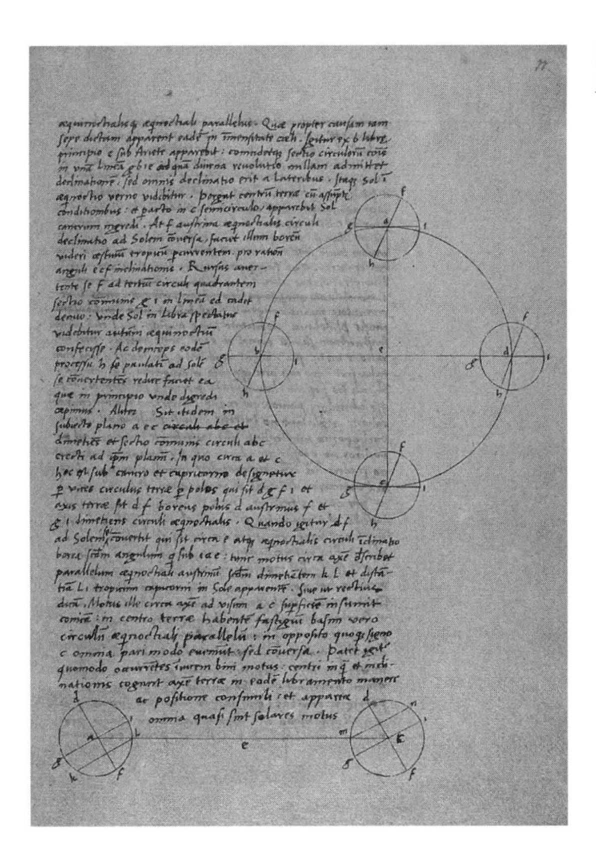

FIGURE 2.3 Page from Copernicus's *On the Revolutions of the Heavenly Spheres.*

We will look briefly at the old vision of the Universe in order to see what was new about Copernicus's new vision.

2.4 THE GEOCENTRIC VIEW

Since we live on Earth and observe the motions of objects in the heavens from Earth, we naturally tend to think of ourselves as being at rest on Earth while everything else is moving around us and that we are at the center of the Universe. Since the heavenly objects appear to move around us on circular paths centered on Earth, we naturally tend to think of ourselves as being at the center of the Universe. This is called the *geocentric view* of the Universe, from the Greek word "geo" for "earth" (from which we also have the words "geology" and "geography"). If you have had an opportunity to study relative motion in the laboratory or to observe the motion of the Sun or Moon, you will know that it is not possible through direct observation to decide if it is we who are at rest and the stars and Sun that are moving,

FIGURE 2.4 Time exposure showing stars' trail around the north celestial pole. The diagonal line represents the path of an artificial earth satellite.

or if it is we who are moving and the stars and Sun that are at rest. For instance, when the Sun "sets" on the horizon, is it the Sun that is moving down below the stationary horizon, or is it the horizon that is moving up to the stationary Sun as the Earth rotates?

No wonder Aristotle's very plausible and well-constructed theory of the Universe was so well received and widely accepted for so long (see the Prologue). But there was one problem with Aristotle's cosmology (theory of the Universe): it was not *quantitative*; it was only *qualitative*. It did not provide a precise, mathematical account of the observed positions and motions of the Sun and Moon and planets, and that is what astronomers really wanted—since astronomy at that time was a considered a branch of mathematics, and astronomers were employed in calculating celestial phenom-

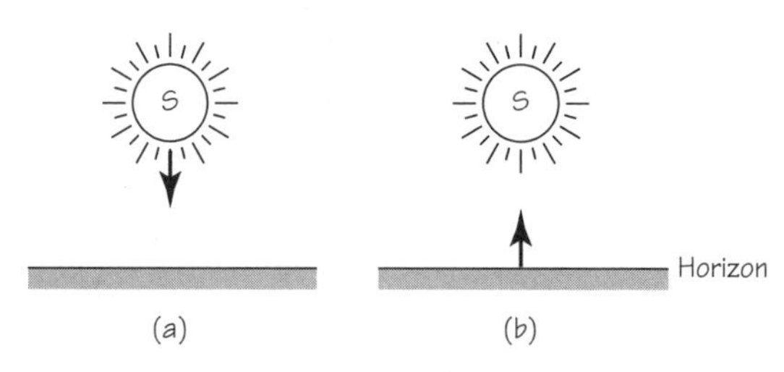

FIGURE 2.5 Sun vs. Earth: which is moving?

ena, reforming the calendar, and performing astrological calculations, which involved a mathematical study of the planets. (However, because it also assumes the influence of the planets on human affairs, an assumption that is not confirmed by the evidence, astrology is not considered a science.)

The Observations

What do you actually see when you look up at the sky?

You see a lot of different celestial objects. Every object in the sky can be located by angles. The two most common angles are the angle above the horizon, called the altitude, and the angle clockwise from due north, the azimuth.

Figure 2.6 depicts what the sky looks like to anyone even today who stands outside, at night for a while, away from lights and buildings: the sky is seemingly a large dome, centered on us. The part we see is part of a *celestial sphere*. If you wait long enough, or come back in an hour, the stars and Sun and Moon (when visible) appear to move slowly from east to west, where they then "set" (go over the horizon).

If you try looking at the positions of the Moon and Sun, each at the same time every day, you will notice something strange. From the Earth, they both seem to circle overhead more slowly than the stars. The Sun seems to fall about 1° behind where it was the day before at the exact same time, while the Moon falls about 12° behind. The next day they are again, respectively, another 1° and 12° behind; after a week the Sun is 7° behind, and so on. How long will it take for the Sun and Moon to slide all the way around a 360° circle and be in line again with the same stars? For the Sun it would take about 360 days, moving at about 1° per day (more precisely,

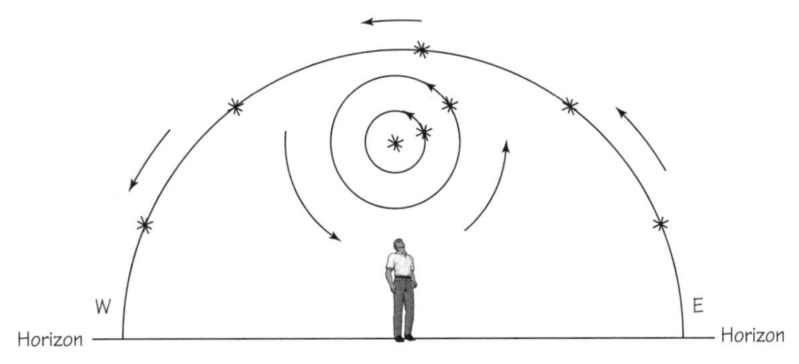

FIGURE 2.6 Seen from the Earth on a clear night, the stars appear to lie on a large dome that rotates slowly from east to west around a fixed star, Polaris.

it is 365.24220 days). In other words, it would take one full year for the Sun to be "lapped" by the stars; indeed, this is how we *define* our year. As seen from the Earth the Moon would take nearly 30 days (more precisely, 29.57333 days) for a complete lunar phase cycle, (or about 27.32152 days as seen from the stars)—about 1 month.

The slipping of the Sun and Moon around the 360° of the celestial sphere suggested to early observers that the dome of the sky is closed, and that it is in fact a sphere. Seen from the Earth, such as outside your building, this dome appears to circle around us once a day from east to west. To the ancients it seemed that the celestial objects *are* actually on the celestial sphere, which rotates around us once every 24 hr, with the Sun and Moon slipping behind in their own way.

The ancient picture of the Universe is presented in Figure 2.7. This shows a closed celestial sphere centered on a spherical Earth with north and south poles and an equator in line with the celestial poles and equator. You may be surprised to see that they thought of the Earth, not as a flat disk, but as a round sphere, just as we do today.

The Round Earth

Contrary to common opinion, it was not Columbus who discovered that the Earth is round when he landed in America in 1492. It was already known to the ancient Greeks, as well as in Columbus's day, that the Earth is round, although the common folk among the sailors on his ships may have believed otherwise. (What Columbus did not know, but which was also known

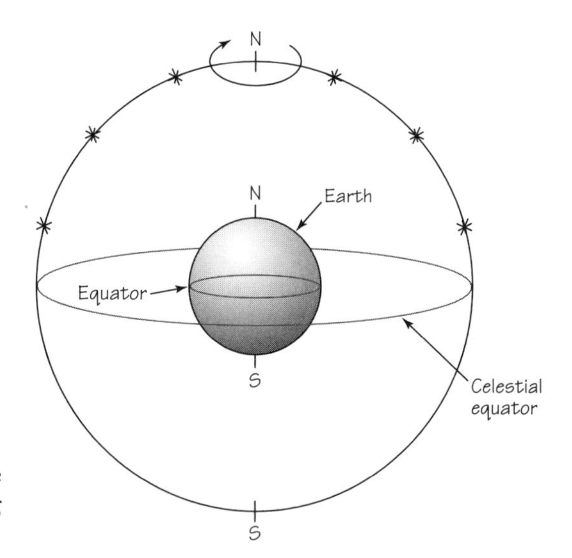

FIGURE 2.7 The celestial sphere and Earth as conceived starting about 500 B.C.

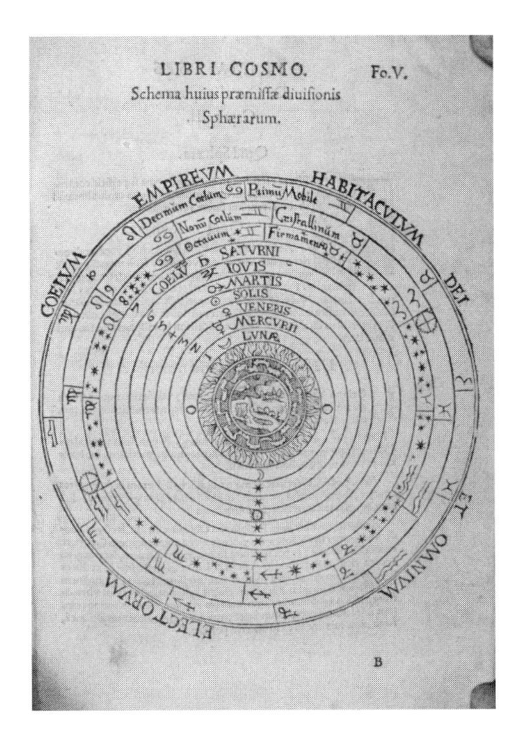

FIGURE 2.8 Geocentric scheme of Petrus Apianus (from his *Cosmographia* of 1551). The Earth is fixed at the center of concentric rotating spheres. The sphere of the Moon ("lune") separates the terrestrial region (composed of concentric shells of the four elements Earth, Water, Air, and Fire) from the celestial region. In the celestial region are the concentric spheres carrying Mercury, Venus, Sun, Mars, Jupiter, Saturn, and the stars.

to the Greeks, is how big the Earth is. Columbus thought it was so small that by sailing west he could reach India more quickly than by traveling east, and thereby find a shorter route for the lucrative spice trade.) Today we know from air travel, satellites, and the impressive pictures of Earth sent back to us from the Moon and outer space that the Earth, our home, can be thought of as a beautiful, round blue ball. But even without airplanes and space flight or sailing trips around the world, we can tell that the Earth really is round—and in the way known to the Greeks over 2000 years ago.

As seafarers, the Greeks knew that when a ship leaving a harbor reaches the horizon, it does not fall over the horizon, but rather it sails out further into the ocean. From the harbor the ship appears to "sink" into the ocean, suggesting that the ship simply goes over the convex curvature of the Earth (see Figure 2.10). Those who traveled in the north–south direction by land or sea noticed that as they traveled north from the equator, the north star (Polaris), which remains fixed over the north pole, seemed to rise in the sky. As Figure 2.11 shows, this could happen only if the Earth is curved outward in the north–south direction. Taking all of these observations into account, people concluded that the Earth must be spherical in shape. These same observations can be made today.

Even though the stars, Sun, and Moon appear to circle around us on the round Earth, someone could still argue that this is just an illusion. The Sun

FIGURE 2.9 Earthrise over moon.

and Moon and stars are really fixed, and it is *we* who are rotating on a north–south axis through the spherical Earth once every day. But there were strong arguments against this interpretation (even though we now know it is the correct one). For instance, common sense seems to tell us that it's impossible for the Earth to be rotating. If the Earth really is a sphere of about 4000 miles in radius (as the Greeks knew), then you can figure out (details in Chapter 3) that people at the equator must be spinning at about 1000 mi/hr through space! Why aren't they hurled off into space, along with assorted animals, rocks, trees, houses, etc? No, common sense seems to agree with the observation that the Earth really is stationary at the center of the Universe and that a celestial sphere carrying the stars and other objects is rotating around it. Anyone who argues differently

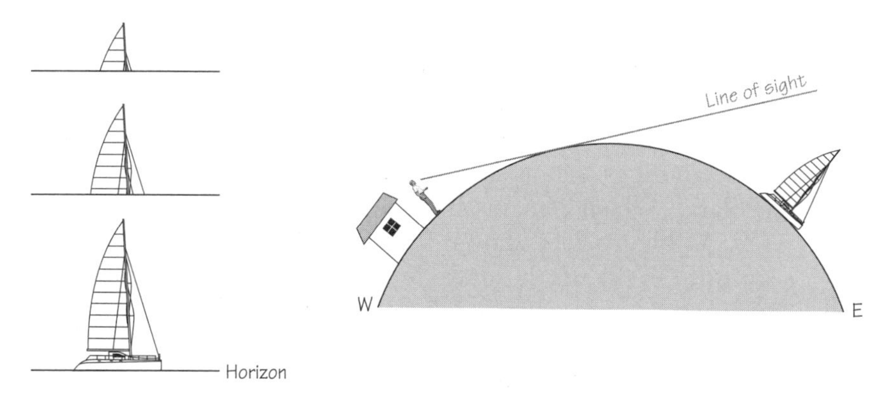

FIGURE 2.10 Ship "sinking" off horizon.

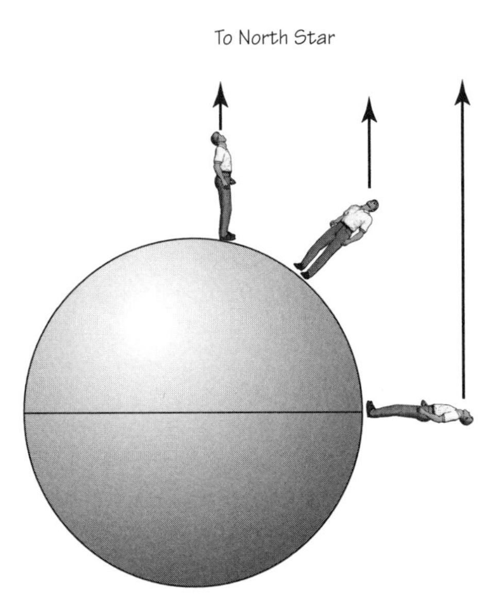

FIGURE 2.11 Observer looking upward towards North Star at three different positions.

had better have some very good arguments, if they want to convince people that it is really the Earth that is moving and the Sun that is stationary.

The Sun and Seasons

Since each day the position of the Sun, compared with the position of the stars just before sunrise or sunset, appears to have slipped behind the stars 1° per day (actually this observation arises from the Earth's motion around the Sun), the Sun appears to move around the celestial sphere once every year in the backward direction, that is, from the west to the east. This backward motion (relative to the stars) is called "eastward drift." The path—called the *ecliptic*—that the Sun traces out on the celestial sphere is not aligned with anything. Rather, the Sun follows a circle tilted at an odd angle, 23.5°, to the circle formed by our equator and the equator of the celestial sphere, as shown in Figure 2.12. The existence of this tilt is extremely important. It is the origin of the seasons.

Let's look at the motion of the Sun from a place on the northern hemisphere of the Earth (the seasons described here will be reversed for the southern hemisphere). For simplicity we will briefly adapt the Earth-centered view. We will follow the apparent path of the Sun along the path shown in Figure 2.13. At position VE, the Sun is just crossing above the equator, and its rays are hitting equally the northern and southern hemispheres of the Earth. There are exactly 12 hr of daylight and 12 hr of night. The time of year when this happens is called an *equinox*, meaning "equal

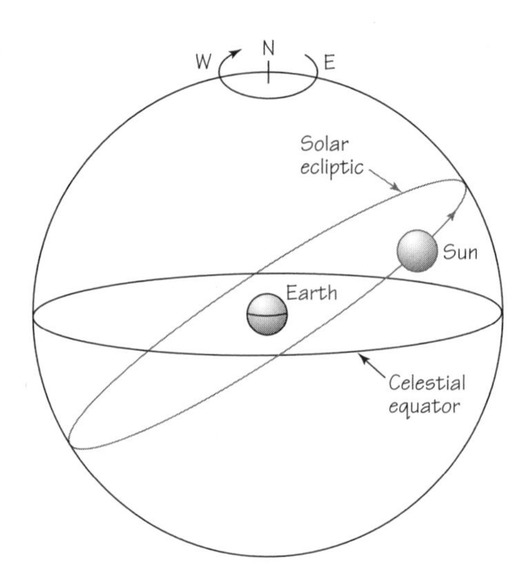

FIGURE 2.12 Solar ecliptic vs. celestial equator.

night." Since the Sun is headed higher in the sky above, it passes through the position VE on the celestial sphere; the position VE is called the *vernal (or spring) equinox*. It is the first day of spring, and the beginning of the old astrological calendar year, according to which nature is born anew every spring.

As the Sun appears to move gradually toward the point labeled SS, the days get longer and the nights shorter. The angle of incidence of the Sun's rays on the ground below become more direct and thus cause more heat. When the Sun reaches the position labeled SS, that moment is called the *summer solstice*; it is the first day of summer. From there the days start to get shorter and the nights longer as the Sun moves toward the position AE, the *autumn equinox*. Finally, winter arrives as the Sun travels to its lowest point in the sky at WS, the *winter solstice*, the first day of winter. Now the rays of the Sun at noon come in at

> This description is for observers in the northern hemisphere. For observers south of the equator, exchange "north" and "south."

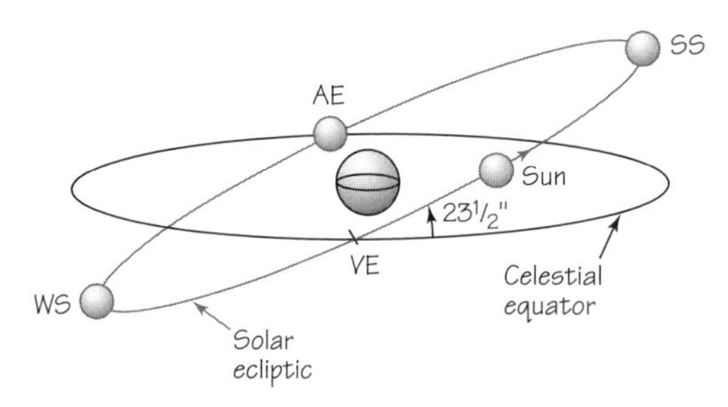

FIGURE 2.13 Diagram showing equinoxes.

the most oblique angle, and leave the ground far less hot than before. The ancient tribes made urgent sacrifices to the Sun at this time in order to induce the sun god not to let the Sun disappear altogether. Apparently the sun god was pleased enough every year to allow the Sun to continue on its journey back to the point VE, where all of nature burst forth once again into a new annual cycle of birth, life, and death.

Notice from Figure 2.13 that at the summer solstice, the point SS, the Sun would be directly overhead at noon to anyone who is 23.5° north of the equator. At the winter solstice, WS, it would be directly overhead at noon to anyone at 23.5° south latitude. The region between these two latitudes is called the *tropic zone*. For those who do not live within the tropic zone, the Sun is never directly overhead at any time during the year. Instead, seen from the northern hemisphere, the Sun reaches its highest point in the sky at noon on the summer solstice, SS, and its lowest point in the sky at noon on the winter solstice, WS. (Your laboratory and demonstration activities may involve your making some of these observations yourself.)

Eclipses and the Phases of the Moon

As discussed earlier, the Moon shares the general east-to-west daily motion of the Sun and stars. But (owing to its own motion around the Earth) the Moon also slips eastward against the background of the stars. It does so much faster than the Sun does, rising each night nearly 1 hr later. When the Moon rises in the east at sunset (opposite the setting Sun in the west), it is a bright, full disk—the "full Moon." Each day after that, it rises later and appears less round. Finally, after about 14 days, it has waned to a thin crescent low in the sky at dawn; when the pale Moon is passing near the bright Sun in the sky and rising with it, you cannot see the Moon at all. At this point it is called "new Moon." After new Moon, you first see the Moon as a thin crescent low in the western sky at sunset. As the Moon moves eastward from the Sun, it gradually fattens until it reaches full Moon again. After full Moon, the cycle repeats itself.

As early as 380 B.C., Plato recognized that the phases of the Moon could be explained by thinking of the Moon as a globe reflecting sunlight and moving around the Earth in about 29 days.

The Moon's path around the sky is very close to the yearly path of the Sun; that is, the Moon is always near the ecliptic. But the Moon's path tilts slightly with respect to the Sun's path. If it did not, the Moon would come exactly in front of the Sun at every new Moon, causing an eclipse of the Sun, a "solar eclipse." In addition, it would be exactly opposite the Sun at every full Moon, moving into the Earth's shadow and causing an eclipse of the Moon, a "lunar eclipse."

(a) (b) (c)

FIGURE 2.14 The Moon as it looks 26 days after New Moon (left); 17 days after New Moon (middle); and three days after New Moon (right). Courtesy of Mount Wilson Palomar Observatory.

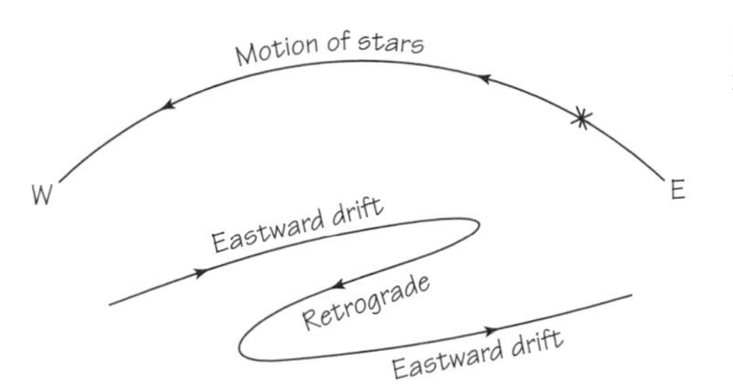

FIGURE 2.15 Diagram of retrograde motion of a planet.

The Planets

Five other objects were also observed on the celestial sphere in antiquity, in addition to the Sun, Moon, and fixed stars. These objects seemed to be stars since, to the naked eye, they appeared as pinpoints of light with no visible disk. But even though they looked like stars, they did not act like stars. As observed over a number of days, a planet, like all other celestial objects, does rise in the east and sets in the west each day, but each day in its rising and setting it falls a few degrees to the east behind the fixed stars, as do the Sun and Moon. This is called "eastward drift." Moreover, occa-

FIGURE 2.16 Babylonian clay tablet recording the observed positions of Jupiter.

FIGURE 2.17 Stonehenge, England.

sionally something very strange happens; the planet appears to speed up relative to the stars and begins moving faster than the stars toward the west, before it settles back into drifting once again toward the east each day. This sudden motion to the west is called "retrograde motion," since the planet seems to be regressing from its eastward drift. In this strange type of motion, the planet forms a looping or S-shaped path against the background of the stars. This apparent motion is now understood, in our present system, in which the Earth and planets orbit the Sun, as an optical illusion (see Figure 2.19b). It occurs whenever the Earth in its orbit passes an outer planet, or an inner planet passes the Earth in its orbit. Because of this seemingly strange behavior, these five objects were called *planets*, which is Greek for "wandering stars." We know today that they are not stars at all but large masses orbiting the sun just like the Earth.

Since the planets seemed to perform their puzzling motion entirely on their own, which only living beings can do, the ancients believed them to be living gods (since they also seemed to be eternal). They also believed that these planet-gods were influential on human affairs. According to the appearance and motions of each planet, the Greeks named the five visible planets for the gods Mercury, Venus, Mars, Jupiter, and Saturn. (There are three other planets, in addition to Earth—Uranus, Neptune, and Pluto. They are not visible without powerful telescopes. Some astronomers do not regard Pluto to be a planet.)

Although the planets were called "wandering stars," they did not wander at random all over the place. They stayed within a narrow band of constellations along the Sun's annual path along the celestial sphere. The ancient Babylonians, who first identified all of our present constellations, also

FIGURE 2.18 Aztec stone calendar.

identified those that lay at that time on the Sun's annual path, the ecliptic. The Greeks called the set of these constellations the *zodiac*. Since the planets stayed within this narrow band of constellations, the zodiac had special significance for ancient believers in astrology. Beginning with our March 21, the first day of nature's "rebirth" in spring, they divided the zodiac into the 12 astrological signs, or constellations, of the zodiac, corresponding to 12 lunar months of about 30 days each. An astrologer could then relate the positions of the planets as they moved within the zodiac to the position of the Sun within each sign of the zodiac, and then pronounce conclusions about human affairs based upon astrological beliefs. As you can see, an astrologer had to be adept at mathematical astronomy, and he would want to see improvements in mathematical astronomy in order to predict the positions of the planets more precisely. But, as noted earlier, there is absolutely no evidence that the planets actually have any effect at all on human affairs. For this reason, the science of astronomy eventually split from the superstition of astrology during the course of the Scientific Revolution.

FIGURE 2.19 Retrograde motion: Ptolemaic (a) and Copernican (b) views.

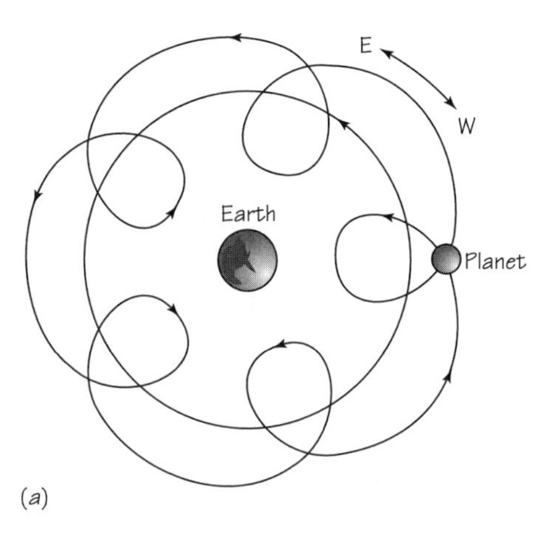

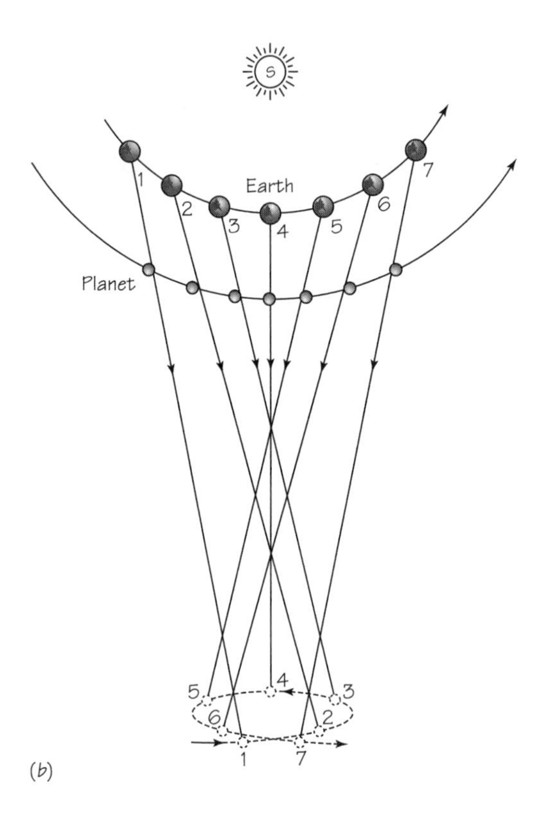

2.5 COPERNICUS VERSUS PTOLEMY

Long before the time of Copernicus, the Greek astronomer Claudius Ptolemy, who lived in Egypt during the second century A.D. and was related to the Ptolemy dynasty of pharaohs who ruled Egypt at that time, created a mathematical theory, or model, of all of the observed celestial motions outlined in the previous section. In fact, he nearly succeeded in reproducing the exact observations of the celestial motions on the basis of a geocentric model of the Universe. Ptolemy saw his work as a solution to Plato's problem, discussed in the Prologue: to provide a model of the Universe in which all of the observed motions can be explained by referring to combinations of perfect circles rotating with uniform speed.

For instance, Ptolemy's explanation of the so-called retrograde motion of the planets is shown in Figure 2.20. He explained the apparent "looping motion" of the planets by placing the center of one rotating circle, called the "epicycle," which carried the planet, on another rotating circle, called the "deferent," so that together the motions of the two circles, the *epicycle-deferent*, produced the observed looping motion of the planet. By choosing the proper circles and speeds, Ptolemy could reproduce these and other observed motions almost exactly! Moreover, the model accounted for the observation that each planet looks nearer to us (bigger and brighter) while in retrograde motion than when it's not in retrograde, since it is closer to us when it is on the inside of the larger circle.

But think about Ptolemy's theory, indicated in Figure 2.20, as a theory of nature. Is it what you would expect for the actual motion of a planet? Would anyone actually believe that this is the way the planets *really* move? Isn't this just an invention to reproduce the observations without worrying if these are actual motions? Also, aren't there some unresolved problems with this theory, such as: What holds the planet on the epicycle, and what holds the epicycle on the deferent? How can the epicycle cut through the deferent

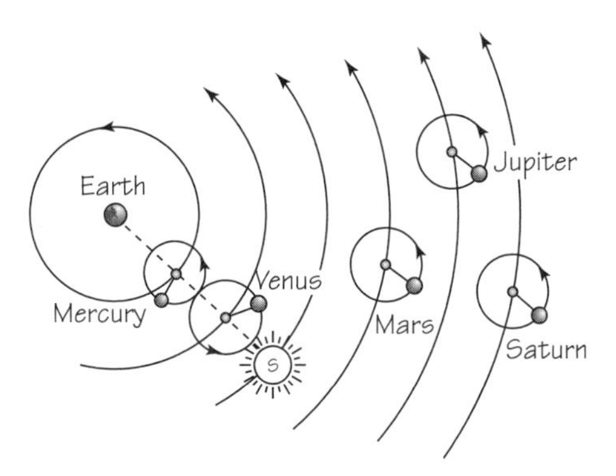

FIGURE 2.20 Simplified representation of the Ptolemaic system.

without breaking it? Moreover, shouldn't all of the circles actually be centered on the Earth? Is this really a satisfactory solution to Plato's problem?

Copernicus did not think so. Raising some of these very same questions in his book, he wrote: "Hence a system of this sort seemed neither sufficiently absolute nor sufficiently pleasing to the mind." Even though it worked, it didn't seem "real" (absolute) to him. Nor was there the harmony, simplicity, or "beauty" to Ptolemy's system that Plato had believed to exist in nature. It wasn't "pleasing to the mind." So Copernicus began looking for an alternative. Apparently he found encouragement in an alternative that had already been considered and rejected in ancient times because it was not fully worked out. Copernicus revived an old proposal (attributed to the Greek Aristarchus, ca. 281 B.C.) in which it is the Earth that moves, while the Sun and stars remain fixed. And he fully developed it into the first viable alternative to Ptolemy's system in 1400 years! In its basic features, it is the system we use today.

Copernicus's Alternative

Instead of interpreting the motions of the celestial objects as revolving around the fixed Earth each day from east to west, Copernicus realized that these observed motions would appear exactly the same for an observer on the Earth *if the Sun and stars are stationary and it is the Earth that rotates on its axis once a day from west to east.*

Copernicus had discovered an ambiguity in the concept of relative motion. Think about this for a moment. As suggested earlier, we can see the Sun rise in the east every morning, but is it the Sun that is rising from the fixed eastern horizon, or is it the horizon that is falling away as the Earth rotates, while the Sun remains fixed? Ptolemy and most people believed the former, Copernicus argued the latter. The Earth, he stated, is rotating west to east while the Sun remains stationary in space. Because the apparent motion of the Sun is an example of relative motion—motion that is relative to an object that itself may in fact be in motion—all of the celestial observations would be exactly the same in either view.

In addition, instead of placing the Sun on a circle centered on the Earth, he realized that the annual path of the Sun against the background of the stars could be obtained equally well by *placing the rotating Earth on a circular orbit that revolves around the stationary Sun once every year*. Our seasons, as caused by the annually changing position of the Sun in the sky, he said, are due to the tilt of the Earth's axis from the perpendicular to the plane of its orbit—not the tilt of the Sun's path (the ecliptic) from the perpendicular to the plane of the Earth's equator. The Earth's tilted axis gives us the seasons. As the Earth orbits the Sun each year, the north pole always

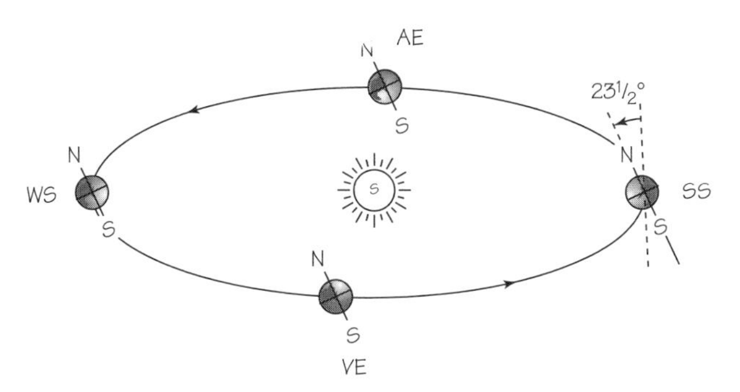

FIGURE 2.21 Copernican view of Earth's revolution about the Sun.

points toward the star Polaris—the "north star"—no matter where the Earth is on its orbit (see Figure 2.21).

In Copernicus's system, only the Moon orbits the Earth, as the Earth orbits the Sun. The Sun remains stationary. (We now know that the Sun too is moving as part of our rotating galaxy, which, in addition, is moving away from other galaxies as the Universe expands. But this does not change the observations within the solar system of planets.) The other planets then also orbit the Sun in the order they are known to have today: Mercury, Venus, Earth, Mars, Jupiter, Saturn (and, as later discovered, Uranus, Neptune, and Pluto). This entire system is called a *heliocentric system* or a *solar system* ("sol" means "sun" in Latin), because the Sun is at the center.

To see how well the new perspective accounts for the observations of the seasons, let's compare the seasons as observed from the moving Earth with the seasons as observed from the perspective of a stationary Earth (review the description in "the Sun and seasons" in Section 2.4). We will be in the northern hemisphere of the Earth and observe the Sun at noon every day as the Earth orbits the Sun once a year, starting from the point labeled VE in Figure 2.13. At this point the Sun's rays are hitting equally the northern and southern hemispheres of the Earth, so this is the Vernal Equinox. In this part of the orbit, as the Earth turns each day, the Sun remains longer in view, and the Sun's rays hit the northern hemisphere more directly. The Earth is heading into summer.

As the Earth moves on its orbit toward the position labeled SS, the north pole of the Earth always remains pointed toward Polaris, so the northern hemisphere begins to tilt toward the Sun: the days grow longer and the nights grow shorter. The position SS is the Summer Solstice, or the first day of summer. From there the days start to get shorter and the nights longer as the Earth moves to AE, the Autumn Equinox. Finally, as the Earth moves into the part of its orbit where the northern hemisphere is tilted away from the Sun, the days get shorter, the nights longer, the rays of the Sun are less direct, and the Sun does not rise as high in the sky as seen by an observer on the Earth. This is winter, and the point WS is the Winter Solstice. After that, the Earth continues on its journey to the Vernal Equinox when the entire cycle is repeated for another year.

You can see from this that, as far as the seasons are concerned, Copernicus's heliocentric theory is just as capable of explaining the observations as is Ptolemy's geocentric theory.

Copernicus's Solar System of planets, including the Earth as the "third rock" from the sun, also provided a wonderfully simple explanation for the puzzling motion of the planets, and without Ptolemy's fictitious circles upon circles. In accounting for retrograde motion, Copernicus replaced all five of the planets' major epicycles with a single circle: the Earth's orbit around the Sun. The numbered lines in Figure 2.19b indicate the lines of sight from the Earth to an outer planet. Most of the time, the outer planet appears to be moving normally on its path against the background of the stars (eastward drift). But the path that we see from the Earth gradually begins to change as the Earth catches up to and passes the outer planet. If you

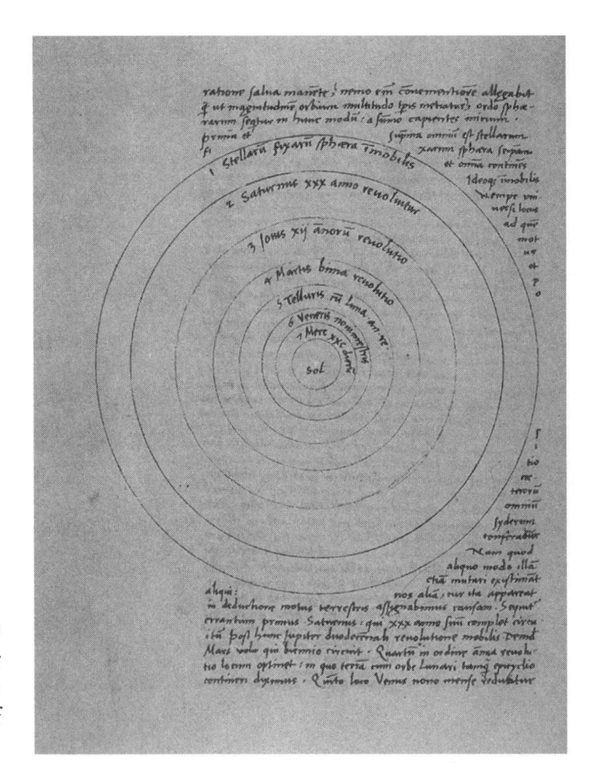

FIGURE 2.22 Page from *On the Revolutions of the Heavenly Spheres* showing the solar system according to Copernicus. Courtesy of Jagiellonian Library, Krakow.

follow the progression of the lines as the Earth passes the planet, you will see a simple explanation of the observed retrograde motion. The planet *seems* to "regress" (move west) for a while then proceed again to the east, creating the appearance of a looping motion, simply because we are passing the planet as the Earth moves along in a different orbit. Since the Earth and the planet are at their closest approach to each other when retrograde occurs, the planet looks bigger and brighter because it is nearer to us. The same thing occurs with the observed retrograde motion of an inner planet, Venus or Mercury, only in this case it is the inner planet that is passing us. (However, it should be noted that, in the end, Copernicus found it necessary to reintroduce small epicycles for each of the planets in order to account for the nonuniformity of each planet's motion around the Sun.)

Once again, the two perspectives involved in relative motion are at work: either the planets can be considered to be moving, while the Earth is at rest, or the Earth can be considered to be moving, while the regressing planet is also moving but at a different speed as seen from the fixed Sun. Without further information, both views are equally valid. As a result, we now have two radically different yet equally capable explanations for one set of observations: Ptolemy's geocentric theory and Copernicus's heliocentric theory. Both of these are equally viable, and so far neither one has a clear advantage over the other.

2.6 ARGUMENTS FOR THE HELIOCENTRIC SYSTEM

You can see how revolutionary Copernicus's ideas were. The centuries-old geocentric system was now, for the first time, under serious challenge by the completely different heliocentric system proposed by Copernicus. Most new theories in science arise from new experimental evidence indicating that the current theory needs to be drastically revised or even replaced by a new theory. However, in this case, there was no new observational evidence leading Copernicus to suggest his new theory. Instead, for him, and for the few others who followed him at first, the most important arguments in its favor were those emphasized centuries earlier by Plato and the Pythagoreans—beauty, harmony, and simplicity.

Like Ptolemy and other mathematical astronomers, Copernicus followed Plato in attempting to account for the observed data in terms of simple mathematical (geometrical) relationships underlying the observations. Copernicus was trying to solve Plato's problem on the basis of motions that were perfectly circular and with perfectly uniform speed, but without resorting to some of the imaginary circles-upon-circles that Ptolemy had introduced to make his theory work. The resulting alternative to Ptolemy's theory could do everything that Ptolemy's theory could do in reproducing the quantitative observations—but no better, nor no worse. Most times scientists have only one theory to account for the observed data; this time they had two equally viable, yet incompatible contenders for the same set of data!

But Copernicus did not stop there. He pointed out what he believed to be the simplicity, harmony, and "beauty" that Plato required of any theory of nature. The theory was simple because Ptolemy's circles-upon-circles and similar contrivances were no longer necessary (except when he got to the finer details of the system); just the rotation of the tilted Earth and the orbits of the Earth and planets were all that were needed to obtain the main observations.

The heliocentric theory had the advantage for Copernicus that it placed the Sun—the symbol of truth and divinity, the giver of light, warmth, and life—in a privileged position. In a statement worthy of a true follower of Plato, some of whose ideas bordered on Sun worship, Copernicus proclaimed in his book:

> In the midst of all, the Sun reposes, unmoving. Who, indeed, in this most beautiful temple would place the light-giver in any other part than whence it can illumine all other parts? . . . So we find underlying this ordination an admirable symmetry in the Universe and clear bond of the harmony in the motion [period] and magnitude [radius] of the spheres, such as can be discovered in no other wise.

The Sun at the center provided a focus to the entire system. As the quotation indicates, there was also a numerical harmony that Copernicus saw in his system when he considered that not only the Earth but all of the other five visible planets also revolved around the Sun. Assuming that all of the observed motions involve primarily perfect circles and uniform speeds, Copernicus used the simple geometry of circles and tangents to determine the radii and periods (time for one revolution) of the orbits of the planets around the Sun as seen from a rotating, orbiting Earth. Analyzing the data on the planets' orbits that had been gathered for centuries, Copernicus used the Earth's day and year as measures of time and the Earth's orbital radius as the unit of distance. Since he didn't know how large the Earth's orbital radius actually is, it could be called one astronomical unit, or 1 AU. This is still used today as a convenient measure for distances within the Solar System. Not only was Copernicus the first to obtain the present-day order of the planets—which Ptolemy did not know—but his results were very close to the modern values for the orbital periods and rel-

FINDING THE RELATIVE RADIUS OF A PLANET'S ORBIT

Let angle *SEP* be the angle of maximum "elongation" of the planet, say Venus, away from the Sun, as viewed from the Earth. Line segment *PE* is then a tangent to the circle of the orbit of Venus. So angle *SPE* must be a right angle. Hence the ratio *SP/SE* must be equal to the sine of the angle *SEP*. Angle *SEP* was observed to be 46°, so

$$\sin 46° = 0.72 = SP/SE.$$

Copernicus defined the distance from the Sun to Earth, *SE*, to be 1 AU. So the distance from the Sun to Venus, *SP*, is 0.72 AU.

A direct observation of the maximum elongation of Mercury or Venus can therefore be used to compute the relative radius of each of these planets. The relative distances of the outer planets from the Sun can be found by a similar, but somewhat more complicated, method.

Modern tables of planetary positions as seen from the Earth are computed from a geocentric model using mathematical techniques that might be considered the equivalent of Ptolemy's epicycles. (See, for example, T.S. Kuhn, *The Copernican Revolution*, pp. 175–176.)

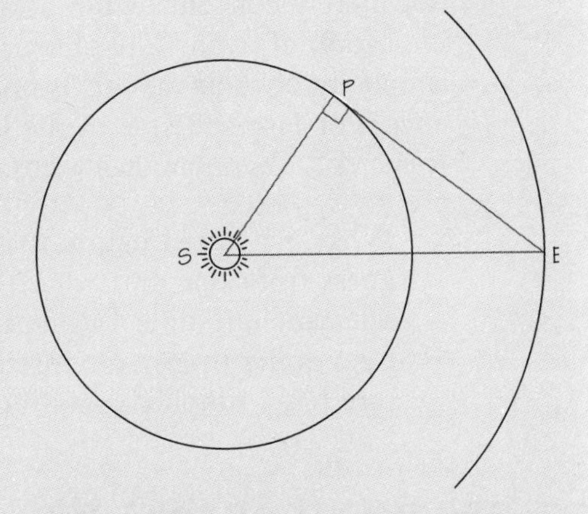

FIGURE 2.23 Method for computing distances.

ative radii of the planets (as well as the Sun and stars at the two extremes), which are given in the table below. (See the insert for an example of how Copernicus obtained the radius of Venus in relation to the radius of the Earth's orbit.)

Object	*Period*	*Radius of Orbit (in AU)*
Sun	0.00	0.00
Mercury	87.97 d	0.39
Venus	224.70 d	0.72
Earth	1.00 yr	1.00
Mars	1.88 yr	1.52
Jupiter	11.86 yr	5.20
Saturn	29.46 yr	9.54

The above table exhibits the harmony that Copernicus so admired in the heliocentric system. Notice that, starting from the Sun, as the relative radii get larger so do the periods for all of the planets, right out to the stars.

This may seem obvious to us today, but for Copernicus, who was the first to notice this relationship, such harmony and simplicity of his system were not merely convenient but also "pleasing to the mind," and therefore an indication of truth. The pleasure which scientists find in the harmony and simplicity of their models is one of the most powerful experiences in science. For instance, one of the leading physicists of the twentieth century, Richard P. Feynman, has written:

> What is it about nature that lets this happen, that it is possible to guess from one part what the rest is going to do? That is an unscientific question: I do not know how to answer it, and therefore I am going to give an unscientific answer. I think it is because nature has a simplicity and therefore a great beauty.*

2.7 ARGUMENTS AGAINST THE HELIOCENTRIC SYSTEM

Not everyone prized the harmony, beauty, and simplicity that Copernicus saw in his new theory. Such aesthetic qualities were not enough to convince anyone who did not think this way. What was also needed was hard evidence about a crucial difference, yet there was no outstanding hard observational evidence that pointed to the heliocentric theory and away from the geocentric theory. Even worse, there were extremely powerful arguments against such a revolutionary idea as a rotating and revolving Earth.

First of all, as noted earlier, simple common sense seems to tell us that the Earth cannot be moving. And a moving Earth would raise a host of excellent questions, such as, if the Earth is spinning on its axis once a day, then why don't objects fly off it? Why aren't there perpetual hurricane-force winds sweeping across the surface of the Earth due to its rotation? Why aren't birds swept from the sky? We now have answers for all of these questions (mainly by using Newtonian mechanics, as we shall see). Copernicus himself attempted to answer some of them, but he worried greatly that, because of such "common-sense" questions, when he published his new theory people "will immediately shout to have me and my opinion hooted off the stage. For my own works do not please me so much that I do not weigh what judgments others will pronounce concerning them." Perhaps for this reason he waited until he was in failing health before publishing his book.

---

* R.P. Feynman, *The Character of Physical Law* (Cambridge, MA: MIT Press, 1967), p. 173.

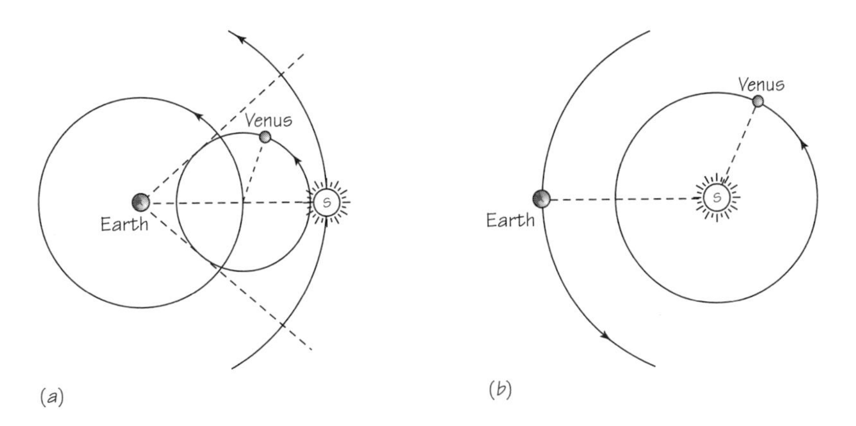

FIGURE 2.24 Changing the frame of reference from Earth to Sun: (a) Venus as seen from a stationary Earth; (b) Venus as seen from a moving Earth in a heliocentric system.

Second, the heliocentric model raised questions about certain natural phenomena. For instance, how does a falling stone know where to land, if the Earth is in constant motion? One of the most cogent of such questions concerned what is known as *stellar parallax*. Why don't the stars appear to shift in the sky if we see them from different angles as we orbit the Sun, much as the angle to an object in the distance shifts as we move our position? If the Earth is indeed revolving around the Sun, as in Figure 2.25c, then the angle representing the position of a star in the sky as seen from the Earth should gradually shift during the course of a year, as the Earth moves along its orbit. However, careful observations made at that time yielded no such shift in the angle of any star; no stellar parallax could be observed. Hence, many concluded, the Earth cannot be orbiting the Sun. Of course, we know today that the Earth is indeed orbiting the Sun, and Copernicus and his followers would have responded with the answer we have today: the stars are so far away from the Earth and Sun that this effect cannot be observed without powerful telescopes. In other words, if Figure 2.25c were drawn to scale, the orbit of the Earth would appear as a dot on the page, and the parallax effect would be impossible to discern. To this argument, Copernicus's contemporaries responded by wondering why God would waste his creative power in creating so much empty space between the Earth and the stars, when much less will do.

Third, in addition to these arguments, there were also major philosophical objections to the heliocentric model. Ptolemy's theory had 1400 years of tradition behind it and Aristotle's Earth-centered cosmology to back it up. Aristotle's entire cosmology, which was the "world view," or

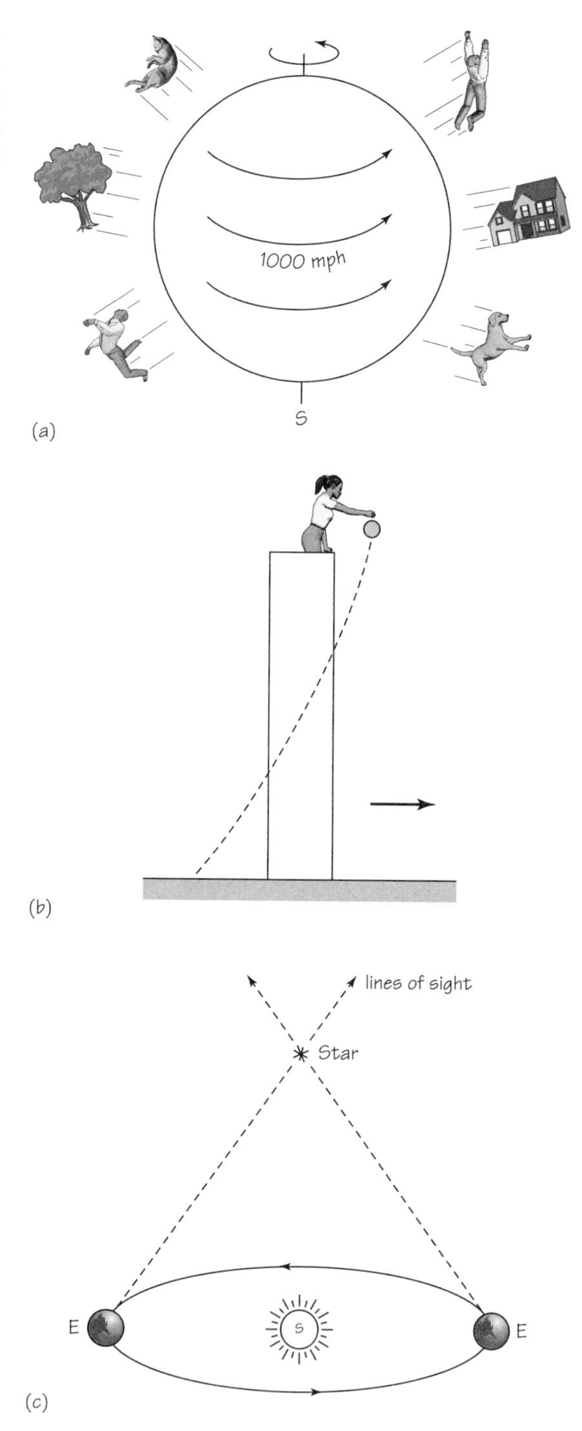

FIGURE 2.25 Arguments against the heliocentric system. (a) refers to the rapidly spinning Earth; (b) what was believed to happen if a ball is dropped from a tower as the Earth moves; (c) the supposed shifting position of a star as seen from the moving Earth.

philosophical outlook, of the day, would collapse if the Earth is not at the center of the Universe and if the celestial sphere of stars and planets is not moving. How will the four elements know their proper place if the Earth is not at the center? Moreover, if the Earth is also a planet, then should the realm of the eternal heavens now be considered imperfect, as is the Earth?

Fourth, there were no clear scientific advantages to Copernicus's theory over Ptolemy's—nor was Ptolemy's theory any better at explaining the observations than Copernicus's theory. No known observation or prediction of planetary position was explained by one system and not by the other. Copernicus had a different viewpoint, as equally well-argued as Ptolemy's, but it was no better than Ptolemy's in predicting the precise positions of the planets, and precision was necessary for the astronomers, astrologers, and calendar-makers of the day. Nevertheless, Copernicus's theory was easier to use, so some astronomers used Copernicus for their calculations but, nevertheless, they continued to believe in the systems of Ptolemy and Aristotle.

Finally, Copernicus's challenge to Aristotle's cosmology also brought his theory sooner or later into conflict with the religious authorities, not only regarding the vastness of outer space. Most learned Europeans at that time recognized the Bible and the Church Fathers on the one hand, and the writings of Aristotle and his followers on the other hand, as the two supreme sources of authority, and both seemed to agree that we humans living on the Earth are at the center of Creation. Copernicus was now attempting to displace the Earth and human beings from the center, rendering the Earth merely the third of six visible planets orbiting a fiery object, the Sun. As a result, many scholars and all of the religious faiths in Europe, including the new Protestants, sooner or later opposed Copernicus. They used biblical quotations (e.g., Joshua 10:12–13) to assert that the Sun and Moon are moving and that the Divine Architect must have worked from a Ptolemaic blueprint. In 1616 (during the Counter Reformation), the Inquisition placed Copernicus's book on the *Index of Forbidden Books* as "false and altogether opposed to Holy Scriptures." Some Jewish communities also prohibited the teaching of Copernicus's theory. Since one of the results of science is a change in knowledge about ourselves and our place in the Universe, it sometimes happens that we don't like what we discover. In fact, often the more we learn about our world, the humbler our place in it seems to be. Many times this results in cultural and religious upheaval and even in attempts to suppress the new theories.

In short, the Sun-centered Copernican scheme was scientifically equivalent to the Ptolemaic scheme in accounting for the observed motions of celestial objects in terms of perfect circles turning at uniform speeds. It had the advantage over the Ptolemaic system in its simplicity, harmony, and beauty, but little else. At the same time, important arguments of science, common sense, and belief were raised against it. The bottom line: There were now two equally viable yet incompatible theories for one set of data; and one theory (Ptolemy's) was in accord with tradition, common sense, and everything people believed at the time about themselves and the Universe in which we live.

Nevertheless, as we have long since discovered, and by arguments we shall encounter later, Copernicus was indeed right:

> The Sun *is* at the center of the solar system, the Earth *does rotate* on its axis once a day, and the Earth and planets *do orbit* the Sun.

But at that time the theory seemed false, absurd, and even dangerous.

2.8 CARRYING FORTH THE REVOLUTION

Following Copernicus's death, many astronomers regarded his heliocentric model as a useful hypothesis for calculations, but most people rejected it

FIGURE 2.26 Tycho Brahe's observatory on the island of Hven.

FIGURE 2.27 The hall containing Tycho Brahe's great quadrant. The rest of the laboratory is shown, with Brahe making an observation (aided by assistants).

as a physical reality. There was no instant revolution. In fact, it took nearly a century and a half of further hard work and brilliant discoveries to dismantle the geocentric theories of Aristotle and Ptolemy and to build up the heliocentric system of today, and with it the physics of today, piece by piece.

Looking back on the period after Copernicus, we can see that Copernicus's ideas about truth and beauty, although important then and now, were simply not enough to induce people to switch suddenly to a radically different point of view, even if, as we now know, it was the correct one. What was needed to accomplish the transition to the heliocentric system was further *scientific evidence*. First, much more precise astronomical data were needed that would enable astronomers to decide between these two theories. Up to that point in time, the available data that had been used for centuries were too imprecise to enable a decision one way or the other.

Fortunately, Tycho Brahe, the greatest observer before the invention of the telescope, was able to provide such data within a few decades after Copernicus's death.

Second, mathematical analysis of Brahe's data was necessary in light of the two competing celestial theories. It was again fortunate that Brahe's data came into the hands of one of the greatest mathematical astronomers, Johannes Kepler. Kepler's lifelong analyses of Brahe's data resulted in new laws of planetary motion that went far beyond Plato's simple assumptions of perfect circles turning at constant speeds.

Third, a new understanding of motion on the Earth was also needed, and this was provided in large part by Galileo and Newton.

Fourth, since science is also a cultural phenomenon, the educated public needed to be brought behind the new theory, and new generations educated into the new theory, and this began to occur through Galileo's popular writings on his sensational discoveries with the telescope.

Finally, we can see that all of these aspects, and many other elements, had to come together into a new alternative for a unified view of nature, to replace the Aristotelian world view. This Isaac Newton provided in one of the greatest scientific books ever written, *Philosophiae Naturalis Principia Mathematica* (1687), usually refered to as the *Principia* or the *Mathematical Principles*, in which he presented a unified theory of the physical world united by the law of universal gravitation. Looking back over his work and that of his predecessors, Newton once wrote to a friend that if he had been able to see farther than others it was because he had been standing on the shoulders of giants.

Let's look closer at some of these giants.

2.9 NEW DATA

Tycho Brahe was born in 1546 of a noble, but not particularly wealthy, Danish family. By the time he was 14 he had become intensely interested in astronomy. Although he was studying law, the studious lad secretly spent his allowance on astronomical tables and books. Later he observed a bright nova, or new star (actually it was a supernova, which is an exploding star). It appeared in the constellation Cassiopeia, then faded over several years. He also observed a comet that he determined to be at least several times farther away from the Earth than the Moon. Aristotle and the ancients had taught that all change must occur below the Moon, while the region beyond the

> Although there were precision sighting instruments, all observations were with the naked eye. The telescope was not to be invented for another 50 years.

FIGURE 2.28 Tycho Brahe (1546–1601).

FIGURE 2.29 A comet like the one Brahe observed in his student days.

Moon is perfect, therefore unchanging. Yet here were two events that indicated that the heavens do change! Evidently at least one assumption of the ancients was wrong; perhaps other assumptions were wrong, too.

As a student, Brahe had read the works of both Ptolemy and Copernicus. Soon he discovered that both men had relied upon tables of observed planetary positions that were not very accurate. He concluded that astronomy needed a complete new set of observations of the highest possible precision, gathered over many years. Only then could a satisfactory theory of planetary motion be created. He decided to devote his life to the task.

With the support of the King of Denmark, Brahe set up the first state-sponsored astronomical observatory, from which he made daily readings of the positions of all of the planets and all visible celestial objects. With state funding, able assistants, and his mechanical skill, Brahe greatly improved the available astronomical instruments, chiefly by making them much larger. This was before the invention of the telescope, so observations did not involve magnifying the observed objects. Instead, the goal was to mea-

FIGURE 2.30 One of Tycho Brahe's sighting devices. Unfortunately, all of Brahe's instruments were destroyed in 1619 during the Thirty Years' War.

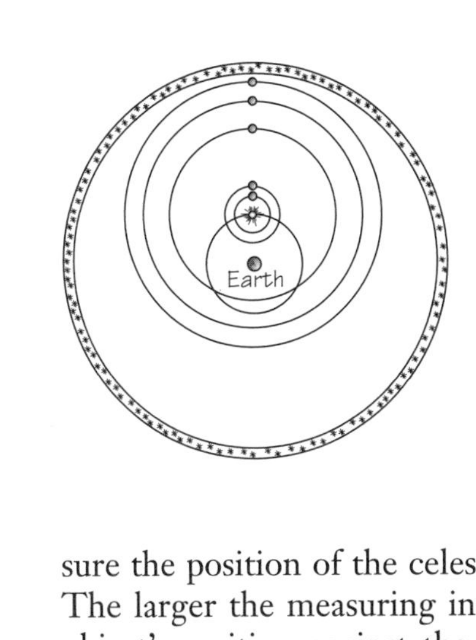

FIGURE 2.31 Main orbits in Tycho Brahe's system of the Universe. The Earth was fixed and was at the center of the Universe. The planets revolved around the Sun, while the Sun, in turn, revolved around the fixed Earth.

sure the position of the celestial objects in the sky as precisely as possible. The larger the measuring instrument, the more precise the angles of the object's position against the background of the celestial sphere could be read. Some of Brahe's instruments were huge. For instance, one of his early instruments was so large that it took several workers to set it into position; but readings with it were accurate to within two minutes of arc. The data Brahe compiled during his decades of nightly observations with these instruments constituted the most accurate measurements of planetary positions ever assembled up to that time.

Four years before his death in 1601, Brahe moved from Denmark to Prague, where the Emperor of Bohemia promised him new support. Not being a mathematician himself, since he was trained in law, Brahe hired a recently graduated German mathematical astronomer, Johannes Kepler, to begin analyzing his years of data in order to decide among the opposing theories of the solar system. There were by then *three* candidates: Ptolemy's geocentric theory, Copernicus's heliocentric theory, and a hybrid "compromise theory" proposed by Brahe himself.

According to Brahe's proposed theory, the planets orbit the Sun, but the Sun orbits the stationary Earth (see Figure 2.31). It was a lawyer's brilliant compromise: it maintained the harmony of the planetary orbits and a privileged position for the Sun, but it also maintained the stationary Earth. Many researchers who liked the harmony of Copernicus's theory, but could not yet accept the moving Earth, welcomed Brahe's compromise as an intermediate theory. It eventually served as a stepping stone from Ptolemy to Copernicus.

When Brahe died unexpectedly in Prague in 1601, a court battle ensued between Kepler and Brahe's heirs over possession of Brahe's lifelong data. Kepler won out in the end. Fortunately for the future of science the data compiled by the world's greatest naked-eye observer now fell into the hands of the world's greatest mathematical astronomer of the day.

2.10 NEW ORBITS

Johannes Kepler (1571–1630) was born into a Protestant German family known equally for its dysfunction as for its mysticism. His mother was accused of being a witch. His father narrowly escaped hanging and later abandoned the family. Kepler attended the University of Tübingen where he studied mathematics, which at that time encompassed mathematical astronomy. Greatly moved by Plato as well as Protestant theology, Kepler became convinced of the Copernican system, in part because he believed that the Sun is the symbol of God and must therefore be at the center of the Universe. Moreover, like Plato, he believed not only that God had used mathematical principles to create the Universe, but also that God and mathematics were in fact identical: "Why waste words?" he wrote. "Geometry existed before the Creation, is coeternal with the mind of God, is God Himself." Kepler became the first astronomer to support Copernicus publicly.

After graduating, Kepler worked as a mathematics teacher, calendar maker, and (to make ends meet) as a court astrologer. He devoted himself to an attempt to discover what he called the "cosmic mystery," the geometrical blueprint according to which God had created the solar system. He wanted to know the mathematical reasons why there are six, and only six, visible planets (the three others being discovered later), and why the planets are in the precise orbits they currently occupy. There must be an undiscovered harmony that accounted for this, he reasoned, since God does nothing by chance. Kepler thought he found the answer to his questions in the five regular solids, also known as the "Platonic solids." Look back at the five regular solids (shown in the Prologue) that Plato had used to account for the five elements. Kepler believed that there are six planets, and only six planets, because God had set up the solar system so that these five solids fit within the five spaces between the six planets. By trial and error, Kepler found that within about 5% accuracy the six planetary orbits could be fit within and around the five solids if they are taken in order of size, as shown in Figure 2.32.

Kepler's idea of using Plato's solids to explain why there are six, and only six, planets (published in 1597) sounds eccentric today, but it was an ingenious one for its time, and it demonstrated Kepler's mathematical capabilities. In fact, Brahe was so impressed by Kepler's abilities that he hired the young man as his assistant. At Brahe's request, Kepler set to work on a careful analysis of the orbit of just one planet, Mars, searching for the geometrical figure

In keeping with Aristotelian physics, Kepler believed that force was necessary to drive the planets along their circles, not to hold them in circles.

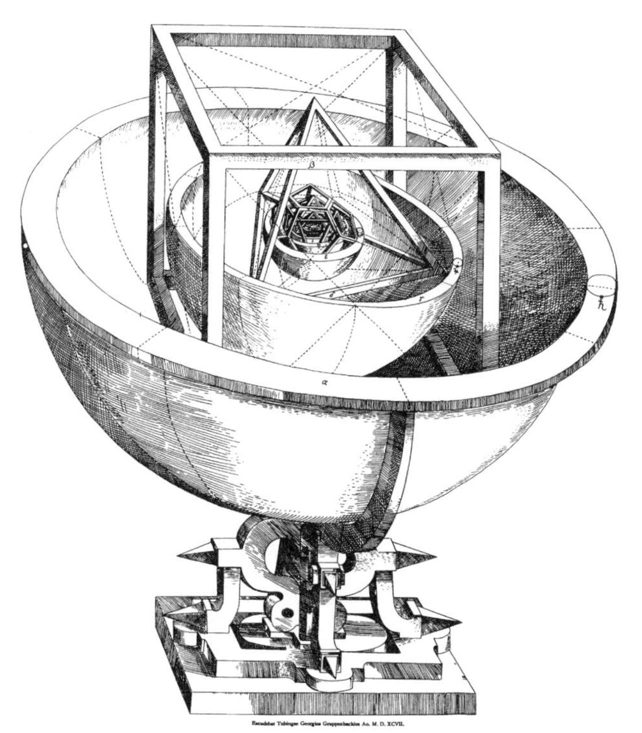

FIGURE 2.32 Kepler's model from *Mysterium Cosmographicum* explaining the spacing of the planetary orbits by means of the regular geometrical solids. Notice that the planetary spheres were thick enough to include the small epicycle used by Copernicus.

that best represented its orbit. Once he found the best fit, he might then be able to apply this geometrical figure to describe better the orbits of the other planets. This would ultimately lead, he hoped, to an understanding of the hidden harmony behind the orbits of the entire solar system.

Kepler naturally began by attempting to fit a circle to the orbit of Mars as seen from the supposed circular orbit of the Earth. He displayed extraordinary tenacity in the work, for after 70 attempts spanning five years, all done by tedious pen-and-paper calculations (of course, no hand calculators or computers in his day!), he was still no closer to the solution. In a book on the eventual solution, titled *New Astronomy*, in 1609, he painfully described every dead-end for his readers, then he wrote:

> If thou, dear reader, art bored with this wearisome method of calculation, take pity on me who had to go through with at least seventy repetitions of it, at a very great loss of time; nor wilst thou be surprised that by now the fifth year is nearly past since I took on Mars.

Fortunately, Kepler had made a major discovery earlier that was crucial to his later work. He found that the orbits of the Earth and other planets were in planes that passed through the Sun. Ptolemy and Copernicus required special explanations for the motion of planets north and south of the ecliptic, but Kepler found that these motions were simply the result of the orbits lying in planes tilted to the plane of the Earth's orbit.

The difficulty for Kepler was that he could make the orbit almost fit a circle—*but not quite*. It fit a circle to within an accuracy of about 8′ of arc. This was better than the fit to any previous data, but Kepler knew that Brahe's data were even more accurate than this, to within 2′ of arc. To his credit, Kepler reluctantly gave up his commitment to circular orbits, and with it over twenty centuries of tradition. After further tedious calculations, Kepler finally arrived at a result that was so universal that it is called a "law of nature."

A *law of nature* is different from a *theory*, since a theory encompasses data and assumptions and hypotheses that can in principle be altered and improved or abandoned as new data and ideas become available. A scientific law is a statement about nature. It can be accepted or rejected, and sometimes expanded to include other circumstances, but it does not contain hypotheses or assumptions. Kepler's laws of planetary motion are still valid today for planets, moons, satellites, and any object orbiting any other object under the action of gravity. But these laws must be modified slightly to take into account additional effects due to gravity and relativity theory.

The First Law

What Kepler discovered from his analysis of Brahe's data is that the planets do not orbit on circles but on *ellipses*. Kepler's result was the first of what we now call Kepler's three laws of planetary motion. They are still valid today.

> Kepler's first law of planetary motion, the law of ellipses: *The planets orbit the sun on ellipses, with the sun at one focus and nothing at the other focus.*

Kepler was fortunate to have chosen Mars for analysis. Its orbit is the second most elliptical among the planets then known. If he had chosen Venus, which is nearly circular, he would not have obtained the First Law.

What is an ellipse? You may have learned in geometry that an ellipse can be drawn when a string is attached to two points and then stretched taut by a pencil. As the pencil moves, it traces out an ellipse (Figure 2.33). The two points, F_1 and F_2 are called the foci (singular: focus). If there are walls around the edge of the ellipse, sound or light emanating from one focus will bounce around the ellipse and

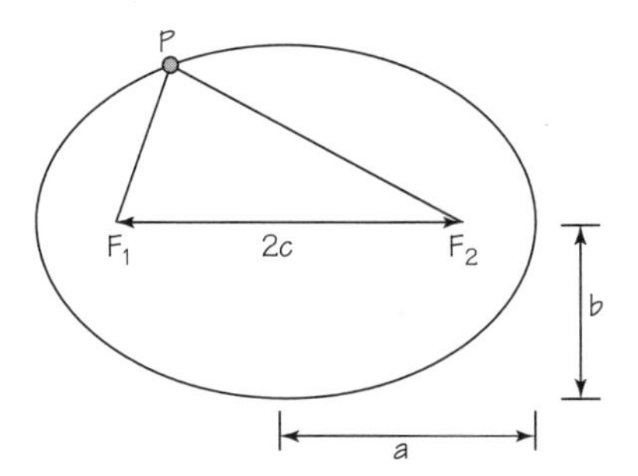

FIGURE 2.33 An ellipse.

eventually converge at the other focus. The closer the foci are to each other, the closer the ellipse comes to being a circle. The farther apart they are, the flatter the ellipse, until finally a straight line is formed. The amount of "flatness" of an ellipse is known as its eccentricity. A measure of the eccentricity of the ellipse is the ratio of the distance F_1F_2 and the long axis of the ellipse. If we call the distance between F_1 and F_2 $2c$, and if we call the length of the long axis $2a$, then the eccentricity e is defined by the equation $e = c/a$. If c is zero, the foci are together and the eccentricity is zero; the ellipse is a perfect circle. Also note that the greatest possible eccentricity for an ellipse is $e = 1.0$. In this case, the ellipse becomes so flat that it approaches a straight line.

Eccentricities of the visible planets	
Mercury	0.206
Venus	0.007
Earth	0.017
Mars	0.093
Jupiter	0.048
Saturn	0.056

Note that all of the orbits are nearly circular except for Mercury's.

We have since discovered that all objects on closed orbits in space travel on ellipses. This includes not only all comets and planets, including Earth, but also the Moon, and satellites in orbit around the Earth or other planets.

The Second Law

Further analyzing Brahe's data, Kepler examined the speed of the planets as they orbited the Sun on their ellipses. He discovered that, here too,

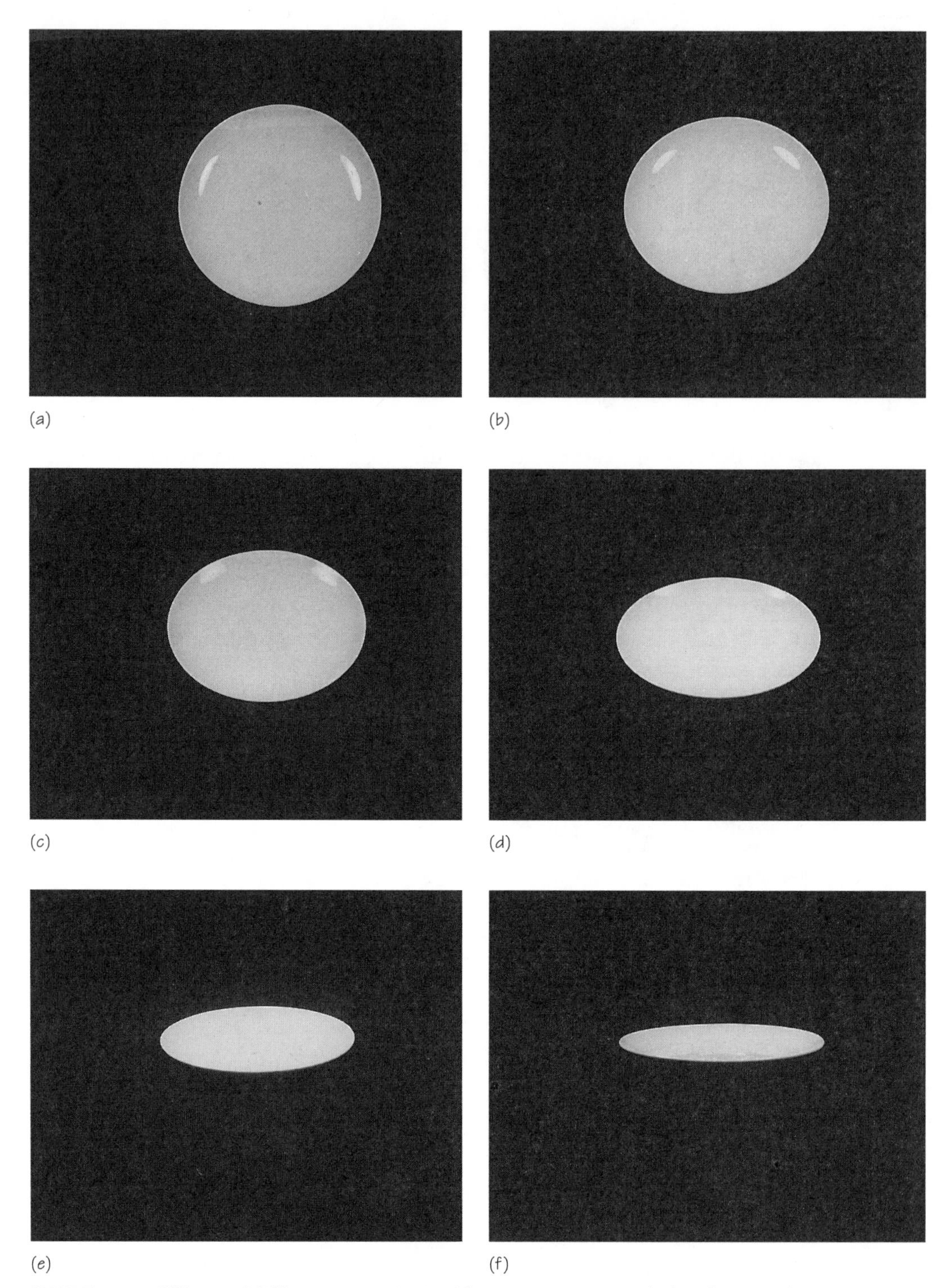

FIGURE 2.34 Ellipses of different eccentricities (the pictures were made by photographing a saucer at different angles).

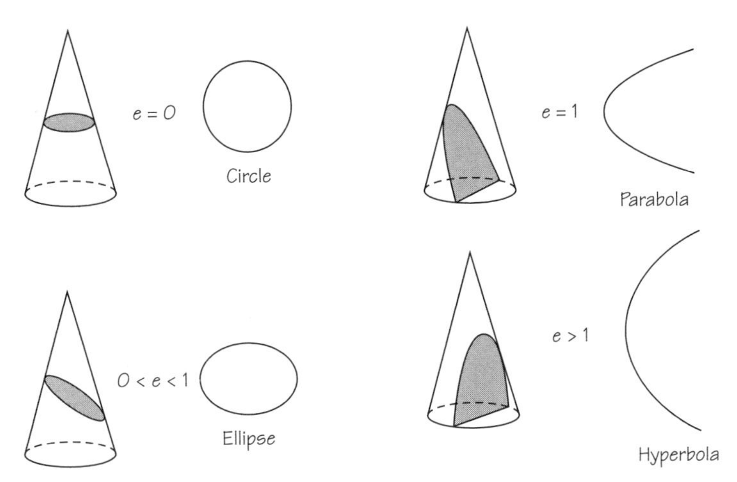

FIGURE 2.35 The conic sections, as shown in the diagram, are figures produced by cutting a cone with a plane. The eccentricity of a figure is related to the angle of the cut. In addition to circles and ellipses, parabolas and hyperbolas are conic sections, with eccentricities greater than ellipses. Newton eventually showed that all of these shapes are possible paths for a body moving under the gravitational attraction of the Sun.

Plato's assumption about planetary motions and centuries of tradition had to be given up. The planets do not travel at uniform speed, but at changing speeds in accordance with a new law, the law of areas:

> Kepler's second law of planetary motion, the Law of Areas: *An imaginary line from the sun to the moving planet sweeps out equal areas in equal amounts of time.*

To see what this means, refer to Figure 2.36. Both of the shaded parts cover equal areas, and the times for a planet to move between the two points on the orbit (AB and CD) are equal. For example, suppose it takes a planet 1 month to travel from point A to point B on the right side of the orbit, when it is nearest the Sun. It also takes the planet 1 month to travel from point C to point D on the other side of the orbit, when it is farthest from the sun. Since the times are equal in each case, this law says that the areas swept out will also be equal to each other. The only way to have both the times and the areas equal to each other is if the distances between these pairs of points are different in each case. As if moving along a giant elliptical pizza pie, the planet sweeps out a short and fat slice of area when nearest the Sun, and a long, thin slice of area when farthest from the Sun. The

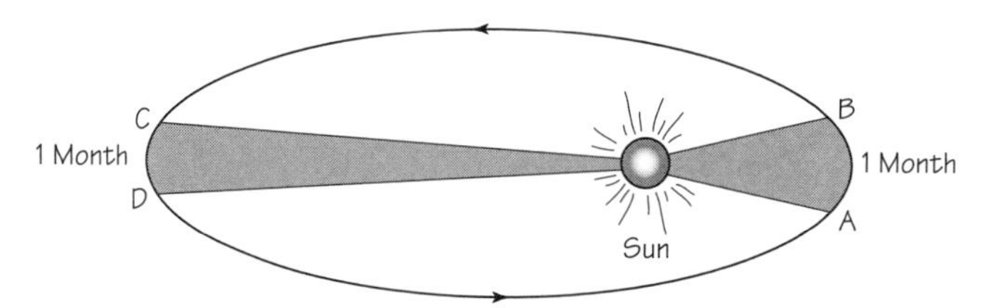

FIGURE 2.36 Kepler's law of areas. A planet (shown here with exaggerated eccentricity) moves along its orbit at such a rate that the line from the Sun to the planet sweeps over areas which are equal for equal time intervals. The time taken to cover *AB* is the same as that for *CD*.

amount of "crust," orbital distance, in each case is different—but the time of travel is the same. As you know, speed is equal to the distance over the time taken. So a larger distance of travel involves a faster speed, while a shorter distance of travel in the same time involves a slower speed. The net result? In moving from A to B near the Sun, the planet moves the fastest. In moving from C to D farthest from the Sun, the planet moves the slowest. And in between, it constantly changes speed as an imaginary line from the Sun to the planet sweeps out equal areas in equal times.

We can now see better why the seasons on Earth differ a little in length. It is not because the Sun is speeding up and slowing down as it orbits the Earth, but because it is the Earth that is doing so as it orbits the Sun on its ellipse. At the same time, the motion of the planets on ellipses with changing speeds put an end to over 2000 years of Plato's problem. In the end, there simply was no problem, because there are no perfect circles and no uniform speeds in the heavens!

Although these Kepler's laws of planetary motion had destroyed Plato's problem, Plato would probably not be too upset, because they provided a much simpler mathematical account of the observed motions of the planets than even Copernicus had provided. With just the simple ellipse and the variable speeds of the planets according to the law of areas, Kepler could do everything that Ptolemy and Copernicus could do. Yet Kepler still sought another harmony that would answer his old question why the planets move as they do in the orbits in which they are found.

The Third Law

Further years of analysis finally led Kepler to his third law of planetary motion, known as the harmonic law. This law did not give the reason "why," but it did reveal a new quantitative relationship between the periods of the planets—the period being the time it takes to complete one orbit—and their average distances from the Sun. (He could use the average distance, since all of the orbits are nearly circular.) It also tied together all of the planets, including the Earth, despite their different ellipses. Kepler's third law is an arithmetic law, not a geometric law:

> Kepler's third law of planetary motion, the harmonic law: *The squares of the periods of the planets are proportional to the cubes of their average distances from the Sun, for all of the planets.*

If we call the period of a planet T, and its average distance from the Sun R_{av}, this law can be expressed in symbols in several different, but equivalent ways, as follows:

$$T^2 \propto R_{av}^3,$$

or

$$T^2 = kR_{av}^3,$$

or

$$\frac{T^2}{R_{av}^3} = k.$$

Here k is a constant, the same constant for all of the planets. This law applies to all the planets as well as to all comets, asteroids, and any other bodies in a closed orbit around the Sun. It also applies to objects orbiting the Earth or any other planet, but in those cases there is a different value for k for each planet.

Let's see what k would be for the planets orbiting the Sun. Of course, the value of k depends upon the units chosen for T and R_{av}. Following standard practice since the days of Copernicus and Kepler, we'll use the Earth's average radius and period as our unit of measure. For the Earth, the period is 1 yr and the average radius is defined as 1 astronomical unit, or 1 AU. So, we can find the value of k in these units by substituting into the expression above:

$$\frac{T^2}{R_{av}^3} = k,$$

$$\frac{(1\ \text{yr})^2}{(1\ \text{AU})^3} = k.$$

Using this value of k and all periods and radii in these units, all of the other planets should have this same ratio. Let's see if we can confirm this, using the data in the table of periods and radii obtain by Copernicus, Section 2.6. For example, for Saturn, $T = 29.46$ yr, $R_{av} = 9.54$ AU:

$$\frac{T^2}{R_{av}^3} = \frac{(29.46\text{ yr})^2}{(9.54\text{ AU})^3} = \frac{867.9\text{ yr}^2}{868.3\text{ AU}^3} = 0.999\text{ yr}^2/\text{AU}^3.$$

For Mercury, $T = 87.97$ d $= 0.24$ yr, $R_{av} = 0.39$ AU:

$$\frac{T^2}{R_{av}^3} = \frac{(0.24\text{ yr})^2}{(0.39\text{ AU})^3} = \frac{0.058\text{ yr}^2}{0.059\text{ AU}^3} = 0.983\text{ yr}^2/\text{AU}^3.$$

So within the limit of uncertainty, k for Saturn and Mercury are equal. Hence, if you know the average radius of a planet, you can find its period, and vice versa.

Evaluating Kepler's Work

Kepler's system was vastly simpler and more precise than the multitude of geometrical devices in the planetary theories of Ptolemy, Copernicus, and even Brahe. Kepler's three laws are so simple that their great power may be overlooked. Combined with his discovery that each planet moves in a plane passing through the Sun, their value is greater still. It is almost as if the solar system is like a gigantic mechanical machine, perhaps a clock, ticking away in precise, predictable fashion according to Kepler's laws of planetary motion. Kepler was among the first astronomers to suggest such a simile. In fact, he was the first to call this a "clockwork universe," a universe, he thought, wound up by God in the beginning and allowed to run like a mechanical clock according to a few laws of motion until the end of time. This was a powerful image, and one which we will encounter again and again.

Although Kepler believed that a magnetic "force" emanating from the Sun was the underlying origin of his laws of planetary motion, these laws are grounded in his painstaking analysis of the data. Data that are obtained from experimental research are often called empirical data, and laws obtained from such data are often called *empirical laws.* They are an important step toward obtaining a theory, but usually they cannot form a theory themselves, since we want a theory to do much more. We want an explanation why these laws occur in the data as they do.

Kepler did try to provide such an explanation by speculating about the action of the supposed magnetic force emanating from the Sun. In doing

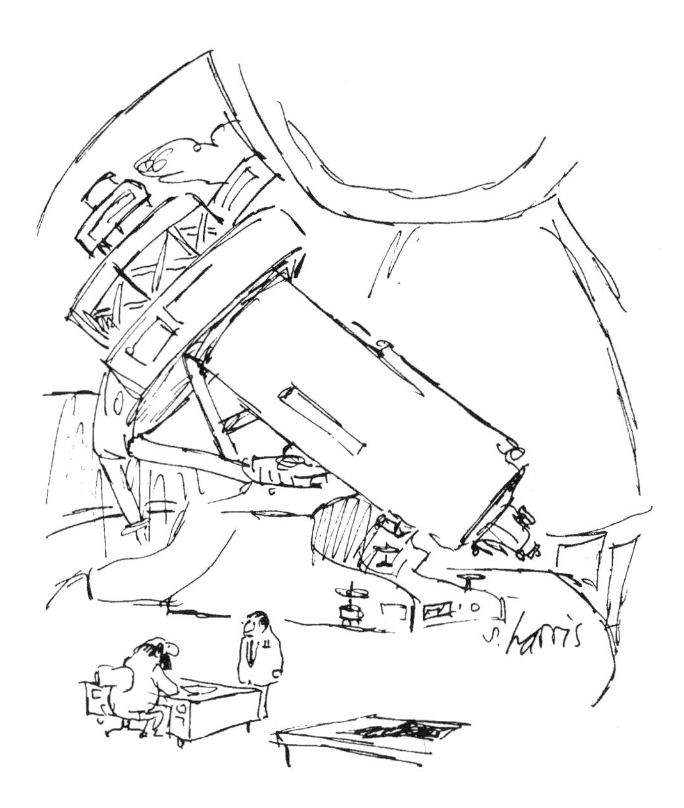

"It's always the same thing — the sun, a few clouds, and that's it.
I'd like a transfer to the night shift."

FIGURE 2.37

this he was the first mathematical astronomer to go beyond an analysis of the observations in an attempt to create a *physics* of the planetary motions—obtaining not just a mathematical description of the motions but the *cause* of the motions. In fact, the full title of his book reporting his work and discoveries was *New Astronomy: A Celestial Physics*.

In obtaining his second law of planetary motion, Kepler realized that the planets move fastest when they are closest to the Sun and slowest when farthest away. He reasoned from this that there might be a force from the Sun that causes the planets to speed up as they move closer. Kepler thought that this force might be a magnetic force of attraction, since recent discoveries had shown that the Earth is a large magnet, and that the strength of a magnet's effect increases as the distance to it decreases. Perhaps the Sun exerts a magnetic attraction on the Earth and other planets as they revolve around the Sun, speeding them up as they approach.

Kepler was almost right! There is an attractive force between the Sun and the planets that does account for his laws. In one of the most important theories ever developed, Newton showed that the attractive force is not magnetism but another force, the force of gravitation between all matter, whether on the Earth, in the solar system, or across the Universe (further discussed in Chapter 4). But Kepler was not able to carry his prescient idea beyond the qualitative stage.

As you will see in Chapter 4, when Newton obtained his theory of universal gravitation, he used Kepler's empirical laws as a guide. His derivation of these laws from the new theory helped to confirm the theory. The power of empirical laws such as Kepler's is in their ability to help guide to general theories from which these laws can be derived. Only then do we believe we have an understanding of the physical processes that give rise to these empirical laws. One may say empirical laws tell us "how"; theories tell us "why."

2.11 NEW OBSERVATIONS

One of the scientists with whom Kepler corresponded was his Italian colleague Galileo. Like Kepler, Galileo was opposed by scholars who believed that the heavens were eternal and could not change. Galileo therefore took special interest in the sudden appearance in 1604 of a new star, a nova. Where there had been nothing visible in the sky, there was now a brilliant star that gradually faded away. Like Brahe and Kepler, Galileo realized that such events conflicted with the then current idea that the stars could not change. Similar to the experiences of Brahe and many future scientists at a young age, this nova awakened in Galileo an interest in astronomy that lasted his entire life.

Four or five years later, as Galileo tells it, he learned that a Dutch lens maker "had constructed a spy glass by means of which visible objects, though very distant from the eye of the observer, were distinctly seen as if nearby." Galileo worked out some of the optical principles involved. Having established a scientific instrument shop in order to supplement his meager income as a professor, he set to work to grind the lenses and to build such an instrument himself. While others used the telescope primarily as a military instrument, for sighting enemy ships and invading armies, Galileo was the first to turn the instrument to the heavens. What he saw there astonished him and the public to whom he reported his observations, as it will amaze you if you have the opportunity to observe the night sky through a telescope.

FIGURE 2.38 Two of Galileo's telescopes.

Imagine being the first person ever to look at the stars through a telescope! For thousands of years people could learn about the heavens only from what they could see with their own eyes, and then only if they had good eyesight. Suddenly a whole new world opened to human eyes for exploration and study. Within a few short weeks in 1609 and 1610 Galileo used his telescope to make a series of major discoveries. First, he pointed his telescope at the Moon. Here is what he saw:

> . . . the surface of the Moon is not smooth, uniform, and precisely spherical as a great number of philosophers believe it (and other heavenly bodies) to be, but is uneven, rough, and full of cavities and prominences, being not unlike the face of the Earth, relieved by chains of mountains and deep valleys.

He did not stop with that simple observation; he supported his conclusions with several kinds of evidence, including ingenious measurements of the heights of the lunar mountains.

Next Galileo looked at the stars. To the naked eye about 3000 stars are visible in the night sky (if you are away from city lights), while the Milky

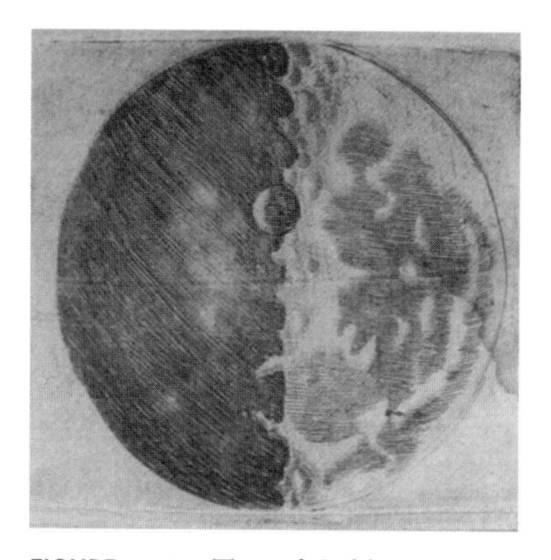

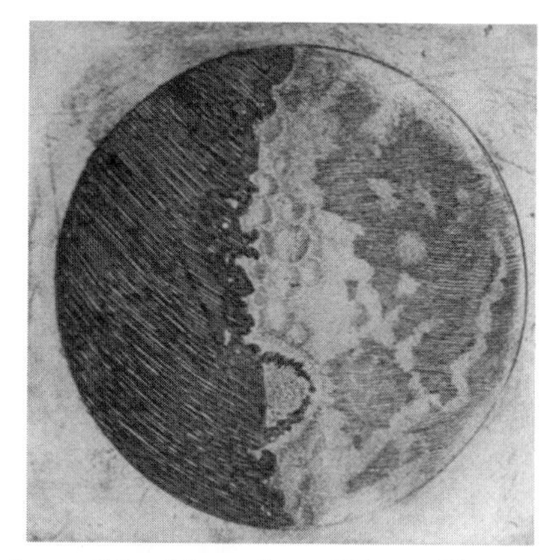

FIGURE 2.39 Two of Galileo's early drawings of the Moon from *Siderius Nuncius (The Starry Messenger).*

Way (now understood to be the major part of our galaxy) seems to be a continuous blotchy band of faint light, almost directly overhead and to either side. Wherever he pointed the telescope, Galileo saw many more stars than had ever been seen before, and he observed the Milky Way to consist of thousands of faint stars. Today, with powerful telescopes such as the Hubble Space Telescope, astronomers can see many billions of stars and other objects—and there is as yet no end in sight.

By projecting an image of the Sun on a screen in order to protect his eyes (never look directly at the Sun!), Galileo observed dark spots on the Sun. These seemed to indicate that the Sun, like the Moon, was not perfect in the Aristotelian sense. He also noticed that the sunspots moved

FIGURE 2.40 Image of stars from Hubble Space Telescope.

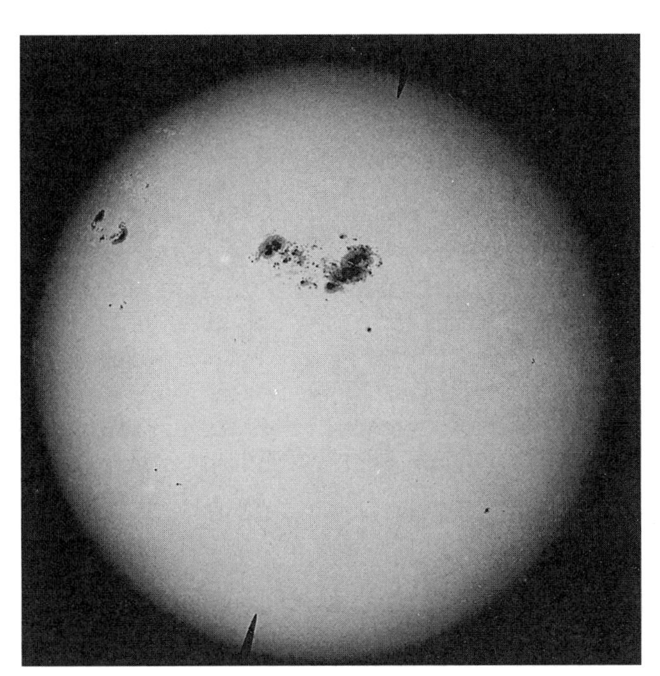

FIGURE 2.41 Solar disk with sunspots.

across the face of the Sun in a regular pattern. He concluded from further study that this motion indicated that the Sun itself rotated on its axis with a period of about 27 days. If the Sun can rotate, he asked, why can't the Earth?

Writing in Italian for the general educated public, Galileo reported on these and his many other discoveries over the following years. Among the most important was, in his words, "the disclosure of four *Planets* never seen from the creation of the world up to our time." He was referring to his discovery of four of the moons that orbit Jupiter. Here, before his eyes, was a miniature solar system, with its own center of revolution—a model for the entire solar system. He named these moons the Medician Planets, in honor of his benefactor in Florence, Cosimo dé Medici. Centuries later, in the 1990s, the first satellite sent to Jupiter for long-term study of the planet and its moons was named *Galileo* in honor of the moons' discoverer.

Galileo also observed that Saturn seemed to carry mysterious "bulges" or "ears" around its equator. The magnification of his telescopes was not large enough to show that these were really the rings of Saturn. In photographing the rings of Saturn during space missions centuries later, the *Voyager* spacecraft revealed these rings to be among the most beautiful objects in the solar system. Yet in some ways—for instance, their fine structure and delicately preserved equilibrium—they still remain mysterious.

So far, none of Galileo's observations was a clear contradiction of the Ptolemaic theory, although they did raise serious doubts. The greatest threat came with Galileo's discovery of the phases of Venus, which are also visible only through a telescope. Like the Moon, Venus shows all phases, and they are of different sizes depending upon the phase, full phase occurring with the smallest size, as shown in Figure 2.45. Galileo pointed out that the full phase should not be observed if Ptolemy's theory is valid, because in it Venus moves always between the Earth and the Sun. But Copernicus's theory can account for all of these phases, and also the different sizes, as indicated in Figure 2.46. This, he told his audience, was clear proof that Ptolemy was wrong and Copernicus was right. But in his eagerness to convince the public, he failed to mention that Brahe's "compromise theory" (see Section 2.9) also accounted for the observations of Venus, and this was still a geocentric theory (although a lopsided one), since the Earth was stationary at the center and all celestial objects orbited about it.

Having collected an impressive array of new information about the heavens with his telescopes, Galileo used it to maximum advantage. He had become convinced of Copernicus's heliocentric system earlier in his career because, like Copernicus, he found it simpler and more pleasing. Now he had observations as well as conviction. A masterful writer and debater, Galileo portrayed his observations to his Italian audience in his *Dialogue Concerning Two Chief World Systems* (1632) and earlier writings as providing irrefutable evidence in favor of Copernicus and against Ptolemy. In fact, while they certainly caused problems for Aristotle's cosmology, they were

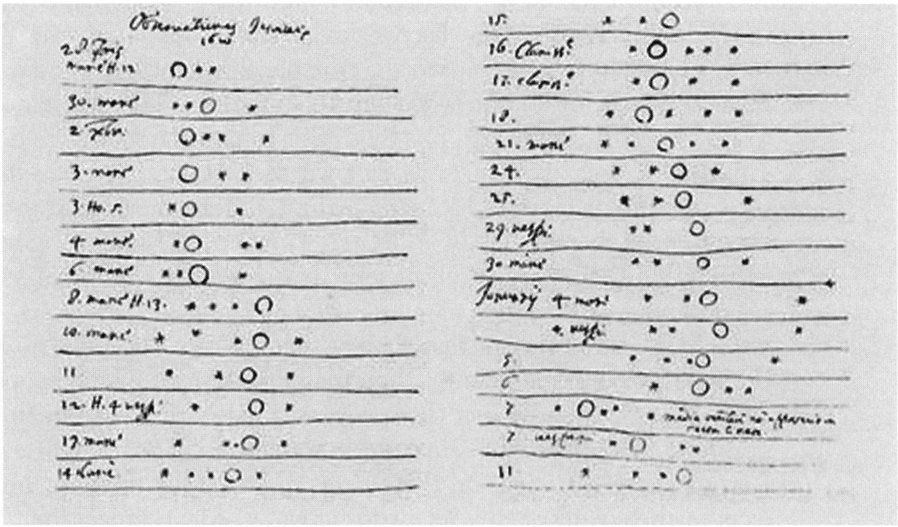

FIGURE 2.42 Galileo's drawings of Jupiter (the large circle) and its moons (the dots) on different days (from *The Starry Messenger*).

FIGURE 2.43 Palomar Observatory, located on Palomar Mountain in southern California, houses the 200-inch Hale reflecting telescope.

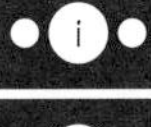

FIGURE 2.44 Saturn seen at different times, as it would have appeared to observers in the seventeenth century (reproduction of sketches).

FIGURE 2.45 Photographs of Venus at various phases with a constant magnification.

not "proof" but rather, at best, circumstantial evidence against the systems of Ptolemy and Aristotle and for Copernicus's heliocentric system. They could lead us to doubting Ptolemy and Aristotle, but not necessarily to rejecting their theories entirely. Only the phases of Venus provided a clear challenge to Ptolemy's theory—but not to all geocentric theories, since Brahe's geocentric theory could still account for the observed phases.

Apparently Galileo expected that his discoveries with the telescope would immediately demolish the deep-seated assumptions and beliefs that prevented widespread acceptance of the Copernican theory. But no matter

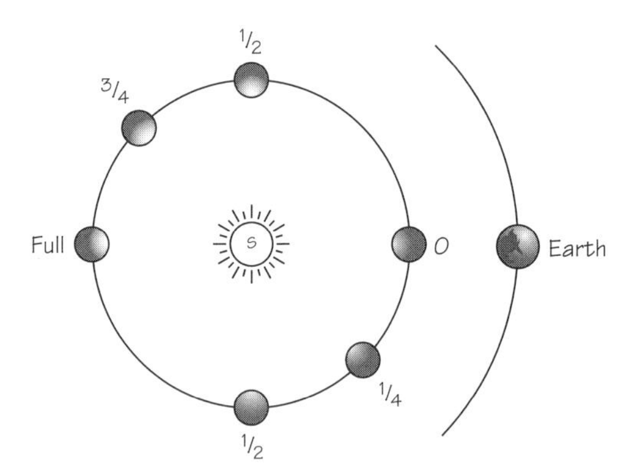

FIGURE 2.46 Explanation of phases of Venus as seen from Earth, based on the heliocentric theory.

what the evidence, people often cannot believe what they are not ready to believe. The Aristotelians firmly continued to believe that the heliocentric theory was absolutely false and contrary to direct, naked-eye observation, common sense, and religious belief. Lest Galileo gain influence among the general public and threaten cherished, pious beliefs, the religious authorities decided to forbid his teachings on this matter.

2.12 GALILEO CONDEMNED

The political and personal tragedy that struck Galileo is described in many books. Many of the documents pertaining to this case have been recently released by the Vatican and are available in English translation. We know that the period was one of turmoil for the Roman Catholic Church. In the wake of the Protestant Reformation, the Church was in the midst of its own reassessment, called the Counter Reformation, to win back some of its members. In addition, there were many intrigues occurring among various factions within the Church. These factions saw Galileo's vulnerability as an opportunity to enhance their own prestige and power within the Church hierarchy.

For several centuries the Church had supported the Inquisition, a theological court established to investigate and stamp out heresy by every means. In 1616, the same year in which the Inquisition placed Copernicus's book on the *Index of Forbidden Books*, the Inquisitors, mindful of Galileo's recent publication of some of his discoveries with the telescope, warned Galileo to cease teaching the Copernican theory as truth. He could continue to teach it only as just one of several possible hypotheses or methods for computing the planetary motions, but not as a literally true model of the Universe. Although a devoutly religious man, Galileo deliberately ruled out questions of religious faith from scientific discussions. This was a fundamental break with the past. But he was ordered "henceforth not to hold, teach, or defend it in any way whatever, either orally or in writing."

In 1632, having obtained permission of the Church censors, Galileo published his *Dialogue Concerning Two Chief World Systems*. Galileo's enemies in the Church flew into a rage when they found their views represented in the book by a fictitious Aristotelian named Simplicio, who in the end was portrayed as persuaded of the heliocentric theory. Galileo's long-time enemies, incensed by his lack of tact, argued that he had directly violated the warning of 1616. These and related motivations marked Galileo for punishment.

Among the many factors in this complex story it is important to remember that Galileo was always religiously faithful, as were most of the scientists of that era. In earlier letters Galileo wrote that God's mind contains all the natural laws. Consequently, the occasional glimpses of these laws that scientists might gain are direct revelations about God, just as true in their way as those in the Bible. He believed scientific research could be considered the retracing of God's thoughts as He created nature long ago. Today, similarly, some view science as one way to contemplate God's creation, whether they are scientists or not. Few people think of scientific findings about the world as conflicting with religion. In Galileo's time, however, such ideas were regarded as symptoms of pantheism, the belief that God is no more (nor less) than the forces and laws of nature. Pantheism was one of the religious "crimes," or heresies, for which the Dominican monk Giordano Bruno, who proclaimed the existence of other worlds, had been burned at the stake in 1600. The Inquisition, alarmed by Galileo's seeming denial of the Bible as the only literal source of knowledge about Nature and of God, ordered him to Rome to stand trial for heresy.

Although old and in ill-health, Galileo was confined in Rome, interrogated, threatened with torture, forced to make a formal confession for holding and teaching forbidden ideas, and finally forced to deny the Copernican theory as heresy. In return for his confession and denial, Galileo was sentenced only to house arrest for the remainder of his life. He never wrote again on the Copernican theory, but he managed to produce perhaps his

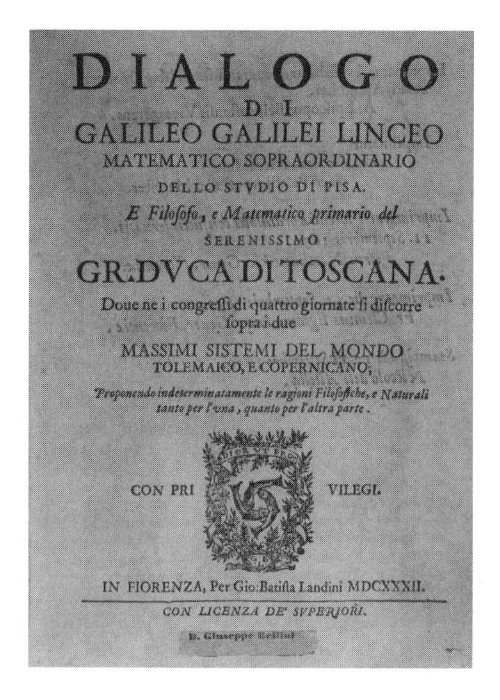
DIALOGO
DI
GALILEO GALILEI LINCEO
MATEMATICO SOPRAORDINARIO
DELLO STVDIO DI PISA.
E Filosofo, e Matematico primario del
SERENISSIMO
GR.DVCA DI TOSCANA.
Doue ne i congressi di quattro giornate si discorre
sopra i due
MASSIMI SISTEMI DEL MONDO
TOLEMAICO, E COPERNICANO;
Proponendo indeterminatamente le ragioni Filosofiche, e Naturali
tanto per l'una, quanto per l'altra parte.

CON PRI VILEGI.

IN FIORENZA, Per Gio:Batista Landini MDCXXXII.

CON LICENZA DE' SVPERIORI.

FIGURE 2.47 Title page from Galileo's *Dialogue on Two Chief World Systems* (1632).

best work, *Dialogue Concerning Two New Sciences*, in which he presented his findings on the science of motion and mechanics. In the end, ironically, this new work—by leading to Newton's work—eventually demolished Aristotle's cosmology more thoroughly than could any polemical writings or dialogues debating the merits of different models for the solar system.

The Inquisition also placed Galileo's *Dialogue on Two Chief World Systems* on the *Index of Forbidden Books*. It remained there, along with Copernicus's book and one by Kepler, until 1835—a warning to all that demands for spiritual conformity also required intellectual conformity. The result was the noticeable decline of science in Italy for nearly two centuries. But science cannot be extinguished. Less than 50 years after the condemnation of Galileo, Newton published his great work, the *Principia*, an achievement that would not have been possible without the work of Galileo and Kepler.

In 1979, during the worldwide celebration of the 100th birthday of Albert Einstein, Pope John Paul II announced that the Vatican would reopen the case against Galileo. In 1984 it released the documents pertaining to the case, and in 1992 a papal commission acknowledged the Vatican's error in condemning Galileo for heresy.

SOME NEW IDEAS AND CONCEPTS

altitude
azimuth
celestial sphere
eastward drift
empirical laws
epicycle-deferent
equinox
geocentric theory
heliocentric theory
law of nature
nova
Renaissance
retrograde motion
Scientific Revolution
solstice
stellar parallax
Zodiac

FURTHER READING

J.R. Christianson, *On Tycho's Island: Tycho Brahe and His Assistants* (New York: Cambridge University Press, 2000).

G. Galilei, *Galileo on the World Systems—A New Abridged Translation and Guide*, M.A. Finocchiaro, transl. and ed. (Berkeley, CA: University of California Press, 1997).

G. Holton and S.G. Brush, *Physics, The Human Adventure* (Piscataway, NJ: Rutgers University Press, 2001), Chapters 1–5.

T.S. Kuhn, *The Copernican Revolution: Planetary Astronomy in the Development of Western Thought* (Cambridge, MA: Harvard University Press, 1982).

J.R. Jacob, *The Scientific Revolution: Aspirations and Achievements, 1500–1700* (Humanities Press, 1998).

M. Caspar, *Kepler*, C.D. Hellman and O. Gingerich, transl. (New York: Dover, 1993).

O. Gingerich, *The Eye of Heaven: Ptolemy, Copernicus, Kepler* (Woodbury, NY: AIP Press, 1993).

A. Koestler, The Watershed. In: *The Sleepwalkers* (New York: Macmillan, 1959), a biography of Kepler.

D. Sobel, *Galileo's Daughter: A Historical Memoir of Science, Faith, and Love* (New York: Walker, 2000).

H. Thurston, *Early Astronomy* (New York: Springer-Verlag, 1994).

STUDY GUIDE QUESTIONS

1. This chapter concerns the development, debates, and impact regarding an important new scientific theory. Describe the new theory and the old theory that it challenged.
2. What was the Scientific Revolution, and what is the importance of astronomy and mechanics to it?

2.4 The Geocentric View

1. Until about A.D. 1700 most people believed that the Earth is stationary at the center of the Universe. We now know that the Earth is rotating and the stars are more or less stationary. Why would anyone believe that the Earth is stationary?
2. Looking at a sunrise, or a sunset, is it possible to decide whether the Sun is moving and the Earth is stationary or the Earth is moving and the Sun is stationary?
3. What is relative motion?
4. As seen from the Earth, briefly describe the motions of the stars, Sun, Moon, and planets.
5. Explain the origin of the seasons in the northern hemisphere in terms of the geocentric theory. (Refer to Discovery Question 4 for further inquiry on this topic.)
6. Explain the origin of the seasons in the northern and southern hemispheres in terms of the heliocentric theory.
7. How would we know that the Earth is round if we didn't have satellites and photographs from the Moon?
8. Is the sun ever directly overhead at your latitude on the Earth? Explain.

2.5 Copernicus versus Ptolemy

1. Who was Copernicus, what did he do, and why did he do it?
2. In what ways was Copernicus's theory a sharp break with the past?
3. How did Ptolemy account for the observed motion of a planet as seen from the Earth?
4. Evaluate Ptolemy's theory, including both its positive contributions as well as its problems.
5. What did Copernicus find wrong with Ptolemy's explanation?
6. How did Copernicus account for the observed motions of the planets?

2.6 Arguments for the Heliocentric System

1. Give some of the arguments in favor of the heliocentric system.
2. What reasons did Copernicus have for proposing this theory?
3. What was the numerical harmony that Copernicus found in the table of relative radii and periods of the planets?

2.7 Arguments against the Heliocentric System

1. Why wasn't Copernicus's idea immediately accepted? After all, he was right!
2. List some of the arguments at that time against his theory.
3. What was the problem of stellar parallax? How did Copernicus respond? And how did his opponents respond to his response?

2.8 Carrying Forth the Revolution

1. At the time of Copernicus's death, how did most astronomers view Copernicus's theory?
2. Comparing the abilities of the alternative theories to account for observations, did either theory have an advantage? Explain why or why not.
3. List some of the work that was done in the century and a half after Copernicus's death, and how it strengthened the case for the new theory.

2.9 New Data

1. Why would the observations of comets and new stars (novae) beyond the Moon pose a challenge for Aristotle's system?
2. What were Brahe's contributions to the debate over a model for the planets?
3. What compromise did he offer, and what were its advantages for either side of the debate?

2.10 New Orbits

1. Who was Kepler? What was he attempting to discover?
2. What did Kepler believe to be the discovery in Question 1?
3. How does a law of nature differ from a theory?
4. State each of Kepler's three laws of planetary motion in your own words.

5. Where in its orbit does a planet move the fastest? the slowest? How does this motion account for the varying lengths of the seasons on Earth?
6. The radius of the planet Venus is given in the table in Section 2.6. Use Kepler's third law to calculate the radius of its orbit, then compare with the value given in the table. Does this support Kepler's third law?
7. How did Kepler's work affect Plato's age-old problem?
8. What are empirical laws, and why are they usually not sufficient to form a theory?
9. How did Kepler attempt to account for his laws?

2.11 New Observations

1. List and briefly describe Galileo's observations with the telescope.
2. Did any of Galileo's observations completely disprove the geocentric theory?
3. What did Galileo observe about the phases of Venus, and what did he claim they proved? Was he right? Explain.

2.12 Galileo Condemned

1. Why did the Church authorities decide to try Galileo for heresy?
2. What was Galileo's position on the relationship between science and religion?
3. What was the Church's position?
4. What was the outcome of the case, in the short term and in the long term?

DISCOVERY QUESTIONS

1. Fundamental new ideas about the world are often hard to accept. Why do you think this is so? Why don't most people gladly accept new and challenging ideas if there is evidence for them?
2. This is supposed to be a physics course, so why is this chapter about astronomy?
3. Why does this chapter have a lot of material on the geocentric model, when we now know that it's wrong?
4. Section 2.4 contained a description of the seasons during 1 year as seen from a position on the northern hemisphere of the Earth. Referring to Figure 2.12, describe what happens to the seasons during 1 year as seen by an observer:
   (a) in the southern hemisphere;
   (b) at the north pole;
   (c) at the south pole.
5. What is so important about the Scientific Revolution and Copernicus's theory?
6. The relationship between science and religion has always been a hot issue. Both deal with nature and our relationship to it, but they have different approaches. Consider what are your own ideas on this controversial issue, but

feel free to discuss without having to reveal your personal beliefs one way or the other.

7. Think of some examples of beliefs and scientific theories. What is the difference, if any, between a scientific theory and a belief? Can there be some overlap? How could a scientist like Galileo also be devoutly religious?
8. In what ways has our place in nature and the Universe become even more humble than it was in Copernicus's day?
9. Set up a debate in class over the heliocentric versus the geocentric theories of the solar system. Give specific arguments for and against each side, and decide upon a winner. Then introduce a compromise and let each side evaluate its acceptability.
10. Instead of an in-class debate, write a brief dialogue between Copernicus, Ptolemy, Galileo, and a modern person, and act it out before the class.
11. Look at the reasons Copernicus gave for his new theory. Do you think they were enough to convince everyone eventually? If not, what other reasons or evidence were needed?
12. Think about or look up some of the other great theories that you know about in science, and compare some of the features of their acceptance with those of the heliocentric theory.
13. Looking back over this chapter, outline the steps in the formation, debate, and acceptance of a new, fundamental theory in physical science. Use this outline later to compare with other theories you will encounter in this course.
14. Many of the astronomers of that day were also astrologers. Why do you think this was so? What are some of the fundamental assumptions in astrology and in astronomy, and how do these two differ from each other? In what ways are they similar?
15. Many of the astronomers then were also greatly concerned with revising the calendar. Look up the history of our modern calendar in an encyclopedia or other reference work and report on it to the class.
16. Do you think a second scientific revolution, with the same implications as the first one, could occur today? Make up an imaginary theory that provides a sharp break with what we understand today about some aspect of nature or the Universe. How would political leaders, religious authorities, scientists, students, the general public, react to this imaginary theory? How does the situation in the United States today compare with the conditions that helped or hindered new theories during the Renaissance?

Quantitative

1. Pluto orbits the Sun with a period of 248.4 yr. What is its average distance from the Sun?
2. A satellite is launched into circular orbit around the Sun at a distance of 3 AU. What is the period of its orbit?
3. Assuming the Earth is a perfect sphere, what would be the altitude of the Sun at solar noon on the Summer Solstice at your latitude?

4. On the day of the Winter Solstice, the Sun's elevation at noon in the northern hemisphere is the lowest it attains throughout the year. At some northern latitudes the Sun never rises on that day. What is the highest northern latitude at which the Sun is still visible at noon on the Winter Solstice?
5. The Moon orbits the Earth at an average distance of 384,403 km with a period (as seen from the stars) of 27.3 days. With this information, find the period of a satellite launched into a circular orbit around the Earth of radius 10,000 km.

CHAPTER 3

Understanding Motion

A. THE THREE LAWS OF MOTION

3.1 NATURAL MOTION

You saw in the previous chapter that the proposal that the Earth moves caused all sorts of problems for the Earth-centered view of the Universe—as well as for theology, philosophy, and just plain common sense. If the Earth really is moving around the Sun, then what keeps the heavy Earth on its orbit? The ground beneath our feet must be moving very fast, so why do heavy objects fall straight down, instead of hitting behind the spot over which they are released?

You probably know that these and other excellent questions can be answered by referring to the law governing the force of gravity. But understanding what gravity is and how it answers these questions requires a further understanding of motion and its causes. We could just list the results for you here in a few paragraphs, but (as usual) we think you will get a much clearer idea of what these concepts really mean, and what goes into them, by seeing where they came from and how people managed to provide the understanding we have today of this complicated thing we so easily call "motion." This understanding, too, began with Galileo, which he discussed in his book *Two New Sciences*, published after his condemnation for religious heresy. At least two other scientists of that era, Descartes and Newton, further enhanced his work and put it into more modern form after his death.

The story begins with what seems to be a very simple concept we have already encountered, the concept of "natural motion"—how an object moves "naturally" without outside help. In Aristotle's physics there were two types of motion: "natural" motion and "violent" motion. Natural motion was the motion of an object on a vertical line directly to the place where it "belonged." For example, a stone falls straight down through the air and further through any water to reach the earth below. On the other hand, an air bubble trapped under water moves upward through the water until it reaches the air above. There was no need to explain why these motions occur—they're just as natural as the movement of an animal making straight for its "home."

In contrast to natural motion Aristotle also defined "violent" motion. This was any motion that is not directly to the place where the object "belonged." Examples are the lifting of a stone upward from the earth, or the motion of a cart across a horizontal tabletop, or the flight of an arrow through the air. This type of motion had to be explained, since it was not natural for an object to move sideways by itself. Something has to *cause* or *force* the object to perform such motion. This was why he called it "vio-

FIGURE 3.1 Examples of Aristotle's notions of natural and violent motion.

FIGURE 3.2

lent" motion. Therefore, according to Aristotle, to maintain these violent motions a *force* had to be applied continuously. As soon as the force ceased, the violent motion ceased. Again, this seems very reasonable at first sight. Anyone who has pushed a heavy cart over a rough pavement knows that it stops moving as soon as you stop pushing. But Galileo showed for the first time that there is more going on here than we realize. The causes and inhibitors to motion are not obvious, and it takes the careful reasoning and experimentation of modern physics that Galileo and others developed to sort this out.

As noted earlier, Galileo was one of the first to realize that he should dispense with less important complications in order to focus on the essential aspects of any motion. He realized that the role of friction and air resistance could be neglected in his thinking in order that the essential properties of the motion itself can be studied. You may recall that when he neglected friction and air resistance for a ball rolling down an inclined plane, he obtained the idea of uniform acceleration. The uniform acceleration increased as the small angle of inclination increased. From this he inferred that the extreme case of a 90° "incline" would also be uniform acceleration. In that case the ball is no longer rolling down the incline, but falling freely to the ground. From this he concluded that free fall must also be an example of uniform acceleration and that all bodies will fall freely at the same uniform acceleration if air resistance is neglected. As a result, heavier objects do *not* fall faster than lighter ones. They all reach the ground at the same time when dropped from the same height (neglecting air resistance).

Now Galileo decided to look at the other extreme. What happens when the angle of the incline becomes smaller and smaller, down to zero? As

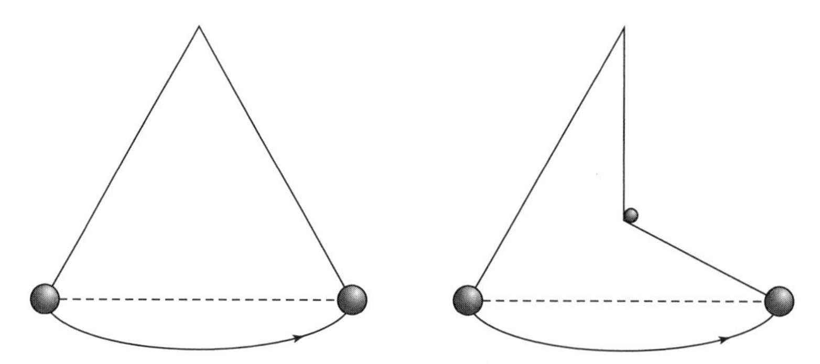

FIGURE 3.3 A swinging pendulum bob returns to its initial height, even if there is an impediment.

usual, he decided to try a "thought experiment" (which can also be done as a real experiment). This thought experiment is based on an actual observation. If a pendulum bob on the end of a string is pulled back and released from rest, it will swing through an arc and rise to very nearly the same height. Indeed, as Galileo showed, the pendulum bob will rise almost to its starting level even if a peg is used to change the path, as shown in the illustration above. Try it yourself!

From this observation Galileo went on to his thought experiment. He predicted that a ball released from a height on a ramp would roll up to the same height on a similar ramp, neglecting all resistances. Consider the following diagram. As the ramp on the right is changed from position A to B and then to C, the ball must roll farther along the upward incline in each case in order to reach its original height.

Now let's look at the extreme case, the limit in which the angle of inclination of the second ramp is zero, and the path is level, as shown in D. In this case the ball never reaches its original height. Therefore, Galileo reasoned, the moving ball on this surface, with no air resistance and no barriers in its way, would continue to roll on and on at uniform speed forever. (Actually, he thought that if the ball continued on long enough, it would

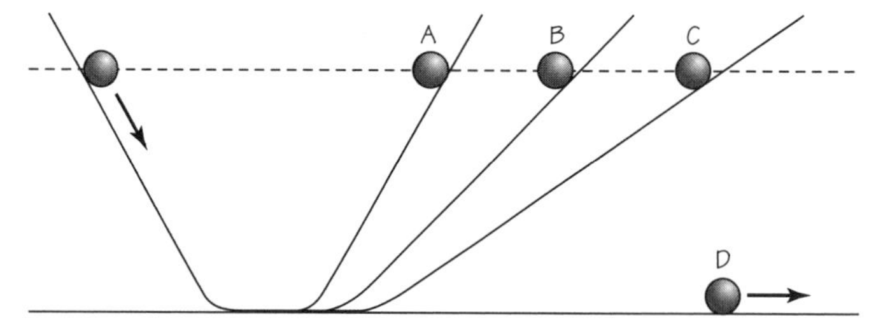

FIGURE 3.4 Ball rolling up ramps of different inclination returning each time to the height, except in D.

For a ball rolling without friction, there is no detectable difference between motion in a straight (horizontal) line in a laboratory on Earth or at a constant height above the Earth. But on a larger scale, Galileo held that eternal rolling would become motion in a circle around the Earth. Newton made clear what is really more important: In the absence of the Earth's gravitational pull or other external forces, the ball's undisturbed path would extend straight out into space without end.

eventually go into a circular orbit around the Earth. However, without the gravity of the Earth, the ball would continue in a straight line forever.) *What Galileo discovered is that, within any reasonable large space—say, a laboratory—it is "natural" for the ball to continue to roll horizontally in a straight line at uniform speed until it is forced by a push or other means to change its motion.* Even though it is not moving vertically, no force is required to keep the ball rolling horizontally, as Aristotle had believed. A force is needed only to stop it from rolling, or to divert it in another direction. Contrary to Aristotle's arguments, the motion of a moving ball at uniform velocity (constant speed in a straight line) is quite "natural." It needs no explanation. What must be explained is why it ever *changes* that motion. And *this* is where the action of a force comes in. The departure from motion with uniform velocity, now called "accelerated motion," occurs only when an object is acted on by a force.

The same holds when objects are at rest. They stay at rest, which can be thought of as a state where the velocity is zero, until they are "forced" to move.

Since objects stay in uniform motion or at rest until they are forced to do otherwise, they behave as if they are "inert," literally unable to change their state by themselves. It takes a force applied from outside themselves to get them to change their motion, either to speed up, or to slow down, or to change direction. When they are at rest, it takes a force to get them to move; otherwise they "naturally" tend to maintain their state of rest or motion. This tendency of all material objects to maintain their state of motion is called *inertia*. Newton elevated this tendency of material objects into the first of his three basic laws of nature governing all motion. These laws are so universal that Newton applied them to all motion everywhere, including throughout the entire Universe. And they are part of the foundation of modern physics. So far, no contradiction to these laws has been found, either on Earth or anywhere else in the Universe—although, as you will see in Chapter 9, modifications become increasingly necessary for events occurring at extremely high speeds approaching the speed of light.

Newton's first law of motion is also called the law of inertia. Expressed in modern terms, it states:

> Every object continues in its state of rest or of uniform velocity (motion at uniform speed in a straight line) unless acted upon by

an unbalanced force (a net force). Conversely, if an object is at rest or in motion with uniform velocity, all forces that may be acting on it must cancel so that the net force is zero.

Notice that Newton is not saying that there are no forces acting on an object at rest or moving with uniform velocity. Rather, he is staying that if there are any forces they must all cancel out if the object is to be in a state of rest or uniform velocity.

For Aristotle, an object rolling horizontally on a tabletop was thought to have a constant force acting on it to keep it moving; otherwise it would stop as soon as the force is removed. Newton's first law of motion makes the situation just the reverse. An object moving with uniform velocity or at rest needs no net force to keep it in that state of motion; instead, any "unbalanced force" (such as friction, air resistance, or a push or pull) will change its state of moving with uniform velocity. Remove all forces and obstacles, and the object will continue on forever. In real life it is very difficult to come close to a situation where all forces are absent or balanced out. But a hockey puck moving on flat ice, after it has been pushed, is a close example of such an object.

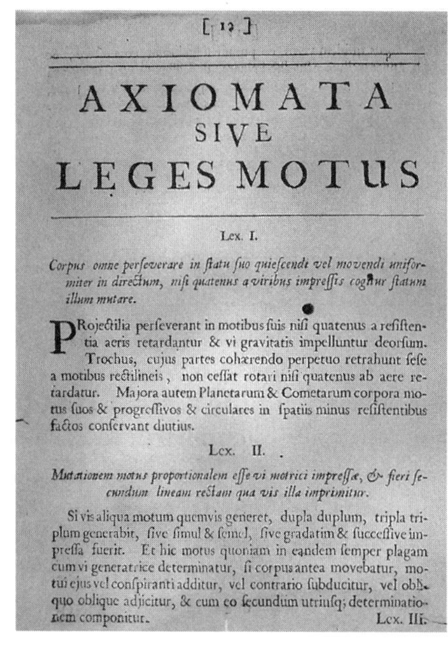

[12]

AXIOMATA
SIVE
LEGES MOTUS

Lex. I.

Corpus omne perſeverare in ſtatu ſuo quieſcendi vel movendi uniformiter in directum, niſi quatenus a viribus impreſſis cogitur ſtatum illum mutare.

PRojectilia perſeverant in motibus ſuis niſi quatenus a reſiſtentia aeris retardantur & vi gravitatis impelluntur deorſum. Trochus, cujus partes cohærendo perpetuo retrahunt ſeſe a motibus rectilineis, non ceſſat rotari niſi quatenus ab aere retardatur. Majora autem Planetarum & Cometarum corpora motus ſuos & progreſſivos & circulares in ſpatiis minus reſiſtentibus factos conſervant diutius.

Lex. II.

Mutationem motus proportionalem eſſe vi motrici impreſſæ, & fieri ſecundum lineam rectam qua vis illa imprimitur.

Si vis aliqua motum quemvis generet, dupla duplum, tripla triplum generabit, ſive ſimul & ſemel, ſive gradatim & ſucceſſive impreſſa fuerit. Et hic motus quoniam in eandem ſemper plagam cum vi generatrice determinatur, ſi corpus antea movebatur, motui ejus vel conſpiranti additur, vel contrario ſubducitur, vel obliquo oblique adjicitur, & cum eo ſecundum utriuſq; determinationem componitur.

Lex. III.

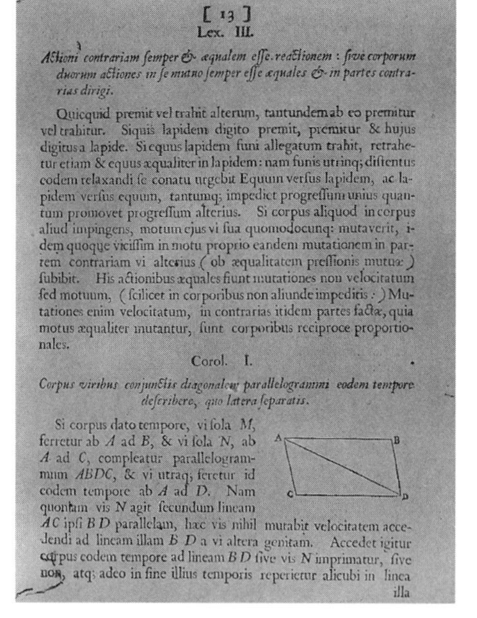

[13]

Lex. III.

Actioni contrariam ſemper & æqualem eſſe reactionem: ſive corporum duorum actiones in ſe mutuo ſemper eſſe æquales & in partes contrarias dirigi.

Quicquid premit vel trahit alterum, tantundem ab eo premitur vel trahitur. Siquis lapidem digito premit, premitur & hujus digitus a lapide. Si equus lapidem funi allegatum trahit, retrahetur etiam & equus æqualiter in lapidem: nam funis utrinq; diſtentus eodem relaxandi ſe conatu urgebit Equum verſus lapidem, ac lapidem verſus equum, tantumq; impediet progreſſum unius quantum promovet progreſſum alterius. Si corpus aliquod in corpus aliud impingens, motum ejus vi ſua quomodocunq; mutaverit, idem quoque viciſſim in motu proprio eandem mutationem in partem contrariam vi alterius (ob æqualitatem preſſionis mutuæ) ſubibit. His actionibus æquales fiunt mutationes non velocitatum ſed motuum, (ſcilicet in corporibus non aliunde impeditis:) Mutationes enim velocitatum, in contrarias itidem partes factæ, quia motus æqualiter mutantur, ſunt corporibus reciproce proportionales.

Corol. I.

Corpus viribus conjunctis diagonalem parallelogrammi eodem tempore deſcribere, quo latera ſeparatis.

Si corpus dato tempore, vi ſola *M*, ferretur ab *A* ad *B*, & vi ſola *N*, ab *A* ad *C*, compleatur parallelogrammum *ABDC*, & vi utraq; feretur id eodem tempore ab *A* ad *D*. Nam quoniam vis *N* agit ſecundum lineam *AC* ipſi *BD* parallelam, hæc vis nihil mutabit velocitatem accedendi ad lineam illam *BD* a vi altera genitam. Accedet igitur corpus eodem tempore ad lineam *BD* ſive vis *N* imprimatur, ſive non, atq; adeo in fine illius temporis reperietur alicubi in linea

illa

FIGURE 3.5 Pages from the original (Latin) edition of Newton's *Philosophiae Naturalis Principia Mathematica (The Mathematical Principles of Natural Philosophy)* presenting the three laws of motion and the parallelogram rule for the addition of forces (See Secs. 3.3 and 3.4).

We must now define exactly what is meant by an "unbalanced force." In fact, what is a force?

3.2 FORCES IN EQUILIBRIUM

The word "force" still has the connotation of violence that Aristotle gave it. In physics, it refers to a push or a pull. You know without having to think about it that forces can make things move. Forces can also hold things still. A cable supporting the main span of the Golden Gate Bridge is under the influence of mighty forces, yet it remains at rest.

This is not surprising. Think of two children quarreling over a toy. If each child pulls with equal determination in the opposite direction, the toy goes nowhere. In this case we say that the forces are "balanced." Or think of a grand chandelier hanging from a ceiling. The pull down by gravity is exactly balanced by the pull upward on the chandelier by the cord and chain on which it hangs. In all such cases we can think of forces in terms of arrows which represent the sizes and directions of the forces in a given situation. As with velocity, forces have both a magnitude (size) and a direction. As discussed in Section 1.5, we call any physical quantity that has both a

FIGURE 3.6 The Golden Gate Bridge in San Francisco—one example of an exquisite arrangement of balanced forces on the structure.

FIGURE 3.7 Children playing tug-of-war. Until one side pulls the other with acceleration, the net force is zero.

magnitude and a direction a "vector," and represent it in a diagram by an arrow. These arrows, representing vectors, can be added up graphically to obtain the resultant, for example, the "net force." (Vectors are discussed further in the next section.) As you can see from Figure 3.8, the two equally strong but opposing forces are represented by arrows equal in size but opposite in direction. The force pulling to the left is "balanced" exactly by the force pulling to the right. So these two forces cancel. The "net force" on the puppet is zero in this case.

Now suppose the child on the left suddenly makes an extra effort and pulls even harder. At that point the forces of course no longer balance each other. There is a net force to the left, and the toy starts to go from rest into motion to the left (Figure 3.8b).

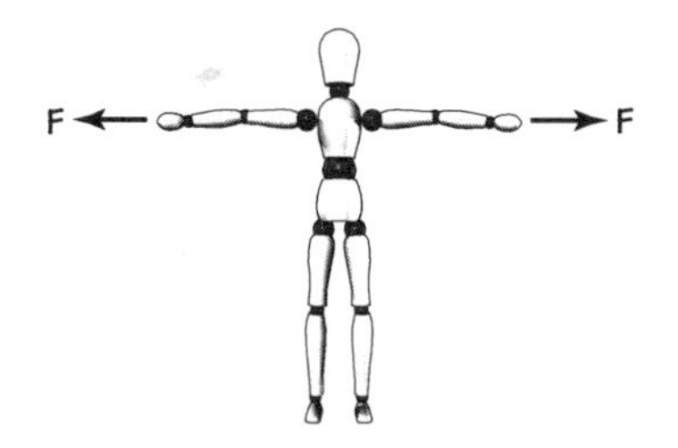

FIGURE 3.8a Diagram of equal forces on a doll.

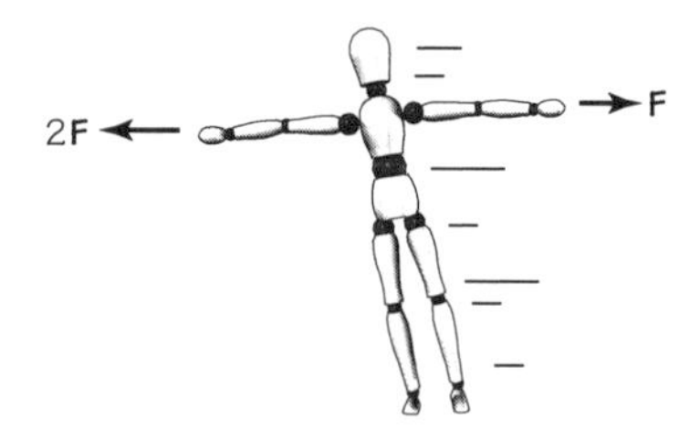

FIGURE 3.8b Diagram of unequal forces on a doll.

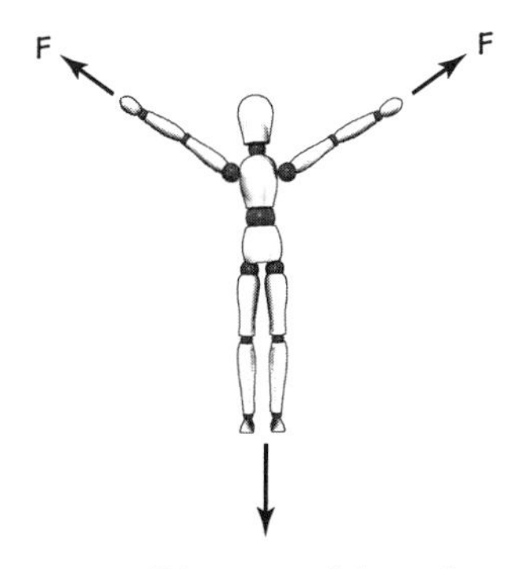

FIGURE 3.8c Diagram of three forces on a doll (non-parallel).

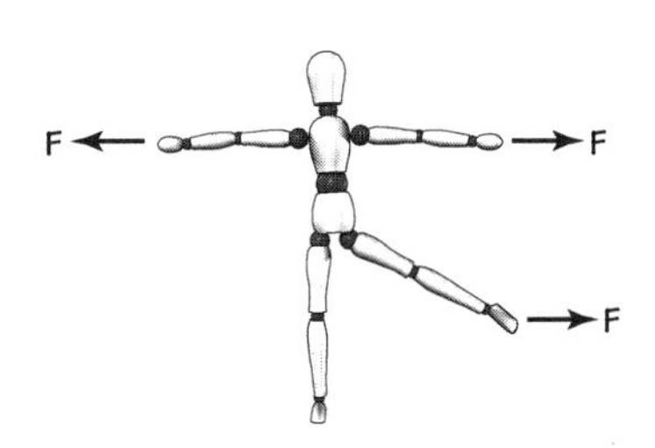

FIGURE 3.8d Diagram of three forces on a doll (parallel).

Suppose now that a third child eyes the toy and tries to grab it from the other two. Perhaps the three children pull in the directions shown in Figure 3.8c. Again there are forces on the toy, but they all balance out, the net force is zero, so the toy stays at rest. Finally, suppose that the original two children decide to "join forces" against the intruder and pull together parallel to each other. As shown in Figure 3.8d, their two forces add together, and the net result of all three forces is in their direction. The result is: they "win" (if the toy is not already torn apart).

The point to keep in mind most of all here is that you can have many forces all acting on an object at once and in different directions; but the net force is zero if the forces are balanced, and in that case the object stays in its original state of motion, either at rest or in uniform velocity. This is the meaning of Newton's first law of motion.

On the other hand, whenever there is an unbalanced or net force on an object, such as a toy, a car, or the space shuttle taking off, a change in its velocity must occur.

This leads to the next step: as you know from Chapter 1, whenever there is a change in uniform velocity there is, by definition, an *acceleration*. So, net forces are related to accelerations. "How" they are related is the subject of one of the mightiest laws of physics, Newton's second law of motion. But first, let's look a little closer at vectors.

3.3 MORE ABOUT VECTORS

As just discussed, forces belong in a class of concepts called *vector quantities*, or just *vectors* for short. Some characteristics of vectors are easy to represent by arrows. It is not obvious that forces should behave like arrows. But arrows drawn on paper happen to be useful for calculating how forces

FIGURE 3.9 Vector sums.

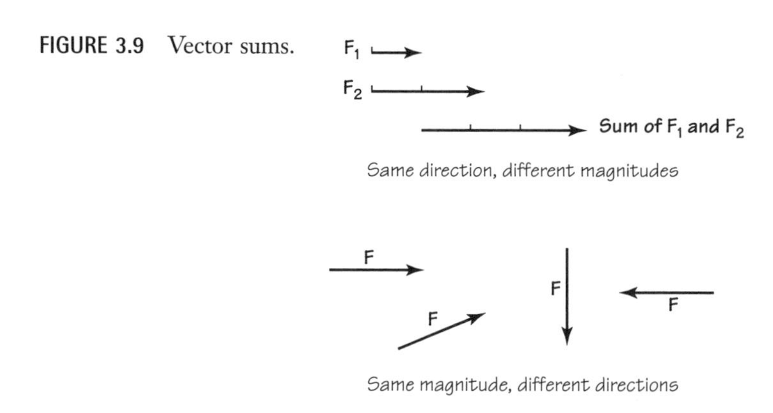

add. (If arrows weren't useful, we simply would look for other symbols that did work.) In particular, vector quantities have *magnitude*, which can be represented by the length of an arrow drawn to scale. They also have *direction*, which can be shown by the direction of an arrow. By experiment, we find that vectors can be *added* in such a way that the total effect of two or more vectors can be represented by the addition of the corresponding arrows placed tail to head. The total effect is represented by a new arrow, from the tail of the first to the head of the last, and is called the *vector resultant*, or the vector sum.

One example of a vector sum involves the distance traveled in a round trip over several days. Each leg of the trip undertaken each day may be represented by a displacement vector, each of which has both a scalar distance traveled and a direction. Since each leg of the trip begins at the point where the previous day left off, the displacement during the course of the trip may be represented by a series of arrows arranged tail to head. The length of each arrow represents the net distance traveled each day, the direction of the arrow represents the direction of the finishing point from the starting point. The net vector displacement from the start at the end of each day can be represented by a vector from the start to the head of the last vector. This is how vectors are added together to form the vector resultant.

One surprising result of this definition is that by the end of the round trip the vector resultant is zero! In other words, while the scalar distance traveled may have been many kilometers, the net displacement was zero. In terms of vectors, the traveler has gone nowhere.

Similar results occur in the example of the tug-of-war between two tots over a toy, where we determined the total effect of equal but opposing vec-

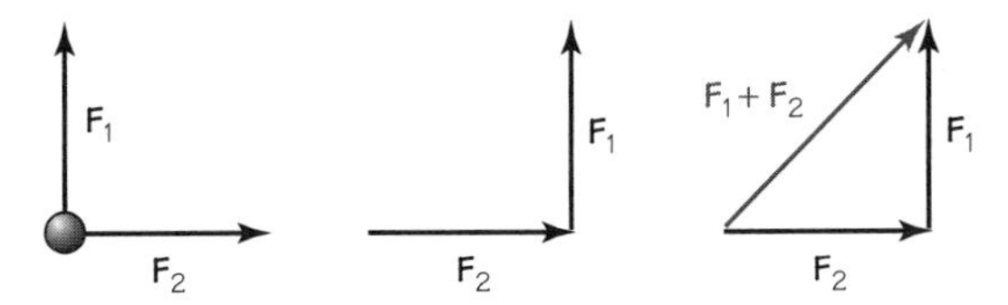

FIGURE 3.10 Adding vectors $\mathbf{F_1}$ and $\mathbf{F_2}$ head to tail to achieve the resultant $\mathbf{F_1} + \mathbf{F_2}$.

tor forces. If two forces act in the *same* direction, the resultant force is found in much the same way as two displacement vectors.

If two forces act at some angle to each other, the same type of sketch is still useful. For example, suppose two forces of equal magnitude are applied to an object at rest but free to move. One force is directed due east and the other due west. The resultant is found by moving each of the vectors parallel to each other and placing them tail to head. The resultant is the vector drawn from the tail of the first vector to the head of the last vector. In this case, this results in a vector of zero length. The object remains in equilibrium, without any acceleration, even though it is under the stress of enormous forces.

If one force on the object is directed due east and the other is directed due north, the object will accelerate in the northeast direction, the direction of the resultant force (see Figure 3.10.) The magnitude and direction of the resultant force vector are represented by the length and direction of the arrow representing the resultant.

The same addition procedure applies to forces of any magnitude and acting at any angle to each other. Suppose one force is directed due east and a somewhat larger force is directed northeast. The resultant vector sum can be found as shown in Figure 3.11.

To summarize, a vector quantity has both direction and magnitude. Vectors can be added by constructing a head-to-tail arrangement of vector arrows (graphical method). An equivalent technique, known as the parallel-

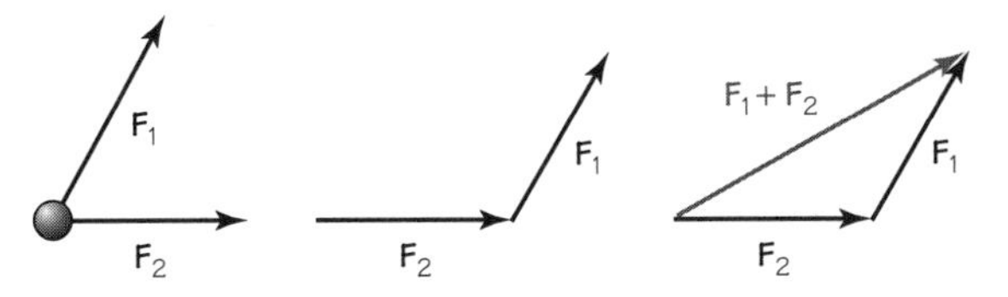

FIGURE 3.11 Adding non-perpendicular vectors ("head-to-tail" method).

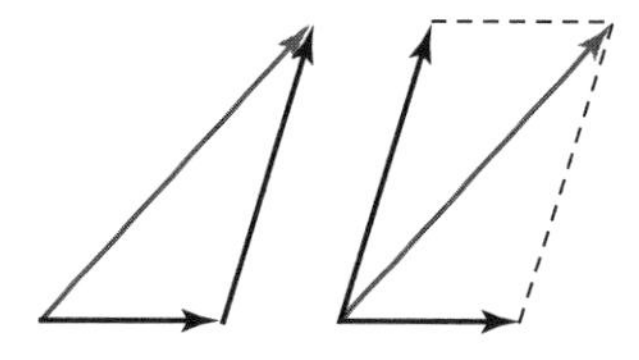

FIGURE 3.12 You can use equally well a graphical construction called the "parallelogram method." It looks different from the "head-to-tail" method, but it is actually equivalent. In the parallelogram construction, the vectors to be added are represented by arrows joined tail-to-tail instead of head-to-tail, and the resultant is obtained by completing the diagonal of the parallelogram.

ogram method, is briefly explained in Figure 3.12. Vectors also have other properties which you will study if you take further physics courses.

3.4 NEWTON'S SECOND LAW OF MOTION

You saw in Chapter 1 that Galileo made tremendous progress in describing motion, speed, acceleration, and free fall, but that he refrained from attempting to uncover the causes of these motions, leaving that to "the attention of other minds." One of the foremost "other minds" was Isaac Newton, who lived from 1642 to 1727. His second law of motion goes far beyond Galileo in attributing the cause of all changes of motion to the action of forces. Once again, it is important to know that Newton's laws of motion are *universal* laws of nature. This means that they apply not just to falling balls and inclined planes and toys pulled by tots, but to all matter everywhere and always—on Earth, in the solar system, and throughout the Universe. (However, at very high speeds, close to the speed of light, relativity theory introduces a modification of the kinematic quantities that enter into these laws.)

As noted earlier, the description of motion, without looking at its causes, is called *kinematics*. The study of motion that involves the causes of changes of motion is called *dynamics*. Together, kinematics and dynamics form the science of *mechanics*, and at the center of this new science are Newton's three laws of motion.

Newton's second law of motion has to do with force and acceleration. We know what acceleration is (see Chapter 1) and we know that force has to do with pushes and pulls generated either by people or by objects, such as magnets or springs.

Newton's second law provides an answer to the question: *What happens when an unbalanced force acts on an object?* Since we think of the force as *causing* the resulting motion, this law is the fundamental law of dynamics.

In qualitative terms, the second law of motion says little more than this: *The reason for a deviation from "natural" motion (i.e., motion with a constant or zero velocity) is that a nonzero net force acts on the object.* However, the law tells us much more: it provides a simple *quantitative* relationship between the change in the state of motion—the acceleration—and the net force. As usual in science, the best way to discern quantitative relationships between two concepts is to head for the laboratory. We will first consider a situation in which different forces act on the same object; then we will consider a situation in which the same force acts on different objects; finally, we will combine the two results into a general relationship. You might do these experiments yourself. We will describe only briefly the procedures and the results here.

Different Forces Acting on the Same Object

In our first experiment we accelerated an object with a steady force so that it accelerated continuously. The object was a cart on wheels but it could be a dry-ice disk or any other nearly frictionless object on a flat table. The force was produced by the horizontal pull provided by a string attached over a pulley to a descending weight. We determined the acceleration ($\mathbf{a}$) by measuring the distance between the dots made every 0.1 s on a tape pulled by the cart.

Repetitions of this experiment with different amounts of force ($\mathbf{F}$) on the same cart produced by different amounts of weight yielded the following results, in words and symbols:

The force $\mathbf{F}$ caused an acceleration of $\mathbf{a}$.

The force $2\mathbf{F}$ caused an acceleration of $2\mathbf{a}$.

The force $\frac{1}{2}\mathbf{F}$ caused an acceleration of $\frac{1}{2}\,\mathbf{a}$.

The force $5.2\mathbf{F}$ caused an acceleration of $5.2\mathbf{a}$.

As you can see, as the force increased or decreased, the acceleration increased or decreased by the exact same amounts. What can you conclude from these results about the general relationship between the force and the acceleration? Here is our conclusion:

> The acceleration of an object is directly proportional to, and in the same direction as, the net force acting on the object.

If $\mathbf{a}$ stands for the acceleration of the object and $\mathbf{F}_{\text{net}}$ stands for the net force on it, this statement may be expressed in symbols as follows:

$$\mathbf{F}_{\text{net}} \propto \mathbf{a}.$$

Notice that the force and the acceleration are both *vector* quantities. In our experiment, the force provided by the horizontal pull of the string and the acceleration of the cart to which the string is attached were both in the same direction. It is understood that when *vectors* are proportional, they must point in the same direction as well as have proportional magnitudes. We used the net force in this conclusion because there was still some friction opposing the force with which we pulled the cart.

To say that two quantities are proportional means that if one quantity is doubled (or multiplied by any number), the other quantity is also doubled (or multiplied by the same number). Thus, as our data indicated, if a certain force produces a certain acceleration, twice the force (on the same object) will produce twice as great an acceleration in the same direction.

The Same Force Acting on Different Objects

Once again we tried an experiment, this time to see what happens when the same force acts on different objects. We can select different objects in different ways, such as by size, shape, density, or weight. In terms of force and motion, one of the most important differences among objects is the amount of matter, or mass, they contain. We can see this when we attempt to apply the same force to a child's wagon and to a parked car in neutral. Each contains a different amount of matter, which is called mass. (Mass is further discussed below.)

In this experiment the different objects are the same cart on wheels but loaded with objects of different masses. In each case we applied the same horizontal force to the cart, by using the same system of a string, a pulley, and a suspended weight. (You can perhaps try this with other setups.) Again we determined the acceleration ($\mathbf{a}$) by measuring the distance between the dots made every 0.1 s on a tape pulled by the cart.

Repetitions of this experiment with different total masses m for the cart and its load yielded the following results, in words:

The mass m experienced an acceleration of $\mathbf{a}$.

The mass $2m$ experienced an acceleration of $\frac{1}{2}\,\mathbf{a}$.

The mass $\frac{1}{3}m$ experienced an acceleration of $3\mathbf{a}$.

The mass $5.2m$ experienced an acceleration of $(1/5.2)\mathbf{a}$.

This time you can see that, as the mass changed, the acceleration changed by exactly the inverse amount. We can conclude from this:

> The acceleration of an object is inversely proportional to the mass of the object. The larger the mass of an object, the smaller will be its acceleration if a given net force is applied to it.

This relationship between acceleration and mass may also be written in symbols. Let a stand for the acceleration (the *magnitude* of the acceleration vector $\mathbf{a}$) and m stand for mass. Then,

$$a \propto 1/m$$

as long as the same net force is acting. Notice that an object with twice (or three times, or whatever) the mass of another object will experience one-half (or one-third, or one-whatever) the acceleration if subjected to the same net force.

Notice also that it is the mass of an object that determines how large a force is required to change its motion. In other words, the mass of an object is a measure of its inertia. It is sometimes called *inertial mass*, to emphasize that it measures inertia. Mass is a *scalar* quantity (it has no direction), and it does not affect the direction of the acceleration.

The General Relationship Between Force, Mass, and Acceleration

We concluded from our two experiments that the acceleration of an object is proportional to:

(1) the net force (for constant m); and
(2) to $1/m$, the reciprocal of the mass (for constant $\mathbf{F}_{\text{net}}$).

It follows from these two proportions that the acceleration is proportional to the product of the net force and the reciprocal of the mass:

$$\mathbf{a} \propto \mathbf{F}_{\text{net}} \cdot \frac{1}{m}.$$

Are there any other quantities (other than net force and mass) on which the acceleration depends? Newton proposed that the answer is no. Only the *net force on the object* and the *mass of the object* being accelerated affect the acceleration. All experience since then indicates that he was right.

Since there are no other factors to be considered, we can make the proportionality into an equality; that is, we can write

$$\mathbf{a} = \frac{\mathbf{F}_{\text{net}}}{m}.$$

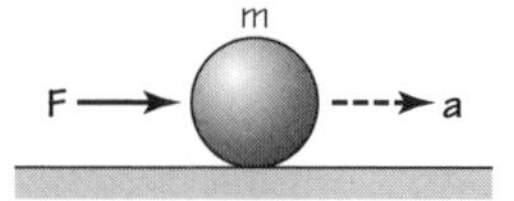

FIGURE 3.13 Ball accelerated by an applied force.

This relationship can also be written as the famous equation

$$\mathbf{F}_{\text{net}} = m\mathbf{a}.$$

In both these equations we have again written $\mathbf{F}_{\text{net}}$ to emphasize that it is the *net* force that determines the acceleration.

This relationship is probably the most basic equation in mechanics and, therefore, in physics. Without symbols we can state it as follows:

> Newton's second law of motion: The net force acting on an object is numerically equal to, and in the same direction as, the acceleration of the object multiplied by its mass. In symbols: $\mathbf{F} = m\mathbf{a}$.

It does not matter whether the forces that act are magnetic, gravitational, simple pushes and pulls, or any combination of these; whether the masses are those of electrons, atoms, stars, or cars; whether the acceleration is large or small, in this direction or that. The law applies universally. And if you think about this and other "universal" laws of science, it is remarkable that there are only two disciplines—physics and mathematics—that have found laws or statements that apply everywhere in the Universe!

3.5 MEASURING MASS AND FORCE

How should we define the units of mass and force that are used in the Newton's laws of motion? The answer depends on the meaning of these two concepts. As you just saw, the *mass* of an object is its inertia, its resistance to changes in its motion. A net force, on the other hand, is what pushes or pulls against the mass to produce an acceleration. If you apply a force to a small mass, you would produce a larger acceleration than if you applied the same force to a larger mass. For instance, pushing a new shopping cart over a smooth surface would produce a larger acceleration than pushing a truck in neutral over the same surface with the same force.

Measuring mass (or inertia) and defining its units are quite different from measuring force and defining its units. When you think of measuring the mass of an object, you might first think of weighing it on a scale—and some laboratory scales are calibrated to convert the weight of the object into a measure of its mass. *But mass and weight are two different concepts.* In the most basic terms: *mass* is the inertia of an object; *weight* is the gravitational force exerted on the object.

The stretch of a spring can be used as a measure of force, since the amount of stretch is proportional to the force exerted. (You may have observed this in the laboratory, as suggested in the *Student Guide*.) If you take apart a typical household scale—a bathroom scale, for instance—you will find that it usually just consists of a spring that compresses as you stand on it. This indicates that it measures your weight, not your inertia or mass, when you stand on it, since the contraction of the spring inside the scale exerts on you an upward force that balances the gravitational force you experience downward. The amount of contraction is also proportional to the force exerted and the scale gives a reading of the corresponding amount of force, which is your weight at your location on the Earth. However, if you stand on your bathroom scale on the Moon, it will show a much smaller *weight*, since the gravitational pull on you downward is less on the Moon than on the Earth. Nevertheless, you will have just the same mass, or inertia.

Since the mass of an object is the same for that object *everywhere*, while the weight depends on the location, the wise thing to do is to choose the unit of mass first, and let the unit of force follow later. The simplest way to define a unit of mass is to choose some convenient object as the *universal standard of mass*, and then compare the masses of all other objects with that one. What is selected to serve as the standard object is arbitrary. In Renaissance England the standard used was a grain of barley ("from the middle of the ear"). The original metric commission, established in France in 1799 in the wake of the French Revolution, proposed the mass of a cubic centimeter of water as the standard mass. Today, for scientific purposes, the standard mass is, by international agreement, a cylinder of platinum–

FIGURE 3.14 Person on spring scale.

iridium alloy kept at the International Bureau of Weights and Measures near Paris. The mass of this cylinder is *defined* to be 1 kilogram (kg), or 1000 grams (g). Accurately made copies of this cylinder are used in various standards laboratories throughout the world to calibrate precision equipment. Further copies are made from these for distribution. (The mass of 1 cm^3 of water is still exactly 1 g, but only under very specific conditions of temperature and pressure. The mass of 1 liter (l) of water under standard conditions is 1000 g, or 1 kg.)

The international agreements that have established the kilogram as the unit of mass also established units of length and time. The meter (m) was originally defined in terms of the circumference of the Earth, but modern measurement techniques make it more precise to define the meter in terms of the wavelength of light, generated in a specific way. The second of time (s) was also originally defined with respect to the Earth (as a certain fraction of the year), but it, too, is now more precisely defined in terms of light waves emitted by a specific group of atoms. The meter and second together determine the units of speed (m/s) and acceleration (m/s^2). The crash of a Mars probe due to a mix-up over the use of the metric or American system of units in guiding the probe is a good example of why it is so important to define and use only one set of units for all measurements.

With the metric units defined so far, we can now go back and calibrate the spring balances used for measuring force. While in the American system of units, the unit of force is the pound or ounce, in the metric system

FIGURE 3.15 The standard kilogram, kept at the Bureau des Poids et Mesures.

the unit of force is called, appropriately the "newton" (N), in honor of Isaac Newton. Again by international agreement, 1 N of force is *defined* to be the amount of net force required to accelerate a mass of 1 kg at the uniform acceleration of one meter per second every second, or, in symbols, 1 m/s/s, which may be written 1 m/s^2. Because of Newton's second law ($\mathbf{F}_{net} = m\mathbf{a}$), we have

$$\mathbf{F}_{net} = m\mathbf{a}$$

$$1\ \text{N} = 1\ \text{kg} \times 1\ \text{m/s}^2,$$

$$1\ \text{N} = 1\ \text{kg m/s}^2.$$

If we are working with the smaller metric units of centimeter (cm) and grams (g), then the unit of force is no longer a newton, since the definition of the newton involves meters, kilograms, and seconds. For centimeters, grams, and seconds another unit of force has been defined. It is called the "dyne" with the symbol D. In this case,

$$1\ \text{D} = 1\ \text{g cm/s}^2.$$

The kilogram (kg), meter (m), and second (s) are the fundamental units of the "mks" system of measurement. The gram (g), centimeter (cm), and second (s) are the fundamental units of the "cgs" system of measurement. Together with units for light and electricity, the mks and cgs units form the International System of units, known as SI (SI stands for *Système Internationale*). Other systems of units are possible. But since the ratios between related units are more convenient to use in a decimal (metric) system, all scientific and technical work, and most industrial work, is now done with SI units in most countries, including the United States.

Example. A 900-kg car accelerates from rest to 50 km/hr in 10 s (50 km/hr is equal to 13.9 m/s). What is the net force on the car?

We need to find the acceleration first

$$a = \frac{\Delta v}{t} = \frac{13.9\ \text{m/s}}{10\ \text{s}} = 1.39\ \text{m/s}^2.$$

We can now find the force

$$F = ma = (900\ \text{kg})(1.39\ \text{m/s}) = 1251\ \text{kg m/s}^2 = 1251\ \text{N}.$$

Example. A 5-g ping-pong ball at rest is hit by a paddle, propelling it to 50 cm/s in 0.1 s. What is the force on the ball?

Again we need to find the acceleration first

$$a = \frac{\Delta v}{t} = \frac{50 \text{ cm/s}}{0.1 \text{ s}} = 500 \text{ cm/s}^2,$$

$$F = ma = (5 \text{ g})(500 \text{ cm/s}^2) = 2500 \text{ g cm/s}^2 = 2500 \text{ D}.$$

3.6 MORE ABOUT WEIGHT, AND WEIGHTLESSNESS

Whenever you observe an object in acceleration, you know there is a force acting on the object. Forces need not be exerted by contact only (often called "mechanical forces"). They can also result from gravitational, electric, magnetic, or other actions originating from objects that are not touching the object that is being accelerated. Newton's laws of motion hold for all of these forces.

The force of gravity acts between objects, attracting them to each other, even if they are not in direct contact. As you know, the gravitational force exerted by the Earth on an object is called the *weight* of the object. Your weight is the downward force that the Earth exerts on your mass, whether you stand or sit, fly or fall, or merely stand on a scale to "weigh" yourself.

The symbol $\mathbf{F}_g$ is often used for gravitational force. The size, or magnitude, of the gravitational pull $\mathbf{F}_g$ on a given mass is roughly the same everywhere on the surface of the Earth. This is because, for objects on the surface, the magnitude of the force of gravity depends on only the mass of the object and the distance from the center of the Earth. Since the Earth is nearly a sphere, the magnitudes of the gravitational force on a given mass at different places on the Earth's surface are nearly the same. Since the gravitational force is a vector, it also has a direction. It is always to the center of the Earth, which means that it is perpendicular to the level ground at every point on the Earth's surface. Scientists since Newton's time have known that an object having a mass of 1.0 kg experiences a gravitational force of about 9.8 N everywhere on the surface of the Earth. An object of mass 2 kg would experience a force twice that of a 1-kg object, or about 19.6 N; a 10-kg object would have a force of 98 N, and so on.

However, these weights are not exactly the same everywhere on the surface of the Earth, since the Earth is not a perfect sphere, nor is it homogeneous. For instance, an object having a mass of 1 kg will experience a gravitational force of 9.812 N in London, but only 9.796 N in Denver, Colorado. Geologists make use of minute variations in locating oil and mineral deposits.

You can sense the magnitude of the gravitational force on an object, for example, the weight of a 1-kg ball, by holding it stationary in your hand. Why doesn't the ball accelerate if there is a gravitational force on it? . . . Remember what Newton's first law of motion tells us. If you hold the ball stationary in your hand, your hand is exerting an upward force that exactly balances the force of gravity downward. It is the upward force that you provide which equals (numerically) the weight of the ball. Since all forces on the ball are balanced, it remains stationary in your hand.

Now release the ball. What happens? The gravitational force is no longer balanced. Rather, it acts as a net force on the 1-kg ball of 9.8 N downward, so the ball begins to accelerate. What is the rate of acceleration? We can obtain this from Newton's second law:

$$\mathbf{F}_g = m\mathbf{a},$$

so

$$\mathbf{a} = \frac{\mathbf{F}_g}{m} = \frac{9.8 \text{ N}}{1.0 \text{ kg}} = 9.8 \text{ N/kg} = 9.8 \text{ m/s}^2 \quad \text{downward.}$$

Now try this with a 2.0-kg ball. As given above, there is a larger force of gravity of 19.6 N downward. What is the acceleration in this case?

$$\mathbf{a} = \frac{\mathbf{F}_g}{m} = \frac{19.6 \text{ N}}{2.0 \text{ kg}} = 9.8 \text{ N/kg} = 9.8 \text{ m/s}^2 \quad \text{downward.}$$

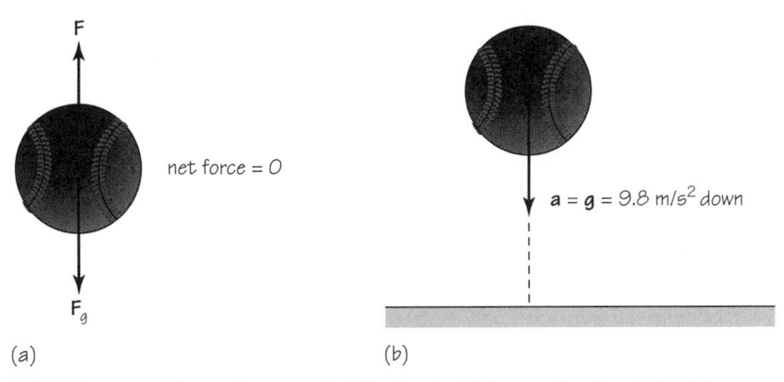

FIGURE 3.16 (a) net force on ball being held in gravitational field is zero; (b) the ball falls with acceleration g if it is not held in the gravitational field.

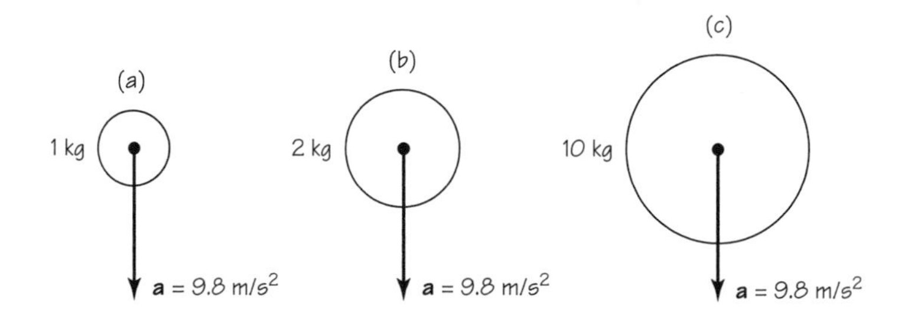

FIGURE 3.17 Acceleration is constant for freely falling balls of different mass.

Finally, the 10-kg ball:

$$\mathbf{a} = \frac{\mathbf{F}_g}{m} = \frac{98 \text{ N}}{10 \text{ kg}} = 9.8 \text{ m/s}^2 \quad \text{downward.}$$

You see the pattern that is emerging. In each case the mass was different and the force of gravity was different, *but the acceleration was the same.*

Now we can understand more clearly the results of Galileo's experiments on falling objects. You may recall (or see Chapter 1) that he found that, in the absence of friction, all objects fall at the same location with the same uniform acceleration. Different masses have different weights, but because they are about the same distance from the center of the Earth, the ratio of weight (force of gravity) to mass always gives the same result at the same location on the surface of the Earth, a uniform acceleration of about 9.8 m/s^2 (give or take a fraction of a percent at different locations around the Earth). This acceleration is called the *acceleration of gravity*. It is often given the symbol **g**. It is a vector quantity with a magnitude 9.8 m/s^2 (or 980 cm/s^2 or 32 ft/s^2) and a direction downward toward the center of the Earth at each place on the surface of the Earth. Free fall is an example of uniform acceleration. All objects fall with the same acceleration, $\mathbf{g} = 9.8$ m/s^2 downward, at the same location on the surface of the Earth. The weight, mass, and acceleration of all objects are related to each other by Newton's second law: $\mathbf{F} = m\mathbf{a}$.

What about weightlessness, such as experienced by astronauts orbiting in the space shuttle? How does that arise? Gravity constantly pulls us and every material object down toward the center of the Earth, but our body hardly "feels" the gravitational force itself. What you can experience directly is the *acceleration* due to gravity as, for instance, when you fall through the air when you dive into a pool. If you stand on a scale to weigh yourself, the spring in it compresses, as noted earlier, until it exerts an upward force on you strong enough to balance the downward force of gravity. The compression of the spring registers on the scale, which indicates your weight in pounds or in newtons. So, what you actually experience as "weight" when you stand on a scale, sit in a chair, or walk on a sidewalk is not the force of gravity but the opposing force of the scale or chair or ground.

Other bodies in the solar system exert the same kind of gravitational forces on objects on their surfaces. For instance, the Moon exerts a gravitational force on objects on or near it, but that force is much smaller than the pull of the Earth at its surface. The astronauts who landed on the surface of the Moon felt the effects of the pull of the Moon, which they experienced as weight, but, since that force was much smaller than on the surface of the Earth, they felt much "lighter" than on Earth and had a grand time bouncing around on the surface.

From the discussion so far you can see that true weightlessness can occur only when there is no net gravitational force pulling on an object—for example, in space, far from any sun or planet. In that case there would be no opposing force that you would experience as your weight. In such a spaceship, when you stand on a scale, it would register zero. But astronauts orbiting the Earth *seem* to be weightless, floating in their spaceship. Does this mean that they are beyond the Earth's pull of gravity and are therefore really weightless? The answer is NO. They are still being pulled by gravity, so they do have weight. But they cannot experience this weight because, while they orbit the Earth or the Moon, they are in free fall! (We will come to explain this curious phenomenon in Section 3.11.)

FIGURE 3.18 Astronauts in "weightless" environment.

In order to get a start on what happens with the apparent weightlessness during orbiting, try taking a bathroom scale into an elevator. While standing on it, notice what happens as the elevator briefly accelerates upward and downward. If you do so, you will notice that the scale registers more than your normal weight when accelerating upward, and less when accelerating downward.

Now imagine for a moment a ridiculous but instructive thought experiment. Suppose you are weighing yourself on a scale in an elevator or room on Earth. As you stand on the scale, the floor suddenly gives way. Like Alice in Wonderland, you and the scale drop into a deep well in free fall. At every instant, your increasing speed of fall and the scale's increasing speed of fall are equal, since all objects fall on Earth with the same acceleration. In fact, any objects that fell into the well with you will fall with you at the same increasing speed as you are experiencing. But to you they appear to float with or around you. Your feet now touch the scale only barely (if at all). Before you awake from this nightmare, you look down at the dial on the scale and see that the scale registers zero because the spring inside is not being compressed. This does not mean that you have lost your weight (the force of gravity on you). Gravity still acts on you as before. In fact, it is the dominant force on you, accelerating you downward. But since the scale is accelerating with you, you are no longer pushing down on it, nor is it pushing up on you. You *feel* weightless, but you still have weight; you just cannot experience it while you are falling. What you do experience is an apparent, but not real, weightlessness.

3.7 NEWTON'S THIRD LAW OF MOTION

In his first law, Newton described the behavior of objects when there are no forces acting on them or when the forces all balance, yielding a net force of zero. His second law explained how the motion of objects changes when the net force is not zero. Newton's third law added an original, new, and surprising insight into forces.

Consider this problem: In a 100-m dash, an athlete goes from rest to nearly top speed in less than 3 s. We could measure the runner's mass before the dash, and we could use high-speed photography to obtain the acceleration. With the mass and acceleration known, we could find the net force acting on the sprinter during the initial acceleration. But where does this force come from? Obviously it must have something to do with the runner himself. But is it possible for him to exert a force on himself as a whole?

Newton's *third law of motion*, also called the *law of action and reaction*, helps to explain just such puzzling situations. In modern language it states:

> If one object exerts a force on another object, the second object at the same time exerts a force on the first object. These two forces, each acting on one of the two objects, are equal in magnitude and opposite in direction.

The startling idea in this statement is that forces always act in pairs, one force acting on one object, the other acting on another object. A single force acting alone, without another force acting back on something else, does not exist in nature. For example, consider the sprinter. When the gun goes off to start the dash, his act of pushing with his feet *back* against the starting blocks (call it the "action") involves simultaneously a push by the starting blocks of an equal amount acting on him in the *forward* direction (call it the "reaction"). It is the reaction by the blocks that propels him forward. The action does not "cause" the reaction; the two forces simply coexist. If somehow the starting blocks came loose from the ground so that they cannot push back on his feet, they would just slide away when he tried to give them a big push, rather than providing the reacting force and the acceleration he needs to get started on the sprint.

A common mistake is to think that these action and reaction forces can balance each other to zero, and give equilibrium, as in the first law of motion. But in fact the two forces do not act on the same object; *each acts on a different object*, so they can't balance out. It is like debt and credit. One is impossible without the other; they are equally large but of opposite sign, and they happen to two different accounts.

Every day you see many examples of Newton's third law of motion at work. A car is set in motion by the push of the ground on the tires forward

FIGURE 3.19 Force diagram on feet of sprinter and on blocks.

■ NOW YOU TRY IT

Michael Johnson, the American sprinter, set a new world record in 1999 of 43.18 s for 400 m, breaking the previous record by 0.11 s.

(a) What was his average speed for the entire run?
(b) Assuming that he was able to accelerate to the average speed in 2.5 s, what was his average acceleration at the start?
(c) Assuming that his mass is about 75 kg, how large was the force on his legs while he accelerated? (Ignore air resistance.)
(d) Explain why the force produced by his legs on the starting block was *not* the force that accelerated him during the first 2.5 s.

FIGURE 3.20 Michael Johnson just off the blocks.

in reaction to the push of the tires on the ground backward. When friction is not sufficient, as when trying to start the car moving on ice, the car just spins its wheels in place because there is no reacting forward push of the ground on the tires. A tennis racket hits a tennis ball, accelerating the ball forward, even while the tennis ball exerts a force backward on the racket, causing tennis elbow in some cases, or even broken arm bones. The Earth exerts a force on an apple and the apple exerts an equally large force on the Earth. When the apple falls, pulled down by the gravitational pull of the Earth (the weight of the apple), the Earth, in turn, is pulled upward by the equal but opposite attraction of the Earth to the apple. Hence, during the apple's fall the Earth accelerates upward—though by only an infinitesimal amount. Of course, we don't notice this motion of the Earth because of the difference in mass between the Earth and the apple, but the effect is there and in all similar situations. Similarly, after the sprinter has left the starting block behind and runs forward (owing to the force the ground exerts *on* his feet), the Earth moves a little in the opposite direction because of the force applied to it *by* his feet. On a small enough planet, this might become noticeable!

The universal nature of Newton's third law of motion makes it, like the other two laws of motion, an extremely valuable and useful law in physics.

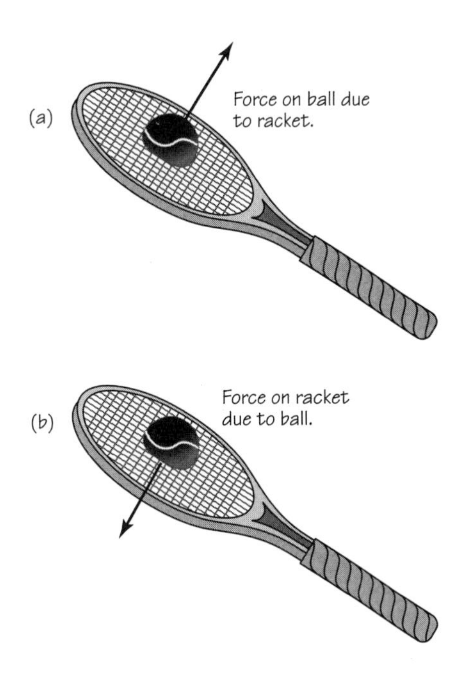

FIGURE 3.21 Tennis racquet: (a) Force on ball and (b) reaction force on racquet.

B. THE THREE LAWS IN ACTION

3.8 PROJECTILE MOTION

You have already seen various types of motion, and you have seen that forces account for changes in motion. Now let's look at some common motions that have major significance for physics, and for our understanding of events around us. The first type is projectile motion: the flight of an arrow shot from a bow, a baseball batted into the outfield, a golf ball hit horizontally from a platform, or an elementary particle shot out into an evacuated tube of an "accelerator." Once again, Galileo was the first person to comprehend fully what is happening here, and to use the results in making some very important conclusions about the heliocentric theory of the solar system.

Returning once again to Galileo's inclined plane (and again neglecting friction and air resistance), you may recall that he obtained two very important results. First, by extrapolation to a 90° angle for the incline, he showed that free fall near the Earth's surface is an example of uniform acceleration, with the acceleration of gravity for all freely falling objects. Second, at the other extreme, for zero-angle inclination for a second inclined plane, he concluded that once in motion a ball will maintain its uniform velocity forever, until something gets in the way or alters its course. Now he considered what happens when something *does* alter a ball's course—namely, what happens when the ball rolls so far in the laboratory that it eventually falls off the table? Galileo realized that a ball rolling with uniform velocity on a flat tabletop will maintain its uniform velocity (constant speed and direction) in the horizontal direction even after it leaves the edge of the table, since there is no net force in the horizontal direction to change its horizontal motion. However, once the ball leaves the tabletop the unbalanced force of gravity also sets it into free fall in the vertical direction. Therefore the ball has two motions simultaneously after it leaves the tabletop:

(1) uniform velocity in the horizontal direction;
(2) uniform acceleration downward, in the vertical direction.

These two motions are completely independent of each other. What happens to one of them has no effect on the other. As the ball flies off the table, the two motions are "compounded" together to form the curved motion of a projectile, or any other object flying through space. The motion of a projectile is the result of two independent motions, put together to form the observed "trajectory" of the object. That is what we now would call our new understanding of "natural" motion.

A real experiment using a stroboscopic camera will help clarify this idea better. In the famous photograph in Figure 3.22, a ball is fired horizontally. At the same moment a second ball is allowed to fall freely down to the same floor. Although the motion of the objects may be too rapid to follow with the eye, if you tried it your ears will tell you that they do in fact

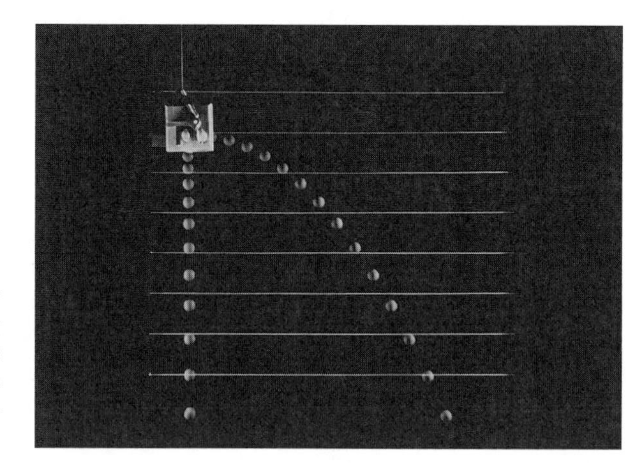

FIGURE 3.22 The two balls in this stroboscopic photograph were released simultaneously. The one on the left was simply dropped from a rest position; the one on the right was given an initial velocity in the horizontal direction.

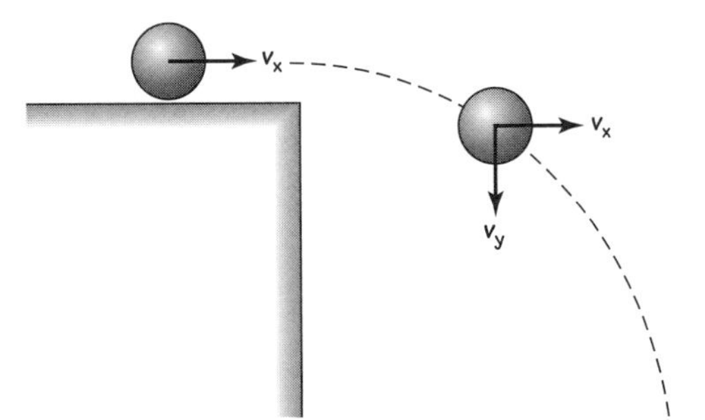

FIGURE 3.23 Schematic of ball on parabolic path.

impact the floor at the same time. This could happen only if they both fell with the exact same vertical motion, even though they had different horizontal motions. To see if this is indeed the case, with equally spaced horizontal lines were introduced into Figure 3.22 to aid the examination of the two motions. Look first at the ball on the left, the one that was released without any horizontal motion. You can see that it accelerates downward because it moves a greater distance downward between successive flashes of the strobe's light in equal time intervals. Now compare the vertical positions of the other ball, the one that was fired horizontally, with the vertical positions of the first ball falling freely. The horizontal lines show that the distances of fall are almost exactly the same for corresponding time intervals (air resistance has a tiny effect). Neglecting the effect of air resistance, the two balls are undergoing the same acceleration in the vertical direction. This experiment supports the idea that the vertical motion is the same whether or not the ball has a horizontal motion. *The horizontal motion does not affect the vertical motion*—something we recall that the Aristotelians did not believe possible.

But does the vertical motion affect the horizontal motion? To answer this question, measure the horizontal distance of the projected ball between each image. You will find that the horizontal distances are practically equal. Since the time intervals between producing each image are equal, you can conclude that the horizontal velocity is constant during the curved descent. *Therefore, the vertical motion does not affect the horizontal motion.* Together, the vertical and horizontal motions make up the path of the projectile that we see. What is this path? Obviously, it's some kind of a curve, as shown in the photograph. On examination it turns out that the actual path, or trajectory, is what is known as a parabola. (See the *Student Guide* for a derivation showing that it is indeed a parabola.)

You can see that the trajectory of a projectile is a parabola more clearly for a ball thrown into the air. Here you can see both halves of the parabola. The second half, the downward side, is equivalent to the motion of the ball

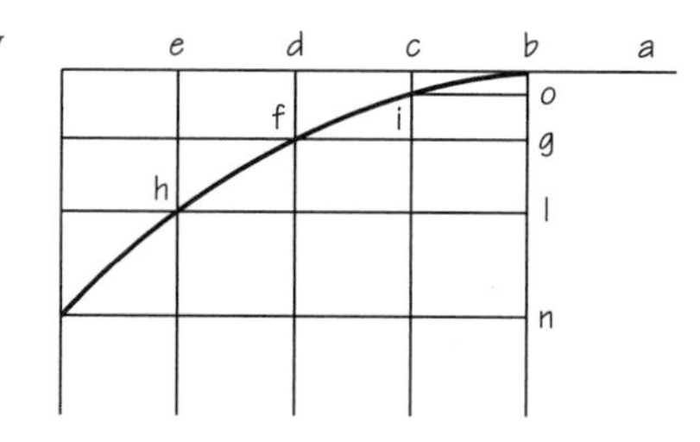

FIGURE 3.24 Drawing of a parabolic trajectory from Galileo's *Two New Sciences.*

projected straight out horizontally. On the upward side of the trajectory, the ball has uniform velocity in the horizontal direction and uniform acceleration downward due to gravity, as before, but it also has a vertical velocity, straight up. Gravity works against the vertical velocity, slowing it down until the vertical velocity finally reaches zero at the top of the arc. (*Note:* The vertical *velocity* reaches zero, but *not* the gravitational acceleration, which stays constant throughout.) At the top of the arc, the ball still has the same horizontal velocity as at the start, since gravity cannot affect motion in the horizontal direction. So the ball continues after reaching the top and begins the downward side of the arc. The vertical velocity now starts accelerating at the acceleration of gravity; the projectile reaches greater and greater speeds until it reaches the initial level with about the same vertical speed and horizontal velocity with which it started (assuming air resistance plays only a minor role).

3.9 THE EARTH CAN MOVE!

As the previous section indicates, understanding the motion of a projectile is not easy to grasp by simple observation. After much thought and study, Galileo concluded that projectile motion is a single motion compounded of two independent motions: uniform velocity in the horizontal direction and changing velocity in the vertical direction owing to the effect of the uniform acceleration of gravity. Once he understood this, Galileo was able to answer at last one of the major objections against the moving-Earth hypothesis, raised at that time and given at the beginning of this chapter. If the Earth really is rotating on its axis once a day, then the surface of the Earth must be moving very rapidly (about 1000 mi/hr at the equator). If that's the case, some argued, then a stone dropped from a high tower would

FIGURE 3.25 Stroboscopic photograph of a ball thrown into the air.

not land directly at the tower's base. During the time the stone is falling, the tower would move forward many meters. The stone would be left behind while it is falling and so would land far away from the base of the tower. But this is *not* what happens. Except for very slight variations, the stone always lands directly below the point of release. Therefore, many of Galileo's critics argued that the tower and the Earth could not possibly be in motion.

Galileo answered this objection showing that the tower experiment can be understood as an example of projectile motion. While the stone lies or is held at the top of the tower, it has the same horizontal velocity as the tower. During the time the stone falls after it is released, Galileo said, the stone continues to have the same horizontal velocity, and it gains vertical

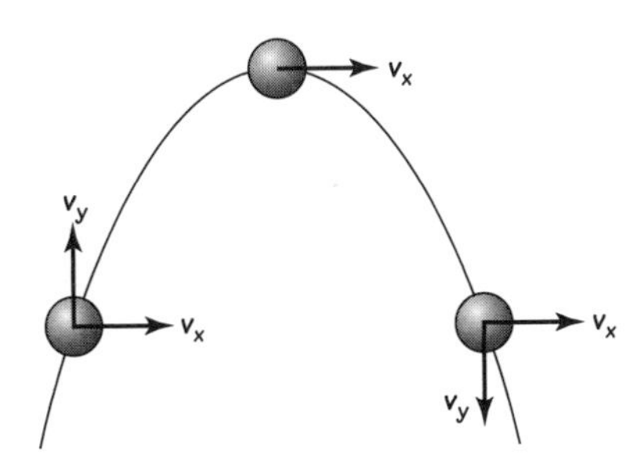

FIGURE 3.26 Components of velocity of a ball in flight.

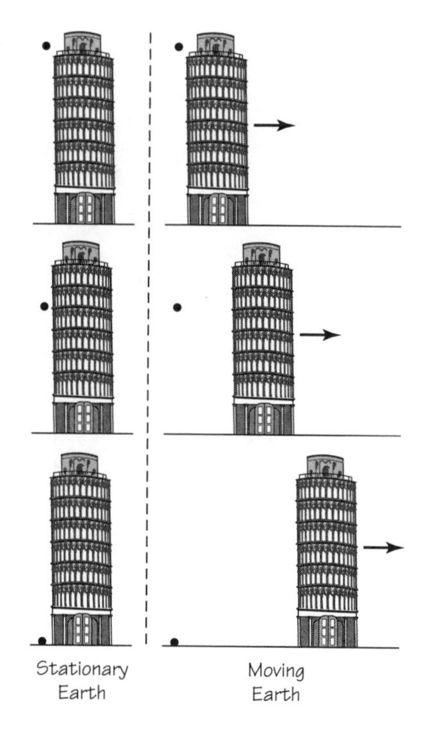

FIGURE 3.27 Galileo's critics' view of how a ball would drop to Earth.

speed as it falls freely under the force of gravity. But during the brief time of the stone's fall, the stone and the tower and the ground under them all continue to move forward with the same uniform horizontal velocity as before; the falling stone behaves like any other projectile. The horizontal and vertical components of its motion are independent of each other. Since the tower and the stone continue to have the same horizontal velocity throughout, the stone will not be left behind as it falls. Therefore, no matter what the speed of the Earth, the stone will always land at the base of the tower. So the fact that falling stones are not left behind can *not* be used to argue that the Earth is standing still. The Earth can be in motion.

Similarly, Galileo said, a stone dropped from the crow's nest at the top of a ship's mast lands at the foot of the mast, whether the ship is standing still or moving with constant velocity in calm water. This was actually tested by experiment in 1642, and many observations today support this view. For example, if you drop or toss a book in a car or bus or train that is moving with constant velocity (no bumps in the road!), you see the book move just as it would if the vehicle were standing still. Similarly, a person walking at constant speed and tossing a ball into the air and catching it, will see the ball move straight up and straight down, as if he were standing still. Try it!

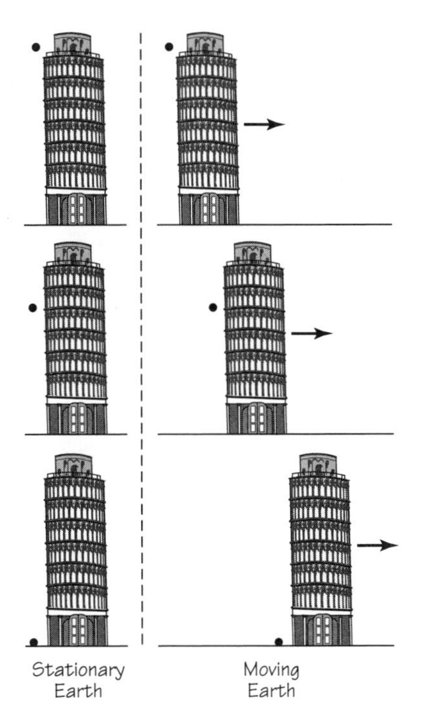

FIGURE 3.28 Galileo argued that while falling, the ball would continue its initial horizontal motion so that it would fall to the base of the tower. Therefore, an observer on Earth could not tell whether the Earth moved by watching the path of the ball.

3.10 GALILEAN RELATIVITY

The observations in the previous section all indicated that objects in a system, such as the Earth, or a ship, or a car, that is moving at constant velocity, will move within that system exactly in the same way as if they were in a system that is at rest. Here is how Galileo himself expressed it in a beautiful thought experiment recounted in 1632 in his *Dialogue on Two Chief World Systems*:

> Shut yourself up with some friend in the main cabin below decks on some large ship, and have with you some flies, butterflies, and other small flying animals. Have a large bowl of water with some fish in it; hang up a bottle that empties drop by drop into a wide vessel beneath it. With the ship standing still, observe carefully how the little animals fly with equal speed to all sides of the cabin. The fish swim indifferently in all directions; the drops fall into the vessel beneath; and, in throwing something to your friend, you need throw it no more strongly in one direction than another, the dis-

> tances being equal; jumping with your feet together, you pass equal spaces in every direction. When you have observed all these things carefully (though there is no doubt that when the ship is standing still everything must happen this way), have the ship proceed with any speed you like, so long as the motion is uniform and not fluctuating this way and that. You will discover not the least change in all the effects named, nor could you tell from any of them whether you were moving or standing still.

We can perform the same experiment today with butterflies or baseballs in a boat, or in a car or train or plane. Today, such a "system" is usually called a *reference frame*, since it is the frame of reference for our observations. The fact that the motions of baseballs and butterflies stay the same, regardless of whether or not the reference frame is moving at constant velocity, indicates that Newton's laws of motion are the same—in fact, that all mechanics (science of motion) is the same—for all reference frames at rest or moving with uniform velocity relative to each other. This conclusion has been called the *Galilean relativity principle*. It can be stated as follows:

> The laws of mechanics are exactly the same for every observer in every reference frame that is at rest or is moving with uniform velocity.

Since objects move in a reference frame that is at rest or in uniform velocity—such as a boat, or the Earth's surface for reasonably short periods—as they were in a frame at rest, there is no way to find out the speed of one's *own* reference frame from any mechanical experiment done *within* that frame. Nor can one pick out any one reference frame as the "true" frame, the one that is, say, absolutely at rest. Thus, there can be no such thing as the "absolute" velocity of an object. All measured velocities are *relative*.

Three centuries later, Albert Einstein expanded upon the Galilean relativity principle in formulating his theory of relativity, as you will see in Chapter 9.

3.11 ORBITING SATELLITES

To recapitulate: After obtaining his new concept of "natural motion" Galileo investigated what happens when a rolling ball moving with uniform velocity reaches the end of the plane on which it is rolling. He observed that it flies off the table and goes into projectile motion. He discovered that this motion is really a compound of independent motions: uniform velocity in

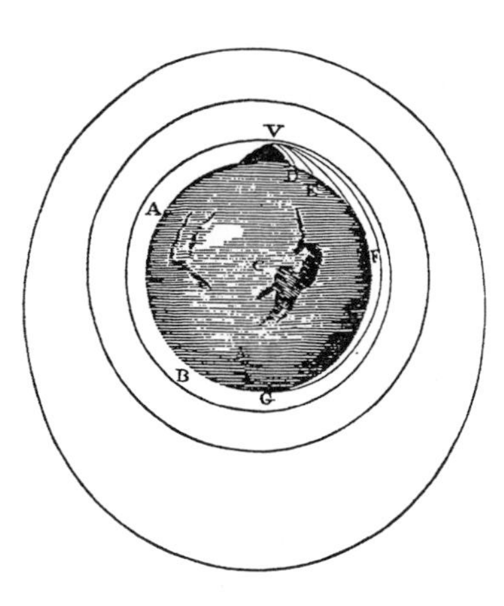

FIGURE 3.29 ". . . the greater the velocity . . . with which [a stone] is projected, the farther it goes before it falls to the Earth. We may therefore suppose the velocity to be so increased that it would describe an arc of 1, 2, 5, 10, 100, 1000 miles before it arrived at the Earth, till at last, exceeding the limits of the Earth, it should pass into space without touching it."—Newton, *Principia*.

the horizontal direction and free fall (with uniform acceleration of gravity) in the vertical direction.

Newton took Galileo's ideas one step further. He considered what happens when a projectile is launched horizontally from a high tower or a mountain on the Earth. Three factors determine where the object strikes the Earth. These are the height of the mountain, the acceleration due to the force of gravity, and the horizontal speed of the projectile. Keeping the first two constant (by starting from the top of the same high mountain), as the horizontal launch speed of the projectile is increased, it will strike the Earth at points farther and farther from the base of the mountain. This is shown in the diagram, taken directly from Newton's *Principia*. Eventually, if it is fast enough at launch, the projectile goes so far that the curvature of the Earth comes into play. Since gravity always pulls objects toward the center of the Earth, the trajectory of the projectile is no longer a parabola because gravity pulls it from slightly different directions as it moves over the curve of the Earth.

If the launch speed is increased even more, the projectile will strike the Earth at points farther and farther around the Earth. Finally, at a certain speed the projectile would not fall fast enough in order to strike the Earth before it returns to its original launch location. At that point, if it is not slowed by air resistance, the projectile would just keep going around and around the Earth, never actually reaching the Earth even though it "keeps falling" toward the Earth. In other words, it would go into orbit: always falling toward the Earth in free fall but moving too fast in the orbit's direction to hit the curved Earth. The amount of fall of the projectile away from its original horizontal motion is roughly matched by the curvature of the Earth's surface. Therefore, the projectile stays in orbit at roughly a constant distance above the surface. (Actually, as Kepler's first law tells us, the orbits may be ellipses, of which the circle is a special case.)

FIGURE 3.30 Space Shuttle blast-off.

The "projectile" could be a research satellite, launched into orbit by a rocket, instead of a ball projected from a high mountain. Putting Newton's third law of motion to work, the rocket, with the satellite fastened to it, lifts off from the launch pad. During its motion upward it is made to tilt in the horizontal direction, moving faster and faster and further from the Earth, until it reaches orbital velocity at the desired height above the surface of the Earth. At that point the main engines shut down and are jettisoned, and the satellite is in orbit.

A common question about the motion of a satellite in orbit is, "What is holding it up"? The answer is of course that it is not being held up. In fact, the tons of metal and electronics are constantly falling toward the Earth, pulled down by gravity. The only difference from direct fall downward is that the satellite also has such a tremendous orbital speed, as much as 18,000 mi/hr (or 29,000 km/hr). This prevents it from ever reaching the curved Earth as it falls downward because the curvature of the satellite's trajectory is about the same as the Earth's curvature. The satellite will stay in orbit as long as it maintains sufficient speed. This can change of course if, following Newton's second law, there acts on it an unbalanced force, say

FIGURE 3.31 An early manned space capsule in orbit.

in the horizontal direction. (One already exists in the vertical direction—gravity.) Such a horizontal force can arise from air resistance, or if the pilot in a space shuttle turns on the forward thrusters to slow the shuttle down and begin the long spiral inward for a landing.

Special Satellites

Three special types of satellites are: communications satellites, weather satellites, and global positioning satellites. Communications satellites are placed into orbits for which the horizontal speed exactly matches the rotational speed of the Earth—this is called a "geosynchronous orbit." Hence the satellite stays over the same location on the Earth, appearing to hang stationary in the sky. This is very useful for television transmission and wireless reception because ground-based satellite dishes can then be aimed at a fixed point in the sky to send and receive signals.

Some weather satellites are placed in polar orbits, in which they orbit the Earth in the north–south direction. As the Earth rotates west to east, the entire Earth passes under the orbit of the satellite. The satellite's cameras can then be turned on at the appropriate time to record events on any desired portion of the Earth.

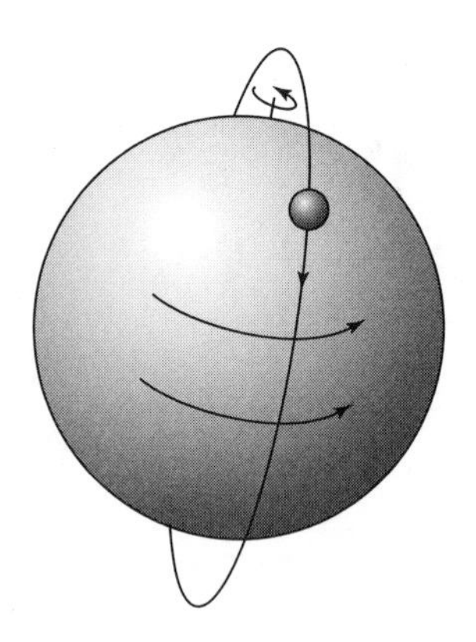

FIGURE 3.32 Path of satellite in polar orbit.

In the Global Positioning System 24 satellites are in orbit around the Earth. A signal is emitted by a satellite and is picked up by a receiver on Earth. From the timing of the sending and receiving of the signal, three of the satellites are used to "triangulate" the location. The fourth satellite corrects for timing errors.

FIGURE 3.33 Space Shuttle with Solar Max satellite (in background) to be launched.

The Space Shuttle

Some might think an orbiting space vehicle stays in orbit because it is far beyond the pull of gravity. After all, they argue, the astronauts and any equipment not fastened down can float around as if "weightless," which "proves" there is no gravity on them and the space shuttle itself. But, as you saw earlier, in orbital motions gravity continues to act. When a space shuttle is about 500 km above the surface of the Earth (whose radius is about 6370 km), the Earth's gravitational force there is still about 90% of its value at the Earth's surface. In fact, if there were no gravity acting on the shuttle, there would be no net force on it, in which case Newton's first law of motion would come into play. According to the first law of motion, the shuttle would fly off into outer space on a straight line at uniform velocity and never go into orbit. As you will see below (Section 3.12), we need a net force owing to gravity to keep any object in orbit; without it there would be no orbit!

So, how can the astronauts feel weightless inside the space shuttle even though the force of gravity is acting on them? The answer is that they are in "free fall," when everything behaves as if it were weightless. For instance, if you

FIGURE 3.34 Astronaut during space walk.

were in free fall and released a bunch of keys taken from your pocket, they would stay with you, falling alongside you, as if gravity had been "turned off."

In order to answer the further objections that people after Copernicus raised against the idea that the Earth is in motion, we have to look a little closer at circular motion.

3.12 CIRCULAR MOTION

The simplest kind of circular motion is *uniform circular motion*. This is motion in a circle at constant speed. If you are in a test car that goes around a perfectly circular track so that at every instant the speedometer reading is, say, 60 mi/hr, you are in uniform circular motion. This is not the case if the track is of any shape other than circular, or if your speed changes at any point.

How can you find out if an object in circular motion is moving at constant speed? You can apply the same test used earlier to decide whether or not an object traveling in a straight line does so with constant average speed: measure the distances traveled along the circumference in given time intervals and determine if the ratios $\Delta d/\Delta t$ are equal. For circular motion, it is often convenient to use one complete circumference of the circle as the distance traveled. If we know the radius r, this distance C is given by the formula: $C = 2\pi r$. Here π is the Greek letter "pi." It represents the value, sufficient for all our calculations, of approximately 3.14. (More precisely, pi represents the value 3.141592653. . . .) The time required for an object to complete one revolution around the circle is the *period* (T) of the mo-

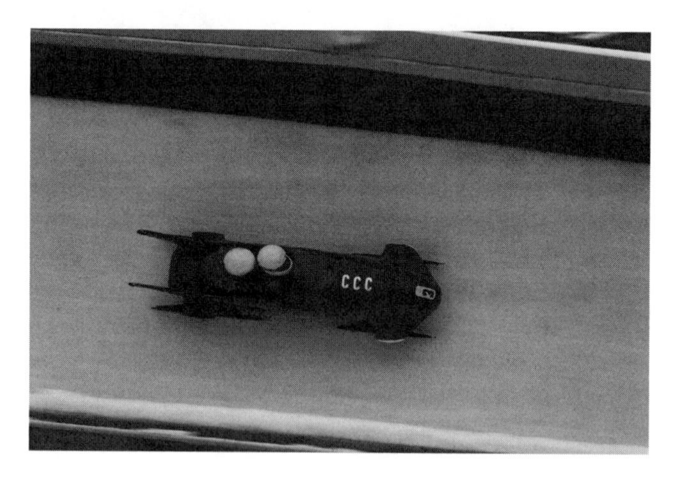

FIGURE 3.35 Bobsled rounding curve.

tion. The number of revolutions completed by an object in a unit of time is called the *frequency* (f) of the motion. If the unit of time is chosen to be one period, then one revolution is completed in one period. So, the frequency of revolution is just 1/period, or $f = 1/T$.

(There is another measure of speed used in circular motion, called *angular speed.* Instead of distance traveled per time interval, which is called *linear speed*, angular speed refers to the angle of the circular segment traversed per time interval. An analogue music record turns at a uniform angular speed, but a modern-day CD turns at a constantly changing rate so that the data, which are arranged along a tight spiral, pass over a laser beam at constant linear speed. In this text we will use only linear speed in uniform circular motion.)

Using the radius of the circle and the frequency of revolution, we can obtain the average linear speed of an object in uniform circular motion as follows:

$$\text{average speed} = \frac{\text{distance traveled on circumference}}{\text{elapsed time}},$$

$$v_{\text{av}} = \frac{2\pi r}{T}$$

or, dropping the "av," if we can assume that there is no change of speed during the elapsed time, we have

$$v = \frac{2\pi r}{T}.$$

This expression is essentially the definition of the speed of an object in uniform circular motion.

What about the *velocity*, which is speed *with direction*? To find out the direction of the velocity of a circulating object, carefully whirl a small object on the end of a string. Being careful not to hit anyone, release the object while it is whirling on the string. In what direction does it move? You will notice that it does not move straight outward from the circle, as one might intuitively believe, nor does it keep spiraling for a while as it moves outward. Instead, upon release it immediately moves in a straight line along a tangent to the circle at the point where you released it. Try this at different points on the circle. You will notice that the object always flies away on a tangent at that point. *We conclude from this that the velocity vector is tangent to the circle at every point on the circle.*

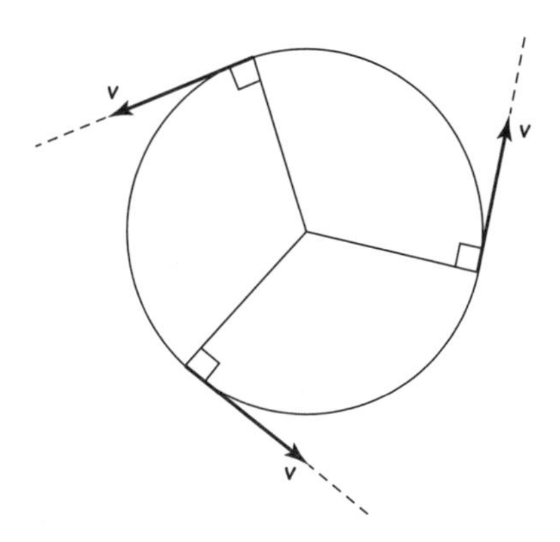

FIGURE 3.36 Velocity vectors for uniform circular motion at three points.

This is a very important conclusion. Since the tangent to a circle is different for every point on the circle, this conclusion means: *the direction of the velocity is different at each point on the circle, even though the speed is the same over the entire circle.* Since the velocity is constantly changing in direction—even though not in magnitude (speed)—there is a uniform acceleration. But it is different from the accelerations we have encountered so far. It is an acceleration that changes only the direction but not the speed of the moving object.

Where there is an acceleration there is a net force. *What is the direction of the vector representing the net force?* Remember that a vector can be represented by an arrow in the direction of the vector. The magnitude of the vector is represented by its length. Since the magnitude of the velocity vector, or the speed, is not changing in uniform circular motion, the lengths of the arrows in the above diagram are all equal. But their directions are all different. Since the force does not help or hinder the speed, it must be perpendicular to the velocity vectors. Since a tangent to a circle is always perpendicular to the radius at that point, the force must lie along the radius, either straight in or straight out.

Now let's look at what happens when arrow **A**, representing the veloc-

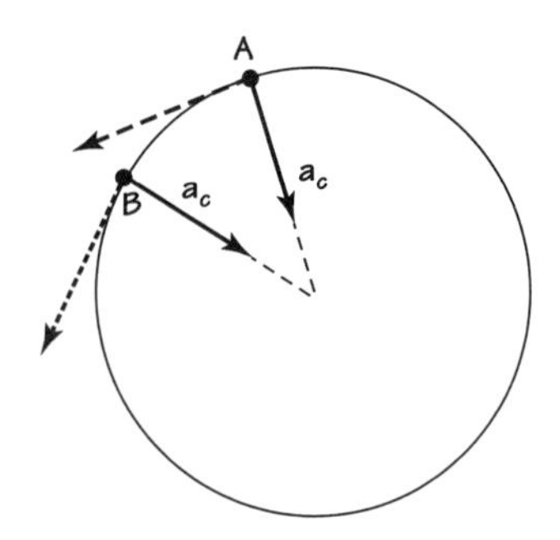

FIGURE 3.37 Force and acceleration vectors on a uniformly circulating object, at two points on its path.

ity vector of a ball in uniform circular motion at point A, turns into arrow **B** at point B. If there is no net force, Newton's first law tells us that the ball will fly off on a straight line in the direction of arrow **A**. But there is a net force, which pulls the ball off the straight-line course, changing the velocity at A, represented by arrow **A**, into a velocity at B in a new direction, represented by the direction of arrow **B**. To force the ball off course and around the curve, you can almost feel how you must pull it inward, not outward. This force must be an inward force along the radius and pointing toward the center. Since it is a pull toward the center, it is often called the *centripetal force*, literally "moving, or directed, toward the center."

The centripetal force acting on a mass causes an acceleration toward the center, the *centripetal acceleration*. These are both vectors and they both point inward toward the center of the circle along the radius at each point on the circle. The magnitude of each is related to the speed of the object and the radius of the circle. You can almost feel that the faster a ball whirls on a string, the harder you must pull on it to keep it in circular motion. As shown in the *Student Guide*, the magnitude of the centripetal acceleration, symbol a_c, is given by

$$a_c = \frac{v^2}{r}.$$

The magnitude of the centripetal force F is given by $F_c = ma_c$, or

$$F_c = \frac{mv^2}{r}.$$

From this you can see that the force and acceleration increase rapidly as you increase speed, while they go down more slowly as you increase the radius of the circle.

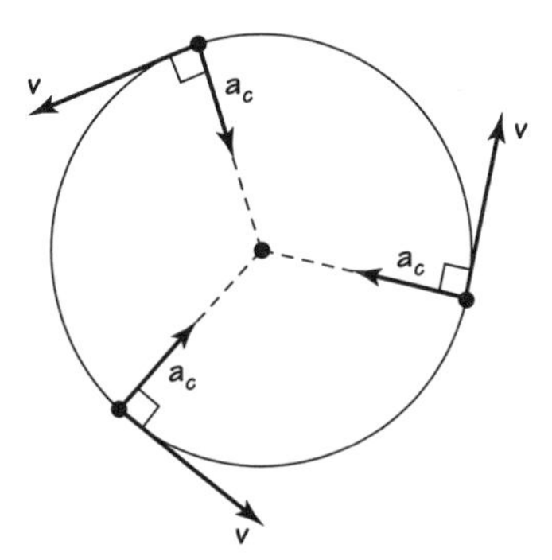

FIGURE 3.38 Velocity and acceleration vectors at three points on the path of a uniformly circulating object.

Some Examples

We have learned now that without a centripetal force there cannot be a circular motion. Unless acted upon by a net, unbalanced force, all objects move in a straight line at constant speed. To make objects change their direction and move in a circle at uniform speed, something must provide a force toward the center of the circle. A stone whirling on a string requires a pull inward on the string. As soon as the inward pull ceases or the string breaks, the ball ceases to be in circular motion.

Another example involves cars traveling at high speed on a curved highway. Why are these highways usually banked on the curves? The reason is to use part of the weight of the car itself to help provide the centripetal force. The faster the expected speeds of the cars, the steeper the bank (according to the square of the speeds). Even with a bank, there must be enough friction between the tires and the road to provide additional centripetal force to enable the cars to perform the circular turn. If the friction is lost on a rainy or icy day, the road might not supply enough centripetal friction force to enable the cars to make the turn. The cars will skid off on a straight line, with possibly disastrous results. This is why speed limits are (or should be) reduced in inclement weather.

An example of nearly uniform circular motion around the Earth was the flight of the space shuttle *Endeavor* in January 1996. Let's find out how fast it was going and what its centripetal acceleration was. To calculate these using the formulas above, we need to know the radius and period of its orbit. According to information supplied by NASA's site on the World Wide Web (http://spacelink.nasa.gov), the average altitude of the *Endeavor* shuttle above the Earth's surface was 288 mi. This is equivalent to 463.4 km. But this is not the radius of its orbit! The orbit is centered on the center of the Earth, which we can assume to be a sphere of radius 6370 km. So the radius of the orbit was

$$R = 6370 \text{ km} + 463 \text{ km} = 6833 \text{ km}.$$

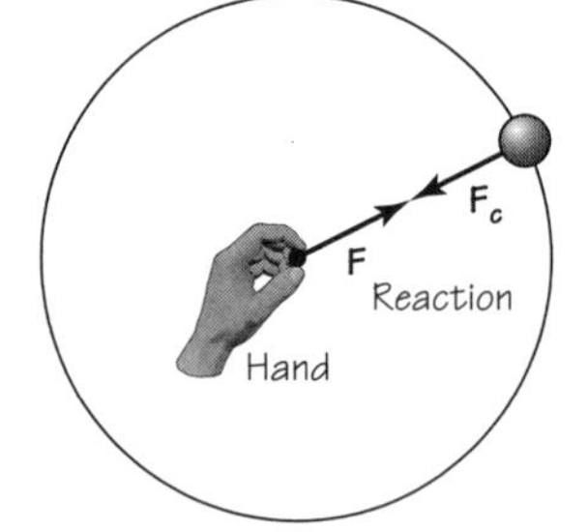

FIGURE 3.39 The centripetal force F_c on the object is equal in magnitude and opposite in direction to the reaction force felt by the hand.

FIGURE 3.40 Space Shuttle in orbit around the Earth.

NASA's Web site also reports that *Endeavor* made 142 orbits of the Earth in approximately 8 days and 22 hr, or 214 hr total. So the period for one orbit is

$$T = \frac{214 \text{ hr}}{142 \text{ orbits}}$$

$$= 1.51 \text{ hr} \times 3600 \text{ s/hr}$$

$$= 5425 \text{ s}.$$

We are now ready to obtain the speed of the shuttle in its orbit

$$v = \frac{2\pi R}{T} = 2\pi \times \frac{6833 \text{ km}}{5425 \text{ s}} = 7.91 \text{ km/s}.$$

This is equivalent to a horizontal speed of 17,798 mi/hr. Some projectile! However, at any speed slower than this, the shuttle would not be able to stay in this orbit, but would spiral down to a lower orbit. If it still decreased in speed, it could not stay in orbit at all and would have to land—which is exactly what it does intentionally when it returns to Earth.

With the speed and the radius of the shuttle in its orbit, we are now ready to obtain the centripetal acceleration

$$a_c = \frac{v^2}{R} = \frac{(7.91 \text{ km/s})^2}{6833 \text{ km}}$$

$$= 0.0092 \text{ km/s}^2 = 9.2 \text{ m/s}^2.$$

What is the origin of the force that gives rise to this acceleration? You already know that it must be due to the Earth's gravitational attraction. Evidently the centripetal acceleration on the shuttle *Endeavor* was just the acceleration of gravity at that height, 9.2 m/s^2. This is in fact the uniform acceleration at that height of a freely falling object. However, it is about 13.6% less than the gravitational acceleration at the surface of the Earth, which is 9.8 m/s^2. It appears, then, that the gravitational force does extend into outer space and that it accounts for the centripetal force acting on orbiting objects, but it appears to decrease with distance.

Is it possible that the same situation occurs for other orbiting objects, such as the Moon around the Earth, the moons of Jupiter observed by Galileo orbiting Jupiter, and perhaps even for the Earth and other planets orbiting the Sun? Is there a centripetal force holding the Earth and the other planets in orbits around the Sun, or is the Earth really physically different from all the other planets, as people believed for centuries? If the former case, what is the exact nature of this centripetal force? For answers to these and other fundamental questions we turn, in the next chapter, to the work of Isaac Newton.

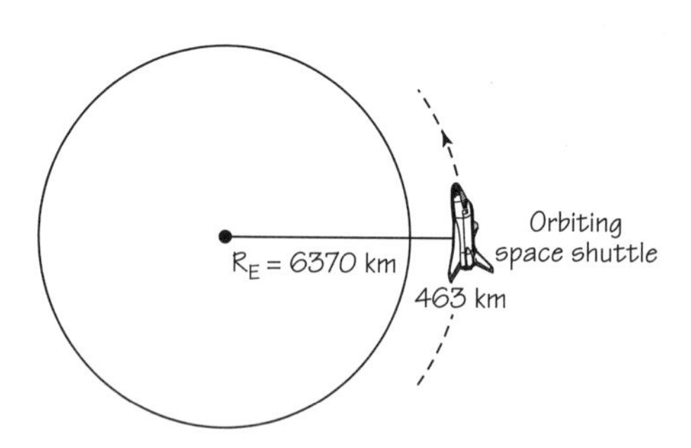

FIGURE 3.41 Diagram of Space Shuttle in orbit.

SUMMARY

Newton's Three Laws of Motion

1. The law of inertia:
   Every object continues in its state of rest or of uniform velocity (motion at uniform speed in a straight line) unless acted upon by an unbalanced force (a net force). Conversely, if an object is at rest or in motion with uniform velocity, all forces that may be acting on it must cancel so that the net force is zero.
2. The force law:
   The net force acting on an object is numerically equal to, and in the same direction as, the acceleration of the object multiplied by its mass. In symbols: $\mathbf{F} = m\mathbf{a}$.
3. The law of action and reaction:
   If one object exerts a force on another object, the second object at the same time exerts a force on the first object. These two forces, each acting on one of the two objects, are equal in magnitude and opposite in direction.

SOME NEW IDEAS AND CONCEPTS

centrifugal force
centripetal force
force
force, unbalanced
Galilean relativity principle
inertia
mass
natural motion
net force
projectile motion
reference frame
uniform circular motion
violent motion
weight
weightlessnes

FURTHER READING

G. Holton and S.G. Brush, *Physics, The Human Adventure* (Piscataway, NJ: Rutgers University Press, 2001), Chapters 8–10.

R.S. Westfall, *The Construction of Modern Science: Mechanisms and Mechanics* (New York: Cambridge University Press, 1978).

T.S. Kuhn, *The Copernican Revolution: Planetary Astronomy in the Development of Western Thought* (Cambridge, MA: Harvard University Press, 1982).

STUDY GUIDE QUESTIONS

1. Look at every equation, graph, table, law, and definition in this chapter and be sure that you understand their meaning to the extent that you are able to explain each of them in your own words to another person.
2. List all the different types of motion discussed in this chapter. Then attempt to see how Newton's laws of motion apply to each one.

A. THE THREE LAWS OF MOTION

3.1 Natural Motion and Newton's First Law

1. What were the so-called natural motion and violent motion? Why did they seem reasonable?
2. How is the old idea of natural motion contradicted by the observation that two objects of different mass fall at the same rate when dropped from the same height?
3. What did Galileo discover about a ball moving on a tabletop without friction?
4. Express the law of inertia in your own words.
5. A roller-blader coasts for a distance along a flat stretch of a street. How would Aristotle and Galileo each account for his motion?

3.2 Forces in Equilibrium

1. What are balanced forces?
2. Give some examples of unbalanced forces.
3. Sketch the examples in Question 2 and represent the forces by vectors.
4. What are two possible states of motion for objects on which balanced forces act?

3.3 More about Vectors

1. Draw arrows to represent two or more forces acting on an object. Then describe how you would find the net vector force acting on the object.
2. A swimmer tries to swim across a fast-moving stream flowing west to east at 4 mi/hr. The swimmer sets out to cross the stream by swimming 3 mi/hr due north. What is his resultant velocity?

3.4 Newton's Second Law of Motion

1. What is a universal law?
2. How do we know that Newton's laws of motion are universal laws?
3. Describe an experiment that explored the relationship between force and acceleration. What was the conclusion?
4. Describe an experiment that explored the relationship between mass and acceleration. What was the conclusion?

5. State Newton's second law of motion in your own words.
6. A net force of 10 N gives an object a constant acceleration of 4 m/s^2. What is the mass of the object?
7. Does Newton's second law hold only when frictional forces are absent? Explain.
8. An unbalanced force can cause acceleration, yet you can push hard on a broken shopping cart on a rough pavement and it still moves only at constant velocity. Explain.
9. What does it mean to say that mass is a scalar quantity?

3.5 Measuring Mass and Force

1. What is weight, and how is it different from mass?
2. How are the kilogram, meter, and second defined?
3. What exactly is inertial mass?
4. What is a "newton"?
5. What are the cgs and mks systems of units? Are they both metric systems?

3.6 More about Weight, and Weightlessness

1. You are standing on a platform on Earth that is part of a "weighing scale."
   (a) Using the metric conversion (to kilograms), how much do you weigh?
   (b) What is your mass?
   (c) In a sketch, draw the forces on you and on the scale.
   (d) If the net force on you is zero, why does the scale register your "weight"?
   (e) The ground suddenly gives way and you fall freely into a deep well. Before you panic, you look down at the scale. What is it reading as your weight?
   (f) What is your mass as you fall in free fall?
2. An astronaut walking on the Moon, which has one-sixth the gravitational pull of the Earth, has a certain amount of weight and mass. Are they the same or different from his weight and mass on the Earth, and if different, by how much?
3. If an object is accelerating, what can we say about the forces acting on it?
4. Name some common types of forces.
5. Different masses on Earth have different weights, yet they have the same acceleration in free fall. How can this be?
6. How do Newton's second law and the force of gravity account for the fact that two objects of different mass hit the ground together when dropped from the same height? (Neglect air resistance.)
7. Why does a skydiver in free fall feel weightless? Is he/she really weightless? Explain.
8. An astronaut is orbiting on the International Space Station, where he feels weightless. At that altitude, the acceleration due to gravity is 13% less than its value on the surface of the Earth. Which of the following is/are true?
   (a) The astronaut's weight is zero.
   (b) The astronaut's mass is zero.

(c) The astronaut's weight is 13% less than its value on Earth.
(d) The astronaut's mass is 13% less than its value on Earth.

3.7 Newton's Third Law of Motion

1. State Newton's third law of motion in your own words.
2. Give some examples of the third law in operation.
3. Why don't action and reaction "cancel out"?

B. THE THREE LAWS IN ACTION

3.8 Projectile Motion

1. A golfer tees off on a fairway, hitting the green some distance away.
   (a) What is the shape of the trajectory of the ball?
   (b) How would Galileo account for the motion of the ball? Would Aristotle agree?
2. What are the two motions, acting at the same time, that make up projectile motion?
3. Does the vertical motion affect the horizontal motion, and vice versa? How do you know?

3.9 The Earth Can Move!

1. What was the main argument against the moving Earth in this section of the text?
2. How did Galileo use the concept of projectile motion to demolish that argument?

3.10 Galilean Relativity

1. What conclusion can be drawn from Galileo's examples of moving objects in systems at rest or in uniform motion?
2. State the Galilean relativity principle in your own words.
3. How does the principle support the argument that the Earth can be in motion?

3.11 Orbiting Satellites

1. Explain how a projectile can become an orbiting satellite.
2. The Space Shuttle weighs hundreds of tons, yet it can orbit the Earth without falling down. How does it stay in orbit if nothing is "holding it up"?
3. Jumbo jets also weigh many tons. What can you say about the forces on them as they are flying at cruising altitude at a constant velocity? What "keeps them up"?
4. Name two types of specialty satellites and describe their orbits.

5. Why do astronauts in the orbiting Space Shuttle feel weightless? Are they really weightless?

3.12 Circular Motion

1. An object is moving around a circle with constant speed. Is the object accelerating, even though the speed is constant? Explain.
2. If an object moving in uniform circular motion is accelerating, then there must be a force on it. Describe the nature of this force, and draw arrows to represent the force vector at different points around a circle depicting the object's motion.
3. What is centripetal acceleration?
4. In which of these cases is an object accelerating?
   (a) Moving with constant speed in a straight line.
   (b) Moving with constant speed in a circle.
   (c) Moving on the trajectory of a projectile.
5. Classify each of the following as either vectors or scalars:
   (a) 4 s;
   (b) 3 m/s eastward;
   (c) 600 g;
   (d) centripetal acceleration;
   (e) 5 N to the right;
   (f) 5 N.

DISCOVERY QUESTIONS

1. During the course of 1 day, observe some of the different motions that you see around you and make a list of them. Examples might be a car accelerating, a bus turning a corner, a rotating fan, falling rain, a basketball arcing toward a basket, even the Sun rising and setting. Now attempt to answer the following:
   (a) What type of motion is each of the motions that you observed?
   (b) If some of these involve combinations of two or more motions, can you break them down into simpler motions?
   (c) Using Newton's laws of motion, try to account for each of these motions.
2. Why do you think that Galileo and Newton found the study of motion to be so difficult? Do you find it difficult?
3. List some examples of action–reaction forces.
4. List some examples of uniform circular motion and describe the centripetal force involved in each case.
5. How would you use a "balance scale" and a set of known masses to measure the mass of an unknown object? How does this make use of Newton's laws of motion?

6. If you know the period and speed of a satellite, you can easily find the acceleration of gravity at the height of the satellite. How would you do this?
7. A person is sitting in a car at rest. In an unfortunate accident, the car is hit from behind. Which way is the person inside thrown, backward or forward? Explain.
8. Have someone drive you for a short distance in a car that includes a number of stops, starts, turning corners and curves, and bumps. Carefully observe and write down all the forces you feel. Then try to explain each one from the material in this chapter. Which of these forces are fictitious forces and which are real forces?
9. While you are sitting in the back of a car on the right side, the car makes a sudden turn to the left. You feel a push on you from the right side of the car. Why?
10. Try the experiment described in Section 3.6. Take a bathroom scale into an elevator and record the reading on the scale as the elevator accelerates upward and downward. Compare the result with the reading when the elevator is at rest. What would the scale read if the acceleration downward increased to the acceleration of gravity, g? What would the scale read if the acceleration upward increased by the same amount?

Quantitative Questions

1. Consider a system consisting of a 1-kg ball and the Earth. The ball is dropped from a short distance above the Earth and falls freely to the Earth (whose mass is 6.0×10^{24} kg).
   (a) Make a vector diagram illustrating the important forces acting on each member of the system during the ball's descent.
   (b) Calculate the acceleration upward of the Earth while the ball accelerates downward.
   (c) Make a vector diagram as in (a), but showing the situation when the ball has come to rest after hitting the ground.
2. A projectile is launched horizontally with a speed of 8 m/s from the edge of a precipice.
   (a) How much time will the projectile take to hit the ground 80 m below?
   (b) How does this time change if the horizontal speed is doubled?
   (c) What is the vertical speed of the projectile when it hits the ground?
   (d) What is the horizontal speed of the projectile when it hits the ground?
3. What is the period of the minute hand of an ordinary clock? If the hand is 3.0 cm long, what is the speed of the tip of the minute hand?
4. Go to the NASA site on the World Wide Web (http://spacelink.nasa.gov) and find data on the orbit of a satellite or space shuttle. From the data given, obtain the speed, period, velocity, and centripetal acceleration of the orbiting object.
5. Do the same as in Question 4 for the International Space Station.
6. What are the velocity and centripetal acceleration of a person standing on the equator of the Earth, owing to the rotation of the Earth?

7. What are the velocity and the centripetal acceleration of the Earth as it orbits the Sun each year? What is the centripetal force on the Earth (mass 6.0×10^{24} kg)?
8. A satellite with a mass of 500 kg completes a circle around the Earth every 380 min in an orbit 18,000 km from the center of the Earth. What are the magnitude and direction of the force vector holding the satellite in orbit? What produces this force?
9. How fast would you have to throw a baseball in order to put it into orbit at the surface of the Earth, assuming that there are no mountains and buildings in the way?

CHAPTER 4

Newton's Unified Theory

4.1 NEWTON AND SEVENTEENTH-CENTURY SCIENCE

Forty-five years passed between the death of Galileo in 1642 and the publication in 1687 of Newton's greatest work, the *Philosophiae Naturalis Principia Mathematica*, or the *Principia* (*Principles*) for short. In those years, major changes occurred in the social organization of scientific studies. The new philosophy of experimental science, applied with enthusiasm and imagination, produced a wealth of new results. Scholars began to work together and to organize scientific societies in Italy, France, and England. One of the most famous, the Royal Society of London for Improving Natural Knowledge, was founded in 1662. Through these societies, scientific experimenters exchanged information, debated new ideas, argued against opponents of the new experimental activities, and published technical papers. Each society sought public support for its work and published studies in widely read scientific journals. Through the societies, scientific activities became well defined, strong, and international.

FIGURE 4.1 Isaac Newton (1642–1727).

This development was part of the general cultural, political, and economic change occurring throughout the sixteenth and seventeenth centuries (1500s and 1600s). Artisans and people of wealth and leisure became involved in scientific studies. Some sought to improve technological methods and products. Others found the study of nature through experiment a new and exciting hobby. However, the availability of money and time, the growing interest in science, and the creation of organizations are not enough to explain the growing success of scientific studies. This rapid growth also depended upon able scientists, well-formulated problems, and good experimental and mathematical tools.

Many well-formulated problems appeared in the writings of Galileo and Kepler. Their studies showed how useful mathematics could be when combined with experimental observation. Furthermore, their works raised exciting new questions. For example, what forces act on the planets and cause the paths that are actually observed? Why do objects fall as they do near the Earth's surface?

Good experimental and mathematical tools were also becoming available in that era. As scientists applied mathematics to physics, studies in each field stimulated development in the others. Similarly, the instrument maker and the scientist aided each other. Another factor of great importance was the rapid buildup of scientific knowledge itself. From the time of Galileo, scientists had reported repeatable experiments in books and journals. Theories could now be tested, modified, and applied. Each new study built upon those done previously.

Newton, who lived during this bustling new scientific age, is the central person in this chapter. However, in science as in any other field, many workers made useful contributions. The structure of science depends not only upon recognized geniuses, but also upon many lesser-known scientists. As Ernest Rutherford, one of the founders of twentieth-century nuclear physics, once said (updating his language slightly):

> It is not in the nature of things for any one person to make a sudden violent discovery; science goes step by step, and every person depends upon the work of his or her predecessors. . . . Scientists are not dependent on the ideas of a single person, but on the combined wisdom of thousands. . . .

In order to tell the story properly, we should trace each scientist's debt to others who worked previously and in the same age, and we should trace each scientist's influence upon future scientists. Within the space available, we can only briefly hint at these relationships.

4.2 ISAAC NEWTON

Isaac Newton was born on Christmas Day, 1642, according to the Julian calendar—4 January 1643, according to the Gregorian calendar now in use—in the small English village of Woolsthorpe in Lincolnshire, north of Cambridge. His father had died before his birth, and Isaac, initially a weak infant, then a quiet farm boy, became very dependent upon his mother. He never married. Like young Galileo, Newton loved to build mechanical gadgets and seemed to have a liking for mathematics. With financial help from an uncle, he went to Trinity College of Cambridge University in 1661, where he enrolled in the study of mathematics and was a successful student.

In 1665, the Black Plague swept through England. The officials closed the college, and Newton went home to Woolsthorpe. Drawing upon what he had learned and his own independent work, there he made spectacular discoveries. In mathematics, he developed the binomial theorem and differential calculus. In optics, he worked out a theory of colors. In mechanics, he had already formulated a clear concept of the first two laws of mo-

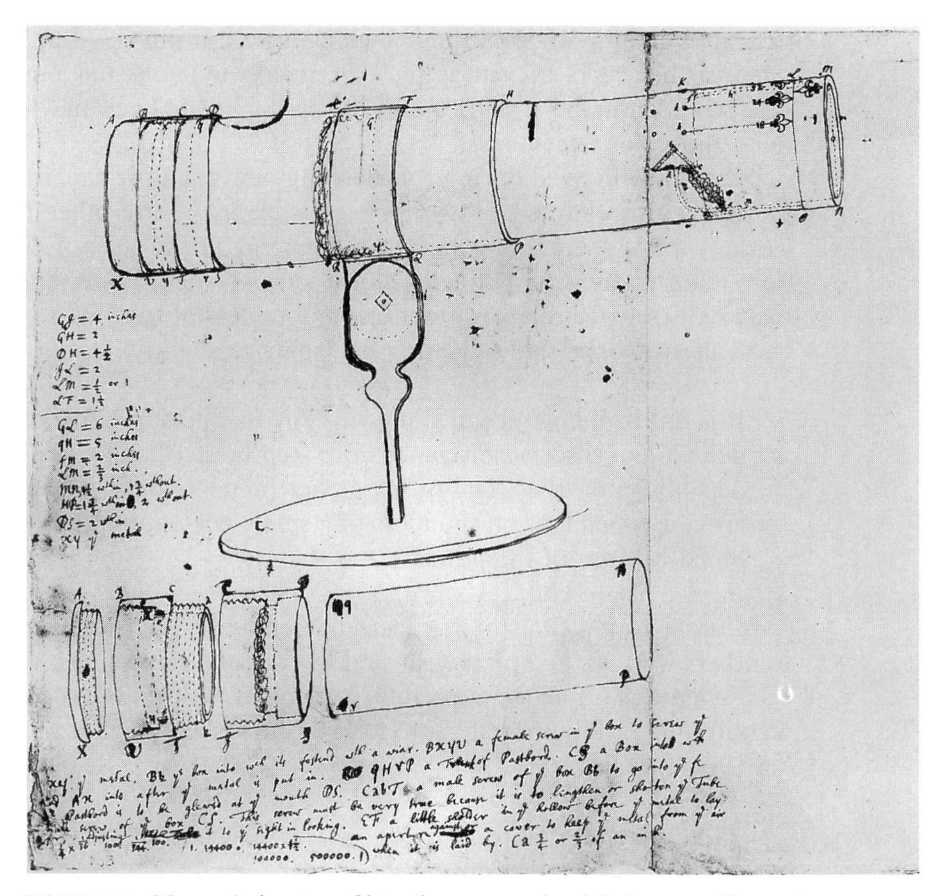

FIGURE 4.2 Newton's drawing of his telescope, made while he was still a student.

tion and the law of gravitational attraction. He also discovered the equation for centripetal acceleration. However, Newton did not announce this equation until many years after Christiaan Huygens' equivalent statement.

This period at Woolsthorpe must have been the time of the famous, though still disputed, fall of the apple. (An ancient apple tree, said to be a descendant of the one from Newton's time, is still on the grounds of the Woolsthorpe home.) One version of the apple story appears in a biography of Newton, written by his friend William Stukeley. In it we read that on a particular occasion Stukeley was having tea with Newton. They were sitting under some apple trees in a garden, and (wrote Stukeley) Newton said that

> he was just in the same situation, as when formerly, the notion of gravitation came into his mind. It was occasion'd by the fall of an apple, as he sat in a contemplative mood. Why should that apple always descend perpendicularly to the ground, thought he to himself. Why should it not go sideways or upwards, but constantly to the Earth's centre?

The main emphasis in this story probably should be placed on the "contemplative mood" and not on the apple. Newton himself later wrote that he had begun at that time to think of the force on the apple as a force that extends out to the Moon, and that this same type of force might also act between the planets and the Sun. "All this was in the two plague years of 1665 and 1666," he later wrote, "for in those days [at age 21 or 22] I was in the prime of my age for invention, and minded mathematics and philosophy more than at any time since." You have seen this pattern before: A great puzzle (here, that of the forces acting on planets) begins to be solved when a clear-thinking person contemplates a familiar event (here the fall of an object on Earth). Where others had seen no relationship, Newton did.

Soon after he returned to Cambridge, Newton succeeded his former teacher as professor of mathematics. Newton taught at the university and contributed papers to the Royal Society. At first his contributions were mainly on optics. His *Theory of Light and Colors*, published in 1672, fired a long and bitter controversy with other scientists who disagreed with his work. Newton, a private and complex man, resolved never to publish again.

In 1684, Newton's devoted friend Edmund Halley, a noted astronomer who later discovered the comet named after him, came to ask Newton's advice. Halley was involved in a discussion with Christopher Wren and Robert Hooke about the force needed to cause a body to move along an ellipse in accord with Kepler's laws. This was one of the most debated and interesting scientific problems of the time. Halley was pleasantly surprised to learn that Newton had already solved this problem "and much other matter," Halley wrote. Halley persuaded his friend to publish these important studies. To encourage Newton, Halley took on responsibility for all the costs of publication. Less than 2 years later, Newton had the *Principia* ready for the printer. Publication of the *Principia* in 1687 quickly established Newton as one of the greatest thinkers in history.

Several years afterward, Newton apparently suffered a nervous breakdown. He recovered, but from then until his death 35 years later, Newton made no major scientific discoveries. He rounded out his earlier studies on heat, optics, chemistry, and electricity, and turned increasingly to theological studies and politics. During those years, he received many honors. In 1699 Newton was appointed Master of the Mint, partly because of his great knowledge of the chemistry of metals. In this position, he helped to reestablish the value of British coins, in which lead and copper had been introduced in place of silver and gold. In 1689 and 1701 Newton represented Cambridge University in Parliament. Queen Anne knighted Newton in 1705. He was president of the Royal Society from 1703 until his death in 1727. Newton is buried in Westminster Abbey.

4.3 NEWTON'S *PRINCIPIA*

Newton's *Principia*, written in scholarly Latin, contained long geometrical arguments, and is very difficult to read, even today in English translation. Happily, as was the case later with the relativity and quantum theories, several gifted writers wrote excellent nontechnical summaries in the vernacular languages that allowed a wide circle of educated readers to learn of Newton's arguments and conclusions. The French philosopher Voltaire published one of the most popular of these books in 1736.

The *Principia* begins with the definitions of mass, momentum, inertia, and force. Next come the three laws of motion and the principles of addition for forces and velocities. In a later edition of the *Principia* Newton also included a remarkable passage on "Rules of Reasoning in Philosophy." The four rules, or assumptions, reflect Newton's profound faith in the uniformity of all nature. Newton intended the rules to guide scientists in making hypotheses. He also wanted to make clear to the reader his own philosophical assumptions. These rules had their roots in ancient Greece and are still useful today. The first has been called a principle of parsimony (simplicity or economy), the second and third, principles of unity. The fourth rule expresses a faith needed to use the process of logic.

In a brief form, and using some modern language, Newton's four rules of reasoning are:

1. *"Nature does nothing . . . in vain, and more is in vain when less will serve."* In short, nature is simple. Therefore, scientists ought not to introduce more hypotheses than are needed to explain observed facts. This fundamental faith of all scientists has a long history, going back to Aristotle, Medieval thought, and Galileo. An example: if a falling apple and the orbiting Moon can be explained by the hypothesis of one force, there is no need for two separate forces, one celestial and one terrestrial.

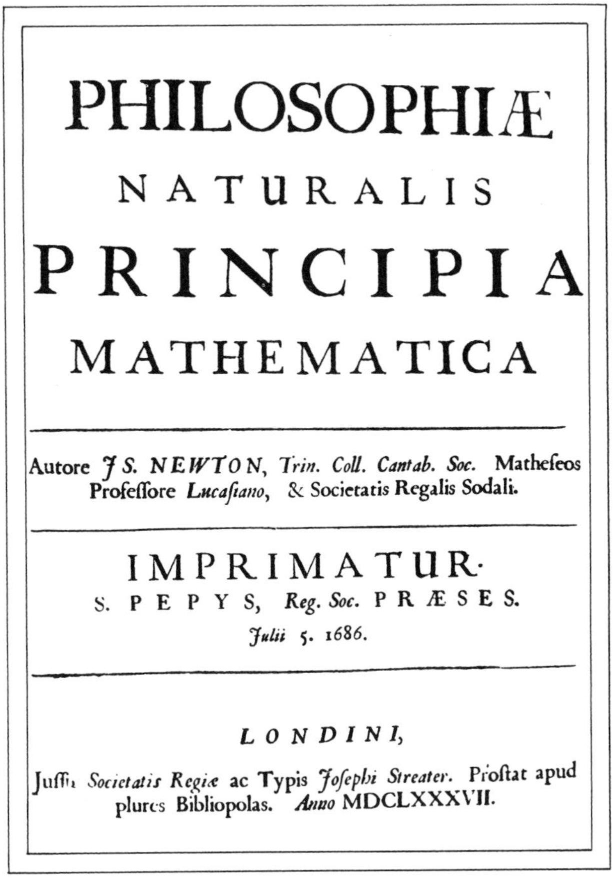

PHILOSOPHIÆ
NATURALIS
PRINCIPIA
MATHEMATICA

Autore *J S. NEWTON*, *Trin. Coll. Cantab. Soc.* Matheſeos Profeſſore *Lucaſiano*, & Societatis Regalis Sodali.

IMPRIMATUR·
S. PEPYS, *Reg. Soc.* PRÆSES.
Julii 5. 1686.

LONDINI,

Juſſu *Societatis Regiæ* ac Typis *Joſephi Streater*. Proſtat apud plures Bibliopolas. *Anno* MDCLXXXVII.

FIGURE 4.3 Title page of Newton's *Principia Mathematica*. The Royal Society sponsored the book, and the title page therefore includes the name of the Society's president, Samuel Pepys (famous for his diary describing life during the seventeenth century).

2. *"Therefore to the same natural effects we must, as far as possible, assign the same causes."* For instance, we can assume, at least initially, that the force attracting the Moon to the Earth is the same type of force attracting the Earth and planets to the Sun.
3. *Properties common to all bodies within reach of experiments are assumed (until proved otherwise) to apply to all bodies in general.* For example, all physical objects known to experimenters had always been found to have mass. So, by this rule, Newton proposed—until proved otherwise by new information—that *every* physical object be considered to have mass, even those objects beyond our reach in the celestial region.
4. *In "experimental philosophy" scientists should accept hypotheses or generalizations based on experimental evidence as being "accurately or very nearly true, notwithstanding any contrary hypotheses that may be imagined."* That is, scientists should accept such experimentally based hypotheses until

> they have additional evidence by which the hypotheses may be made more accurate or revised.

The *Principia* is an extraordinary document. Its three main sections contain a wealth of mathematical and physical discoveries. Overshadowing everything else is the theory of universal gravitation, with the geometrical proofs and arguments leading to it. We shall now restate many of the steps Newton used in his proofs in modern terms in order to make them more accessible.

The central idea of universal gravitation is breathtaking in its generality but can be simply stated: *Every object in the Universe attracts every other object.* Moreover, the amount of attraction in every instance depends in a simple way on the masses of the objects and the distance between them. (We shall patiently develop the steps taken toward finding the equation in Section 4.8, which, in a few symbols, expresses that grand law.)

Newton's great synthesis was boldly to combine terrestrial laws of force and motion with astronomical laws of motion. Gravitation is a *universal* force. It applies to the Earth and apples, to the Sun and planets, and to all other bodies (such as comets) moving in the solar system. Celestial and terrestrial physics were united in one grand system dominated by the law of universal gravitation. The general astonishment and awe were reflected in the famous words of the English poet Alexander Pope:

> Nature and Nature's laws lay hid in night:
> God said, Let Newton be! and all was light.

Readers of Newton's work must have been excited and perhaps puzzled by the new approach and assumptions. For 2000 years, from the time of the ancient Greeks until well after Copernicus, people believed that the world is separated into two distinct realms, the celestial and the terrestrial. They had used the ideas of natural place and natural motion to explain the general position and movements of the planets. From the time of the Greeks, scholars had widely believed that the planets' orbits were their "natural motion." Therefore, their orbital motion required no explanation; it was natural. However, to Newton the natural motion of a body was continuing to go at a uniform rate along a straight line, in the absence of forces. Motion in a curve showed that a net force was continuously accelerating the planets away from their natural straight-line motion, causing them to move in an orbit. The force acting on the planets was entirely natural and acted between all bodies in heaven and on Earth. Furthermore, it was the same force that caused bodies on the Earth to fall. What a reversal of the old assumptions about what was "natural"!

4.4 THE INVERSE-SQUARE LAW

Newton believed that the influence of the Sun forced the natural straight-line path of a planet into a curve. He demonstrated that Kepler's law of areas could be true if, and only if, forces exerted on the planets are always directed toward a single point. Such a force is termed a *central force*. Newton also showed that the single point is the location of the Sun. Planets obey the law of areas no matter what *magnitude* the force has, as long as the force is always directed to the same point. Newton still had to show that a central gravitational force would cause the exact relationship observed between the orbital radii and the periods of the planets, as given by Kepler's third law. How great was the gravitational force and how did it differ for different planets?

The combination of Kepler's three laws of planetary motion with Newton's three laws of motion in general provides a fine example of the power of logical reasoning. Compare these different sets of laws (you may want to review Section 2.10 and Chapter 3):

Newton's Laws of Motion

1. Every object continues in its state of rest or of uniform velocity (motion at uniform speed in a straight line) unless acted upon by an unbalanced force (a net force). Conversely, if an object is at rest or in motion with uniform velocity, all forces that may be acting on it must cancel so that the net force is zero.
2. The net force acting on an object is numerically equal to, and in the same direction as, the acceleration of the object multiplied by its mass. In symbols: $\mathbf{F}_{\text{net}} = m\mathbf{a}$.
3. If one object exerts a force on another object, the second object at the same time exerts a force on the first object. These two forces, each acting on one of the two objects, are equal in magnitude and opposite in direction.

Kepler's Laws of Planetary Motion

1. The planets orbit the Sun on ellipses, with the Sun at one focus and nothing at the other focus.
2. An imaginary line from the Sun to the moving planet sweeps out equal areas in equal amounts of time.
3. The squares of the periods of the planets are proportional to the cubes of their average distances from the Sun, for all of the planets.

Now we will see how these different laws can be put to more general use. According to Newton's first law, a change in motion, either in direction or in magnitude (speed), requires the action of a net force. According to Kepler, the planets move in orbits that are ellipses, that is, curved orbits. Therefore, a net force must be acting to change their motion. Notice that this conclusion alone does not yet specify the type or direction of the net force.

Combining Newton's second law with the first two laws of Kepler clarifies the direction of the force. According to Newton's second law, the net force is exerted in the direction of the observed acceleration. What is the direction of the force acting on the planets? Newton's geometrical analysis indicated that a body moving under a central force will, when viewed from the center of the force, move according to Kepler's law of areas. Kepler's law of areas invokes the distance of the planets from the Sun. Therefore, Newton could conclude that the Sun at one focus of each ellipse was the source of the central force acting on each of the corresponding planets.

Newton then found that motion in an elliptical path would occur only when the central force was an inverse-square force

$$\mathbf{F} \propto \frac{1}{R^2}.$$

Thus, only an inverse-square force exerted by the Sun would result in the observed elliptical orbits described by Kepler. Newton then proved the argument by showing that such an inverse-square law would in fact result in Kepler's third law, the law of periods, $T^2 = kR^3_{\text{av}}$.

From this analysis, Newton concluded that one general law of universal gravitation applied to all bodies moving in the solar system. This is the central argument of Newton's great synthesis.

Early in his thinking, Newton considered the motions of the six then-known planets in terms of their centripetal acceleration toward the Sun. By Newton's proof, mentioned above, this acceleration decreases inversely as the square of the planets' average distances from the Sun. For the spe-

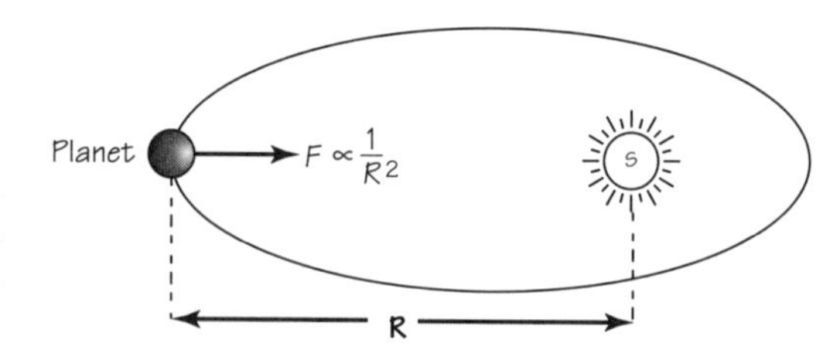

FIGURE 4.4 Planet orbiting Sun, which exerts a force F on the planet that is inversely proportional to the square of the distance to the Sun's center. The ellipse is greatly exaggerated here.

cial case of an ellipse that is just a circle, the proof is very simple and short. The expression for centripetal acceleration a_c of a body moving uniformly in a circular path may be found in terms of the radius R and the period T:

$$a_c = \frac{v^2}{R},$$

$$v = \frac{2\pi R}{T},$$

so

$$a_c = \frac{4\pi^2 R^2}{T^2} \cdot \frac{1}{R} = \frac{4\pi^2 R}{T^2}.$$

Kepler's law of periods stated a definite relationship between the orbital period of every planet and its average distance from the Sun, i.e.,

T^2 is proportional to R^3,

or

$$\frac{T^2}{R^3} = \text{constant}.$$

Using the symbol k for constant

$$T^2 = kR^3_{av}.$$

For circular orbits, R_{av} is just R. Substituting kR^3 for T^2 in the centripetal force equation gives

$$a_c = \frac{4\pi^2 R}{kR^3} = \frac{4\pi^2}{kR^2}.$$

Since $4\pi^2/k$ is also a constant, we have

$$a_c \propto \frac{1}{R^2}.$$

As you can see from the derivation, this conclusion follows necessarily from Kepler's law of periods and the definition of acceleration. If Newton's second law, $\mathbf{F}_{net} \propto \mathbf{a}$, holds for planets as well as for bodies on Earth, then there must be a centripetal force $\mathbf{F}_c$ acting on a planet. Furthermore, this force must decrease in proportion to the square of the distance of the planet from the Sun

$$F_c \propto \frac{1}{R^2}.$$

Newton showed that the same result holds for all ellipses. Indeed, this proportionality holds for any object moving in an orbit around a center from which the force on the body is applied (even in Bohr's model for hydrogen, in which an electron is thought to orbit around the nucleus). The center of force (in this case the Sun) acts upon any such object by a centripetal force that varies inversely with the square of the distance from the center of force.

Newton had still more evidence from telescopic observations of Jupiter's satellites and Saturn's satellites. The satellites of each planet obeyed Kepler's law of areas around the planet as a center. For Jupiter's satellites, Kepler's law of periods, T^2/R^3 = constant, held. But the *value* of the constant was different from that for the planets around the Sun. The law held also for Saturn's satellites, but with still a different constant. The reason is that Jupiter's satellites were acted on by a central force directed toward Jupiter and decreasing with the square of the distance from Jupiter. The same held true for Saturn's satellites and Saturn. These observed interactions of astronomical bodies supported Newton's proposed central attractive force obeying the $1/R^2$ rule.

4.5 THE LAW OF UNIVERSAL GRAVITATION

Subject to further evidence, you can now accept the idea that a central force is holding the planets in their orbits. Furthermore, the strength of this central force changes inversely with the square of the distance from the Sun. This strongly suggests that the Sun is the origin of the force, but it does not necessarily require this conclusion. Newton's results so far describe the force in mathematical terms, but they do not provide any mechanism for its transmission.

The French philosopher Descartes (1596–1650) had proposed that all space was filled with a thin, invisible fluid. This ethereal fluid carried the planets around the Sun in a huge whirlpool-like motion. This interesting idea was widely accepted at the time. However, Newton proved by a pre-

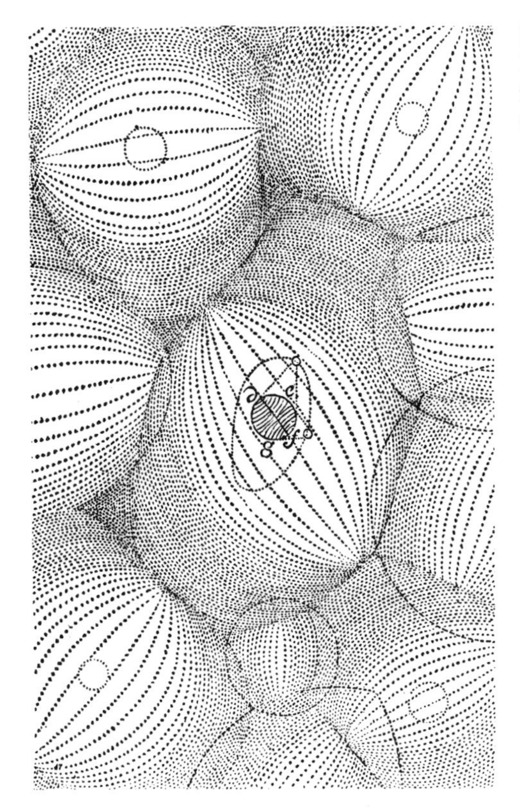

FIGURE 4.5 Drawing by Descartes (1596–1650) illustrating his theory of a space filled with whirlpools of matter that drive the planets along their orbits.

cise argument that this mechanism could not explain the details of planetary motion summarized in Kepler's laws.

Kepler had made a different suggestion some years earlier. He proposed that a magnetic force from the Sun kept the planets moving. Kepler was the first to regard the Sun as the controlling mechanical agent behind planetary motion. But Kepler's magnetic model was inadequate. The problem remained: Was the Sun actually the source of the force? If so, on what properties of the Sun or planets did the amount of the force depend?

As you read in Section 4.2, Newton had begun to think about planetary force while living at home during the Black Plague. There an idea came to him (perhaps when he saw an apple fall). Newton's idea was that the force between the Sun and the planets was the same as the force that caused objects near the Earth's surface to fall. He first tested this idea on the Earth's attraction for the Moon. It was an ingenious argument. The data available to him fixed the distance between the center of the Earth and the center of the Moon at nearly 60 times the radius of the Earth. Newton believed that the attractive force varies as $1/R^2$. Therefore, the gravitational accel-

FIGURE 4.6 An apple on Earth's surface.

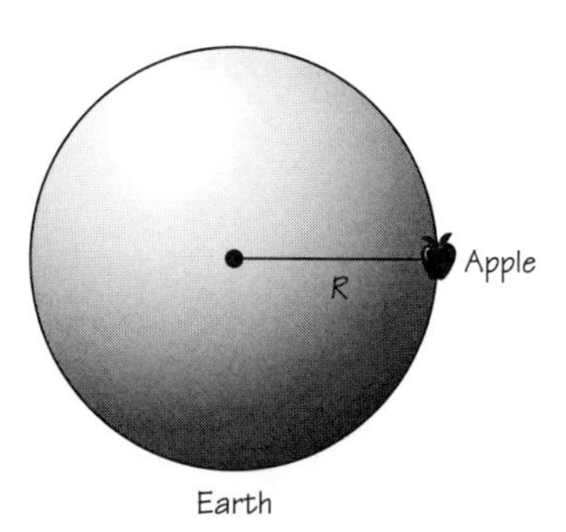

eration the Earth exerts on the Moon should be only $1/(60)^2 = 1/3600$ of that exerted upon objects at the Earth's surface. Observations of falling bodies had long established gravitational acceleration at the Earth's surface as about 9.80 m/s/s, that is, 9.8 $\mathrm{m/s^2}$. Therefore, the Moon *should* fall at 1/3600 of that acceleration value:

$$9.80\ \mathrm{m/s^2} \times (1/3600) = 2.72 \times 10^{-3}\ \mathrm{m/s^2}.$$

It was also common knowledge that the orbital period of the Moon is very nearly 27.33 days. The centripetal acceleration a_c of a body moving uniformly with period T in a circle of radius r is $a_c = 4\pi^2 R/T^2$, as indicated earlier. When you insert the observed values for the quantities R and T (in meters and seconds) for the case of the Moon orbiting the Earth and do the arithmetic, you find that the centripetal acceleration of the Moon, based upon observed quantities, should be

$$a_c = 2.74 \times 10^{-3}\ \mathrm{m/s^2}.$$

This is in very good agreement with the value of $2.72 \times 10^{-3}\ \mathrm{m/s^2}$ predicted above. From the values available to Newton, which were close to these, he concluded that he had, in his words,

> compared the force requisite to keep the moon in her orbit with the force of gravity at the surface of the earth, and found them to answer pretty nearly.
>
> Therefore, the force by which the moon is retained in its orbit becomes, at the very surface of the earth, equal to the force of gravity which we observe in heavy bodies there. And, therefore, (by rules of reasoning 1 and 2) the force by which the moon is retained in its orbit is that very same force which we commonly call gravity. . . .

This was really a triumph. The same gravity that brings apples down from trees also keeps the Moon in its orbit, i.e., making it constantly fall away from a straight-line motion into space, just enough to make it move around the Earth. This assertion is an aspect of what is known as the law of universal gravitation: *Every object in the Universe attracts every other object with a gravitational force*. If this is so, there must be gravitational forces not only between a rock and the Earth, but also between the Earth and the Moon, between Jupiter and its satellites, and between the Sun and each of the planets. And by Newton's third law, we should expect the Moon to exert an equal (but oppositely directed) gravitational force on the Earth, and the planets on the Sun. Indeed that is so, and we shall study this in the next section in detail. But for now, note that because the Earth is so much more massive than the Moon, the "wobble" of the Earth caused by the pull of the Moon is tiny. Similarly, our Sun, enormously more massive than the planets of our solar system, is acted on by them, but this action produces only very small motions of the Sun. This effect has led to a remarkable finding: other stars than our Sun have been observed to exhibit a small wobbling motion that is attributed to the presence of one or more (not directly observable) planets orbiting these stars, thus revealing that other planetary systems exist throughout our Universe.

Newton did not stop at saying that a gravitational force exists between the Earth and Moon, the planets and the Sun. He further claimed that the force is exactly the right magnitude and direction to explain *completely* the motion of every planet. No other mechanism (whirlpools of invisible fluids or magnetic forces) is needed. Gravitation, and gravitation alone, underlies the dynamics of the heavens.

Recently indications have been found that some galaxies experience a repulsion, not predominantly a gravitational attraction, from one another. Hence the expansion of the universe is proceeding at an accelerating rate!

4.6 NEWTON'S SYNTHESIS

The concept of gravitation is now so commonplace that you might be in danger of passing it by without really appreciating what Newton was claiming. First, he proposed a truly universal physical law. Following his rules of reasoning, Newton extended to the whole Universe what he found true for its observable parts. He excluded no object in the Universe from the effect of gravity.

The idea that terrestrial laws and forces are the same as those that regulate the entire Universe had stunning impact. Less than a century before, it would have been dangerous even to suggest such a thing. Kepler and

Galileo had laid the foundation for combining the physics of the heavens and Earth. Newton carried this work to its conclusion. Today, Newton's extension of the mechanics of terrestrial objects to the motion of celestial bodies is called the *Newtonian synthesis*.

Newton's claim that a planet's orbit is determined by the gravitational attraction between it and the Sun had another effect. It moved science away from geometrical explanations and toward physical ones. Most philosophers and scientists before Newton were occupied mainly with the question "What are the motions?" Newton asked, instead, "What force *explains* the motions?" In both the Ptolemaic and Copernican systems, the planets moved about *points* in space rather than about *objects*. The planets moved as they did because of their "nature" or geometrical shape, not because forces acted on them. Newton, on the other hand, spoke not of points, but of things, of objects, of physical bodies. For example, unless the gravitational attraction to the Sun deflected them continuously from straight-line paths, the planets would fly out into the darkness of deep space. Thus, it was the physical Sun that was important, not the point at which the Sun happened to be located.

Newton's synthesis centered on the idea of gravitational force. By calling it a force of gravity, Newton knew that he was not explaining *why* it existed. When you hold a stone above the surface of the Earth and release it, it accelerates to the ground. The laws of motion tell you that there must be a force acting on the stone to accelerate it. You know the *direction* of the force. You can find the *magnitude* of the force by multiplying the mass of the stone by the acceleration. You know that this force is weight, or gravitational attraction to the Earth. But why such an interaction between bodies exists at all remains a puzzle. It is still an important problem in physics today.

4.7 NEWTON AND HYPOTHESES

Newton's claim that there is a mutual force (gravitational interaction) between a planet and the Sun raised a new question: How can a planet and the Sun act upon each other at enormous distances without any visible connections between them? On Earth you can exert a force on an object by pushing it or pulling it. You are not surprised to see a cloud or a balloon drifting across the sky, even though nothing seems to be touching it. Air is invisible, but you know that it is actually a material substance that you can feel when it moves. Objects falling to the Earth, and pieces of iron being attracted to a magnet are harder to explain, but at least the distances involved are small. However, the Earth is over 144 million kilometers, and

Saturn more than 1 billion kilometers, from the Sun. How could there possibly be any physical contact between such distant objects? How can we account for such "action at a distance"?

In Newton's time and for a long time afterward, scholars advanced suggestions for solving this problem. Most solutions involved imagining space to be filled with some invisible substance, called the "ether," which transmitted force. Newton himself privately guessed that such an ether was involved. But he could find no way to test this belief. Therefore, at least in public, he refused to speculate on possible mechanisms. In a famous passage that he added in the second edition of the *Principia* (1713), Newton declared:

> Hitherto I have not been able to discover the cause of those properties of gravity from phenomena, and I frame no hypotheses; for whatever is not deduced from the phenomena is to be called an hypothesis; and hypotheses, whether metaphysical or physical, whether of occult qualities or mechanical, have no place in experimental philosophy. . . . And to us it is enough that gravity does really exist, and acts according to the laws which we have explained, and abundantly serves to account for all the motions of the celestial bodies, and of our sea.

Newton is quoted at length here because one particular phrase is often taken out of context and misinterpreted. The original Latin reads: *hypotheses non fingo*. This is translated above as "I frame no hypotheses." The sense is, "I do not make untestable, possibly *false*, hypotheses." Newton in fact made many hypotheses in his publications that led to testable results. Also, his letters to friends contain many speculations which he did not publish. So his stern denial of "framing" hypotheses must be properly interpreted.

The fact is that there are two main kinds of hypotheses or assumptions. The most common hypothesis is a proposal of some hidden mechanism to explain observations. For example, you observe the moving hands of a watch. You might propose or imagine some arrangement of gears and springs that causes the motion. This would be a *hypothesis that is directly or indirectly testable, at least in principle, by reference to phenomena*. The hypothesis about the watch, for example, can be tested by opening the watch or by making an X ray film of it. In this context, consider the invisible fluid that supposedly transmitted gravitational force, the so-called "ether." Newton and others thought that certain direct tests might establish the presence of this substance. Many experimenters tried to "catch" the ether. A common approach involved pumping the air from a bottle. Then tests were made to see if any wind, pressure, or friction due to the ether remained to affect objects in the bottle. Nothing of this sort worked (nor has it since).

So Newton wisely avoided making public any hypothesis for which he could not also propose a test.

A quite different type of assumption is often made in published scientific work. It involves a hypothesis which everyone knows is not directly testable, but which still is necessary just *to get started on one's work*. An example is such a statement as "nature is simple," or any other of Newton's four rules of reasoning. Acceptance of either the heliocentric system or the geocentric system is another example. In choosing the heliocentric system, Copernicus, Kepler, and Galileo made the hypothesis that the Sun is at the center of the Universe. They knew that this hypothesis was not directly testable and that either system seemed to explain the observed behavior of the celestial objects equally well. Yet they adopted the point of view that seemed to them simpler, more convincing, and more "pleasing to the mind." It was this kind of initial hypothesis that Newton used without apology in his published work, and it turned out to be right to do so since it did lead to testable results in the end.

Every scientist's work involves both kinds of hypothesis. One popular image of the scientist is of a person who uses only deliberate, logical, objective thoughts, and immediately tests them by definitive experiments. But, in fact, the working scientist feels quite free to entertain any guess, imaginative speculation, or hunch, provable or not, that might be helpful in the early stages of research. (Sometimes these hunches are dignified by the phrase "working hypotheses." Without them there would be little progress!) As Einstein once said, the initial ideas and concepts used in starting research may be considered "free conventions. They appear to be a priori only insofar as thinking without the positing of categories and of concepts in general would be as impossible as is breathing in a vacuum." However, most scientists today do not like to publish something that is still only an unproven hunch. It has to prove its usefulness to be accepted in the final theory.

4.8 THE MAGNITUDE OF THE GRAVITATIONAL FORCE

The general statement that gravitational forces exist universally must now be turned into a quantitative law, as indeed Newton did. An expression is needed for both the *magnitude* and the *direction* of the forces any two objects exert on each other. It was not enough for Newton to assert that a mutual gravitational attraction exists between the Sun and Jupiter. To be convincing, he had to specify what quantitative factors determine the mag-

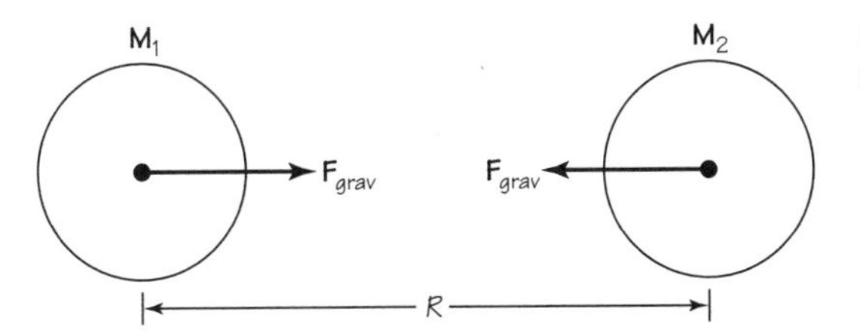

FIGURE 4.7 Mutual gravitational forces.

nitudes of those mutual forces. He had to show how they could be measured, either directly or indirectly.

The first problem was defining precisely the distance R. Should it, for example, be taken as the distance between the surface of the Earth and the surface of the Moon? For many astronomical problems, the sizes of the interacting bodies are extremely small compared to the distances between them. In such cases, the distance between the surfaces is practically the same as the distance between the centers. (For the Earth and the Moon, the distance between centers is only about 2% greater than the distance between surfaces.) Yet, some historians believe Newton's uncertainty about a proper answer to this problem led him to drop the study for many years.

Eventually, Newton solved the problem. His calculation showed that the gravitational force exerted *by* a spherical body is the same as if all its mass were concentrated at its center. The gravitational force exerted *on* a spherical body by another body is the same as would be exerted on it if all its mass were concentrated at its center. Therefore, *the distance* R *in the law of gravitation is the distance between centers.*

This was a very important discovery. The gravitational attraction between spherical bodies can be considered as though their masses were concentrated at single points. Thus, in thought, the objects can be replaced by *mass points.*

Newton's third law states that action equals reaction. If this is universally true, the amount of force the Sun exerts on a planet must exactly equal the amount of force the planet exerts on the Sun. For such a very large mass and such a relatively small mass, this may seem contrary to common sense. But the equality is easy to prove. First, assume only that Newton's third law holds between small pieces of matter. For example, a 1-kg piece of Jupiter pulls *on* a 1-kg piece of the Sun as much as it is pulled *by* it. Now consider the total attraction between Jupiter and the Sun, whose mass is about 1000 times greater than Jupiter's. You can consider the Sun as a globe containing about 1000 Jupiters. Define one unit of force as the force that two Jupiter-sized masses exert on each other when separated by the distance of Jupiter from the Sun. Then Jupiter pulls on the *Sun* (a globe of 1000 Jupiters) with a total force of 1000 units. Each of the 1000 parts of the Sun also pulls *on* the planet Jupiter with 1 unit. Therefore, the total pull of the Sun on Jupiter is also 1000 units. Each part of the massive Sun not only pulls *on* the planet, but is also pulled upon *by* the planet. The more mass there is to *attract*, the more there is to be *attracted*. But note: Although the mutual attractive forces on the two bodies are equal in magnitude, the resulting *accelerations* of these are not. Jupiter pulls on the Sun as hard as the Sun pulls on Jupiter, but the Sun *responds* to the pull with only 1/1000 of the acceleration, because its *inertia* is 1000 times Jupiter's.

We saw earlier why bodies of different mass fall with the same acceleration near the earth's surface. The greater the inertia of a body, the more strongly it is acted upon by gravity; that is, near the Earth's surface, the gravitational force on a body is directly proportional to its mass. Like Newton, extend this earthly effect to all gravitation. You then can assume that the gravitational force exerted on a planet by the *Sun* is proportional to the mass of the planet. Similarly, the gravitational force exerted on the Sun by the *planet* is proportional to the mass of the Sun. You have just seen that the forces the sun and planet exert on each other are equal in magnitude (though opposite in direction). It follows that the magnitude of the gravitational force is proportional to the mass of the Sun *and* to the mass of the planet; that is, the gravitational attraction between two bodies is proportional to the *product* of their masses. That is, if the mass of either body is tripled, the force each experiences is tripled. If the masses of both bodies are tripled, the force is increased by a factor of 9. Using the symbol F_{grav} for the magnitude of the forces

$$F_{\text{grav}} \propto m_{\text{planet}}\, m_{\text{Sun}}.$$

The conclusion is that the amount of attraction between the Sun and a planet is proportional to the product of their masses. Earlier you saw that the gravitational attraction also depends on the square of the distance between the centers of the bodies.

$$F_{\text{grav}} \propto \frac{1}{R^2}.$$

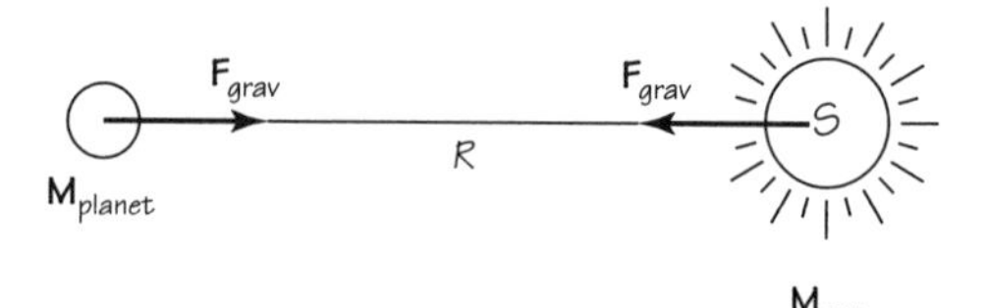

FIGURE 4.8 Gravitational attraction between a planet and the Sun.

Combining these two proportionalities gives *one* force law, which now includes both mass and distance

$$F_{grav} \propto \frac{m_{planet}\, m_{Sun}}{R^2}.$$

Such a proportionality can be written as an equation by introducing a constant. (The constant allows for the units of measurement used.) Using G as a symbol of the proportionality constant, the law of planetary forces can be written as an equation

$$F_{grav} = \frac{G m_{planet}\, m_{Sun}}{R^2}.$$

This equation asserts that the force between the Sun and any planet depends *only* upon three factors. These factors are the masses of the Sun and planet and the distance between them. The equation seems unbelievably simple when you remember how complex the observed planetary motions seemed. Yet every one of Kepler's empirical laws of planetary motion agrees with this relation. In fact, we can *derive* Kepler's empirical laws from this force law and Newton's second law of motion. More important still, details of planetary motion not obtainable with Kepler's laws alone can be calculated using this force law.

Newton's proposal that this simple equation describes completely the forces between the Sun and planets was not the final step. He saw nothing to restrict this mutual force to the Sun and planets, or to the Earth and apples. Rather, again in line with his rules of reasoning, Newton insisted that an identical relation should apply *universally* to all matter everywhere throughout the Universe. This relation would hold true for *any two bodies* separated by a distance that is large compared to their dimensions. It would apply equally to two atoms or two stars. In short, Newton proposed a *general law of universal gravitation*

$$\boxed{F_{grav} = \frac{G \cdot m_1 m_2}{R^2}}$$

Here m_1 and m_2 are the masses of the two bodies and R is the distance between their centers. The numerical constant G is called the *constant of universal gravitation*. Newton assumed it to be the same for all gravitational interactions, whether between two grains of sand, two members of a solar system, or two stars in different parts of the sky. As you will see, the very great successes made possible by this simple relationship have borne out Newton's assumption. In fact, scientists have come to assume that this equation applies everywhere and at all times, past, present, and future.

Even before we consider more supporting evidence, the sweeping majesty of Newton's law of universal gravitation should command your wonder and admiration. It also leads to the question of how such a bold universal law can be proved. There is no complete proof, of course, for that would mean examining every interaction between all bodies everywhere in the Universe! But the greater the variety of individual tests made, the greater will be our belief in the validity of the law.

4.9 THE VALUE OF *G*, AND SOME CONSEQUENCES

The masses of small solid objects on the surface of the Earth can be found easily enough from their weights. Measuring the distance between solid objects of spherical shape presents no problem. But how can one measure the tiny mutual gravitational force between relatively small objects in a laboratory? (Remember that each object is also experiencing separately a huge gravitational force toward the tremendously massive Earth.) In addition, how can we find the actual value of G?

This serious technical problem was eventually solved by the English scientist, Henry Cavendish (1731–1810). For measuring gravitational forces, he employed a torsion balance. In this device, the gravitational attraction between two pairs of lead spheres, one pair fixed in the laboratory, the other suspended from a wire on a rod holding up that pair (see the illustration). The force producing a twist of the wire could be calibrated by applying first to the suspended pair of lead spheres small known forces. A typical experiment might test the attraction between one of the fixed spheres of, say, 100 kg and one of the suspended spheres of, say, 1 kg, at a center-to-center distance of 0.1 m. The resulting force would be found to be about one-millionth of a newton (0.000001 N)! These data, inserted into the force law of universal gravitation, allow one to calculate a value for G; it comes out to be about 10^{-10} (N m^2/kg^2). The experiment to find the value of G has been steadily improved ever since, and the accepted value of G is now

$$G = 0.00000000006673 \text{ N m}^2/\text{kg}^2.$$

This can be written more concisely in scientific notation:

$$G = 6.673 \times 10^{-11} \text{ N m}^2/\text{kg}^2.$$

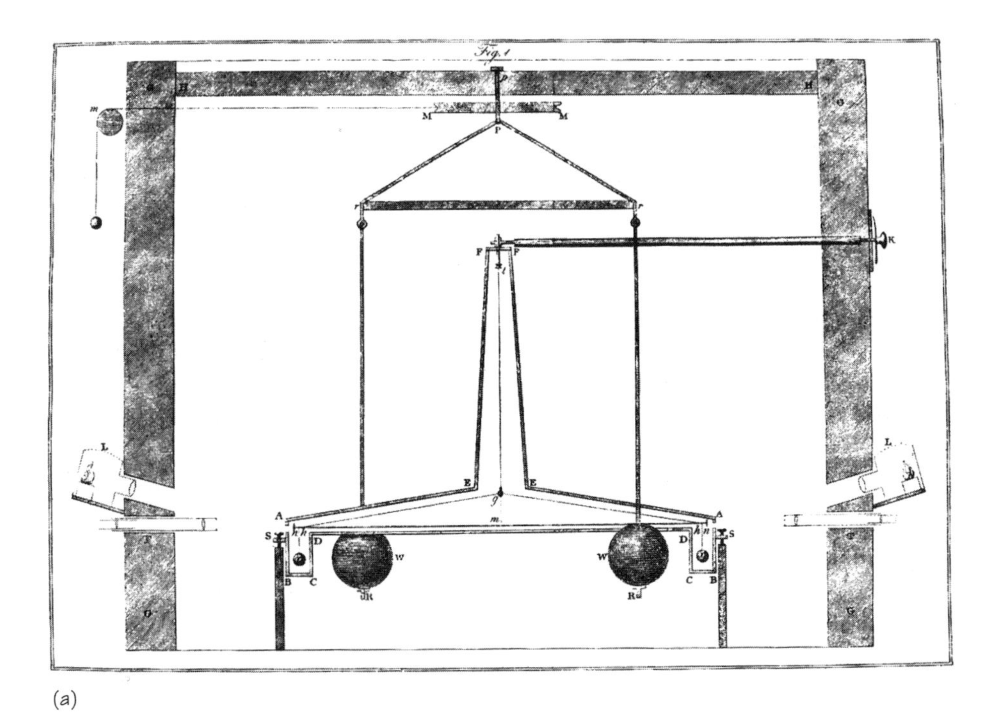

(a)

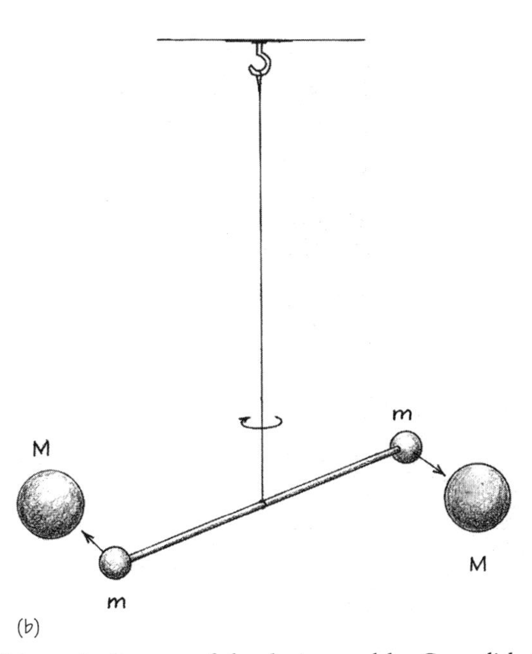

(b)

FIGURE 4.9 (a) Schematic diagram of the device used by Cavendish for determining the value of the gravitational constant G. Large lead balls of masses M_1 and M_2 were brought close to small lead balls of masses m_1 and m_2. The mutual gravitational attraction between M_1 and m_1 and between M_2 and m_2 caused the vertical wire to be twisted by a measurable amount. (b) Diagram of Cavendish's apparatus for determining the value of G. To prevent disturbance from air currents, Cavendish enclosed the apparatus in a sealed case and observed the deflection of the balance rod from outside with telescopes.

Using this result in the formula giving the law of universal gravitation allows one to calculate the force in newtons (N), if the distance R between the centers of the attracting objects is measured in meters (m) and the masses of the objects are measured in kilograms (kg).

G is obviously a very small number. The measurement of G confirms that the gravitational force between everyday objects is indeed very small. Let's see how small. Suppose you have two 1-kg masses separated by a distance of 1 m between their centers. How large is the gravitational attraction between them? To find out, simply substitute into the equation for F_{grav}:

$$
\begin{aligned}
F_{\text{grav}} &= \frac{Gm_1m_2}{R^2} \\
&= (6.67 \times 10^{-11} \text{ N m}^2/\text{kg}^2)(1 \text{ kg} \cdot 1 \text{ kg}/1 \text{ m}^2) \\
&= 6.67 \times 10^{-11} \text{ N}
\end{aligned}
$$

or

$$F_{\text{grav}} = 0.0000000000667 \text{ N}.$$

Obviously this is an extremely small force. Could you ever measure a force to this accuracy in your laboratory? For comparison, it turns out that an average apple (mass of about 0.1 kg) weighs (has the gravitational force on it of) about 1 N—which is only right!

The Acceleration of Gravity g

You know from the previous chapter that all objects on Earth fall at the same rate of acceleration, regardless of their mass. Since this rate of acceleration is nearly the same everywhere on Earth, it is given a special symbol, g. Using Newton's law of gravitation, and the value of G, how can we explain that g is nearly constant over the entire Earth's surface? Consider an apple of mass m. From the previous chapter, the gravitational force on the apple at the Earth's surface is

$$F_{\text{grav}} = m \cdot a = m \cdot g.$$

We found in this chapter that the gravitational force is given also by Newton's equation

$$F_{\text{grav}} = G \cdot \frac{M_{\text{Earth}}\, m}{R_{\text{Earth}}^2}.$$

Therefore, these two expressions must be equal to each other

$$G \cdot \frac{M_{\text{Earth}}\, m}{R_{\text{Earth}}^2} = m \cdot g.$$

Canceling the mass of the apple (m) on both sides of the last equation, we find that g is related to the mass and radius of the Earth

$$G \cdot \frac{M_{\text{Earth}}}{R_{\text{Earth}}^2} = g.$$

This equation helps to explain why the acceleration of gravity g is nearly the same for all objects on the Earth, regardless of their own mass: The value of g is determined by the universal constant G, the mass of the Earth M_{Earth}, which is the same in each case, and the radius of the Earth, R_{Earth}, also near enough the same in each case.

To be sure, g is not *exactly* the same everywhere on the Earth's surface, because the distance to the center of the Earth and its local mass density are slightly different at different places on the surface of the Earth. The Earth is not a perfect, homogeneous sphere. Nevertheless, to one decimal place, the result for g is near enough the same everywhere on the Earth. (For example, at the top of Mount Everest the value of g is 9.7647 m/s^2, while in Toronto, Canada, it is 9.8049 m/s^2.) This means that the gravitational force on objects near the Earth's surface, also known as the weight, is different for each mass; but the acceleration (g) will be about the same for all.

Of course, the acceleration due to gravity will be different from the acceleration on Earth if the attracting object is not the Earth. For instance, on the surface of the Moon, which has a smaller radius and a smaller mass than the Earth's, the acceleration due to gravity is only 1.6 m/s^2. Thus, an astronaut having a mass of 70 kg, would weigh 690 N on the surface of the Earth, but only 110 N on the Moon.

The law of gravitation may also be used to "weigh the Earth," i.e., to find its mass. This amazing calculation may be found in the *Student Guide*.

Revisiting the Space Shuttle

In the previous chapter (Section 3.12) we looked at the flight of the space shuttle *Endeavor* as it orbited the Earth in January 1996. Data available on the NASA Web site indicated that the shuttle flew in a nearly circular orbit of 288 miles average altitude above the Earth's surface with a period of

■ AVIATION TECHNOLOGY

The English physicist George Cayley worked out the essentials of aerodynamic theory in 1799. However, it would take another one hundred years before aviation technology would be ready for the flight of a heavier-than-air machine that seemed to defy gravitation (which it did not). For the next century those studying aerodynamics focused on finding the working ratio between the weight of the craft and the power required to make it fly. Wilbur and Orville Wright, two bicycle manufacturers from Dayton, Ohio, were the first to solve this problem successfully. On December 17, 1903, Orville Wright became the first man ever to fly an airplane, using a 12-horsepower engine to fly a biplane with a 40-ft 4-in wingspan.

FIGURE 4.10 Orville Wright (in prone position) piloting Flyer, December 17, 1903.

After World War I commercial airlines began to be formed in order to harness the increasing number of aviation inventions. In contrast to European government subsidies, the 1920s American government had no coherent national policy concerning the production of aircraft or the operation of airlines. As a result many private American airline companies were created during the 1920s but very few of them lasted more than a couple of years before going out of business. The most successful American airline during this period was Pan American Airlines, whose founder Juan Trippe concentrated on routes to Latin America.

Two central problems still remained to be overcome, however, if air flight was going to challenge the supremacy of rail travel, particularly in America. These problems were safe high-altitude flying and safe night flying. In 1930, the airplane manufacturer Boeing responded to this need by designing and producing the Monomail, which was an all-metal plane with retractable landing equipment. This evolved into the 247 model in 1933, which could cruise at 155 mi/hr carrying ten passengers, twice as fast as its competitors.

Airline operators and military users of these planes, however, wanted still bigger and faster craft. American manufacturers had begun tinkering with various ways to improve the piston engines: in particular, they had been experimenting with turbochargers during the previous decade.

In a turbocharger, exhaust gases from the piston engine drive the turbine. The turbine drives a supercharger, which acts as an air compressor, delivering additional air to run the engine.

The aim of the turbocharger was to maintain power at high altitudes, by compressing the air that goes into the piston engine. The compressor would be driven by a turbine placed in the hot engine exhaust whose gases carried so much power that this system was extremely effective, and cost free. However, regardless of this simplicity and the cost effectiveness of this design, the turbocharger did not attract investors within the aviation industry, largely because the temperatures within the turbine were too high for the steel that was used in its construction.

The next major improvement in aircraft engines was the turbojet; which was developed in Germany by Max Hahn and Hans von Ohain and in Britain by Frank Whittle. The main benefit of the turbojet was that its light weight offered a favorable weight to power ratio. The American company Boeing became one of the first successfully to design and manufacture a passenger plane powered by a turbojet. They introduced their 707 model onto the market in 1958, and it was so economical that it was soon followed by the Boeing 727 that could carry 100 passengers, and therefore rivaled the biggest piston planes on the market. When the Boeing 747, known as the "jumbo jet," came onto the market in 1970, it could carry 500 passengers, and still have room for cargo, freight, and mail.

Further Reading

T.A. Heppenheimer, *Turbulent Skies: The History of Commercial Aviation* (New York: Wiley, 1998).

5425 seconds for each orbit. From these data we calculated that the shuttle exhibited a centripetal acceleration of

$$a_c = 9.2 \text{ m/s}^2$$

which is about 13.6% less than the gravitational acceleration g at the surface of the Earth. We surmised that the centripetal acceleration on the shuttle was provided by the force of gravity at that altitude. From this we concluded that the gravitational force appears to decrease with distance from the Earth.

Now we can affirm this conclusion on the basis of the law of universal gravitation. Earlier in this section we found that the acceleration of gravity at the distance R from the center of a massive object M is

$$a_g = \frac{GM}{R^2}.$$

This holds if the point where the calculation is measured is outside of the object. If the centripetal acceleration experienced by the shuttle Endeavor does indeed arise from the force of gravity at the shuttle's altitude, then the calculated centripetal acceleration, $a_c = 9.2 \text{ m/s}^2$, should be close to the acceleration of gravity calculated from the law of gravitation. So we may ask which is correct:

$$a_c = a_g$$

or

$$9.2 \text{ m/s}^2 = \frac{GM}{R^2} \text{ ?}$$

In order to answer these questions, we need to substitute the values for G, M, and R in the second equation and see if the result is 9.2 m/s^2. We have now found the value for G and for M, which is the mass of the Earth, $M = 6.0 \times 10^{24}$ kg (to one decimal place). The radius R of the shuttle's orbit, which is assumed to be circular, must be measured from the center of the Earth (not from the surface). We found in Section 3.12 that $R = 6833$ km $= 6.8 \times 10^6$ m (to one decimal place). Substituting, we have

$$\frac{GM}{R^2} = \frac{(6.8 \times 10^{-11} \text{ N m}^2/\text{kg}^2)(6.0 \times 10^{24} \text{ kg})}{(6.8 \times 10^6 \text{ m})^2}$$

$$= \left[\frac{(6.8 \times 10^{-11})(6.0 \times 10^{24})}{(46.2 \times 10^{12})}\right] \frac{\text{N m}^2 \text{ kg}}{\text{m}^2 \text{ kg}^2}$$

$$= 8.8 \text{ m/s}^2.$$

The calculated result of 8.8 m/s^2 is within about 4.3% of the expected acceleration of 9.2 m/s^2. Since we have assumed that the orbit of the shuttle was a perfect circle, which it was not, we can be fairly certain that our result is within the limits of accuracy of this assumption. Therefore our conclusion that the centripetal acceleration on the shuttle is just the force of gravity at that altitude is supported. However, we would have to repeat this calculation again for the flights of other satellites, and perhaps even test it for a satellite in an orbit that is as nearly circular as possible. This is how conclusions based upon experimental evidence are often obtained.

4.10 FURTHER SUCCESSES

Newton did not stop with the fairly direct demonstrations described so far, accessible to him at that time. In the *Principia*, he showed that his law of universal gravitation could explain other complicated gravitational interactions. Among these were the tides of the sea and the peculiar drift of comets across the sky.

The Tides

Knowledge of the tides had been vital to navigators, traders, and explorers through the ages, including of course especially Newton's fellow countrymen, the British traders and explorers. But the *cause of* the tides had remained a mystery despite the studies of such scientists as Galileo. By applying the law of universal gravitation, Newton was able to explain the main features of the ocean tides. He found them to result from the attraction of the Moon and the Sun upon the waters of the Earth. Because the Moon is so much nearer to the Earth, its attractive force on the oceans is greater than the Sun's force on the oceans. Each day, as the Earth rotates, two high tides normally occur. Also, twice each month, at full Moon and at new Moon, the Moon, Sun, and Earth are in line with each other. At these times the tidal changes are greater than average.

Two questions about tidal phenomena demand special attention. First, why do high tides occur on both sides of the Earth, including the side away from the Moon? Second, why does high tide occur at a given location some hours after the Moon is highest in the sky?

Newton knew that the gravitational attractions of the Moon and Sun act not only on the ocean but also on the whole solid Earth. These forces accelerate both the fluid water on the Earth's surface and the Earth itself. Newton realized that the tides result from the *difference* in acceleration of the Earth and its waters. The Moon's distance from the Earth's center is about 60 Earth radii. On the side of the Earth nearer the Moon, the distance of the water from the Moon is only 59 radii. On the side of the Earth away from the Moon, the water is 61 Earth radii from the Moon. (See Figure 4.11.) On the side nearer the Moon, the force on the water toward the

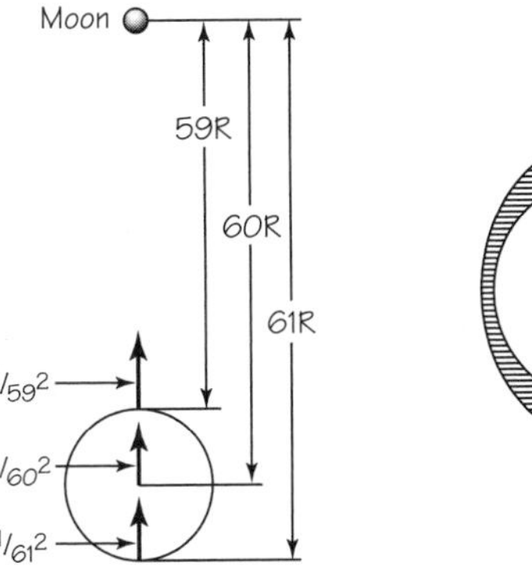

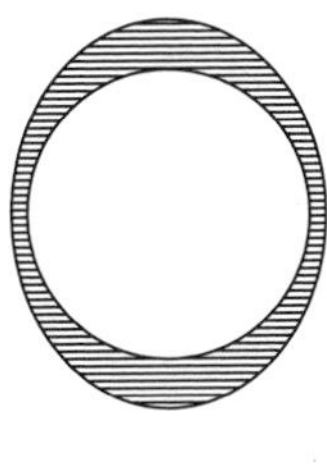

FIGURE 4.11 Relative amounts of tidal forces (note that the Earth-Moon distance indicated in the figure is greatly reduced owing to space limitations).

Moon is slightly greater than the force on the Earth as a whole, because the distance R is slightly smaller. The net effect is that the water is pulled away from the Earth. On the side of the Earth away from the Moon, the pull on the water toward the Moon is less than that of the Earth as a whole since its distance to the Moon is slightly greater. The net result is that the Earth is pulled away from the water there. A point on the ocean shore on the Earth therefore experiences two high tides each day. In between, there are low tides because the water is depleted at those points on the globe between the high tides.

Perhaps you have watched the tides change at the seashore or examined tide tables. If so, you know that high tide occurs some hours *after* the Moon is highest in the sky. To understand this, even qualitatively, you must remember that relative to the size of the Earth the oceans are not very deep. The ocean waters moving in from more distant parts of the oceans in response to the Moon's attraction are slowed by friction with the ocean floors, especially in shallow water. Thus, the time of high tide is delayed. In any particular place, the amount of delay and the height of the tides depend greatly upon how easily the waters can flow. No general theory can account for all the particular details of the tides. Most local predictions in the tide tables are based, in part, on empirical rules using the tidal patterns recorded in the past.

Since there are tides in the seas, you may wonder if the atmosphere and the Earth itself undergo tides. They do. The Earth is not completely rigid, but bends somewhat, like steel. The high tide in the Earth's crust is about 30 cm high. The atmospheric tides are generally masked by other weather changes. However, at altitudes of about 160 km, satellites have recorded considerable rises and falls in the thin atmosphere.

Comets

From earliest history through the Middle Ages, people have interpreted comets as omens of disaster. Halley and Newton showed them to be only shiny, cloudy masses moving around the Sun according to Kepler's laws, just as planets do. They found that most comets are visible only when closer to the Sun than the distance of Jupiter. Today we know that they are composed of frozen water and gases, as well as dirt—a "dirty snowball."

Several very bright comets have orbits that take them well inside the orbit of Mercury. Such comets pass within a few million kilometers of the Sun. Many orbits have eccentricities near 1.0; these comets have periods of thousands or even millions of years. Some other faint comets have periods of only 5–10 years. Occasionally comets collide with planets, as happened when the Schoemaker–Levy comet smashed into Jupiter in 1994. An

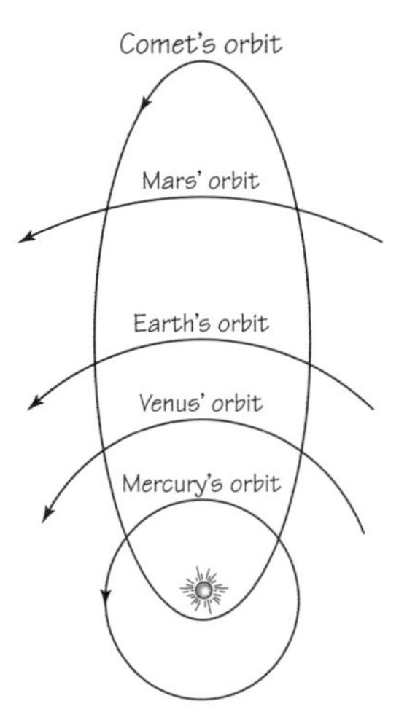

FIGURE 4.12 Schematic diagram of the orbit of a comet in the ecliptic plane. Comet orbits are tilted at all angles.

observation satellite studying Jupiter at that time, appropriately named Galileo, recorded the spectacular effects on the outer gases of Jupiter, providing much valuable information on Jupiter's atmosphere.

Unlike the planets, all of whose orbits lie nearly in a single plane, the planes of comet orbits tilt at all angles. Yet, like all members of the solar system, they obey all the laws of dynamics, including Kepler's laws and the law of universal gravitation.

Edmund Halley (1656–1742) applied Newton's concepts of celestial motion to the motion of bright comets. Among the comets he studied were those that had been observed in 1531, 1607, and 1682. Halley found the orbits for these comets to be very nearly the same. He suspected that all these observations were really of one and the same comet, moving in a closed orbit with a period of about 75 years. He predicted that the comet would return to where people on Earth could see it in about 1757—which it did, although Halley did not live to see it. It was a spectacular verification of Newton's law of gravitation. Halley's comet appeared again in 1833, 1909, and in 1985–86. It is due to return to the vicinity of the Sun in 2061.

With the period of Halley's bright comet known, its dates of appearance could be tracked back in history. Ancient Indian, Chinese, and Japanese documents record all of the expected appearances since 240 B.C. except one. Almost no European records of this great comet exist. This is a sad comment upon the level of culture in Europe during the so-called Dark Ages. One of the few European records is on part of the famous Bayeux tapestry, embroidered with 72 scenes of the Norman Conquest of England in 1066. One scene shows the comet overhead while King Harold of England and his court cower below, taking the appearance to presage a disaster. But with Newtonian science explaining the paths of comets, they came to be seen as regular members of the solar system, instead of unpredictable, fearful events.

Beyond the Solar System

You have seen how Newton's laws explain motions and other physical events on the Earth and in the solar system. Now consider a new and even broader question: Do Newton's laws also apply at greater distances, for example, among the stars?

Over the years following publication of the *Principia*, several sets of observations provided an answer to this important question. One observer was William Herschel, a British musician turned amateur astronomer. In the late 1700s, with the help of his sister Caroline, Herschel made a remarkable series of observations. Using homemade, high-quality telescopes, Herschel hoped to measure the parallax of stars owing to the Earth's motion around the Sun. Occasionally he noticed that one star seemed quite close to another. Of course, this might mean only that two stars happened to lie in the same line of sight. But Herschel suspected that some of these pairs were actually double stars held together by their mutual gravitational attractions.

He continued to observe the directions and distances of one star with respect to the other in such pairs. In some cases, one star moved during a few years through a small arc of a curved path around the other. Other astronomers gathered more information about these so-called *double stars*, far removed from the Sun and planets. Eventually, it was clear that they do move around each other according to Kepler's laws. Therefore, their motions also agree with Newton's law of universal gravitation. Using the same equation as that used above for the planets, see *Student Guide*, astronomers have calculated the masses of these stars. They range from about 0.1 to 50 times the Sun's mass.

Here we stop briefly to note again that the essence of science is that it is not dogmatic, i.e., newly found facts can modify and improve a theory. The-

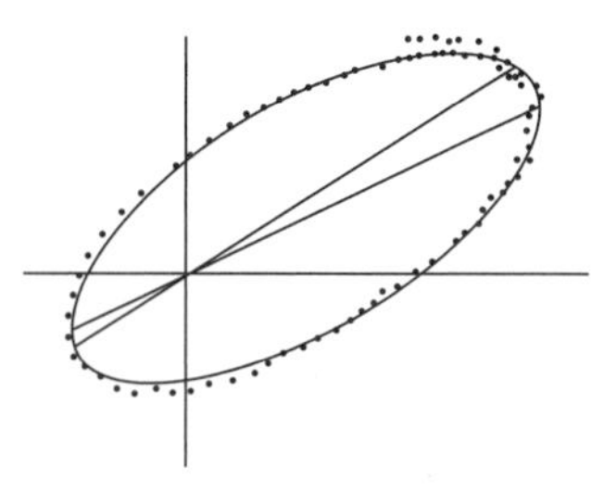

FIGURE 4.13 The motion over many years of one of two components of a binary star system. Each dot indicates the average of observations made over an entire year.

ories become increasingly acceptable as they are found useful over a wider and wider range of problems. No theory has stood this test better than Newton's theory of universal gravitation as applied to the planetary system. It took nearly a century for physicists and astronomers to comprehend, verify, and extend Newton's work on planetary motion. As late as the end of the nineteenth century, most of what had been accomplished in mechanics since Newton's day was but a development or application of his work.

As indicated earlier in reference to Kepler's laws of planetary motion (Section 2.10), a scientific *theory* differs from a *law of nature*, since a theory encompasses data and assumptions and hypotheses that can be altered and improved or abandoned as new experimental data and new ideas become available. A scientific theory is therefore quite different from the common usage of the word "theory," as, for instance, in such an everyday statement as "That's just a theory, which is merely your opinion." On the other hand, a scientific law is a statement about nature. It can be accepted or rejected, and sometimes expanded, but it does not contain hypotheses or assumptions. Examples are Kepler's laws, Newton's laws of motion, and Newton's law of universal gravitation: *Every object in the Universe attracts every other object with a gravitational force given by* $F_{grav} = GM_1M_2/R^2$. Newton's *theory* of gravitation encompasses his law of universal gravitation, but it also encompasses Newton's synthesis of celestial and terrestrial phenomena, as expressed, for instance, in his *Rules of Reasoning*: Not only is there one universal law of gravitation, but it is an expression of the circumstance that there is one universal physics that applies to all physical processes occurring on the Earth and throughout the entire Universe. As we have seen in this chapter, Newton's synthesis and his law of universal gravitation have

FIGURE 4.14 Globular cluster M80. Globular clusters like this one contain tens of thousands of stars held together by gravitational attraction.

been confirmed many times in many different comparisons with experimental evidence.

4.11 THE NATURE OF NEWTON'S WORK

Today, Newton and his system of mechanics are honored for many reasons. The *Principia* formed the basis for the development of much of our physics and technology. Also, the success of Newton's approach has made it a model for all the physical sciences.

Throughout Newton's work, you will find his basic belief that celestial phenomena can be explained by applying quantitative, earthly laws. Newton felt that his laws had real physical meaning, that they were not just mathematical conveniences behind which unknowable laws lay hidden. The natural physical laws governing the universe *are* accessible through human reason and observation, and they can be expressed in simple mathematical forms of the laws.

Newton combined the skills and approaches of both the experimenter and the theoretician. He invented research equipment, such as the first reflecting telescope. He performed skillful experiments, especially in optics. Yet he also applied his great mathematical and logical powers to the creation of specific, testable predictions.

Like all scientists, Newton also had another weapon: the useful concepts developed by earlier scientists and those of his own time. Galileo and Descartes had contributed the first steps leading to a proper idea of inertia, which became Newton's first law of motion. Kepler's planetary laws were central in Newton's consideration of planetary motions. Huygens, Hooke, and others clarified the concepts of force and acceleration, ideas that had been evolving for centuries.

In addition to his own experiments, Newton selected and used data from a large number of sources. Tycho Brahe was only one of several astronomers whose observations of the motion of the Moon he used. When Newton could not complete his own measurements, he knew whom he could ask or where to look.

Last, recall how completely and how fruitfully he used and expanded his own specific contributions. A good example is his theory of universal gravitation. In developing it, Newton used his laws of motion and his various mathematical inventions again and again. Yet Newton was modest about his achievements. He once said that if he had seen further than others "it was by standing upon the shoulders of Giants."

But, today, scientists recognize that Newton's mechanics, while immensely useful still, holds true only within a well-defined region of science. For example, the forces within each galaxy appear to be Newtonian. But this may not be true for forces acting between one galaxy and another or across the entire universe. In addition to gravitation between galaxies, there appears to be an effect so far little understood but first suspected by Albert Einstein and given the name "dark energy," that seems to overcome that attraction and indeed to accelerate the galaxies away from one another.

At the other end of the scale are atoms and subatomic particles. During the past century, non-Newtonian concepts had to be developed to explain the observed motions of these particles, as will be discussed in Chapter 14.

Even within the solar system, there are several small differences between the predictions based on Newtonian gravitation and the observations. The most famous involves the angular motion of the axis of Mercury's orbit around the Sun. This motion is greater than the value predicted from Newton's laws by about one part in 800 per century. What causes this difference? For a while, it was thought that gravitational force might not vary inversely *exactly* with the square of the distance. One of the greatest triumphs of Einstein's theory of general relativity over 200 years later involved an explanation of this effect, without throwing doubt on the accuracy of Newton's equation for the gravitational force for the cases to which they had been generally applied.

Einstein's success suggests that any perceived difficulties with Newtonian gravitation should not be hastily assigned to an imperfection in the theory as a whole. The law of gravitation applies with unquestionable accuracy to all other planetary motions. But when applied to extreme situations, such as the behavior of a massive object like the planet Mercury moving rapidly in the near vicinity of the Sun, some facets of the theory make it too limited. Many studies have shown that there is no way to modify the details of Newtonian mechanics to explain certain observations in such extreme cases. Instead, these observations can be accounted for only by constructing *new* theories, such as *relativity theory* and *quantum mechanics*, based on some very different assumptions. The predictions from these theories, when applied to ordinary, nonextreme situations, are almost identical to those from Newton's laws for familiar phenomena. But they are accurate also in some extreme cases where the Newtonian predictions alone begin to show inaccuracies. Thus, Newtonian science is linked at one end with *relativity theory*, which is important for bodies with very great mass or moving at very high speeds. At the other end, Newtonian science approaches *quantum mechanics*, which is important for particles of extremely small mass and size, for example, atoms, molecules, and nuclear particles. However, for a vast range of problems between these extremes, Newtonian theory gives accurate results and is far simpler to use. Moreover, it is in Newtonian mechanics that relativity theory and quantum mechanics historically have their roots. (For more on Newton's impact see the *Student Guide* discussion for Chapter 4, "Impact and Reaction.")

SOME NEW IDEAS AND CONCEPTS

action at a distance
ether
gravitation
law of universal gravitation
Newton's synthesis

FURTHER READING

G. Holton and S.G. Brush, *Physics, The Human Adventure* (Piscataway, NJ: Rutgers University Press, 2001), Chapters 12–14, is especially helpful on the structure and method in physical science, as is Chapter 11 on Newton's law of universal gravitation.

D.H. Levy, *More Things in Heaven and Earth: Poets and Astronomers Read the Night Sky* (Wolfville, Nova Scotia: Wombat Press, 1997).

I.B. Cohen, *Science and the Founding Fathers* (New York: Norton, 1995.)

I. Newton, *The Principia: Philosophiae Naturalis Principia Mathematica*, a new translation by I.B. Cohen and A. Whitman, assisted by J. Budenz (Berkeley, CA: University of California Press, 1999).

STUDY GUIDE QUESTIONS

1. Write a brief outline of this chapter, including the main ideas contained in each section.
2. State in your own words Newton's law of universal gravitation.
3. What is Newton's equation for the magnitude of the gravitational force? Carefully define every symbol in this equation.

4.1 Newton and Seventeenth-Century Science

List some characteristics of society during Newton's lifetime that fostered scientific progress.

4.3 Newton's *Principia*

1. Explain Newton's concept of the "whole burden of philosophy," that is, the job of the scientist.

2. In your own words, state Newton's four rules of reasoning and give an example of each.
3. What do these rules have to do with the theory of universal gravitation?
4. State, in your own words, the central idea of universal gravitation.
5. How did Newton differ from Aristotle, who believed that the rules of motion on Earth are different from the rules of motion in the heavens?

4.4 The Inverse-Square Law

1. What can be proved from Kepler's law that the planets sweep out equal areas with respect to the Sun in equal times?
2. With what relationship can Kepler's third law, T^2/R^3_{av} = constant, be combined to prove that the gravitational attraction varies as $1/R^2$?
3. What simplifying assumption was made in the derivation given in this section?
4. Did Newton limit his own derivation by the same assumption?
5. How did Newton reach the conclusion that one general law of universal gravitation must apply to all bodies moving in the solar system?
6. If two objects are moved twice as far away from one another, by how much is the gravitational force between them decreased? If they are moved three times as far? By how much is the gravitational force increased if the objects are moved together to one-fourth their original separation?
7. While two objects are held at a fixed distance from each other, what happens to the gravitational force if one of the two masses is doubled; if it is tripled; and if both masses are halved?

4.5 The Law of Universal Gravitation

1. What idea came to Newton while he was thinking about falling objects and the Moon's acceleration?
2. Kepler, too, believed that the Sun exerted forces on the planets. How did his view differ from Newton's?
3. What quantitative comparison did Newton make after (supposedly) seeing the apple fall? What was the result?
4. The Moon is 60 times farther from the center of the Earth than objects at the Earth's surface. How much less is the gravitational attraction of the Earth acting on the Moon than on objects at its surface? Express this value as a fraction of 9.8 m/s^2.
5. The central idea of this and the next chapter is the "Newtonian synthesis." What is a synthesis? What did Newton synthesize?

4.6 Newton's Synthesis

1. What was the nature of Newton's synthesis?

4.7 Newton and Hypotheses

1. If Newton could not test the gravitational attraction of every body in the Universe, how could he dare to formulate a "universal" law of gravitation?

2. What is meant by "action at a distance"?
3. What was the popular type of explanation for "action at a distance"? Why did Newton not use this type of explanation?
4. What are two main types of hypotheses used in science?
5. Newton's claim to "frame no hypotheses" seems to refer to hypotheses that cannot be tested. Which of the following claims are not directly testable?
   (a) Plants need sunlight to grow, even on other planets.
   (b) This bandage is guaranteed to be free from germs unless the package is opened.
   (c) Virtual particles exist for a time that is too short for them to affect anything.
   (d) Life exists in the distant galaxies.
   (e) The Earth really does not move, since you would feel the motion if it did.
   (f) Universal gravitation holds between every pair of objects in the Universe.

4.8 The Magnitude of the Gravitational Force

1. How did Newton define the value of R in the force equation?
2. Two large balls of equal size and mass are touching each other. The radius of each is 1 m. In finding their mutual gravitational attraction, what is the value of R that goes into Newton's equation?
3. Two objects attract each other by gravitation. If the mass of each object is doubled, how does the force between them change?
4. Two people attract each other by gravitation. If the distance between them is doubled, but their masses stay the same, how does the force between them change?
5. Finally, if both masses and the distance between two people are doubled, how does the gravitational force between them change?
6. Can a theory ever be absolutely confirmed or proved? If not, does this throw doubt on the usefulness of a theory? Explain.
7. According to Newton's law of action and reaction, the Earth should experience a force and accelerate toward a falling stone.
   (a) How does the force on the Earth compare with the force on the stone?
   (b) How does the Earth's acceleration compare with the stone's acceleration?

4.9 The Value of *G*, and Some Consequences

1. Which of the quantities in the equation $F_{grav} = Gm_1m_2/R^2$ did Cavendish measure?
2. Knowing a value for G, what other information can be used to find the acceleration of gravity, g?
3. The mass of the Sun is about 1000 times the mass of Jupiter. How does the Sun's acceleration, owing to Jupiter's attraction, compare with Jupiter's acceleration owing to the Sun's attraction?
4. Two young persons you know are standing about 1 ft apart. Stating what assumption you make (e.g., about their respective masses, etc.), calculate the approximate value of the attraction between them.

5. What is the difference between a theory and a law in science? Explain with some examples.

4.10 Further Successes

1. How does the Moon cause the water level to rise on both sides of the Earth?
2. Explain why at a point on the ocean shore two high tides and two low tides are observed each day.
3. In which of the following does the Moon produce tides?
   (a) the seas;
   (b) the atmosphere;
   (c) the solid Earth.
4. Why is the calculation of the Moon's motion so difficult?
5. How are the orbits of comets different from the orbits of the planets?
6. Do these differences affect the validity of Newton's law of universal gravitation as applied to comets?

4.11 The Nature of Newton's Work

1. How did the orbit of Mercury challenge Newton's theory?
2. The theories of relativity and quantum mechanics contradict Newtonian physics in some situations. Does this mean that Newtonian physics must now be rejected? If not, why not?
3. In what ways are relativity theory and quantum mechanics "linked" to Newtonian physics?

DISCOVERY QUESTIONS

1. Write a brief outline of the steps Newton took in arriving at the law of universal gravitation. Include the data, assumptions, empirical laws, and hypotheses he used in constructing his theory.
2. Think of one of the theories you have encountered so far in this book. How did the proposer of the theory obtain it? What were the reasons for its acceptance and/or rejection?
3. An apple is attracted to the Earth by the force of gravitation. What is the reaction force to this action? What happens to the Earth as a result? Do we see this reaction? Why or why not?
4. How would you answer the following question: What keeps the Moon up?
5. Since Newton can be regarded as the "culmination of the Scientific Revolution," how would you answer the following:
   (a) What happened to Plato's problem? Was it solved?
   (b) Why do we believe today in the heliocentric theory? What is the evidence for it?
   (c) Is there any fundamental difference between the timeless heavens and the constantly changing Earth?

6. Is Newton's work only of historical interest, or is it still useful today? Explain.
7. What are some of the major consequences of Newton's work on scientists' views of the world?
8. What is the relationship, if any, between physics and our culture and society?
9. Set up a discussion or debate between a person living in the Aristotelian world view before Newton and a person living in the Newtonian world view.

Quantitative

1. Using the value for the Sun's mass $m_{sun} = 1.98 \times 10^{30}$ kg and the mass of the Earth from the text, find the gravitational attraction of the Sun on the Earth.
2. How could you find the attractive force of the Sun on the Earth if you did not know the mass of the Sun?
3. Using the values for the mass and radius of the Earth in the text, find the acceleration of gravity at the Earth's surface in meters per seconds squared. Assume that the Earth is a homogeneous sphere.
4. One of Jupiter's moons, Europa, has an ocean on its surface whose surface is frozen. Some scientists speculate that some form of life may exist below the surface. Find from tables the radius and period of Europa's orbit around Jupiter. From this information, find the mass of Jupiter. How does the mass of Jupiter, the largest planet, compare with the mass of Earth?
5. How high could an astronaut throw a ball on the Moon, if he can throw it to the height of 30 m on Earth?
6. (a) The gravitational force from Earth extends all the way to the Moon and beyond. How far out into space does it extend? Does the gravitational force ever decline to zero?
   (b) Black holes are thought to contain mass that is concentrated into a space approaching a radius of 0. What would happen to the gravitational force if the distance R between the centers of two masses were 0? Can this ever occur?
   (c) Draw a graph of the gravitational force as a function of R; place F on the y-axis and R on the x-axis. Let Gm_1m_2 remain constant.

CHAPTER 5

Conserving Matter and Motion

5.1 CONSERVATION OF MASS

Newton's success in mechanics altered profoundly the way in which scientists viewed the Universe. The motions of the Sun and planets could now be considered as purely mechanical, that is, governed by the laws of mechanics, much like a machine. As for any machine, whether a clock or the solar system, the motions of the parts were completely determined once the system had been put together.

This model of the solar system is called the *Newtonian world machine*. As is true of any model, certain things are left out. The mathematical equations that govern the motions of the model cover only the main properties of the real solar system. The masses, positions, and velocities of the parts

FIGURE 5.1 Newtonian physics inspired a mechanistic view of the universe as a self-contained "clock" designed by God to run on its own according to discernible principles and without any further need for Divine intervention (except, Newton thought, for occasional "fine tuning").

of the system, and the gravitational forces among them, are well described. But the Newtonian model neglects the internal structure and chemical composition of the planets, as well as heat, light, and electric and magnetic forces. Nevertheless, it serves splendidly to deal with observed motions in mechanics, and to this day is in constant use, in physics, engineering, sports, etc. Moreover, Newton's approach to science and many of his concepts became useful later in the study of those aspects he had to leave aside.

The idea of a world machine does not trace back only to Newton's work. In his *Principia Philosophiae* (1644), René Descartes, the most influential French philosopher of the seventeenth century, had written:

> I do not recognize any difference between the machines that artisans make and the different bodies that nature alone composes, unless it be that the effects of the machines depend only upon the adjustment of certain tubes or springs, or other instruments, that, having necessarily some proportion with the hands of those who make them, are always so large that their shapes and motions can be seen, while the tubes and springs that cause the effects of natural bodies are ordinarily too small to be perceived by our senses.

Robert Boyle (1627–1691), a British scientist, is known particularly for his studies of the properties of air. Boyle, a pious man, expressed the "mechanistic" viewpoint even in his religious writings. He argued that a God who could design a universe that ran by itself, as an ideal machine would, was more wonderful than a God who simply created several different kinds of matter and gave each a natural tendency to behave as it does. Boyle also thought it was insulting to God to believe that the world machine would be so badly designed as to require any further divine adjustment once it had been created. He suggested that an engineer's skill in designing "an elaborate engine" is more deserving of praise if the engine never needs supervision or repair. Therefore, if the "engine" of the Universe is to keep running unattended, the amounts of matter and motion in the Universe must remain constant over time. Today we would say that they must be *conserved*.

The idea that despite ever-present, obvious change all around us the total amount of material in the Universe does not change is really very old. It may be found, for instance, among the ancient atomists (see Prologue).

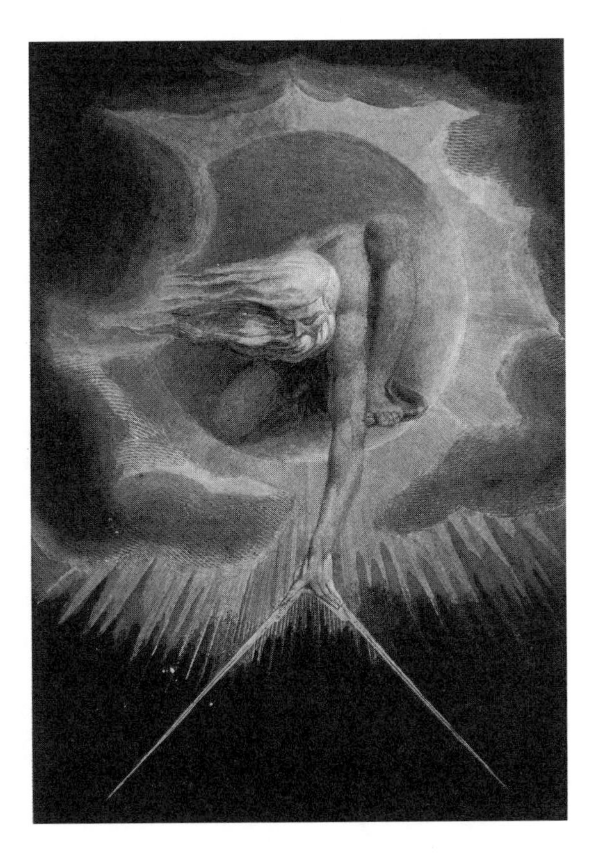

FIGURE 5.2 *The Ancient of Days* by William Blake (1757–1827), an English poet and artist who had little sympathy with the Newtonian style of "Natural Philosophy."

And just 24 years before Newton's birth, the English philosopher Francis Bacon included the following among his basic principles of modern science in *Novum Organum* 1620):

> There is nothing more true in nature than the twin propositions that "nothing is produced from nothing" and "nothing is reduced to nothing" . . . the sum total of matter remains unchanged, without increase or diminution.

This view agrees with everyday observation to some extent. While the form in which matter exists may change, in much of our ordinary experience matter appears somehow indestructible. For example, you may see a large boulder crushed to pebbles and not feel that the amount of matter in the Universe has diminished or increased. But what if an object is burned to ashes or dissolved in acid? Does the amount of matter remain unchanged even in such chemical reactions? What of large-scale changes such as the forming of rain clouds or seasonal variations?

In order to test whether the total quantity of matter actually remains constant, you must know how to measure that quantity. Clearly, it cannot be measured simply by its volume. For example, you might put water in a container, mark the water level, and then freeze the water. If you try this, you will find that the volume of the ice is greater than the volume of the water you started with. This is true even if you carefully seal the container

FIGURE 5.3 In some open-air chemical reactions, the mass of objects seems to decrease, while in others it seems to increase

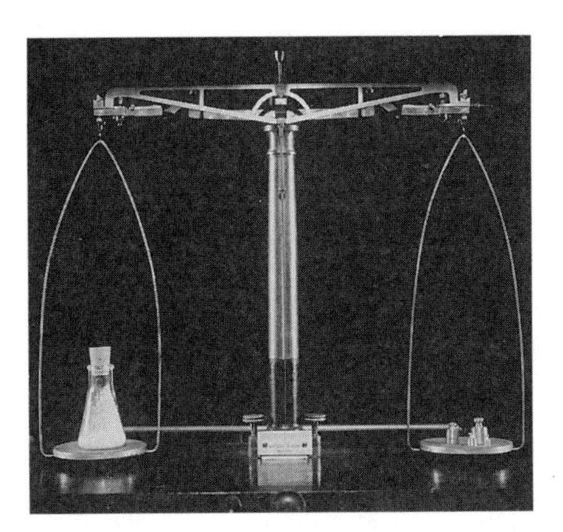

FIGURE 5.4 Conservation of mass was first demonstrated in experiments on chemical reactions in closed flasks.

so that no water can possibly come in from the outside. Similarly, suppose you compress some gas in a closed container. The volume of the gas decreases even though no gas escapes from the container.

Following Newton, we regard the *mass* of an object as the proper measure of the amount of matter it contains. In all the examples in previous chapters, we assumed that the mass of a given object does not change. However, a burnt match has a smaller mass than an unburnt one; an iron nail increases in mass as it rusts. Scientists had long assumed that something escapes from the match into the atmosphere and that something is added from the surroundings to the iron of the nail. Therefore, nothing is really "lost" or "created" in these changes. Not until the end of the eighteenth century was sound experimental evidence for this assumption provided. The French chemist Antoine Lavoisier produced this evidence.

Lavoisier (1743–1794), who is often called the "father of modern chemistry," closely examined chemical reactions that he caused to occur in *closed* flasks (a "closed system"). He carefully weighed the flasks and their contents before and after each reaction. For example, he burned iron in a closed flask. He found that the mass of the iron oxide produced equaled the sum of the masses of the iron and oxygen used in the reaction. With experimental evidence like this at hand, he could announce with confidence in *Traité Elémentaire de Chimie* (1789):

> We may lay it down as an incontestable axiom that in all the operations of art and nature, nothing is created; an equal quantity of matter exists both before and after the experiment . . . and nothing takes place beyond changes and modifications in the combinations of these elements. Upon this principle, the whole art of performing chemical experiments depends.

THE FATHER OF MODERN CHEMISTRY

Antoine Laurent Lavoisier showed the decisive importance of quantitative measurements, confirmed the principle of conservation of mass in chemical reactions, and helped develop the present system of nomenclature for the chemical elements. He also showed that organic processes such as digestion and respiration are similar to burning.

To earn money for his scientific research, Lavoisier invested in a private company which collected taxes for the French government. Because the tax collectors were allowed to keep any extra tax which they could collect from the public they became one of the most hated groups in France. Lavoisier was not directly engaged in tax collecting, but he had married the daughter of an important executive of the company, and his association with the company was one of the reasons why Lavoisier was guillotined during the French Revolution.

Also shown in the elegant portrait by J.L. David is Madame Lavoisier. She assisted her husband by taking data, translating scientific works from English into French, and making illustrations. About 10 years after her husband's execution, she married another scientist, Count Rumford, who is remembered for his experiments which cast doubt on the caloric theory of heat.

TRAITÉ
ÉLÉMENTAIRE
DE CHIMIE,
PRÉSENTÉ DANS UN ORDRE NOUVEAU
ET D'APRÈS LES DÉCOUVERTES MODERNES;
Avec Figures :

Par M. LAVOISIER, de l'Académie des Sciences, de la Société Royale de Médecine, des Sociétés d'Agriculture de Paris & d'Orléans, de la Société Royale de Londres, de l'Institut de Bologne, de la Société Helvétique de Basle, de celles de Philadelphie, Harlem, Manchester, Padoue, &c.

A PARIS,
Chez CUCHET, Libraire, rue & hôtel Serpente.

M. DCC. LXXXIX.
Sous le Privilège de l'Académie des Sciences & de la Société Royale de Médecine.

FIGURE 5.6 Title page from Lavoisier's *Traite*.

FIGURE 5.5 The Lavoisiers.

Lavoisier knew that if he put some material in a well-sealed bottle and measured its mass, he could return at any later time and find the same mass. It would not matter what had happened to the material inside the bottle. It might change from solid to liquid or liquid to gas, change color or consistency, or even undergo violent chemical reactions. At least one thing would remain unchanged: the *total* mass of all the different materials in the bottle.

In the years after Lavoisier's pioneering work, a vast number of similar experiments were performed with ever-increasing accuracy. The result was always the same. As far as scientists now can measure with sensitive balances (having a precision of better than 0.000001%), mass is *conserved*, that is, remains constant, in chemical reactions.

To sum up, despite changes in location, shape, chemical composition, and so forth, *the mass of any closed system remains constant*. This is the statement of the *law of conservation of mass*. This law is basic to both physics and chemistry.

5.2 COLLISIONS

Looking at moving things in the world around us easily leads to the conclusion that everything set in motion eventually stops. Every actual machine, left to itself, eventually runs down. It appears that the amount of motion in the Universe must be decreasing. This suggests that the Universe, too, must be running down, though, as noted earlier, many philosophers of the seventeenth century could not accept such an idea. Some definition of "motion" was needed that would permit one to make the statement that "the quantity of motion in the Universe is constant."

Is there a constant "quantity of motion" that keeps the world machine going? To suggest an answer to this question, you can do some simple laboratory experiments (Figure 5.7). Use a pair of carts with equal mass and nearly frictionless wheels; even better are two dry-ice disks or two air-track gliders. In a first experiment, a lump of putty is attached so that the carts will stick together when they collide. The carts are each given a push so that they approach each other with equal speeds and collide head-on. As you will see when you do the experiment, both carts stop in the collision; their motion ceases. But is there anything related to their motions that does not change?

The answer is yes. If you add the velocity $\mathbf{v}_A$ of one cart to the velocity $\mathbf{v}_B$ of the other cart, you find that the *vector sum* does not change. The vector sum of the velocities of these oppositely moving equally massive carts is zero *before* the collision. It is also zero for the carts at rest *after* the collision.

Does this finding hold for all collisions? In other words, is there a "law of conservation of velocity"? The example above was a very special circumstance. Carts with equal masses approach each other with equal speeds. But suppose the mass of one of the carts is twice the mass of the other cart. We let the carts approach each other with equal speeds and collide, as before. This time the carts do *not* come to rest. There is some motion remaining. Both objects move together in the direction of the initial velocity of the more massive object. So the vector sum of the velocities is not conserved in all collisions. (See Figure 5.7.)

Another example of a collision will confirm this conclusion. This time let the first cart have twice the mass of the second, but only half the velocity. When the carts collide head-on and stick together, they stop. The vector sum of the velocities is equal to zero *after* the collision. But it was not equal to zero *before* the collision. Again, there is no conservation of velocity; the total "quantity of motion" is not always the same before and after a collision.

The problem was solved by Newton. He saw that the mass played a role in such collisions. He redefined the "quantity of motion" of a body as the product of its mass and its velocity, $m\mathbf{v}$. This being a vector, it includes the idea of the *direction* of motion as well as the speed. For example, in all

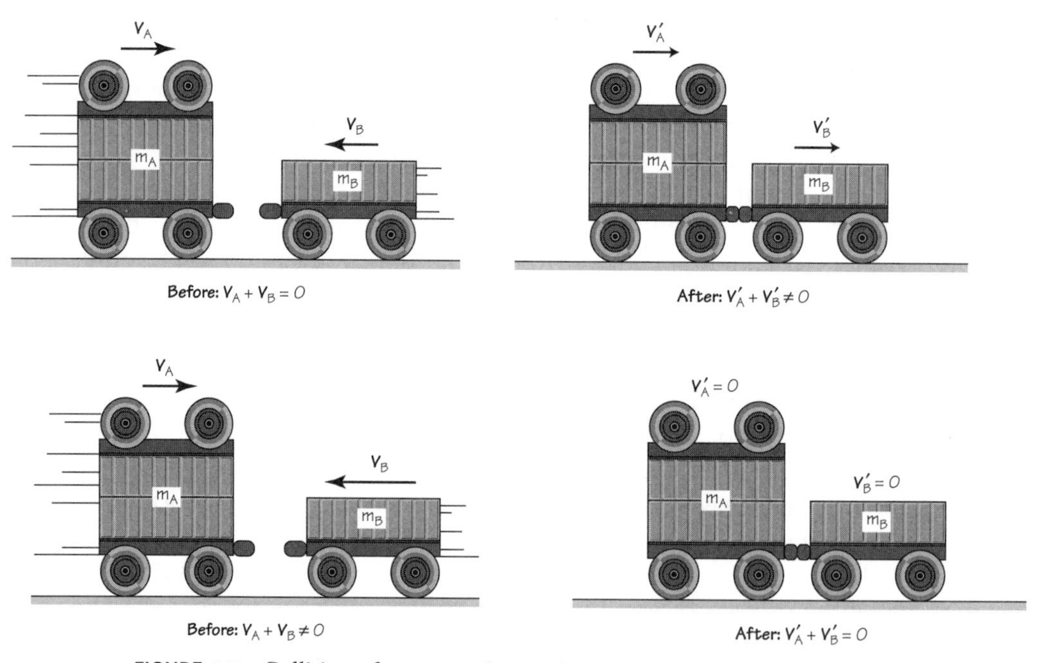

FIGURE 5.7 Collision of two carts (see text).

three collisions we have mentioned above, the motion of the carts before and after collision is described by the equation

$$\underbrace{m_A \mathbf{v}_A + m_B \mathbf{v}_B}_{\text{before collision}} = \underbrace{m_A \mathbf{v}'_A + m_B \mathbf{v}'_B}_{\text{after collision}}$$

Here m_A and m_B (which remain constant) represent the respective masses of the two carts, $\mathbf{v}_A$ and $\mathbf{v}_B$ represent their velocities before the collision, and $\mathbf{v}'_A$ and $\mathbf{v}'_B$ represent their velocities after the collision. Earlier, we represented initial and final velocities by $\mathbf{v}_i$ and $\mathbf{v}_f$. Here they are represented by $\mathbf{v}$ and $\mathbf{v}'$ because we now need to add subscripts, such as A and B.

In words, the above equation states:

> The *vector sum of the quantities mass × velocity before the collision is equal to the vector sum of the quantities mass × velocity after the collision. The vector sum of these quantities is constant, or conserved, in all these collisions.*

The above equation is very important and useful, leading directly to a powerful law, and of course is useful in allowing us to predict, at least qualitatively, the motions after collisions of the two colliding carts in the above examples.

5.3 CONSERVATION OF MOMENTUM

The product of mass and velocity often plays an important role in mechanics. It therefore has been given a special name. Instead of being called "quantity of motion," as in Newton's time, it is now called *momentum.* The total momentum of a system of objects (e.g., the two carts) is the vector sum of the momenta of all objects in the system. Consider each of the collisions examined. The momentum of the system as a whole, that is, the vector sum of the individual parts, is the same before and after collision. Thus, the results of the experiments can be summarized briefly: The momentum of the system is conserved.

This rule (or law, or principle) follows from observations of special cases, such as that of collisions between two carts that stuck together after colliding. But in fact, this *law of conservation of momentum* (often abbreviated LCM) turns out to be a completely general, universal law. The momentum of *any* system is conserved *if one condition is met:* that no net force is acting on the system—or, to put it in other words, that the system of objects can be considered closed to any effect from outside the system.

To see just what this condition means, let's examine the forces acting on one of the carts in the earlier experiment. Each cart, on a level track, experiences three main forces. There is, of course, a downward pull $\mathbf{F}_{\text{grav}}$ exerted by the Earth, and an equally large upward push $\mathbf{F}_{\text{table}}$ exerted by the table. (See Figure 5.8.) During the collision, there is also on each a push $\mathbf{F}_{\text{from other cart}}$ exerted by the other cart. The first two forces evidently cancel, since the cart is not accelerating up or down while on the tabletop. Thus, the net force on each cart is just the force exerted on it by the other cart as they collide. (We assume that frictional forces exerted by the table and the air are small enough to neglect. That was the reason for using dry-ice disks, air-track gliders, or carts with "frictionless" wheels. This assumption makes it easier to discuss the law of conservation of momentum. Later, you will see that the law holds whether friction exists or not.)

The two carts form a *system* of bodies, each cart being a part of the system. The force exerted by one cart on the other cart is a force exerted by one part of the system on another part. It is *not* a force on the system as a whole. The outside forces acting on the carts (by the Earth and by the table) exactly cancel. Thus, there is no *net* outside force. The system is "isolated." If this condition is met, the total momentum of all parts making up the system stays constant, it is "conserved." This is the *law of conservation of momentum* for systems of bodies that are moving with linear velocity **v**.

The Universality of Momentum Conservation

So far, you have considered only cases in which two bodies collide directly and stick together. The remarkable thing about the law of conservation of momentum is how universally it applies. For example:

1. It holds true no matter what *kind* of forces the bodies exert on each other. They may be gravitational forces, electric or magnetic forces,

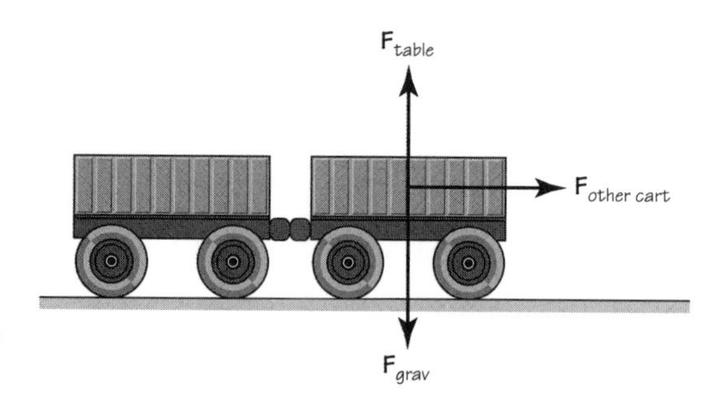

FIGURE 5.8 Forces on one of the carts during collision.

tension in strings, compression in springs, attraction or repulsion. The sum of the (mass × velocity) before is equal to the sum of the (mass × velocity) of all parts after any interaction.

2. The LCM also holds true even when there are friction forces present. If a moving object is slowed or stopped by frictional forces, for example, a book sliding to a stop on a tabletop, then the Earth, to which the table is attached, will take up the initial momentum of the book. In general, the object producing friction becomes part of the system of bodies to which the LCM applies.
3. It does not matter whether the bodies stick together or scrape against each other or bounce apart. They do not even have to touch. When two strong magnets repel or when a positively charged alpha particle is repelled by a nucleus (which is also positive), conservation of momentum still holds in each of those systems.
4. The law is not restricted to systems of only two objects; there can be any number of objects in the system. In those cases, the basic conservation equation is made more general simply by adding a term for each object to both sides of the equation.
5. The size of the system is not important. The law applies to a galaxy as well as to atoms.
6. The angle of the collision does not matter. All of the examples so far have involved collisions between two bodies moving along the same straight line. They were "one-dimensional collisions." If two bodies make a glancing collision rather than a head-on collision, each will move off at an angle to the line of approach. The law of conservation of momentum applies to such "two-dimensional collisions" also. (Remember that momentum is a vector quantity.) The law of conservation of momentum also applies in *three* dimensions. The vector sum of the momenta is still the same before and after the collision.

In general symbols, for n objects, this law may be written:

$$\sum_{i=1}^{n} (m_i\mathbf{v}_i)_{\text{before}} = \sum_{i=1}^{n} (m_i\mathbf{v}_i)_{\text{after}},$$

where $\sum_i$ represents the sum of the quantities in parentheses.

In the *Student Guide* for this chapter you will find a worked-out example of a collision between a spaceship and a meteorite in outer space that will help you become familiar with the law of conservation of momentum. On p. 225, the sidebar, "A Collision in Two Dimensions," shows an analysis of a two-dimensional collision. There are also short VHS or DVD videos of colliding bodies and exploding objects. These include collisions and explosions in two and three dimensions. The more of them you analyze, the more convinced you will be that the law of conservation of momentum applies to *any* isolated system.

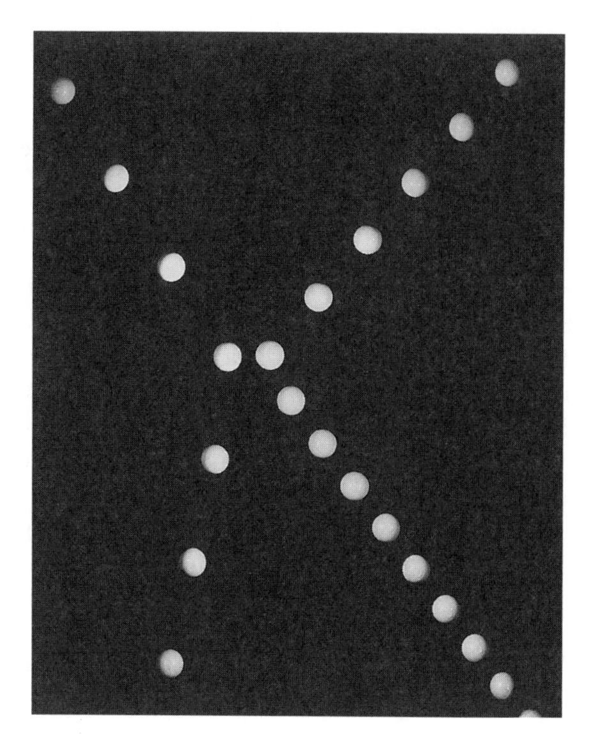

FIGURE 5.9 Stroboscopic photographs of two balls colliding. A ball enters from left top at a higher speed than the one from the right top. They collide near the center of the picture and then separate at different speeds.

These worked-out examples display a characteristic feature of physics: again and again, physics problems are solved by applying the expression of a *general* law to a specific situation. Both the beginning student and the veteran research physicist find it helpful, but also awesome, that one can *do* this. It seems strange that a few general laws enable one to solve an almost infinite number of specific individual problems. As Einstein expressed it in a letter to a friend:

> Even though the axioms of a theory are posed by human beings, the success of such an enterprise assumes a high degree of order in the objective world which one is not at all authorized to expect a priori. This is the wonder which is supported more and more with the development of our knowledge.*

Everyday life seems so very different. There you usually cannot calculate answers from general laws. Rather, you have to make quick decisions, some based on rational analysis, others based on "intuition." But the use of general laws to solve scientific problems becomes, with practice, quite natural also.

---

* A. Einstein to M. Solvine, letter of March 30, 1952.

5.4 MOMENTUM AND NEWTON'S LAWS OF MOTION

Earlier in this chapter we developed the concept of momentum and the law of conservation of momentum by considering experiments with colliding carts. The law was an "empirical" law; that is, it was discovered (perhaps "invented" or "induced" are better terms) as a generalization from experiment.

We can show, however, that the law of conservation of momentum also follows directly from Newton's laws of motion. It takes only a little algebra; that is, we can *deduce* the law from an established theory! Conversely, it is also possible to derive Newton's laws from the conservation law. Which of these is the fundamental law and which the conclusion drawn from it is therefore a bit arbitrary. Newton's laws used to be considered the fundamental ones, but since about 1900 the conservation law has been assumed to be the fundamental one.

Newton's second law expresses a relation between the net force $\mathbf{F}_{net}$ acting on a body, the mass m of the body, and its acceleration $\mathbf{a}$. We wrote this as $\mathbf{F}_{net} = m\mathbf{a}$. We can also write this law in terms of *change of momentum* of the body. Recalling that acceleration is the rate-of-change of velocity, $\mathbf{a} = \Delta\mathbf{v}/\Delta t$, we can write

$$\mathbf{F}_{net} = m\mathbf{a} = \frac{\Delta \mathbf{v}}{\Delta t},$$

or

$$\mathbf{F}_{net}\,\Delta t = m\,\Delta\mathbf{v}.$$

If the mass of the body is constant, the change in its momentum, $\Delta(m\mathbf{v})$, is the same as its mass times its change in velocity, $m(\Delta\mathbf{v})$, since only the velocity changes. Then we can write

$$\mathbf{F}_{net}\,\Delta t = \Delta(m\mathbf{v}).$$

That is, *the product of the net force on a body and the time interval during which this force acts equals the change in momentum of the body*. (The quantity $\mathbf{F}\,\Delta t$ is called the "impulse.")

This statement of Newton's second law is more nearly how Newton expressed it in his *Principia*. Together with Newton's third law, it enables us to derive the law of conservation of momentum for the cases we have studied. The details of the derivation are given in the *Student Guide*, "Deriving Conservation of Momentum from Newton's Laws." Thus, Newton's laws

A COLLISION IN TWO DIMENSIONS

The stroboscopic photograph shows a collision between two wooden disks on a frictionless horizontal table photographed from straight above the table. The disks are riding on tiny plastic spheres which make their motion nearly frictionless. Body B (marked ×) is at rest before the collision. After the collision it moves to the left, and Body A (marked −) moves to the right. The mass of Body B is known to be twice the mass of Body A: $m_B = 2m_A$. We will analyze the photograph to see whether momentum was conserved. (*Note:* The size reduction factor of the photograph and the [constant] stroboscopic flash rate are not given here. But as long as all velocities for this test are measured in the same units, it does not matter here what those units are.)

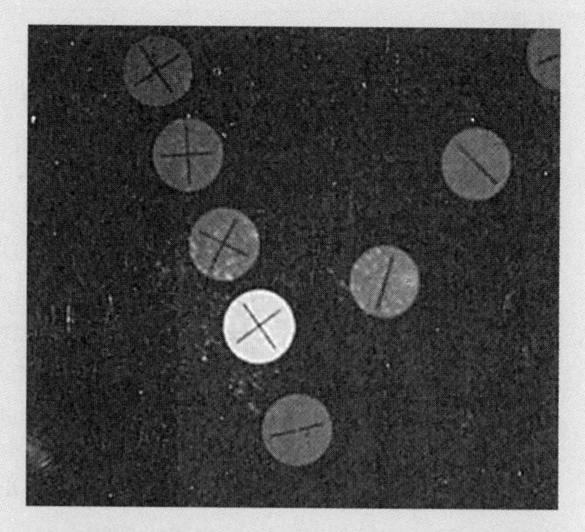

FIGURE 5.10

In this analysis, we will measure in centimeters the distance the disks moved on the photograph. We will use the time between flashes as the unit of time. Before the collision, Body A (coming from the lower part of the photograph) traveled 36.7 mm in the time between flashes: $\mathbf{v}_A = 36.7$ speed-units. Similarly, we find that $\mathbf{v}'_A = 17.2$ speed-units, and $\mathbf{v}'_B = 11.0$ speed units.

The total momentum before the collision is just $m_A\mathbf{v}_A$. It is represented by an arrow 36.7 momentum-units long, drawn at right.

The vector diagram shows the momenta $m_A\mathbf{v}'_A$ and $m_B\mathbf{v}'_B$ after the collision; $m_A\mathbf{v}'_A$ is represented by an arrow 17.2 momentum-units long. Since $m_B = 2m_A$, the $m_B\mathbf{v}'_B$ arrow is 22.0 momentum-units long.

The dotted line represents the vector sum of $m_A\mathbf{v}'_A$ and $m_B\mathbf{v}'_B$, that is, the total momentum after the collision. Measure-

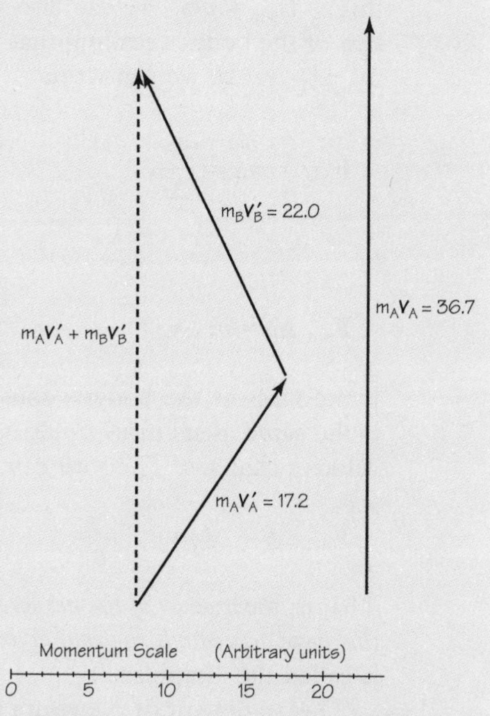

FIGURE 5.11 Momentum diagram of the two-dimensional collision pictured in Fig. 5.10.

ment shows it to be 34.0 momentum-units long. Thus, our measured values of the total momentum before and after the collision differ by 2.7 momentum-units. This is a difference of about 7%. We can also verify that the direction of the total is the same before and after the collision to within a small uncertainty.

Have we now demonstrated that momentum was conserved in the collision? Is the 7% difference likely to be due entirely to measurement inaccuracies? Or is there reason to expect that the total momentum of the two disks after the collision is really a bit less than before the collision?

and the law of conservation of momentum are not separate, independent laws of nature.

In all the examples considered so far and in the derivation above, we have considered each piece of the system to have a constant mass. But the definition of momentum permits a change of momentum to arise from a change of mass as well as from a change of velocity. In many cases, the mass of the object involved is in fact changing. For example, as a rocket spews out exhaust gases, its mass is decreasing; conversely, the mass of a train of coal cars increases as it rolls past a hopper that drops coal into the cars. The LCM remains valid for cases such as these, where the masses of the objects involved are not constant, as long as no net forces act on the system as a whole, and the momenta of all parts (including, say, that of the rocket's exhaust, are included).

One great advantage of being able to use the LCM is that it is a law of the kind that simply says "before = after." Thus, it applies in cases where you do not have enough information to use Newton's laws of motion during the whole interval between "before" and "after." For example, suppose a cannon that is free to move fires a shell horizontally. Although it was initially at rest, the cannon is forced to move while firing the shell; it *recoils.* The expanding gases in the cannon barrel push the cannon backward just as hard as they push the shell forward. You would need a continuous record of the magnitude of the force in order to apply Newton's second law separately to the cannon and to the shell to find their respective accelerations during their movement away from each other. A much simpler way is to use the LCM to calculate the recoil. The momentum of the system (cannon plus shell) is zero initially. Therefore, by the LCM, the momentum will also be zero after the shell is fired. If you know the masses of the shell and the cannon, and the speed of the emerging shell after firing, you can calculate the speed of the recoil (or the speed of the shell, if you measure the cannon's recoil speed). Moreover, if both speeds can be measured after the separation, then the ratio of the masses of the two objects involved can be calculated.

5.5 ISOLATED SYSTEMS

There are important similarities between the conservation law of mass and that of momentum. Both laws are tested by observing systems that may be considered to be isolated from the rest of the Universe. When testing or using the law of conservation of *mass*, an isolated system such as a sealed flask is used. Matter can neither enter nor leave this system. When testing or using the law of conservation of *momentum*, another kind of isolated system, one which experiences no net force from outside the system, is used.

Consider, for example, two frictionless carts colliding on a smooth horizontal table, or two hockey pucks colliding on smooth ice. The very low friction experienced by the pucks allows us to think away the ice on which they move, and to consider just the pucks to form a very nearly closed or isolated system. The table under the carts and the ice under the pucks do not have to be included since their individual effects on each of the objects cancel. That is, each puck experiences a downward gravitational force exerted by the Earth, while the ice on the Earth exerts an equally strong upward push.

Even in this artificial example, the system is not entirely isolated. There is always a little friction with the outside world. The layer of gas under the puck and air currents, for example, provide some friction. All outside forces are not *completely* balanced, and so the two carts or pucks do not form a truly isolated system. Whenever this is unacceptable, one can expand or extend the system so that it *includes* the bodies that are responsible for the external forces. The result is a new system on which the unbalanced forces are small enough to ignore.

For example, picture two automobiles skidding toward a collision on an icy road. The frictional forces exerted by the road on each car may be several hundred newtons. These forces are very small compared to the immense force (thousands of newtons) exerted by each car on the other when they collide. Thus, for many purposes, the action of the road can be ignored. For such purposes, the two skidding cars *before, during, and after the collision* are nearly enough an isolated system. However, if friction with the road (or the table on which the carts move) is too great to ignore, the law of conservation of momentum still holds, if we apply it to a larger system—one which includes the road or table. In the case of the skidding cars or the carts, the road or table is attached to the Earth. So the entire Earth would have to be included in a "closed system."

DERIVING CONSERVATION OF MOMENTUM FROM NEWTON'S LAWS

Suppose two bodies with masses m_A and m_B exert forces on each other (by gravitation or by magnetism, etc.). $\mathbf{F}_{AB}$ is the force exerted on body A by body B, and $\mathbf{F}_{BA}$ is the force exerted on body B by body A. No other unbalanced force acts on either body; they form an isolated system. By Newton's third law, the forces $\mathbf{F}_{AB}$ and $\mathbf{F}_{BA}$ are at every instant equal in magnitude and opposite in direction. Each body acts on the other for exactly the same time Δt. Newton's second law, applied to each of the bodies, says

$$\mathbf{F}_{AB}\,\Delta t = \Delta(m_A \mathbf{v}_A)$$

and

$$\mathbf{F}_{BA}\,\Delta t = \Delta(m_B \mathbf{v}_B).$$

By Newton's third law,

$$\mathbf{F}_{AB} = -\mathbf{F}_{BA}$$

so that

$$\mathbf{F}_{AB}\,\Delta t = -\mathbf{F}_{BA}\,\Delta t.$$

Therefore,

$$\Delta(m_A \mathbf{v}_A) = -\Delta(m_B \mathbf{v}_B).$$

Suppose that the masses m_A and m_B are constant. Let $\mathbf{v}_A$ and $\mathbf{v}_B$ stand for the velocities of the two bodies at some instant,

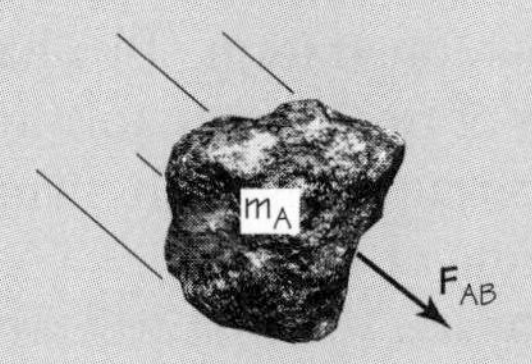

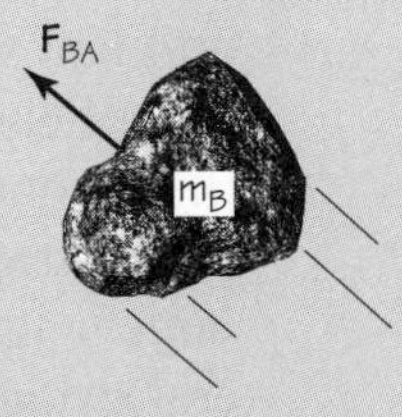

FIGURE 5.12 Collision between two rocks.

and let $\mathbf{v}'_A$ and $\mathbf{v}'_B$ stand for their velocities at some later instant. Then we can write the last equation as

$$m_A \mathbf{v}'_A - m_A \mathbf{v}_A = -(m_B \mathbf{v}'_B - m_B \mathbf{v}_B).$$

and

$$m_A \mathbf{v}'_A - m_A \mathbf{v}_A = -m_B \mathbf{v}'_B + m_B \mathbf{v}_B$$

A little rearrangement of terms leads to

$$m_A \mathbf{v}'_A + m_B \mathbf{v}'_B = m_A \mathbf{v}_A + m_B \mathbf{v}_B.$$

You will recognize this as our original expression of the law of conservation of momentum.

Here we are dealing with a system consisting of two bodies. This method works equally well for a system consisting of any number of bodies.

5.6 ELASTIC COLLISIONS

In 1666, members of the recently formed Royal Society of London witnessed a demonstration. Two hardwood balls of equal size were suspended at the ends of two strings, forming two pendula. One ball was released from rest at a certain height. It swung down and struck the other, which had been hanging at rest.

After impact, the first ball stopped at the point of impact while the second ball swung from this point, as far as one could easily observe, to the same height as that from which the first ball had been released. When the second ball returned and struck the first, it was now the second ball which stopped at the point of impact as the first swung up to almost the same height from which it had started. This motion repeated itself for several swings. (You can repeat it with a widely available desk toy.)

This demonstration aroused great interest among members of the Royal Society. In the next few years, it also caused heated and often confusing arguments. Why did the balls rise each time to nearly the same height after each collision? Why was the motion "transferred" from one ball to the

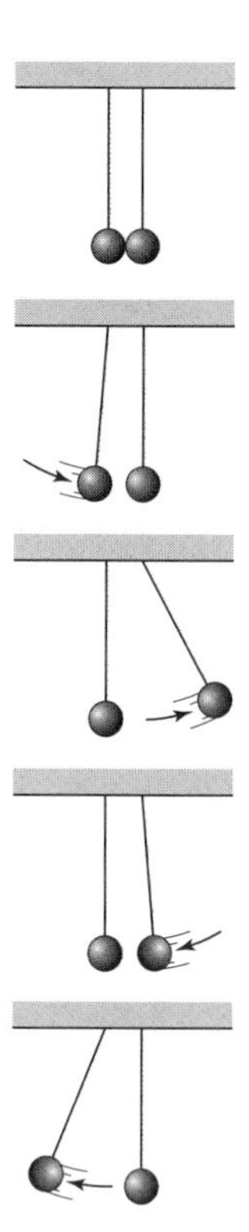

FIGURE 5.13 Demonstration with two pendula (similar to the demonstration witnessed by Royal Society members in 1666)

other when they collided? Why did the first ball not bounce back from the point of collision, or continue moving forward after the second ball moved away from the collision point?

The LCM explains what is observed, but it would also allow quite different results for different cases. The law says only that the momentum of ball A just before it strikes the resting ball B is equal to the total momentum of A and B just after collision. It does not say how A and B share the momentum. The actual result is just one of infinitely many different outcomes that would all agree with conservation of momentum. For example, suppose (though it has never been observed to happen) that ball A bounced back with ten times its initial speed. Momentum would still be conserved *if* ball B went on its way at 11 times A's initial speed.

In general symbols,
$\Delta \sum_i (\frac{1}{2} m_i \mathbf{v}_i^2) = 0.$

In 1668, three men reported to the Royal Society on the whole matter of impact. The three were the mathematician John Wallis, the architect and scientist Christopher Wren, and the physicist Christian Huygens. Wallis and Wren offered partial answers for some of the features of collisions; Huygens analyzed the problem in complete detail.

Huygens explained that in such collisions *another conservation law*, in addition to the law of conservation of momentum, also holds. Not only is the vector sum of the values of (mass × velocity) conserved, but so is the ordinary arithmetic sum—as we would now express it—of the values of $\frac{1}{2}mv^2$ for the colliding spheres! In modern algebraic form, the relationship he discovered can be expressed as

$$\tfrac{1}{2}m_A v_A^2 + \tfrac{1}{2}m_B v_B^2 = \tfrac{1}{2}m_A v_A'^2 + \tfrac{1}{2}m_B v_B'^2.$$

The quantity $\frac{1}{2}mv^2$—a scalar, not a vector—has come to be called *kinetic energy*, from the Greek word *kinetos*, meaning "moving." (The origin of the $\frac{1}{2}$, which does not really affect the rule here, is shown in the *Student Guide* discussion for this chapter, "Doing Work on a Sled.") The equation stated above, then, is the mathematical expression of *the conservation of kinetic energy*. This relationship holds for the collision of two "perfectly hard" objects similar to those observed at the Royal Society meeting. There, ball A stopped and ball B went on at A's initial speed. This is the *only* result that agrees with *both* conservation of momentum and conservation of kinetic energy, as you can demonstrate yourself.

But is the conservation of kinetic energy as general as the law of conservation of momentum? Is the total kinetic energy present conserved in *any* interaction occurring in *any* isolated system?

It is easy to see that it is not, that it holds only in special cases such as that observed at the Royal Society test (or on making billiard ball colli-

FIGURE 5.14 Christian Huygens (1629–1695) was a Dutch physicist and inventor. He devised an improved telescope with which he discovered a satellite of Saturn and saw Saturn's rings clearly. Huygens was the first to obtain the expression for centripetal acceleration (v^2/R); he worked out a wave theory of light; and he invented a pendulum-controlled clock. Huygens' reputation would undoubtedly have been greater had he not been overshadowed by his contemporary, Newton.

sions). Consider the first example of Section 5.2. Two carts of equal mass (and with putty between the bumping surfaces) approach each other with equal speeds. They meet, stick together, and stop. The kinetic energy of the system after the collision is 0, since the speeds of both carts are zero. Before the collision the kinetic energy of the system was $\frac{1}{2}m_A v_A^2 + \frac{1}{2}m_B v_B^2$. Both $\frac{1}{2}m_A v_A^2$ and $\frac{1}{2}m_B v_B^2$ are always positive numbers. Their sum cannot equal zero (unless both v_A and v_B are zero, in which case there would be no collision and not much of a problem). The kinetic energy of the system is *not* conserved in this collision in which the bodies stick together, while momentum *is* conserved. In fact, *no* collision in which the bodies stick together will show conservation of kinetic energy. It applies only to the collision of "perfectly hard" bodies that bounce back from each other.

The law of conservation of kinetic energy, then, is *not* as general as the law of conservation of momentum. If two bodies collide, the kinetic energy may or may not be conserved, depending on the type of collision. It *is* conserved if the colliding bodies do not crumple or smash or dent or stick together or heat up or change physically in some other way. Bodies that rebound without any such change are called *perfectly elastic*, whether they are billiard balls or subatomic particles. Collisions between them are called *perfectly elastic collisions*. In perfectly elastic collisions, *both* momentum and kinetic energy are conserved.

But most collisions are not perfectly elastic, and kinetic energy is not conserved. Thus, the sum of the $\frac{1}{2}m\mathbf{v}^2$ values after the collision is *less* than that before the collision. Depending on how much kinetic energy is "lost," such collisions might be called "partially elastic" or "perfectly inelastic." The loss of kinetic energy is greatest in perfectly inelastic collisions, when the colliding bodies remain together.

Collisions between steel ball bearings, glass marbles, hardwood balls, billiard balls, or some rubber balls (silicone rubber) are almost perfectly elastic, if the colliding bodies are not damaged in the collision. The total kinetic energy after the collision might be as much as, say, 96% of this value before the collision. Examples of perfectly elastic collisions are found only in collisions between atoms or subatomic particles. But all is not lost—we shall see how to deal with inelastic collisions also.

5.7 LEIBNIZ AND THE CONSERVATION LAW

Gottfried Wilhelm Leibniz (1646–1716) extended conservation ideas to phenomena other than collisions. For example, when a stone is thrown straight upward, its kinetic energy decreases as it rises, even without any

FIGURE 15.15 Gottfried Wilhelm Leibniz (1646–1716), a contemporary of Newton, was a German philosopher and diplomat and advisor to Louis XIV of France and Peter the Great of Russia. Independently of Newton, Leibniz invented the method of mathematical analysis called calculus. A long public dispute resulted between the two great men concerning the priority of ideas.

WHAT IS CONSERVED? THE DEBATE BETWEEN DESCARTES AND LEIBNIZ

René Descartes believed that the total quantity of motion in the Universe did not change. He wrote in his *Principles of Philosophy*:

> It is wholly rational to assume that God, since in the creation of matter He imparted different motions to its parts, and preserves all matter in the same way and conditions in which He created it, so He similarly preserves in it the same quantity of motion.

Descartes proposed to define the quantity of motion of an object as the product of its mass and its speed. As you saw in Section 5.3, this product is a conserved quantity only if there are no outside forces.

Gottfried Wilhelm Leibniz was aware of the error in Descartes' ideas on motion. In a letter in 1680 he wrote:

> M. Descartes' physics has a great defect; it is that his rules of motion or laws of nature, which are to serve as the basis, are for the most part false. This is demonstrated. And his great principle, that the quantity of motion is conserved in the world, is an error.

FIGURE 5.16 René Descartes (1596–1650) was the most important French scientist of the seventeenth century. In addition to his early contribution to the idea of momentum conservation, he is remembered by scientists as the inventor of coordinate systems and the graphical representation of algebraic equations. Descartes' system of philosophy, which used the deductive structure of geometry as its model, is still influential.

collision. At the top of the trajectory, kinetic energy is zero for an instant. Then it reappears and increases as the stone falls. Leibniz wondered whether something applied or given to a stone at the start is somehow *stored* as the stone rises, instead of being lost. His idea would mean that kinetic energy is just one part of a more general and really conserved quantity.

It was a hint that was soon followed up, with excellent results—once more an illustration of how science advances by successive innovators improving on partial truths.

5.8 WORK

In everyday language, pitching, catching, and running on the baseball field are "playing," while using a computer, harvesting in a field, or tending to an assembly line are "working." However, in the language of physics, "work" has been given a rather special definition, one that involves physical concepts of force and displacement instead of the subjective ones of reward or accomplishment. It is more closely related to the simple sense of effort or labor. The work done on an object is defined as the *product of the force exerted on the object times the displacement of the object along the direction of the force*. (You will see in Chapter 6 one origin of this definition in connection with the steam engine.)

When you move the hand and arm to throw a baseball, you exert a large force on it while it moves forward for about 1 m. In doing so, you (i.e., your muscles) do a large amount of work, according to the above definition. By contrast, in writing or in turning the pages of a book you exert only a small force over a short distance. This does not require much work, as the term "work" is understood in physics.

Suppose you have to lift boxes from the floor straight upward to a table

FIGURE 5.17 Major League baseball pitcher Mike Hampton.

at waist height. Here the language of common usage and that of physics both agree that you are doing work. If you lift two identical boxes at once, you do twice as much work as you do if you lift one box. If the table were twice as high above the floor, you would do twice as much work to lift a box to it. The work you do depends on both the *magnitude* of the force you must exert on the box and the *distance* through which the box moves in the direction of the force. Note that the work you do on a box does not depend on how long it takes to do your job.

We can now define the work W done on an object by a force $\mathbf{F}$ more precisely as the product of the magnitude F of the force and the distance d that the object moves *in the direction of* F while the force is being exerted; in symbols,

$$W = Fd$$

Note that work is a scalar quantity; it has only a magnitude but not a direction. As an example, while you are lifting a box weighing 100 N upward through 0.8 m you are applying a force of 100 N to the box. The work you

FIGURE 5.18

More generally, the definition of mechanical work is $W = Fd \cos \theta$, where θ is the angle between the vectors **F** and **d**. So, if $\theta = 90°$, $\cos \theta = 0$, and $W = 0$; if $\theta = 0°$, $\cos \theta = 1$, and $W = Fd$.

have done on the box to move it through the distance is $100 \text{ N} \times 0.8 \text{ m} = 80 \text{ N} \cdot \text{m}$.

From the definition of work, it follows that no work is done if there is no displacement. No matter how hard you push on a wall, no work is done if the wall does not move. Also, by our definition, no work is done if the only motion is perpendicular to the direction of the force. For example, suppose you are carrying a book bag. You must pull up against the downward pull of gravity to keep the bag at a constant height. But as long as you are standing still you do no work on the bag. Even if you walk along with it steadily in a horizontal line, the only work you do is in moving it forward against the small resisting force of the air.

5.9 WORK AND KINETIC ENERGY

Work is a useful concept in itself. The concept is most useful in understanding the concept of *energy*. There are a great many forms of energy, in addition to kinetic energy discussed in Section 5.6. A few of them will be discussed in this and succeeding chapters. We will define them, in the sense of describing how they can be measured and how they can be expressed algebraically. We will also discuss how energy changes from one form to another. The *general* concept of energy is difficult to define. But to define some *particular* forms of energy is easy enough. The concept of work helps greatly in making such definitions.

The chief importance of the concept of work is that work represents an amount of energy transformed from one form to another. For example, when you throw a ball you do work on it. While doing so, you transform chemical energy, which your body obtains from food and oxygen, into energy of motion of the ball. When you lift a stone (doing work on it), you transform chemical energy into what is called gravitational potential energy (discussed in the next section). If you release the stone, the Earth pulls it downward (does work on it); gravitational potential energy is transformed into kinetic energy. When the stone strikes the ground, it compresses the ground below it (does work on it), and its kinetic energy is transformed into heat and into work done to deform the ground on which it lands. In each case, the work is a measure of how much energy is transferred.

The form of energy called kinetic energy is the simplest to deal with. We can use the definition of work, $W = Fd$, together with Newton's laws of motion, to get an expression of this form of energy. Imagine that you exert a constant net force **F** on an object of mass m. This force accelerates the object over a distance **d** in the same direction as **F** from rest to a speed v. Using Newton's second law of motion, we can show that

$$Fd = \tfrac{1}{2}mv^2.$$

(The details of this derivation are given in the *Student Guide*, "Doing Work on a Sled.")

Fd is the expression for the work done on the object by whatever agency exerted the force **F**. The work done on the object equals the amount of energy transformed from some form into the energy of motion, the kinetic energy, of the object. The symbol KE is often used to represent kinetic energy. By definition, then

$$KE = \tfrac{1}{2}mv^2.$$

The expression $\tfrac{1}{2}mv^2$ relates directly to the concept of work and so provides a useful expression for the energy of motion.

The equation $Fd = \tfrac{1}{2}mv^2$ was obtained by considering the case of an object initially at rest. In other words, the object had an initial kinetic energy of zero. The relation can be extended to hold also for an object already in motion when the net force is applied (e.g., a bat hitting a moving ball). In that case, the work done on the object still equals the change in its kinetic energy from its initial to its final value

$$Fd = \Delta(KE).$$

The quantity $\Delta(KE)$ is, by definition, equal to $(\tfrac{1}{2}mv^2)_{\text{final}} - (\tfrac{1}{2}mv^2)_{\text{initial}}$.

Work is defined as the product of a force and a distance. Therefore, its units in the mks system are *newtons* × *meters* or newton-meters: A newton-meter is given a special name. It is also called a *joule* (symbol J) in honor of James Prescott Joule, the nineteenth-century physicist famous for his experiments showing that heat is a form of energy (Chapter 6). The joule is the unit of work or of energy, when force is measured in newtons and distance in meters. When force is measured in dynes and distance in centimeters, the unit of work or energy is *dynes* × *centimers*. A dyne-centimeter is also given a special name: *erg*.

5.10 POTENTIAL ENERGY

As you saw in the previous section, doing work on an object can increase its kinetic energy. Work also can be done on an object *without* increasing its kinetic energy. For example, while you lifted that box up to the table in

Section 5.8 at a small, constant speed, kinetic energy remains constant. But you were doing work on the box. By doing work you are using your body's store of chemical energy. Into what form of energy is it being transformed?

The answer, as Leibniz suggested, is that there is "energy" somehow associated with height above the Earth. This energy is now called *gravitational potential energy*. Lifting a box or a book higher and higher increases the gravitational potential energy associated with the lifted object. You can see clear evidence of this effect when, say, you pick up a book from the floor, lift it to a certain height, and then let it drop. The gravitational potential energy is transformed rapidly into the kinetic energy of a fall. In general terms, suppose a force F is used to displace an object upward a distance d, without changing its KE. Then, the increase in gravitational potential energy, symbolized by $\Delta(PE)_{\text{grav}}$, is

$$\Delta(PE)_{\text{grav}} = F_{\text{applied}} \cdot d$$

$$= -F_{\text{grav}} \cdot d.$$

Potential energy can be thought of as *stored* energy. As the book falls, its gravitational potential energy decreases while its kinetic energy increases correspondingly. When the book reaches its original level, all of the gravitational potential energy stored during the lift will have been transformed into kinetic energy.

Many useful applications follow from this idea of potential or stored energy. For example, a steam hammer used by construction crews is driven upward by high-pressure steam (thus gaining potential energy). When the steam supply stops, the hammer drops, the gravitational potential energy is converted into kinetic energy. Another example is the use of energy from electric power plants during low-demand periods to pump water into a high reservoir. When there is a large demand for electricity later, the water is

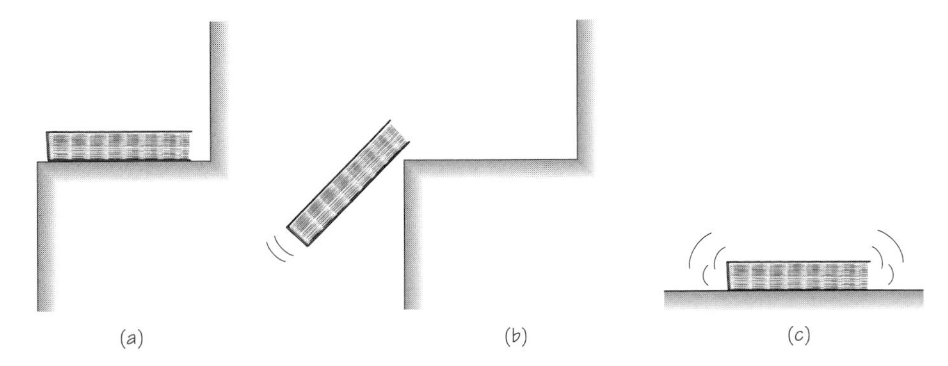

FIGURE 5.19 Object (a book) at rest, falling, and coming to rest again.

FIGURE 5.20 Two forms of potential energy (in an extended spring and in a pair of magnets).

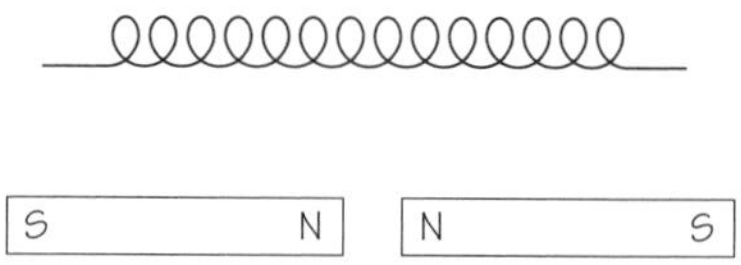

allowed to run down and drive the electric generators, recouping the stored energy for use.

There are forms of potential energy other than gravitational. For example, if you stretch a rubber band or a spring, you increase its *elastic potential energy*. When you release the rubber band, it can deliver the stored energy to a projectile in the form of kinetic energy. Nearly all of the work done in blowing up an elastic balloon is also stored as potential energy.

Other forms of potential energy are associated with other kinds of forces. In an atom, the negatively charged electrons are attracted by the positively charged nucleus. If an externally applied force pulls an electron *away* from the nucleus, the *electric potential energy* increases. If, then, the electron is freed of the applied forces and moves back *toward* the nucleus, the potential energy decreases as the electron's kinetic energy increases. Or, if two magnets are pushed together with north poles facing, the *magnetic potential energy* increases. When released, the magnets will move apart, gaining kinetic energy as they lose potential energy.

Where is the potential energy located in all these cases? It is easy to think at first that it "belongs" to the body that has been moved. But without the presence of the other object (the Earth, the nucleus, the other magnet) neither would work be needed for steady motion, nor would there be

FIGURE 5.21 Chemical energy (ultimately from the Sun) is stored in rice and converted into work by the farmer who cultivates the crop.

any increase in potential energy. Rather, action on the object would increase only the kinetic energy of the object on which work was done. We must conclude that the potential energy belongs not to *one* body, but to the whole system of interacting bodies involved! This is evident in the fact that the potential energy gained is available to any one or to all of these interacting bodies. For example, either of the two magnets would gain all the kinetic energy just by releasing it and holding the other in place. Or suppose you could fix the book somehow to a hook in space that would hold it fixed there. The Earth would then "fall" up toward the book, being attracted to it. Eventually the Earth would gain just as much kinetic energy at the expense of stored potential energy as the book would if it were free to fall to the Earth.

The increase in gravitational potential energy "belongs" to the Earth–book *system*, not the book alone. The work is done by an "outside" agent (you), increasing the total energy of the Earth–book system. When the book falls, it is responding to forces exerted by one part of the system on another. The *total energy* of the system does not change; it is converted from *PE* to *KE*. This is discussed in more detail in the next section.

In fact, during the fall of the book, as initially postulated, the Earth would be moving a little toward the falling book. Why don't we observe this?

5.11 CONSERVATION OF MECHANICAL ENERGY

In Section 5.9, you learned that the amount of work done on an object is *equal* to the amount of energy transformed from one form to another. For example, the chemical energy of muscles is transformed into the kinetic energy of a thrown ball. The work you have done while throwing the ball is equal to the energy you have given up from your store of chemical energy. This statement in the first sentence implies that the *amount* of energy involved during an interaction does not change; only its *form* changes. This is particularly obvious in motions where no "outside" force is applied to a mechanical system.

While a stone falls freely, for example, the gravitational potential energy of the stone–Earth system is continually transformed into kinetic energy. Neglecting air friction, the *decrease* in gravitational potential energy is, for any portion of the path, equal to the *increase* in kinetic energy. Consider a stone thrown upward, which we will call the positive direction. Between any two points in its path, the *increase* in gravitational potential energy equals the *decrease* in kinetic energy. Consider a stone rising upward after being thrown in the upward direction. After it leaves your hand, the only force applied is F_{grav} (neglect external forces such as friction). The gravitational force is pointing downward in the negative direction. The work done by this force (with d positive for upward displacements) is

$$-F_{grav}\, d = \Delta(PE)_{grav}$$

$$= -\Delta KE.$$

In words, these equations state that the work done by gravity as the stone rises against it results in an increase in potential energy but a decrease in kinetic energy—the stone slows down. This relationship can be rewritten as

$$\Delta(KE) + \Delta(PE)_{grav} = 0,$$

or, still more concisely, as

$$\Delta(KE + PE_{grav}) = 0.$$

If $(KE + PE_{grav})$ represents the *total mechanical energy* of the system (consisting here of the stone and the Earth), then the *change* in the system's total mechanical energy is *zero, provided there is no outside work added to the system (e.g., a strong wind acting on the stone).* In other words, when outside work is zero, the total mechanical energy of a system, $(KE + PE_{grav})$, remains constant; it is *conserved.*

A similar statement can be made for a vibrating guitar string. While the string is being pulled away from its unstretched position, the string–guitar system gains elastic potential energy. When the string is released, the elastic potential energy decreases while the kinetic energy of the string increases. The string coasts through its unstretched position and becomes stretched in the other direction. Its kinetic energy then decreases as the elastic potential energy increases. As it vibrates, there is a repeated transformation of elastic potential energy into kinetic energy and back again. The string loses some mechanical energy; for example, sound waves radiate away. Otherwise, the decrease in elastic potential energy over any part of the string's motion would be accompanied by an equal increase in kinetic energy, and vice versa:

The equations in this section are valid only if friction is negligible. We shall extend the range later to include friction, which can cause the conversion of mechanical energy into heat energy.

$$\Delta(PE)_{elastic} = -\Delta(KE),$$

FIGURE 5.22

or

$$-\Delta(PE)_{\text{elastic}} = \Delta(KE).$$

In such an ideal case, the total mechanical energy ($KE + PE_{\text{elastic}}$) remains constant; it is conserved.

Galileo's experiment with the pendulum (Section 3.1) can also be described in these terms. The gravitational potential energy is determined by the height to which the pendulum was originally pulled. That potential energy is converted to kinetic energy at the bottom of the swing and back to potential energy at the other side. Since the pendulum retains its initial energy, it will stop temporarily ($KE = 0$, $PE = \text{max}$) only when it returns to its initial height.

Adding Work

You have seen that the potential energy of a system can be transformed into the kinetic energy of some part of the system, and vice versa. Suppose that an amount of work W is done on part of the system by some external force. Then the energy of the system is increased by an amount equal to W. Consider, for example, a suitcase–Earth system. You must do work on the suitcase to pull it away from the Earth up to the second floor. This work increases the total mechanical energy of the Earth–suitcase system. If

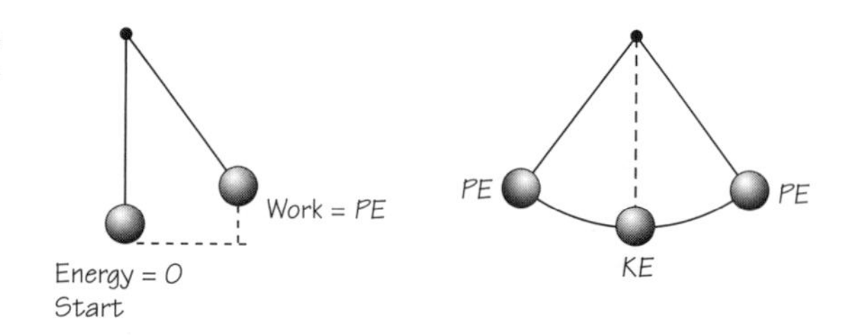

FIGURE 5.23 Work, kinetic, and potential energy in a swinging pendulum.

you yourself are included in the system, then your internal chemical energy decreases in proportion to the work you do. Therefore, the *total* energy of the lifter + suitcase + Earth system does not change, though of course the energy is redistributed among the parts of the system.

Relationship to Newton's Laws of Motion

The law of conservation of energy can be derived from Newton's laws of motion (although Newton himself did not do this). Therefore, it tells nothing that could not, in principle, be computed directly from Newton's laws of motion. However, there are situations where there is simply not enough information about the forces involved to apply Newton's laws, or where it would be inconvenient to try to measure them. It is in these cases that the law of conservation of mechanical energy demonstrates its maximum usefulness.

A perfectly elastic collision is a good example of a situation where we often cannot simply apply Newton's laws of motion. For such collisions, we cannot easily measure the force that one object exerts on the other. We do know that during the actual collision among perfectly elastic bodies, the objects distort one another. The distortions are produced against elastic forces. Thus, some of the combined kinetic energy of the objects is transformed into elastic potential energy as they distort one another. Then elastic potential energy is transformed back into kinetic energy as the objects separate. In an ideal case, both the objects and their surroundings are exactly the same after colliding as they were before.

However, the law of conservation of mechanical energy involves only the *total* energy of the objects before and after the collision. It does not try to give the kinetic energy of each object separately. So it is incomplete knowledge, but useful nevertheless. You may recall that the law of conservation of momentum also supplies incomplete but useful knowledge. It can be used to find the *total* momentum of elastic objects in collision, but not the *individual* momentum vectors. In Section 5.6 you saw how conservation of momentum and conservation of mechanical energy *together* limit the possible outcomes of perfectly elastic collisions. (As often in physics, two or

FIGURE 5.24 During its contact with a golf club, a golf ball is distorted. As the ball moves away from the club, and while the ball recovers its normal spherical shape, its elastic potential energy is transformed into (additional) kinetic energy.

more laws together show the limits of what science can say about phenomena.) For two colliding objects, these two restrictions are enough to give an exact solution for the two velocities after collision. For more complicated systems, conservation of energy remains important. Scientists usually are not interested in the detailed motion of every part of a complex system. They are not likely to care, for example, about the motion of every molecule in a rocket exhaust. Rather, they may want to know only about the exhaust's overall thrust and temperature. These can be found from the overall conservation laws.

5.12 FORCES THAT DO NO WORK

In Section 5.8, the *work* done on an object was defined as the product of the magnitude of the force **F** applied to the object and the magnitude of the distance **d** in the direction of **F** through which the object moves while the force is being applied. In all the examples so far, the object moved in the same direction as that of the force vector.

Usually, the direction of motion and the direction of the force are *not* the same. Let us revisit the example we briefly mentioned before (end of Section 5.8). Suppose you carry a book bag at constant speed horizontally, so that its kinetic energy does not change. Since there is no change in the book bag's energy, you are doing no work on it (by the definition of work). You do apply a force on the book bag, and the bag does move through a distance. But here the applied force and the distance the bag moves are at right angles. You exert a vertical force on the bag upward to balance its weight. But it moves horizontally, with you. Here, an applied force **F** is ex-

FIGURE 5.25 A balanced rock (Mojave Desert): potential energy waiting to be converted.

erted on an object while the object moves at right angles to the direction of the force. Therefore, **F** has no component in the direction of **d** and so the force *does no work*. This statement agrees entirely with the idea of work in physics as *energy being transformed from one form to another*. Since the book bag's speed is constant, its kinetic energy is constant. Since its distance from the Earth is constant, its gravitational potential energy is constant. Therefore, there is no transfer of mechanical energy. (Nevertheless, your arm does tire as you carry the book bag horizontally. The reason for this is that muscles are not rigid. They are constantly relaxing and tightening up again. This requires chemical energy, even though no work is done on the bag.)

A similar reasoning, but not so obvious, applies to a satellite in a circular orbit. The speed and the distance from the Earth are both constant. Therefore, the kinetic energy and the gravitational potential energy are both constant, and there is no energy transformation. For a circular orbit, the centripetal force vector is perpendicular to the tangential direction of motion at any instant. No work is being done on the satellite. But to put an artificial satellite into a circular orbit to start with requires work. Once it is in orbit, however, the *KE* and *PE* stay constant, and no further work is done on the satellite.

However, if the satellite's orbit is eccentric, the force vector is generally not perpendicular to the direction of motion. In such cases, energy is continually transformed between kinetic and gravitational potential forms. The total energy of the system (satellite and Earth) remains of course constant.

Situations where the net force is exactly perpendicular to the motion are as rare as situations where the force and motion are in exactly the same direction. What about the more usual case, involving some angle between the force and the motion?

In general, the work done on an object depends on how far the body moves *in the direction of the force*. As stated before, the equation $W = Fd$ properly defines work only if d is the distance the body moves in the direction of the force. The gravitational force is directed *down*. So only the distance *down* determines the amount of work done by $\mathbf{F}_{grav}$. Change in gravitational potential energy depends *only* on change in height, at least near the Earth's surface. For example, consider raising a suitcase from the first floor to the second floor of a building. The same increase in PE_{grav} of the suitcase–Earth system occurs regardless of the path by which the suitcase is raised. Also, each path requires the same amount of work.

More generally, change in PE_{grav} depends only on change of position. The details of the path followed in making the change make no difference at all. The same is true for changes in elastic potential energy, electric potential energy, etc. The changes depend only on the initial and final positions, and not on the path taken between these positions.

An interesting conclusion follows from the fact that change in PE_{grav} depends only on change in height. For example, consider a child on a slide. Starting from the top position, the gravitational potential energy decreases as his/her altitude decreases. If frictional forces are vanishingly small, all the work the Earth's pull does on him/her goes into transforming PE_{grav} into KE. Therefore, the increases in KE depend only on the decreases in altitude. In other words, the child's speed when he/she reaches the ground will be the same (absent friction) whether he/she slides down or jumps off the top.

A similar principle holds for satellites in orbit and for electrons in TV tubes. In the absence of losses to parts outside the system, the change in kinetic energy depends only on the initial and final positions, and not on

FIGURE 5.26 The change in kinetic energy depends only on the initial and final positions, and not on the path taken between them.

the path taken between them. This principle gives great simplicity to the use of physical laws, as you will see when you study gravitational and electric fields in Chapter 10.

SOME NEW IDEAS AND CONCEPTS

conservation law
elastic collision
isolated system
joule
kinetic energy
law of conservation of kinetic energy
law of conservation of mass
law of conservation of mechanical energy
law of conservation of momentum (LCM)
mechanical energy
momentum
Newtonian world machine
potential energy
system
work

FURTHER READING

F.L. Holmes, *Lavoisier and the Chemistry of Life* (Madison: University of Wisconsin Press, 1989).

G. Holton and S.G. Brush, *Physics, The Human Adventure* (Piscataway, NJ: Rutgers University Press, 2001), Chapters 15–18.

R.S. Westfall, *The Construction of Modern Science: Mechanisms and Mechanics* (New York: Cambridge University Press, 1978).

STUDY GUIDE QUESTIONS

5.1 Conservation of Mass

1. Why did scientists look for conservation laws?
2. True or false: Mass is conserved in a closed system only if there is no chemical reaction in the system.
3. If 50 cm^3 of alcohol is mixed with 50 cm^3 of water, the mixture amounts to only 98 cm^3. An instrument pack weighs much less on the Moon than on Earth. Are these examples of contradictions to the law of conservation of mass?
4. Which one of the following statements is true?
   (a) Lavoisier was the first person to believe that the amount of material in the Universe does not change.
   (b) Mass is measurably increased when heat enters a system.
   (c) A closed system was used to establish the law of conservation of mass experimentally.

5. Five grams (5 g) of a red fluid at 12°C having a volume of 4 ml are mixed in a closed bottle with 10 g of a blue fluid at 5°C having a volume of 8 ml. On the basis of this information only, what can you be sure of about the resulting mixture?

5.2 Collisions

1. Descartes defined the quantity of motion of an object as the product of its mass and its speed. Is his quantity of motion conserved as he believed it was? If not, how would you modify his definition so the quantity of motion would be conserved?
2. Two carts collide head-on and stick together. In which of the following cases will the carts be at rest immediately after the collision?

	Cart A		*Cart B*	
	Mass (kg)	*Speed before (m/s)*	*Mass (kg)*	*Speed before (m/s)*
(a)	2	3	2	3
(b)	2	2	3	3
(c)	2	3	3	2
(d)	2	3	1	6

5.3 Conservation of Momentum

1. State the law of conservation of momentum in terms of
   (a) a change in the total momentum of a system;
   (b) the total initial momentum and final momentum;
   (c) the individual parts of a system.
2. Under what condition is the law of conservation of momentum valid?
3. Which of the following has the least momentum? Which has the greatest momentum?
   (a) a pitched baseball;
   (b) a jet plane in flight;
   (c) a jet plane taxiing toward the terminal.
4. A girl on ice skates is at rest on a horizontal sheet of smooth ice. As a result of catching a rubber ball moving horizontally toward her, she moves at 2 cm/s. Give a rough estimate of what her speed would have been:
   (a) if the rubber ball were thrown twice as fast;
   (b) if the rubber ball had twice the mass;
   (c) if the girl had twice the mass;
   (d) if the rubber ball were not caught by the girl, but bounced off and went straight back with no change of speed.
5. A boy and a girl are on ice skates at rest near each other. The boy throws a ball to the girl in a straight line. Does he move? If so, in what way and why?

After she catches the ball, she throws it back to him in a high arc. Does she move? If so, in what way and why?

5.4 Momentum and Newton's Laws of Motion

1. Since the law of conservation of momentum can be derived from Newton's laws, what good is it?
2. Explain why a cannon shooting a cannon ball must experience a recoil.
3. What force is required to change the momentum of an object by 50 kg m/s in 15 s?

5.5 Isolated Systems

1. Define what is meant by a "closed" or "isolated" system for the purpose of the law of conservation of mass; for the purpose of the law of conservation of momentum.
2. Explain whether or not each of the following can be considered an isolated system:
   (a) a baseball thrown horizontally, after it leaves the thrower's hand;
   (b) a space shuttle orbiting the Earth;
   (c) the Earth and the Moon.
3. Three balls in a closed system have the following masses and velocities:
   ball A: 4 kg, 8 m/s left;
   ball B: 10 kg, 3 m/s up;
   ball C: 8 kg, 4 m/s right.
   Using the principles of mass and momentum conservation, what can you discover about the final condition of the system after the balls have come to rest in the system? What cannot be discovered?

5.6 Elastic Collisions

1. Which phrases correctly complete the statement? Kinetic energy is conserved:
   (a) in all collisions;
   (b) whenever momentum is conserved;
   (c) in some collisions;
   (d) when the colliding objects are not too hard.
2. Under what condition does the law of conservation of kinetic energy hold?
3. Explain why the conservation laws of kinetic energy and momentum are both sometimes needed to describe the outcome of a collision of two bodies.
4. Is the law of conservation of kinetic energy as general as the law of conservation of momentum? Explain.
5. Kinetic energy is never negative because:
   (a) scalar quantities are always positive;
   (b) it is impossible to draw vectors with negative length;
   (c) speed is always greater than zero;
   (d) kinetic energy is proportional to the square of the speed.

5.7 Leibniz and the Conservation Law

1. How would Leibniz have explained the apparent disappearance of the quantity $\frac{1}{2}m\mathbf{v}^2$ in the following situations?
   (a) during the upward motion of a thrown object;
   (b) when the object strikes the ground.
2. Give an example of a situation in which momentum is conserved but kinetic energy is not conserved.

5.8 Work

1. If a force of magnitude F is exerted on an object while the object moves a distance d in the direction of the force, the work done on the object is:
   (a) F; (b) Fd; (c) F/d; (d) $\frac{1}{2}Fd^2$.
2. Give two examples of situations in which a force is exerted on an object but no work is done.

5.9 Work and Kinetic Energy

1. The kinetic energy of a body of mass m moving at a speed v is given by the expression:
   (a) $\frac{1}{2}mv$; (b) $\frac{1}{2}mv^2$; (c) mv^2; (d) $2mv^2$; (e) m^2v^2.
2. What is the general relationship between work and energy?
3. You lift a book from the floor and put it on a shelf.
   (a) What happens to the work that you put into lifting the book?
   (b) Can the work ever be turned into kinetic energy? Explain how.

5.10 Potential Energy

1. Name some forms of potential energy. Describe a situation for each of these in which the potential energy would be greater than zero.
2. A stone of mass m falls a vertical distance d, pulled by its weight $F_{grav} = mg$, where g is the acceleration of gravity. The decrease in gravitational potential energy during the fall is:
   (a) md; (b) mg; (c) mgd; (d) $\frac{1}{2}md^2$; (e) d.
3. When you compress a coil spring, you do work on it. The elastic potential energy:
   (a) disappears; (b) breaks the spring; (c) increases; (d) decreases.
4. Two electrically charged objects repel one another. To increase the electric potential energy, you must:
   (a) make the objects move faster;
   (b) move one object in a circle around the other object;
   (c) attach a rubber band to the objects;
   (d) pull the objects farther apart;
   (e) push the objects closer together.
5. A pendulum bob is swinging back and forth. Where is the kinetic energy of the bob the greatest? Where is it the least?

5.11 Conservation of Mechanical Energy

1. As a stone falls frictionlessly:
   (a) its kinetic energy is conserved;
   (b) its gravitational potential energy is conserved;
   (c) its kinetic energy changes into gravitational potential energy;
   (d) no work is done on the stone;
   (e) there is no change in the total energy.
2. In which position is the elastic potential energy of a vibrating guitar string greatest? In which position is its kinetic energy greatest?
3. If a guitarist gives the same amount of elastic potential energy to a bass string and to a treble string, which one will gain more speed when released? (The mass of 1 m of bass string is greater than that of 1 m of treble string.)
4. Describe the changes in kinetic energy and potential for the system of the two colliding pendula observed at the Royal Society, as described in Section 5.6.

5.12 Forces That Do No Work

1. How much work is done on a satellite during each revolution if its mass is m, its period is T, its speed is v, and its orbit is a circle of radius R?
2. Two skiers were together at the top of a hill above a ski jump. While one skier skied down the slope and went off the jump, the other had a change of mind and rode the ski lift back down. Compare their changes in gravitational potential energy.
3. A third skier went directly down a straight slope next to the ski jump. How would this skier's speed at the bottom compare with that of the skier who went off the jump?
4. No work is done (select one):
   (a) on a heavy box when it is pushed at constant speed along a rough horizontal floor;
   (b) on a nail when it is hammered into a board;
   (c) when there is no component of force parallel to the direction of motion;
   (d) when there is no component of force perpendicular to the direction of motion.

DISCOVERY QUESTIONS

1. In the examples in this chapter we carefully neglected the effects of friction and air resistance. How would friction and air resistance affect the mechanical energy of a pendulum or a flying baseball? How would they affect the conservation of momentum?
2. A child is swinging on a swing. She asks you to push her higher.
   (a) Using the concepts in this chapter, explain why pushing her makes her go higher.
   (b) What is the best part of the swing to exert a push? Why?

3. Using the laws of conservation of momentum and energy, explain why cyclists should wear helmets and football players should wear padding.
4. Discuss the conversion between kinetic and potential forms of energy in the system of a planet orbiting the Sun.
5. Furniture movers are paid for their work in moving furniture. In which of the following cases is work, as defined in physics, *not* done by them or the truck on the furniture?
   (a) A piano is carried downstairs.
   (b) The piano is lifted onto a truck.
   (c) The truck accelerates to 50 mi/hr in 30 s.
   (d) The truck hits an ice patch and skids at constant speed for 2 s.
   (e) The truck hits an obstacle and comes to rest.
6. (a) Why can ocean liners or airplanes not turn corners sharply?
   (b) In the light of your knowledge of the relationship between momentum and force, comment on reports about so-called unidentified flying objects (UFOs) turning sharp corners in full flight.
7. The philosopher John Locke (1632–1704) proposed a science of human nature that was strongly influenced by Newton's physics. In Locke's atomistic view, elementary ideas ("atoms") are produced by elementary sensory experiences and then drift, collide, and interact in the mind. Thus, the formation of ideas was only a special case of the universal interactions of particles.

   Does such an approach to the subject of human nature seem reasonable to you? What argument for and against this sort of theory can you think of?

Quantitative

1. A person who lifts a 10-N book a distance d of 0.8 m straight up does 8 J of work. How much work would the person do if the book is lifted up 1.6 m?
2. The kinetic energy of a ball on a tabletop increases from 10 J to 20 J. How much work is done on it?
3. One joule (1 J) of work is put into lifting a pendulum bob from the zero-energy position. After it is let go, the kinetic energy at one point was found to be 0.25 J. What is the potential energy at that point?
4. A freight car of mass 10^5 kg travels at 2.0 m/s and collides with a motionless freight car of mass 1.5×10^5 kg on a horizontal track. The two cars lock and roll together after impact.
   (a) Using the law of conservation of momentum, find the velocity of the two cars after collision.
   (b) Using the result from (a), find the total kinetic energy of the two cars after the collision, and compare it with the total kinetic energy before the collision. Is this an example of an elastic collision?
   (c) Which quantities are conserved in this collision, and which are not?
5. A 1-kg billiard ball moving on a pool table at 0.8 m/s collides head-on with the cushion along the side of the table. The collision can here be regarded as perfectly elastic. What is the momentum of the ball:
   (a) before impact?

(b) after impact?
(c) What is the change in momentum of the ball?
(d) Is the momentum of the ball conserved?
(e) Is the kinetic energy of the ball conserved?
(f) If the duration of the collision is 0.01 s, what average force does the cushion experience?

6. A system consists of three elastic bodies with masses of 4 g, 6 g, and 8 g. They are squeezed together at rest at a single point and then released. They fly away from each other under the influence of the elastic forces, which are assumed equal for each. The 4-g body is moving with a velocity of 20 cm/s north, and the 6-g body is moving at 3 cm/s east. What is the velocity of the 8-g body?
7. Calculate the kinetic energy of a car and driver traveling at 100 km/hr (about 60 mi/hr). The mass of the car and driver is about 1000 kg. How did they obtain this kinetic energy? Where does it go when the driver puts on the brakes and comes to a stop? Where does it go if the car collides with another?

CHAPTER 6

The Dynamics of Heat

6.1 HEAT AS A FORM OF ENERGY

Consider a book sent sliding across a tabletop. If the surface is rough, it will exert a fairly large frictional force on the book, and the book will soon come to a stop as its kinetic energy rapidly disappears. No corresponding increase in potential energy will occur, since there is no change in height. It appears that, in this example, mechanical energy is not conserved.

However, close examination of the book and the tabletop would show that they are slightly warmer than before. The disappearance of kinetic energy of the book is accompanied by the appearance of *heat*. This suggests, though by no means proves, that the kinetic energy of the book was transformed into heat. If so, heat must be one form of energy. This section deals with how the idea of heat as a form of energy gained acceptance. You will see how theory was aided by practical knowledge of the relationship between heat and work.

Until about the middle of the nineteenth century, heat was generally thought to be some kind of fluid, called *caloric fluid*. No heat is lost or gained overall when hot and cold bodies are mixed; for example, mixing equal amounts of boiling water (100°C) and nearly freezing water (0°C) produces water at just about 50°C. One could therefore conclude that the caloric fluid is conserved in that kind of experiment. Some substances, like wood or coal, seemed to have locked up the caloric fluid, which is then released during combustion.

Although the idea that the heat content of a substance is represented by a quantity of conserved fluid was an apparently useful one, it does not adequately explain some phenomena involving heat. Friction, for example, was known to produce heat (e.g., just rub your hands together rapidly). But it was difficult to understand how the conservation of caloric fluid applied to friction.

In the late eighteenth century, while boring cannon for the Elector of Bavaria, Benjamin Thompson (Count Rumford) observed that a great deal of heat was generated. Some of the cannon shavings—provided by the work done on the metal by a drill—were hot enough to glow. Rumford made some careful measurements by immersing the cannon in water and measuring the rate at which the temperature rose. His results showed that so much heat evolved that the cannon would have melted had it not been cooled. From many such experiments, Rumford concluded that heat is not

FIGURE 6.1 Benjamin Thompson was born in Woburn, Massachusetts, in 1753. After several years as a shopkeeper's apprentice, he married a rich widow and moved to Concord (then called Rumford). During the Revolution, Thompson was a Tory; he left with the British army for England when Boston was taken by the rebels. In 1783, Thompson left England and ultimately settled in Bavaria, where he designed fortifications and built munitions and served as an administrator. The King of Bavaria was sufficiently impressed to make him a Count in 1790, and Thompson took the name Rumford. In 1799, he returned to England and continued to work on scientific experiments. Rumford was one of the founders of the Royal Institution. In 1804 he married Lavoisier's widow; the marriage was an unhappy one, and they soon separated. Rumford died in France in 1814, leaving his estate to institutions in the United States.

a conserved fluid but is generated when work is done, and it continues to appear without limit as long as work is done. His estimate of the ratio of heat produced to work performed was within an order of magnitude ("power of ten") of the presently accepted value.

Rumford's experiments and similar work by Humphry Davy at the Royal Institution in London did not convince many scientists at the time. The reason may have been that Rumford could give no clear suggestion of just what heat is, at least not in terms that were compatible with the accepted models for matter at that time.

Nearly a half-century later, James Prescott Joule repeated on a smaller scale some of Rumford's experiments. Starting in the 1840s and continuing for many years, Joule refined and elaborated his apparatus and his techniques. In all cases, the more careful he was, the more exact was the proportionality of the quantity of heat, as measured by a change in temperature, and the amount of work done. Here, Joule, like others, made the assumption that the quantity of heat produced, say, in water, symbolized by ΔQ (Q is the usual symbol for heat), is equal to the mass of the water times the change of its temperature, ΔT:

$$\Delta Q = m\,\Delta T.$$

Today, we know that the amount of heat corresponding to a given temperature change is different for different substances being heated. In order to take this into account, the constant c, called the *specific heat*, is introduced into the above equation. In the metric system of units, the specific heat c is the amount of heat, measured in the units of calories, required to raise 1 g of the substance by 1°C under standard conditions (i.e., at prescribed temperature and pressure) and without any loss of heat to the surroundings. So the relationship between heat and temperature may be written

$$\Delta Q = m\,c\,\Delta T.$$

In order to define a *calorie* of heat, the specific heat of water under standard conditions is defined as $c = 1$ cal/g°C. In other words, 1 cal is defined as the amount of heat required to raise the temperature of 1 g of water by 1°C under standard conditions. So, in these units, and with water as the material being heated, we have

$$\Delta Q = m\,\Delta T.$$

For one of his early experiments on the relationship between heat and work, Joule constructed a simple electric generator, which was driven by a falling weight. The electric current that was generated heated a wire. The

FIGURE 6.2 James Prescott Joule (1818–1889). Joule was the son of a wealthy Manchester brewer. His arduous experiments were initially motivated by the desire to develop more efficient engines for the family brewery.

wire was immersed in a container of water, which it heated. From the distance that the weight descended Joule calculated the work done (the decrease in gravitational potential energy). The product of the mass of the water and its temperature rise allowed him to calculate the amount of heat produced. In another experiment, Joule compressed gas in a bottle immersed in water, measuring the amount of work done to compress the gas. He then measured the amount of heat given to the water as the gas grew hotter on compression.

Joule's most famous experiments involved an apparatus in which slowly descending weights turned a paddle wheel in a container of water. Owing to the friction between the wheel and the liquid, the wheel performed work on the liquid, raising its temperature.

All these experiments, some repeated many times with improved apparatus, led Joule to announce two very important, quantitative results in 1849. As expressed in modern terms and units, they are as follows:

- The quantity of heat produced by the friction of bodies, whether solid or liquid, is always proportional to the quantity of energy expended.
- The quantity of heat (in calories) capable of increasing the temperature of 1 kg of water by 1°C requires for its evolution the change of me-

chanical energy represented by the fall of a weight of 4180 N through the distance of 1 m.

Joule's first statement in the above quote is the evidence that heat is a form of energy, contrary to the caloric theory that heat is a fluid. The second statement gives the numerical magnitude of the ratio he had found of mechanical energy to the equivalent heat energy. The ratio of the mechanical energy, E, to the equivalent amount of heat energy, Q, is generally called the *mechanical equivalent of heat*. Its value, by the most recent measurements, is 4.18×10^3 joules/Calorie in the mks system of units, where Calorie (with a capital C) is 1000 calories (often abbreviated "kcal"). In the cgs system of units, the mechanical equivalent of heat is 4.18×10^5 erg/calorie. (See the insert for review.)

By the time Joule performed his famous experiments, the idea that heat is a form of energy was slowly gaining acceptance. Joule's experiments served as a strong argument in favor of that idea.

UNITS

A reminder: In the metric system used today in most countries, quantities are measured using either grams, centimeters, and seconds (cgs units) or kilograms, meters, and seconds (mks units). (See the discussion in the *Student Guide*.) You saw in Chapter 5 that mechanical energy in mks units is expressed as joules (J) and in cgs units as ergs:

$$1\ \text{J} = 1\ \text{kg m}^2/\text{s}^2 = 1\ \text{N} \cdot \text{m (newton-meter)},$$

$$1\ \text{erg} = 1\ \text{g cm}^2/\text{s}^2 = 1\ \text{D} \cdot \text{cm (dyne-centimeter)}.$$

As indicated on page 255, heat energy in cgs units is expressed as calories (abbreviated as cal), while in mks units heat energy is expressed in kilocalories (1000 calories), which is often written either as kcal or with an uppercase initial C, Calorie. (The Calorie, abbreviated as Cal, is also the measure used to express the energy content in food.) These units are summarized in the table below, along with the modern values for the mechanical equivalent of heat.

Units	*Mechanical energy*	*Heat energy*	$J = W/Q$
mks	joule (J)	Calorie, kcal	4.18×10^3 J/Cal
cgs	erg	calorie	4.18×10^7 erg/cal
English	foot-pound (ft-lb)	Btu	1.29×10^{-3} ft-lb/Btu

6.2 THE STEAM ENGINE AND THE INDUSTRIAL REVOLUTION

The development of the modern science of heat was closely tied to the development of the modern technology of engines designed to perform useful work. For millennia until about two centuries ago, most work was done by people or by animals. Wind and water also provided mechanical work, but these were generally unreliable as sources of energy. For one thing, they were not always available when and where they were needed.

In the eighteenth century, miners began to dig deeper and deeper in search of a greater coal supply. Water tended to seep in and flood these deeper mines. The need arose for an economical method of pumping water out of these mines. The steam engine was developed initially to meet this very practical need.

The steam engine is a device for converting the heat energy of heat-producing fuel into mechanical work. For example, the chemical energy of wood, coal, or oil, or the nuclear energy of uranium, can be converted into heat. The heat energy in turn is used to boil water to form steam, and the energy in the steam is then turned into mechanical energy. This mechanical energy can be used directly to perform work, as in a steam locomotive, used to pump water, or to transport loads, or is transformed into electrical energy. In typical industrial societies today, most of the energy used in fac-

FIGURE 6.3 Camel driving a water wheel.

FIGURE 6.4 Woodcut from Georgius Agricola's *De Re Metallica* (1556), a book on mining techniques in the sixteenth century.

tories and in homes comes from electrical energy. Falling water is used to generate electricity in some areas, but steam-powered generators still generate most of the electrical energy used in the United States today. (This is further discussed in Chapter 11.) There are other devices that convert fuel to heat energy for the production of mechanical energy, such as internal combustion engines used in cars and trucks, for example. But the steam engine remains a good model for the basic operation of the whole family of so-called *heat engines*, and the chain of processes from heat input to work output and heat exhaust is a good model of the basic *cycle* involved in all heat engines.

Since ancient times, people knew that heat can be used to produce steam, which can then do mechanical work. One example was the "aeolipile," in-

(a) (b)

FIGURE 6.5 Old windmill and new wind turbine.

vented by Heron of Alexandria about A.D. 100. (See Figure 6.6.) It worked on the principle of the rotating lawn sprinkler, except that the driving force was steam instead of water pressure. Heron's aeolipile was a toy, meant to entertain rather than to do any useful work. Perhaps the most "useful" application of steam in the ancient world was another of Heron's inventions. This steam-driven device astonished worshipers in a temple by causing a door to open when a fire was built on the altar.

Not until late in the eighteenth century, however, did inventors develop commercially successful steam engines. Thomas Savery (1650–1715), an English military engineer, invented the first such engine. It could pump water out of a mine by alternately filling a tank with high-pressure steam, driving the water up and out of the tank, and then condensing the steam, drawing more water into the tank.

Unfortunately, inherent in the Savery engine's use of high-pressure steam was a serious risk of boiler or cylinder explosions. Thomas Newcomen (1663–1729), another English engineer, remedied this defect. Newcomen invented an engine that used steam at lower pressure (see Figure 6.7). His engine was superior in other ways also. For example, it could raise loads other than water. Instead of using the steam to force water into and out of

FIGURE 6.6 A model of Heron's aeolipile. Steam produced in the boiler escapes through the nozzles on the sphere, causing the sphere to rotate.

a cylinder, Newcomen used the steam to force a piston forward and air pressure to force it back. The motion of the piston could then be used to drive a pump or other engine. It is the back-and-forth force provided by the motion of the piston in a steam engine that is one origin of the definition of mechanical work, W, as force × distance, $W = Fd$.

The Newcomen engine was widely used in Britain and other European countries throughout the eighteenth century. By modern standards, it was not a very good engine. It burned a large amount of coal but did only a

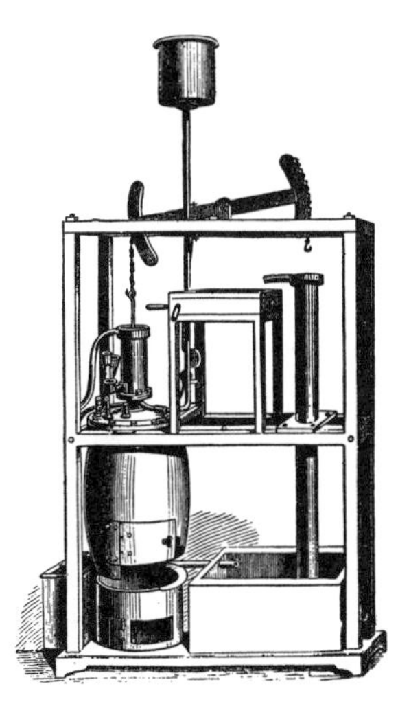

FIGURE 6.7 Model of the Newcomen engine, which inspired Watt to conceive of the separation of condenser and piston.

small amount of work at a slow, jerky rate. But the great demand for machines to pump water from mines produced a good market, even for that uneconomical engine.

The work of a Scotsman, James Watt, led to a greatly improved steam engine and one that had profound economic consequences. Watt's father was a carpenter with a successful business selling equipment to shipowners. James Watt was in poor health much of his life and gained most of his early education at home. In his father's attic workshop, he developed considerable skill in using tools. He wanted to become an instrument maker and went to London to learn the trade. Upon his return to Scotland in 1757, he obtained a position as instrument maker at the University of Glasgow.

In the winter of 1763–1764, Watt was asked to repair a model of Newcomen's engine that was used for demonstration lectures at the university. In acquainting himself with the model, Watt was impressed by how much steam was required to run the engine. He undertook a series of experiments on the behavior of steam and found that a major problem was the temperature of the cylinder walls. Newcomen's engine wasted most of its heat in warming the walls of its cylinder, since the walls were cooled on each cycle as cold water was injected to condense the steam, forcing the piston back under air pressure.

Early in 1765, Watt remedied this wasteful defect by devising a modified type of steam engine. (See Figure 6.8.) In retrospect, it sounds like a simple idea. After pushing the piston up, the steam was admitted to a *separate* container, called the *condenser*, where the steam condensed at a low temperature. With this system, the *cylinder* containing the piston could be kept hot all the time, and the condenser could be kept cool all the time.

The use of the separate condenser allowed huge fuel savings. Watt's engine could do twice as much work as Newcomen's with the same amount of fuel. Watt also added many other refinements, such as automatically controlled valves that were opened and closed by the reciprocating action of the piston itself, as well as a *governor* that controlled the amount of steam reaching the engine, to maintain a constant speed for the engine (see Figure 6.9). The latter idea of using part of the output of the process to regulate the process itself, is called *feedback*. It is an essential part of the design of many modern mechanical and electronic systems.

Like Thomas Edison later, or successful computer technologists in our day, Watt and his associates were good businessmen as well as good engineers. They made a fortune manufacturing and selling the improved steam engines. Watt's inventions stimulated the development of engines that could do many other jobs. Steam drove machines in factories, railway locomotives,

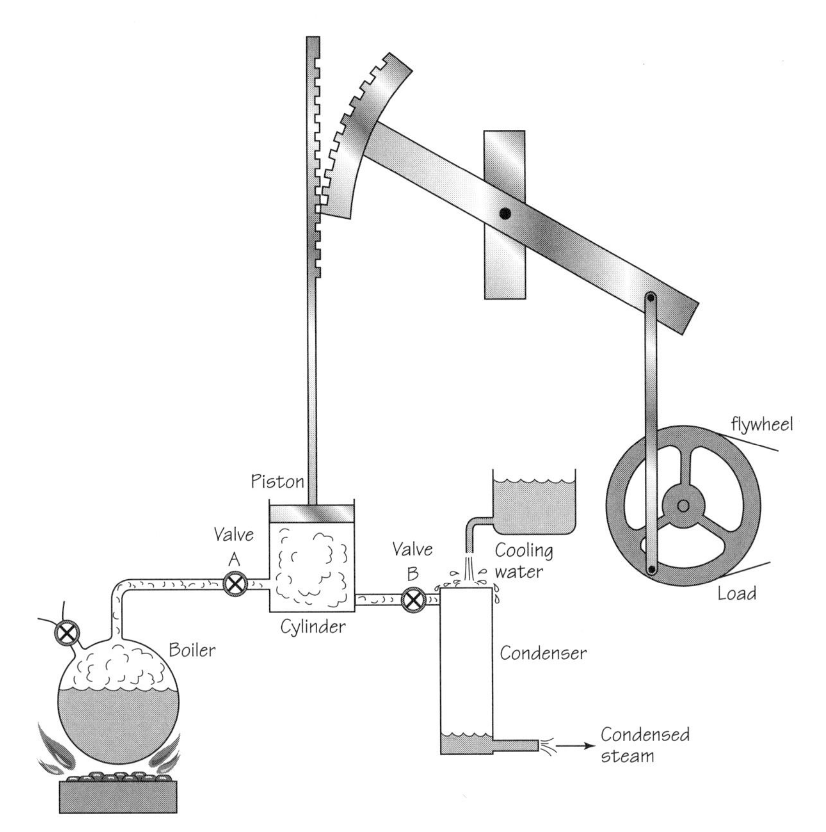

FIGURE 6.8 Schematic diagram of Watt's steam engine. With valve A open and valve B closed, steam under pressure enters the cylinder and pushes the piston upward. When the piston nears the top of the cylinder, valve A is closed to shut off the steam supply. Then valve B is opened, so that steam leaves the cylinder and enters the condenser. The condenser is kept cool by water flowing over it, so the steam condenses. As steam leaves the cylinder, the pressure there decreases. Atmospheric pressure (helped by the inertia of the flywheel) pushes the piston down. When the piston reaches the bottom of the cylinder, valve B is closed, and valve A is opened, starting the cycle again. Valves A and B are connected to the piston directly, so that the motion of the piston itself operates them.

steamboats, and even early steam cars. Watt's engine gave an enormous stimulus to the growth of industry in Europe and America. It thereby helped transform the economic and social structure of industrial civilization.

The widespread development of engines and machines revolutionized the mass production of consumer goods, construction, and transportation. The average standard of living in Western Europe and the United States

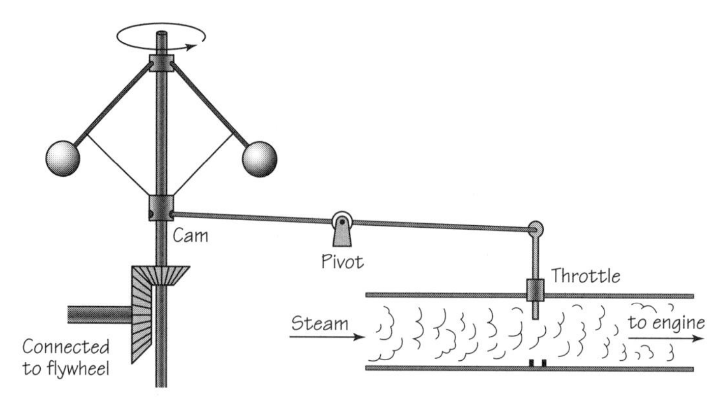

FIGURE 6.9 Watt's "governor." If the engine speeds up for some reason, the heavy balls swing out to rotate in larger circles. They are pivoted at the top, so the sleeve below is pulled up. The cam that fits against the sleeve is therefore also pulled up; this forces the throttle to move down and close a bit. The reduction in steam reaching the engine thus slows it down again. The opposite happens when the engine starts to slow down. The net result is that the engine tends to operate at nearly a stable level.

rose sharply. It is difficult for most people in the industrially "developed" countries to imagine what life was like before this "Industrial Revolution." But not all the effects of industrialization have been beneficial. The nineteenth-century factory system provided an opportunity for some greedy and cruel employers to treat workers almost like slaves. With no labor laws or even child protection laws, those employers made huge profits while keeping employees and their families on the edge of starvation.

FIGURE 6.10 Steam-powered locomotive.

FIGURE 6.11 The *Charlotte Dundas*, the first practical steamboat, built by William Symington, an engineer who had patented his own improved steam engine. It was tried out on the Forth and Clyde Canal in 1801.

This situation, which was especially serious in England early in the nineteenth century, led to demands for reform. New laws eventually eliminated the worst excesses.

More and more people left the farms, either voluntarily or forced by poverty and new land laws, to work in factories. Conflict grew intense between the working class, made up of employees, and the middle class, made up of employers and professionals. At the same time, some artists and intellectuals, many of the Romantic movement, began to attack the new tendencies of their society. They saw this society becoming increasingly dom-

FIGURE 6.12 Engraving of an early steam-powered factory. Matthew Boulton (Watt's business partner) proclaimed to Boswell (biographer of Samuel Johnson): "I sell here, sir, what all the world desires to have: POWER!"

■ AGRICULTURAL STEAM TECHNOLOGY

There are three distinct operations in the harvesting of grain. Reaping, which is cutting the stem from the ground, threshing which is separating the grain from the kernel, and winnowing which cleans the grain from the chaff.

Hand-reaping methods, using first sickles and then scythes for millennia, were only replaced by new methods employing mechanical machinery in the first half of the nineteenth century. In the 1830s Hiriam Moore, a farmer in Michigan, began to design a machine known as a combine that would use horsepower to cut, thresh, and clean.

The next important breakthrough came from George Berry, a wheat farmer in the Sacramento Valley. Although he had been impressed by the reduction in labor costs resulting from the introduction of the combine, Berry had seen many of his horses die in the intense heat of July and August harvesting. Berry decided to use the steam-traction engines that had begun to appear on some farms to power his own combine. The combine was moved forward by a 26-horsepower steam-traction engine, while a smaller, stationary 6-horsepower steam-traction engine powered the cutter, thresher, and separator.

The steam-traction engines were used not only to power combines, but also for plowing, planting, and cultivation as well. However, a different engine was used for each different job, and few farmers could justify this kind of expenditure on machinery. In 1921, Alexander Legge, general manager of International Harvester Company, authorized his company to begin developing an all-purpose engine. The new tractor became known as the Farmall, and it was the first engine that truly achieved George Berry's vision of replacing horses in farming.

Further Reading

C. Canine, *Dream Reaper* (Chicago, University of Chicago Press, 1997).

FIGURE 6.13 Nineteenth-century French steam cultivator

FIGURE 6.14 The Farmall: "It would be tall and maneuverable enough to cultivate row crops, yet it would still have the power to plow, pull implements, and deliver belt power to threshers, grinders, and crop elevators."

inated by commerce, machinery, and an emphasis on material goods. In some cases, they confused research science itself with its technical applications (as is still done today). Sometimes they accused scientists of explaining away all the awesome mysteries of nature. These artists denounced both science and technology, while often refusing to learn anything about them. A poem by William Blake contains the questions:

> And did the Countenance Divine
> Shine forth upon our clouded hills?
> And was Jerusalem builded here
> Among these dark Satanic mills?

Elsewhere, Blake advised his readers "To cast off Bacon, Locke, and Newton." John Keats was complaining about science when he included in a poem the line:

> Do not all charms fly
> At the mere touch of cold philosophy?

These attitudes are part of an old tradition, going back to the ancient Greek opponents of Democritus' atomism. As noted in Chapter 4, many of the Romantic writers and artists attacked Galilean and Newtonian physics for supposedly distorting moral values. The same type of accusation can still be heard today.

6.3 POWER AND EFFICIENCY OF ENGINES

The usefulness of an engine for many tasks is given by the rate at which it can deliver energy. The rate at which an engine delivers energy is called its *power*. By definition, the power (P) is the amount of energy (E) delivered per unit of time (t):

$$P = \frac{E}{t}.$$

As with energy, there are many common units of power with definitions rooted in tradition. Before the steam engine, the standard source of power was the workhorse. Watt, in order to rate his engines in a unit people could understand, measured the power output of a horse. He found that a strong horse, working steadily, could lift an object of 75-kg mass, which weighed about 750 N, at a speed of about 1 m/s (of course, Watt used the units of pounds and feet). The "horsepower" unit is still used today, but its value is now given by definition, not by experiment.

In metric units, the unit of power is appropriately named the watt, symbol W, which is not to be confused with the symbol W for "work." (You can usually tell from the context which unit is intended.) One watt is defined as one joule of energy per second, or in symbols, 1 W = 1 J/s. Thus, Watt's horse had a power rating of about 750 W. This means that in this case one horsepower was about 750 W. It is a curious case of the persistence of ancient habits that the unit "horsepower" is still used today, for example, for rating car engines and electric motors.

A further example: A light bulb rated at 100 W is using energy at the rate of 100 J/s. To find the total energy the bulb uses in a specific case, we need to specify the time interval during which it is on. Once the time is specified, and if the power usage is known, the energy can be found (from $P = E/t$) by multiplying the time and the power. In a typical case, the energy E used by a 100-W bulb during a period of, say, 10 hr is

$$E = P \cdot t = (100\text{ W})\ (10\text{ hr})$$

$$= 1000\text{ W} \cdot \text{hr} = 1\text{ kWhr},$$

$$1\text{ kWhr} = (1000\text{ J/s})(1\text{ hr})(3600\text{ s/hr})$$

$$= 3.6 \times 10^6\text{ J}.$$

The answer is over three million joules! Since the amount of energy consumed is so large, the commercial energy used by a typical home is billed in units of kilowatt-hours (kW-hr). Look at the monthly bill for your home's use of electricity and see how much electric energy is used, and what it costs effectively per kW-hr. (Then perhaps consider how to cut down on the home's use of electricity.)

Efficiency

Section 6.1 showed that the amount of mechanical energy corresponding to a unit of heat energy is known as the "mechanical equivalent of heat." Joule's finding a value for the mechanical equivalent of heat made it possible to describe engines in a new way. The concept of *efficiency* can be applied to an engine or to any device that transforms energy from one form to another, such as from heat energy into mechanical energy. *Efficiency is defined as the ratio of the useful output energy to the amount of input energy*. If E_{in} is input energy and E_{out} is the useful output, then efficiency (eff), can be defined in symbols

$$\text{eff} = \frac{\text{useful } E_{out}}{E_{in}}.$$

Efficiency may also be expressed as a percentage

$$\text{eff (\%)} = \left(\frac{\text{useful } E_{out}}{E_{in}}\right) \times 100.$$

Since energy cannot be lost, the greatest possible efficiency of any engine would be 100%, which would occur when *all* of the input energy appears as useful output. Obviously, efficiency must be considered as seriously as power output in designing engines. Fuel is, after all, a part of the cost of running an engine, and the more efficient an engine is, the cheaper it is to run.

Watt's engine was more efficient than Newcomen's, which in turn was more efficient than Savery's. Is there any limit to improvements in efficiency? The upper limit, 100%, is of course imposed by the law of energy conservation. That is, no engine can put out more mechanical energy than the energy put into it. But even before that law had been formulated, a young French engineer, Sadi Carnot, established that there is in practice a much lower limit. The reasons for this limit are just as fundamental as the law of energy conservation.

6.4 CARNOT AND THE BEGINNINGS OF THERMODYNAMICS

Carnot was one of a group of French engineers who had set out to study the scientific principles underlying the operation of the steam engine with the goal of achieving maximum power output at maximum efficiency. As a result of their studies, Carnot and others helped to establish the physics of heat, which is known as *thermodynamics*.

Carnot started with the experimentally obtained fact that heat does not by itself flow from a cold body to a hot one. It then follows that if, in a given situation, heat is made to flow from cold to hot, some other change

FIGURE 6.15 Sadi Carnot (1796–1832). Son of one of Napoleon's most trusted generals, Sadi Carnot was one of the new generation of expert administrators who hoped to produce a new enlightened order in Europe. He died of cholera in Paris at the age of 36.

must take place elsewhere. Some work must be done. Using an elegant argument, which is summarized in the materials for this chapter in the *Student Guide*, Carnot showed that no engine can be more efficient than an ideal, reversible engine, and that all such engines have the same efficiency. A *reversible engine is one in which the cycle from input energy to output work and exhausted energy, then back to input energy, can be run in reverse without any loss or gain of heat or other forms of energy*. For example, a refrigerator or an air conditioner is also a "heat engine," but its cycle operates in reverse fashion to a steam engine or automobile engine. It *takes in* work (in the form of electrical or mechanical power) to pump heat from a cold body (from inside the cold compartment or room) to a hotter one (the outside room or outside air). Naturally, because of friction and other outside effects, a truly reversible engine cannot be realized in practice, but it can be approached.

Since all reversible engines have the same efficiency, one has only to choose a simple version of a reversible engine and calculate its efficiency for one cycle to find an upper limit to the efficiency of any engine. Such a simple engine is depicted schematically in the diagram below. During one cycle of operation, the engine, represented by the small rectangle, takes in heat energy Q_1 from the hot body, produces useful work W, and exhausts some wasted heat energy Q_2 to the cold body. The cycle may then be repeated many times.

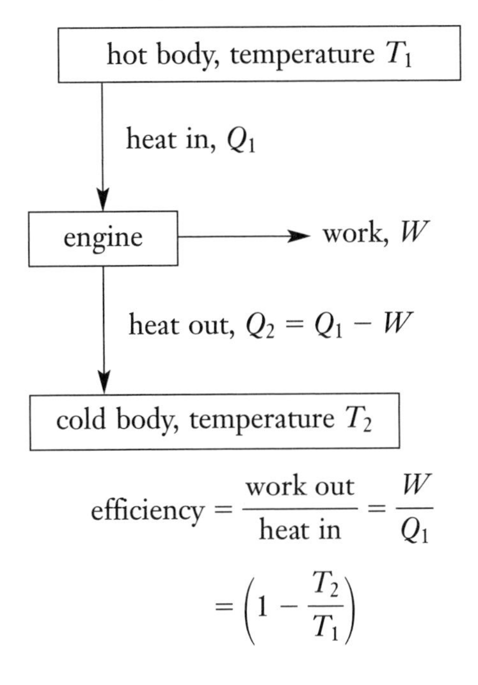

Carnot calculated the efficiency of this schematic engine cycle and found that the ratios of heat and work in such a reversible engine depend only on the temperature of the hot substance from which the engine obtains its heat and on the temperature of the cold substance that extracts the waste heat from the engine. The temperatures used in this case are called *absolute*, or *Kelvin*, temperatures (named for Lord Kelvin who first introduced this scale). On the absolute scale, temperature measurements are equal to temperatures (t) on the Celsius scale (°C) plus 273. (Following current standard practice, no degree sign, °, is used for degrees Kelvin; the symbol used is K.)

$$T \text{ (absolute, in K)} = t \text{ (Celsius, in °C)} + 273.$$

Thus, on the Kelvin scale, water freezes at 273 K, while "absolute zero," $T = 0$ K, is

$$t = -273°\text{C}.$$

The expression found by Carnot for the efficiency of reversible engines, in modern terms, is

$$\text{efficiency} = \frac{\text{work out}}{\text{heat in}} = \frac{W}{Q_1} = \left(1 - \frac{T_2}{T_1}\right)$$

Although Carnot did not write the formula this way, we are making use of the fact that heat and energy are equivalent.

Notice that unless T_2, the temperature of the cold body that receives the exhaust from the engine, is 0 K—an unattainably low temperature—no engine can have an efficiency of 1 (or 100%). *This means that every engine must exhaust some "waste" heat to the outside before returning to get more energy from the hot body.*

In steam engines, the "hot body" is the steam fresh from the boiler, and the waste heat is extracted at the condenser. The *cycle* starts with the piston in the cylinder at rest. Steam is let into the cylinder at high temperature and pressure. The steam expands against the piston, performing work on the piston and ultimately on the outside surroundings. As the space occupied by the steam increases, the steam cools down. The steam cools further as it is let into the condenser, where it condenses into water. Air pressure then works on the piston, pushing it back into the cylinder, until it returns to where it started, thus completing one cycle.

As you see, the engine could not operate without a condenser to remove the heat from the steam after it had caused the piston to move forward, thus enabling the cylinder to be filled once again with steam. In an internal combustion engine (a car engine, for example), the "hot body" is the gasoline vapor inside the cylinder just as it explodes, and the cold substance is the exhaust. Any engine that derives its mechanical energy from heat must also be cooled to remove the "waste" heat at a lower temperature. If there is any friction or other inefficiency in the engine, it will add further heat to the waste and reduce the efficiency to below the theoretical limit of an ideal engine.

However, despite the inefficiencies of all real engines, it is important to know that none of the total energy is destroyed. Rather than being destroyed, the part of the energy that is extracted at the exhaust is unavailable for doing work. For instance, the exhausted heat cannot be recycled as input energy to run the engine to produce more useful work and thus increase the efficiency of the engine, by reducing the amount of waste energy because the input reservoir of heat is at a higher temperature than the output, and heat does not flow of its own accord from cold to hot.

The generalization of Carnot's finding is now known as one expression of the *second law of thermodynamics*. It is equivalent to Carnot's earlier observation that *heat does not by itself flow from a cold body to a hot one*. The need for air conditioners and refrigerators makes this abundantly clear.

The second law of thermodynamics, of which more will be said in Section 6.6, is recognized as one of the most powerful laws of physics. Even in simple situations it can help explain natural phenomena and the fundamental limits of technology.

Some Examples of Carnot's Result

If you burn oil to heat your home, the furnace requires some inefficiency to burn cleanly, so some heat is lost out the chimney. But recent advances in boiler technology have resulted in boilers with rated efficiencies as high as 0.86, or 86%.

If you install "flameless electric heat," which uses electric heating elements placed along the floor where it meets the wall, the electric power company still has to burn oil, coal, or natural gas in a boiler, use the steam to generate electricity, and deliver the electricity to your home. Because metals melt above a certain temperature and because the cooling water never gets below freezing, Carnot's finding makes it impossible to make the efficiency of electrical generation greater than about 60%. Since the power company's boiler also loses some of its energy out the chimney, and since the electricity loses some of its energy on the way from the power plant, only about one-quarter to one-third of the energy originally in the fuel actually makes it to your home. Obviously, electric heating wastes a lot of irreplaceable fossil fuel.

Because of the limits placed by Carnot's finding on heat engines, it is sometimes important not only to give the actual efficiency of a heat engine but also to specify how close it comes to the maximum possible. The more carefully you look at a process, the more information is seen to be important. Home-heating apparatus and many large electrical heat-engine devices, such as refrigerators and air conditioners, now come with an "energy guide" sticker indicating the efficiency of the apparatus and the potential annual savings in electricity costs. Some states may even reward consumers with a rebate for making an energy-efficient purchase.

> For steam engines the coldest temperature feasible for T_2 is about 280 K. (Why?) The hottest possible temperature for T_1 is about 780 K. So the maximum efficiency is 0.64.

6.5 ARRIVING AT A GENERAL CONSERVATION LAW

The law of conservation of *mechanical* energy was presented in Section 5.11. This law applies only to "closed systems," i.e., to situations where no work is done on or by the system, and where no mechanical energy is transformed into heat energy or vice versa. Early in the nineteenth century, developments in science, engineering, and philosophy suggested new ideas about energy. It appeared that all forms of energy (including heat) could be transformed into one another with no loss. Therefore, it appeared that the total amount of energy in nature, that is, the Universe, must remain constant.

In 1800 Italian scientist Alessandro Volta invented the electric battery, demonstrating that chemical reactions could produce electricity. It was soon found that electric currents could produce heat and light, as in passing through a thin wire. In 1820, Hans Christian Oersted, a Danish physicist, discovered that an electric current produces magnetic effects. In 1831, Michael Faraday, the English scientist, discovered electromagnetic induction. When a magnet moves near a coil or a wire, an electric current is produced in the coil or wire. To some thinkers, these discoveries (discussed further in Chapter 11) suggested that all of the phenomena of nature were somehow united. Perhaps all natural events result from the same basic "force." This idea, though vague and imprecise, eventually bore fruit in the form of the *law of conservation of energy*, one of the most important laws in all of science:

> Natural events may involve a transformation of energy from one form to another; but the total quantity of energy does not change during the transformation.

Joule began his long series of experiments by investigating the "duty" of electric motors. In this case, duty was measured by the work the motor could do when a certain amount of zinc was used up in the battery that ran the motor. Joule's interest was to see whether motors could be made economically competitive with steam engines.

The invention and use of steam engines helped in establishing the law of conservation of energy (often abbreviated LCE) by showing how to measure energy changes. For example, Joule used the work done by descending weights driving a paddle wheel in a tank of water as a measure of the amount of gravitational potential energy transformed into heat energy in the water by its friction with the paddles.

In 1843, Joule stated that in such experiments, whenever a certain amount of mechanical energy seemed to disappear, a definite amount of heat always appeared. To him, this was an indication of the conservation of what we now call energy. Joule said that he was

> . . . satisfied that the grand agents of nature are by the Creator's fiat *indestructible;* and that, wherever mechanical [energy] is expended, an exact equivalent of heat is *always* obtained.

Joule was basically a practical man who had little time to speculate about a deeper philosophical meaning of his findings. But others, though using speculative arguments, were also concluding that the total amount of energy in the Universe is constant. Before going into the detailed uses of the LCE (as we shall in the next section), it will be interesting to look briefly at an example of the interaction of science and other cultural trends of the time.

Nature Philosophy

A year before Joule's remark, Julius Robert Mayer, a German physician, had also proposed a general law of conservation of energy. Mayer had done no quantitative experiments; but he had observed body processes involving heat and respiration. He had also used other scientists' published data on the thermal properties of air to calculate the mechanical equivalent of heat, obtaining about the same value that Joule had.

Mayer had been influenced by the German philosophical school now known as *Naturphilosophie* or "Nature Philosophy." This school, related to the Romantic movement, flourished during the late eighteenth and early nineteenth centuries. According to Nature Philosophy, the various phenomena and forces of nature—such as gravity, electricity, and magnetism—are not really separate from one another but are all manifestations of some unifying "basic" natural force. This philosophy therefore encouraged experiments searching for that underlying force and for connections between different kinds of forces observed in nature.

FIGURE 6.16 Friedrich von Schelling (1775–1854), one of the founders of German *Naturphilosophie*.

The most influential thinkers of the school of Nature Philosophers were Johann Wolfgang von Goethe and Friedrich Wilhelm Joseph von Schelling. Neither of these men is known today as a scientist, although Goethe did write extensively on geology and botany, and did develop a theory of colors that differed from Newton's. Goethe is generally considered Germany's greatest poet and dramatist, while Schelling was a philosopher. Both men had great influence on the generation of European scientists educated in the first decades of the nineteenth century.

The Nature Philosophers were closely associated with the Romantic movement in literature, art, and music. As noted Section 6.2 and in Chapter 4, the Romantics protested against the idea of the Universe as a great machine, the "Newtonian world machine." This idea seemed to them morally empty and artistically worthless. They refused to believe that the richness of natural phenomena, including human intellect, emotions, and hopes, could be understood as the result of the motions of particles—an opinion which in fact almost no scientists then did, or now do, hold or defend.

The Nature Philosophers claimed that nature could be understood as it really is only by direct observation, or "experience." No complicated, "artificial" apparatus must be used, only the senses, feelings, and intuitions. For Goethe the goal of his philosophy was "that I may detect the inmost force which binds the world, and guides its course."

Although its emphasis on the unity of nature led followers of Nature Philosophy to some very useful insights—such as the general concept of the conservation of energy—its romantic and antiscientific bias made it less and less influential. Scientists who had previously been influenced by it, including Mayer, now strongly opposed it. In fact, some hard-headed scientists at first doubted the law of conservation of energy simply because of their distrust of Nature Philosophy. For example, William Barton Rogers, founder of the Massachusetts Institute of Technology, wrote in 1858:

> To me it seems as if many of those who are discussing this question of the conservation of force [we would now call it energy] are plunging into the fog of mysticism.

However, the law was so quickly and successfully put to use in physics that its philosophical origins were soon forgotten. Yet, this episode is a reminder of a familiar lesson: In the ordinary day-to-day work of physical scientists, experiment and mathematical theory are the usual guides. But in making a truly major advance in science, philosophical speculation may also play an important role.

Mayer and Joule were only two of at least a dozen people who, between 1832 and 1854, proposed in some form the idea that energy is conserved.

Some expressed the idea vaguely; others expressed it quite clearly. Some arrived at the belief mainly through philosophy; others from a practical concern with engines and machines or from laboratory investigations; still others from a combination of factors. Many, including Mayer and Joule, worked quite independently of one another. The idea of energy conservation was somehow "in the air," leading to essentially simultaneous, separate discoveries.

The initial wide acceptance of the LCE owed much to the long-range influence of a paper published in 1847, 2 years before Joule published the results of his most precise experiments. The author, a young German physician and physicist named Hermann von Helmholtz, entitled his work "On the Conservation of Force." Helmholtz (using "force" in the modern sense of "energy"), boldly asserted the idea that others were only vaguely expressing, namely, "that it is impossible to create a lasting motive force out of nothing." He restated this theme even more clearly many years later in one of his popular lectures:

> We arrive at the conclusion that Nature as a whole possesses a store of force [energy] which cannot in any way be either increased or diminished, and that, therefore, the quantity of force in Nature is

FIGURE 6.17 Hermann von Helmholtz (1821–1894).

> just as eternal and unalterable as the quantity of matter. Expressed in this form, I have named the general law "The Principle of the Conservation of Force."

Any machine or engine that does work (provides energy) can do so only by drawing from some source of energy. The machine cannot supply more energy than it obtains from the source. When the source runs out, the machine will stop working. Machines and engines can only *transform* energy; they cannot create it or destroy it.

6.6 THE TWO LAWS OF THERMODYNAMICS

Two laws summarize many of the ideas in this chapter. Both of these laws are called laws of thermodynamics. They may be stated in completely analogous fashion as statements of impossibility.

The First Law

The first law of thermodynamics is a general statement of the conservation of energy in thermal processes. It is based on Joule's finding that heat and energy are equivalent. It would be pleasingly simple to call heat "internal" energy associated with temperature. We could then add heat to the potential and kinetic energies of a system, and call this sum the total energy that is conserved. In fact, this solution works well for a great variety of phenomena, including the experiments of Joule. Difficulties arise with the idea of the heat "content" of a system. For example, when a solid is heated to its melting point, further heat input causes melting *but without increasing the temperature*. Simply regarding heat energy measured by a rise in temperature as a part of a system's total energy will not give a complete general law.

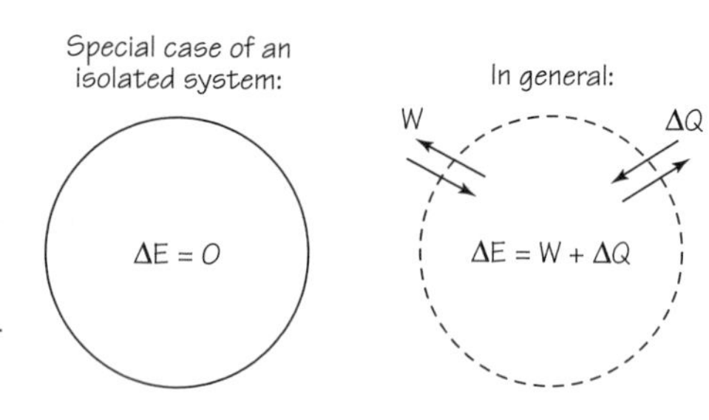

FIGURE 6.18 Diagram of a thermodynamic system.

Instead of "heat," we can use the idea of an *internal energy*—energy in the system that may take forms not directly related to temperature. We can then use the word "heat" to refer only to a *transfer* of energy between a system and its surroundings. (In a similar way, the term *work* is not used to describe something contained in the system. Rather, it describes the transfer of energy from one system to another.)

Even these definitions do not permit a simple statement such as "Heat input to a system increases its internal energy, and work done on a system increases its mechanical energy." Heat input to a system can have effects other than increasing internal energy. In a steam engine, for example, heat input increases the mechanical energy of the piston. Similarly, *work* done on a system can have effects other than increasing mechanical energy. In rubbing your hands together on a cold day, for example, the work you do increases the internal energy of the skin of your hands. In short, a general conservation law of energy must include *both* work *and* heat transfer. Further, it must deal with change in the *total energy* of a system, not with a "mechanical" part and an "internal" part.

In an isolated system, that is, a system that does not exchange energy with its surroundings, the total energy must remain constant. If the system exchanges energy with its surroundings, it can do so in only one of two ways: Work can be done on or by the system, or heat can be passed to or from the system. In the latter case, the change in energy of the system must equal the net energy gained or lost by the surroundings. More precisely, let W stand for the *work* done on or by the system (such as the cylinder in a steam engine). If the work is done by the system, W will be positive; if the work is done on the system, W will be negative. Similarly, let ΔQ represent the net *heat transfer* to or from the system. If the net heat transfer is to the system, ΔQ will be positive; if the net transfer is from the system, ΔQ will be negative.

With these definitions, *the first law of thermodynamics states that the change in the total energy of the system, ΔE, is given by the sum of the work done on or by the system and the net heat transfer to or from the system, or in symbols*

$$\Delta E = W + \Delta Q.$$

This general expression includes as special cases the preliminary versions of the energy-conservation law given earlier in the chapter. If there is no heat transfer at all, then $\Delta Q = 0$, and so $\Delta E = W$. In this case, the change in energy of a system equals the work done on or by it. On the other hand, if work is done neither on nor by a system, then $W = 0$, and $\Delta E = \Delta Q$. Here the change in energy of a system is equal to the net heat transfer.

The equation above is enormously useful. But we still need a description of that part of the total energy of a system called "heat" (or better, "internal" energy). So far, we have seen only that an increase in internal energy is sometimes associated with an increase in temperature. We also mentioned the long-held suspicion that internal energy involves the motion of the "small parts" of bodies. We will take up this problem in detail in the next chapter.

The Second Law

The second law of thermodynamics is a general statement of the limits of the heat engine and is based on Carnot's work. We indicated in Section 6.4 that a reversible engine is the most efficient engine. Any other engine is not as efficient. In order to formulate that idea generally and precisely, a new concept, *entropy*, must be introduced.

The change in entropy of a system, ΔS, is defined as the net heat, ΔQ, gained or lost by the system, divided by the temperature (in Kelvin) of the system, T:

$$\Delta S = \frac{\Delta Q}{T}.$$

This equation defines only changes of entropy, S, rather than the absolute value of entropy. But this is similar to what we encountered in defining potential energy. In both cases what interests us is only the *change*. Once a standard state for the system for which $S = 0$ is chosen, the total entropy for any state of the system can be determined.

We introduced the concept of an ideal, reversible engine in Section 6.4. Such an engine, working in a cycle between hot and cold bodies (as any heat engine does), must have the same entropy at the end of a cycle as it does at the start. This is because, at the end of the cycle, T is back to its initial value, and as much heat and work energy as has been given up in one part of the cycle as has been gained in the rest of the cycle; so ΔQ on the whole during the entire cycle is zero. Since the change of entropy is defined as $\Delta S = \Delta Q/T$, the change in entropy during one cycle is also zero, $\Delta S = 0$.

What about an engine that is not reversible and thus less than ideal, such as an actual steam engine? You know it must be less efficient than a perfectly reversible engine, which would have 100% efficiency. So, for such an engine, the heat transfers must be greater than those for an ideal engine.

At the end of each work cycle, ΔQ within the engine will not be zero but positive, and ΔS, correspondingly, will have a positive value. In short, though the total energy inside and outside the engine will, by the first law, be unchanged, the *entropy* of the system will have *increased*. Note that this will happen again and again as this or any other engine of this sort repeats its work cycle. So the result is that the entropy of the universe will constantly increase while the less-than-ideal engine is running.

We can summarize our results for the change in entropy of the universe resulting from the operation of simple heat engines as follows:

$$\Delta S_{\text{universe}} = 0 \quad \text{(reversible processes)},$$

$$\Delta S_{\text{universe}} > 0 \quad \text{(any other process)}.$$

Although proven here only for these simple heat engines, these results are general ones. In fact, these apply to all thermal processes. For simplicity, these two expressions may be joined together by using the greater than or equal to sign, $\geqq$ or more simply $\geq$:

$$\Delta S_{\text{universe}} \geq 0,$$

where the = sign refers to reversible processes; the > sign refers to any other process. *The last expression is, in fact, a mathematical formulation expressing the second law of thermodynamics.*

Rudolf Clausius, who first formulated the second law in the form given here, paraphrased the two laws of thermodynamics in 1850, as follows: "*The energy of the Universe remains constant, but its entropy seeks to reach a maximum.*"

THE "THIRD" LAW

Some physicists include a third law among the laws of thermodynamics. The third law states that no system can be cooled to absolute zero.

If we include the third law, a lighthearted synopsis of the three laws is:

1. you cannot win; you can only break even;
2. you can break even only at absolute zero;
3. you cannot reach absolute zero.

6.7 FAITH IN THE LAWS OF THERMODYNAMICS

For over a century, the law of conservation of energy has stood as one of the most fundamental laws of science. You will encounter it again and again in this course, in studying electricity and magnetism, the structure of the atom, and nuclear physics. Throughout the other sciences, from chemistry to biology, and throughout engineering studies, the same law applies. Indeed, no other law so clearly brings together the various scientific fields, giving all natural scientists and engineers a common set of concepts.

The law of conservation of energy has been immensely successful. It is so firmly believed that it seems almost impossible that any new discovery could disprove it. However, in some experiments, energy does seem to appear or disappear in a system, without being accounted for by changes in known forms of energy. For instance, as heat is added to a melting ice cube, the temperature of the ice cube does not increase. In such cases, physicists preferred to assume that the added heat takes the form of a kind of energy not yet investigated, rather than to consider seriously the possibility that energy is not conserved. The German philosopher Leibniz once proposed that energy could be dissipated among "the small parts" of bodies. He advanced this idea specifically in order to maintain the principle of conservation of energy in inelastic collisions and frictional processes. Leibniz's faith in energy conservation was justified. Other evidence showed that "internal energy," stored by the motion of submicroscopic particles in the material being experimented on, changed by just the right amount to explain observed changes in external energy, such as the case of a melting ice cube.

Another similar example is the "invention" of the neutrino by the physicist Wolfgang Pauli in 1930. Experiments had suggested that energy disappeared in certain nuclear reactions. Pauli proposed that a previously unsuspected and then undetectable subatomic particle, which Enrico Fermi named the "neutrino," was produced in these reactions. He proposed that the neutrino carried off some of the energy. Physicists accepted the neutrino theory for more than 20 years even though neutrinos by themselves could not be shown to exist. Finally, in 1956, neutrinos were indeed detected, in experiments using the radiation from a nuclear reactor. (The experiment could not have been done in 1930, since no nuclear reactor existed until over a decade later.) Again, faith in the law of conservation of energy turned out to be justified.

The theme of "conservation" is so powerful in science that scientists believe it will always be justified. Any apparent exceptions to the law will sooner or later be understood in a way which does not require us to give up the law. At most, these exceptions may lead us to discover new forms of matter or energy, making the law even more general and powerful.

The French mathematician and philosopher Henri Poincaré expressed this idea in 1903 in his book *Science and Hypothesis:*

> . . . the principle of conservation of energy signifies simply that there is *something* which remains constant. Indeed, no matter what new notions future experiences will give us of the world, we are sure in advance that there will be something which will remain constant, and which we shall be able to call *energy*.

Today, it is agreed that the discoveries of various conservation laws with cosmic reach we have discussed (and others to be mentioned later) were among the most important achievements of science. These laws are powerful and valuable tools of analysis. All of them basically affirm that, whatever happens within a system of interacting bodies, certain measurable quantities will remain constant as long as the system remains isolated.

The list of known conservation laws has grown in recent years. The area of fundamental (or "elementary") particles has yielded much of this new knowledge. Some of the newer laws are imperfectly and incompletely understood. Others are on uncertain ground and are still being argued. Below is a list of conservation laws to date. This list is not complete or eternal, but it does include the conservation laws that make up the working tool-kit of physicists today.

Conservation Laws

1. Linear momentum.
2. Angular momentum (including spin).
3. Energy (including mass).
4. Electric charge.
5. Electron-family number.
6. Muon-family number.
7. Baryon-family number.
8. Strangeness number.
9. Isotopic spin.

Number 3 is a generalized form of the first law of thermodynamics; the inclusion of mass will be explained in the chapter on relativity theory. Numbers 4–9 result from work in nuclear physics, high-energy physics, or elementary or fundamental particle physics.

The Status of the Second Law

The second law of thermodynamics has a status rather different from the conservation laws. It, too, is extremely successful and powerful. It, too, has continued to stand as one of the fundamental laws of science. Unlike the conservation laws or Newton's laws of motion, the second law of thermodynamics is less concerned with giving precisely measurable results than with saying that certain processes or phenomena are impossible. For example, it is impossible to make the entropy of the Universe (or of an isolated system) decrease; it is impossible to make heat flow from a cold body to a hot one without doing work on something; it is impossible to invent a heat engine with efficiency greater than 100%. In other words, the processes involving heat happen in one direction only: The entropy increases; heat flows from hot objects to cold ones. Thus, the second law is connected in some fundamental way with the notion that time proceeds in one direction only. To word it differently, when a movie taken of real events is run backward, what you see cannot, in detail, be found to occur in the real world. For example, while two liquids can quickly mix together by themselves; they cannot spontaneously un-mix themselves. These ideas will be examined in more detail in the next chapter.

The second law states that during reversible processes—such as those involving ideal, frictionless heat engines—the entropy of the Universe will remain constant. However, all other processes, all of which would not be reversible—that is, they are *irreversible*—the entropy of the Universe would increase. In no case will the net entropy of the Universe decrease. For example, as the second law requires, heat will not flow by itself from cold bodies to hot bodies, because that would involve a decrease in entropy. A ball lying on the floor will not somehow gather energy from its surroundings and suddenly leap up. An egg will not unscramble itself. An ocean liner cannot be powered by an engine that takes heat from the ocean water and ejects ice cubes. All these and many other events could occur without violating any principles of Newtonian mechanics, including the law of conservation of energy. But they do not happen; they are "forbidden" by the second law of thermodynamics. (They are "forbidden" in the sense that such things do not happen in nature.)

All familiar processes are to some degree irreversible, and thus contribute to an increase in the entropy of the Universe. As this happens, the usefulness of the heat available for work in engines will decline. Lord Kelvin predicted that eventually all bodies in the Universe would reach the same temperature by exchanging heat with one another. When this happened, it would be impossible to produce any useful work from heat, since work can only be done by means of heat engines when heat flows from a hot body

FIGURE 6.19 An illustration of the "heat death" of the universe, from Camille Flammarion's 1893 book: *La Fin du Monde (The End of the World).*

to a colder body. Finally, the Sun and other stars would cool, all life on Earth would cease, and the Universe would be dead.

This idea of a "heat-death" of the Universe, based on predictions from thermodynamics, aroused some popular interest at the end of the nineteenth century. The idea later appeared in several books, such as H.G. Wells' *The Time Machine*, and in Isaac Asimov's short story "The Last Question."

SOME NEW IDEAS AND CONCEPTS

absolute temperature scale
absolute zero
caloric fluid
calorie
Calorie
Carnot's finding
condenser
cycle
cylinder
efficiency
entropy
feedback
governor
heat-death of the Universe
law of conservation of energy (LCE)
mechanical equivalent of heat
Nature Philosophy (*Naturphilosophie*)
power
reversible engine
specific heat
thermodynamics, first law
thermodynamics, second law

SOME IMPORTANT EQUATIONS

$$Q = mc\,\Delta T,$$

$$\text{eff} = \frac{\text{useful } E_{\text{out}}}{E_{\text{in}}},$$

$$P = \frac{E}{t},$$

$$\Delta E = W + \Delta Q,$$

$$\Delta S_{\text{universe}} \geq 0,$$

$$\Delta S = \frac{\Delta Q}{T}.$$

DEFINITIONS OF IMPORTANT UNITS

$$1\ \text{W} = 1\ \text{J/s},$$

$$T\,(\text{K}) = t\,(^\circ\text{C}) + 273.$$

STUDY GUIDE QUESTIONS

6.1 Heat as a Form of Energy

1. When a book slides to a stop on the horizontal rough surface of a table:
   (a) the kinetic energy of the book is transformed into potential energy;
   (b) heat is transformed into mechanical energy;
   (c) the kinetic energy of the book is transformed into heat energy;
   (d) the momentum of the book is conserved.
2. The kilocalorie is:
   (a) a unit of temperature;
   (b) a unit of energy;
   (c) 1 kg of water at 1°C.
3. In Joule's paddle-wheel experiment, was all the change of gravitational potential energy used to heat the water?
4. How are heat and temperature related to each other?
5. How are heat and mechanical energy related to each other?

6.2 The Steam Engine and the Industrial Revolution

1. Describe in your own words, possibly with a drawing, the operation of Watt's steam engine.
2. The purpose of the separate condenser in Watt's steam engine is to:
   (a) save the water so it can be used again;
   (b) save fuel by not having to reheat the cylinder after cooling;
   (c) keep the steam pressure as low as possible;
   (d) make the engine more compact.
3. The history of the steam engine suggests that the social and economic effects of technology are:
   (a) always beneficial to everyone;
   (b) mostly undesirable;
   (c) unimportant one way or another;
   (d) none of the above.

6.3 Power and Efficiency of Engines

1. A strong horse, working steadily, is able to lift a 75-kg mass at a speed of 1 m/s. What is the power output of the horse in watts?
2. During a certain period, a home consumes 1 kW-hr of energy. How many joules is this? How long could this amount of energy light a 100-W light bulb?
3. The efficiency of a heat engine is the ratio of:
   (a) the work output to the heat input;
   (b) the work output to the heat output;
   (c) the heat output to the heat input.
4. What would be the efficiency, in percent, of a steam engine that took in the equivalent of 3000 J of heat and produced 1200 J of work?

6.4 Carnot and the Beginnings of Thermodynamics

1. What does Carnot's finding say about the efficiency of heat engines?
2. A heat engine is most efficient when it works between objects that have:
   (a) a large temperature difference;
   (b) a small temperature difference;
   (c) a large size.
3. In what way is a refrigerator or an air conditioner also a heat engine?
4. How are the Kelvin and Celsius temperature scales related to each other? What would be the temperatures of freezing water, boiling water, and absolute zero on these two scales?

6.5 Arriving at a General Conservation Law

1. The significance of German Nature Philosophy in the history of science was that it:
   (a) was the most extreme form of the mechanistic viewpoint;
   (b) was a reaction against excessive speculation;

(c) stimulated speculation about the unity of natural phenomena;
(d) delayed progress in science by opposing Newtonian mechanics.

2. Discoveries in electricity and magnetism early in the nineteenth century contributed to the discovery of the law of conservation of energy because:
   (a) they attracted attention to the transformation of energy from one form to another;
   (b) they made it possible to produce more energy at less cost;
   (c) they revealed what happened to the energy that was apparently lost in steam engines;
   (d) they made it possible to transmit energy over long distances.
3. The development of steam engines helped the discovery of the law of conservation of energy because:
   (a) steam engines produced a large amount of energy;
   (b) the caloric theory could not explain how steam engines worked;
   (c) the precise idea of work was developed to rate steam engines;
   (d) the internal energy of a steam engine was always found to be conserved.

6.6 The Two Laws of Thermodynamics

1. Define the meaning of the symbols ΔE, ΔQ, W, and ΔS for a system.
2. State the first law of thermodynamics in your own words.
3. What is the difference between the first law and the law for energy conservation?
4. The value of ΔQ can be positive, negative, or zero. Using an example, such as the steam engine cylinder, explain what is happening in each case and exactly why ΔQ is positive, negative, or zero.
5. Give an example of an imaginable situation that would violate the law regarding entropy increase.
6. The first law of thermodynamics is:
   (a) true only for steam engines;
   (b) true only when there is no friction;
   (c) a completely general statement of conservation of energy;
   (d) the only way to express conservation of energy.
7. What two ways are there for changing the total energy of a system?
8. The second law of thermodynamics says that the entropy of the Universe:
   (a) cannot increase;
   (b) cannot decrease;
   (c) must increase;
   (d) must decrease.
9. The presumed "heat-death of the Universe" refers to a state in which:
   (a) all mechanical energy has been transformed into heat energy;
   (b) all heat energy has been transformed into other forms of energy;
   (c) the temperature of the Universe decreases to absolute zero;
   (d) the supply of coal and oil has been used up.

6.7 Faith in the Laws of Thermodynamics

1. Give some examples of situations in which energy seems not to be conserved.
2. How was the law of energy conservation confirmed in the situations in 1?
3. How does the status of the second law of thermodynamics differ from that of the conservation laws?
4. How might the second law of thermodynamics be connected with the forward motion of time?
5. What is meant by the "heat-death" of the universe?

DISCOVERY QUESTIONS

1. The introduction of steam technology transformed the economic and social lives of people in the industrial world. In what ways has the introduction of computer technology changed our economic and social lives?
2. Give some examples of the first and second laws of thermodynamics in actual cases.
3. Give an example of a hypothetical situation in which the first law of thermodynamics would not be violated but the second law is.
4. Give an example of a hypothetical situation in which the second law of thermodynamics would not be violated but the first law is.
5. What is the difference between work and power?
6. Suppose the temperatures of the "hot" and "cold" bodies in a heat engine were the same. What would be the efficiency of a heat engine that attempted to operate between these two bodies? What would be the work output?
7. The introduction of the steam engine had both positive and negative effects, although all of these effects were not predicted at the time.
   (a) List several *actual* effects on society, both beneficial and undesirable ones, of the steam engine and of the gasoline internal combustion engine.
   (b) List several *predicted* effects of nuclear power and of solar power by its inventors and the general public at the time of its invention. List both beneficial and undesirable ones.
8. Explain why all ideal reversible engines have the same efficiency and why this efficiency is the maximum possible for an engine. What is an ideal engine? What is a reversible engine?
9. Assuming that no real engine can be perfectly reversible, why does the formula for the maximum efficiency of an engine imply that absolute zero can never be reached in practice?
10. Any of the terms in the equation $\Delta E = W + \Delta Q$ can have negative values.
    (a) What would be true of a system for which:
        (1) ΔE is negative?
        (2) ΔQ is negative?
        (3) W is negative?

(b) Which terms would be negative for the following systems?
   (1) a person digging a ditch;
   (2) a car battery while starting a car;
   (3) an electric light bulb just after it is turned on;
   (4) an electric light bulb an hour after it is turned on;
   (5) a running refrigerator;
   (6) an exploding firecracker.

11. In each of the following, trace the chain of energy transformations from the Sun to the energy in its final form.
    (a) A pot of water is boiled on an electric stove.
    (b) An automobile accelerates from rest on a level road, climbs a hill at constant speed, and comes to a stop at a traffic light.
    (c) A windmill pumps water out of a flooded field.
12. Why is it that despite centuries of attempts no perpetual motion machine has been constructed? Is it likely that one will be in the future?
13. Why can a block of ice at 0°C and water at 0°C coexist in an insulated bucket without changes either way in the amount of ice?
14. Why can an ocean liner not run all its engines simply by drawing heat from the ocean, thereby making the ocean a bit colder?
15. Since there is a tendency for heat to flow from hot to cold, will the Universe eventually reach absolute zero?

Quantitative

1. You walk up a flight of stairs to the second floor, which is 10 m above the first floor, in 15 s. At the same time, a friend runs up the stairs in 5 s. Both of you weigh 80 N. Find the amount of work and the amount of power that each of you exerts in getting up the stairs.
2. A skier of 70-kg mass experiences a pull on a ski lift from an engine transmitting 140 W to the cable. Neglecting friction, how high can the engine pull the skier in 500 s?
3. One hundred joules (100 J) of heat is put into two engines. Engine A can lift 5 N a distance of 10 m in 10 s. Engine B pulls with a force of 2 N for 5 s a distance of 20 m. Calculate the efficiency and power of each engine.
4. While traveling in Switzerland on his honeymoon, James Prescott Joule attempted to measure the difference in temperature of the water at the top and at the bottom of a waterfall. Assuming that the amount of heat produced in the water when it is stopped at the bottom is equal to the decrease in its gravitational potential energy, calculate roughly the temperature difference you would expect to observe between the top and bottom of Niagara Falls, which is 50 m in height. Does it matter how much water goes down the waterfall?
5. If you place a hot body and a cold one in thermal contact, heat will flow from the first to the second spontaneously. Suppose an amount of heat Q flows from a body at temperature T_1 to a body at T_2. What is the entropy change of the universe?

6. An ice cube (10 g) at 0°C melts in a glass of water (100 g) at a temperature just above 0°C. Melting the ice requires 3.4×106 J of energy (which comes from cooling the water). Neglecting temperature changes, what is the entropy change of the ice? of the water? of the universe?
7. In Section 6.3 you saw that a 100-W light bulb burning for 10 hr consumes 1 kWhr of energy. The production of 1 kWhr of electricity to run the light bulb requires the burning of about 0.8 lb of coal in an electric generating plant.
   (a) If the average citizen consumes energy at the rate of 1.5 kJ/s, how many pounds of coal do you consume per year? during an expected life span of 75 yr?
   (b) The United States population is about 280 million. For their energy needs, how many pounds of coal would they consume in 1 day?
8. On average, a person emits as much heat from his/her body as a 100-watt bulb. How come?

CHAPTER 7

Heat—A Matter of Motion

A. THE KINETIC THEORY

7.1 AN IDEAL GAS

During the 1840s, many scientists recognized that heat is not a substance but a form of energy that can be converted into other forms. James Prescott Joule and Rudolf Clausius went a step further. Heat can produce mechanical energy, and mechanical energy can produce heat; therefore, they reasoned, the "heat energy" of a substance is simply the kinetic energy of its atoms and molecules. This idea, which forms the basis of the *kinetic-molecular theory of heat*, is largely correct.

However, the idea of atoms and molecules was not completely accepted in the nineteenth century. (Molecules, as you know, are the smallest pieces of a substance. They may themselves be composed of atoms of simpler substances.) If such small bits of matter really existed, they would be too small to observe even under the most powerful microscopes. Since scientists could not observe molecules, they could not confirm directly the hypothesis that heat is molecular kinetic energy. So they resorted to the *indirect* way of checking the plausibility of the hypothesis, which is always useful: They derived from this hypothesis predictions about the behavior of measurably large samples of matter, and then tested these predictions by experiment.

For reasons that will be explained, it is easiest to test such hypotheses by observing the properties of gases. The resulting development of the kinetic theory of gases in the nineteenth century led to one of the last major triumphs of Newtonian mechanics. This chapter deals mainly with the kinetic theory as applied to gases. We start with the observed properties of a simple gas; then, in the following sections, we will attempt to account for these properties and to understand origins of the laws of thermodynamics on the basis of the kinetic theory.

One of the most easily measured characteristics of a confined gas is its pressure. Experience with balloons and tires makes the idea of air pressure seem obvious. But there are important subtleties. One involves the pressure exerted by the air around us.

Galileo, in his book on mechanics, *Two New Sciences* (1638), noted that a lift-type pump cannot raise water more than the equivalent of 10 m. This fact was well known, and such pumps were widely used to obtain drinking water from wells and to remove water from mines. Already you have seen one important consequence of this limited ability of pumps to lift water out of deep mines: the initial stimulus for the development of steam engines. But why only 10 m? Why did the lift pumps work at all?

Air Pressure

These puzzles were solved as a result of experiments by Torricelli (a student of Galileo), as well as by Guericke, Pascal, and Boyle. By 1660, it was fairly clear that the operation of a "lift" pump depends on the pressure of the air. By removing some air above the water in the pump, it merely reduces the pressure at the top of the water in the pipe. It is the pressure exerted by the atmosphere on the pool of water below which forces water up the pipe. A good pump can reduce the pressure at the top of the pipe to nearly zero. Then the atmospheric pressure can force water up above the pool. But atmospheric pressure at sea level is not great enough to support a column of water any higher than 10 m. Mercury is almost 14 times as dense as water. Thus, ordinary pressure on a pool of mercury can support a column only one-fourteenth as high, or about 0.76 m (76 cm). This is a convenient height of an instrument for measuring atmospheric pressure. Therefore, much of the seventeenth-century research on air pressure was

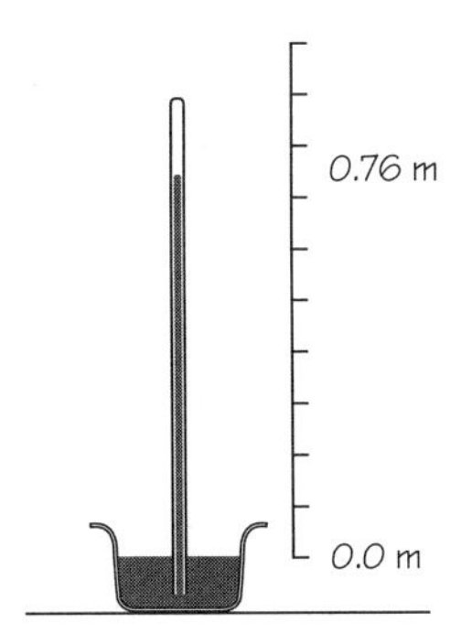

FIGURE 7.1 Torricelli's barometer is a glass tube standing in a pool of mercury. The topmost part of the tube is empty of air. The air pressure on the pool supports the weight of the column of mercury in the tube up to a height of about 0.76 m. The unit of pressure in the metric system is N/m^3, which has been given the name "pascal" (symbol: Pa).

done with a column of mercury, a mercury *barometer*. Torricelli designed the first of these barometers.

The height of the mercury column that can be supported by air pressure does not depend on the diameter of the tube; that is, it depends not on the total amount of mercury, but only on its height. This may seem strange at first. To understand it, you must consider the difference between *pressure* and *force*. *Pressure is defined as the magnitude of the force acting perpendicularly on a surface divided by the area of that surface:*

$$P = \frac{F_{\perp}}{A} \quad \text{(in units of N/m}^2\text{).}$$

Thus, a large force may produce only a small pressure if it is spread over a large area. For example, you can walk on snow without sinking in it if you wear snowshoes. On the other hand, a small force can produce a very large pressure if it is concentrated on a small area. Women's spike-heel shoes have dented many a wooden floor or carpet. The pressure exerted by a spike heel can be greater than that under an elephant's foot!

In short, the pressure measurement is not affected by the cross-sectional area of the barometer tube or by the total weight of the column it supports, because pressure is the ratio of the two. Twice the weight owing to twice the size of the tube still provides the same value for the ratio that defines pressure.

The Relationship Between Pressure and Volume

In 1661, two English scientists, Richard Towneley and Henry Power, discovered an important basic relation. They found that—e.g., in a thin balloon—*the pressure exerted by a gas is inversely proportional to the volume occupied by that gas*. Doubling the pressure (by letting in more gas) will dou-

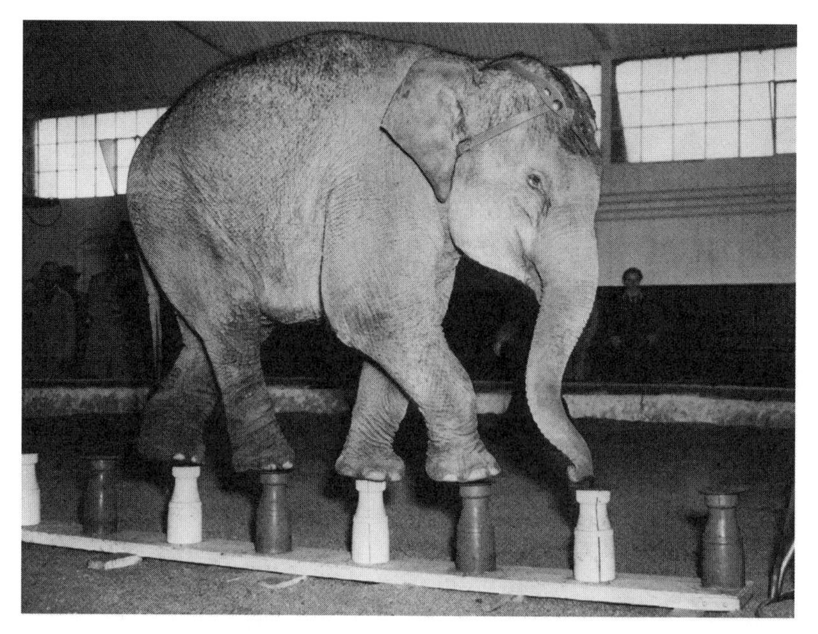

FIGURE 7.2 Because the force acts on a very small surface, the pressure under a thin, high heel is greater than under an elephant's foot (which acts on a larger surface).

ble the volume of the balloon. Using P for pressure and V for volume, this relationship is

$$P \propto \frac{1}{V}$$

or

$$P = \frac{a}{V}$$

or

$$PV = a,$$

where a is some constant. For example, if the volume of a given quantity of air in a balloon is halved (say, by compressing it), the pressure exerted by the gas inside doubles. On the other hand, if you double the volume of the closed container with a certain amount of air inside, the pressure inside is halved. Robert Boyle confirmed this relation by extensive experiments. It is an empirical rule, now generally known as *Boyle's law*. However, the law holds true exactly only under special conditions, as given below.

The Effect of Temperature on Gas Pressure and Volume

Boyle recognized that if the temperature of a gas changes during an experiment, the relation $P = a/V$ no longer applies. For example, the pressure exerted by a gas in a closed container of fixed size increases if the gas is heated, even though its volume remains constant. However, if the temperature of a gas is held constant, then Boyle's law does apply. Thus we modify the rule as follows:

$$P = \frac{a}{V} \quad \text{if } T \text{ is constant.} \tag{a}$$

Many scientists throughout the eighteenth century also investigated how, say in a thin balloon, gases expand when heat is supplied, even though the pressure remains the same. Eventually, evidence for a surprisingly simple general law appeared. The French chemist Joseph-Louis Gay-Lussac (1778–1850) found that all the gases he studied (air, oxygen, hydrogen, carbon dioxide, etc.) changed their volume in the same way. *If the pressure remained constant, then the change in volume was proportional to the change in temperature*. This may be expressed in symbols:

$$\Delta V \propto \Delta T, \quad \text{if } P \text{ is constant.} \tag{b}$$

On the other hand, *if the volume was kept constant (say, in using a rigid container), the change in pressure of the gas inside was proportional to the change in temperature:*

$$\Delta P \propto \Delta T \quad \text{if } V \text{ is constant.} \tag{c}$$

The experimental data obtained by Boyle, Gay-Lussac, and many other scientists are expressed in the three proportionalities, (a), (b), and (c). These relate the three main characteristics of a fixed amount of a gas—the pressure, volume, and temperature, each measured from zero—when any one variable is held constant. Introducing a new constant, k, these three proportionalities may be joined together into a single, general equation known as the *ideal gas law*.

$$\boxed{PV = kT}$$

This equation is one of the most important discoveries about gases, but in use one must be careful about the units! The proportionality constant k depends on the kind of gas; T, the temperature of the gas, has to be given on the absolute, or Kelvin, scale, where, as defined earlier,

$$T\,(\text{K}) = t\,(^\circ\text{C}) + 273.15.$$

(T is temperature in the same units as used for the definition of entropy given earlier in Chapter 6: $S = \Delta Q/T$.) The pressure P is always the *total* pressure (in units of N/m^2, which is given the name Pascal, abbreviation Pa), *including* the so-called ambient pressure of the atmosphere. Thus a car tire blown up to 32 lb/in^2 above atmospheric pressure has an actual gas pressure inside of 32 lb/in^2 + 15 lb/in^2 = 47 lb/in^2, or in mks units 2.2×10^5 Pa + 1.0×10^5 Pa = 3.2×10^5 Pa.

The equation relating P, V, and T is called the *ideal* gas law because it is not completely accurate for real gases at very low pressures. It also does not apply when pressure is so high, or temperature so low, that the gas

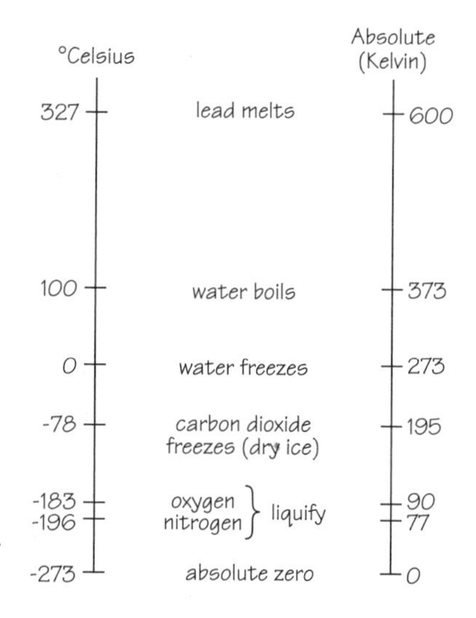

FIGURE 7.3 Comparison of the Celsius and Absolute temperature scales.

FIGURE 7.4 Lord Kelvin (William Thomson) (1824–1907).

nearly changes to a liquid. Thus, it is not a law of physics in the same sense as the majestic law of conservation of energy, which is valid under all circumstances. Rather, it simply gives an experimentally based and useful but approximate summary of the observed properties of real gases.

Consider what would happen if we tried to lower the temperature of the gas to absolute zero; that is, $T = 0$ K (or, in Celsius, $t = -273.15$°C). In this extreme case, the entire factor involving temperature would be zero. According to the ideal gas law, the PV term must also fall to zero at this temperature. At constant pressure, the volume would shrink to zero. In fact, all real gases become liquid before that temperature is reached. Both experiment and thermodynamic theory indicate that it is impossible actually to cool anything—gas, liquid, or solid—down to exactly this temperature. However, a series of cooling operations can produce temperatures closely approaching this limit.

In view of the unique meaning of the lowest temperature for a gas obeying the ideal gas law, Lord Kelvin (William Thomson) proposed the *absolute temperature scale* and put its zero at -273.15°C. This is why the absolute scale is sometimes called the Kelvin scale, and why temperatures on this scale are measured in kelvins.

The ideal gas law, $PV = kT$, summarizes *experimental facts* about gases. The kinetic theory of gases offers a *theoretical* explanation for these facts. To prepare for that, we need to develop a kinetic model of a gas.

7.2 A MODEL FOR THE GASEOUS STATE

What are the differences between a gas and a liquid or solid? You know by observation that, if not compressed, liquids and solids have definite volume. Even if their shapes change, they still take up the same amount of space. A gas, on the other hand, will spontaneously expand to fill any container (such as a room). If not confined, it will leak out and spread in all directions. Gases have low densities compared to those of liquids and solids, typically about 1000 times smaller. Therefore gas molecules are usually relatively far apart from one another. In the model of a gas we are constructing here, we can reasonably assume that the forces between molecules act only over very short distances. In other words, gas molecules are considered to be moving freely most of the time. In liquids, the molecules are closer together; forces act among them continually, and keep them from flying apart. In solids, the molecules are usually even closer together, and the forces between them keep them in a definite orderly arrangement.

> Gases can be confined without a container. A star, for example, is a mass of gas confined by gravitational force. Another example is the Earth's atmosphere.

Our initial model of a gas is thus very simple, following Newton's advice to start with the simplest hypotheses. We will assume that the molecules behave like extremely small balls or marbles; they are tiny spheres or clumps of spheres that exert no force at all on each other except when they happen to make contact. Moreover, all the collisions of these spheres are assumed to be perfectly elastic. Thus, the total kinetic energy of two spheres is the same before and after they collide.

Note that the word "model" is used in two different senses in science. Earlier we mentioned the model of Newcomen's engine which James Watt was given to repair. That was a *working model*. It actually did function, although it was much smaller than the original engine, and contained some parts made of different materials. Now we are discussing a *theoretical model* of a gas. This model exists only in the imagination. Like the points, lines, triangles, and spheres studied in geometry, this theoretical model can be discussed mathematically. The results of such a discussion is intended to understand the real world although, of course, the model will eventually have to be tested to see if it approximates reality.

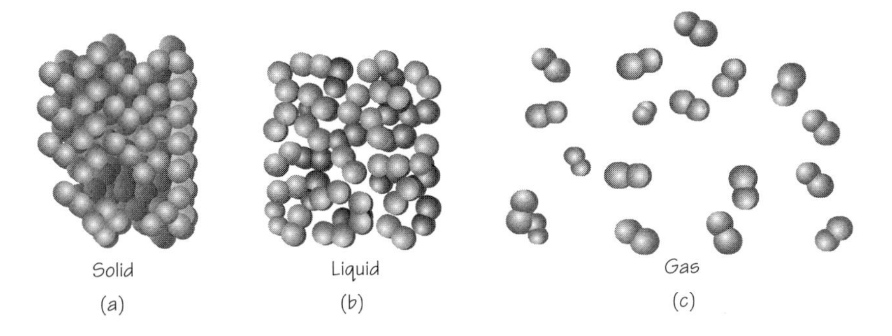

FIGURE 7.5 A very simplified "model" of the three states of matter: (a) solid, (b) liquid, and (c) gas.

Our theoretical model represents the gas as consisting of *a large number of very small particles in rapid, disordered motion.* "*A large number*" means something like a billion billion (10^{18}) or more particles in a sample as small as a bubble in a soft drink. "*Very small*" means a diameter about a hundred-millionth of a centimeter (10^{-8} cm). "*Rapid motion*" means an average speed at normal temperatures of a few hundred meters per second.

What is meant by "*disordered*" motion? Nineteenth-century kinetic theorists assumed that each individual molecule moved in a definite way, determined by Newton's laws of motion. Of course, in practice it is impossible to follow billions upon billions of particles at the same time. They move in all directions, and each particle changes its direction and speed during collisions with other particles or with the wall of the container. Therefore, we cannot make a definite prediction of the motion of any one *individual* particle. Instead, we must be content with describing the *average* behavior of large collections of particles. From moment to moment, each individual molecule behaves according to the laws of motion. But it is easier to describe the *average* behavior, and to assume complete ignorance about any *individual* motions.

The word "gas" was originally derived from the Greek word *chaos*; it was first used by the Belgian chemist Jan Baptista van Helmont (1580–1644).

To see why this is so, consider the results of flipping a large number of coins all at once. If you assume the coins behave randomly, you can confidently predict that flipping a million coins will give approximately 50% heads and 50% tails. The same principle applies to gas molecules bouncing around in a container. You can safely assume, for example, that about as many are moving in one direction as in another. Furthermore, at a given moment the same number of molecules is equally likely to be found in any one cubic centimeter of space inside the container as in any other. "Disordered," then, means that velocities and positions are distributed *randomly*. Each molecule is just as likely to be moving to the right as to the left (or in any other direction). It is just as likely to be near the center as near the edge (or any other position).

7.3 THE SPEEDS OF MOLECULES

The basic idea of the kinetic theory of matter is that heat energy is related to the kinetic energy of moving molecules. As we shall see, this is right. This idea had been frequently suggested in the past. However, many difficulties stood in the way of its general acceptance. Some of these difficulties are well worth mentioning. They show that not all good ideas in science (any more than outside science) are immediately successful.

In 1738, the Swiss mathematician Daniel Bernoulli showed how a kinetic model could explain a well-known property of gases. This property is described by a variant of Boyle's law: As long as the temperature does not change, the pressure of a gas is proportional to its density.

$$P \propto D \quad \text{if } T \text{ is constant.}$$

Here density D is defined as the amount of mass (m) per unit volume (V) of the gas, $D = m/V$.

FIGURE 7.6 Daniel Bernoulli (1700–1782).

Bernoulli assumed that the pressure of a gas is simply a result of the impacts of individual molecules striking the wall of the container. If the density of the gas were twice as great, there would be twice as many molecules per cubic centimeter. Thus, Bernoulli said, there would be twice as many molecules striking the wall per second and hence twice the pressure. Bernoulli's proposal seems to have been the first step toward the modern kinetic theory of gases. Yet it was generally ignored by other scientists in the eighteenth century. One reason for this was that Newton had proposed a different theory in his *Principia* (1687). Newton showed that Boyle's law *could* be explained by a model in which particles *at rest* exert forces that repel neighboring particles. Newton did not claim that he had proved that gases *really are* composed of such fixed, mutually repelling particles. But most scientists, impressed by Newton's discoveries, simply assumed that his treatment of gas pressure was also right. (It was not.)

The kinetic theory of gases was proposed again in 1820 by English physicist John Herapath. Herapath, on his own, rediscovered Bernoulli's findings on the relations between pressure and density, or volume, of a gas and the speeds of the particles. Herapath's work was also ignored by most other scientists.

James Prescott Joule, however, did see the value of Herapath's work. In 1848, he read a paper to the Manchester Literary and Philosophical Society in which he tried to revive the kinetic theory. This paper, too, was ignored by other scientists. For one thing, physicists do not generally look in the publications of a "literary and philosophical society" for scientifically important papers. However, evidence for the equivalence of heat and mechanical energy continued to mount. Several other physicists independently worked out the consequences of the hypothesis that the heat energy in a gas is explained and given by the kinetic energy of its molecules. Rudolf Clausius in Germany published a paper in 1856 on "The Nature of the Motion We Call Heat." This paper established the basic principles of kinetic theory essentially in the form accepted today. Soon afterward, James Clerk Maxwell in Britain and Ludwig Boltzmann in Austria set forth the full mathematical details of the theory.

Maxwell's Velocity Distribution

It did not seem likely that at a given moment all molecules in a gas would have the same speed. In 1859, Maxwell applied the mathematics of probability to this problem. He suggested that the speeds of molecules in a gas are distributed over all possible values. Most molecules have speeds not very far from the average speed. Some have much lower speeds and some much higher speeds.

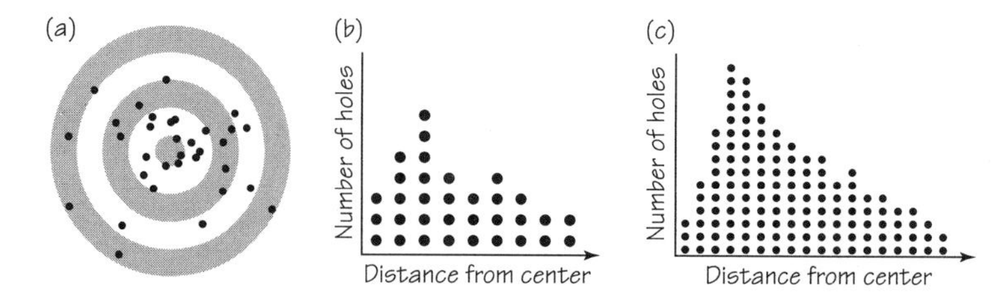

FIGURE 7.7 Target practice experiment: (a) scatter of holes in target; (b) graph showing number of holes in each ring from the bull's-eye; (c) graph showing that the distribution becomes smooth for a very large number of shots and for very narrow rings.

A simple example helps to understand Maxwell's distribution of molecular speeds. Suppose a person shoots arrows at a practice target. Some arrows will probably hit the bull's-eye. Others will miss by smaller or larger amounts, as shown in (a) in the sketch above. The number of arrows scattered at various distances from the center are counted. A graph of the results is shown in (b). This graph shows the distribution of the holes made by arrow hits for one set of shots. If you plot the distribution of hits for a very large number of shots, you will get a distribution like the one in (c). For still larger numbers of arrow shots the *spread* between the holes you see in (c) will become smaller, ultimately too small to be noticed. By analogy, the number of molecules in a gas being very large indeed, a graph showing the distribution of molecular *speeds* is smooth at any scale that can be drawn.

The actual shape of the curve shown in (c) was determined by many things about the bow, the arrows, the person, and so on. Other processes give rise to other shapes of curves. The speeds of the molecules in a gas are determined by the collisions they have with each other. Maxwell used

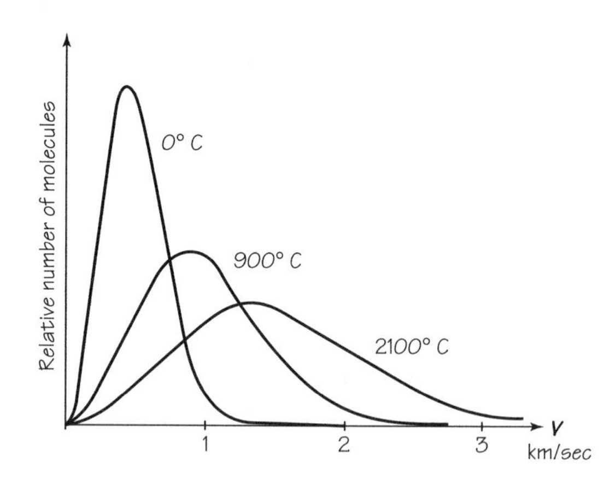

FIGURE 7.8 Maxwell's distribution of speeds in gases at different temperatures.

a clever mathematical argument to deduce what the distribution of molecular speeds should be.

Maxwell's proposed distribution for molecular speeds in a gas is shown in the diagram in graphical form for three different temperatures. For a gas at any given temperature, the "tail" of each curve is much longer on the right (high speeds) than on the left (low speeds). As the temperature increases, the peak of the curve shifts to higher speeds, and the speed distribution becomes more broadly spread out.

All this had to be tested, to see if the theoretical model is of real use, or not. The best test was a direct one, by experiments in the 1920s. Otto Stern in Germany devised an ingenious method for measuring the speeds in a beam of molecules. Stern and others found that molecular speeds are indeed distributed according to Maxwell's theory. This gave a most direct proof of the kinetic-molecular model of a gas. (See next page.)

7.4 THE SIZES OF MOLECULES

On the way to that proof, there were reasonably skeptical questions to wrestle with. Was it reasonable to suppose that gases consist of molecules moving at speeds up to several hundred meters per second at room temperature? If that model were correct, gases should mix with each other very rapidly. But anyone who has studied chemistry knows that they do not. Suppose someone opens a bottle of perfume or a container containing ammonia gas in a corner of the classroom. Several minutes may pass before the odor is noticed at the other end. But according to Maxwell's speed distribution, each of the gas molecules should have crossed the room hundreds of times by then. Therefore, something must be wrong with the kinetic-theory model.

Rudolf Clausius recognized this as a valid objection to his own version of the kinetic theory. His 1856 paper had assumed that the particles are so small that they can be treated like mathematical points. If this were true, particles would almost never collide with one another. However, the observed *slowness* of diffusion and mixing convinced Clausius to change his model. He thought it likely that the molecules of a gas are not vanishingly small, but of a finite size. Particles of finite size moving very rapidly would often collide with one another. An individual molecule might have an instantaneous speed of several hundred meters per second, but it changes its direction of motion every time it collides with another molecule. The more often it collides with other molecules, the less likely it is to move very far in any one direction. How often collisions occur depends on their size and

DIRECT MEASUREMENT OF MOLECULAR SPEEDS

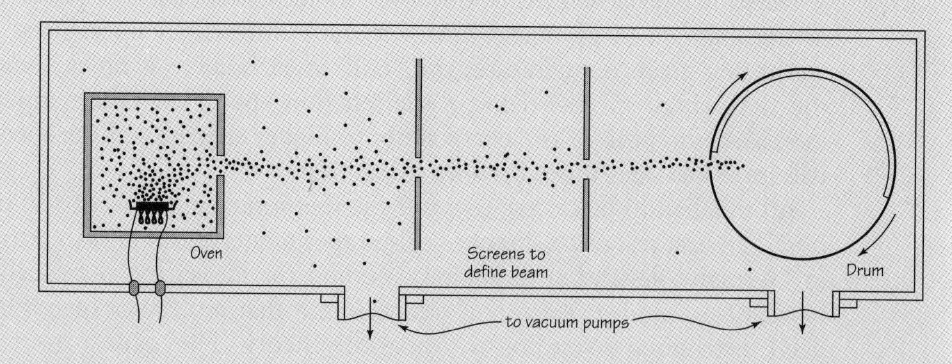

A narrow beam of molecules is formed by letting molecules of a hot gas pass through a series of slits. In order to keep the beam from spreading out, collisions with randomly moving molecules must be avoided. Therefore, the source of gas and the slits are housed in a highly evacuated chamber. The molecules are then allowed to pass through a slit in the side of a cylindrical drum that can be spun very rapidly. The general scheme is shown in the drawing above.

As the drum rotates, the slit moves out of the beam of molecules. No more molecules can enter until the drum has rotated through a whole revolution. Meanwhile, the molecules in the drum continue moving to the right, some moving quickly and some moving slowly.

Fastened to the inside of the drum is a sensitive film that acts as a detector. Any molecule striking the film leaves a mark. The faster molecules strike the film first, before the drum

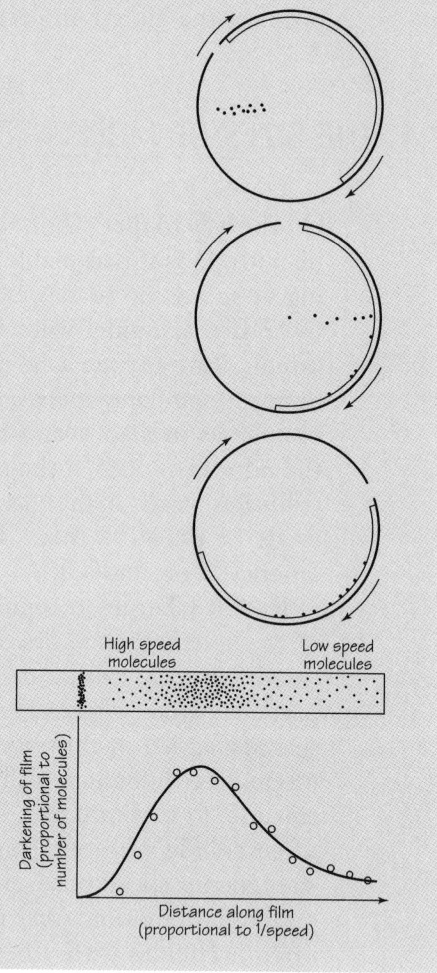

FIGURE 7.9 Schematic presentation of Otto Stern's experiment on the distribution of the speeds of gas molecules.

has rotated very far. The slower molecules hit the film later, after the drum has rotated farther. In general, molecules of different speeds strike different parts of the film.

The darkness of the film at any point is proportional to the number of molecules that hit it there. Measurement of the darkening of the film shows the relative distribution of molecular speeds. The speckled strip represents the unrolled film, showing the impact position of molecules over many revolutions of the drum. The heavy band indicates where the beam struck the film before the drum started rotating. (It also marks the place to which infinitely fast molecules would get once the drum was rotating.)

A comparison of some experimental results with those predicted from theory is shown in the graph. The dots show the experimental results, and the solid line represents the predictions from the kinetic theory.

on how crowded the molecules are. For most purposes, you can think of molecules as being relatively far apart and of very small size. But they are just large enough and crowded enough to get in one another's way. Realizing this, Clausius was able to modify his model to explain why gases mix so slowly, a process known as *diffusion*.

Clausius now was faced with a problem that plagues every theoretical physicist. If a simple model is modified to explain better the observed properties, it becomes more complicated. Some plausible adjustment or approximation may be necessary in order to make any predictions from the model. If the predictions disagree with experimental data, is this because of a flaw in the model or a calculation error introduced by the approxima-

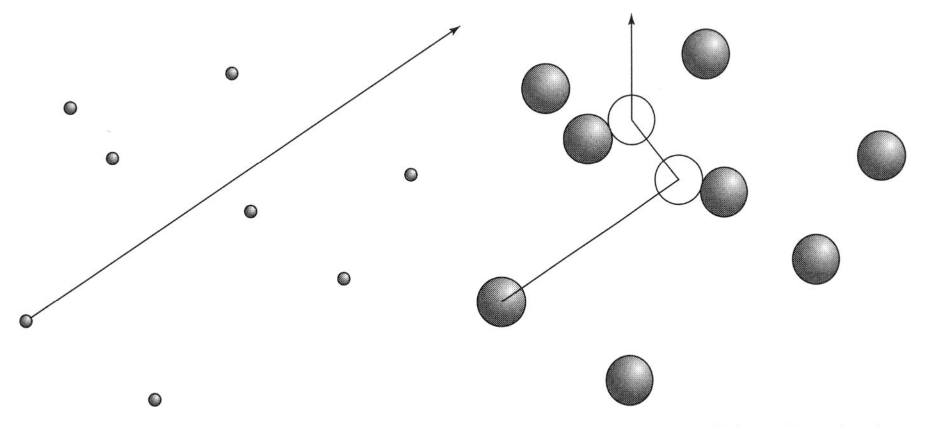

FIGURE 7.10 The larger the molecules, the more likely they are to collide with each other, thus delaying the progress of a molecule through the gas.

tions? The development of a theory often involves a compromise between adequate explanation of the data and mathematical convenience.

Nonetheless, it soon became clear that the new model was a great improvement over the old one. It turned out that certain other properties of gases also depend on the size of the molecules. By combining data on several such properties, it was possible to work backward and find fairly reliable values for molecular sizes. Here, only the result of these calculations is reported. Typically, the diameter of gas molecules came out to be of the order of 10^{-10} to 10^{-9} m. This is not far from the modern values—an amazingly good result. After all, no one previously had known whether a molecule was many times smaller or bigger than that. In fact, as Lord Kelvin remarked:

> The idea of an atom has been so constantly associated with incredible assumptions of infinite strength, absolute rigidity, mystical actions at a distance and indivisibility, that chemists and many other reasonable naturalists of modern times, losing all patience with it, have dismissed it to the realms of metaphysics, and made it smaller than "anything we can conceive."

Kelvin showed that other methods could also be used to estimate the size of atoms and molecules. None of these methods gave results as reliable as did the kinetic theory. But it was encouraging that they all led to the same order of magnitude (power of ten).

One early attempt to obtain the order of magnitude of the size of a molecule was made by Benjamin Franklin. Dropping a spoonful of oil on a pond, he was able to estimate the width of a single molecule by assuming that the oil formed a slick one molecule thick, then comparing the original volume of the oil to the area covered by the oil on the pond. A similar experiment is provided in the *Student Guide*, "Avogadro's Number and the Size and Mass of a Molecule."

B. APPLYING THE KINETIC THEORY

7.5 KINETIC-THEORY EXPLANATION OF THE IDEAL GAS LAW

As we now know, according to the kinetic theory, the pressure of a gas results from the continual impacts of gas particles against the container wall. This explains why pressure is inversely proportional to the volume and di-

If the pressure were kept constant, then according to the ideal gas law, the *volume* of a sample of gas would shrink to zero at −273°C.

rectly proportional to density: the smaller the volume or the greater the density, the greater the number of particles colliding with the wall.

But pressure also depends on the *speed* of the individual particles (hence, on the kinetic energy, $\frac{1}{2}mv^2$ of them). This speed determines the force exerted on the wall during each impact and the frequency of the impacts. If the collisions with the wall are perfectly elastic, the law of conservation of momentum will describe the results of the impact. (The detailed reasoning for this procedure is worked out in the *Student Guide*.) An atom bouncing off a wall undergoes a change of momentum. As you know from Section 5.4, whenever there is a change of momentum, there must be a force exerted on the object, an atom in this case. At the same time, there must be a reaction force on the wall that repelled the atom, which contributes to what we observe as pressure. This is a beautifully simple application of Newtonian mechanics.

Applying Newtonian mechanics to the kinetic-molecular model of gases, the result of the actual calculation (in three dimensions of motion) leads to the conclusion that pressure P is related to the average of the squared speed of the atoms $(v^2)_{av}$ by the expression

$$P = \frac{\frac{1}{3}m(v^2)_{av}}{V},$$

where V is the volume of the gas, and m is the mass of an individual gas atom. The steps in the derivation are quite straightforward (see the derivation in the *Student Guide*) and a beautiful case of applying Newton's laws in a region Newton himself never did.

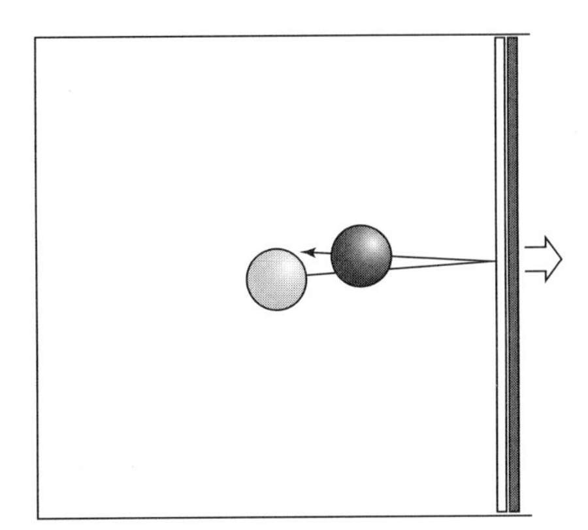

FIGURE 7.11 Model of gas particle in a container with a moveable wall.

We now have two expressions for the pressure of a gas. One summarizes the experimental facts, $PV = kT$. The other is derived by Newton's laws from a theoretical model, $PV = \frac{1}{3}m(v^2)_{av}$. The *theoretical* expression will agree with the *experimental* expression only if $kT = \frac{1}{3}m(v^2)_{av}$. This would mean that the temperature of a gas is proportional to $(v^2)_{av}$. The mass m of each molecule is a constant, so the temperature T is also proportional to $m(v^2)_{av}$. But this allows us also to write $T \propto \frac{1}{2}m(v^2)_{av}$, which immediately tells us that according to *the kinetic theory the temperature of a gas is proportional to the average kinetic energy of its molecules!* We already had some idea that raising the temperature of a material somehow affected the motion of its "small parts." We were aware that the higher the temperature of a gas, the more rapidly its molecules are moving. But the conclusion $T \propto \frac{1}{2}m(v^2)_{av}$ is a precise quantitative relationship derived from the kinetic model and empirical laws. At last we know for certain, for a gas, that heat is not some fluid (caloric) or other substance. It is just the kinetic energy of the particles (atoms) making up the material (or, as we shall see later, radiant energy).

FIGURE 7.12 Balloon for carrying weather forecasting apparatus.

Kinetic-Theory Explanation of the Temperature–Volume Relationship

The kinetic theory makes possible other quantitative predictions. We know by experience (e.g., by inflating a bicycle tire quickly) that when a gas is compressed or condensed rapidly, its temperature changes. The general gas law ($PV = kT$) applies. Can our model explain this result?

In the model used in the previous subsection, atoms or molecules were bouncing back and forth between the walls of a box. Every collision with the wall was perfectly elastic, so the particles rebounded with no loss in speed (or kinetic energy). Suppose the outside force that holds one wall in place is suddenly increased. What will happen to the wall? The wall moves inward, compressing the gas. As it compresses the gas, it does work on the particles, increasing their kinetic energy. As kinetic energy goes up, the temperature of the gas should rise—which is just what happens when a gas is compressed quickly.

If the outside force on the wall is decreased instead of increased, just the opposite happens. Again, what we learned earlier in mechanics comes in handy. As long as the wall was stationary, the particles did no work on it, and the wall did no work on the particles. Now if the wall is free to move outward, in the same direction as the force exerted on it by the particles as they smash into the wall, the picture changes. Since the particles, by their collisions, exert a force on the wall and the wall moves in the direction of the force, the particles must be doing work on the wall. The energy needed to do this work must come from somewhere. The only available source of energy here is part or all of the kinetic energy ($\frac{1}{2}mv^2$) of the particles. In fact, it can be shown that molecules colliding perfectly elastically with a receding wall rebound with slightly less speed. Therefore, the kinetic energy of the particles must decrease. The

> This phenomenon can be demonstrated by means of the expansion cloud chamber, or cooling of CO_2 fire extinguisher while in use. In the last case, the "wall" is the air mass being pushed away.

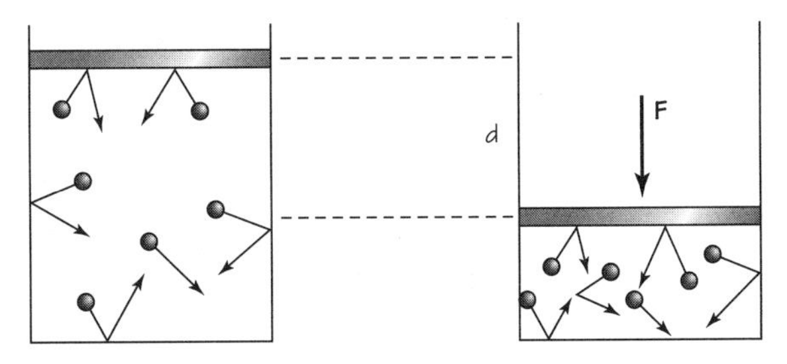

FIGURE 7.13 Particles in a cylinder being compressed.

relationship $T \propto \frac{1}{2}m(v^2)_{av}$ implies that the temperature of the gas will drop. This is exactly what happens when a container holding a gas expands!

Brownian Motion

Many different kinds of experimental evidence support these conclusions and therefore support the kinetic-theory model of a gas. Perhaps the most sophisticated evidence is the motion of very small particles seen through a microscope, when they are suspended in a gas or liquid. The gas or liquid molecules themselves are too small to be seen directly, but their effects on a larger particle (e.g., a particle of smoke) can be observed through the microscope. At any instant, swarms of molecules moving at very different speeds are striking the larger particle from all sides. So many molecules are taking part that their total effect *nearly* cancels. Any remaining effect changes in magnitude and direction from moment to moment. Therefore, the impact of the *invisible* molecules makes the visible *particles* appear to "dance" or jitter randomly in the view field of the microscope. The hotter the gas or liquid, the more lively the motion, as the relationship $T \propto \frac{1}{2}m(v^2)_{av}$ predicts.

This observation is known as *Brownian motion*. It was named after the English botanist, Robert Brown, who in 1827 observed the phenomenon

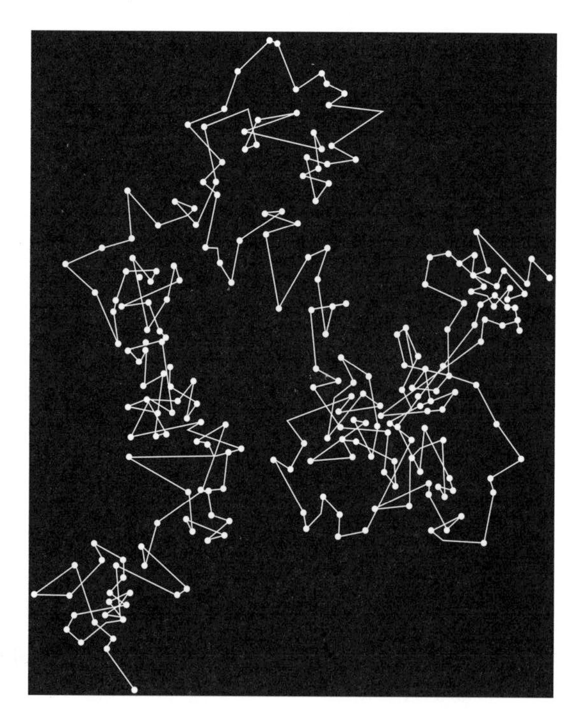

FIGURE 7.14 Track of a particle in Brownian motion. Successive positions, recorded every 20 seconds, are connected by straight lines. The actual paths between recorded positions would be as erratic as the overall path.

while looking at a suspension of the microscopic grains of plant pollen. The same kind of motion of suspended particles ("thermal motion") also exists in liquids and solids, although there the particles are far more constrained than in gases.

The origin of Brownian motion remained a mystery for many years, until in 1905 Albert Einstein, using the kinetic theory, predicted that such motion should occur. Comparison between his detailed predictions and the observations of Brownian motion helped to convince most of the remaining sceptics at the time about the reality of atoms. This phenomenon, which is simple to set up and fascinating to watch, gives striking evidence that the smallest parts of all matter in the Universe are indeed in a perpetual state of lively, random motion.

7.6 KINETIC-THEORY EXPLANATION OF THE SECOND LAW

The kinetic-theory model can explain the behavior of a gas when it is compressed or expanded, warmed or cooled. In the late nineteenth century, the model was refined to take into account many effects we have not discussed. There proved to be limits beyond which the model breaks down. For example, radiated heat comes from the Sun through the vacuum of space, or from an electric grill. This is not explainable in terms of the thermal motion of particles; rather, that thermal radiation is a form of electromagnetic waves. But in most cases the model worked splendidly, explaining the phenomenon of heat in terms of the ordinary motions of submicroscopic particles. It fulfilled much of the hope Newton had expressed in the *Principia* and in the *Opticks* that all phenomena of nature could be explained in terms of the motion of the small parts of matter (atoms).

As we noted earlier, a basic philosophical theme of the Newtonian cosmology is the idea that the world is like a machine whose parts never wear out and which never runs down. This idea inspired the search for conservation laws applying to matter and motion. So far in this text, you have seen that this search has been successful. We can measure "matter" by mass, and "motion" by momentum or by kinetic energy. By 1850, the law of conservation of mass had been firmly established in chemistry. In physics, the laws of conservation of momentum and of energy had been equally well established.

Yet these successful conservation laws could not banish the suspicion that somehow the world *is* running down, the parts of the machine *are* wearing out. Energy may be conserved in burning fuel, but it loses its *usefulness* as

> Our life runs down in sending up the clock.
> The brook runs down in sending up our life.
> The sun runs down in sending up the brook.
> And there is something sending up the sun.
> It is this backward motion toward the source,
> Against the stream, that most we see ourselves in
> The tribute of the current to the source.
> It is from this in nature we are from.
> It is most us.
>
> Excerpt from "West-Running Brook," (by Robert Frost).

the heat goes off into the atmosphere. The coal burned in a steam engine can never be recovered once it is burned. Mass may be conserved in scrambling an egg, but the organized *structure* of the egg is lost forever. In these transformations, something is conserved, but something is also lost. Some processes are irreversible; that is, they will not run backward. There is no way to unscramble an egg, although such a change would not violate mass conservation. There is no way to draw smoke and hot fumes back into a blackened stick, forming a new, unburned match. There is no way to run a steam engine backward and obtain all of the heat originally obtained from the burning of the fuel.

Section 6.4 discussed one type of irreversible process, that involving heat engines, which is governed by *the second law of thermodynamics*. As we saw, that law can be stated in several equivalent ways:

- Heat will not by itself flow from a cold body to a hot one.
- It is impossible to convert fully a given amount of heat into work.
- The entropy of an isolated system, and therefore of the Universe, tends to increase.

The processes of scrambling an egg, of mixing smoke and air, or of wearing down a piece of machinery do not, at first sight, seem to obey the same laws as do heat engines. However, these processes are also governed by the second law. Heat, as you have seen, is represented by the *disordered* motions

FIGURE 7.15 Waterfall.

FIGURE 7.16 Stroboscopic photograph of a bouncing ball.

of atoms and molecules. Converting ordered mechanical work into heat—say, the push of a piston straight into a cylinder full of gas—thus leads to an increase in disordered motion in the heated material. In fact, entropy can be defined mathematically as a measure of the disorder of a system (though it is not necessary to go into the mathematics here). In sum, irreversible processes are processes for which entropy increases, and the increase in entropy is a measure of the increase in the disorder of the atoms, molecules, and any other components making up the system.

7.7 MAXWELL'S DEMON AND THE STATISTICAL VIEW OF THE SECOND LAW

The Austrian physicist Ludwig Boltzmann, thinking about "irreversible" phenomena, detected a loophole in the rather pessimistic picture of a Universe running down as entropy increased. He concluded that the tendency toward dissipation of energy is not an *absolute* law of physics that holds in every situation. Rather, when it concerns the behavior of many particles, it is only a *statistical* law.

Think of a balloon filled with air containing billions upon billions of molecules. Boltzmann agreed that, of all conceivable arrangements of the gas molecules at a given instant, their motion would nearly always be completely "disordered," as we have noted (see the meaning of "disorder" in Section 7.2). Yet, it is conceivable that at some moment most of the molecules happen, by chance, to be moving in the same direction. In any ran-

FIGURE 7.17 Ludwig Boltzmann (1844–1906).

Consider also a pool table. The ordered motion of a cue ball moving into a stack of resting ones soon gets "randomized."

dom arrangement fluctuations from complete disorder are bound to occur. But the greater the fluctuation toward order, the less likely it is to occur. For collections of particles as large as 10^{23}, the chance of a fluctuation large enough to be measurable is vanishingly small, but not zero. By the same argument, it is *conceivable* that a cold kettle of water will heat up on its own after being struck by only the most energetic molecules that happen to be in the surrounding air. It is also *conceivable* that for a brief moment air molecules will "gang up" and strike only one side of a rock, pushing it uphill. Such events, while conceivable, are *utterly improbable*.

For *small* collections of particles, however, it is a different story. Just as it is quite probable that the average height of the people on a particular bus will be considerably greater or less than the national average, it is probable in the same way that more molecules will hit one side of a microscopic particle than the other side. That is just what causes the observable, Brownian motion of microscopic particles in a gas or liquid. Fluctuations, virtually undetectable for any large collection of molecules familiar in the everyday world, are an important aspect of the world of very small particles.

One outcome of these considerations is that the second law of thermo-

To illustrate Boltzmann's argument, consider a pack of cards when it is shuffled. Most possible arrangements of the cards after shuffling are fairly disordered. If you start with an ordered arrangement, for example, the cards sorted by suit and rank, then shuffling would almost certainly lead to a more disordered arrangement. (Nevertheless, it does occasionally happen that a player is dealt 13 spades, even if no one has stacked the deck.)

dynamics is different in character from all the other fundamental laws of physics presented so far. The difference is that it deals with probabilities, not certainties. For example, it says it is highly probable that when ice is dropped into hot water, the ordered structure of the cube will break apart and the cube will melt, forming water. But this law does not rule out the extremely unlikely possibility that the slowest molecules of hot water will join together for an instant to form an ice cube. Such an event has never been observed, and probably never will be, but it is possible in principle. The second law is thus a statistical law, giving the statistical outcome of a huge number of individual events (collisions of molecules).

Testing the limits of conceivable consequences of the second law, Maxwell proposed an interesting "thought experiment." Suppose a container of gas is divided by a thin membrane into two parts, the gas in one part being hotter than in the other. "Now conceive of a finite being," Maxwell suggested, "who knows the paths and velocities of all the molecules but who can do no work except open and close a hole in the diaphragm." This "finite being," now referred to as "Maxwell's demon," can make the hot gas hotter and the cold gas cooler, just by letting fast molecules move in only one direction through the hole (and slow molecules in the other), as is shown in the diagram.

Of course, there exists no such fanciful demon (even in machine form) that can observe and keep track of all the molecules in a gas—hence no such procedure for violating the second law can be realized in practice. If somehow such a "demon" could be made to exist, one might find that the demon's own entropy is affected by its actions. For example, its entropy might increase enough to compensate for the decrease in entropy of the gas. This is what happens in other systems where local order is created, such as in a tray of water freezing into ice cubes in a freezer; the entropy must increase somewhere else in the universe, such as in the room outside the freezer, where waste heat from the freezer engine is exhausted.

Some biologists have suggested that certain large molecules, such as enzymes, may function as "Maxwell's demons." Large molecules may influence the motions of smaller molecules to build up the ordered structures of living systems. This result is different from that of lifeless objects and is in apparent violation of the second law of thermodynamics. This suggestion, however, shows a misunderstanding of the law. The second law does not say that the order can *never* increase in any system. It makes that claim *only for closed or isolated systems*. Any "open" system, one that can exchange

FIGURE 7.18 How Maxwell's "demon" could use a small, massless door to increase the order of a system and make heat flow from a cold gas to a hot gas.

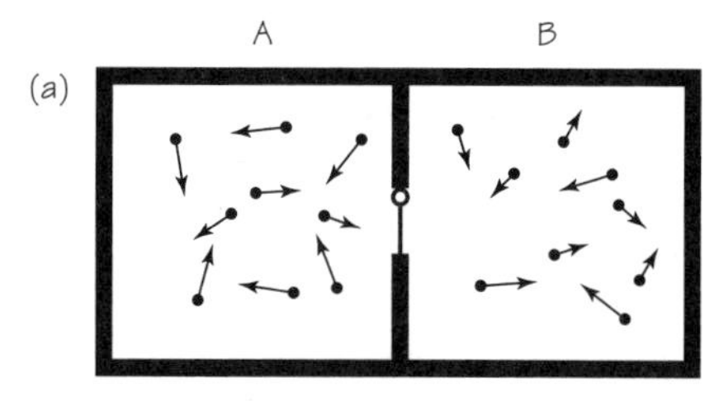

Initially the average KE of molecules is equal in A and B.

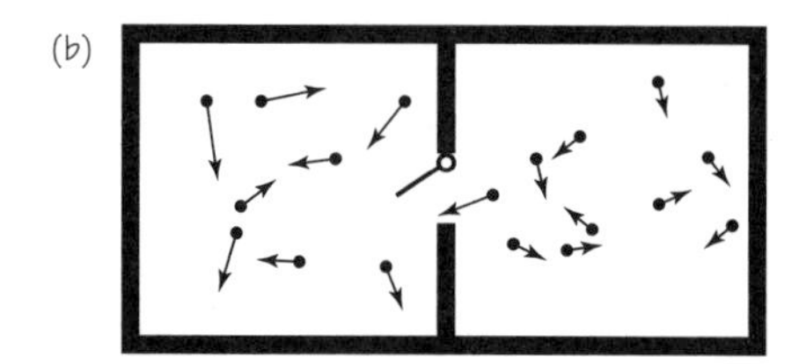

Only fast molecules are allowed to go from B to A.

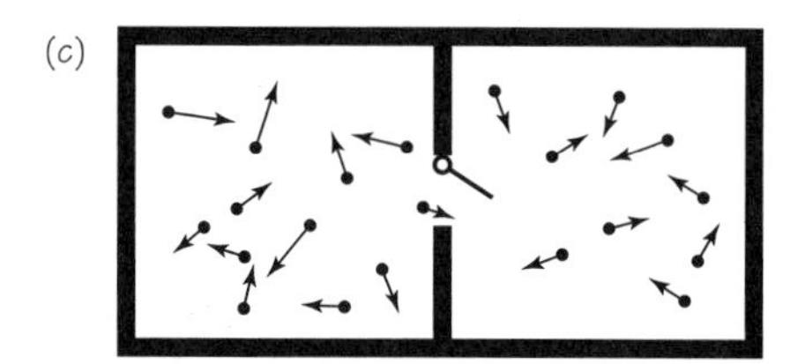

Only slow molecules are allowed to go from A to B.

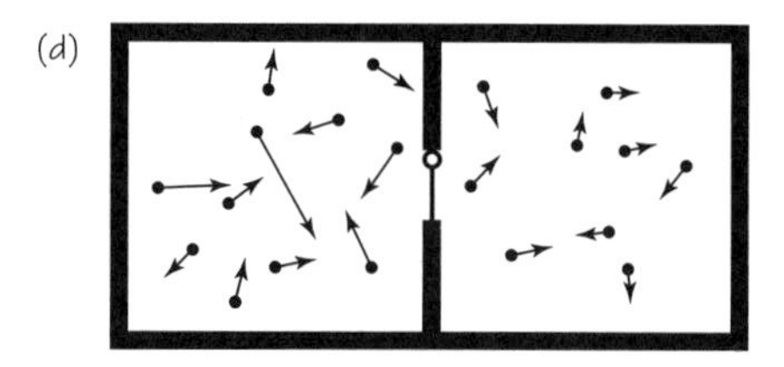

As this continues, the average KE in A increases and the average KE in B decreases.

energy with its surroundings, can increase its own order without violating the second law, for the entropy of the surroundings will increase.

The flow of energy through a system that is not closed may tend to increase the order in the system. The existence of highly organized structures that have life may be a result of supplying energy to them, as from the Sun. But the organizing phenomenon resulting in life does have its cost

FIGURE 7.19 A living system appears to contradict the second law of thermodynamics by bringing order out of disorder.

in terms of the effect on the rest of the total system. This point is expressed vividly in the following passage from a UNESCO document on environmental pollution:

> Some scientists used to feel that the occurrence, reproduction, and growth of order in living systems presented an exception to the second law. This is no longer believed to be so. True, the living system may increase in order, but only by diffusing energy to the surroundings and by converting complicated molecules (carbohydrates, fats) called food into simple molecules (CO_2, H_2O). For example, to maintain a healthy human being at constant weight for one year requires the degradation of about 500 kilograms (one-half ton) of food, and the diffusion into the surroundings (from the human and the food) of about 500,000 kilocalories (two million kilojoules) of energy. The "order" in the human may stay constant or even increase, but the order in the surroundings decreases much, much more.

7.8 TWO CHALLENGES

Late in the nineteenth century, a small but influential group of scientists began to question the basic philosophical assumptions of Newtonian mechanics. They even questioned the very idea of atoms. The Austrian physicist Ernst Mach argued that scientific theories should not depend on assuming the existence of things (such as atoms) which could not be directly observed.

Critics of kinetic theory pointed to two apparent contradictions between kinetic theory and the second law of thermodynamics. These contradictions are known as the *reversibility paradox* and the *recurrence paradox*. Both paradoxes were based on possible exceptions to the second law; both were thought to cast doubt on the kinetic theory.

The Reversibility Paradox

Lord Kelvin and Josef Loschmidt, both of whom supported atomic theory, discovered the reversibility paradox during the 1870s. It was not regarded as a serious objection to the kinetic theory until the 1890s. The paradox is based on the simple fact that Newton's laws of motion are reversible in time. For example, if you watch a video of a bouncing ball, it is easy to tell whether the video tape is being run forward or backward: You know that the collisions of the ball with the floor are inelastic and that the ball rises less high after each bounce. (See Figure 7.16.) If, however, the ball made perfectly elastic bounces, it would rise to the same height after each bounce. Then you could not tell whether the tape was being run forward or backward. In the kinetic theory, molecules *are* assumed to make perfectly elastic collisions. Imagine that you could make a video recording of gas molecules colliding elastically according to this assumption. (See Figure 7.20.) When showing this video, there would be no way to tell whether it was being run forward or backward. Either way would show valid sequences of collisions. But here is the paradox: Consider videos of interactions involv-

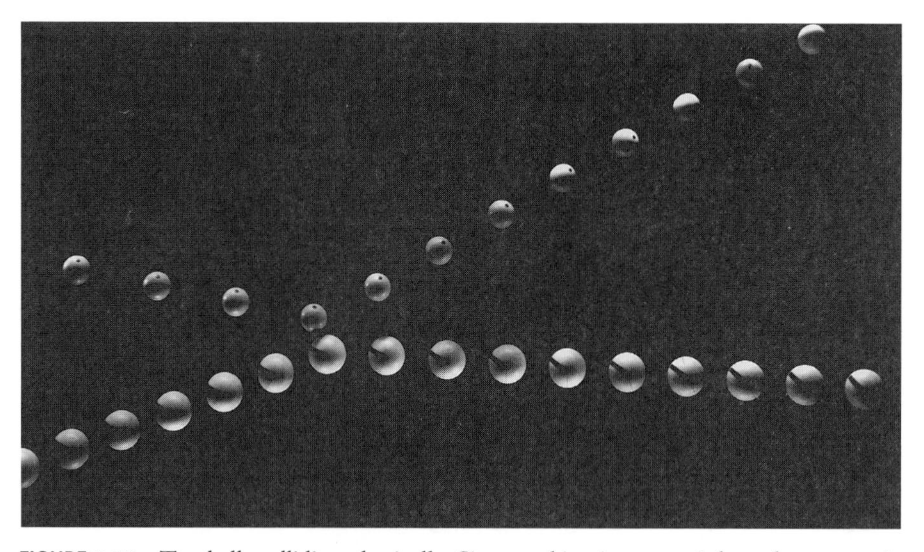

FIGURE 7.20 Two balls colliding elastically. Since no kinetic energy is lost, the process is reversible. The reversibility paradox: Can a model based on reversible events explain a world in which so many events are irreversible?

FIGURE 7.21 Example of an irreversible process.

ing large objects, containing many molecules. You can immediately tell the difference between forward (true) and backward (impossible) time direction. For example, a smashed light bulb does not reassemble itself in real life, although a video run backward can make it appear to do so.

The kinetic theory is based on laws of motion that assume the motions are reversible for each individual molecular interaction. How, then, can it explain the existence of *irreversible* processes on a large scale, involving many molecular collisions? The existence of such processes seems to indicate that time flows in a definite direction, that is, from past to future. This contradicts the possibility, implied in Newton's laws of motion, that, when it comes to observing physical phenomena, it does not matter whether we think of time as flowing forward or backward. As Kelvin expressed the paradox:

> If . . . the motion of every particle of matter in the Universe were precisely reversed at any instant, the course of nature would be simply reversed for ever after. The bursting bubble of foam at the foot of a waterfall would reunite and descend into the water; the thermal motions would reconcentrate their energy, and throw the mass up the fall in drops reforming into a close column of ascending water. Heat which had been generated by the friction of solids and dissipated by conduction, and radiation with absorption, would come again to the place of contact, and throw the moving body back against the force to which it had previously yielded. . . . But the real phenomena of life infinitely transcend human science; and speculation regarding consequences of their imagined reversal is utterly unprofitable.

Kelvin himself, and later Boltzmann, used statistical probability to explain why we do not observe such large-scale reversals. There are almost

FIGURE 7.22 A bonfire—another example of an irreversible process.

infinitely many possible disordered arrangements of water molecules at the bottom of a waterfall. Only an extremely small number of these arrangements would lead to the process described by Kelvin. Reversals of this kind are possible *in principle*, but for all practical purposes they are out of the question.

The Recurrence Paradox

Another small possibility allowed in kinetic theory leads to a situation that seems unavoidably to contradict the second law. The *recurrence paradox* revived an idea that appeared frequently in ancient philosophies and is present also in Hindu philosophy to this day: the myth of the "eternal return." According to this myth, the long-range history of the world is cyclic. All historical events eventually repeat themselves, perhaps many times. Given enough time, even the matter that people were made of will eventually reassemble by chance. Then people who have died may be born again and go through the same life. The German philosopher Friedrich Nietzsche was convinced of the truth of this idea.

The recurrence paradox begins with the fact that the number of molecules in the world is finite, hence there is only a finite number of possible arrangements of molecules. Therefore, if time continues infinitely, the same combination of molecules is bound to come up again. At some instant, all

> The World's great age begins
> anew,
> The golden years return.
> The earth cloth like a snake
> renew
> His winter weeds outworn . . .
> Another Athens shall arise
> And to remoter time
> Bequeath, like sunset to the
> skies,
> The splendour of its prime . . .
>
> [Percy Bysshe Shelley,
> "Hellas" (1822)]

the molecules in the Universe would reassemble exactly the same arrangement they had at some previous time. All events following this point would then be exactly the same as the events that followed it before. That is, if any single instant in the history of the Universe is ever *exactly* repeated, then the entire history of the Universe will be repeated from that instant on. As a little thought shows, it would then be repeated over and over again to infinity. Thus, energy would *not* endlessly become dissipated as required by the second law. Nietzsche claimed that this view of the eternal return disproved the "heat-death" theory.

At about the same time, in 1889, the French mathematician Henri Poincaré published a theorem on the possibility of recurrence in mechanical systems. According to Poincaré, even though the Universe might undergo a heat death, it would ultimately come alive again:

> A bounded world, governed only by the laws of mechanics, will always pass through a state very close to its initial state. On the other hand, according to accepted experimental laws (if one attributes absolute validity to them, and if one is willing to press their consequences to the extreme), the Universe tends toward a certain final state, from which it will never depart. In this final state, from which will be a kind of death, all bodies will be at rest at the same temperature.
>
> . . . the kinetic theories can extricate themselves from this contradiction. The world, according to them, tends at first toward a state where it remains for a long time without apparent change; and this is consistent with experience; but it does not remain that way forever; . . . it merely stays there for an enormously long time, a time which is longer the more numerous are the molecules. This state will not be the final death of the Universe, but a sort of slumber, from which it will awake after millions of centuries.

According to this theory, to see heat pass from a cold body to a warm one, it will not be necessary to have the acute vision, the intelligence, and the dexterity of Maxwell's demon; it will suffice to have a little patience.

Poincaré was willing to accept the possibility of a violation of the second law after a very long time has passed. Others refused to admit even this possibility.

FIGURE 7.23 The ruins of the pyramid of Zosher (a step pyramid in Saggara, Egypt) testify to the continual encroachment of disorder.

The outcome of the dispute between the defenders and the critics of the kinetic theory was that both sides were partly right and partly wrong. Mach and his followers were correct in believing that Newton's laws of mechanics cannot fully describe molecular and atomic processes. We will come back to this subject when we discuss quantum mechanics in Chapter 15. For example, it is only approximately valid to describe gases in terms of collections of frantic little balls. But Boltzmann was right in defending the usefulness of the molecular model. The kinetic theory is very nearly correct except for those properties of matter that involve the structure of molecules themselves.

The new success of the kinetic theory in Einstein's account of Brownian motion, along with discoveries in radioactivity and atomic physics, persuaded almost all the critics that atoms and molecules do, in fact, exist. But the problems of irreversibility and of whether the laws of physics must distinguish between past and future survived. In a new form, these issues still interest physicists today.

SOME NEW IDEAS AND CONCEPTS

Boyle's law
Brownian motion
density
diffusion
entropy
ideal gas
ideal gas law
irreversible processes
kinetic-molecular theory of heat
Maxwell's "demon"
model
open system
pressure
recurrence paradox
reversiblity paradox
reversibile processes

IMPORTANT EQUATIONS

$$P = \frac{F_{\perp}}{A},$$

$$D = \frac{m}{V},$$

$$PV = kT.$$

FURTHER READING

G. Holton, and S.G. Brush, *Physics, The Human Adventure* (Piscataway, NJ: Rutgers University Press, 2000), Chapters 18–22.

H.C. von Baeyer, *Maxwell's Demon: Why Warmth Disperses and Time Passes* (New York: Random House, 1998). Softcover: *Warmth Disperses and Time Passes: A History of Heat* (New York: Modern Library, 1999).

STUDY GUIDE QUESTIONS

A. THE KINETIC THEORY

7.1 An Ideal Gas

1. Early forms of the kinetic-molecular theory were based on the assumption that heat energy is (select one of the following):
   (a) a liquid;
   (b) a gas;
   (c) the kinetic energy of molecules;
   (d) made of molecules.
2. The relationship between the volume and the pressure of a gas expressed by Boyle's law, $P = a/V$, holds true
   (a) for any gas under any conditions;
   (b) for some gases under any conditions;
   (c) only if the temperature is kept constant;
   (d) only if the volume is constant.
3. How is it possible for the pressure on the floor under a spiked heel to be greater than the pressure under an elephant's foot?
4. State the "ideal gas law" in your own words.
5. What is "ideal" about the ideal gas law?

6. The text states that the three proportionalities among P, V, and T can be combined into one equation, the ideal gas law. Confirm that by holding one of the variables constant, the equation does indeed yield each of the three proportions.
7. Explain why the ideal gas law is not a law of physics in the same sense as the law of conservation of energy.
8. Assuming the ideal gas law holds to the lowest temperatures, what happens to a gas at absolute zero?

7.2 A Model for the Gaseous State

1. What is meant by a "model of a gas"?
2. What kind of a model is a test model of a bridge made of balsa wood? a computer program that simulates the forces acting on a bridge? What are the differences between "theoretical" and "working" models? How are theoretical models tested?
3. What are some of the assumptions in the kinetic theory of gases?
4. Read the following description of a model of a gas and give a suitable numerical estimation for each underlined phrase: "a <u>large number</u> of <u>small particles</u> in <u>rapid, disordered motion</u>."
5. In the kinetic theory, particles are thought to exert significant forces on one another:
   (a) only when they are far apart;
   (b) only when they are close together;
   (c) all the time;
   (d) never.
6. Why was the kinetic theory first applied to gases rather than to liquids or solids?
7. Why do we have to use statistics in studying gases? Why can't we just follow the motion of each atom?

7.3 The Speeds of Molecules

1. In the kinetic theory of gases, it is assumed that the pressure of a gas on the walls of the container is due to:
   (a) gas molecules colliding with one another;
   (b) gas molecules colliding against the walls of the container;
   (c) repelling forces exerted by molecules on one another.
2. The idea of speed distribution for gas molecules means that:
   (a) each molecule always has the same speed;
   (b) there is a wide range of speeds of gas molecules;
   (c) molecules are moving fastest near the center of the container of the gas.
3. What happens to the speed distribution as the temperature increases? As the temperature reaches absolute zero?
4. The average speed of a molecule in a gas at 900°C is about 1 km/s. Is it possible that there are molecules in the gas at speeds of 10 km/s? With zero speed?

7.4 The Sizes of Molecules

1. What was the objection raised against Clausius' kinetic theory for a gas?
2. In his revised kinetic-theory model Clausius assumed that the particles have a finite size, instead of being mathematical points, because:
   (a) obviously everything must have some size;
   (b) it was necessary to assume a finite size in order to calculate the speed of molecules;
   (c) the size of a molecule was already well known before Clausius' time;
   (d) a finite size of molecules could account for the slowness of diffusion.

B. APPLYING THE KINETIC THEORY

7.5 Kinetic-Theory Explanation of the Ideal Gas Law

1. Using the kinetic theory of gases, explain in your own words each of the three empirical relationships between P, V, and T for an ideal gas.
2. Does your response to Question 1 prove that the kinetic theory of gases is valid? Explain.
3. Using the concept of work and the kinetic theory of gases, explain why the temperature of a gas and the kinetic energy of its molecules both increase if a piston is suddenly pushed into the container holding the gas.
4. What are the limits under which the ideal gas law describes the behavior of real gases?

7.6 Kinetic-Theory Explanation of the Second Law

1. Which of the following statements agrees with the second law of thermodynamics?
   (a) Heat does not naturally flow from cold bodies to hot bodies.
   (b) Energy tends to transform itself into less useful forms.
   (c) No engine can transform all its heat input into mechanical energy.
   (d) Most processes in nature are reversible.
2. If the water in a pot placed on a hot stove suddenly froze, Newton's laws would not have been violated. Why would this event violate the second law of thermodynamics? If an extremely small volume of the total water in the pot cooled for a moment, would this violate the second law?
3. What is a reversible process?

7.7 Maxwell's Demon and the Statistical View of the Second Law

1. In each of the following pairs, which situation is more ordered?
   (a) an unbroken egg; a scrambled egg;
   (b) a glass of ice and warm water; a glass of water at uniform temperature;
   (c) a falling ball; a ball that has hit a tabletop, bounced, and come to rest.
2. What is Maxwell's "demon" and what is it supposed to do?

3. True or false?
   (a) Maxwell's demon was able to get around the second law of thermodynamics.
   (b) Scientists have made a Maxwell's demon.
   (c) Maxwell believed that his demon actually existed.
4. How did Boltzmann account for the entropy version of the second law?
5. How does the kinetic theory account for the fact that heat does not flow by itself from cold objects to hot objects?
6. Is it ever possible for a scrambled egg to unscramble itself? Why don't we observe this happening occasionally?
7. A growing plant takes in nutrients from the soil to create the increasingly complete structure of its stem, roots, leaves, flowers, etc. Order seems to be emerging by itself out of disorder. Is this a violation of the second law of thermodynamics? Explain.

7.8 Two Challenges

1. What is the "reversibility paradox" and how did Boltzmann resolve it?
2. What is the "recurrence paradox" and how was it finally resolved?
3. The kinetic energy of a falling stone is transformed into heat when the stone strikes the ground. Obviously, this is an irreversible process; you never see the heat transform into kinetic energy of the stone, so that the stone rises off the ground. Scientists believe that the process is irreversible because:
   (a) Newton's laws of motion prohibit the reversed process;
   (b) the probability of such a sudden ordering of molecular motion is extremely small;
   (c) the reversed process would not conserve energy;
   (d) the reversed process would violate the second law of thermodynamics.
4. What is Brownian motion? How did it provide convincing evidence for the existence of atoms and molecules?

DISCOVERY QUESTIONS

1. Using the kinetic theory, explain why when exposed to room temperature, hot water near boiling cools down faster than cold water, initially near freezing, warms up?
2. An ice cube is dropped into an insulated cup of hot water. Describe what happens and why, based on the laws of thermodynamics.
3. Explain the results in Question 2 using the kinetic theory, including the fact that this process does not reverse itself.
4. Using material in this chapter, how would you answer this question: "How do we know that atoms really do exist?"
5. The idea of randomness can be used in predicting the results of flipping a large number of coins. Give some other examples where randomness is useful.

6. A drop in barometric pressure is often a signal that cold, wet weather is on the way. A rise in air pressure indicates warm, dry weather. Why do you think this is so?
7. At sea level, water boils at 100°C, while at higher elevations it boils at slightly lower temperature. Using the kinetic theory, how would you explain this?
8. Why is the melting of an ice cube considered to be an irreversible process, even though it could easily be refrozen?
9. Explain why you feel cold on getting out of a shower, even if the room is not cold.
10. Consider these aspects of the curves showing Maxwell's distribution of molecular speeds:
    (a) all show a peak;
    (b) the peaks move toward higher speed at higher temperatures.
    Explain these characteristics on the basis of the kinetic model.
11. State the ideal gas law. What three proportionalities are contained in this law? What are the limitations of this law?
12. Starting from the definition of density, $D = m/V$ (where m is the mass of a sample and V is its volume), write an expression relating pressure P, temperature T, and density D of a gas.
13. Distinguish between two uses of the word "model" in science.
14. What would you expect to happen to the temperature of a gas that was released from a container into empty space (i.e., with nothing to push back)?
15. List some of the directly observable properties of gases.
16. What aspects of the behavior of gases can the kinetic-molecular theory be used to explain successfully?
17. When a gas in an enclosure is compressed by pushing in a piston, its temperature increases. Explain this fact in two ways:
    (a) by using the first law of thermodynamics;
    (b) by using the kinetic theory of gases.
18. Since all the evidence is that molecular motions are random, you might expect that any given arrangement of molecules will recur if you just wait long enough. Explain how a paradox arises when this prediction is compared with the second law of thermodynamics, viewed as an absolute law of nature.
19. (a) Explain what is meant by the statement that Newton's laws of motion are time-reversible.
    (b) Describe how a paradox arises when the time-reversibility of Newton's laws of motion is compared with the second law of thermodynamics.
20. Where did Newtonian mechanics run into difficulties in explaining the behavior of molecules?
21. What are some advantages and disadvantages of theoretical models?

Quantitative

1. Benjamin Franklin observed in 1773 that a teaspoonful of oil would spread out to form a film of about 22,000 ft^2 on a pond near London. This helped to give him an estimate of the upper limit of the size of a molecule. Suppose that

1 cm^3 of oil forms a continuous layer one molecule thick that just covers an area on water of 1000 m^2.

(a) How thick is the layer?

(b) What is the size of a single molecule of the oil (considered to be a cube for simplicity)? (See the Laboratory Exploration "Avogadro's Number and the Size and Mass of a Molecule," in the *Student Guide*.)

2. How high could water be raised with a lift pump on the Moon?
3. (a) The pressure of a gas in a container is 100 N/m^2. If the temperature is doubled while the volume is made to shrink to one-third, what is the new pressure?

   (b) The temperature of a gas is 100°C. If the pressure is doubled and the volume is also doubled, what is the new temperature?

CHAPTER 8

Wave Motion

A. WAVES

8.1 WHAT IS A WAVE?

The world is continually criss-crossed by waves of all sorts. Water waves, whether giant rollers in the middle of the ocean or gently formed rain ripples on a still pond, are sources of wonder or pleasure. If the Earth's crust shifts, violent waves in the solid Earth cause tremors thousands of kilome-

FIGURE 8.1 Waves crashing on the shore.

ters away. A musician plucks a guitar string, and sound waves pulse against the ears. Wave disturbances may come in a concentrated bundle, like the shock front from an airplane flying at supersonic speeds. Or the disturbances may come in succession like the train of waves sent out from a steadily vibrating source, such as a bell or a string.

All of these examples are *mechanical* waves, in which bodies or particles physically move back and forth. There are also wave disturbances in electric and magnetic fields. Such waves are responsible for what we experience as X rays, visible light, or radio waves. In all cases involving waves, however, the effects produced depend on the flow of energy, not matter, as the wave moves forward. Waves are cases of energy transfer without matter transfer.

So far in this text, you have considered motion in terms of individual particles or other objects. In this chapter, you will study the cooperative motion of collections of particles in "continuous media," oscillating back and forth as the mechanical waves pass by. You will see how closely related are the ideas of particles and waves used to describe events in nature. Then we shall deal with the properties of light and other electromagnetic waves.

8.2 THE PROPERTIES OF WAVES

To introduce some necessary terms to discuss the fascinating world of waves, suppose that two people are holding opposite ends of a taut rope. Suddenly one person snaps the rope up and down quickly once. That "disturbs" the rope and puts a hump in it which travels along the rope toward the other person. The traveling hump is one kind of a wave, called a *pulse*.

Originally, the rope was motionless. The height above ground of each point on the rope depended only upon its position along the rope and did not change in time. But when one person snaps the rope, a rapid change is created in the height of one end. This disturbance then moves away from its source, down the rope to the other end. The height of each point on the rope now depends also upon time, as each point eventually oscillates up and down and back to the initial position, as the pulse passes.

The disturbance is thus a pattern of *displacement* moving along the rope. The motion of the displacement pattern from one end of the rope toward the other is an example of a *wave*. The hand snapping one end is the *source* of the wave. The rope is the *medium* in which the wave moves.

Consider another example. When a pebble falls into a pool of still liquid, a series of circular crests and troughs spreads over the surface. This moving displacement pattern of the liquid surface is a wave. The pebble is the source; the moving pattern of crests and troughs is the wave; and the liquid surface is the medium. Leaves or other objects floating on the surface of the liquid bob up and down as each wave passes. But they do not experience any net displacement on the average. No *material* has moved from the wave source along with the wave, either on the surface or among the particles of the liquid—only the energy and momentum contained in the disturbance have been transmitted. The same holds for rope waves, sound waves in air, etc.

As any one of these waves moves through a medium, the wave produces a changing displacement of the successive parts of the medium. Thus, we can refer to these waves as *waves of displacement*. If you can see the medium and recognize the displacements, then you can easily see waves. But waves also may exist in media you cannot see, such as air; or they may form as disturbances of a state you cannot detect with your unaided eyes, such as pressure or an electric field.

You can use a loose spring coil (a Slinky) to demonstrate three different kinds of motion in the medium through which a wave passes. First, move

FIGURE 8.2 The transverse disturbance moves in the horizontal plane of the ground, rather than in the vertical plane.

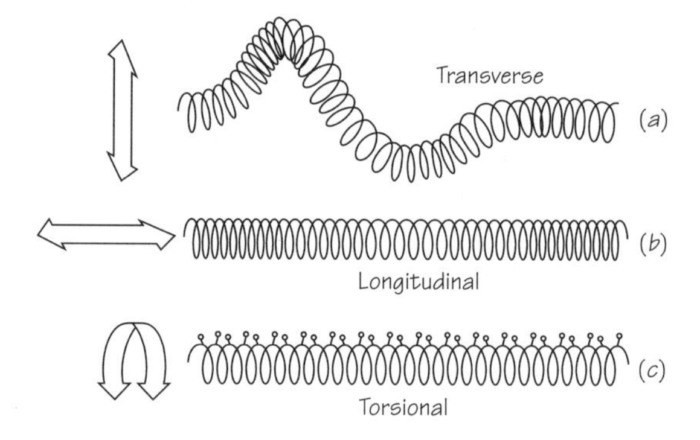

FIGURE 8.3 "Snapshots" of three types of waves on a spring. In (c), the small markers have been put on the top of each coil in the spring.

the end of the spring from side to side, or up and down as in Figure 8.3 (a). A wave of side-to-side or up-and-down displacement will travel along the spring. Now push the end of the spring back and forth, along the direction of the spring itself, as in sketch (b). A wave of back-and-forth displacement will travel along the spring. Finally, twist the end of the spring quickly clockwise and counterclockwise, as in sketch (c). A wave of angular displacement will begin to travel along the spring. (See also the suggested laboratory exploration on waves in the *Student Guide*.)

Waves like those in (a), in which the displacements are perpendicular to the direction the wave travels, are called *transverse* waves. Waves like those in (b), in which the displacements are in the direction the wave travels, are called *longitudinal* waves. Waves like those in (c), in which the displacements are twisting in a plane perpendicular to the direction the wave travels, are called *torsional* waves.

All three types of wave motion can be set up in solids. In fluids, however, transverse and torsional waves die out very quickly and usually cannot be produced at all, except on the surface. Therefore, sound waves in air and water are longitudinal. The molecules of the medium are displaced back and forth along the direction in which the sound energy travels.

It is often useful to make a graph on paper, representing the wave patterns in a medium. This is of course easy to do for transverse waves, but not for longitudinal or torsional waves. But there is a way out. For example, the graph in Figure 8.4 represents the pattern of *compressions* at a given moment as a (longitudinal) sound wave goes through the air. The graph line goes up and down because the graph represents a snapshot of the increase and decrease in density of the air along the path of the wave. It does *not* represent an up-and-down motion of the molecules in the air themselves.

To describe completely transverse waves, such as those in ropes, you must specify the *direction* of displacement. When the displacement pattern of a transverse wave is along one line in a plane perpendicular to the direction

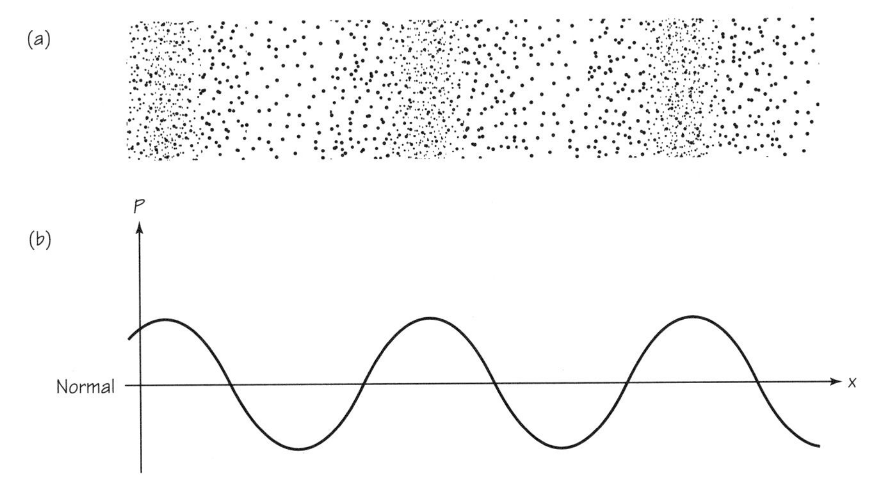

FIGURE 8.4 (a) "Snapshot representation of a sound wave progressing to the right. The dots represent the density of air molecules. (b) Graph of air pressure, P, versus position, x, at the instant of the snapshot.

of motion of the wave, the wave is said to be *polarized*. See the diagrams in Figure 8.5. For waves on ropes and springs, you can observe the polarization directly. In Section 8.18 you will see that for light waves, for example, polarization can have important effects.

All three kinds of waves—longitudinal, transverse, and torsional—have an important characteristic in common. The disturbances move away from their sources through the media and *continue on their own* (although their amplitude may diminish owing to energy loss to friction and other causes). We stress this particular characteristic by saying that these waves *propagate*. This means more than just that they "travel" or "move." An example will clarify the difference between waves that propagate and those that do not. You may have seen one of the great wheat plains of the Middle West,

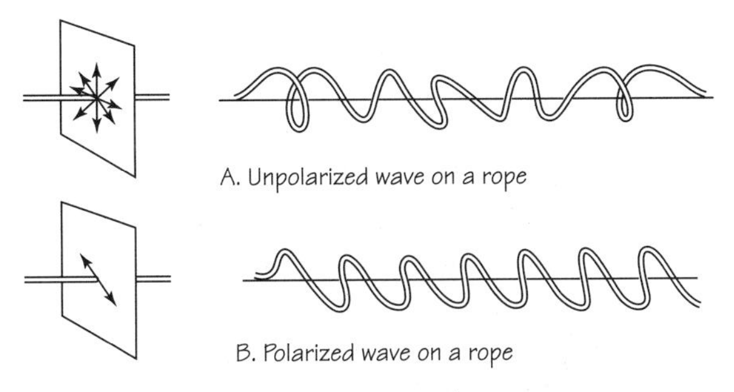

FIGURE 8.5 Polarized/unpolarized waves on rope.

Canada, or Central Europe. Such descriptions usually mention the "beautiful, wind-formed waves that roll for miles across the fields." The medium for such a wave is the wheat, and the disturbance is the swaying motion of the wheat. This disturbance does indeed travel, but it does *not* propagate; that is, the disturbance does not originate at a source and then go on *by itself.* Rather, it must be continually fanned by the wind. When the wind stops, the disturbance does not roll on, but stops, too. The traveling "waves" of swaying wheat are not at all the same as rope and water waves. This chapter will concentrate on waves that originate at sources and propagate themselves through the medium. For the purposes of this chapter, *waves are disturbances which propagate in a medium.*

8.3 WAVE PROPAGATION

Waves and their behavior are perhaps best studied by beginning with large mechanical models and focusing our attention on pulses. Consider, for example, a freight train, with many cars attached to a powerful locomotive, but standing still. If the locomotive starts abruptly, its pull on the next neighboring car sends a displacement wave running down the line of cars.

FIGURE 8.6 A displacement.

The shock of the starting displacement proceeds from the locomotive, clacking through the couplings one by one. In this example, the locomotive is the source of the disturbance, while the freight cars and their couplings are the medium. The "bump" traveling along the line of cars is the wave. The disturbance proceeds all the way from end to end, and with it goes *energy* of displacement and motion. Yet no particles of matter are transferred that far; each car only jerks ahead a bit.

How long does it take for the effect of a disturbance created at one point to reach a distant point? The time interval depends of course on the speed with which the disturbance or wave propagates. This speed, in turn, depends upon the type of wave and the characteristics of the medium. In any case, the effect of a disturbance is never transmitted instantly over any distance. Each part of the medium has inertia, and each portion of the medium is compressible. So time is needed to transfer energy from one part to the next.

The same comments also apply to transverse waves. The series of sketches in the accompanying diagram (Figure 8.7) represents a wave on a rope. Think of the sketches as frames of a motion picture film, taken at equal time intervals. We know that the material of the rope does *not* travel

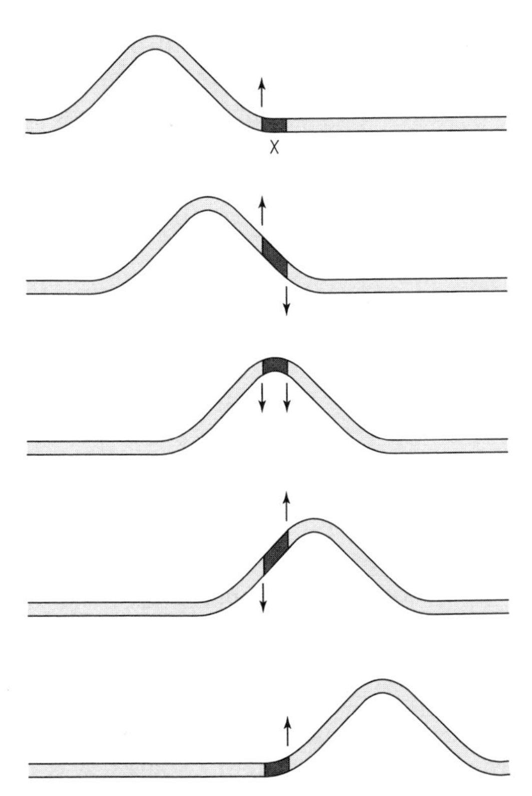

FIGURE 8.7 A rough representation of the forces at the ends of a small section of rope as a transverse pulse moves past.

along with the wave. But each bit of the rope goes through an up-and-down motion as the wave passes. Each bit goes through exactly the same motion as the bit to its left, except a little later.

Consider the small section of rope labeled X in the first diagram. When the pulse traveling on the rope first reaches X, the section of rope just to the left of X exerts an upward force on X. As X is moved upward, a restoring downward force is exerted by the next section. The further upward X moves, the greater the restoring forces become. Eventually, X stops moving upward and starts down again. The section of rope to the left of X now exerts a restoring (downward) force, while the section to the right exerts an upward force. Thus, the trip down is similar, but opposite, to the trip upward. Finally, X returns to the equilibrium position when both forces have vanished.

The time required for X to go up and down, that is, the time required for the pulse to pass by that portion of the rope, depends on two factors. These factors are the *magnitude of the forces* on X and the *mass* of X. To put it more generally: The speed with which a wave propagates depends on the *stiffness* and on the *density* of the medium. The stiffer the medium, the greater will be the force each section exerts on neighboring sections. Thus, the greater will be the propagation speed. On the other hand, the greater the density of the medium, the less it will respond to forces. Thus, the slower will be the propagation. In fact, the speed of propagation depends on the *ratio* of the stiffness factor and the density factor. The exact meaning of stiffness and density factors is different for different kinds of waves and different media. For tight strings, for example, the stiffness factor is the tension T in the string, and the density factor is the mass per unit length, m/l. The propagation speed v is given by

$$v = \sqrt{\frac{T}{m/l}}.$$

8.4 PERIODIC WAVES

Many of the disturbances we have considered so far have been sudden and short-lived, set up by a brief motion like snapping one end of a rope or suddenly displacing one end of a train. In each case, you see a single wave running along the medium with a certain speed. As noted, this kind of wave is called a pulse.

Now consider *periodic waves*, continuous regular rhythmic disturbances in a medium, resulting from *periodic vibrations* of a source. A good example of an object in periodic vibration is a swinging pendulum. Neglecting the effects of air resistance, each swing is virtually identical to every other swing, and the swing repeats over and over again in time. Another example is the up-and-down motion of a weight at the end of a coiled spring. In each case, the maximum displacement from the position of equilibrium is called the *amplitude*, A, as shown in the diagram below for the case of the spring. The time taken to complete one vibration is called the *period*, T, usually given in seconds. The number of vibrations per second is called the *frequency*, f. Note that T and f are reciprocals, in the sense that $T = 1/f$.

What happens when a periodic vibration is applied to the end of a rope? Suppose that the left end of a taut rope is fastened to the oscillating (vibrating) weight on a spring in Figure 8.8. As the weight vibrates up and down, you observe a wave propagating along the rope (see the illustration). The wave takes the form of a series of moving crests and troughs along the length of the rope. The source executes "simple harmonic motion" up and down. Ideally, every point along the length of the rope executes simple harmonic motion in turn. The wave travels to the right as crests and troughs follow one another. Each point or small segment along the rope simply os-

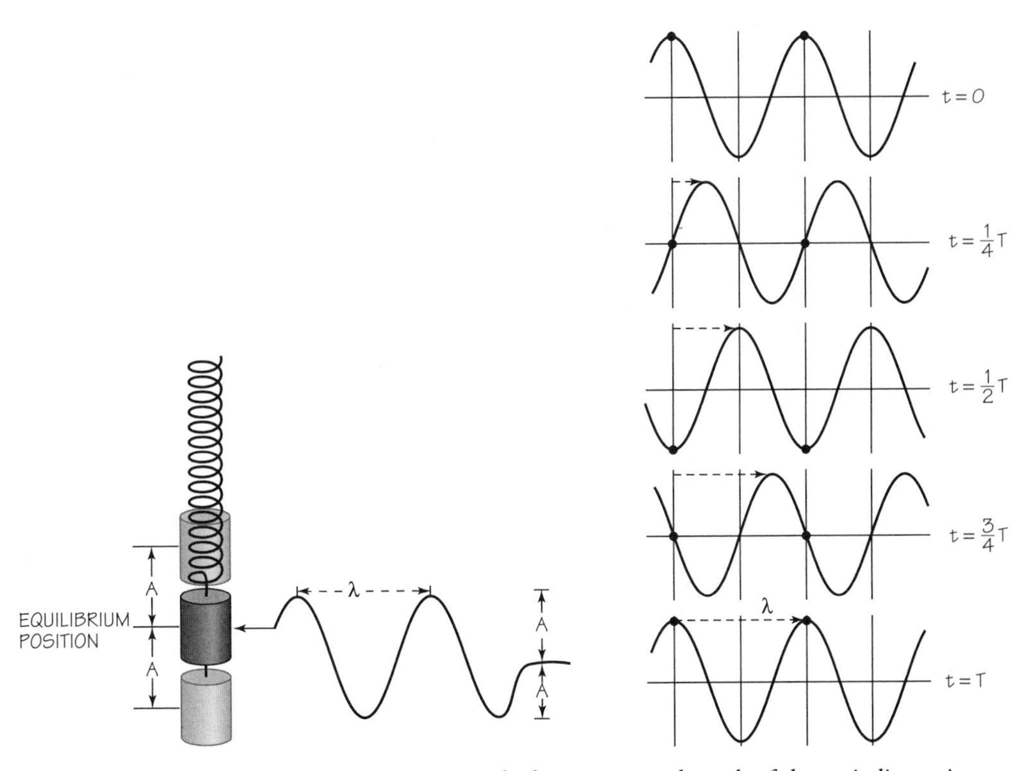

FIGURE 8.8 Spring-mass system attached to a rope, and graph of the periodic motion.

cillates up and down at the same frequency as the source. The amplitude of the wave is represented by A. The distance between any two consecutive crests or any two consecutive troughs is the same all along the length of the rope. This distance, called the *wavelength* of the periodic wave, is conventionally represented by the Greek letter λ (lambda).

If a single pulse or a wave crest moves fairly slowly through the medium, you can easily find its *speed*. In principle, all you need is a clock and a meter stick. By timing the pulse or crest over a measured distance, you can get the speed.

To be sure, it is not always so simple to observe the motion of a pulse or a wave crest. But the speed of a periodic wave can be found indirectly, if one can measure both its frequency and its wavelength. Here is how this works. Using the example of the rope wave, we know that as the wave progresses, each point in the medium oscillates with the frequency and period of the source. The diagram in Figure 8.8 illustrates a periodic wave moving to the right, as it might look in snapshots taken every one-quarter period. Follow the progress of the crest that started out from the extreme left at time $t = 0$. The time it takes this crest to move a distance of one wavelength is equal to the time required for one complete oscillation of the source, or equally of any point on the rope; that is, the crest moves one wavelength λ during one period of oscillation T. The speed v of the crest is therefore given by the equation

$$v = \frac{\text{distance moved}}{\text{corresponding time interval}}$$

$$= \frac{\lambda}{T}.$$

All parts of the wave pattern propagate with the same speed along the rope. Thus, the speed of any one crest is the same as the speed of the wave as a whole. Therefore, the speed v of the wave is also given by

$$v = \frac{\text{wavelength}}{\text{period of oscillation}}$$

$$= \frac{\lambda}{T}.$$

But $T = 1/f$, where f = frequency. Therefore,

$$v = f\lambda$$

or

wave speed = frequency × wavelength.

We can also write this relationship as

$$\lambda = \frac{v}{f}$$

or

$$f = \frac{v}{\lambda}.$$

These expressions show that, for waves of the same speed, the frequency and wavelength are inversely proportional; that is, a wave of twice the frequency would have only half the wavelength, and so on. This inverse relationship of frequency and wavelength will turn out to be very useful in later chapters.

We now go to the last of the definitions that will help to understand how waves behave. The diagram below represents a periodic wave passing through a medium. Sets of points are marked that are moving "in step" as the periodic wave passes. The crest points C and C′ have reached maximum displacement positions in the upward direction. The trough points D and D′ have reached maximum displacement positions in the downward direc-

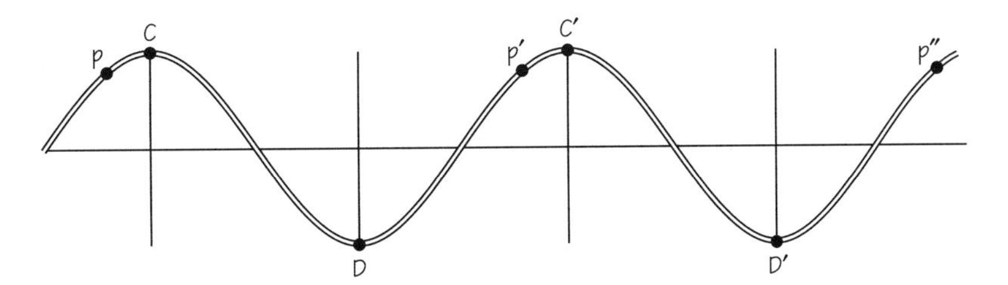

FIGURE 8.9 A "snapshot" of a periodic wave moving to the right. Letters indicate sets of points with the same phase.

tion. The points C and C′ have identical displacements and velocities at any instant of time. Their vibrations are identical and in unison. The same is true for the points D and D′. Indeed there are infinitely many such pairs of points along the medium that are vibrating identically when this wave passes. Note that C and C′ are a distance λ apart, and so are D and D′.

Points that move "in step," such as C and C′, are said to be *in phase* with one another. Points D and D′ also move in phase. Indeed, points separated from one another by distances of λ, 2λ, 3λ, . . . , and $n\lambda$ (n being any whole number) are all in phase with one another. These points can be anywhere along the length of the wave. They need not correspond with only the highest or lowest points. For example, points such as P, P′, P″, are all in phase with one another. Each such point is separated by a distance λ from the next one in phase with it.

On the other hand, we can also see that some pairs of points are exactly *out* of step. For example, point C reaches its maximum upward displacement at the same time that D reaches its maximum downward displacement. At the instant that C begins to go down, D begins to go up (and vice versa). Points such as these are one-half period *out of phase* with respect to one another. C and D′ also are one-half period out of phase. Any two points separated from one another by distances of $\frac{1}{2}\lambda$, $\frac{3}{2}\lambda$, $\frac{5}{2}\lambda$, etc., are one-half period out of phase.

8.5 WHEN WAVES MEET

With the above definitions in hand, we can explore a rich terrain. So far, we have considered single waves. What happens when two waves encounter each other in the same medium? Suppose two waves approach each other on a rope, one traveling to the right and one traveling to the left. The series of sketches in Figure 8.10 shows what would happen if you made this experiment. The waves pass through each other without being modified. After the encounter, each wave looks just as it did before and is traveling onward just as it did before. (How different from two particles meeting head-on!) This phenomenon of waves passing through each other unchanged can be observed with all types of waves. You can easily see that this is true for surface ripples on water. It must be true for sound waves also, since two conversations can take place across a table without distorting each other.

What happens during the time when the two waves overlap? The displacements they provide add together at each point of the medium. The

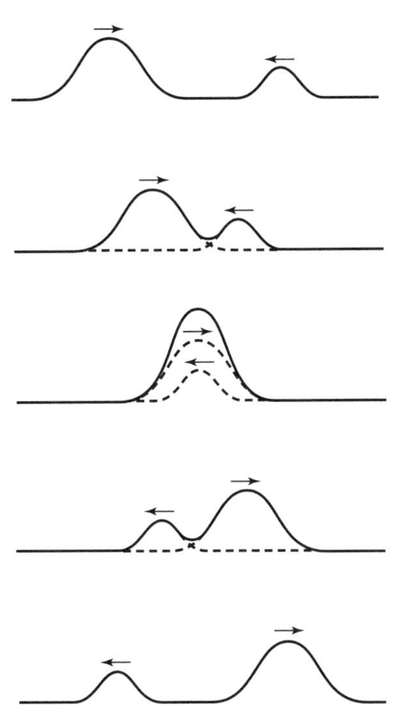

FIGURE 8.10 The superposition of two rope pulses at a point.

displacement of any point in the overlap region is just the *sum* of the displacements that would be caused at that moment by each of the two waves separately, as shown in Figure 8.10. Two waves travel toward each other on a rope. One has a maximum displacement of 0.4 cm upward and the other a maximum displacement of 0.8 cm upward. The total maximum upward displacement of the rope at a point where these two waves pass each other is 1.2 cm.

What a wonderfully simple behavior, and how easy it makes everything! Each wave proceeds along the rope making its own contribution to the rope's displacement no matter what any other wave is doing. This property of waves is called *superposition*. Using it, one can easily determine ahead of time what the rope will look like at any given instant. All one needs to do is to add up the displacements that will be caused by each wave at each point along the rope at that instant. Another illustration of wave superposition is shown in Figure 8.11. Notice that when the displacements are in opposite directions, they tend to cancel each other.

The *superposition principle* applies no matter how many separate waves or disturbances are present in the medium. In the examples just given, only

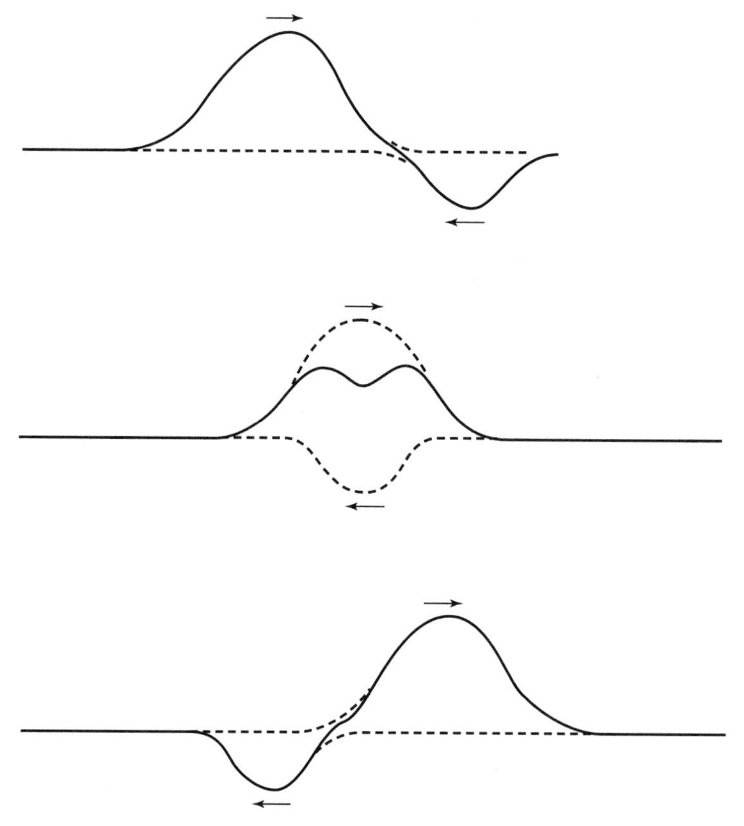

FIGURE 8.11 Superposition of two pulses on a rope.

two waves were present. But you would find by experiment that the superposition principle works equally well for three, ten, or any number of waves. Each makes its own contribution, and the net result is simply the sum of all the individual contributions (see Figure 8.12).

If waves add as just described, then you can think of a complex wave as the sum of a set of simple, sinusoidal waves. In 1807, the French mathematician Augustin Jean Fourier advanced a very useful theorem. Fourier stated that any continuing periodic oscillation, however complex, could be analyzed as the sum of simpler, regular wave motions. This, too, can be demonstrated by experiment. The sounds of musical instruments can be analyzed in this way also. Such analysis makes it possible to "imitate" instruments electronically, by combining and emitting just the right proportions of simple vibrations, which correspond to pure tones.

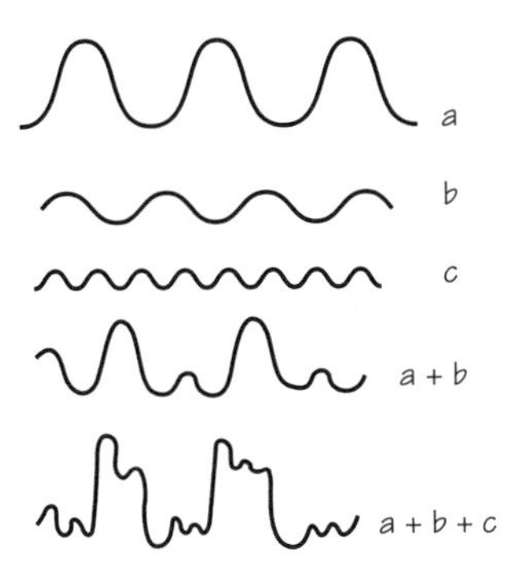

FIGURE 8.12 Sketch of complex waves as addition of two or three waves.

8.6 A TWO-SOURCE INTERFERENCE PATTERN

The figures on page 346 show ripples spreading from a vibrating source touching the water surface in a "ripple tank." The drawing shows a "cutaway" view of the water level pattern at a given instant. The image on the right introduces a phenomenon that will play an important role in later parts of the course. It shows the pattern of ripples on a water surface disturbed by *two* vibrating sources. The two small sources go through the up-and-down motions together, that is, they are in phase. Each source creates its own set of circular, spreading ripples. The image captures the pattern made by the overlapping sets of waves at one instant. This pattern is called an *interference pattern*.

You can interpret what you see here in terms of what you already know about waves. You can predict how the pattern will change with time. First, tilt the page so that you are viewing the interference pattern from a glancing direction. You will see more clearly some nearly straight gray bands. One can explain this feature by the superposition principle.

To start with, suppose that two sources produce identical pulses at the same instant. Each pulse contains one crest and one trough. (See Figure 8.16.) In each pulse the height of the crest above the undisturbed or average level is equal to the depth of the trough below. The sketches show the patterns of the water surface after equal time intervals. As the pulses spread out, the points at which they overlap move too. In the figure, a completely darkened small circle indicates where a crest overlaps another crest. A half-darkened small circle marks each point where a crest overlaps a trough. A blank small circle indicates the meeting of two troughs. According to the superposition principle, the water level should be highest at the completely darkened circles (where the crests overlap). It should be lowest at the blank

WAVES IN A RIPPLE TANK

When something drops in the water, it produces periodic wave trains of crest and troughs, somewhat as shown in the "cut-away" drawing at the left below.

Figure 8.13 is an instantaneous photograph of the shadows of ripples produced by a vibrating point source. The crests and troughs on the water surface show up in the image as bright and dark circular bands. In the photo below right, there were two point sources vibrating in phase. The overlapping waves create an interference pattern.

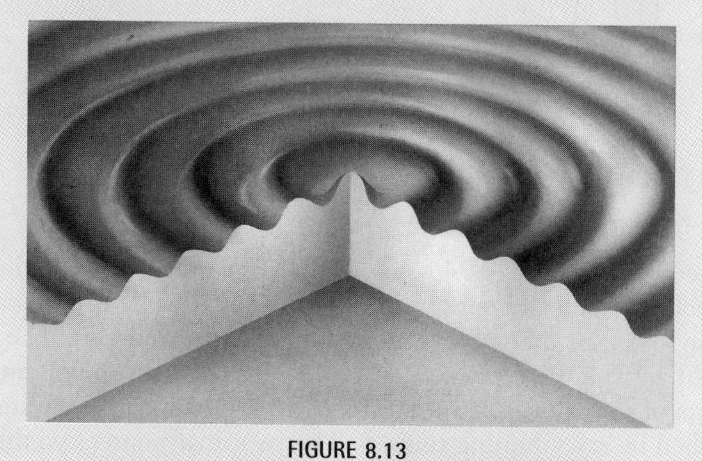

FIGURE 8.13

FIGURE 8.14

FIGURE 8.15

FIGURE 8.13–8.15 When an object drops in the water, it produces periodic wave trains of crests and troughs, somewhat as shown in the "cut-away" drawing here. Also represented here are two ripple patterns produced by one vibrating point source (left) and two point sources vibrating in phase (right). The overlapping waves create an interference pattern.

circles, and at average height at the half-darkened circles. Each of the sketches in Figure 8.16 represents the spatial pattern of the water level at a given instant.

At the points marked with darkened circles in the figure, the two pulses arrive in phase. At the points indicated by open circles, the pulses also ar-

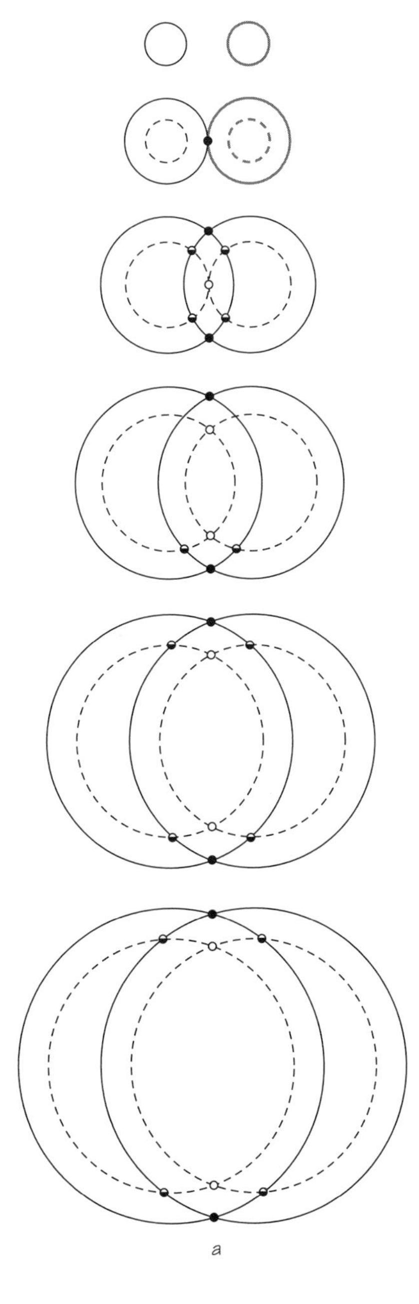

FIGURE 8.16 Pattern produced when two circular pulses, each of a crest and a trough, spread through each other. The very small circles indicate the net displacement at those points (dark circle = double height peak; half-dark circle = average level; blank circle = double depth trough).

rive in phase. In either case, the waves reinforce each other, causing a *greater* amplitude of either the crest or the trough. Thus, the waves are said to *interfere constructively*. In this case, all such points are at the same distance from each source. As the ripples spread, the region of maximum disturbance moves along the central dotted line in (a). At the points marked with half-darkened circles, the two pulses arrive completely out of phase. Here the waves cancel and so are said to interfere *destructively*, leaving the water surface undisturbed.

When two periodic waves of equal amplitude are sent out instead of single pulses, overlap occurs all over the surface, as is also shown in Figure 8.17. All along the central dotted line in Figure 8.17, there is a doubled disturbance amplitude. All along the lines labeled N, the water height remains undisturbed. Depending on the wavelength and the distance between the sources, there can be many such lines of constructive and destructive interference.

Now you can interpret the ripple tank interference pattern shown in the previous drawings (Figures 8.14 and 8.15). The gray bands are areas where waves cancel each other at all times; they are called *nodal lines*. These bands correspond to lines labeled N in the drawing above. Between these bands are other bands where crest and trough follow one another, where the waves reinforce. These are called *antinodal* lines.

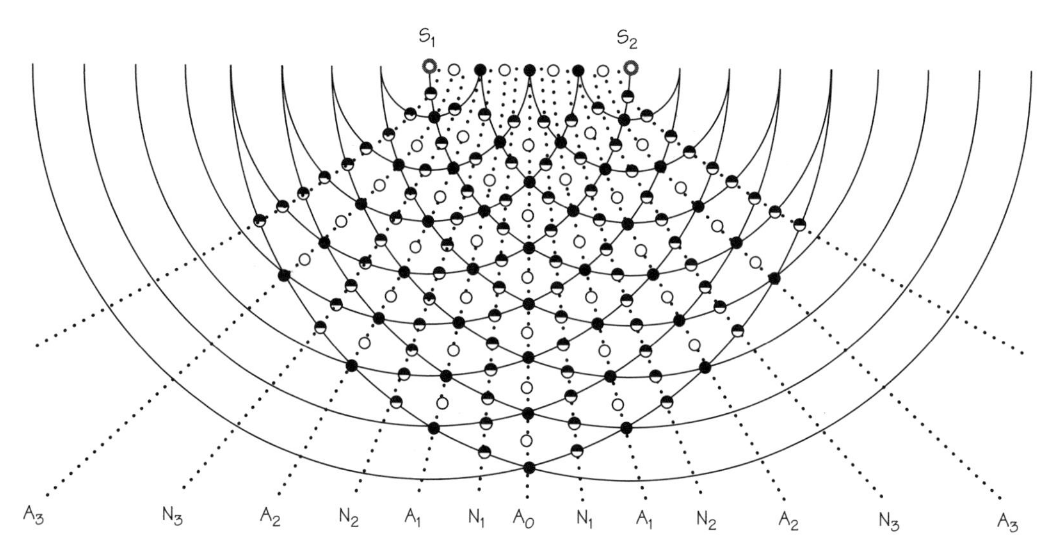

FIGURE 8.17 Analysis of interference pattern. The dark circles indicate where crest is meeting crest, the blank circles where trough is meeting trough, and the half-dark circles where crest is meeting trough. The other lines of maximum constructive interference are labeled A_0, A_1, A_2, etc. Points on these lines move up and down much more than they would because of waves from either source alone. The lines labeled N_1, N_2, etc. represent bands along which there is maximum destructive interference. Points on these lines move up and down much less than they would because of waves from either source alone.

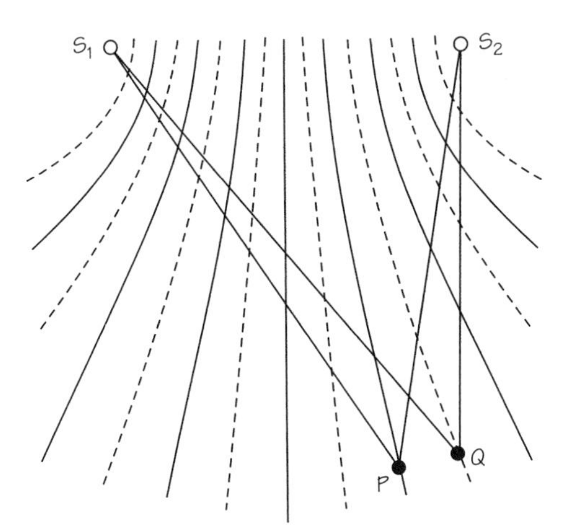

FIGURE 8.18 Detail of interference pattern.

Such an interference pattern is set up by overlapping waves from two sources. For water waves, the interference pattern can be seen directly. But whether visible or not, all waves, including earthquake waves, sound waves, or X rays, can set up interference patterns. For example, suppose two loudspeakers powered by the same receiver are working at the same frequency. By changing your position in front of the loudspeakers, you can find the nodal regions where destructive interference causes only a little sound to be heard. You also can find the antinodal regions where a strong signal comes through.

The beautiful symmetry of these interference patterns is not accidental. Rather, the whole pattern is determined by the wavelength λ and the source separation S_1S_2. From these, you could calculate the angles at which the nodal and antinodal lines spread out to either side of A_0. Conversely, you might know S_1S_2, and might have found these angles by probing around in the two-source interference pattern. If so, you can calculate the wavelength even if you cannot see the crests and troughs of the waves directly. This is very useful, for most waves in nature cannot be directly seen. Their wavelength has to be found by letting waves set up an interference pattern, probing for the nodal and antinodal lines, and calculating λ from the geometry.

The above figure shows part of the pattern of the diagram in Figure 8.17. At any point P on an *antinodal* line, the waves from the two sources arrive *in phase*. This can happen only if P is equally far from S_1 and S_2, or if P is some whole number of wavelengths farther from one source than from the other. In other words, the difference in distances $(S_1P - S_2P)$ must equal $n\lambda$, λ being the wavelength and n being zero or any whole number. At any point Q on a *nodal* line, the waves from the two sources arrive exactly *out of phase*. This occurs because Q is an odd number of half-wavelengths ($\frac{1}{2}\lambda$, $\frac{3}{2}\lambda$, $\frac{5}{2}\lambda$, etc.) farther from one source than from the other. This condition can be written $S_1Q - S_2Q = (n + \frac{1}{2})\lambda$.

The distance from the sources to a detection point may be much larger than the source separation d. In that case, there is a simple relationship between the node position, the wavelength λ, and the separation d. The wavelength can be calculated from measurements of the positions of nodal lines. (The details of the relationship and the calculation of wavelength are described in the *Student Guide* for this chapter.)

This analysis allows you to calculate from simple measurements made on an interference pattern the wavelength of any wave. It applies to water ripples, sound, light, etc. You will find this method very useful later. One important thing you can do now is find λ for a real case of interference of waves in the laboratory. This practice will help you later in finding the wavelengths of other kinds of waves.

8.7 STANDING WAVES

If you and a partner shake both ends of a taut rope with the same frequency and same amplitude, you will observe an interesting result. The interference of the identical waves coming from opposite ends results in certain points on the rope not moving at all! In between these nodal points, the entire rope oscillates up and down. But there is no apparent propagation of wave patterns in either direction along the rope. This phenomenon is called a *standing wave* or a *stationary wave*. The remarkable thing behind this phenomenon is that the standing oscillation you observe is really the effect of two *traveling* waves.

To see this, let us start with a simpler case. To make standing waves on a rope (or Slinky), there do not have to be two people shaking the opposite ends. One end can be tied to a hook on a wall or to a door knob. The train of waves sent down the rope by shaking one end back and forth will reflect back from the fixed hook. These reflected waves interfere with the new, oncoming waves, and it is this interference that can produce a standing pattern of nodes and oscillation. In fact, you can go further and tie both ends of a string to hooks and pluck (or bow) the string. From the plucked point a pair of waves go out in opposite directions, and are then reflected from the ends. The interference of these reflected waves that travel in opposite directions can produce a standing pattern just as before. The strings of guitars, violins, pianos, and all other stringed instruments act in just this fashion. The energy given to the strings sets up standing waves. Some of

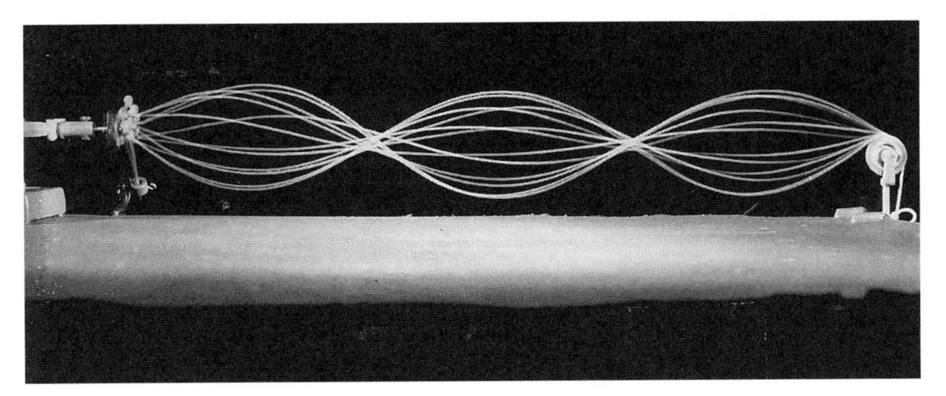

FIGURE 8.19 Time exposure: A vibrator at the left produces a wave train that runs along the rope and reflects from the fixed end at the right. The sum of the oncoming and reflected waves is a standing wave pattern.

the energy is then transmitted from the vibrating string to the body of the instrument; the sound waves sent forth from there are at essentially the same frequency as the standing waves on the string.

The vibration frequencies at which standing waves can exist depend on two factors. One is the speed of wave propagation along the string. The other is the length of the string. A connection between the length of string and the musical tone it can generate was recognized over 2000 years ago, and contributed indirectly to the idea that nature is built on mathematical principles. Early in the development of musical instruments, people learned how to produce certain pleasing harmonies by plucking a string constrained to different lengths by stops. Harmonies result if the string is plucked while constrained to lengths in the ratios of small whole numbers. Thus, the length ratio 2:1 gives the octave, 3:2 the musical fifth, and 4:3 the musical fourth. This striking connection between musical harmony and simple numbers (integers) encouraged the Pythagoreans to search for other numerical ratios or harmonies in the Universe. This Pythagorean ideal strongly affected Greek science, and many centuries later inspired much of Kepler's work. In a general form, the ideal flourishes to this day in many beautiful applications of mathematics to physical experience.

The physical reason for the appearance of harmonious notes and the relation between them were not known to the Greeks. But using the superposition principle, we can understand and define the harmonic relationships much more precisely. First, we must stress an important fact about standing wave patterns produced by reflecting waves from the boundaries of a medium. One can imagine an unlimited variety of waves traveling back and forth. But, in fact, *only certain wavelengths (or frequencies) can produce*

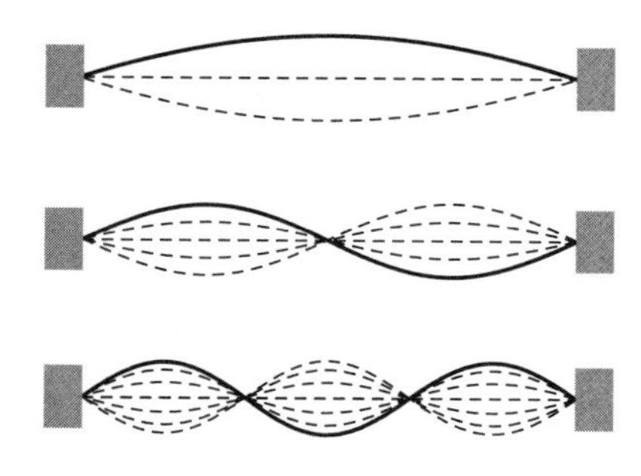

FIGURE 8.20 Standing wave patterns: first three nodes.

standing waves in a given medium. In the example of a stringed instrument, the two ends are fixed and so must be nodal points. This fact puts an upper limit on the length of standing waves possible on a fixed rope of length l. Such waves must be those for which one-half wavelength just fits on the rope ($l = \lambda/2$). Shorter waves also can produce standing patterns, having more nodes. But *always*, some whole number of one-half wavelengths must just fit on the rope, so that $l = n\lambda/2$. For example, in the first of the three illustrations in Figure 8.20, the wavelength of the interfering waves, λ_1, is just $2l$. In the second illustration, λ_2 is $\frac{1}{2}(2l)$; in the third, it is $\frac{1}{3}(2l)$. The general mathematical relationship giving the expression for all possible wavelengths of standing waves on a fixed rope is thus

$$\lambda_n = \frac{2l}{n},$$

where n is a whole number representing the harmonic. Or we can write simply,

$$\lambda_n \propto \frac{1}{n}.$$

That is, if λ_1 is the longest wavelength possible, the other possible wavelengths will be $\frac{1}{2}\lambda_1$, $\frac{1}{3}\lambda_1$, . . . $(1/n)\lambda_1$. Shorter wavelengths correspond to higher frequencies. Thus, *on any bounded medium, only certain frequencies of standing waves can be set up*. Since frequency f is inversely proportional to wavelength, $f \propto 1/\lambda$, we can rewrite the expression for all possible standing waves on a plucked string as

$$f_n \propto n.$$

VIBRATION OF A DRUM

FIGURE 8.21 A marked rubber "drumhead" vibrating in several of its possible modes. Here we see side-by-side pairs of still photographs from three of the symmetrical modes and from an anti-symmetrical mode.

In other circumstances, f_n may depend on n in some other way. The lowest possible frequency of a standing wave is usually the one most strongly present when the string vibrates after being plucked or bowed. If f_1 represents this lowest possible frequency, then the other possible standing waves would have frequencies $2f_1, 3f_1, \ldots, nf_1$. These higher frequencies are called "overtones" of the "fundamental" frequency f_1. On an "ideal" string, there are in principle an unlimited number of such frequencies, but each being a simple multiple of the lowest frequency.

In real media, there are practical upper limits to the possible frequencies. Also, the overtones are not exactly simple multiples of the fundamental frequency; that is, the overtones are not strictly "harmonic." This effect is still greater in systems more complicated than stretched strings. In a flute, saxophone, or other wind instrument, an *air column* is put into standing wave motion. Depending on the shape of the instrument, the overtones produced may not be even approximately harmonic.

As you might guess from the superposition principle, standing waves of different frequencies can exist in the same medium at the same time. A strongly plucked guitar string, for example, oscillates in a pattern which is the superposition of the standing waves of many overtones. The relative oscillation energies of the different instruments determine the "quality" of the sound they produce. Each type of instrument has its own balance of overtones. This is why a violin sounds different from a trumpet, and both sound different from a soprano voice, even if all are sounding at the same fundamental frequency.

8.8 WAVE FRONTS AND DIFFRACTION

Unlike baseballs, bullets, and other pieces of matter in motion, waves can go around corners. For example, you can hear a voice coming from the other side of a hill, even though there is nothing to reflect the sound to you. You are so used to the fact that sound waves do this that you scarcely notice it. This spreading of the energy of waves into what you might expect to be "shadow" regions is called *diffraction*.

Once again, water waves will illustrate this behavior most clearly. From among all the arrangements that can result in diffraction, we will concentrate on two. The first is shown in the second photograph in Figure 8.22. Straight water waves (coming from the bottom of the second picture) are diffracted as they pass through a narrow slit in a straight barrier. Notice that the slit is less than one wavelength wide. The wave emerges and spreads in all directions. Also notice the *pattern* of the diffracted wave. It is basi-

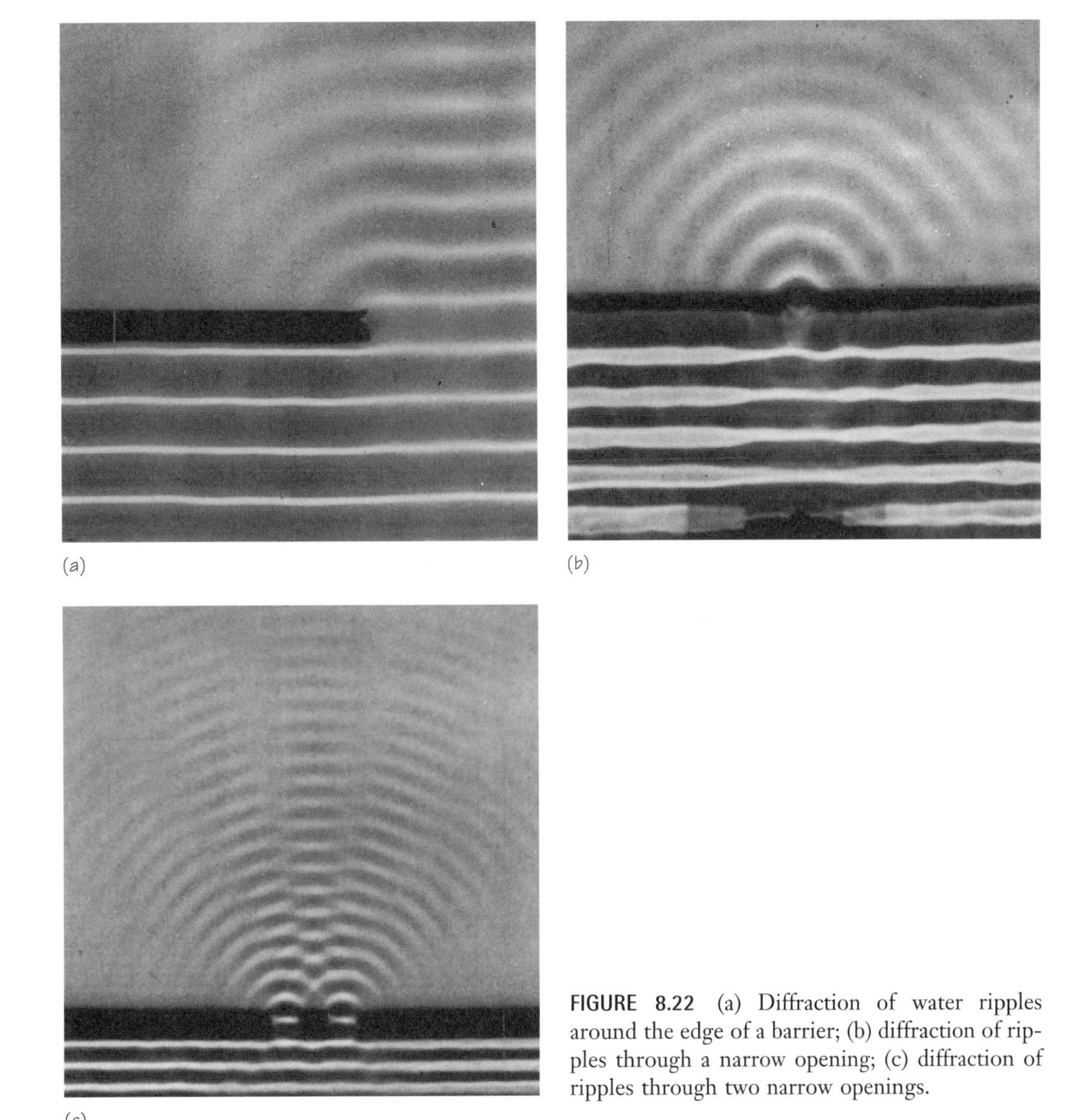

FIGURE 8.22 (a) Diffraction of water ripples around the edge of a barrier; (b) diffraction of ripples through a narrow opening; (c) diffraction of ripples through two narrow openings.

cally the same pattern a vibrating point source would set up if it were placed where the slit is.

The bottom photograph shows a second barrier arrangement. Now there are two narrow slits in the barrier. The pattern resulting from superposition of the diffracted waves from both slits is the same as that produced by two point sources vibrating in phase. The same kind of result is obtained when many narrow slits are put in the barrier; that is, the final pattern just matches that which would appear if a point source were put at the center of each slit, with all sources in phase.

One can describe these and all other effects of diffraction if one understands a basic characteristic of waves. This characteristic was first stated by Christiaan Huygens in 1678 and is now known as *Huygens' principle*. To understand it one first needs the definition of a *wave front*.

For a water wave, a wave front is an imaginary line along the water's surface, with every point along this line in exactly the same stage of vibration; that is, all points on the line are *in phase*. For example, crest lines are wave fronts, since all points on the water's surface along a crest line are in phase. Each has just reached its maximum displacement upward, is momentarily at rest, and will start downward an instant later.

Since a sound wave spreads not over a surface but in three dimensions, its wave fronts form not lines but surfaces. The wave fronts for sound waves from a very small source are very nearly spherical surfaces, just as the wave fronts for ripples, made by a very small source of waves on the surface of water, are circles.

Huygens' principle, as it is generally stated today, is that *every point on a wave front may be considered to behave as a point source for waves generated in the direction of the wave's propagation*. As Huygens said:

> There is the further consideration in the emanation of these waves, that each particle of matter in which a wave spreads, ought not to communicate its motion only to the next particle which is in the straight line drawn from the [source], but that it also imparts some of it necessarily to all others which touch it and which oppose themselves to its movement. So it arises that around each particle there is made a wave of which that particle is the center.

The diffraction patterns seen at slits in a barrier are certainly consistent with Huygens' principle. The wave arriving at the barrier causes the water in the slit to oscillate. The oscillation of the water in the slit acts as a source for waves traveling out from it in all directions. When there are two slits and the wave reaches both slits in phase, the oscillating water in each slit acts like a point source. The resulting interference pattern is similar to the pattern produced by waves from two point sources oscillating in phase.

Consider what happens behind the breakwater wall as in the aerial photograph of the harbor. By Huygens' principle, water oscillation near the end of the breakwater sends circular waves propagating into the "shadow" region.

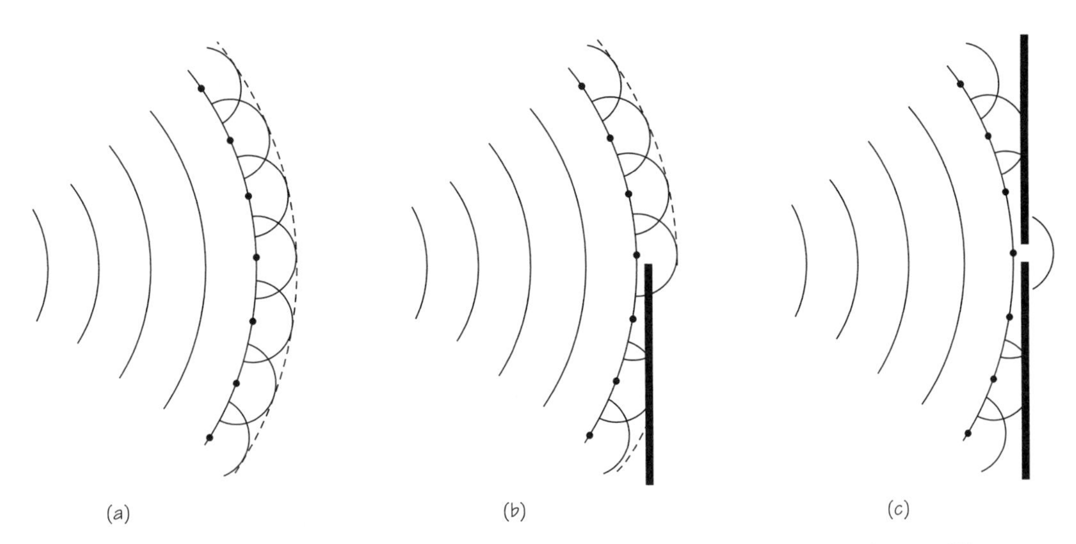

FIGURE 8.23 (a) Each point on a wave front can be thought of as a point source of waves. The waves from all the point sources interfere constructively only along their envelope, which becomes the new wave front. (b) When part of the wave front is blocked, the constructive interference of waves from points on the wave front extends into "shadow" region. (c) When all but a very small portion of a wave front is blocked, the wave propagating away from that small portion is nearly the same as that from a point source.

FIGURE 8.24 Reflection, refraction, and diffraction of water waves around an island.

You can understand all diffraction patterns if you keep both Huygens' principle and the superposition principle in mind. For example, consider a slit wider than one wavelength. In this case, the pattern of diffracted waves contains no nodal lines unless the slit width is about λ (see the series of images in Figure 8.25).

Figure 8.26 helps to explain why nodal lines appear. There must be points like P that are just λ farther from side A of the slit than from side B; that is, there must be points P for which distance AP differs from distance BP by exactly λ. For such a point, AP and OP differ by one-half wavelength, $\lambda/2$. By Huygens' principle, you may think of points A and O as in-phase point sources of circular waves. But since AP and OP differ by $\lambda/2$, the two waves will arrive at P completely out of phase. So, according to the superposition principle, the waves from A and O will cancel at point P.

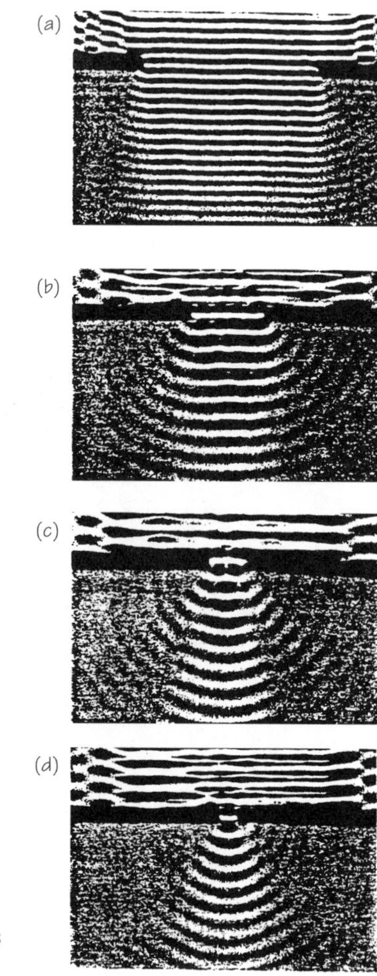

FIGURE 8.25 Single-slit diffraction of water waves with slits of different sizes.

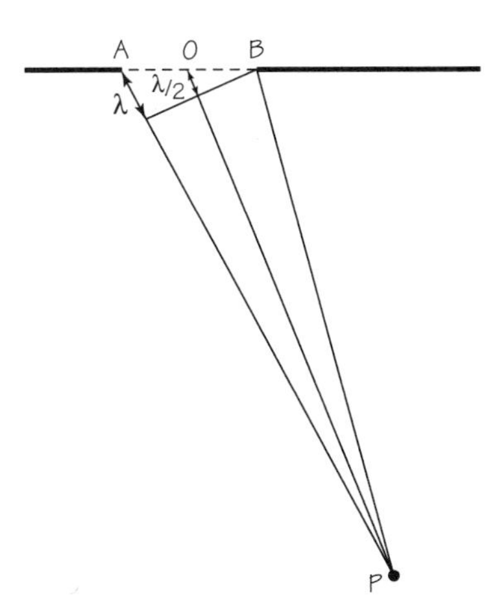

FIGURE 8.26 Diagram of a single slit showing how nodal lines appear (see text).

This argument also holds true for the pair of points consisting of the first point to the right of A and the first to the right of O. In fact, it holds true for *each* such matched pair of points, all the way across the slit. The waves originating at each such pair of points all cancel at point P. Thus, P is a nodal point, located on a nodal line. On the other hand, if the slit width is less than λ, then there can be *no* nodal point. This is obvious, since no point can be a distance λ farther from one side of the slit than from the other. Slits of widths less than λ behave nearly as point sources. The narrower they are, the more nearly their behavior resembles that of point sources.

One can compute the wavelength of a wave from the interference pattern set up where diffracted waves overlap. (See the *Student Guide* for such a calculation.) This is one of the main reasons for interest in the interference of diffracted waves. By locating nodal lines formed beyond a set of slits, you can calculate λ even for waves that you cannot see. Moreover, this

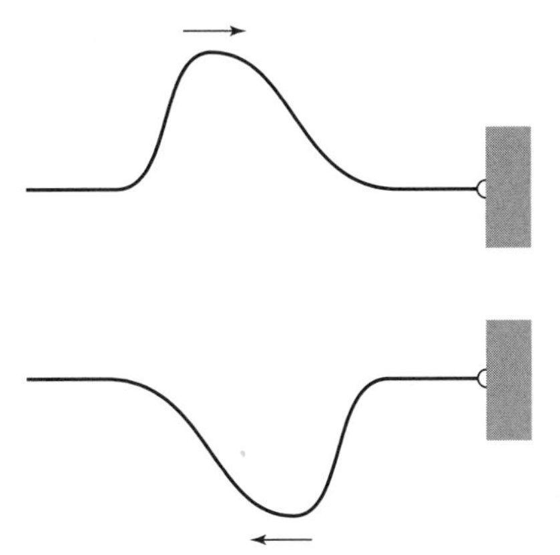

FIGURE 8.27 Wave on rope reflected from a wall to which it is attached.

is one very important way of identifying a series of unknown rays as consisting of either particles or waves.

For two-slit interference, the larger the wavelength compared to the distance between slits, the more the interference pattern spreads out. That is, as λ increases or d decreases, the nodal and antinodal lines make increasingly large angles with the straight-ahead direction. Similarly, for single-slit diffraction, the pattern spreads when the ratio of wavelength to the slit width increases. In general, diffraction of longer wavelengths is more easily detected. Thus, when you hear a band playing around a corner, you hear the bass drums and tubas better than the piccolos and cornets, even if they actually are playing equally loudly.

8.9 REFLECTION

You have seen that waves can pass through one another and spread around obstacles in their paths. Waves also are reflected, at least to some degree, whenever they reach any boundary of the medium in which they travel. Echoes are familiar examples of the reflection of sound waves. All waves share the property of being capable of reflection. Again, the superposition principle will help understand what happens when reflection occurs.

Suppose that one end of a rope is tied tightly to a hook securely fastened to a massive wall. From the other end, a pulse wave is sent down the rope toward the hook. Since the hook cannot move, the force exerted by the rope wave can do no work on the hook. Therefore, the energy carried in the wave cannot leave the rope at this fixed end. Instead, the wave bounces back, is *reflected*, ideally with the same energy.

What does the wave look like after it is reflected? The striking result is that the wave seems *to flip upside down* on reflection. As the wave comes in from left to right and encounters the fixed hook, it pulls up on it. By Newton's third law, the hook must exert a force on the rope in the opposite direction while reflection is taking place. The details of how this force varies in time are complicated, but the net effect is that an inverted wave of the same form is sent back down the rope.

The three sketches in Figure 8.28 show the results of reflection of water waves from a straight wall. You can check whether the sketches are accurate by trying to reproduce the effect in a sink or bathtub. Wait until the water is still, then dip your fingertip briefly into the water, or let a drop fall into the water. In the upper part of the sketch, the outer crest is approaching the barrier at the right. The next two sketches show the po-

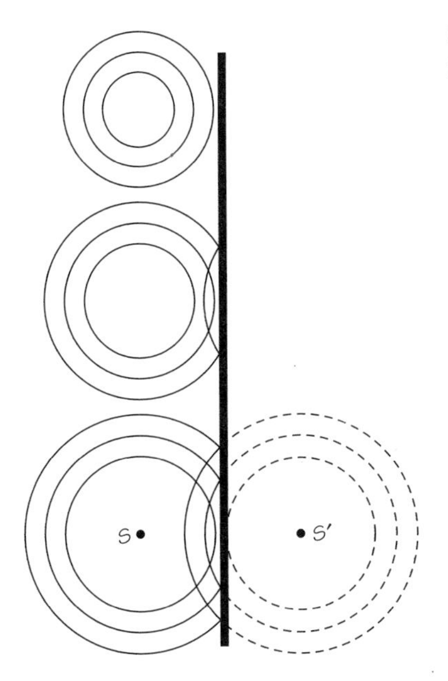

FIGURE 8.28 Two-dimensional circular wave reflecting from a wall.

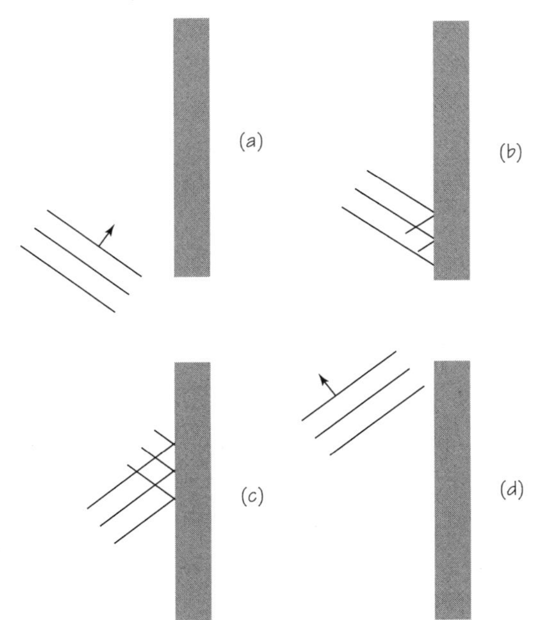

FIGURE 8.29 Two-dimensional plane wave reflecting from a wall.

FIGURE 8.30 Angles of incidence and reflection.

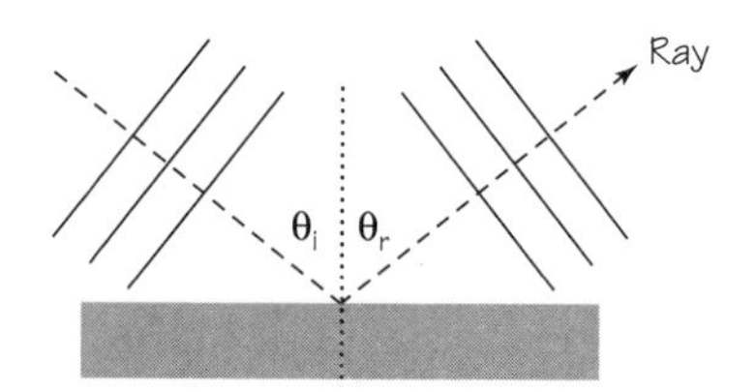

sitions of the crests after first one and then two of them have been reflected. Notice the dashed curves in the last sketch. They show that the reflected wave appears to originate from a point S′ that is as far behind the barrier as S is in front of it. The imaginary source at point S′ is called the *image* of the source S.

Reflection of circular waves is studied first, because that is what you usually notice first when studying water waves. But it is easier to see a general principle for explaining reflection by observing a straight wave front, reflected from a straight barrier. The ripple-tank photograph (Figure 8.32a) shows one instant during such a reflection. (The wave came in from the upper left at an angle of about 45°.) The sketches below indicate in more detail what happens as the wave crests reflect from the straight barrier.

The description of wave behavior is often made easier by drawing lines perpendicular to the wave fronts. Such lines, called *rays*, indicate the direction of propagation of the wave. Notice Figure 8.30 for example. Rays have been drawn for a set of wave crests just before reflection and just after reflection from a barrier. The straight-on direction, perpendicular to the reflecting surface, is shown by a dotted line. The ray for the *incident*

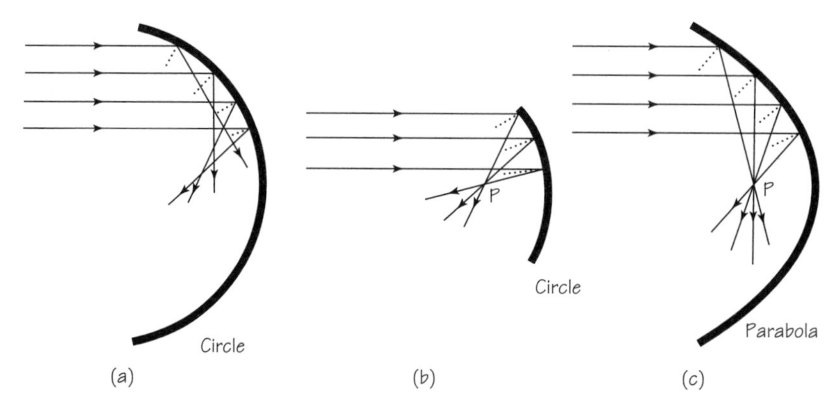

FIGURE 8.31 Rays reflecting from concave surfaces (circular and parabolic).

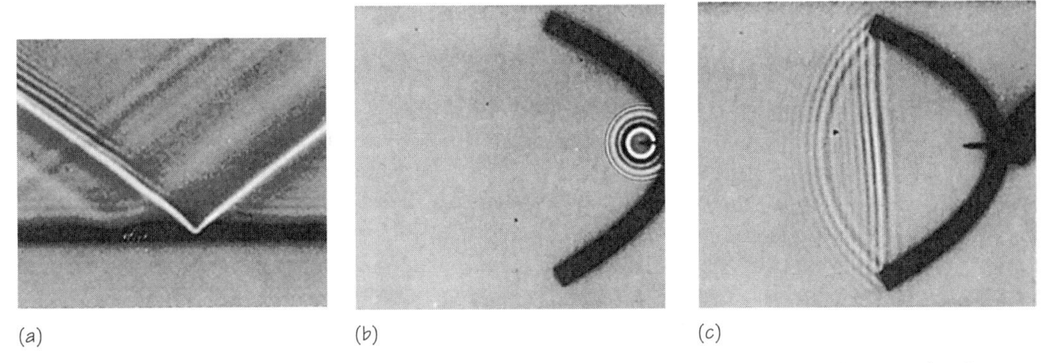

(a) (b) (c)

FIGURE 8.32 (a) Reflection of a water wave from a wall; (b) and (c) ripple tank photographs showing how circular waves produced at the focus of a parabolic wall are reflected from the wall into straight waves.

crests makes an angle θ_i with the straight-on direction. The ray for the *reflected* crests makes an angle θ_r with it. The *angle of reflection* θ_r is equal to the *angle of incidence* θ_i; that is,

$$\theta_r = \theta_i.$$

This is an experimental fact, which you can easily verify.

Many kinds of wave reflectors are in use today. One can find them in radar antennae or infrared heaters. Figure 8.31 (a) and (b) shows how straight-line waves reflect from two circular reflectors. A few incident and reflected rays are shown. (The dotted lines are perpendicular to the barrier surface.) Rays reflected from the half-circle (a) head off in all directions. However, rays reflected from a small segment of the circle (b) come close to meeting at a single point. A barrier with the shape of a parabola (c) focuses straight-line *rays*, quite precisely at a point—which is to say that a parabolic surface reflects *plane waves* to a sharp focus. An impressive example is a radio telescope. Its huge parabolic surface reflects faint radio waves from space to focus on a detector. Another example is provided by the dish used for satellite TV reception.

The wave paths indicated in the sketches could just as well be reversed. For example, spherical waves produced at the focus become plane waves when reflected from a parabolic surface. The flashlight and automobile headlamp are familiar applications of this principle. In them, white-hot wires placed at the focus of parabolic reflectors produce almost parallel beams of light.

RADAR WAVES AND TECHNOLOGY

One of the highly useful wave phenomena we shall encounter in Chapter 12 is the propagation and reflection of electromagnetic waves, such as light and microwaves. An example of the latter, as a preview of the general usefulness of the idea of wave propagation, is Radar.

Radar (an acronym for Radio Detection and Ranging) is an electromagnetic sensor for detecting, locating, tracking, and identifying various kinds of objects at various distances. The most popular form of radar signal is made up of short continual pulses, and the shorter the width of this pulse, the more accurate the radar is at locating the target

Radar, developed during World War II, is credited with being the key technology that prevented a victory by the German air force, especially during its bombing campaign on cities in England. It shifted the course of the war, and forever changed the face of military, astronomical, and weather technology.

Radar is now used for innumerable tasks, all of which require some form of detection at distance: police detection of speeders, air traffic controllers following the path of aircraft, satellite detection of topography on Earth and on other bodies in the solar system—all utilize the basic principles of radar. These principles rest on two fundamental effects of radio waves, the echo effect and the Doppler shift.

The echo is a very familiar phenomenon—shout out your name in a large empty room and the walls seem to shout it back. This type of echo results from the reflection of sound waves off a surface. The same effect can be obtained with radio waves, which travel at the speed of ordinary (visible) light in space. A pulse of radio waves sent out from an antenna will reflect from any object it hits, and part of the wave will return to where it originated. The time it takes between the emission of the pulse and the reception of the reflected part of the pulse can be used to determine the distance between the point and the reflecting surface. Thus the radar stations in England could be alerted that a hostile aircraft was present even if it was still far away, and fighter planes could scramble to fight off the expected attack.

The Doppler shift, also common in everyday life, occurs when waves of any kind are emitted or reflected by a moving body. (Everyone has experienced it as the shift in frequency of a car horn or a train whistle while in motion.) If waves sent out from a point are reflected by a moving body, the returning waves will appear to have a higher frequency as the object moves toward the original point, and a lower frequency as the object moves away from it. Therefore measuring the Doppler shift of reflected waves can be used to determine quickly the speed and direction of the reflecting surface.

Since these principles apply to sound waves, it is possible to make a "sound radar," or Sonar. That device works well enough in water, and has been used with great success to detect and trace submarines. But sonar is impractical for use in the air, because there, ordinary sound waves travel less far, and their echo would be too faint to be useful in precise detections.

After a radar transmitter has sent out a short burst, or pulse, of radio waves, the transmitter shuts off and a receiver is turned on, measuring the time and Doppler shift of the detected reflection. From monitoring the movement of cars on Route 1 to mapping the surface of Venus, radar has given us a new way to see the world.

In a standard radar system there is a transmitter, which produces a signal in the form of electromagnetic energy that is sent to the

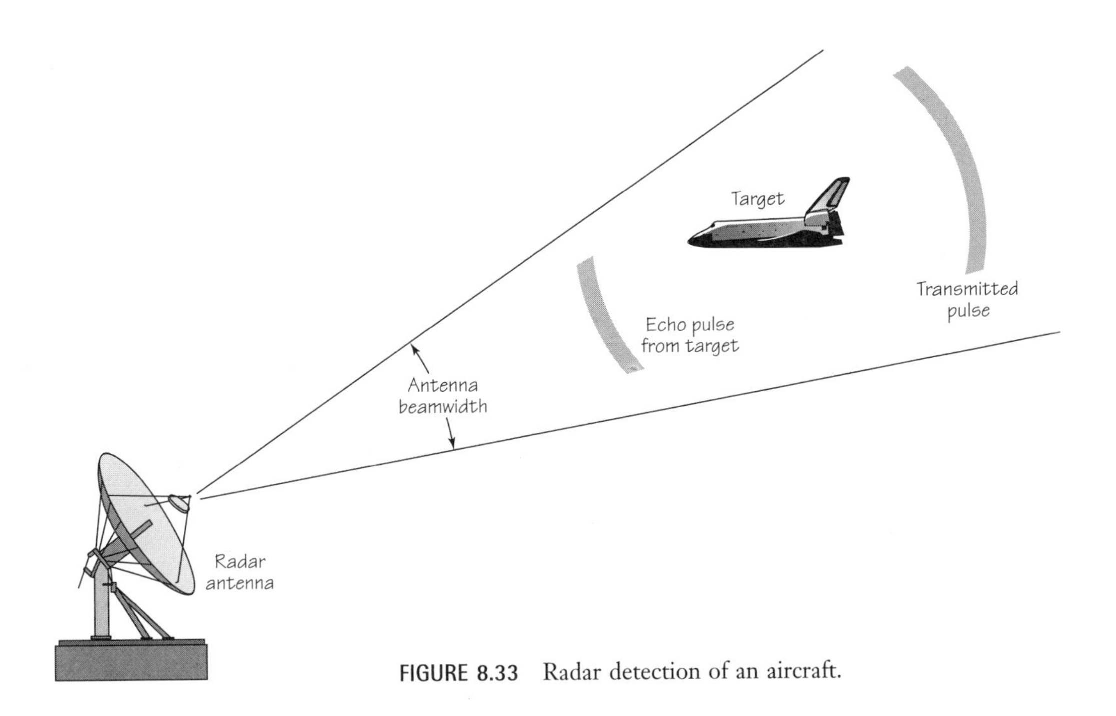

FIGURE 8.33 Radar detection of an aircraft.

antenna. A radar antenna is commonly a parabolic reflector, with a small antenna placed at the center of the parabola to illuminate the surface of the reflector. The electromagnetic energy is radiated from this surface in the form of a narrow beam. When this electromagnetic beam passes over a target, the object reflects an amount of the radiated energy back to the radar, where a receiver filters and amplifies the echoes. The single processor then differentiates the signals obtained from a target from those produced by clutter, such as atmospheric effects. A computer then processes this information and the output is displayed on a monitor. With the information that is provided by the radar it is possible to calculate the location of the target in terms of range and angular direction.

Development of Radar

The scientific origins of radar can be found in the work of the German physicist Heinrich Hertz. Hertz showed, as Maxwell's equations had predicted, that radio waves exist, and in the same way as light waves, are reflected from metallic objects. However, not until the 1930s did radar become a focus for scientific research, largely in response to the fear of war. The most important needs of a radar transmitter are often conflicting. For example, to guarantee the greatest accuracy a transmitter must be powerful and have a wide bandwidth, but at the same time it cannot be too heavy or too large, to fit into an aircraft or ship. One of the main objectives for researchers working on radar during World War II was to solve these conflicts by developing a system that was small enough to fit in fighter aircraft but worked at higher frequencies.

The benefits of using microwaves, waves of wavelength 1 m or less, in radar technology had been recognized for some time. They included greater accuracy, increased efficiency at reducing clutter, and an expanded potential

■ RADAR WAVES AND TECHNOLOGY (*Continued*)

for discriminating between targets. The cavity magnetron, which was invented in Britain in 1939 at the University of Birmingham, opened up the possibility of using microwaves and was small enough to fit in the palm of a hand. In 1940, as part of a transatlantic scientific exchange, a prototype of the magnetron was sent to America where it soon became the basis for some of the most important work on radar during World War II. As a direct spinoff of this exchange, a new laboratory, known as the Radiation Laboratory, was founded at the Massachusetts Institute of Technology that helped to develop more than 150 radar systems between 1940 and 1950.

Current Uses of Radar Technology

Radar technology is utilized today in many different ways. Armed forces all over the world continue to use radar as a detector of aircraft and ships, as they did in World War II. However, it is now also used to distinguish many different kinds of targets, to control and guide weapons, and to provide information on the damage caused by these weapons. Sophisticated weather forecasting techniques are also highly dependent on radar technology in the form of remote sensing. Radar technology is also crucial to civilian air traffic control, where it provides information on air traffic and weather conditions, as well as a tool for guiding pilots in unfavorable weather conditions.

Further Reading

R. Buderi, *The Invention that Changed the World* (New York: Touchstone Books, 1998).

8.10 REFRACTION

What happens when a wave propagates from one medium to another medium in which its speed of propagation is different? Look at the simple situation pictured in Figure 8.35. Two one-dimensional pulses approach a boundary separating two media. The speed of the propagation in medium 1 is greater than it is in medium 2. Imagine the pulses to be in a light rope (medium 1) tied to a relatively very heavy rope (medium 2). Part of each pulse is reflected at the boundary. This reflected component is flipped upside down relative to the original pulse. (Recall the inverted reflection at a hook in a wall discussed earlier. The heavier rope here tends to hold the boundary point fixed in just the same way.) But what happens to that part of the wave that continues into the second medium?

As shown in the figure, the transmitted pulses are closer together in medium 2 than they are in medium 1. The reason is that the speed of the pulses is less in the heavier rope. So the second pulse, while still in the light

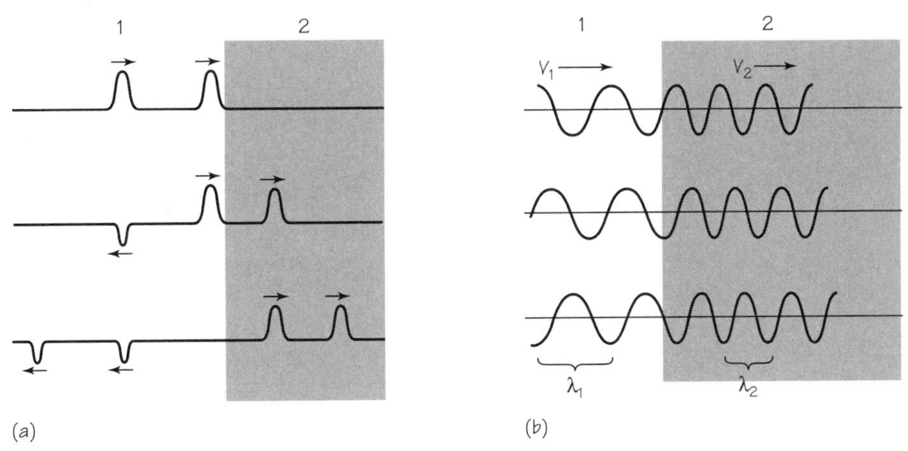

FIGURE 8.34 (a) Pulses encountering a boundary between two different media. (b) continuous wave train crossing the boundary between two different media. In both cases, the speed of propagation is less in medium 2.

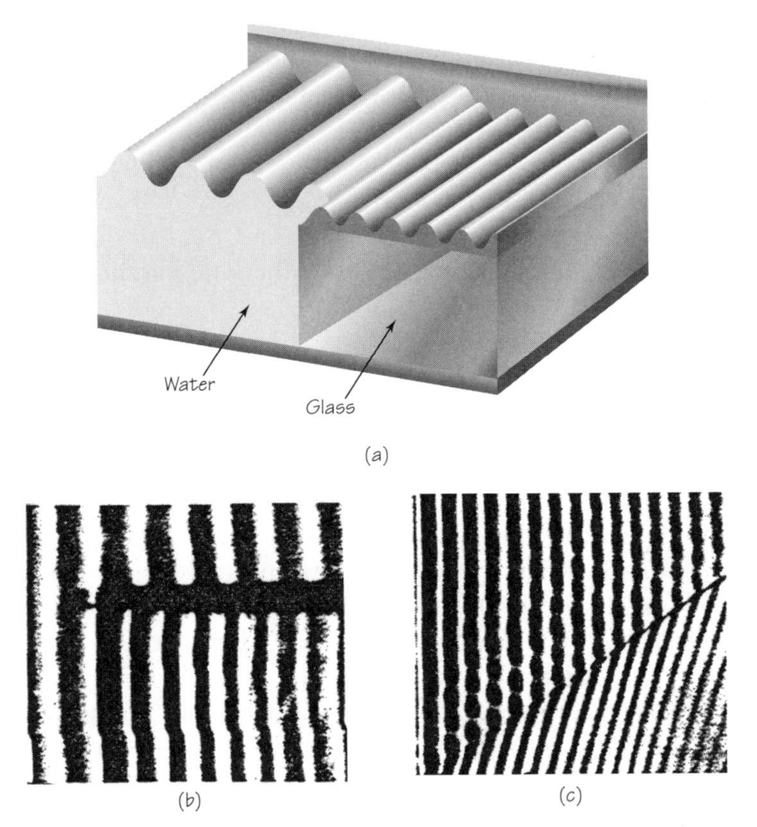

FIGURE 8.35 (a) Cross section of waves in a ripple tank; (b) ripples on water (coming from the left) encounter the shallow region over the corner of a submerged glass plate; (c) ripples on water (coming from the left) encounter a shallow region over a glass plate placed at an angle to the wave fronts.

rope, is catching up with the one that is already in the heavy rope. For the same reason, each separate pulse is itself squeezed into a narrower form; that is, when the front of a pulse has entered the region of less speed, the back part of it is still moving ahead with greater speed.

Something of the same sort happens to a periodic wave at such a boundary. This situation is pictured in Figure 8.35b. For the sake of simplicity, assume that all of the wave is transmitted and none of it is reflected. Just as the two pulses were brought closer and each pulse was squeezed and narrowed, the periodic wave pattern is squeezed together, too. Thus, the wavelength λ_2 of the transmitted wave is shorter than the wavelength λ_1 of the incoming, or incident, wave.

Although the wavelength changes when the wave passes across the boundary, the frequency of the wave cannot change. If the rope is unbroken, the pieces immediately on either side of the boundary must go up and down together. The frequencies of the incident and transmitted waves must, then, be equal. We can simply label both of them f.

The wavelength, frequency, and speed relationship for both the incident and transmitted waves can be written separately

$$\lambda_1 f = v_1 \qquad \text{and} \qquad \lambda_2 f = v_2.$$

Dividing one of these equations by the other and eliminating f:

$$\frac{\lambda_1}{\lambda_2} = \frac{v_1}{v_2}.$$

This equation tells us that the ratio of the wavelengths in the two media equals the ratio of the speeds.

The same sort of thing happens when water ripples cross a boundary. Experiments show that the ripples move more slowly in shallower water. A piece of plate glass is placed on the bottom of a ripple tank to make the water shallower there. This creates a boundary between the deeper and shallower part (medium 1 and medium 2). Figure 8.35a shows the case where this boundary is parallel to the crest lines of the incident wave. As with rope waves, the wavelength of water waves in a medium is proportional to the speed in that medium.

Water waves offer a possibility not present for rope waves. The crest lines can approach the boundary at any angle, not only head-on. Photograph (c) shows such an event. A ripple-tank wave approaches the boundary at an angle. The wavelength and speed, of course, change as the wave passes across the boundary. The *direction* of the wave propagation also changes. As each

FIGURE 8.36 Aerial photograph of the refraction of ocean waves approaching shore.

part of a crest line in medium 1 enters medium 2, its speed decreases, and it starts to lag behind. In time, the directions of the whole set of crest lines in medium 2 are changed from their directions in medium 1.

This phenomenon is called *refraction*. Refraction occurs whenever a wave passes into a medium in which the wave velocity is different. In this case, the wave fronts are turned (refracted) so that they are more nearly parallel to the boundary. (See Figures 8.35a and 8.35b.) This accounts for something that you may have noticed if you have been at an ocean beach. No matter in what direction the waves are moving far from the shore, when they come near the beach that slopes gently into the deeper water, their crest lines are nearly parallel to the shoreline. A wave's speed is steadily reduced as it moves into water that gets gradually more shallow. So the wave is refracted continuously as if it were always crossing a boundary between different media, as indeed it is. The refraction of sea waves, coming from one direction, can be so great that wave crests can curl around a very small island with an all-beach shoreline and provide surf on all sides.

The slowing of starlight by increasingly dense layers of the atmosphere produces refraction that changes the apparent position of the star.

8.11 SOUND WAVES

Sound waves are mechanical disturbances that propagate through a medium, such as the air. Typically, sound waves are *longitudinal* waves, producing changes of density and pressure in the medium through which they travel. The medium can be a solid, liquid, or gas. If the waves strike the ear, they cause the ear drum to vibrate, which produces a signal in the acoustic nerve that can produce the sensation of hearing. The biology and psychology of hearing, as well as the physics of sound, are important to the science of acoustics. Here, of course, we will concentrate on sound as an example of wave motion. Sound has all the properties of wave motion considered so far. It exhibits reflection, refraction, diffraction, and the same relations among frequency, wavelength, and propagation speed and interference. Only the property of polarization is missing, because sound waves are longitudinal, not transverse. In addition, sound waves travel faster in cold air than in hot air because of the increased density of the medium, air, when it is cold.

Vibrating sources for sound waves may be as simple as a tuning fork or as complex as the human larynx with its vocal cords. Tuning forks and some electronic devices produce a steady "pure tone." Most of the energy in such a tone is in simple harmonic motion at a single frequency. Frequency is often measured in units of hertz (Hz), where 1 Hz is one cycle (or oscillation) per second; 1 Hz = 1/s. (There is no unit for "cycle.")

The normal human ear can hear sound waves with frequencies between about 20 Hz and 15,000 Hz. Dogs can hear over a much wider range (15 Hz–50,000 Hz). Bats, porpoises, and whales generate and respond to frequencies up to about 120,000 Hz.

Loudness (or "volume") of sound is, like pitch, a psychological variable. Loudness is strongly related to the *intensity* of the sound. Sound intensity is a physical quantity. It is defined in terms of the energy carried by the wave and is usually measured in the number of watts per square centimeter transmitted through a surface perpendicular to the direction of motion of a wave front. The human ear can perceive a vast range of intensities of sound. Figure 8.37 illustrates this range. It begins at a level of 10^{-16} W/cm^2 (relative intensity = 1). Below this "threshold" level, the normal ear does not perceive sound. It is customary to measure loudness in decibels (db). The number of decibels is 10 times the exponent in the relative intensity of the sound. Thus, a jet plane at takeoff, making a noise of 10^{14} relative intensity, is said to emit noise at the 140-db level.

Levels of noise intensity about 10^{12} times threshold intensity can be felt as an uncomfortable tickling sensation in the normal human ear. Beyond

Relative intensity	Sound
1	Threshold of hearing
10^1	Normal breathing
10^2	Leaves in a breeze
10^3	
10^4	Library
10^5	Quiet restaurant
10^6	Two-person conversation
10^7	Busy traffic
10^8	Vacuum cleaner
10^9	Roar of Niagara Falls
10^{10}	Subway train
10^{11}	
10^{12}	Propeller plane at takeoff
10^{13}	Machine-gun fire
10^{14}	Small jet plane at takeoff
10^{15}	
10^{16}	Wind tunnel
10^{17}	Space rocket at lift-off

FIGURE 8.37 Loudness chart.

that, the sensation changes to pain and may damage the unprotected ear. Since many popular music concerts produce, in an auditorium, steady sound levels at this intensity (and above it for the performers), there are many cases of impaired hearing among people extensively exposed to such sound.

Often the simplest way of reducing noise is by *absorbing* it after it is produced but before it reaches your ears. Like all sound, noise is the energy of back and forth motion of the medium through which the noise travels. Noisy machinery can be muffled by padded enclosures in which the energy of noise is changed to heat energy, which then dissipates. In a house, a thick rug on the floor can absorb 90% of room noise. (A foot of fresh fluffy snow is an almost perfect absorber of noise outdoors. Cities and countrysides are remarkably hushed after a snowfall.)

It has always been fairly obvious that sound takes time to travel from source to receiver. By timing echoes over a known distance, the French mathematician Marin Mersenne in 1640 first computed the speed of sound in air. It took another 70 years before William Derham in England, comparing the flash and noise from cannons across 20 km, came close to the modern measurements. Sound in air at sea level at 20°C moves at about 344 m/s. As for all waves, the speed of sound waves depends on the properties of the medium: the temperature, density, and elasticity. Sound waves generally travel faster in liquids than in gases, and faster still in solids. In seawater, their speed is about 1500 m/s; in steel, about 5000 m/s; in quartz, about 5500 m/s.

Interference of sound waves can be shown in a variety of ways. In a large hall with hard, sound-reflecting surfaces, there will be "dead" spots. At these spots, sound waves coming together after reflection tend to cancel each other. Acoustic engineers must consider this in designing the shape, position, and materials of an auditorium. Another interesting and rather dif-

FIGURE 8.38 Boston Symphony concert hall.

ferent example of sound interference is the phenomenon known as *beats*. When two notes of slightly different frequency are heard together, they interfere. This interference produces beats, a rhythmic pulsing of the sound. Piano tuners and string players use this fact to tune two strings to the same pitch. They simply adjust one string or the other until the beats disappear.

Refraction of sound by different layers of air explains why you sometimes cannot hear the thunder after seeing lightning. Similar refraction of sound occurs in layers of water of different temperatures. Geologists use the refraction of sound waves to study the Earth's deep structure and to locate fossil fuels and minerals. Very intense sound waves are set up in the ground (as by dynamite blasts). The sound waves travel through the Earth and are received by detection devices at different locations. The path of the waves, as refracted by layers in the Earth, can be calculated from the relative intensities and times of sound received. From knowledge of the paths, estimates can be made of the composition of the layers.

As mentioned, diffraction is a property of sound waves. Sound waves readily bend around corners and barriers to reach the listener within range.

Sound waves reflect, as do rope or water waves, wherever they encounter a boundary between different media. The architectural features called "whispering galleries" show vividly how sound can be focused by reflection from curved surfaces. All these effects are of interest in the study of acoustics. Moreover, the proper acoustical design of public buildings is now recognized as an important function by all good architects.

So far in this chapter, you have studied the basic phenomena of mechanical waves, ending with the theory of sound propagation. The explanations of these phenomena were considered the final triumph of Newtonian mechanics as applied to the transfer of energy of particles in motion. Most of the general principles of acoustics were discovered in the 1870s. Since then, perhaps its most important influence on modern physics has been its effect on the imagination of scientists. The successes of acoustics encouraged them to take seriously the power of the wave viewpoint, even in fields far from the original one—the mechanical motion of particles that move back and forth or up and down in a medium.

We now turn to an especially important type of wave phenomenon—light.

B. LIGHT

8.12 WHAT IS LIGHT?

The conviction that the world and everything in it consists of *matter in motion* drove scientists prior to the twentieth century to search for mechanical models for light as well as heat; that is, they tried to imagine how the effects of light, heat, and other phenomena could be explained in detail as the action of material objects. For example, consider the way light bounces off a mirror. A model for this effect might picture light as consisting of particles of matter that behave somewhat like tiny ping-pong balls. On the other hand, light exhibits interference and diffraction, suggesting a model involving waves. Such mechanical models were useful for a time, but in the long run proved far too limited. Still, the search for these models led to many new discoveries, which in turn brought about important changes in science, technology, and society.

In most basic terms, light is a form of energy. The physicist can describe a beam of light by stating measurable values of its speed, wavelength or frequency, and intensity. But to scientists, as to all people, "light" also means brightness and shade, the beauty of summer flowers and fall foliage, of red sunsets, and of the canvases painted by masters. These are different ways

FIGURE 8.39 Roman temple in Evora, Portugal. Light beams travel in straight lines, as shown by the shadow lines.

of appreciating light. One way concentrates on light's measurable aspects, an approach enormously fruitful in physics and technology. The other way concerns aesthetic responses to viewing light in nature or art. Still another way of considering light deals with the biophysical process of vision.

These aspects of light are not easily separated. Thus, in the early history of science, light presented more subtle and more elusive problems than did most other aspects of physical experience. Some Greek philosophers believed that light travels in straight lines at high speed and contains particles that stimulate the sense of vision when they enter the eye. For centuries after the Greek era, this particle model survived almost intact. Around 1500, Leonardo da Vinci, noting a similarity between sound echoes and the reflection of light, speculated that light might have a wave character.

A decided difference of opinion about the nature of light emerged among scientists of the seventeenth century. Some, including Newton, favored a model largely based on the idea of light as a stream of particles. Others, including Huygens, supported a wave model. By the late nineteenth century, there appeared to be overwhelming evidence in support of the wave model. This part of the chapter will deal with the question: *How accurate is a wave model in explaining the observed behavior of light?* The wave model will be taken as a hypothesis, and the evidence that supports it examined. Remember that any scientific model, hypothesis, or theory has two chief functions: to explain what is known, and to make predictions that can be tested experimentally. Both of these aspects of the wave model will be discussed. The result will be rather surprising. The wave model turns out to work splendidly to this day for all properties of light known before the twentieth century. But in Chapter 13 you will find that for some purposes a particle model must be used. Then in Chapter 15 *both* models will be combined, merging two apparently conflicting theories.

The ancient opinion, later proved by experiment, that light travels in straight lines and at high speed has been mentioned. The daily use of mirrors shows that light can also be reflected. Light can also be refracted, and it shows the phenomena of interference and diffraction, as well as other phenomena characteristic of waves, such as dispersion, polarization, and scattering. All of these characteristics lent strong support to the wave model of light.

8.13 PROPAGATION OF LIGHT

There is ample evidence that light travels in straight lines. A shadow cast by an object intercepting sunlight has well-defined outlines. Similarly, sharp shadows are cast by smaller sources closer by. The distant Sun and the nearby small source are approximate *point* sources of light. Such point sources produce sharp shadows.

Images can also demonstrate that light travels in straight lines. Before the invention of the modern camera with its lens system, a light-tight box with a pinhole in the center of one face was widely used. As the *camera obscura* (meaning "dark chamber" in Latin), the device was highly popular in the Middle Ages. Leonardo da Vinci probably used it as an aid in his sketching. In one of his manuscripts he says that "a small aperture in a window shutter projects on the inner wall of the room an image of the bodies which are beyond the aperture." He includes a sketch to show how the straight-line propagation of light explains the formation of an image.

It is often convenient to use a straight line to represent the direction in which light travels. The pictorial device of an infinitely thin *ray* of light is useful for thinking about light. But no such rays actually exist. A light beam

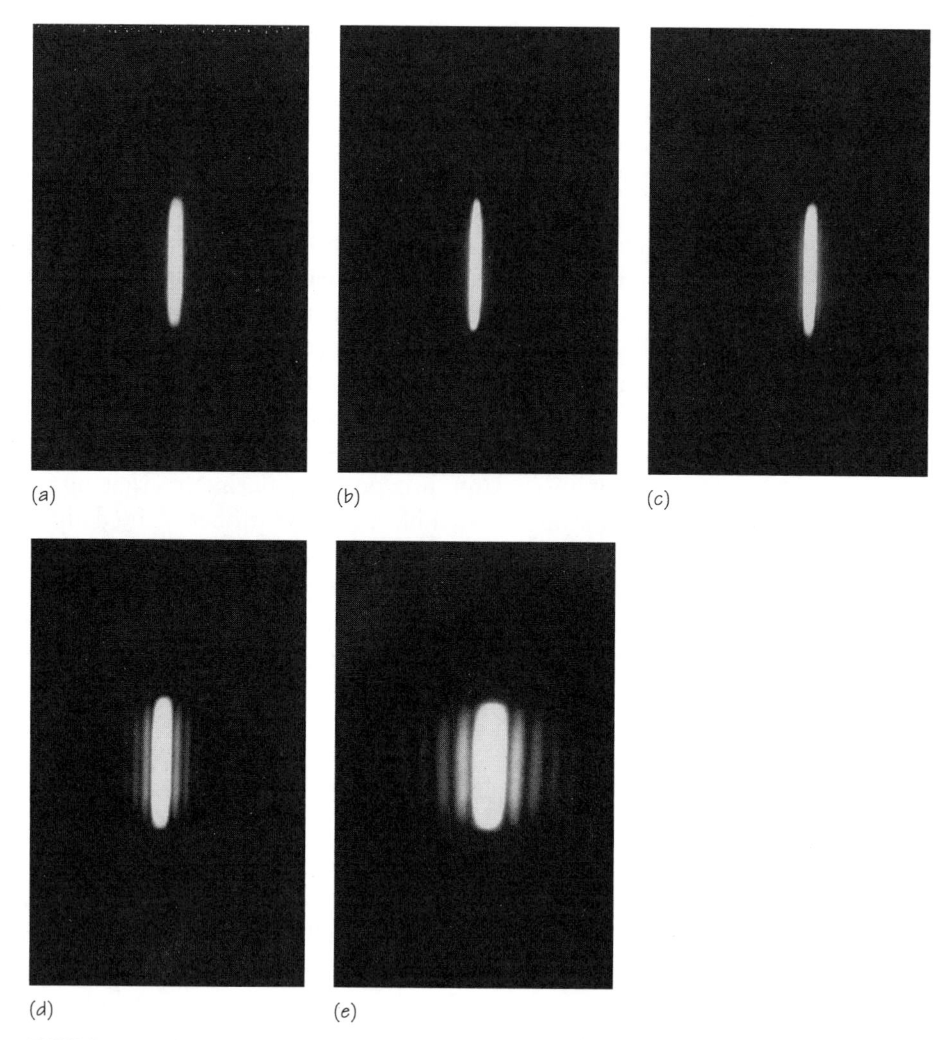

FIGURE 8.40 An attempt to produce a "ray" of light. To make the pictures, a parallel beam of red light was directed through increasingly narrow slits to a photographic plate. Of course, the narrower the slit, the less light gets through. This was compensated for by longer exposures in these photographs. The slit widths were (a) 1.5 mm; (b) 0.7 mm; (c) 0.4 mm; (d) 0.2 mm; and (e) 0.1 mm.

emerging from a good-sized hole in a screen is as wide as the hole. You might expect that if you made the hole extremely small, you would get a very narrow beam of light, ultimately just a single ray. This is not the case. Diffraction effects, such as those observed for water and sound waves, appear when the beam of light passes through a small hole. So an infinitely thin ray of light, although it is pictorially useful, cannot be produced in practice. But the idea can still be used in order to *represent the direction* in which a train of waves in a beam of light is traveling.

The beam of light produced by a laser comes as close as possible to the ideal case of a thin, parallel bundle of rays. As you will find in Chapter 14, light is often produced by the action of electrons within the atoms of its source. Lasers are designed in such a way that their atoms produce light in unison with one another, rather than individually and at random, as in other sources of light. As a result, light from a laser can yield a total beam of considerable intensity, and one that is much more nearly monochromatic—that is, of a single color—than light from any conventional source. In addition, since the individual wavelets from the atoms of a laser are produced simultaneously, they are able to interfere with each other constructively to produce a beam of light that is narrow and very nearly parallel. In fact, such light spreads out so little that thin beams emitted by lasers on Earth, when directed at the surface of the Moon 400,000 km away, have been found to produce spots of light only 1 m in diameter on the Moon.

Given that light can be considered to travel in straight lines, can we tell how fast it goes? Galileo discussed this problem in his *Two New Sciences* (published in 1638). He pointed out that everyday experiences might lead one to conclude that light propagates instantaneously. But these experiences, when analyzed more closely, really show only that light travels much faster than sound. For example, "when we see a piece of artillery fired, at a great distance, the flash reaches our eyes without lapse of time; but the sound reaches the ear only after a noticeable interval."

But how do you really know whether the light moved "without lapse of time" unless you have some accurate way of measuring the lapse of time? Galileo went on to describe an experiment by which two people standing on distant hills flashing lanterns might measure the speed of light. He concluded that the speed of light is probably finite, not infinite. Galileo, however, was not able to estimate a definite value for it.

Experimental evidence was first successfully related to a finite speed for light by a Danish astronomer, Ole Rœmer. Detailed observations of Jupiter's satellites had shown an unexplained irregularity in the times recorded between successive eclipses of the satellites by the planet. Such an eclipse was expected to occur at 45 s after 5:25 a.m. on November 9, 1676 (Julian calendar). In September of that year, Rœmer announced to the Academy of Sciences in Paris that the observed eclipse would be 10 min late. On November 9, astronomers at the Royal Observatory in Paris carefully studied the eclipse. Though skeptical of Rœmer's mysterious prediction, they reported that the eclipse did occur late, just as he had foreseen.

Later, Rœmer revealed the theoretical basis of his prediction to the baffled astronomers at the Academy of Sciences. He explained that the originally expected time of the eclipse had been calculated from observations made when Jupiter was near the Earth. But now Jupiter had moved farther away. The delay in the eclipse occurred simply because light from the area around Jupiter takes time to reach the Earth. Obviously, this time interval must be greater when the relative distance between Jupiter and the Earth in their orbits is greater. In fact, Rœmer estimated that it takes about 22 min for light to cross the Earth's own orbit around the Sun.

Shortly after this, the Dutch physicist Christian Huygens used Rœmer's data to make the first calculation of the speed of light. Huygens combined Rœmer's value of 22 min for light to cross the Earth's orbit with his own estimate of the diameter of the Earth's orbit. (This distance could be estimated for the first time in the seventeenth century, in good part as a result of the advances in astronomy described in Chapter 2.) Huygens obtained a value for the speed of light in space which, in modern units, is about 2×10^8 m/s. This is about two-thirds of the currently accepted value. The error in Huygens' value was due mainly to Rœmer's overestimate of the time interval. Scientists now know that it takes light only about 16 min to cross the Earth's orbit.

The speed of light has been measured in many different ways since the seventeenth century. The development of electronic devices in the twentieth century allowed very precise measurements, making the speed of light one of the most precisely measured physical constants known today. Because of the importance of the value of the speed of light in modern physical theories, physicists continue to improve their methods of measurement.

The most precise recent measurements indicate that the speed of light in vacuum is 2.99792458×10^8 m/s. The uncertainty of this value is thought to be about 1 m/s, or 0.000001%, the precision being limited only by the precision to which lengths can be measured (times can be measured to several orders of magnitude greater precision). The speed of light is usually represented by the symbol c; for most purposes it is sufficient to use the approximate value $c = 3 \times 10^8$ m/s. Indeed, there has been general agreement not to pursue endlessly the search for greater accuracy, and some physicists have accepted the speed of light (in vacuum) to be *by definition* 2.9979×10^8 m/s.

8.14 REFLECTION AND REFRACTION

What does each model of light predict will happen when light traveling in one medium (e.g., air) hits the boundary of another medium (e.g., glass)? The answers to this question depend on whether a particle or a wave the-

The incident, reflected, and refracted rays are all in the same plane, a plane perpendicular to the surface.

ory of light is used. Here is an opportunity to test which theory is better.

Reflection and refraction from the wave viewpoint were discussed in Sections 8.9 and 8.10. Recall the results obtained there and apply them to light:

1. A ray may be taken as the line drawn perpendicular to a wave's crest lines. Such a ray represents the direction in which a train of waves is traveling.
2. In reflection, the angle of incidence (θ_i) is equal to the angle of reflection (θ_r).
3. Refraction involves a change of wavelength and speed of the wave as it passes into another medium. When the speed decreases, the wavelength decreases, and the ray bends in a direction toward a line perpendicular to the boundary. This bending toward the perpendicular is observed, for example, when a ray of light passes from air to glass.

What about explaining the same observations by means of the particle model? To test this model, first consider the nature of the surface of glass. Though apparently smooth, it is actually a wrinkled surface. A powerful microscope would show it to have endless hills and valleys. If particles of light were at all similar to little balls of matter, then on striking such a wrinkled surface they would scatter in all directions. They would not be reflected and refracted as noted above. Therefore, Newton argued, there must actually be "some feature of the body which is evenly diffused over its surface and by which it acts upon the ray without immediate contact." Obviously, in the case of reflection, the acting force would have to be one that

FIGURE 8.41 Two narrow beams of light, coming from the upper left, strike a block of glass. Can you account for all the observed effects?

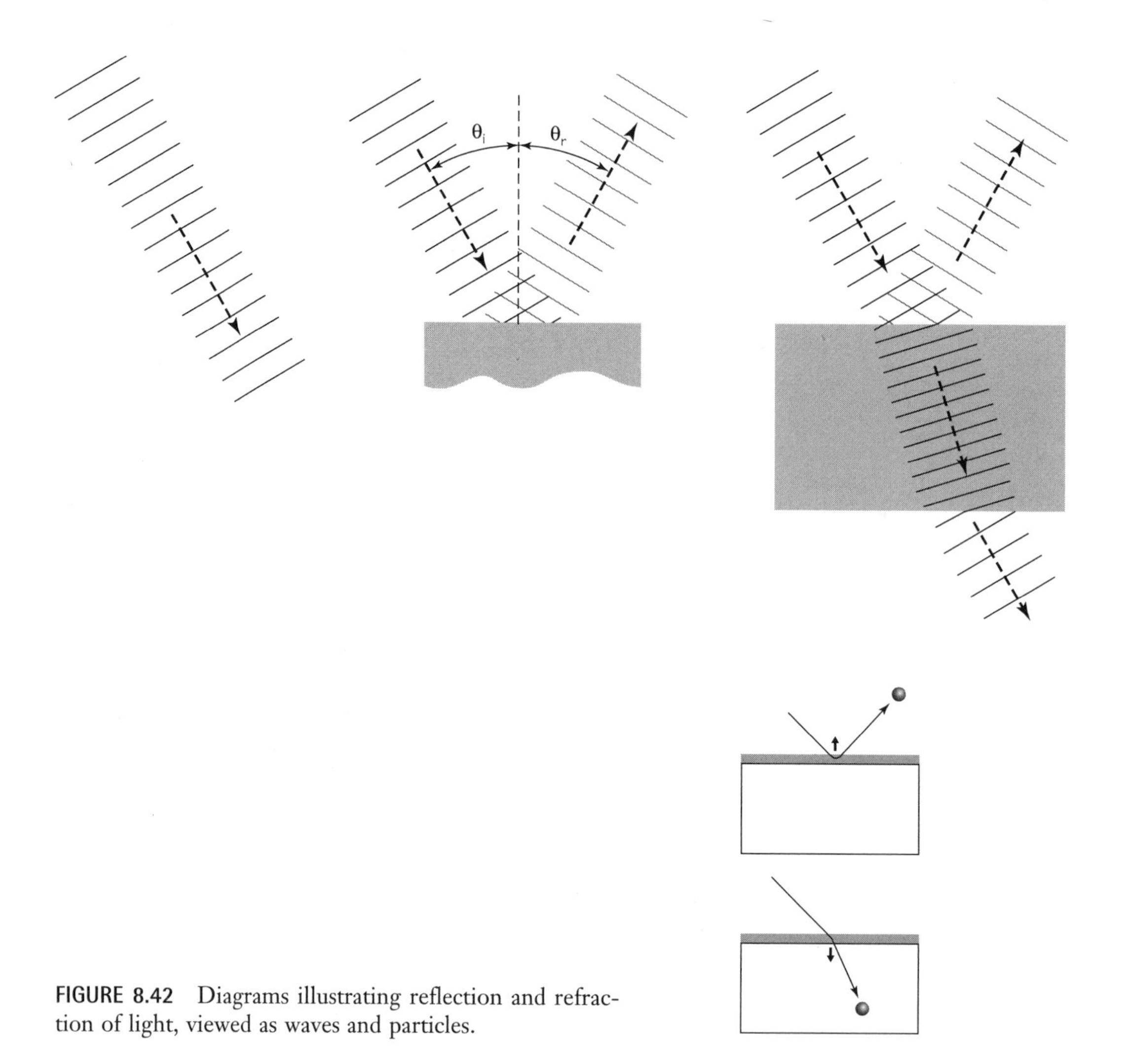

FIGURE 8.42 Diagrams illustrating reflection and refraction of light, viewed as waves and particles.

repelled the particles of light. Similarly, a force that attracted light particles instead of repelling them could explain refraction. As a particle of light approached a boundary of another medium, it would first have to overcome the repelling force. If it did that, it would then meet an attractive force in the medium that would pull it into the medium. Since the attractive force would be a vector with a component in the direction of the particle's original motion, the particle's speed would increase. If the ray of particles were moving at an oblique angle to the boundary, it would change direction as it entered the medium, bending toward the line perpendicular to the boundary. Notice that to make this argument we have had to make an assumption about the size of Newton's light "particles." The particles must be at least as small as the irregularities in the surface of a mirror. Similarly, a concrete wall is quite rough, but a tennis ball rebounds from such a wall almost exactly as light reflects from a mirror.

According to the *particle* model, therefore, you can make the following statements about reflection and refraction:

1. A ray represents the direction in which the particles are moving.
2. In reflection, the angles of incidence and reflection are equal. This prediction can be derived by applying the law of conservation of momentum to particles repelled by a force as shown on the last sketch.
3. Refraction involves a change of speed of the particles as they enter another medium. In particular, when an attractive power acts, *the speed increases*, and the ray is bent into the medium.

Compare these features of the particle model with the corresponding features of the wave model. The only difference is in the predicted speed for a refracted ray. You *observe* that a ray is bent toward the perpendicular line when light passes from air into water. The particle theory *predicts* that light has a *greater* speed in the second medium. The wave theory *predicts* that light has a *lower* speed.

You might think that it would be fairly easy to devise an experiment to determine which prediction is correct. All one has to do is measure the speed of light after it has entered water and compare it with the speed of light in air. But in the late seventeenth and early eighteenth centuries, when Huygens was arguing the wave model and Newton a particle model, no

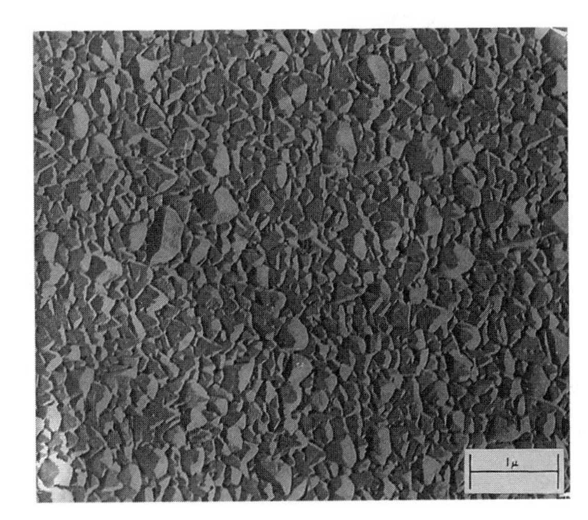

FIGURE 8.43 The surface of a mirror, as shown by a scanning electron microscope. The surface is a 3-micron thick aluminum film. The magnification here is nearly 26,000 times (one micron = 10^{-6} m).

such experiment was possible. The only available way of measuring the speed of light was an astronomical one. Not until the middle of the nineteenth century did Armand H.L. Fizeau and Jean B.L. Foucault measure the speed of light in water. *The results agreed with the predictions of the wave model*: The speed of light is less in water than in air.

The Foucault–Fizeau experiments of 1850 were widely regarded as driving the last nail in the coffin of the Newtonian particle theory of light, for, by the time these experiments were done, most physicists had already accepted the wave model for other reasons. Some of these stemmed from the work of the English scientist Thomas Young, to whom we now turn.

8.15 INTERFERENCE AND DIFFRACTION

Early in the nineteenth century, when Newton's prestige still contributed greatly to the support of the particle theory of light, Thomas Young revived the wave theory of light. In experiments made between 1802 and 1804, Young found that light shows the phenomenon of *interference* de-

FIGURE 8.44 Thomas Young (1773–1829) was an English linguist, physician, and expert in many fields of science. At the age of 14, he was familiar with Latin, Greek, Hebrew, Arabic, Persian, French, and Italian, and later was one of the first scholars successful at decoding Egyptian hieroglyphic inscriptions. He studied medicine in England, Scotland, and Germany. While still in medical school, he made original studies of the eye and later developed the first version of what is now known as the three-color theory of vision. Young also did research in physiology on the functions of the heart and arteries and studied the human voice mechanism, through which he became interested in the physics of sound and sound waves. Young then turned to optics and showed that many of Newton's experiments with light could be explained in terms of a simple wave theory of light. This conclusion was strongly attacked by some scientists in England and Scotland who were upset by the implication that Newton might have been wrong.

scribed in general for transverse waves in Section 8.6. The particle theory of light could not easily explain the interference patterns produced by light. Young's famous "double-slit experiment" provided convincing evidence that light does have properties that are explainable only in terms of waves.

Young's experiment should be done in the laboratory, rather than only talked about; we will describe it only briefly here. Basically, it involves splitting a single beam of light into two beams in order to ensure that they are in phase. The split beams are then allowed to overlap, and the two wave trains interfere, constructively in some places and destructively in others. To simplify the interpretation of the experiment, assume that it is done with light that has a single definite wavelength λ.

Young used a black screen with a small hole punched in it to produce a narrow beam of sunlight in a dark room. In the beam he placed a second black screen with two narrow slits cut in it, close together. Beyond this screen he placed a white screen. The light coming through each slit was diffracted and spread out into the space beyond the screen. The light from each slit interfered with the light from the other, and the interference pattern showed on the white screen. Where interference was constructive, there was a bright band on the screen. Where interference was destructive, the screen remained dark.

It is remarkable that Young actually found, by experiment, numerical values for the very short wavelength of light (see the *Student Guide Calculation* for this chapter). Here is his result:

> From a comparison of various experiments, it appears that the breadth of the undulations constituting the extreme red light must be supposed to be, in air, about one 36 thousandth of an inch [7×10^{-7} m], and those of the extreme violet about one 60 thousandth [4×10^{-7} m].

In announcing his result, Young took special pains to forestall criticism from followers of Newton, who was generally considered a supporter of the particle theory. He pointed out that Newton himself had made several statements favoring a theory of light that had some aspects of a wave theory. Nevertheless, Young was not taken seriously. It was not until 1818, when the French physicist Augustin Fresnel proposed his own mathematical wave theory, that Young's research began to get the credit it deserved. Fresnel also had to submit his work for approval to a group of physicists who were committed to the particle theory. One of them, the mathematician Simon Poisson, tried to refute Fresnel's wave theory of light. If it really did describe the behavior of light, Poisson said, a very peculiar thing ought to happen when a small solid disk is placed in a beam of light. Dif-

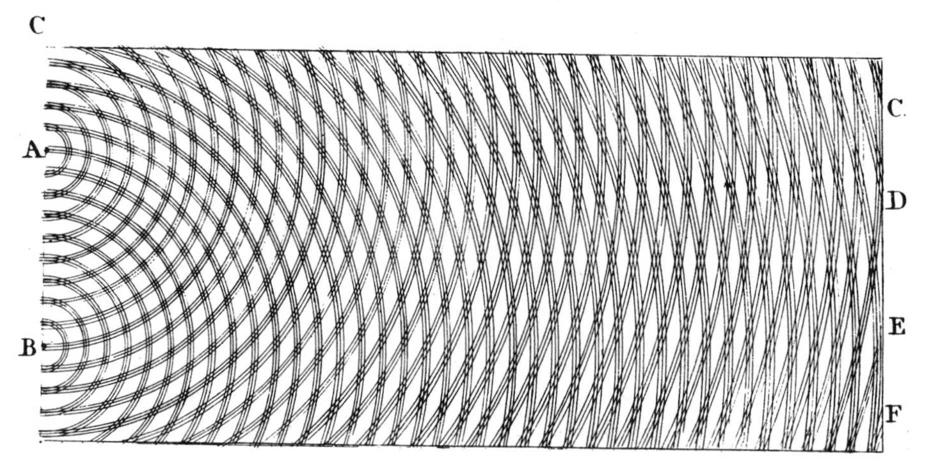

FIGURE 8.45 Thomas Young's original drawing showing interference effects in overlapping waves. The alternate regions of reinforcement and cancellation in the drawing can be seen best by placing your eye near the right edge and sighting at a grazing angle along the diagram.

fraction of some of the light waves all around the edge of the round disk should lead to constructive interference, producing a bright spot in the center of the disk's shadow on a white screen placed behind the disk. But the particle theory of light allowed no room for ideas such as diffraction and constructive interference. In addition, such a bright spot had never been reported, and even the very idea of a bright spot in the center of a shadow seemed absurd. For all of these reasons, Poisson announced that he had refuted the wave theory.

Fresnel accepted the challenge, however, and immediately arranged for Poisson's seemingly ridiculous prediction to be tested by experiment. The result was that a bright spot *did* appear in the center of the shadow!

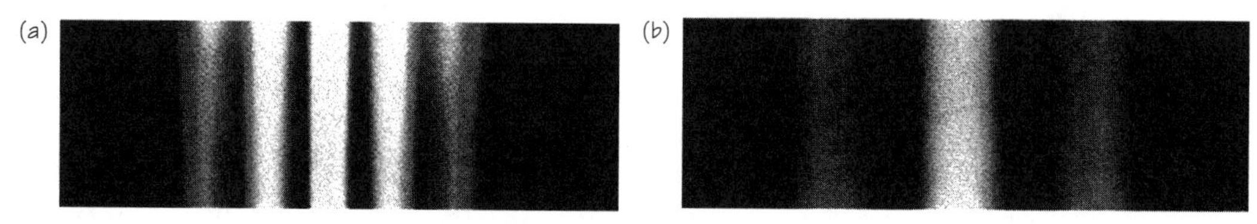

FIGURE 8.46 (a) A double-slit fringe pattern. When white light is used in Young's experiment, each wavelength produces its own fringe pattern slightly shifted from the others. The result is a central white band surrounded by fringes that are increasingly colored. (b) When the separation between slits, a, is decreased, the distance of the fringes from the central axis increases and the fringes broaden. All of this can be seen easily with an ordinary long-filament display light bulb, viewed through the space between two straight fingers.

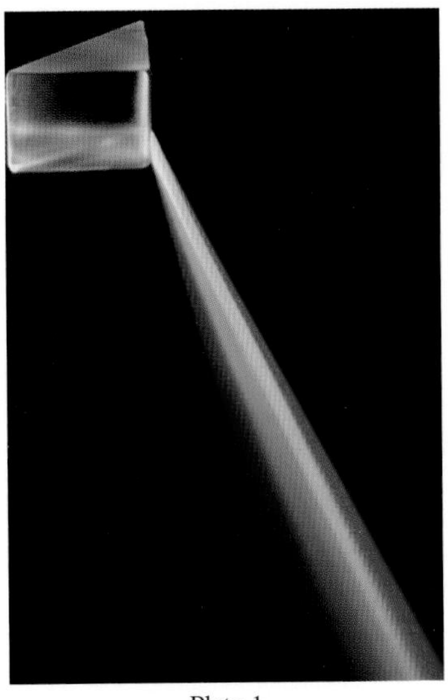

Plate 1

Plate 2

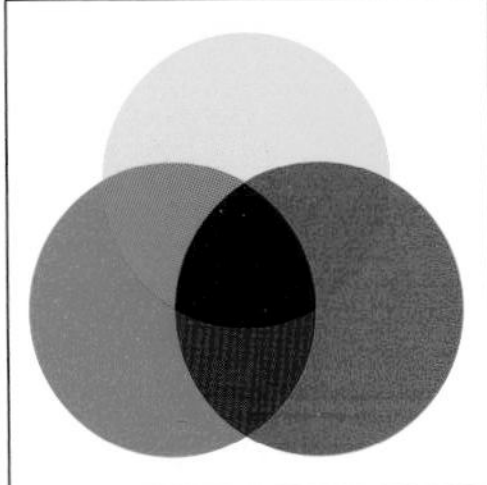

Plate 3

PLATE 1: Light spectrum from glass prism.

PLATE 2: Illustration of the principle of additive coloration.

PLATE 3: Illustration of the principle of subtractive coloration.

PLATE 4: "Entrance to the Harbor" by Georges Seurat (1888). Art historians believe that Seurat's technique of *pointillism*, the use of tiny dots of pure color to achieve all effects in a painting, was based on his understanding of the physical nature of light.

PLATE 5: Emission spectra for five elements undergoing electrical discharge excitation. A. Hydrogen, B. Helium, C. Oxygen, D. Sodium, E. Carbon.

PERIODIC TABLE OF THE ELEMENTS

IA 1	IIA 2	IIIB 3	IVB 4	VB 5	VIB 6	VIIB 7	VIIIB 8	 9	 10	IB 11	IIB 12	IIIA 13	IVA 14	VA 15	VIA 16	VIIA 17	VIIIA 18
1 **H** 1.0080																	2 **He** 4.0026
3 **Li** 6.939	4 **Be** 9.0122											5 **B** 10.811	6 **C** 12.011	7 **N** 14.007	8 **O** 15.999	9 **F** 18.998	10 **Ne** 20.183
11 **Na** 22.990	12 **Mg** 24.312											13 **Al** 26.982	14 **Si** 28.086	15 **P** 30.974	16 **S** 32.064	17 **Cl** 35.453	18 **Ar** 39.948
19 **K** 39.102	20 **Ca** 40.08	21 **Sc** 44.956	22 **Ti** 47.90	23 **V** 50.942	24 **Cr** 51.996	25 **Mn** 54.938	26 **Fe** 55.847	27 **Co** 58.933	28 **Ni** 58.71	29 **Cu** 63.54	30 **Zn** 65.37	31 **Ga** 69.72	32 **Ge** 72.59	33 **As** 74.922	34 **Se** 78.96	35 **Br** 79.909	36 **Kr** 83.80
37 **Rb** 85.47	38 **Sr** 87.62	39 **Y** 88.905	40 **Zr** 91.22	41 **Nb** 92.906	42 **Mo** 95.94	43 **Tc** 98.907	44 **Ru** 101.07	45 **Rh** 102.91	46 **Pd** 106.42	47 **Ag** 107.87	48 **Cd** 112.40	49 **In** 114.82	50 **Sn** 118.69	51 **Sb** 121.75	52 **Te** 127.60	53 **I** 126.90	54 **Xe** 131.30
55 **Cs** 132.91	56 **Ba** 137.34	† 71 **Lu** 174.97	72 **Hf** 178.49	73 **Ta** 180.95	74 **W** 183.85	75 **Re** 186.21	76 **Os** 190.2	77 **Ir** 192.22	78 **Pt** 195.09	79 **Au** 196.97	80 **Hg** 200.59	81 **Tl** 204.38	82 **Pb** 207.19	83 **Bi** 208.98	84 **Po** 208.98	85 **At** 209.99	86 **Rn** 222.02
87 **Fr** 223.02	88 **Ra** 226.03	‡ 103 **Lw** 262.11	104 **Rf** (261)	105 **Db** (262)	106 **Sg** (263)	107 **Bh** (264)	108 **Hs** (265)	109 **Mt** (268)	110 **Uun** (269)	111 **Uuu** (272)	112 **Uub** (277)	113	114 **Uuq** (285)	115	116 **Uuh** (289)	117	118 **Uuo** (293)

† Lanthanide series	57 **La** 138.91	58 **Ce** 140.12	59 **Pr** 140.91	60 **Nd** 144.24	61 **Pm** 144.91	62 **Sm** 150.35	63 **Eu** 151.96	64 **Gd** 157.25	65 **Tb** 158.92	66 **Dy** 162.50	67 **Ho** 164.93	68 **Er** 167.26	69 **Tm** 168.93	70 **Yb** 173.04
‡ Actinide series	89 **Ac** 227.03	90 **Th** 232.04	91 **Pa** 231.04	92 **U** 238.03	93 **Np** 237.05	94 **Pu** 244.06	95 **Am** 243.06	96 **Cm** 247.07	97 **Bk** 247.07	98 **Cf** 251.08	99 **Es** 252.08	100 **Fm** 257.10	101 **Md** 258.10	102 **No** 259.10

PLATE 6: The Periodic Table of the elements. Elements possessing similar properties are highlighted in color. The groups are indicated by the numbered columns.

FIGURE 8.47 Augustin Jean Fresnel (1788–1827) was an engineer of bridges and roads for the French government. In his spare time, he carried out extensive experimental and theoretical work in optics. Fresnel developed a comprehensive wave model of light that successfully accounted for reflection, refraction, interference, and polarization. He also designed a lens system for lighthouses that is still used today.

Thereafter, increasing numbers of scientists realized the significance of the Young double-slit experiment and the "Poisson bright spot." By 1850, the wave model of light was generally accepted; physicists had begun to concentrate on working out the mathematical consequences of this model and applying it to the different properties of light.

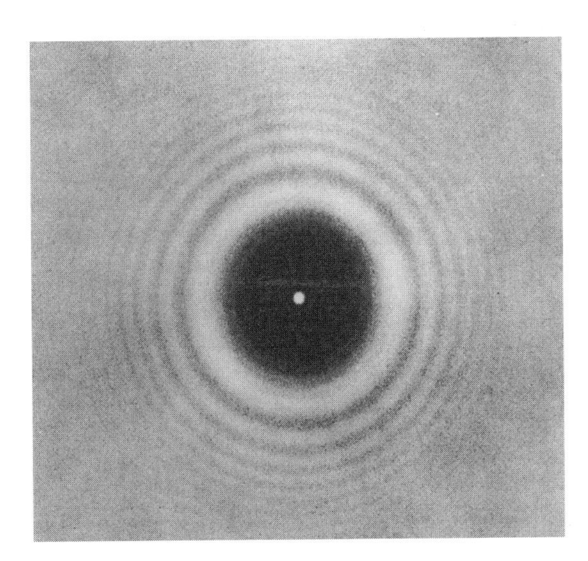

FIGURE 8.48 Diffraction pattern caused by an opaque circular disk, showing the Poisson bright spot in the center of the shadow. Note also the bright and dark fringes of constructive and destructive interference.

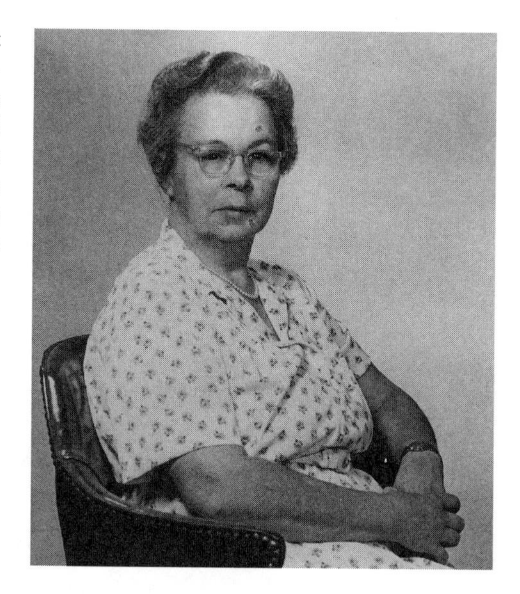

FIGURE 8.49 Katherine Burr Blodgett (1898–1979). Dr. Blodgett developed "invisible" glass by applying 44 layers of a one-molecule thick transparent liquid soap to glass to reduce reflections from its surface. Today, nearly all camera lenses and optical devices have non-reflective coatings on their surfaces which facilitate the efficient passage of light.

8.16 WHAT IS COLOR?

The coloring agents found in prehistoric painting and pottery show that humans have appreciated color since earliest times. But no scientific theory of color was developed before the time of Newton. Until then, most of the accepted ideas about color had come from artist–scientists like da Vinci, who based their ideas on experiences with mixing pigments.

Unfortunately, the lessons learned in mixing pigments rarely apply to the mixing of different-colored light beams. In ancient times, it was thought that light from the Sun was "pure." Color resulted from adding impurity, as was considered to be the case when a beam of "pure light" was refracted in glass and emerged with colored fringes.

Newton became interested in colors even while he was still a student at Cambridge University. In 1672, at the age of 29, Newton published a theory of color in the *Philosophical Transactions* of the Royal Society of London. This was his first published scientific paper. He wrote:

> In the beginning of the Year 1666, at which time I applyed myself to the grinding of Optick glasses of other figures than *Spherical*, I procured me a Triangular glass-Prisme, to try therewith the celebrated *Phaenomena* of *Colours*. And in order thereto haveing darkened my chamber, and made a small hole in my window-shuts, to

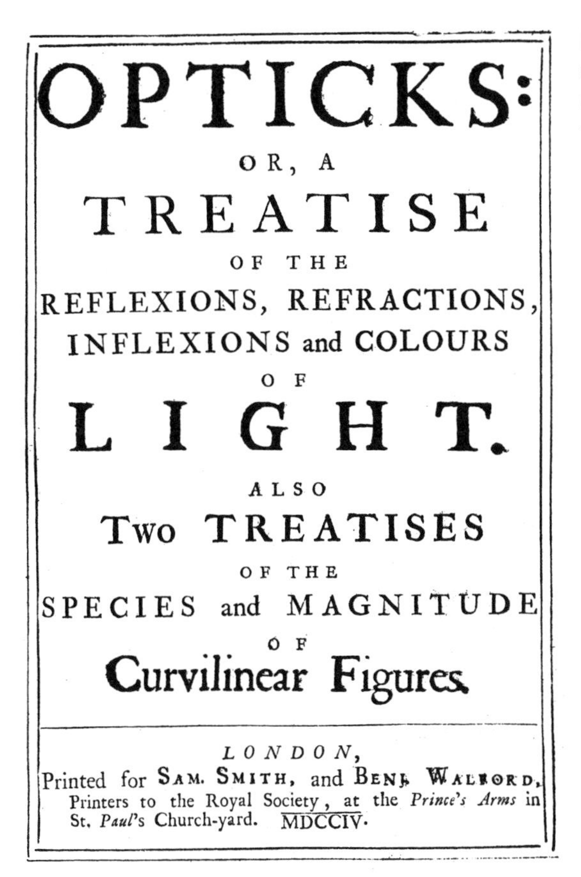
OPTICKS:

OR, A

TREATISE

OF THE

REFLEXIONS, REFRACTIONS, INFLEXIONS and COLOURS

OF

LIGHT.

ALSO

Two TREATISES

OF THE

SPECIES and MAGNITUDE

OF

Curvilinear Figures.

LONDON,

Printed for SAM. SMITH, and BENJ. WALFORD, Printers to the Royal Society, at the *Prince's Arms* in St. *Paul's* Church-yard. MDCCIV.

FIGURE 8.50 Title page from the first edition of Newton's *Opticks* (1704), in which he described his theory of light.

> let in a convenient quantity of the Suns light, I placed my Prisme at his entrance, that it might be thereby refracted to the opposite wall. It was at first a very pleasing divertisement, to view the vivid and intense colours produced thereby. . . .

The cylindrical beam of "white" sunlight from the circular opening passed through the prism and produced an elongated patch of colored light on the opposite wall. This patch was violet at one end, red at the other, and showed a continuous gradation of colors in between. For such a pattern of colors, Newton invented the name *spectrum*.

But, Newton wondered, *where do the colors come from*? And why is the image spread out in an elongated patch rather than circular? Newton passed the light through different thicknesses of the glass, changed the size of the hole in the window shutter, and even placed the prism outside the window.

None of these changes had any effect on the spectrum. Perhaps some unevenness or irregularity in the glass produced the spectrum, Newton thought. To test this possibility, he passed the colored rays from one prism through a second similar prism turned upside down. If some irregularity in the glass caused the beam of light to spread out, then passing this beam through the second prism should spread it out even more. Instead, the second prism, when properly placed, brought the colors *back together* fairly well. A spot of *white* light was formed, as if the light had not passed through either prism.

By such a process of elimination, Newton convinced himself of a belief that he probably had held from the beginning: *White light is composed of colors*. The prism does not manufacture or add the colors; they were there all the time, but mixed up so that they could not be distinguished. When white light passes through a prism, each of the component colors is refracted at a different angle. Thus, the beam is spread into a spectrum.

As a further test of this hypothesis, Newton cut a small hole in a screen on which a spectrum was projected. In this way, light of a single color could be separated out and passed through a second prism. He found that the second prism had no further effect on the color of this beam, though it refracted the beam more. That is, once the first prism had done its job of separating the colored components of white light, the second prism could not change the color of the components.

Summarizing his conclusions, Newton wrote:

> Colors are not *Qualifications of Light* derived from Refraction or Reflection of natural Bodies (as 'tis generally believed) but Original and Connate Properties, which in divers Rays are divers. Some Rays are disposed to exhibit a Red Colour and no other; some a Yellow and no other, some a Green and no other, and so of the rest. Nor are there only Rays proper and particular to the more Eminent Colours, but even to all their intermediate gradations.

Apparent Colors of Objects

So far, Newton had discussed only the colors of rays of light. In a later section of his paper he raised the important question: Why do objects appear to have different colors? Why is the grass green, a paint pigment yellow or red? Newton proposed a very simple answer:

> That the Colours of all Natural Bodies have no other Origin than this, that they . . . Reflect one sort of Light in greater plenty than another.

Most colors observed for real materials are "body" colors, produced by selective absorption of light which penetrates a little beyond the surface before being scattered back. This explains why the light transmitted by colored glass has the same color as the light reflected from it. Thin metallic films, however, have "surface" color, resulting from selective regular reflection. Thus, the transmitted light will be the complement of the reflected light. For example, the light transmitted by a thin film of gold is bluish–green, while that reflected is yellow.

In other words, a red paint pigment looks red to us because when white sunlight falls on it, the pigment absorbs most of the rays of other colors of the spectrum and reflects mainly the red to our eyes.

According to Newton's theory, color is not a property of an object by itself. Rather, color depends on how the object reflects and absorbs the various colored rays that strike it. Newton backed up this hypothesis by pointing out that an object may appear to have a different color when a different kind of light shines on it. For example, consider a pigment that reflects much more red light than green or blue light. When illuminated by white light, it will reflect mostly the red component of the white light, and so will appear red. But if it is illuminated with blue light, there is no red there for it to reflect, and it can reflect only a very little of the blue light. Thus, it will appear to be dark and perhaps dimly blue. (However, Newton was not suggesting that the rays themselves possess color, only that they raise the sensation of color in the eye, or the mind.)

Reactions to Newton's Theory

Newton's theory of color met with violent opposition at first. Other British scientists, especially Robert Hooke, objected on the grounds that postulating a different kind of light for each color was unnecessary. It would be simpler to assume that the different colors were produced from pure white light by some kind of modification. For example, the wave front might be twisted so that it is no longer perpendicular to the direction of motion.

Newton was aware of the flaws in Hooke's theory, but he disliked public controversy. In fact, he waited until after Hooke's death in 1703 to publish his own book, *Opticks* (1704), in which he reviewed the properties of light and his many convincing experiments on light.

Newton's *Principia* was a more important work from a purely scientific viewpoint. But his *Opticks* had also considerable influence on the literary world. This was in part because the work was written in English rather than in Latin and because the book contained little mathematics. English poets gladly celebrated the discoveries of their country's greatest scientist. Most poets, of course, were not deeply versed in the details of Newton's theory of gravity. The technical aspects of the geometric axioms and proofs in the *Principia* were beyond most of its readers. Although some students, including young Thomas Jefferson, learned their physics out of that book, translations and popularized versions soon appeared. But Newton's theory of colors and light provided good material for poetic fancy, as in James Thomson's, "To the Memory of Sir Isaac Newton" (1727):

> . . . First the flaming red,
> Springs vivid forth; the tawny orange next;
> And next delicious yellow; by whose side
> Fell the kind beams of all-refreshing green.
> Then the pure blue, that swells autumnal skies,
> Ethereal played; and then, of sadder hue,
> Emerged the deepened indigo, as when
> The heavy-skirted evening droops with frost;
> While the last gleamings of refracted light
> Died in the fainting violet away.

Leaders of the nineteenth-century Romantic movement in literature and the German "Nature Philosophers" did not think so highly of Newton's theory of color. The scientific procedure of dissecting and analyzing natural phenomena by experiments was distasteful to them. They preferred to speculate about the unifying principles of all natural forces, hoping somehow to grasp nature as a whole. The German philosopher Friedrich Schelling wrote in 1802:

> Newton's *Opticks* is the greatest illustration of a whole structure of fallacies which, in all its parts, is founded on observation and experiment.

The German poet Goethe (mentioned in Chapter 4) rejected Newton's theory of colors and proposed his own theory, based upon his own direct observations as well as passionate arguments. Goethe insisted on the purity of white light in its natural state, rejecting Newton's argument that white light is a mixture of colors. Instead, he suggested, colors may be produced by the interaction of white light and its opposite—darkness. Goethe's observations on the psychology of color perception were of some value to science. But his theory of the physical nature of color could not stand up under further detailed experiment. Newton's theory of the colors of the spectrum remained firmly established.

8.17 WHY IS THE SKY BLUE?

Newton suggested that the apparent colors of natural objects depend on which color is most strongly reflected or scattered to the viewer by the object. In general, there is no simple way of predicting from the surface structure, chemical composition, etc., what colors a substance will reflect or scatter. However, the blue color of the clear sky can be explained by a fairly simple argument.

As Thomas Young showed experimentally (Section 8.15), different wavelengths of light correspond to different colors. The wavelength of light may be specified in units of *nanometers* (abbreviated nm; 1 nm = 10^{-9} m) or, alternatively, in Ångstroms (Å), named after Anders Jonas Ångstrom, a Swedish astronomer who, in 1862, used spectroscopic techniques to detect the presence of hydrogen in the Sun. One angstrom (symbol Å) is equal to 10^{-10} m. The range of the spectrum visible to humans is from about 400 nm (4000 Å) for violet light to about 700 nm (7000 Å) for red light.

Small obstacles can scatter the energy of an incident wave of any sort in all directions, and the amount of scattering depends on the wavelength. This fact can be demonstrated by simple experiments with water waves in

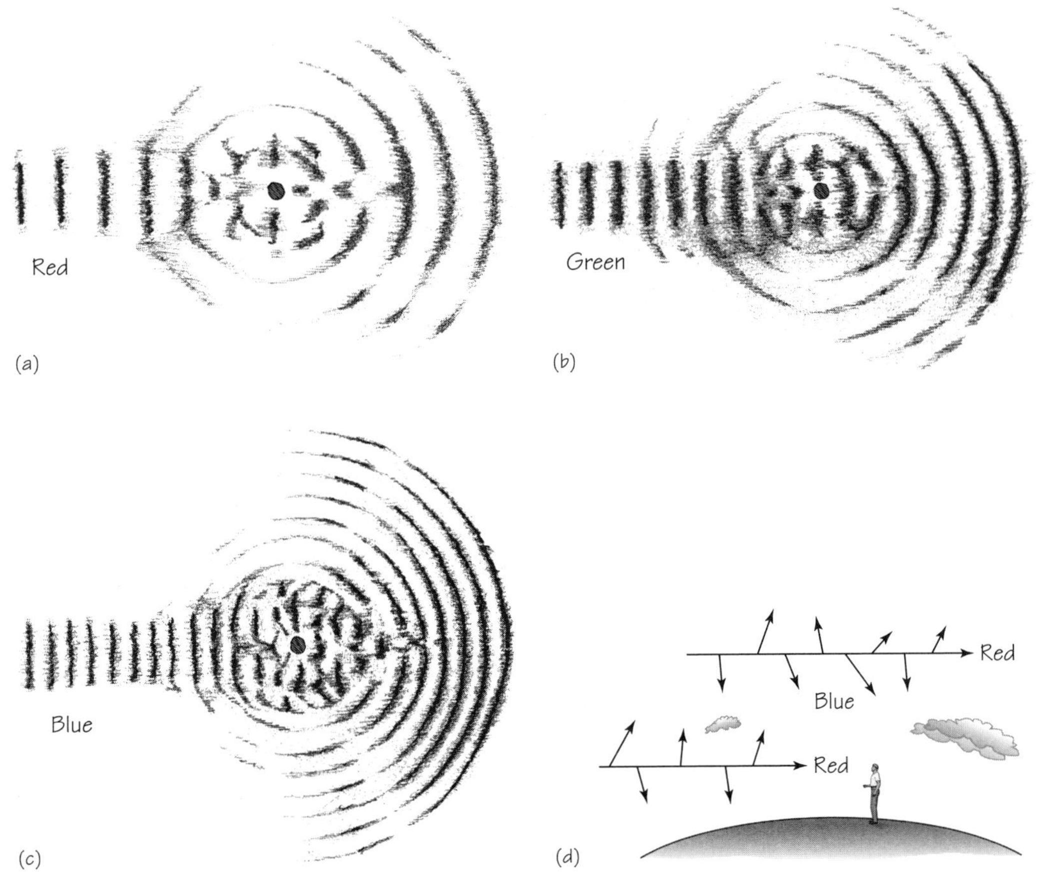

FIGURE 8.51 If you look at a sunset on a hazy day, you receive primarily unscattered colors, such as red, whereas if you look overhead, you will receive primarily scattered colors, the most dominant of which is blue.

a ripple tank. As a general rule, *the larger the wavelength is compared to the size of the obstacle, the less the wave is scattered by the obstacle.* For particles smaller than one wavelength, the amount of scattering of light varies inversely with the fourth power of the wavelength. For example, the wavelength of red light is about twice the wavelength of blue light. Therefore, the scattering of red light is only about one-sixteenth as much as the scattering of blue light.

Now it is clear why the sky is blue. Light from the sun is scattered by separate molecules of vapor, particles of dust, etc., in the air above, all of which are usually very small compared to the wavelengths of visible light. Thus, on a clear day light of short wavelengths (blue light) is much more strongly scattered by the particles than is light of longer wavelengths, and so-to-speak fills the firmament from end to end. When you look up into a clear sky, it is mainly this scattered light that enters your eyes. The range of scattered short wavelengths (and the color sensitivity of the human eye) leads to the sensation of blue. On the other hand, suppose you look at a sunset on a hazy day. You receive directly from the Sun a beam that has had the blue light almost completely scattered out in all directions, while the longer wavelengths have *not* been scattered out. So you perceive the Sun as reddish.

If the Earth had no atmosphere, the sky would appear black, and stars would be visible by day. In fact, starting at altitudes of about 16 km, where the atmosphere becomes quite thin, the sky does look black, and stars can be seen during the day, as astronauts have found.

If light is scattered by particles considerably larger than one wavelength (such as water droplets in a cloud), there is not much difference in the scattering of different wavelengths. So we receive the mixture we perceive as white.

The blue–gray haze that often covers large cities is caused mainly by particles emitted by internal combustion engines (cars, trucks) and by industrial plants. Most of these pollutant particles are invisible, ranging in size from 10^{-6} m to 10^{-9} m. Such particles provide a framework to which gases, liquids, and other solids adhere. These larger particles then scatter light and produce haze. Gravity has little effect on the particles until they become very large by collecting more matter. They may remain in the atmosphere for months if not cleaned out by repeated rain, snow, or winds. The influences of such clouds of haze or smog on the climate and on human health are substantial.

8.18 POLARIZATION

Hooke and Huygens proposed that light is in many ways like sound, that is, that light is a wave propagated through a medium. Newton could not accept this proposal and argued that light must also have some particle-like properties, in addition to its wave nature. He noted two properties of light that, he thought, could not be explained unless light had particle properties. First, a beam of light is propagated in space in straight lines, while waves such as sound spread out in all directions and go around corners. This objection could not be answered until early in the nineteenth century, when Thomas Young measured the wavelength of light and found how extremely small it is. Even the wavelength of red light, the longest wavelength of the visible spectrum, is less than one-thousandth of a millimeter. As long as a beam of light shines on objects or through holes of ordinary size (a few millimeters or more in width), the light will appear to travel in straight lines. As we saw, diffraction and scattering effects do not become strikingly evident until a wave passes over an object or through a hole whose size is about equal to or smaller than the wavelength.

Newton based his second objection on the phenomenon of "polarization" of light. In 1669, the Danish scientist Erasmus Bartholinus discovered that crystals of Iceland spar (calcite) could split a ray of light into two rays. Writing or small objects viewed through the crystal looked double.

Newton thought this behavior could be explained by assuming that light is made up of particles that have different "sides," for example, rectangular cross sections. The double images, he thought, represent a sorting out of light particles that had entered the medium with different orientations.

Around 1820, Young and Fresnel gave a far more satisfactory explanation of polarization, using a modified wave theory of light. Before then, scientists had generally assumed that light waves, like sound waves, must be *longitudinal*. Young and Fresnel showed that if light waves are *transverse*, this could account for the phenomenon of polarization.

In a transverse wave of a mechanical type, the motion of the medium itself, such as a rope, is always perpendicular to the direction of propagation of the wave. This does not mean that the motion of the medium is always in the same direction. In fact, it could be in any direction in a plane perpendicular to the direction of propagation. However, if the motion of the medium *is* mainly in one direction (e.g., vertical), the wave is *polarized*. Thus, a polarized wave is really the simplest kind of transverse wave. An unpolarized transverse wave is more complicated, since it is a mixture of various transverse motions. All of this applies to light waves, which do not need a medium in which to propagate.

Scientific studies of polarization continued throughout the nineteenth century. For instance, the

Ordinary light, when scattered by particles, shows polarization to different degrees, depending on the direction of scattering. The eyes of bees, ants, and other animals are sensitive to the polarization of scattered light from the clear sky. This enables a bee to navigate by the Sun, even when the Sun is low on the horizon or obscured. Following the bees' example, engineers have equipped airplanes with polarization indicators for use in Arctic regions.

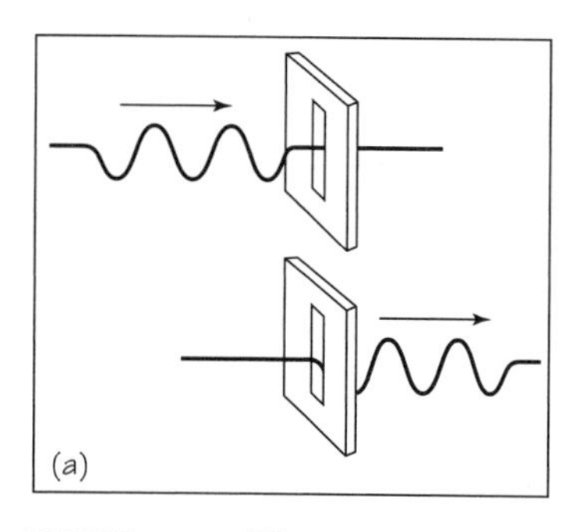

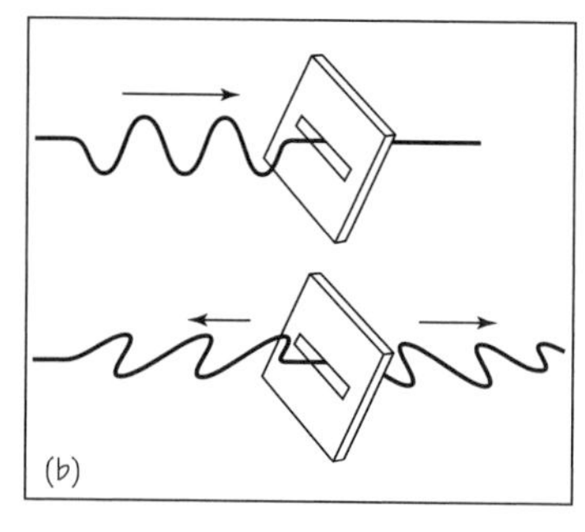

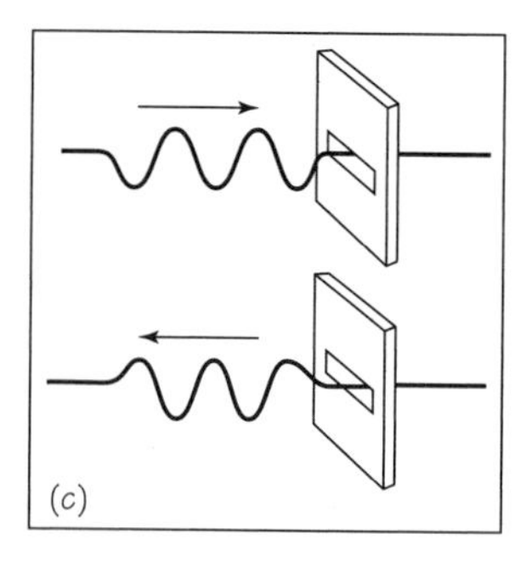

FIGURE 8.52 The same short-wave train on the rope approaches the slotted board in each of the three sketches. Depending on the orientation of the slot, the train of waves (a) goes entirely through the slot; (b) is partly reflected and partly transmitted with changed angles of rope vibration; or (c) is completely reflected.

way in which Iceland spar separates an unpolarized light beam into two polarized beams is sketched in Figure 8.53. Practical applications, however, were frustrated, mainly because polarizing substances like Iceland spar were scarce and fragile. One of the best polarizers was the synthetic crystal "herapathite," or sulfate of iodo-quinine. The needle-like herapathite crystals absorb light that is polarized in the direction of the long crystal axis and absorb very little of the light polarized in a direction at 90° to the long axis.

Herapathite crystals were so fragile that there seemed to be no way of using them. But in 1928, Edwin H. Land, while still a freshman in college, invented a polarizing plastic sheet he called "Polaroid." His first polarizer was a plastic film in which many small crystals of herapathite were embedded. When the plastic was stretched, the needle-like crystals lined up in one direction. Thus, they all acted on incoming light in the same way.

Some properties of a polarizing material are easily demonstrated. For example, you can obtain two polarizing sheets from the lenses of a pair of polarizing sunglasses, or from the "three-dimensional" eyeglasses used in Imax theatres. Hold one of the lenses in front of a light source. Then look at the first lens through the second one. Rotate the first lens. You will notice that, as you do so, the light alternately brightens and dims. You must rotate the sheet through an angle of 90° to go from maximum brightness to maximum dimness.

How can this effect be explained? The light that strikes the first lens, or polarizing sheet, is originally unpolarized, that is, a mixture of waves polarized in different directions. The first sheet transmits only those waves that are polarized in one direction, and it absorbs the rest. The transmitted wave going toward the second sheet is now polarized in one direction. Whenever this direction coincides with the direction of the long molecules

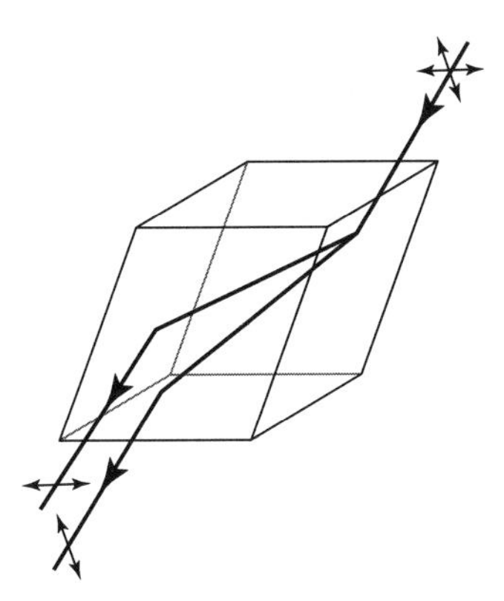

FIGURE 8.53 Double refraction by a crystal of Iceland spar. The "unpolarized" incident light can be thought of as consisting of two polarized components. The crystal separates these two components, transmitting them through the crystal in different directions and with different speeds.

in the second sheet, the wave will not be absorbed by the second sheet (i.e., the wave will set up vibrations within the molecules of the crystals which will transmit most of its energy). However, if the direction is *perpendicular* to the long axis of the crystal molecules, the polarized light will not go through the second sheet but instead will be absorbed.

In conclusion, we see that interference and diffraction effects required a wave model for light. To explain polarization phenomena, the wave model was made more specific; polarization could be explained only if the light waves are transverse waves. Altogether, this model for light explains very satisfactorily all the characteristics of light considered so far.

8.19 THE ETHER

One factor seems clearly to be missing from the wave model for light. Earlier in this chapter we defined waves as disturbances that propagate in some substance or "medium," such as a rope or water. What is the medium for the propagation of light waves?

Is air the medium for light waves? No, because light can pass through airless space, as it does between the Sun or other stars and the Earth. Even before it was definitely known that there is no air between the Sun and the Earth, Robert Boyle had tried the experiment of pumping almost all of the air out of a glass container. He found that the objects inside remained visible.

A wave is a disturbance, and it was difficult to think of a disturbance without specifying what was being disturbed. So it was natural to propose that a medium for the propagation of light waves existed. This hypothetical medium was called the *ether*. The word "ether" was originally the name

FIGURE 8.54 Polarized waves.

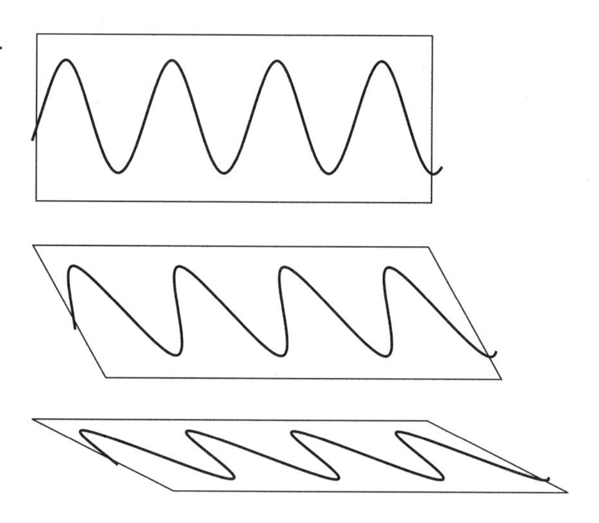

for Aristotle's fifth element, the pure transparent fluid that filled the heavenly sphere and was later called "quintessence." In the seventeenth and eighteenth centuries, the ether was imagined to be an invisible fluid of very low density. This fluid could penetrate all matter and fill all space. It might somehow be associated with the "effluvium" (something that "flows out") that was imagined to explain magnetic and electric forces. But light waves must be transverse in order to explain polarization, and transverse waves usually propagate only in a *solid* medium. A liquid or a gas cannot transmit transverse waves for any significant distance for the same reason that you cannot "twist" a liquid or a gas. So nineteenth-century physicists assumed that the ether must be a solid.

As stated in Section 8.3, the speed of propagation increases with the stiffness of the medium, and decreases with its density. The speed of propagation of light is very high compared to that of other kinds of waves, such as sound. Therefore, the ether was thought to be a very stiff solid with a very low density. Yet it seems absurd to say that a stiff, solid medium (ether) fills all space. The planets move through space without slowing down, so apparently they encounter no resistance from a stiff ether. And, of course, you feel no resistance when you move around in a space that transmits light freely.

Without ether, the wave theory of light seemed improbable. But the ether itself had absurd properties. Until early in this century, this problem remained unsolved, just as it had for Newton. We shall shortly see how, following Einstein's modification of the theory of light, the problem was solved.

SOME NEW IDEAS AND CONCEPTS

diffraction
ether
frequency
Huygens' principle
in phase
interference
longitudinal wave
medium
nodal lines
out of phase
period
polarization
propagation
ray
spectrum
superposition principle
transverse wave
wave
wave front
wavelength

AN IMPORTANT EQUATION

$v = f\lambda.$

AN IMPORTANT UNIT

1 Hz = 1/s.

FURTHER READING

G. Holton and S.G. Brush, *Physics, The Human Adventure* (Piscataway, NJ: Rutgers University Press, 2001), Chapter 23.
D. Park, *The Fire within the Eye: A Historical Essay on the Nature and Meaning of Light* (Princeton, NJ: Princeton University Press, 1997).
J. Hecht, *City of Light: The Story of Fiber Optics.* Sloan Technology Series (New York: Oxford University Press, 1999).

STUDY GUIDE QUESTIONS

A. WAVES

8.1 What Is a Wave?

1. How would you answer the question, What is a wave?

8.2 The Properties of Waves

1. What kinds of mechanical waves can propagate in a solid?
2. What kinds of mechanical waves can propagate in a fluid?
3. What kinds of mechanical waves can be polarized?
4. Suppose that a mouse runs along under a rug, causing a bump in the rug that travels with the mouse across the room. Is this moving disturbance a propagating wave?

8.3 Wave Propagation

1. What is transferred along the direction of wave motion?
2. On what two properties of a medium does wave speed depend?
3. If a spring is heated to make it less stiff, does it carry waves faster or slower? If the boxcars in a train are unloaded and empty, does the longitudinal start-up wave travel faster or slower?

8.4 Periodic Waves

1. Of the variables *frequency*, *wavelength*, *period*, *amplitude*, and *polarization*, which ones describe:
   (a) *space* properties of waves?
   (b) *time* properties of waves?
2. A vibration of 100 Hz (cycles per second) produces a wave:
   (a) What is the wave frequency?
   (b) What is the period of the wave?
   (c) If the wave speed is 10 m/s, what is the wavelength?
3. If points X and Y on a periodic wave are one-half period "out of phase" with each other, which of the following must be true?
   (a) X oscillates at half the frequency at which Y oscillates.
   (b) X and Y always move in opposite directions.
   (c) X is a distance of one-half wavelength from Y.

8.5 When Waves Meet

1. Two periodic waves of amplitudes A1 and A2 pass through a point P. What is the greatest possible displacement of P?
2. What is the displacement of a point produced by two waves together if the displacements produced by the waves separately at that instant are +5 cm and −6 cm, respectively? What is the special property of waves that makes this simple result possible?

8.6 A Two-Source Interference Pattern

1. Are nodal lines in interference patterns regions of cancellation or of reinforcement?
2. What are antinodal lines? antinodal points?

3. Nodal points in an interference pattern are places where:
   (a) the waves arrive "out of phase";
   (b) the waves arrive "in phase";
   (c) the point is equidistant from the wave sources;
   (d) the point is one-half wavelength from both sources.
4. Under what circumstances do waves from two in-phase sources arrive at a point out of phase?

8.7 Standing Waves

1. When two identical waves of the same frequency travel in opposite directions and interfere to produce a standing wave, what is the motion of the medium at:
   (a) the nodes of the standing wave?
   (b) the places between nodes (called antinodes or loops) of the standing wave?
2. If the two interfering waves have the same wavelength λ, what is the distance between the nodal points of the standing wave?
3. What is the wavelength of the longest traveling waves that can produce a standing wave on a string of length l?
4. Can standing waves of *any* frequency higher than that of the fundamental be set up in a bounded medium?

8.8 Wavefronts and Diffraction

1. What characteristic do all points on a wave front have in common?
2. State Huygens' principle in your own words.
3. Can there be nodal lines in a diffraction pattern from an opening less than one wavelength wide? Explain.
4. What happens to the diffraction pattern from an opening as the wavelength of the wave increases?
5. Can there be diffraction without interference? interference without diffraction?

8.9 Reflection

1. What is a "ray"?
2. What is the relationship between the angle at which a wave front strikes a barrier and the angle at which it leaves?
3. What shape of reflector can reflect parallel wave fronts to a sharp focus?
4. What happens to wave fronts originating at the focus of such a reflecting surface?

8.10 Refraction

1. If a periodic wave slows down on entering a new medium, what happens to:
   (a) its frequency?
   (b) its wavelength?
   (c) its direction?

2. Complete the sketch below to show roughly what happens to a wave train that enters a new medium beyond the vertical line in which its speed is greater.

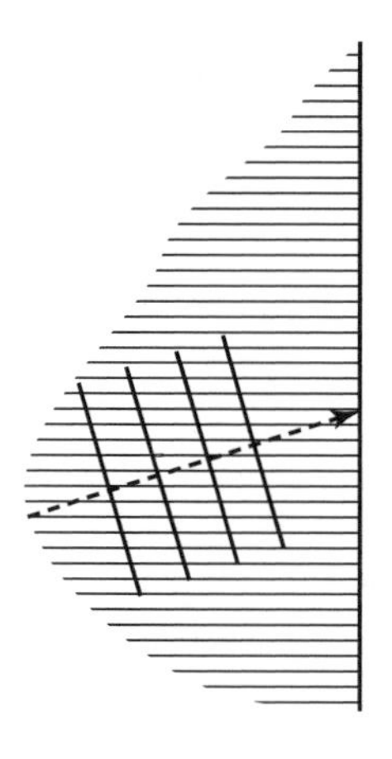

8.11 Sound Waves

1. List five wave behaviors that can be demonstrated with sound waves.
2. Can sound waves be polarized? Explain.

B. LIGHT

8.12 What Is Light?

1. How would you answer the question in the title above?

8.13 Propagation of Light

1. Can a beam of light be made increasingly narrow by passing it through narrower and narrower slits? What property of light does such an experiment demonstrate?
2. What reason did Rœmer have for thinking that the eclipse of a particular satellite of Jupiter would be observed later than expected?
3. What was the most important outcome of Rœmer's work?

8.14 Reflection and Refraction

1. What evidence showed conclusively that Newton's particle model for light could not explain all aspects of refraction?
2. If light has a wave nature, what changes take place in the speed, wavelength, and frequency of light on passing from air into water?

8.15 Interference and Diffraction

1. How did Young's experiments support the wave model of light?
2. In what way is diffraction involved in Young's experiments?

3. What phenomenon did Poisson predict on the basis of Fresnel's wave theory? What was the result?
4. What does the Poisson–Fresnel debate tell about the way science grows?
5. Recall from discussions earlier in the text how difficult it often is for new scientific ideas to be accepted.
   (a) List the cases, just by names.
   (b) Some people object to science as being "too dogmatic" and unchallengeable. Do these cases help or undermine such asssertions?

8.16 What Is Color?

1. How would you answer the question "What is color"?
2. How did Newton show that white light was not "pure"?
3. Why could Newton be confident that, say, green light was not itself composed of different colors of light?
4. How would Newton explain the color of a blue shirt?
5. Why was Newton's theory of color attacked by the Nature Philosophers?

8.17 Why Is the Sky Blue?

1. Why is the sky blue?
2. How does the scattering of light waves by tiny obstacles depend on the wavelength?
3. What would you expect the sky to look like on the Moon? Why?

8.18 Polarization

1. What two objections did Newton have to a pure wave model?
2. Of the phenomena we have discussed, which ones agree with the wave model of light?

8.19 The Ether

1. Why did scientists assume that there existed an "ether" that transmitted light waves?
2. What remarkable property must the ether have if it is to be the mechanical medium for the propagation of light?

DISCOVERY QUESTIONS

1. On the basis of the evidence presented in this chapter, can light be considered to consist of particles or of waves? Give evidence in support of your answer.

2. The drawing below represents a pulse that propagates to the right along a rope. What is the shape of a pulse propagating to the left that could for an instant cancel this one completely?

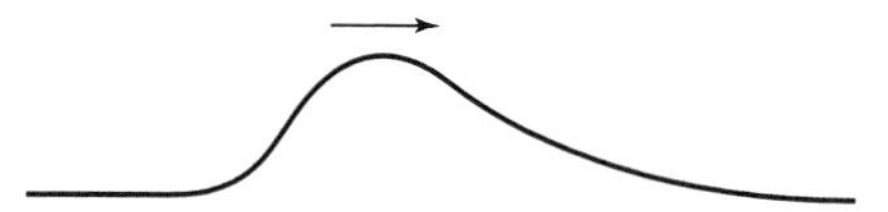

3. What shape would the nodal regions have for sound waves from two loudspeakers that emit the same sound?
4. Explain why it is that the narrower a slit in a barrier is, the more nearly it can act like a point source of waves.
5. If light is also a wave, why have you not seen light being diffracted by the slits of a picket fence, or diffracted around the corner of a house?
6. If the frequency of a wave traveling in a medium is increased, what will happen to its speed? What determines the speed of waves in a medium?
7. How can sound waves be used to map the floors of oceans?
8. Waves reflect from an object in a definite direction only when the wavelength is small compared to the dimensions of the object. This is true for sound waves as well as for any other. What does this tell you about the sound frequencies a bat must generate if it is to catch a moth or a fly? Actually, some bats can detect the presence of a wire about 0.12 mm in diameter. Approximately what frequency would that require?
9. Suppose the reflecting surfaces of every visible object were somehow altered so that they completely absorbed any light falling on them; how would the world appear to you?
10. Because of atmospheric refraction, you see the Sun in the evening some minutes after it is really below the horizon, and also for some minutes before it is actually above the horizon in the morning.
    (a) Draw a simple diagram to illustrate how this phenomenon occurs.
    (b) What would sunset be like on a planet with a very thick and dense (but still transparent) atmosphere?
11. Using the phenomena of diffraction and interference, show how the wave theory of light can explain the bright spot found in the center of the shadow of a disk illuminated by a point source.
12. It is a familiar observation that clothing of certain colors appears different in artificial light and in sunlight. Explain why.
13. To prevent car drivers from being blinded by the lights of approaching automobiles, polarizing sheets could be placed over the headlights and windshields of every car. Explain why these sheets would have to be oriented in the same way on every vehicle and must have their polarizing axis at 45° to the vertical.
14. A researcher has discovered some previously unknown rays emitted by a radioactive substance. She wants to determine if the rays are made up of waves or particles. Design a few experiments that she could use to answer her question.

15. When Wilhelm Roentgen discovered X rays, which we now know to have a wavelength of the order of 10^{-10} m, he could not decide by experiment whether X rays were particles or waves. Why do you think he might have had that difficulty?

Quantitative

1. (a) What is the speed of sound in air if middle C (256 Hz) has a wavelength of 1.34 m?
   (b) What is the wavelength in water of middle C if sound waves travel at 1500 m/s in water?
   (c) What is the period of a wave sounding middle C in air? in water?
2. Assuming that light is a wave phenomenon, what is the wavelength of green light if the first node in a diffraction pattern is found 10 cm from the center line at a distance of 4 m from the slits which have a separation distance of 2.5×10^{-3} cm?
3. A convenient unit for measuring astronomical distances is the light year, defined as the distance that light travels in 1 year. Calculate the number of meters in a light year to two significant figures.
4. Suppose a space vehicle had a speed one-thousandth that of light. How long would it take to travel the 4.3 light years from Earth to the closest known star other than the Sun, alpha Centauri? Compare the speed given for the space vehicle with the speed of approximately 10 km/s maximum speed (relative to the Earth) that a space capsule reaches on an Earth–Mars trip.
5. Calculate how much farther than expected Jupiter must have been from Earth when Rœmer predicted a 10-min delay for the eclipse of 1676.
6. Green light has a wavelength of approximately 5×10^{-7} m (500 nm). What frequency corresponds to this wavelength? Compare this frequency to the carrier frequency of the radio waves broadcast by a radio station you listen to.

CHAPTER 9

Einstein and Relativity Theory

9.1 THE NEW PHYSICS

Following Newton's triumph, work expanded not only in mechanics but also in the other branches of physics, in particular, in electricity and magnetism. This work culminated in the late nineteenth century in a new and successful theory of electricity and magnetism based upon the idea of electric and magnetic fields. The Scottish scientist James Clerk Maxwell, who formulated the new electromagnetic field theory, showed that what we observe as light can be understood as an electromagnetic wave. Newton's physics and Maxwell's theory account, to this day, for almost everything we observe in the everyday physical world around us. The motions of planets, cars, and projectiles, light and radio waves, colors, electric and magnetic effects, and currents all fit within the physics of Newton, Maxwell, and their contemporaries. In addition, their work made possible the many wonders of the new electric age that have spread throughout much of the world since the late nineteenth century. No wonder that by 1900 some distinguished physicists believed that physics was nearly complete, needing only a few minor adjustments. No wonder they were so astonished when, just 5 years later, an unknown Swiss patent clerk, who had graduated from the Swiss Polytechnic Institute in Zurich in 1900, presented five major research papers that touched off a major transformation in physics that is still in progress. Two of these papers provided the long-sought definitive evidence for the existence of atoms and molecules; another initiated the development of the quantum theory of light; and the fourth and fifth papers introduced the theory of relativity. The young man's name was Albert Einstein, and this chapter introduces his theory of relativity and some of its many consequences.

Although relativity theory represented a break with the past, it was a gentle break. As Einstein himself put it:

> We have here no revolutionary act but the natural continuation of a line that can be traced through centuries. The abandonment of certain notions connected with space, time, and motion hitherto treated as fundamentals must not be regarded as arbitrary, but only as conditioned by the observed facts.*

The "classical physics" of Newton and Maxwell is still intact today for events in the everyday world on the human scale—which is what we would expect, since physics was derived from and designed for the everyday world. However, when we get away from the everyday world, we need to use relativity theory (for speeds close to the speed of light and for extremely high densities of matter, such as those found in neutron stars and black holes) or quantum theory (for events on the scale of atoms), or the combination of both sets of conditions (e.g., for high-speed events on the atomic scale). What makes these new theories so astounding, and initially difficult to grasp, is that our most familiar ideas and assumptions about such basic concepts as space, time, mass, and causality must be revised in unfamiliar, yet still understandable, ways. But such changes are part of the excitement of science—and it is even more exciting when we realize that much remains to be understood at the frontier of physics. A new world view is slowly emerging to replace the mechanical world view, but when it is fully revealed

---

* *Ideas and Opinions*, p. 246.

(a)

(b)

(c)

FIGURE 9.1 Albert Einstein (1879–1955). (a) in 1905; (b) in 1912; and (c) in his later years.

it will probably entail some very profound and unfamiliar ideas about nature and our place in it.

9.2 ALBERT EINSTEIN

Obviously to have founded relativity theory and to put forth a quantum theory of light, all within a few months, Einstein had to be both a brilliant physicist and a totally unhindered, free thinker. His brilliance shines throughout his work, his free thinking shines throughout his life.

Born on March 14, 1879, of nonreligious Jewish parents in the southern German town of Ulm, Albert was taken by his family to Munich 1 year later. Albert's father and an uncle, both working in the then new profession of electrical engineering, opened a manufacturing firm for electrical and plumbing apparatus in the Bavarian capital. The firm did quite well in the expanding market for recently developed electrical devices, such as telephones and generators, some manufactured under the uncle's own patent. The Munich business failed, however, after the Einsteins lost a municipal contract to wire a Munich suburb for electric lighting (perhaps similar in our day to wiring fiber-optic cable for TV and high-speed Internet access). In 1894 the family pulled up stakes and moved to Milan, in northern Italy, where business prospects seemed brighter, but they left Albert, then aged 15, behind with relatives to complete his high-school education. The teenager lasted alone in Munich only a half year more. He quit school, which he felt too militaristic, when vacation arrived in December 1894, and headed south to join his family.

Upon arriving in Milan, the confident young man assured his parents that he intended to continue his education. Although underage and without a high-school diploma, Albert prepared on his own to enter the Swiss Federal Polytechnic Institute in Zurich, comparable to the Massachusetts Institute of Technology or the California Institute of Technology, by taking an entrance examination. Deficiencies in history and foreign language doomed his examination performance, but he did well in mathematics and science, and he was advised to complete his high-school education, which would ensure his admission to the Swiss Polytechnic. This resulted in his fortunate placement for a year in a Swiss high school in a nearby town. Boarding in the stimulating home of one of his teachers, the new pupil blossomed in every respect within the free environment of Swiss education and democracy.

Einstein earned high marks, graduated in 1896, and entered the teacher training program at the Swiss Polytechnic, heading for certification as a high-school mathematics and physics teacher. He was a good but not an outstanding student, often carried along by his friends. The mathematics and physics taught there were at a high level, but Albert greatly disliked the lack of training in any of the latest advances in Newtonian physics or Maxwellian electromagnetism. Einstein mastered these subjects entirely by studying on his own.

One of Einstein's fellow students was Mileva Marić, a young Serbian woman who had come to Zurich to study physics, since at that time most other European universities did not allow women to register as full-time students. A romance blossomed between Mileva and Albert. Despite the opposition of Einstein's family, the romance flourished. However, Mileva gave birth to an illegitimate daughter in 1902. The daughter, Liserl, was apparently given up for adoption. Not until later did Einstein's family finally accede to their marriage, which took place in early 1903. Mileva and Albert later had two sons, Hans Albert and Eduard, and for many years were happy together. But they divorced in 1919.

Another difficulty involved Einstein's career. In 1900 and for sometime after, it was headed nowhere. For reasons that are still unclear, probably anti-Semitism and personality conflicts, Albert was continually passed over for academic jobs. For several years he lived a discouraging existence of temporary teaching positions and freelance tutoring. Lacking an academic sponsor, his doctoral dissertation which provided further evidence for the existence of atoms was not accepted until 1905. Prompted by friends of the family, in 1902 the Federal Patent Office in Bern, Switzerland, finally offered Einstein a job as an entry-level patent examiner. Despite the full-time work, 6 days per week, Albert still found time for fundamental research in physics, publishing his five fundamental papers in 1905.

The rest, as they say, was history. As the importance of his work became known, recognized at first slowly, Einstein climbed the academic ladder, arriving at the top of the physics profession in 1914 as Professor of Theoretical Physics in Berlin.

In 1916, Einstein published his theory of general relativity. In it he provided a new theory of gravitation that included Newton's theory as a special case. Experimental confirmation of this theory in 1919 brought Einstein world fame. His earlier theory of 1905 is now called the theory of special relativity, since it excluded accelerations.

When the Nazis came to power in Germany in January 1933, Hitler being appointed chancellor, Einstein was at that time visiting the United States, and vowed not to return to Germany. He became a member of the newly formed Institute for Advanced Study in Princeton. He spent the rest of his life seeking a unified theory which would include gravitation and electromagnetism. As World War II was looming, Einstein signed a letter to President Roosevelt, warning that it might be possible to make an "atomic bomb," for which the Germans had the necessary knowledge. (It was later found that they had a head-start on such research, but failed.) After World War II, Einstein devoted much of his time to organizations advocating world agreements to end the threat of nuclear warfare. He spoke and acted in favor of the founding of Israel. His obstinate search to the end for a unified field theory was unsuccessful; but that program, in more modern guise, is still one of the great frontier activities in physics today. Albert Einstein died in Princeton on April 18, 1955.

9.3 THE RELATIVITY PRINCIPLE

Compared with other theories discussed so far in this book, Einstein's theory of relativity is more like Copernicus's heliocentric theory than Newton's universal gravitation. Newton's theory is what Einstein called a "constructive theory." It was built up largely from results of experimental evidence (Kepler, Galileo) using reasoning, hypotheses closely related to empirical laws, and mathematical connections. On the other hand, Copernicus' theory was not based on any new experimental evidence but primarily on aesthetic concerns. Einstein called this a "principle theory," since it was based on certain assumed principles about nature, of which the deduction could then be tested against the observed behavior of the real world. For Copernicus these principles included the ideas that nature should be simple, harmonious, and "beautiful." Einstein was motivated by similar concerns. As one of his closest students later wrote,

> You could see that Einstein was motivated not by logic in the narrow sense of the word but by a sense of beauty. He was always looking for beauty in his work. Equally he was moved by a profound religious sense fulfilled in finding wonderful laws, simple laws in the Universe.*

Einstein's work on relativity comprises two parts: a "special theory" and a "general theory." The special theory refers to motions of observers and events that do not exhibit any accelerations. The velocities remain uniform. The general theory, on the other hand, does admit accelerations.

Einstein's special theory of relativity began with aesthetic concerns which led him to formulate two fundamental principles about nature. Allowing

---

* Banesh Hoffmann in *Strangeness in the Proportion*, H. Woolf, ed., see Further Reading.

himself to be led wherever the logic of these two principles took him, he then derived from them a new theory of the basic notions of space, time, and mass that are at the foundation of all of physics. He was not constructing a new theory to accommodate new or puzzling data, but deriving by deduction the consequences about the fundamentals of all physical theories from his basic principles.

Although some experimental evidence was mounting against the classical physics of Newton, Maxwell, and their contemporaries, Einstein was concerned instead from a young age by the inconsistent way in which Maxwell's theory was being used to handle relative motion. This led to the first of Einstein's two basic postulates: the Principle of Relativity, and to the title of his relativity paper, "On the Electrodynamics of Moving Bodies."

Relative Motion

But let's begin at the beginning: *What is relative motion?* As you saw in Chapter 1, one way to discuss the motion of an object is to determine its average speed, which is defined as the distance traveled during an elapsed time, say, 13.0 cm in 0.10 s, or 130 cm/s. In Chapter 1 a small cart moved with that average speed on a tabletop, and the distance traveled was measured relative to a fixed meter stick. But suppose the table on which the meter stick rests and the cart moves is itself rolling forward in the same direction as the cart, at 100 cm/s relative to the floor. Then *relative to a meter stick on the floor*, the cart is moving at a different speed, 230 cm/s (100 + 130), while the cart is still moving at 130 cm/s *relative to the tabletop*. So, in measuring the average speed of the cart, we have first to specify what we will use as our reference against which to measure the speed. Is it the tabletop, or the floor, or something else? The reference we finally decide upon is called the "reference frame" (since we can regard it to be as a picture frame around the observed events). *All speeds are thus defined relative to the reference frame we choose.*

But notice that if we use the floor as our reference frame, it is not at rest either. It is moving relative to the center of the Earth, since the Earth is

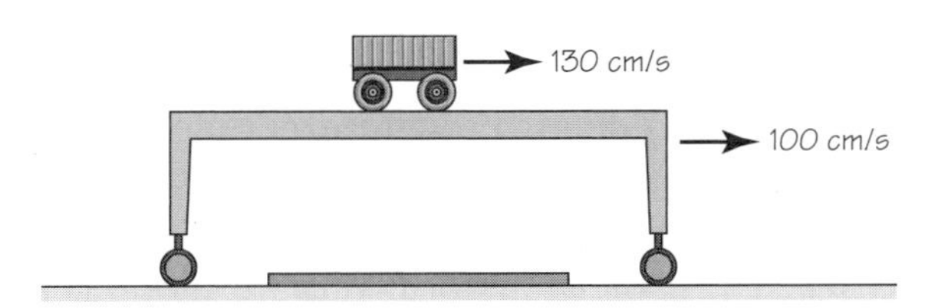

FIGURE 9.2 Moving cart on a moving table.

rotating. Also, the center of the Earth is moving relative to the Sun; and the Sun is moving relative to the center of the Milky Way galaxy, and on and on. . . . Do we ever reach an end? Is there something that is at *absolute rest*? Newton and almost everyone after him until Einstein thought so. For them, it was space itself that was at absolute rest. In Maxwell's theory this space is thought to be filled with a substance that is not like normal matter. It is a substance, called the "ether," that physicists for centuries hypothesized to be the carrier of the gravitational force. For Maxwell, the ether itself is at rest in space, and accounts for the behavior of the electric and magnetic forces and for the propagation of electromagnetic waves (further details in Chapter 12).

Although every experimental effort during the late nineteenth century to detect the resting ether had ended in failure, Einstein was most concerned from the start, not with this failure, but with an inconsistency in the way Maxwell's theory treated relative motion. Einstein centered on the fact that it is only the relative motions of objects and observers, rather than any supposed absolute motion, that is most important in this or any theory. For example, in Maxwell's theory, when a magnet is moved at a speed v relative to a fixed coil of wire, a current is induced in the coil, which can be calculated ahead of time by a certain formula (this effect is further discussed in Chapter 11). Now if the magnet is held fixed and the coil is moved at the same speed v, the same current is induced but a *different* equation is needed to calculate it in advance. Why should this be so, Einstein wondered, since only the relative speed v counts? Since absolutes of velocity, as of space and time, neither appeared in real calculations nor could be determined experimentally, Einstein declared that the absolutes, and on their basis in the supposed existence of the ether, were "superfluous," unnecessary. The ether seemed helpful for imagining how light waves traveled—but it was not needed. And since it could not be detected either, after Einstein's publication of his theory most physicists eventually came to agree that it simply did not exist. For the same reason, one could dispense with the notions of absolute rest and absolute motion. In other words, Einstein concluded, *all motion, whether of objects or light beams, is relative motion*. It must be defined relative to a specific reference frame, which itself may or may not be in motion relative to another reference frame.

The Relativity Principle—Galileo's Version

You saw in Section 3.10 that Galileo's thought experiments on falling objects dropped from moving towers and masts of moving ships, or butterflies trapped inside a ship's cabin, indicated that to a person within a reference frame, whether at rest or in uniform relative motion, there is no

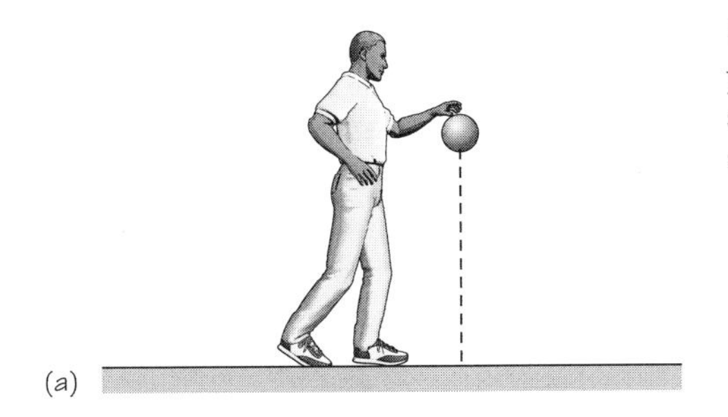

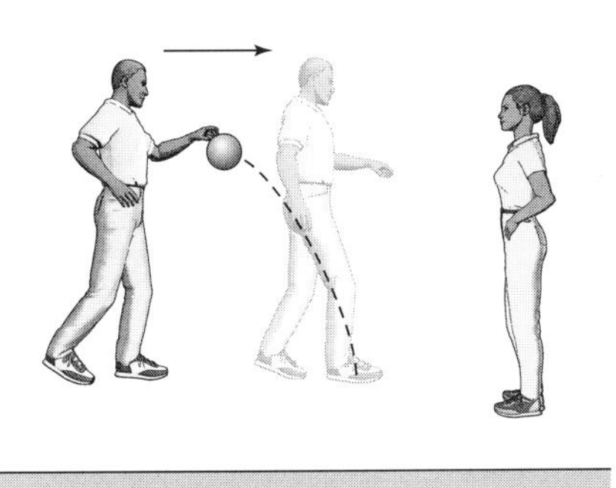

FIGURE 9.3 (a) Falling ball as seen by you as you walk forward at constant speed; (b) falling ball as seen by stationary observer.

way for that person to find out the speed of his *own* reference frame from any mechanical experiment done *within* that frame. Everything happens within that frame as if the frame is at rest.

But how does it look to someone outside the reference frame? For instance, suppose you drop a ball in a moving frame. To you, riding with the moving frame, it appears to fall straight down to the floor, much like a ball dropped from the mast of a moving ship. But what does the motion of the ball look like to someone who is not moving with you, say a classmate standing on the shore as your ship passes by? Or sitting in a chair and watching you letting a ball drop as you are walking by? Try it!

Looking at this closely, your classmate will notice that from her point of view the ball does not fall straight down. Rather, as with Galileo's falling ball from the mast or the moving tower, the ball follows the parabolic trajectory of a projectile, with uniform velocity in the horizontal direction as well as uniform acceleration in the vertical direction.

The surprising result of this experiment is that two different people in two different reference frames will describe the same event in two different ways. As you were walking or sailing past, you were in a reference frame with respect to which the ball is at rest before being released. When you let it go, you see it falling straight down along beside you, and it lands at your feet. But persons sitting in chairs or standing on the shore, in their own reference frame, will report that they see something entirely different: a ball that starts out with you—not at rest but in forward motion—and on release it moves—not straight down, but on a parabola toward the ground, hitting the ground at your feet. Moreover, this is just what they would expect to see, since the ball started out moving horizontally and then traced out the curving path of a projectile.

So who is correct? Did the ball fall straight down or did it follow the curving path of a projectile? Galileo's answer was: *both are correct*. But how can that be? How can there be two different observations and two different explanations for one physical event, a ball falling to someone's feet? The answer is that different observers observe the same event differently when they are observing the event from different reference frames in relative motion. The ball starts out stationary relative to one frame (yours), whereas it is, up to its release, in constant (uniform) motion relative to the other reference frame (your classmate's). Both observers see everything happen as they expect it from Newton's laws applied to their situation. But what they see is different for each observer. Since there is no absolute reference frame (no reference frame in uniform velocity is better or preferred over any other moving with uniform velocity), there is no absolute motion, and their observations made by both observers are equally valid.

Galileo realized that the person who is at rest relative to the ball could not determine by any such mechanical experiment involving falling balls, inclined planes, etc., whether or not he is at rest or in uniform motion relative to anything else, since all of these experiments will occur as if he is simply at rest. A ball dropping from a tower on the moving Earth will hit the base of the tower as if the Earth were at rest. Since we move with the Earth, as long as the Earth can be regarded as moving with uniform velocity (neglecting during the brief period of the experiment that it actually rotates), there is no mechanical experiment that will enable us to determine whether or not we are really at rest or in uniform motion.

Note: The observation of events are frame dependent. But the laws of mechanics are not. They are the same in reference frames that are at rest or in relative uniform motion. All objects that we observe to be moving relative to us will also follow the same mechanical laws (Newton's laws, etc.). As discussed in Section 3.10, this statement applied to mechanical phenomena is known as the *Galilean relativity principle*.

The Relativity Principle—Einstein's Version

In formulating his theory of relativity, Einstein expanded Galileo's principle into the *Principle of Relativity* by including *all of the laws of physics*, such as the laws governing light and other effects of electromagnetism, not just mechanics. Einstein used this principle as one of the two postulates of his theory of relativity, from which he then derived the consequences by deduction. Einstein's *Principle of Relativity* states:

> All the laws of physics are exactly the same for every observer in every reference frame that is at rest or moving with uniform relative velocity. This means that there is no experiment that they can perform in their reference frames that would reveal whether or not they are at rest or moving at uniform velocity.

Reference frames that are at rest or in uniform velocity relative to another reference frame have a technical name. They are called *inertial reference frames* (since Newton's law of inertia holds in them). Reference frames that are accelerating relative to each other are called *noninertial reference frames*. They are *not* included in this part of the theory of relativity. That is why this part of the theory of relativity is called the *theory of special relativity*. It is restricted to inertial reference frames, those which are either at rest or moving with uniform velocity relative to each other.

Notice that, according to Einstein's Relativity Principle, Newton's laws of motion and all of the other laws of physics remain the same for phenomena occurring in any of the inertial reference frames. This principle does *not* say that "everything is relative." On the contrary, it asks you to look for relationships that do *not* change when you transfer your attention from one moving reference frame to another. The physical measurements but not the physical laws depend on the observer's frame of reference.

9.4 CONSTANCY OF THE SPEED OF LIGHT

The Relativity Principle is one of the two postulates from which Einstein derived the consequences of relativity theory. The other postulate concerns the speed of light, and it is especially important when comparing observations between two inertial reference frames in relative motion, since we rely chiefly on light to make observations.

You recall that when Einstein quit high school at age 15 he studied on his own to be able to enter the Swiss Polytechnic Institute. It was probably during this early period that Einstein had a remarkable insight. He asked himself what would happen if he could move fast enough in space to catch up with a beam of light. Maxwell had shown that light is an electromagnetic wave propagating outward at the speed of light. If Albert could

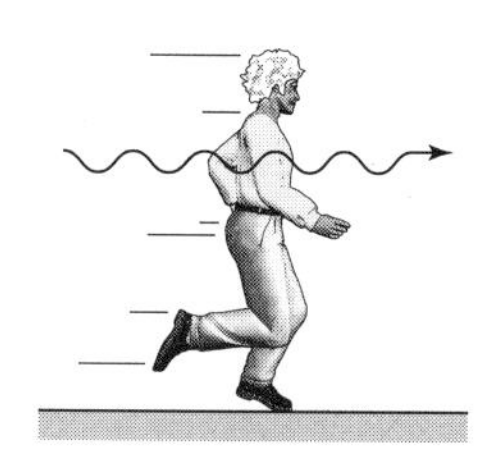

FIGURE 9.4 Running alongside a beam of light.

ride alongside, he would not see a wave propagating. Instead, he would see the "valleys" and "crests" of the wave fixed and stationary with respect to him. This contradicted Maxwell's theory, in which no such "stationary" landscape in free space was possible. From these and other, chiefly theoretical considerations, Einstein concluded by 1905 that Maxwell's theory must be reinterpreted: the speed of light will be exactly the same—a universal constant—for all observers, no matter whether they move (with constant velocity) relative to the source of the light. This highly original insight became Einstein's second postulate of special relativity, the *Principle of the Constancy of the Speed of Light*:

> Light and all other forms of electromagnetic radiation are propagated in empty space with a constant velocity c which is independent of the motion of the observer or the emitting body.

Einstein is saying that, whether moving at uniform speed toward or away from the source of light or alongside the emitted light beam, any observer always measures the exact same value for the speed of light in a vacuum, which is about 3.0×10^8 m/s or 300,000 km/s (186,000 mi/s). (More precisely, it is 299,790 km/s.) This speed was given the symbol c for "constant." If light travels through glass or air, the speed will be slower, but the speed of light in a vacuum is one of the *universal physical constants of nature*. (Another is the gravitational constant G.) It is important to note that, again, this principle holds only for observers and sources that are in inertial reference frames. This means they are moving at uniform velocity or are at rest relative to each other.

In order to see how odd the principle of the constancy of the speed of light really is, let's consider a so-called "thought experiment," an experiment that one performs only in one's mind. It involves two "virtual student researchers." One, whom we'll call Jane, is on a platform on wheels moving at a uniform speed of 5 m/s toward the second student, John, who is standing on the ground. While Jane is moving, she throws a tennis ball to John at 7 m/s. John catches the ball, but before he does he quickly measures its speed (this is only a thought experiment!). What speed does he obtain? . . . The answer is 5 m/s + 7 m/s = 12 m/s, since the two speeds combine.

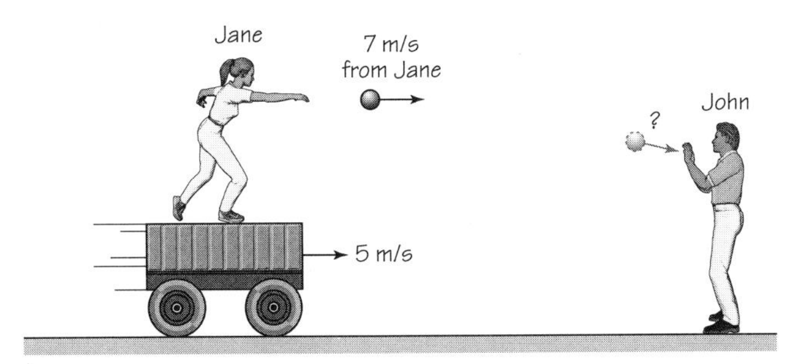

FIGURE 9.5 Ball thrown from a cart moving in the same direction. Jane is moving at 5 m/s, and the ball is thrown to John at a speed of 7 m/s.

Let's try it in the opposite direction. Jane is on the platform now moving at 5 m/s *away* from John. She again tosses the ball to John at 7 m/s, who again measures its speed before catching it. What speed does he measure? . . . This time it's -5 m/s + 7 m/s = 2 m/s. The velocities are subtracted. All this was as expected.

Now let's try these experiments with light beams instead of tennis balls. As Jane moves toward John, she aims the beam from a laser pen at John (being careful to avoid his eyes). John has a light detector that also measures the speed of the light. What is the speed of the light that he measures? . . . Neglecting the minute effect of air on the speed of light, Jane and John are surprised to find that Einstein was right: The speed is exactly the speed of light, no more, no less. They obtain the same speed when the platform moves away from John. In fact, even if they get the speed of the platform almost up to nearly the speed of light itself (possible only in a thought experiment), the measured speed of light is still the same in both instances. Strange as it seems, the speed of light (or of any electromagnetic wave) always has the same value, no matter what the relative speed is of the source and the observer.

FIGURE 9.6 Ball thrown from a cart moving in the opposite direction.

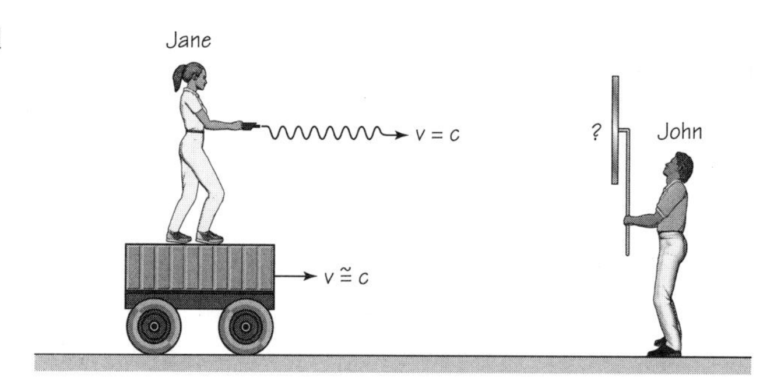

FIGURE 9.7 Light beam directed from a moving cart.

Let's consider some consequences that followed when Einstein put together the two fundamental postulates of special relativity theory, the Principle of Relativity and the Principle of the Constancy of the Speed of Light in space.

9.5 SIMULTANEOUS EVENTS

Applying the two postulates of relativity theory to a situation similar to Galileo's ship, Einstein provided a simple but profound thought experiment that demonstrated a surprising result. He discovered that two events that occur simultaneously for one observer may not occur simultaneously for another observer in relative motion with respect to the events. In other words, the simultaneity of events is a relative concept. (Nevertheless, the laws of physics regarding these events still hold.)

Einstein's thought experiment, an experiment that he performed through logical deduction, is as follows in slightly updated form. An observer, John, is standing next to a perfectly straight level railroad track. He is situated at the midpoint between positions A and B in Figure 9.8. Imagine that he is holding an electrical switch which connects wires of equal length to lights bulbs placed at A and B. Since he is at the midpoint between A and B, if he closes the switch, the bulbs will light up, and very shortly thereafter John will see the light from A and from B arriving at his eyes at the same moment. This is because the light from each bulb, traveling at the constant speed of light and covering the exact same distance to John from each bulb, will take the exact same time to reach his eyes. John concludes from this that the two light bulbs lit up simultaneously.

Now imagine a second observer, Jane, standing at the middle of a flat railroad car traveling along the track at a very high uniform speed to the right. Jane and John have agreed that when she reaches the exact midpoint between A and B, John will instantly throw the switch, turning on the light

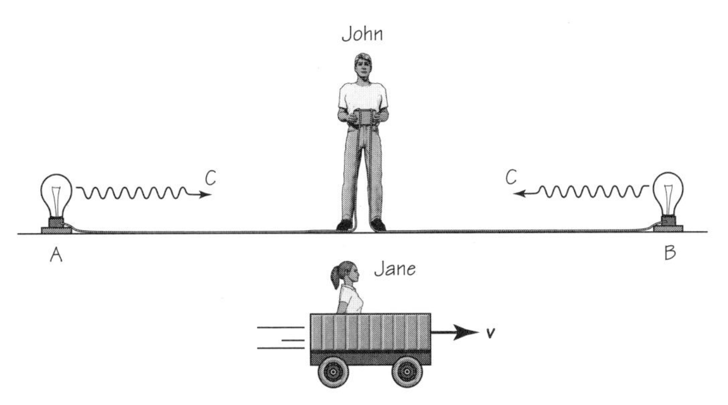

FIGURE 9.8 Einstein's thought experiment demonstrating the relativity of simultaneous events.

bulbs. (Since this is a thought experiment, we may neglect his reaction time, or else he might use a switch activated electronically.)

John and Jane try the experiment. The instant Jane reaches the midpoint position between A and B, the switch is closed, the light bulbs light up, and John sees the flashes simultaneously. But Jane sees something different: to her the flashes do *not* occur simultaneously. In fact, the bulb at B appeared to light up before the bulb at A. Why? Because she is traveling toward B and away from A and, because the speed of light is the same regardless of the motion of the observer, she will encounter the beam from B before the beam from A reaches her. Consequently, she will see the flash at B before she sees the flash at A. The conclusion: The two events that John perceives to occur simultaneously do not occur simultaneously for Jane. The reasons for this discrepancy are that the speed of light is the same for both observers and that each observer is moving in a different way relative to the events in question.

It might be argued that Jane could make a calculation in which she computed her speed and the speed of light, and then very simply find out if the flashes actually occurred as she saw them or as John claimed to see them. However, if she does this, then she is accepting a specific frame of reference: That is, she is assuming that she is the moving observer and that John is the stationary observer. But according to the relativity postulate motions are relative, and she need not assume that she is moving since there is no preferred frame of reference. Therefore she could just as well be the stationary observer, and John, standing next to the track, could be the moving observer! If that is so, then Jane could claim that the flash at B actually did occur before the flash at A and that John perceived them to occur simultaneously only because from her point of view he was moving toward

A and away from B. On the other hand, John could argue just the reverse, that he is at rest and it is Jane who is moving.

Which interpretation is correct? There is no "correct" interpretation because there is no preferred frame of reference. Both observers are moving relative to each other. They can agree on what happened only if they agree on the frame of reference, but that agreement is purely arbitrary.

The conclusion that the simultaneity of two events, such as two flashes from separate light bulbs, depends upon the motion of the observer, led to the possibility that time itself might also be a relative concept when examined in view of the relativity postulates.

9.6 RELATIVITY OF TIME

Let's see what happens to the measurement of time when understood through special relativity.

We'll follow Einstein's original argument and examine another, somewhat updated thought experiment. In this experiment one observer—again we'll call her Jane—is in a spaceship moving at an extremely fast uniform speed relative to the Earth and in the horizontal direction relative to another observer, John, who is stationary on the Earth. In Jane's spaceship (i.e., in her reference frame) there is a clock that measures time in precise intervals by using a laser pulse. The pulse travels straight up from a laser, hits a mirror, and is reflected back down. When the pulse returns to the starting point, it is detected by a photosensor, which then registers the elapsed time Δt, a fraction of a second, say, 10^{-7} s, and emits another pulse upward. Since the speed of light is constant and the distance that it travels is fixed, it takes the second pulse the exact same amount of time to make the round trip. So another 10^{-7} s is registered by the detector. These identical time intervals are used as a clock to keep time.

Since Jane is traveling at uniform velocity, Einstein's Principle of Rela-

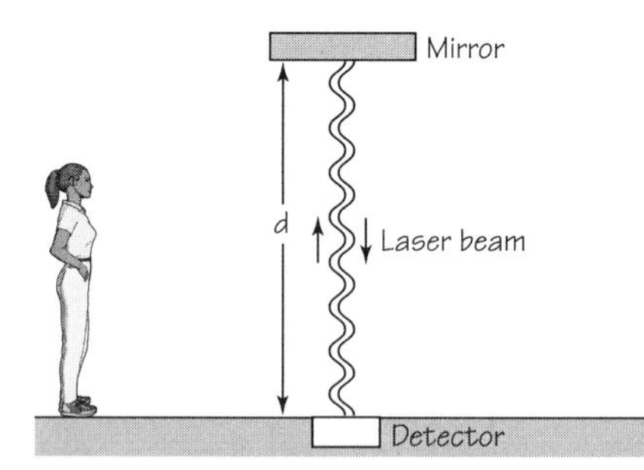

FIGURE 9.9 Laser clock in spaceship (as seen from spaceship frame of reference).

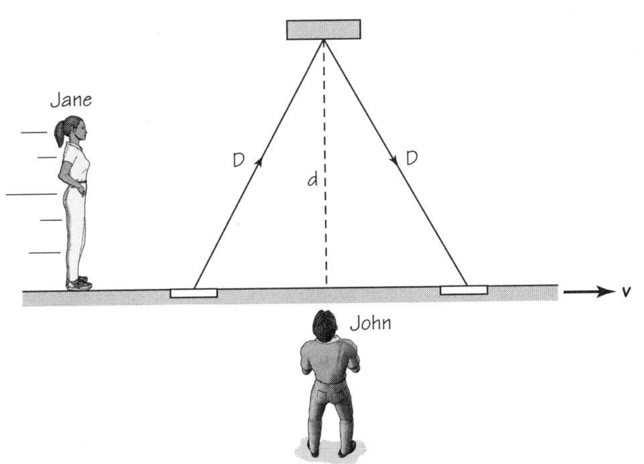

FIGURE 9.10 Laser clock in spaceship (as seen from an outside observer's frame of reference).

tivity tells her that the clock behaves exactly as it would if she were at rest. In fact, according this principle, she could not tell from this experiment (or any other) whether her ship is at rest or moving relative to John, without looking outside the spaceship. But to John, who is not in her reference frame but in his own, she appears to him to be moving forward rapidly in the horizontal direction relative to him. (Of course, it might be John who is moving backward, while Jane is stationary; but the observation and the argument that follows will be the same.)

Observing Jane's laser clock as her spaceship flies past him, what does John see? Just as before, in the experiment with the ball observed to be falling toward the floor when released by a moving person, John sees something quite different from what Jane sees. Because her spaceship is moving with respect to him, he observes that the light pulse follows a diagonal path upward to the upper mirror and another diagonal path downward to the detector. Let us give the symbol t' for the time he measures for the round trip of the light pulse.

Here enters the second postulate: the measured speed of light must be the same as observed by both John and Jane. But the distance the light pulse travels during one round trip, as Jane sees it, is shorter than what John sees. Call the total distance the pulse travels from the emitter to the upper mirror and back d for Jane and d' for John. The speed of light, c, which is the same for each, is

Jane: $$c = \frac{d}{t},$$

John: $$c = \frac{d'}{t'}.$$

DERIVATION OF TIME DILATION: THE LIGHT CLOCK

The "clock" consists of a stick of length l with a mirror and a photodetector P at each end. A flash of light at one end is reflected by the mirror at the other end and returns to the photodetector next to the light source. Each time a light flash is detected, the clock "ticks" and emits another flash.

Diagram (a) below shows the clock as seen by an observer riding with the clock. The observer records the time t between ticks of the clock. For this observer, the total distance traveled by the light pulse during the time t is $d = 2l$. Since the light flash travels at the speed of light c:

$$d = 2l = ct.$$

So

$$l = ct/2.$$

Diagram (b) shows the same clock as seen by an observer who is "stationary" in his or her own framework, with the clock apparatus moving by. This observer observes and records the time t' between ticks of the clock. For this observer, the total distance traveled by the light beam is d' in time t'. Since light travels at the same speed for all observers moving at uniform speed relative to each other, we have

$$d' = ct'.$$

Let's look at the left side of drawing (b). Here the motion of the clock, the vertical distance l, and the motion of the light beam form a right triangle. The base of the triangle is the distance traveled by the clock in time $t'/2$, which is $vt'/2$. The distance the beam travels in reaching the mirror is $d'/2$. Using the Pythagorean theorem, we obtain

$$\left(\frac{d'}{2}\right)^2 = l^2 + \left(\frac{vt'}{2}\right)^2.$$

From the above, we can substitute $d' = ct'$ and $l = ct/2$:

$$\left(\frac{ct'}{2}\right)^2 = \left(\frac{ct}{2}\right)^2 + \left(\frac{vt'}{2}\right)^2.$$

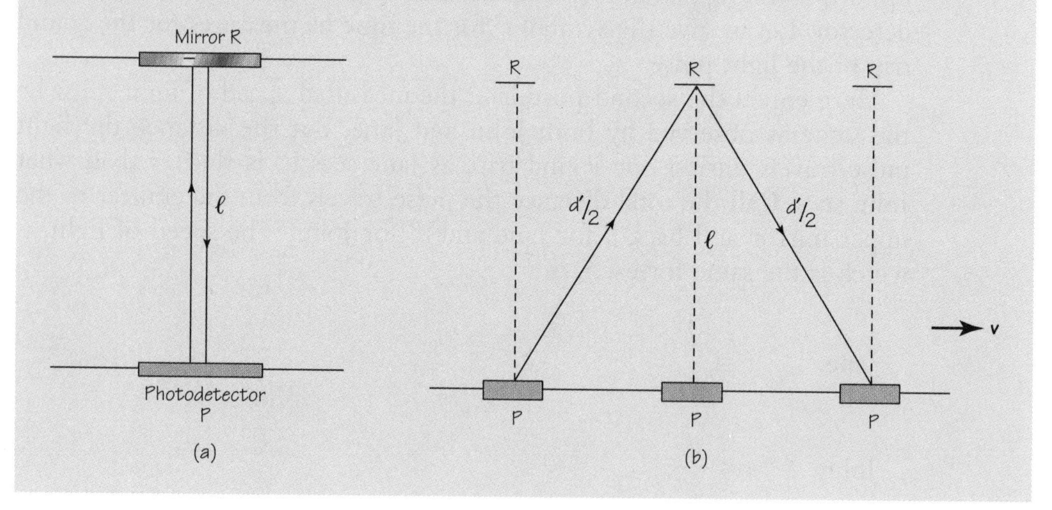

Squaring and canceling like terms, we have

$$c^2t'^2 = c^2t^2 + v^2t'^2.$$

Now, let's solve for t':

$$c^2t'^2 - v^2t'^2 = c^2t^2,$$

$$t'^2\,(c^2 - v^2) = c^2t^2,$$

$$t'^2 = \frac{c^2t^2}{c^2 - v^2},$$

$$t'^2 = \frac{t^2}{1 - v^2/c^2},$$

or

$$t' = \frac{t}{\sqrt{1 - v^2/c^2}},$$

Since $1 - v^2/c^2$ is here always less than 1, the denominator is less than 1, and the fraction is *larger* than t alone. Thus, the time interval t' registered by the clock as seen by the stationary observer is "dilated" compared to the time interval t registered by the clock as seen by the observer riding with the clock. In other words, the moving clock appears to run slower as measured by the stationary observer than when the clock is not moving with respect to the observer. Note also the crucial role of Einstein's second postulate in this derivation.

Since d' is larger than d, t' must be larger than t, in order for the ratios on the right side of both equations to have the same value, c. This means that the time interval ($\Delta t'$) for the round trip of the light pulse, as registered on the clock as John observes it, is longer than the time interval (Δt) registered on the clock as Jane observes it.

The surprising conclusion of this thought experiment (which is really a deduction from the postulates of relativity theory) is:

> Time intervals are not absolute and unchanging, but relative. A clock (such as Jane's), or any repetitive phenomenon which is moving relative to a stationary observer appears to the stationary observer to run slower than it appears to do when measured by the observer moving *with* the clock—and it appears to run slower the faster the clock is moving. This is known as *time dilation.*

Just how much slower does a clock seem when it is moving past an observer? To get the answer, you can use the diagram in Figure 9.10 of John and Jane and apply the Pythagorean theorem. After a bit of basic algebra (see the derivation in the insert), you obtain the exact relationship between the time elapsed interval registered by a clock that is stationary with respect to the observer (as in the case of Jane)—call it now ΔT_s—and the time elapsed interval for the same phenomenon—call it ΔT_{m}—as measured by someone who observes the clock in motion at constant velocity v (as in the case of John). The result is given by the following equation:

$$\Delta T_m = \frac{\Delta T_s}{\sqrt{1 - \frac{v^2}{c^2}}}.$$

In words: ΔT_{m}, John's observation of time elapsed registered by the moving clock, is different from ΔT_{s}, Jane's observation of time elapsed registered on the same clock, which is stationary in her frame, by the effect of the factor $\sqrt{1 - v^2/c^2}$ in the denominator.

9.7 TIME DILATION

What may make the equation for time dilation appear complicated is the term in the square root, which contains much of the physics. Study this equation and all of the symbols in it. The symbol c is the speed of light, and v is the speed of the clock moving relative to the observer measuring the time elapsed interval ΔT_{m}. As shown on page 427, for actual objects v is always less than c. Therefore v/c is always less than one, and so is v^2/c^2. In the equation on this page, v^2/c^2 is subtracted from 1, and then you take the square root of the result and divide it into ΔT_{s}, the time elapsed interval registered by the "stationary" clock.

Before we look at the full meaning of what the equation tries to tell us, consider a case where $v = 0$, for example, when Jane's spaceship has stopped relative to the Earth where John is located. If $v = 0$, then v^2/c^2 will be zero, so $1 - v^2/c^2$ is just 1. The square root of 1 is also 1; so our equation reduces to $\Delta T_{\mathrm{m}} = \Delta T_{\mathrm{s}}$: The time interval seen by John is the same as seen by Jane, when both are at rest with respect to each other, as we of course expect.

Now if v is not zero but has some value up to but less than c, then v^2/c^2 is a decimal fraction; so $1 - v^2/c^2$ and its square root are also decimal fractions, less than 1. (Confirm this by letting v be some value, say $\frac{1}{2}c$.) Dividing a decimal fraction into ΔT_{s} will result in a number larger than ΔT_{s}; so by our equation giving ΔT_{m}, ΔT_{m} will turn out to be larger than ΔT_{s}. In other words, the time interval as observed by the stationary observer watching the moving clock is larger (longer) than it would be for someone who is riding with the clock. The clock appears to the observer to run slower.

What Happens at Very High Speed?

Let's see what happens when the speed of the moving clock (or any repetitive process) is extremely fast, say 260,000 km/s (161,000 mi/s) relative to another inertial reference frame. The speed of light c in vacuum is, as always, about 300,000 km/s. When the moving clock registers a time interval of 1 s in its own inertial frame ($\Delta T_s = 1$ s), what is the time interval for someone who watches the clock moving past at the speed of 260,000 km/s? To answer this, knowing that ΔT_s is 1 s, we can find ΔT_m by substituting the relevant terms into the equation for ΔT_m:

$$\left[\frac{v}{c}\right]^2 = \left[\frac{260{,}000 \text{ km/s}}{300{,}000 \text{ km/s}}\right]^2$$

$$= [0.867]^2 = 0.75.$$

Therefore

$$1 - \sqrt{\frac{v^2}{c^2}} = \sqrt{1 - 0.75}$$

$$= \sqrt{0.25} = 0.5.$$

So

$$\Delta T_m = \frac{1 \text{ s}}{0.5}$$

$$= 2 \text{ s}.$$

This result says that a clock moving at 260,000 km/s that registers an interval of 1 s in its own inertial frame appears to an observer at rest relative to the clock to be greatly slowed down. While the person riding with the clock registers a 1-s interval, the resting observer will measure it (with respect to his own clock) to be 2 s. Note again that the clock does not seem to be slowed down at all to the person moving with the clock; but to the outside observer in this case the time interval has "dilated" to exactly double the amount.

What Happens at an Everyday Speed?

Notice also in the previous situation that we obtain a time dilation effect of as little as two times only when the relative speed is 260,000 km/s, which is nearly 87% of the speed of light. For slower speeds, the effect decreases very rapidly, until at everyday speeds we cannot notice it at all, except in very delicate experiments. For example, let's look at a real-life situation, say a clock ticking out a 1-s interval inside a jet plane, flying at the speed of sound of 760 mi/hr, which is about 0.331 km/s. What is the corresponding time interval observed by a person at rest on the ground? Again we substitute into the expression for time dilation.

$$\left[\frac{v}{c}\right]^2 = \left[\frac{0.331 \text{ km/s}}{300{,}000 \text{ km/s}}\right]^2$$

$$= [1.10 \times 10^{-6}]^2 = 1.22 \times 10^{-12}$$

$$\sqrt{1 - \frac{v^2}{c^2}} = \sqrt{1 - (1.22 \times 10^{-12})}$$

$$= \sqrt{0.99999999999878} = 0.99999999999938.$$

So

$$\Delta T_{\text{m}} = \frac{1 \text{ s}}{0.99999999999938}$$

$$= 1.00000000000061 \text{ s}.$$

With such an incredibly minute amount of time dilation, no wonder this effect was never observed earlier! Because the effect is so tiny, Newton's physics is still fine for the everyday world of normal speeds for which it was designed. This is also why it is false to say (as Einstein never did) that relativity theory proved Newton wrong. Nevertheless, the effect on moving clocks is there, and was in fact confirmed in a famous experiment involving a very precise atomic clock flown around the world on a jet airliner. It has also been tested and confirmed by atomic clocks flown on satellites and on the space shuttle at speeds of about 18,000 mi/hr. But the effect is so small that it can be neglected in most situations. It becomes significant only at relative speeds near the speed of light—which is the case in high-energy laboratory experiments and in some astrophysical phenomena.

What Happens When the Speed Reaches the Speed of Light?

If we were to increase the speed of an object far beyond 260,000 km/s, the time dilation effect becomes more and more obvious, until, finally, we approach the speed of light $v = c$. What happens as this occurs? Examining the time dilation equation, v^2/c^2 would approach 1 as v approaches c, so the denominator in the equation, $\Delta T_m = \Delta T_s/\sqrt{(1 - v^2/c^2)}$, would become smaller and smaller, becoming zero at $v = c$. As the denominator approaches zero, the fraction $\Delta T_s/\sqrt{(1 - v^2/c^2)}$ would grow larger and larger without limit, approaching infinity at $v = c$. And ΔT_m would thus become infinite when the speed reaches the speed of light c. In other words, a time interval of 1 s (or any other amount) in one system would be, by measurement with the clock in the other system, an infinity of time; the moving clock will appear to have stopped!

What Happens If v Should Somehow Become Greater Than c?

If this could happen, then v^2/c^2 would be greater than 1, so $(1 - v^2/c^2)$ would be negative. What is the square root of a negative number? You will recall from mathematics that there is no number that, when squared, gives a negative result. So the square root of a negative number itself has no physical reality. It is often called an "imaginary number." In practice, this means that objects cannot have speeds greater than c. This is one reason that the speed of light is often regarded as the "speed limit" of the Universe. *Neither objects nor information can travel faster in vacuum than does light.* As you will see in Section 9.9, nothing that has mass can even reach the speed of light, since c acts as an asymptotic limit of the speed.

Is It Possible to Make Time Go Backward?

The only way for this to happen would be if the ratio $\Delta T_s/\sqrt{(1 - v^2/c^2)}$ is negative, indicating that the final time after an interval has passed is less than the initial time. As you will also recall from mathematics, the solution of every square root has two values, one positive and one negative. Usually in physics we can ignore the negative value because it has no physical meaning. But if we choose it instead, we would obtain a negative result, suggesting that time, at least in theory, would go backward. But this would also mean that mass and energy are negative. That could not apply to ordinary matter, which obviously has positive mass and energy.

In Sum

You will see in the following sections that the square root in the equation for time dilation also appears in the equations for the relativity of length and mass. So it is important to know its properties at the different values of the relative speed. Because it is so important in these equations, the square root $\sqrt{(1 - v^2/c^2)}$ is often given the symbol γ, the Greek letter gamma.

We summarize the properties of $\gamma \equiv \sqrt{1 - v^2/c^2}$, discussed in this section:

$v = 0,$ $\gamma = 1,$

$0 < v < c,$ $\gamma =$ a fraction between 0 and 1, depending on the value of v^2/c^2

$v = 260{,}000$ km/s, $\gamma = 0.5,$

$v = c,$ $\gamma = 0,$

$v > c,$ $\gamma =$ imaginary.

9.8 RELATIVITY OF LENGTH

The two postulates of relativity theory also lead to the relativity of a second fundamental measured quantity, length. Einstein again applied the two postulates to a thought experiment (not a real experiment) on a simple measuring process. This was one way of deducing the physical consequences from his two fundamental postulates. Again the constant speed of light is the key, while the relativity principle is the underlying assumption.

We'll give Jane and John a rest and ask Alice and Alex, two other virtual researchers, to perform this thought experiment. Let Alice be at rest, while Alex is riding on a platform moving at uniform velocity relative to her. Alex carries a meter stick to measure the length of his platform in the direction it is moving. He obtains exactly 1 m. Alice tries to measure the length of Alex's platform with her meter stick as Alex's platform moves past her at constant velocity. She has to be quick, since she must read the two ends of the meter stick at the exact same instant; otherwise if she measures one end first, the other end will have moved forward before she gets to it. But there is a problem: light from the front and the rear of the platform take a certain amount of time to reach her, and in that brief lapse of time, the platform has moved forward.

Using only a little algebra and an ingenious argument (see the insert "Length Contraction"), Einstein derived an equation relating the measurements made by our two observers. The calculation, which is similar to the one for time dilation, yielded the result that, because the speed of light

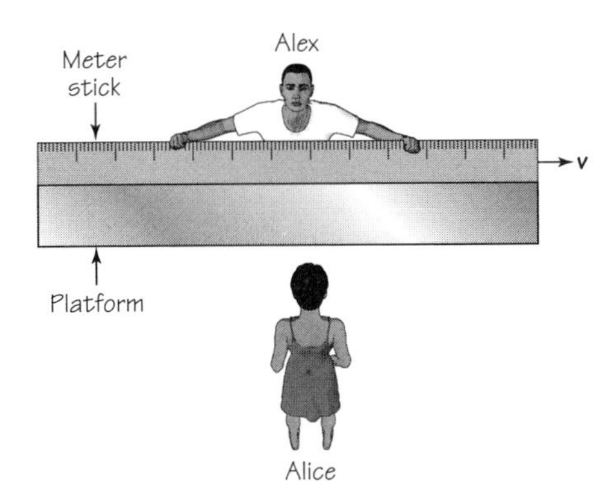

FIGURE 9.11 Length and contraction of a meter stick.

is not infinite, Alice's measurement of the length of the moving platform always turns out to be shorter than the length that Alex measures. The faster the platform moves past her, the shorter it is by Alice's measurement. The lengths as measured by the two observers are related to each other by the same square root as for time dilation. Alex, who is at rest relative to his platform, measures the length of the platform to be l_s, but Alice, who must measure the length of Alex's moving platform from her stationary frame, measures its length to be l_m. Einstein showed that, because of the constant speed of light, these two lengths are not equal but are related instead by the expression

$$l_m = l_s \sqrt{1 - \frac{v^2}{c^2}}.$$

Again the square root appears, which is now multiplied by the length l_s in Alex's system to obtain the length l_m as measured by Alice. Again, you will notice that when $v = 0$, i.e., when both systems are at rest with respect to each other, the equation shows there is no difference between l_m and l_s, as we expect. When the platform moves at any speed up to nearly the speed of light, the square root becomes a fraction with the value less than 1, which indicates that l_m is less than l_s. The conclusion:

> Length measurements are not absolute and unchanging, but relative. In fact, an object moving relative to a stationary observer appears to that observer in that reference frame to be shorter in the direction of motion than when its length is measured by an observer moving with the object—and it appears shorter the faster the object is moving.

LENGTH CONTRACTION

Consider a meter stick in a spaceship moving past you at high speed v. The meter stick is aligned in the direction of motion. Alex is an observer riding on the spaceship. He has a high-speed timing device and a laser emitter. With that equipment, she intends to measure the speed of light by emitting a laser pulse along a meter stick, which is aligned along the direction of motion of his spaceship. He will time the duration required for the light pulse to traverse the length of the meter stick. After performing the measurement, the time interval he measures is T_s and the length of the meter stick is l_s, the s indicating that they are stationary relative to her. Calculating the speed of the light pulse, l_s/T_s, he obtains the speed of light c, as expected.

Meanwhile, Alice is fixed on Earth as Alex's spaceship speeds past. She observes his experiment and makes the same measurements using her own clock—however her result for the time interval T_m registered on Alex's moving clock is different from Alex's measurement because of time dilation. Nevertheless, according to Einstein's second postulate Alice must obtain the exact same value for the speed of the light pulse, c. The only way this is possible is if the length of the meter stick in Alex's moving spaceship as measured by Alice, l_m as measured with her own measuring device, appears to have contracted by the same amount that the time interval she measured on the moving clock has expanded. The moving length l_m must therefore appear to be contracted in the direction of motion according to the relationship

$$l_m = l_s\sqrt{1 - \frac{v^2}{c^2}}.$$

This effect is known as *length contraction*. But note that the object is not actually contracting as it moves—the observed "contraction," which is in the direction of motion only, not perpendicular to it, is an effect of the *measurement* made from another system—as was the effect on the relative observations of elapsed time, the "time delay."

When $v = 0.8c$, for example, the apparent foreshortening seen by Alice of Alex's platform moving to the right, and of Alex himself and everything moving with him, would be about $0.6\ l_s$. Moreover, it is symmetrical! Since Alex can consider his frame to be at rest, Alice seems to be moving fast to the left, and it is she and her platform which seem to Alex to be foreshortened by the same amount.

The apparent contraction continues all the way up to the speed of light, at which point the length of the moving object would appear to the stationary observer to be zero. However, no mass can be made to reach the speed of light, so we can never attain zero length, although in accelerators (colliders) elementary particles come pretty close to that limit.

9.9 RELATIVITY OF MASS

You saw in Section 3.4 that inertial mass is the property of objects that resists acceleration when a force is applied. The inertial mass, or simply "the mass," is the constant of proportionality between force and acceleration in Newton's second law of motion

$$\mathbf{F}_{\text{net}} = m\mathbf{a}.$$

Therefore a constant force will produce a constant acceleration. So, once an object is moving, if you keep pushing on it with the same force, it will keep accelerating, going faster and faster and faster without limit, according to this formula. Newton's second law thus contains no speed limit. But this is inconsistent with the relativity theory, which imposes a speed limit for objects in space of about 300,000 km/s (186,000 mi/s), the speed of light. The way out is to amend Newton's second law. Einstein's way was to note that m, the inertial mass, does not stay constant but increases as the speed increases—as in fact is experimentally observed, for example, for high-speed elementary particles. When the speed increases, it takes more and more force to continue the same acceleration—eventually an infinite force trying to reach the speed of light. Einstein deduced from the two postulates of special relativity theory that the inertia of a moving object increases with speed, and it does so in the same way as the time relation in time dilation. (The derivation is provided in the *Student Guide* for this chapter.) Using our familiar square root factor, we can write

$$m_{\text{m}} = \frac{m_{\text{s}}}{\sqrt{1 - v^2/c^2}}.$$

Here m_{m} is the mass of the object in relative motion, and m_{s} is the mass of the same object before it starts to move. Often m_{s} is called the "rest mass."

Similarly to the measurement of time intervals, as an object's speed increases the mass as observed from a stationary reference frame also increases. It would reach an infinite (or undefined) mass if it reached the speed of light. This is another reason why anything possessing mass cannot actually be made to attain the speed of light; it would require applying an infinite force to accelerate it to that speed. By the same argument, entities that do move at the speed of light, such as light itself, must therefore have zero rest mass.

Following Einstein's result that the mass of an object increases when it is in motion relative to a stationary observer, Newton's equation relating the force and the acceleration can be written as a more general law

$$\mathbf{F}_{\text{net}} = \frac{m_{\text{s}}}{\sqrt{1 - v^2/c^2}}\,\mathbf{a}.$$

THE RELATIVISTIC INCREASE OF MASS WITH SPEED

v/c	m/m_0	v/c	m/m_0
0.0	1.000	0.95	3.203
0.01	1.000	0.98	5.025
0.10	1.005	0.99	7.089
0.50	1.155	0.998	15.82
0.75	1.538	0.999	22.37
0.80	1.667	0.9999	70.72
0.90	2.294	0.99999	223.6

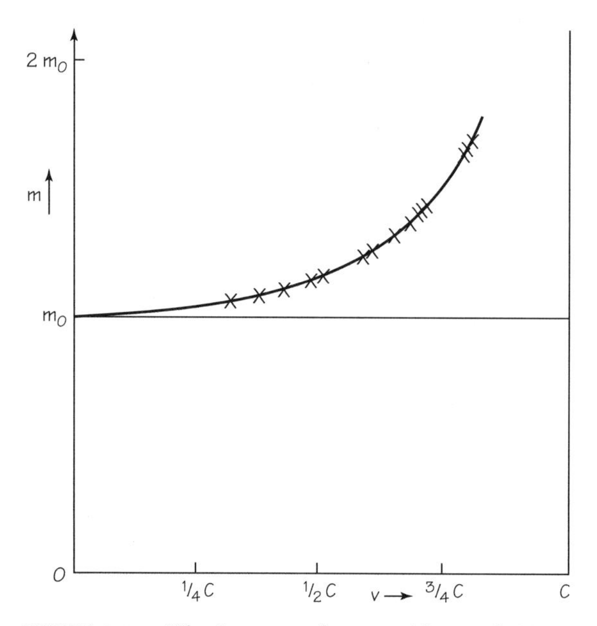

FIGURE 9.12 The increase of mass with speed. Note that the increase does not become large until v/c well exceeds 0.50.

Notice that as the relative speed decreases to zero, this equation transforms continuously into Newton's equation

$$\mathbf{F}_{\text{net}} = \frac{m_s}{\sqrt{1 - v^2/c^2}}\,\mathbf{a} \rightarrow \mathbf{F} = m\mathbf{a}, \text{ as } v \rightarrow 0.$$

This indicates again that Einstein's physics did not break with Newton's physics. Instead Einstein's physics is a continuation of Newton's physics.

9.10 MASS AND ENERGY

After Einstein completed his paper on the special theory of relativity in 1905 he discovered one more consequence of the relativity postulates, which he presented, essentially as an afterthought, in a three-page paper later that year. In terms of the effect of physics on world history, it turned out to be the most significant of all his findings.

We discussed in Chapter 5 that when work is done on an object, say hitting a tennis ball with a racket, the object acquires energy. In relativity theory, the increase in speed, and hence the increase in kinetic energy of a tennis ball or any object, results in an increase in mass (or inertia), although in everyday cases it may be only an infinitesimal increase.

Examining this relation between relative speed and effective mass more closely, Einstein discovered that *any* increase in the energy of an object should yield an increase of its measured mass—whether speeding up the object, or heating it, or charging it with electricity, or merely by doing work by raising it up in the Earth's gravitational field. In short, Einstein discovered that a change in energy is equivalent to a change in mass. Moreover, he found that the equivalence works both ways: An increase or decrease in the energy in a system correspondingly increases or decreases its mass, and an increase or decrease in mass corresponds to an increase or decrease in energy. In other words, mass itself is a measure of an equivalent amount of energy.

To put Einstein's result in symbols and using the delta (Δ) symbol: a change in the amount of energy of an object is directly proportional to a change in its mass, or

$$\Delta E \propto \Delta m.$$

Einstein found that the proportionality constant is just the square of the speed of light, c^2:

$$\Delta E = (\Delta m)c^2,$$

or, expressed more generally,

$$\boxed{E = mc^2.}$$

In its two forms, this is probably the most famous equation ever written. What it means is that an observed change of mass is equivalent to a change of energy, and vice versa. It also means that an object's mass itself, even if it doesn't change, is equivalent to an enormous amount of energy, since the proportionality constant, c^2, the square of the speed of light in vacuum, is a very large number. For example, the amount of energy contained in just 1 g of matter is

$$E = mc^2 = (0.001 \text{ kg})(3 \times 10^8 \text{ m/s})^2$$
$$= (1 \times 10^{-3} \text{ kg})(9 \times 10^{16} \text{ m}^2/\text{s}^2)$$
$$= 9 \times 10^{13} \text{ kg m}^2/\text{s}^2$$
$$= 9 \times 10^{13} \text{ J}.$$

This enormous amount of energy is roughly equivalent to the chemical energy released in 20 tons of TNT, or the amount of energy consumed in the whole United States on average in 30 s. It is the source of the energies released by radioactive substances, our Sun and other stars, by nuclear weapons, and by nuclear reactors producing electrical energy.

Not only are mass and energy "equivalent," we may say *mass is energy*. This is just what Einstein concluded in 1905: "The mass of a body is a measure of its energy content." We can think of mass as "frozen energy," frozen at the time the Universe cooled soon after the Big Bang and energy clumped together into balls of matter, the elementary particles of which ordinary matter is made. Thus any further energy pumped into a mass will increase its mass even more. For instance, as we accelerate protons in the laboratory to nearly the speed of light, their mass increases according to the relativistic formula for m_{m}. This increase can also be interpreted as an increase in the energy content of the protons. These two different deductions of relativity theory—mass increase and energy–mass equivalence—are consistent with each other.

This equivalence has exciting significance. First, two great conservation laws become alternate statements of a single law. In any system whose total mass is conserved, the total energy is also conserved. Second, the idea arises that some of the rest energy might be transformed into a more familiar form of energy. Since the energy equivalent of mass is so great, a very small reduction in rest mass would release a tremendous amount of energy, for example, kinetic energy or electromagnetic energy.

9.11 CONFIRMING RELATIVITY

Einstein's theory is not only elegant and simple, it is extraordinarily far-reaching, although its consequences were and still are surprising when first encountered. By noticing an inconsistency in the usual understanding of

Note: Einstein did not initially call his theory the theory of relativity. That term was given to it by others. Einstein later said he would have preferred calling it the theory of *invariance*. Why? Because, as said before, the laws of physics remain invariant, unchanged, the same for the "stationary" and the "moving" observer. That is extremely important, and makes it obvious why it is so *wrong* to say that Einstein showed that "everything is relative."

Maxwell's theory, and by generalizing Galileo's ideas on relative motion in mechanics, Einstein had been led to state two general postulates. Then he applied these two postulates to a study of the procedures for measuring the most fundamental concepts in physical science—time, length, mass, energy—and, as one does in a geometry proof, he followed these postulates to wherever the logic led him. The logic led him to conclude that the measurements of these quantities can be different for different observers in motion relative to each other. While the laws of physics—properly amended, as in the case of $\mathbf{F}_{\text{net}} = m\mathbf{a}$ becoming $\mathbf{F}_{\text{net}} = (m_s/\sqrt{1 - v^2/c^2})\ \mathbf{a}$—and the speed of light are the same for all observers, these basic quantities that enter into the laws of physics, such as time or mass, are not the same for all, they are relative with respect to the measurement frame. This is why it is called the theory of relativity. More precisely, it is called the theory of *special* relativity, since in this theory the relative velocities of the observers must be uniform (no acceleration), hence applying only to inertial systems.

But, you may object, anyone can come up with a couple of postulates, correctly deduce some strange consequences from them, and claim that they now have a new theory. In fact, this happens all too often, and usually is rejected as poor science. Why do we accept Einstein's theory as good science? The answer is of course eventual experimental confirmation, internal consistency, and consistency with other well-established theories. Every theory in science, whether deduced from a few postulates or induced from experimentally based hypotheses, must pass the rigorous test of experimental examination by various researchers, usually over a long period of time. In fact, as one astronomer recently remarked, the more profound the theory, the more extensive the experimental evidence that is required before it can be accepted. In addition, of course, the derivation of the theory cannot contain any logical mistakes or unfounded violations of accepted laws and principles. And it must be compatible with existing theories, or else it must show how and why these theories must be revised.

Far from being "dogmatic," as some would have it, scientists are always skeptical until the evidence is overwhelming. Indeed, it took more than a decade of research to confirm that relativity theory is indeed internally consistent as well as experimentally sound. The above sections also indicate how and why the classical physics of Newton and Maxwell had to be revised for application to phenomena at high relative speeds. But as the relative speed decreases, all of the results of relativity theory fade smoothly into the classical physics of the everyday world. There is no "incommensurability" between the worlds of Newton and Einstein.

Relativity theory is so well tested that it is now used as a tool for studying related theories and for constructing new experiments. Most of these experiments involve sub-microscopic particles moving at extremely high speeds, such as are found in modern-day accelerators. But some are also at everyday speeds. Here are a few of the most well-known confirmations of the postulates and deductions of special relativity theory.

The Constancy of the Speed of Light

The validity of the two postulates of relativity theory also extends to classical physics (e.g., mechanics), as Galileo showed for the early relativity postulate with the tower experiment, and as Einstein apparently realized as he thought about running alongside a light beam. A direct confirmation of the constancy of the speed of light has been obtained from the study of double stars, which are stars that orbit about each other. If the orbit of one star is close to the line of sight from the Earth, then at one side of the orbit it is moving toward the Earth, on the other side it is moving away. Careful studies of the speed of light emitted by such stars as they move toward and away from us at high speed show no difference in the speed of light, confirming that the speed of light is indeed independent of the speed of the source.

Another of the many experiments involved a high-speed particle in an accelerator. While moving at close to the speed of light, it emitted electromagnetic radiation in opposite directions, to the front and to the rear. Sensitive instruments detected the radiation and measured its speed. Astonishing as it may seem to the uninitiated, the speed of the radiation emit-

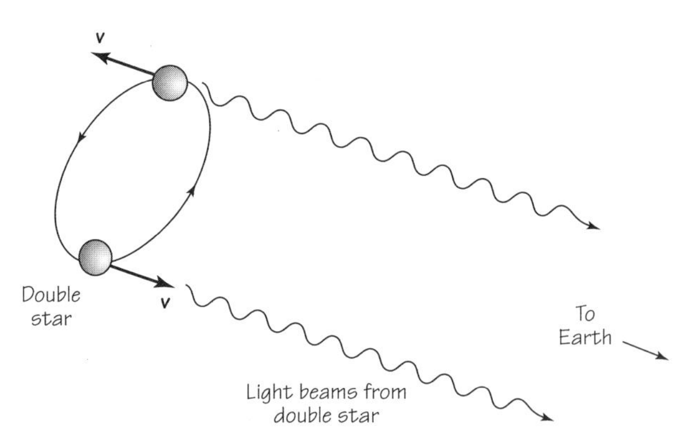

FIGURE 9.13 Light beams from a double-star system.

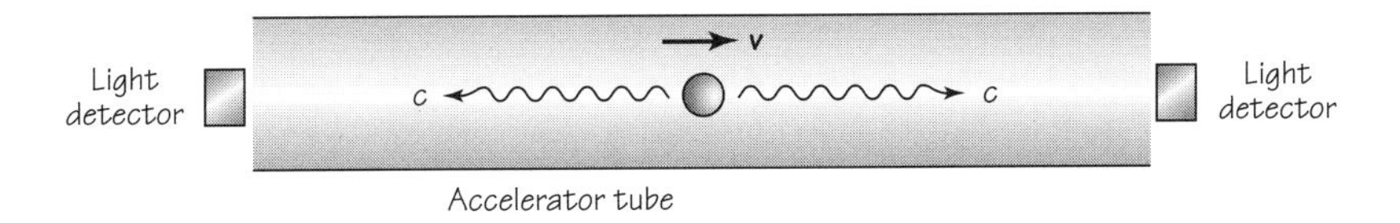

FIGURE 9.14 Particle in an accelerator emitting light beams simultaneously in opposite directions.

ted in both directions turned out to be exactly the speed of light, even though the particle itself was moving close to the speed of light—a striking confirmation of the constant-light-speed postulate, which amounts to a law of nature.

The Relativity of Time

The relativity theory predicts that a moving clock, as seen by a stationary observer, will tick slower than a stationary clock. We noted earlier that this effect has been tested and confirmed using atomic clocks inside airplanes and satellites.

An equally dramatic confirmation of the relativity of time occurred with the solution to a curious puzzle. Cosmic rays are high-speed protons, nuclei, and other particles that stream through space from the Sun and the galaxy. When they strike the Earth's atmosphere, their energy and mass are converted into other elementary particles—a confirmation in itself of the mass–energy equivalence. One of the particles they produce in the atmosphere is the so-called *mu-meson*, or simply the *muon*. When produced in the laboratory, slow muons are found to have a short life. On average they last only about $2.2\ 10^{-6}$ s, at which time there is a 50–50 chance that each one will decay into other elementary particles. (10^{-6} s is a microsecond, symbol: μs.)

The puzzle is that the muons created in the upper atmosphere and moving at high speed were found to "live" longer before they decay than those laboratory-generated ones. They last so long that many more survive the long trip down to the detectors on the ground than should be possible. Considering the speed they are traveling and the distance they have to traverse from the upper atmosphere to sea-level (about 30 km), their average lifetime of 2.2 μs, as measured for slow muons, should not be sufficient for them to survive the journey. Most of them should decay before hitting the ground; but in fact most of them do reach the ground. How can this be? The answer is the time dilation predicted by relativity theory. Relative to the detectors on the ground, the muons are moving at such high speed that their "clock" appears slowed, allowing them to survive long enough to reach the ground. The amount of slowing, as indicated by the number of muons reaching the ground, was found to be exactly the amount predicted by relativity theory.

Relativity of Length

Recall that one of the basic ideas of relativity theory is that all speed is relative to the observer who measures the speed. So let's return to the puzzle of the long-lived muons, only this time we will jump to the perspective of the hapless muon crashing through the Earth's atmosphere. From the muon's point of view, it is at rest, while the Earth is flying up toward it at close to the speed of light. Since the Earth is now moving with respect to the muon's own frame of reference, the distance from the top of the atmosphere to the ground undergoes a length contraction when seen from the point of view of the muon. The contraction is so great that, from the muon's perspective, it has no trouble covering this short distance in the mere 2.2 μs of the short life it has in its "stationary" reference frame. Again, the observations are in complete agreement with the predictions of special relativity theory.

Relativity of Mass

Relativity theory predicts that the observed mass of an object will increase as the relative speed of the object increases. Interestingly, this effect had been observed even before Einstein's theory, when scientists were puzzled to notice an increase in the mass of high-speed electrons in vacuum tubes. This effect is easily observed today in particle accelerators, where elementary charged particles such as electrons or protons are accelerated by electromagnetic fields to speeds as high as 0.9999999 the speed of light. The masses of these particles increase by exactly the amount predicted by Einstein's formula. At that speed the increase of their mass (m_m is about 2236 times the rest mass; $m_m = 2236\ m_s$). In fact, circular accelerators have to be designed to take the mass increase into account. As the particles are accelerated to high speeds by electric fields, they are curved into a circular path by magnetic fields to bring them back and let them undergo repeated accelerations by the fields. You saw in Section 3.12 that an object moving in a circular path requires a centripetal force. This force is given by the equation $F = mv^2/R$. Here R is the radius of the circle, which is fixed; v is the particle's speed, which increases; and m is the moving mass, which also increases according to relativity theory.

If scientists do not take the mass increase into account in their particle

FIGURE 9.15 Fermi National Accelerator Laboratory (Fermilab), in Batavia, Illinois, one of the world's most powerful particle accelerators.

accelerators, the magnetic force would not be enough to keep the particles on the circular track of the accelerator, and they would hit the wall or come out there through a portal. A simple circular accelerator is called a *cyclotron*. But when the increase of the accelerating force is precisely synchronized with the increases in speed and relativistic mass, it is called a *synchrocyclotron*.

Equivalence of Mass and Energy

Einstein regarded the equivalence of mass and energy, as expressed in the equation $E = mc^2$, to be a significant theoretical result of special relativity, but he did not believe it had any practical importance when he announced his finding. The hidden power became most obvious, of course, in the explosion of the atomic (more precisely "nuclear") bombs in 1945. The tremendous energy unleashed in such a bomb is derived from the transformation in the nuclei of a small amount of uranium or plutonium mass into the equivalent, huge amount of energy.

FIGURE 9.16 Trajectories of a burst of elementary particles in the magnetic field inside a buble chamber.

Nuclear bombs and reactors are powered by the splitting of heavy atoms. An opposite process, a fusion reaction, takes place using the joining together of nuclei of light elements. Again, a tiny amount of the mass is converted into energy according to Einstein's formula. Despite much effort, it has not yet been possible to control this fusion process on a scale sufficient to produce electricity for domestic and industrial use; however, the absence of harmful radioactive by-products would make such a device very desirable. But the nuclear fusion process does have a very practical importance: It powers the energy output of the Sun and all other stars in the Universe. Without it, life could not exist on the surface of the Earth. (Nuclear fission and fusion and their applications are further discussed in Chapter 18.)

The conversion of energy into mass can also be observed in the collisions of elementary particles that have been accelerated to enormously high speeds. Photographs of the results, such as the one here, display the creation of new particles.

9.12 BREAKING WITH THE PAST

Although Einstein's theory of special relativity did not represent a major break with classical physics, it did break with the mechanical world view. Our understanding of nature provided by special relativity, together with subsequent advances in quantum mechanics, general relativity theory, and other innovations, will slowly shape the new world view that is emerging.

Special relativity introduced an important break with the mechanical world view concerning the notion of absolute rest and absolute motion, which ceased to exist as a result of Einstein's work. Until that time, most physicists defined absolute rest and motion in terms of the so-called ether, the stuff that filled all of the space and transmitted light and electric and magnetic forces. As noted earlier, Einstein simply ignored the ether as "superfluous," since only relative motions were used in his theory. At the same time, and even before, a large number of careful experiments of different sorts to detect the ether had failed completely. One of these, the most famous one, was a series of experiments, during the 1880s, in which the American scientists Albert A. Michelson and Edward Morley attempted to detect the "wind" of ether experienced by the Earth as it moved through the supposed stationary ether on its orbit around the Sun. If such an ether existed, scientists believed, it should cause an "ether wind" over the surface of the Earth along the direction of motion. Since light was believed to be a wave moving through the ether, somewhat like sound waves through the air, it should be affected by this wind. In particular, a light wave traveling into the wind and back should take longer to make a round trip than a wave traveling the exact same distance at a right angle, that is, across the wind and back. (See the calculation in the *Student Guide*.) Comparing two such waves, Michelson and Morley could find no difference in their times of travel, within the limits of precision of their experiment. Within a few years of Einstein's theory, most physicists had abandoned the notion of an ether. If it could not be detected, why keep it?

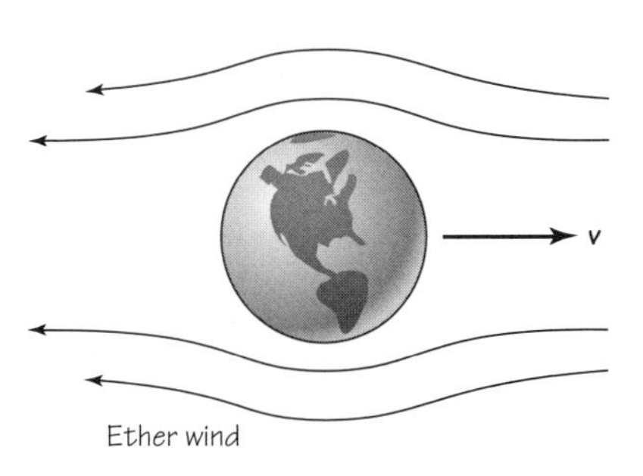

FIGURE 9.17 Earth moving through the stationary ether, according to nineteenth-century concepts.

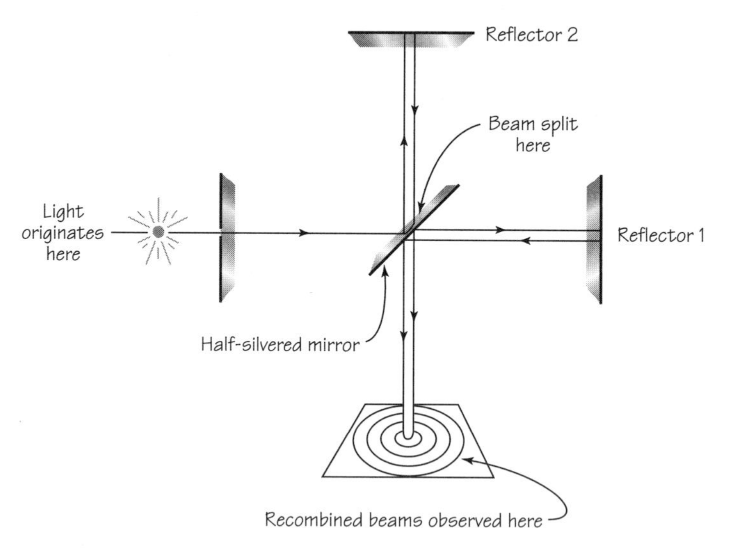

FIGURE 9.18 Schematic diagram of the Michelson–Morley experiment.

Not only did the loss of the ether rule out the concepts of absolute rest and absolute motion, but scientists had to rethink their understanding of how forces, such as electricity, magnetism, and gravity, operate. Ether was supposed to transmit these forces. Suddenly there was no ether; so what are these fields? Scientists finally accepted the idea that fields are independent of matter. There was now more to the world than just matter in motion. There were now matter, fields, and motion, which meant that not everything can be reduced to material interactions and Newton's laws. Nonmaterial fields had also to be included, and be able to carry energy across empty space in the form of light beams. The world was suddenly more complicated than just matter and motion. (You will read more about fields in Chapter 10.)

Another break with the mechanical world view concerned the concepts of space and time. Newton and his followers in the mechanical view had regarded space and time to be absolute, meaning the same for all observers, regardless of their relative motion. Einstein demonstrated in special relativity that measurements of space and time depend upon the relative motion of the observers. Moreover, it turned out that space and time are actually entwined with each other. You can already see this in the problem of making measurements of the length of a moving platform. The ends of the meter stick must be read off at the ends of the platform at the same instant in time. Because of the postulate of the constancy of the speed of light, a person at rest on the platform and a person who sees the platform moving will not agree on when the measurements are simultaneous. In 1908 the German mathematician Hermann Minkowski suggested that in relativity theory time and space can be viewed as joined together to form the four dimensions of a universal four-dimensional world, called *spacetime*. Four-dimensional spacetime is universal because an "interval" measured in this world would turn out to be the same for all observers, regardless of their relative motion at uniform velocity, but the "interval" would include both distance and time.

The space in which we live consists, of course, of three dimensions: length, width, and breadth. For instance, the event of a person sitting down on a chair in a room can be defined, in part, by the person's three coordinates. Starting at one corner of the room, the length along one wall may be 3 m, the length along the other wall may be 4 m, and the height to his chair seat may be 0.5 m. But to specify this event fully, you must also specify the time: say, 10:23 a.m. These four coordinates, three of space and one of time, form the four dimensions of an event in *spacetime*. Events take place not only in space but also in time. In the mechanical world view, space and time are the same for all observers and completely independent of each other. But in relativity theory, space and time are different for different observers moving relative to each other, and space and time are entwined together into a four-dimensional construct, "spacetime," which is the same for all observers.

SOME NEW CONCEPTS AND IDEAS

constructive theory
ether
length contraction
mass–energy equivalence
Michelson–Morley experiment
principle of constancy of speed of light
principle of relativity
reference frame
reference frame, inertial
relative motion
spacetime
theory of special relativity
time dilation

FURTHER READING

D. Cassidy, *Einstein and Our World* (Amherst, NY: Prometheus Books, 1995).
A. Einstein, *Ideas and Opinions* (New York: Bonanza Books, 1988).

A. Einstein, *Relativity: The Special and the General Theory* (New York: Crown, 1995), and many other editions; originally published 1917.
A. Einstein, *The World As I See It* (New York: Citadel Press, 1993).
A. Einstein and Leopold Infeld, *The Evolution of Physics* (New York: Simon and Schuster, 1967).
A. Fölsing, *Albert Einstein: A Biography*, E. Osers, transl. (New York: Penguin, 1998).
P. Frank, *Einstein: His Life and Times*, rev. ed. (New York: Da Capo).
M. Gardner, *Relativity Simply Explained* (New York: Dover, 1997).
B. Hoffmann, *Albert Einstein: Creator and Rebel* (New York: Viking Press, 1972).
G. Holton, *Einstein, History, and Other Passions* (Cambridge, MA: Harvard University Press, 2000).
G. Holton, *Thematic Origins of Scientific Thought: Kepler to Einstein* (Cambridge, MA: Harvard University Press, 1988), Part II: "On Relativity Theory."
G. Holton and S.G. Brush, *Physics, the Human Adventure*. (Piscataway, NJ: Rutgers University Press, 2001), Chapter 30.
E.F. Taylor, and J.A. Wheeler, *Spacetime Physics: Introduction to Special Relativity*, 2nd ed. (New York: Freeman, 1992).
H. Woolf, ed., *Some Strangeness in the Proportion: A Centennial Symposium to Celebrate the Achievements of Albert Einstein* (Reading, MA: Addison-Wesley, 1980).

Web sites

See the course Web site at: http://www.springer-ny.com/
A. Einstein: http://www.aip.org/history/einstein
A. Einstein: http://www.pbs.org/wgbh/nova/einstein

STUDY GUIDE QUESTIONS

1. What is "relative" in the theory of relativity?
2. What is special about the theory of special relativity?
3. Why did Einstein later say he would have preferred if it had been called the theory of invariance (or constancy)?
4. State in your own words the two principles, or postulates, on which special relativity is based.
5. What are four deductions of the theory?
6. Briefly describe an experimental confirmation of each of these four deductions.

9.1 The New Physics

1. What is "classical physics"?
2. What was new about the new physics?

9.2 Albert Einstein

1. Who was Albert Einstein? What did he do?
2. Give a brief synopsis of his life. Give a brief synopsis of his views apart from science.

9.3 The Relativity Principle

1. How does Einstein's theory compare with other theories described in previous chapters?
2. What is relative motion? Give an example.
3. A student measures the speed of a cart on a laboratory table to be 150 cm/s toward the north. The laboratory table is on wheels and is moving forward at 150 cm/s to the north. What is the speed of the cart relative to the floor?
4. What do Einstein's Relativity Principle and the Galilean Relativity Principle have in common? In what ways are they different?
5. A sailing ship is moving at uniform velocity on a calm sea. A ball drops from the mast of the ship.
   (a) Where does the ball land?
   (b) Compare the observation of the falling ball by a person riding on the ship with the observation of another person standing on the shore. Explain any difference.
   (c) If these are different, then which observation is correct?
6. Can the observer on the ship that is moving smoothly forward determine (from a windowless cabin within) if he is really moving or not? Explain.
7. Two people moving relative to each other while observing a falling ball see two different trajectories. Why would they be wrong to say as a result that "everything is relative"?
8. What does the statement mean that physical measurements, but not physical laws, depend on the observer's frame of reference?

9.4 Constancy of the Speed of Light

1. A small asteroid is headed straight toward the Earth at 20 km/s. Suddenly a gas jet on the asteroid fires a chunk of rock toward the Earth with a speed of 3 km/s relative to the asteroid. Scientists on Earth measure the speed of the chunk as it flies toward them. What is the measured speed?
2. An elementary particle is moving toward the Earth at 0.999999 the speed of light. It emits a light wave straight at the Earth. The light wave is detected by equipment on the ground and the speed of the wave is measured. What will the measured speed be? With what speed does the light move away from the elementary particle, as measured by an experimenter moving along with the particle?

9.5–9.7 Relativity of Time and Time Dilation

1. In your own words explain the thought experiment that shows that to an observer who is stationary, a moving clock consisting of a laser and mirror runs slower than it does to an observer riding with the clock.
2. Examine the formula for time dilation and define every symbol in it.
3. Using the formula for time dilation, explain why a stationary observer will measure that time slows down for events in the moving system, for relative speeds greater than zero but less than c.
4. The equation for time dilation refers to any relative speed v up to the speed of light c.
   (a) What happens when $v = 0$?
   (b) Why is it that we don't notice any time dilation at even the fastest speeds we can encounter in the everyday world, for example, a supersonic jet plane?
5. Explain why the time dilation equation indicates that there can be no relative speeds of objects greater than the speed of light.

9.8–9.9 Relativity of Length and Mass

1. In your own words present the thought experiment that shows that to an observer who is stationary, a moving meter stick appears shorter than it does to an observer riding with the meter stick.
2. Examine the formula for length contraction and define every symbol in it.
3. Using the formula for length contraction, show what happens when the meter stick is moving at 260,000 km/s and its length is measured by an observer at rest.
4. Why would it take an infinite force to accelerate any mass up to the speed of light?
5. Why can't anything that possesses mass at zero speed attain the speed of light?
6. Why is it wrong to say that Einstein's relativity represents a sharp break with Newtonian mechanics?
7. Why is it wrong to say that time is really dilated and length is really contracted in a moving system?
8. Why has it been wrong to say, as some did, that relativity theory undermines morality because it shows that "it all depends on your point of view"?

9.10 Mass and Energy

1. What does it mean to say that mass and energy are equivalent?
2. Illogically, one often hears that mass can be "converted" into energy, and energy be "converted" into mass. What would be a more accurate way of expressing the facts summarized by $E = mc^2$?
3. If mass and energy are interchangeable, what happens to the law of conservation of energy?

9.11 Confirming Relativity

1. By using different frames of reference, give two explanations for the fact that numerous high-speed muons generated in the upper atmosphere are able to

survive the trip to the Earth's surface, even though they are so short-lived that only a few should survive.

2. Aside from the increasing speed of the elementary particles, why does the centripetal force in a circular accelerator have to be constantly increased as a bunch of particles moves faster and faster?

9.12 Breaking with the Past

1. In what ways did relativity theory introduce a sharp break with the mechanical world view?
2. How were electric forces, fields, and the ether believed to be related?
3. What was the Michelson–Morley experiment? What were they attempting to detect? How did they attempt to detect it? What was their result?
4. What impact on the mechanical view did the rejection of the ether have?
5. What other break with the past did special relativity introduce?
6. What is four-dimensional spacetime? Give an example of an event in four-dimensional spacetime.

DISCOVERY QUESTIONS

1. Two observers are in uniform relative motion with respect to each other. They are in direct communication by cell phone and are attempting to decide who is really in motion and who is really at rest. What arguments can each one give to claim that he is at rest and the other person is moving?
2. Some people have argued that the theory of relativity supports the idea that "everything is relative." Would you agree or disagree with this conclusion, and how would you support your position, say in a discussion with others in class?
3. A person is exercising on a tread mill in a gym. The speed of the tread mill is set at 3 mi/hr. What is the speed of the exerciser relative to the ground? What is the exerciser's speed relative to the belt of the tread mill?
4. In studying this Part One of the text, you have followed some of the immense transformation of humankind's culture, from the pre-scientific period to current research questions. After thoughtful reflection on this experience, write a page or two summarizing the stages in this adventure of the creative mind.

Quantitative

5. On a piece of graph paper, plot the results, for a few distributed points, of the equation for length contraction, with l_m on the y-axis and the ratio v/c on the x-axis from 0 to 1.
   (a) Study this graph and explain why using this relativistic equation is usually not needed for relative speeds below about 20% of the speed of light.
   (b) From this graph explain why objects cannot attain the speed of light.
   (c) Explain why objects cannot go faster than the speed of light.

6. How fast would a pitcher have to throw a baseball to have its mass increase by 1%?
7. How much would be the rate of your body's "clock," the heart beat, decrease as measured by someone at rest on the ground, if you were flying in a plane at the speed of sound, about 330 m/s?
8. In Question 7, if you weighed yourself on a supersensitive scale during flight, how much would your weight seem to have increased, if at all, as observed from the ground?
9. If you have an opportunity to use the computer program "Space-Time," use the program to take a "trip" to Alpha Centauri, and observe the twin paradox. How is the paradox resolved?

PART TWO

FIELDS AND ATOMS

Prologue to Part Two

1. A REVOLUTION IN SCIENCE

The work of Isaac Newton concluded the scientific revolution that began in large part in A.D. 1549 when Nicholaus Copernicus first argued that the Earth is not stationary at the center of the Universe but rotates on its axis once a day and orbits the Sun once a year, along with the other planets. The revolution in science extended over the work of numerous "giants," as Newton called them—Galileo, Kepler, Descartes, and many others—resulting, by Newton's death in 1724, in the basis for the understanding of the physical world that we have today, what we call modern physics.

Many of the characteristics of modern physics that were established during the scientific revolution actually derived from the work of ancient thinkers, especially the work of the Ancient Greeks. This was because the Greeks were the first influential thinkers to seek explanations of natural events in terms of rational causes, rather than in the actions of supernatural beings. This meant that, for the first time, people regarded nature as accessible to human inquiry and study and governed by humanly understandable, rational principles. Truly scientific research was now possible.

As established by the scientific revolution, scientific research consists of the gathering of data through active experimentation and testing, not just passive observation or no observation at all. Galileo called experimental inquiry the "interrogation" of nature. The experimental evidence and closely related hypotheses are then joined together through rational processes and further experimental testing into a theory about the workings of nature that

Albert Einstein (1879–1955).

is then subject thereafter to revision, or even rejection, as new evidence becomes available.

Two of the main characteristic assumptions about nature that we have today can be traced, through Newton and the scientific revolution, back to the Ancient Greeks. The first is that nature is governed by a few, simple, clear rational laws and principles. From Plato and his followers, Newton and others came to realize that many of these principles can be expressed in mathematical terms and that the basic concepts of nature behave according to the rules of mathematics. Many centuries later, Albert Einstein declared: "the supreme task of the physicist is to arrive at those universal elementary laws from which the cosmos can be built up by pure deduction."* In another statement he declared:

> I am convinced that we can discover by means of purely mathematical constructions the concepts and the laws connecting them with each other, which furnish the key to the understanding of natural phenomena.†

---

* *Ideas and Opinions*, p. 226.

† *Ideas and Opinions*, p. 274.

A second characteristic assumption of contemporary physics goes back even further than Plato to the Greek thinker Democritus. Smelling the baking of bread one day, he reasoned that something must travel from the bread to his nose, even though he could not see anything traveling. He surmised that what traveled were extremely small, invisible "atoms" of bread. These and other considerations led to the hypothesis that the natural world consists of myriads of tiny atoms moving through empty space and clumping together to form the matter we see around us. Although this hypothesis was out of favor for nearly two millennia, it was revived and established as a foundation of physical science during the Scientific Revolution. The reason for the sudden popularity of atoms at that time is that, assuming nature consists of these inert bits of matter, the laws governing the events of nature can be clearly understood by referring to atoms, without having to refer to hidden spirits or other nonmaterial causes.

By assuming that matter consists of inert atoms moving around and clumping together in empty space, the behavior of everyday matter could be easily obtained from the properties of the atoms themselves and the few simple laws that govern their motion. These simple laws were Newton's famous *three laws of motion* (although they are not quite so simple as they may seem). They are:

1. The Law of Inertia:
   Every object continues in its state of rest or of uniform velocity (motion at uniform speed in a straight line) unless acted upon by an unbalanced force (a net force). Conversely, if an object is at rest or in motion with uniform velocity, all forces that may be acting on it must cancel so that the net force is zero.

Isaac Newton (1642–1727).

2. The Force Law:
   The net force acting on an object is numerically equal to, and in the same direction as, the acceleration of the object multiplied by its mass. In symbols: $\mathbf{F}_{net} = m\mathbf{a}$.
3. The Law of Action and Reaction:
   If one object exerts a force on another object, the second object at the same time exerts a force on the first object. These two forces, each acting on one of the two objects, are equal in magnitude and opposite in direction.

These laws are valid for all matter everywhere in the Universe, and they apply to matter in any form it takes, from planets and galaxies to space satellites and moving electrons in computers. They are still valid today.

We will not go into the details of these laws here (see Section 3.3), but you should be aware that for some concepts in these laws, such as force, acceleration, and velocity, the direction of the force or acceleration is as important as how large they are, their "magnitude." Concepts in which the direction and the magnitude are important are given a special name. They are called *vectors*. They are represented in this text in boldface. Other quantities that do not have a direction—such as mass and temperature—are called *scalars*. They are usually given in this text in italics, as are symbols representing only magnitudes of vectors.

You should also be aware that all quantities that can be measured are measured in carefully defined units. Although the American–English system of pounds, feet, and gallons is common in the United States, the metric system is used in science, since it is a decimal system. In this system, distance is usually measured in meters (m) or centimeters (cm), time in seconds (s), and force in newtons (N). Here

$$1 \text{ N} = 1 \text{ kg m/s}^2.$$

(See the *Student Guide* for further details. Your instructor may go over these units with you in class.)

Newton's crowning achievement was his discovery of the law of universal gravitation. According to this law, every massive object in the Universe attracts any other massive object with a force that can be given by a mathematical expression. This expression says that the force is equal to a constant times the product of the two masses, divided by the square of the distance between their centers, or in symbols:

$$F = \frac{Gm_1m_2}{R^2}.$$

This law is still valid today. For instance, it guides the launching of satellites from Earth or the fall of apples to the ground. However, while it is simple, it is not very clear. What exactly is this force and how does it operate? Newton was unable to provide an explanation in terms of only atoms moving through space; and it is still a problem today. But that does not prevent us from using the concept and equation for gravitational force today.

In the end atoms moving through space and time and clumping together by gravitation and other forces formed the basic conception of the entire Universe known as the "mechanical world view."

2. THE MECHANICAL WORLD VIEW

Once scientists realized the usefulness of the combination of the atomic hypothesis with the laws of motion and the force laws, all of which are based on experimental evidence, they were convinced that they had found the ultimate principles by which we can understand events in the physical world. This was an extremely important discovery. Since in physics the science of mechanics has to do with the study of matter in motion, this point of view became known as the "mechanical world view," which dominated physical science until well into the twentieth century. It is still prevalent today for everyday events on the human scales of distance, time, and speed, but not in situations where relativity and the quantum theory are important.

According to the mechanical world view, or the "mechanical philosophy," the physical world—the Earth, planets, and the entire Universe—can be understood in terms of atoms of matter interacting with each other as they move through space and time according to Newton's laws of motion and the pushes and pulls of gravitation and other forces. These mechanical principles were considered to be so fundamental that for centuries most scientists believed that a rational divine being must have established the Universe and everything in it according to mechanical principles.

In the mechanical view, atoms moved around and clumped together against the background of a flat, fixed, infinite space and in an unending linear unfolding of time. In an era when the first mechanical "grandfather" clocks were being invented, the Universe seemed to these scientists a clockwork universe, one that operates like a gigantic mechanical clock, constructed and set in motion by God in the beginning, and running smoothly against the backdrop of space and time ever since.

3. ENERGY AND ATOMS

The idea of a clockwork universe, based on the mechanical world view, enabled physical scientists to make enormous progress toward understanding the physical Universe. With this outlook scientists could be confident that all of the complex phenomena and their numerous unanswered questions about their world and the entire Universe would eventually be understood in terms of the basic, well-known principles of mechanical philosophy. Fired with the new confidence engendered by Newton's success, scientists turned to new areas of research in an attempt to discover the mechanical principles underlying these as yet unexplained phenomena, such as the behavior of heat, light, and electricity and magnetism. The first two of these are discussed in Part One of this text; we will start with electricity and magnetism in this part of the course.

Applying the mechanical world view to these fields led to great scientific progress in the centuries ahead. They also led to such important technological inventions as the steam engine, the electric generator, and the electric motor, which together helped to unleash the Industrial Revolution and the electric age, transforming the economic and social fabric of the world in which we live today. But, in the end, as you will see in the following chapters, these sciences also pointed to the limitations of the mechanical philosophy, leading ultimately to the contemporary theories of relativity theory and quantum mechanics. Part One of this text, which focuses on the study of motion, ended with relativity theory. This part, in which we investigate atoms and electromagnetism, will lead us into the strange and exciting world of quantum mechanics. We include the discussion of electromagnetism with that of atoms, because the structure and behavior of the atom itself involves electrical forces, while many properties of matter in the everyday world arise from the electrical behavior of atoms.

Before we enter the atomic world, you should know about the concepts of energy and atoms, which are more fully treated in Part One.

Energy

Energy is closely tied to forces. It is transported or transformed into different forms by the action of forces. For instance, a force exerted by a hand on a ball can transform the chemical energy in your muscles into the energy of motion of the ball. The energy of motion is called *kinetic energy*. This action is called *work*, and it is in the performance of useful work that machines are used in industrial processes. Like the arm throwing a ball, the work done by a machine represents the transformation of one form of energy, such as heat energy or electrical energy, into another form of energy, such as the spinning of a wheel.

During the 1800s researchers came to recognize many different forms of energy—heat energy, electrical energy, chemical energy, mechanical energy, and others. After careful studies they came to recognize that all of these forms of energy are different manifestations of one underlying entity—

energy—which can appear in the different forms that we observe. Further study, along with the encouragement of some philosophical notions about energy, led to the idea that in all energy-conversion processes occurring anywhere in nature—on Earth, in the Sun, inside an electric generating plant, etc.—the total amount of energy in the Universe, is never changed, it always stays the same—that is, the total amount of energy in the Universe is "conserved," although it can be transferred into different forms. This is the *Law of Conservation Energy*. It is still a fundamental law of nature today. It states:

> All natural events involve a transformation of energy from one form to another, but the total quantity of energy does not change during the transformation.

How Do We Know that Atoms Really Exist?

Since at least the time of the Ancient Greeks, many thinkers have postulated the existence of atoms. By "atoms" the Greeks meant invisibly small, hard, unbreakable balls of matter that make up the different elements constituting the material objects in the physical world. Such an idea was disregarded for many centuries until it was revived during the Scientific Revolution of the seventeenth century by such esteemed thinkers as Descartes, Boyle, Newton, and Gassendi. Studies of heat and chemistry during the nineteenth century lent strong support for this existence of the supposed atoms which, of course, no one could actually see (and which we can only barely see with today's apparatus). But not until about 1916 were the last doubters finally satisfied.

There are now several answers to the question: How do we know that atoms really exist? The *first answer* is that in *chemistry during the nineteenth century the work of John Dalton and many others showed that the elements combine in definite integral proportions to form chemical compounds*. This indicated that the atoms of the different elements combined in whole-number ratios to form molecules. These combinations never occurred in fractions of amounts, suggesting that there are indivisible integral units of each element, which we can call atoms. For instance, 1 unit of sodium combined with 1 unit of chlorine to produce 1 unit of salt, but never 1.5 units of sodium and 1.1 units of chlorine.

Second, *we know that atoms exist because the atomic theory of matter is highly successful in accounting for properties of matter that cannot be easily explained in any other way*. During the late nineteenth century, the properties of gases and the behavior of heat as a form of energy in different situations were all shown to be comprehensible in terms of the motions and interactions of the atoms, according to Newton's laws of motion.

But still some suggested that, although the atomic hypothesis was helpful and convenient, that did not make it necessarily valid. The third answer to the question proved to be decisive: in 1905 Albert Einstein showed that the *atomic theory provides the only detailed account of the purely random motion of large-scale molecules or microscopic pollen grains suspended in a stationary liquid.* The phenomenon, known as "Brownian motion" when pollen is involved, could be explained only by assuming that the suspended grains are subjected to random collisions by the much smaller molecules of the liquid, colliding randomly with each grain from different directions. Careful comparisons over the next few years resulted in complete agreement between Einstein's predictions and the observed random motion of Brown's pollen grains. Since then scientists have never seriously doubted the existence of atoms as the fundamental building blocks of all the forms of matter that we observe around us. Today, scanning tunneling microscopes and other high-resolution processes enable us, at last, to see direct evidence of individual atoms.

SOME NEW IDEAS AND CONCEPTS

atoms
Brownian motion
clockwork universe
conservation of energy
laws of motion
mechanical world view
scalars
vectors

STUDY GUIDE QUESTIONS

1. A Revolution in Science

1. What was the Scientific Revolution, and why is it important for us today?
2. What were some of the consequences of the scientific revolution for physics?
3. What was Newton's "crowning achievement?" How would you express it in your own words?

2. The Mechanical World View

1. Describe the mechanical world view in your own words.
2. What were some of the consequences of this world view for science?

3. Energy and Atoms

1. What is energy?
2. What is meant by the conservation of energy?
3. How do we know that atoms really exist?
4. How does Brownian motion relate to the existence of atoms?

CHAPTER 10

Electricity and Magnetism

10.1 GILBERT'S *MAGNETS*

Two natural substances, amber and lodestone, have awakened curiosity since ancient times. Amber is sap that oozed long ago from certain softwood trees, such as pine. Over many centuries, it hardened into a semitransparent solid akin to model plastics and ranging in color from yellow to brown. It is a handsome ornamental stone when polished, and sometimes contains the remains of insects that were caught in the sticky sap. Ancient Greeks recognized a strange property of amber. If rubbed vigorously against cloth, it can attract nearby objects, such as bits of straw or grain seeds.

Lodestone is a metallic mineral that also has unusual properties. It attracts iron. Also, when suspended or floated, a piece of lodestone always turns to one particular position, a north–south direction. The first known written description of the navigational use of lodestone as a compass in Western countries dates from the late twelfth century, but its properties were known even earlier in China. Today, lodestone would be called magnetized iron ore.

The histories of lodestone and amber are the early histories of magnetism and electricity. The modern developments in these subject areas began in 1600 with the publication in London of William Gilbert's book *De Magnete (On Magnets)*. Gilbert (1544–1603) was an influential physician, who served as Queen Elizabeth's chief physician. During the last 20 years of his life, he studied what was already known of lodestone and amber. Gilbert made his own experiments to check the reports of other writers and summarized his conclusions in *De Magnete*. The book is a classic in scientific literature, primarily because it was a thorough and largely successful attempt to test complex speculation by detailed experiment.

Gilbert's first task in his book was to review and criticize what had previously been written about lodestone. He discussed various theories about the cause of magnetic attraction. When it was discovered that lodestone and magnetized needles or iron bars tend to turn in a north–south direction, many authors offered explanations. But, says Gilbert:

> they wasted oil and labor, because, not being practical in the research of objects of nature, being acquainted only with books . . . they constructed certain explanations on the basis of mere opinions.

FIGURE 10.1 William Gilbert (1544–1603).

As a result of his own researchers, Gilbert himself proposed the real cause of the lining-up of a suspended magnetic needle or lodestone: The Earth itself is a lodestone and thus can act on other magnetic materials. Gilbert performed a clever experiment to show that his hypothesis was a likely one. Using a large piece of natural lodestone in the shape of a sphere, he showed that a small magnetized needle placed on the surface of the lodestone acts just as a compass needle does at different places on the Earth's surface. (In fact, Gilbert called his lodestone the *terrella*, or "little Earth.") If the directions along which the needle lines up are marked with chalk on the lodestone, they form meridian circles. Like the lines of equal longitude on a globe of the Earth, these circles converge at two opposite ends that may be called "poles." At the poles, the needle points perpendicular to the surface of the lodestone. Halfway between, along the "equator," the needle lies along the surface. Small bits of iron wire, when placed on the spherical lodestone, also line up in these same directions.

Nowadays, discussion of the actions of magnets generally involves the idea that magnets set up "fields" all around themselves, as further discussed in Section 10.3. The field can act on other objects, near or distant. Gilbert's description of the force exerted on the needle by his spherical lodestone was a step toward the modern field concept:

> The terrella's force extends in all directions. . . . But whenever iron or other magnetic body of suitable size happens within its sphere of influence it is attracted; yet the nearer it is to the lodestone the greater the force with which it is borne toward it.

Gilbert also included a discussion of electricity in his book. He introduced the word *electric* as the general term for "bodies that attract in the same way as amber." (The word *electric* comes from the Greek word *electron*, which means "amber." Today the word *electron* refers to the smallest free electric charge.) Gilbert showed that electric and magnetic forces are different. For example, a lodestone always attracts iron or other magnetic bodies. An electric object exerts its attraction only when it has been recently rubbed. On the other hand, an electric object can attract small pieces of many different substances. But magnetic forces act only between

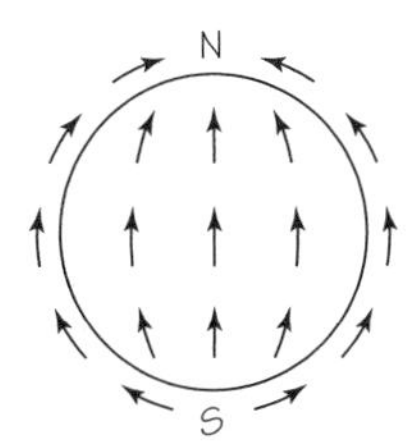

FIGURE 10.2 The Earth as a lodestone showing tiny magnets lined up at different locations on Earth.

a few types of substances. Objects are attracted to a rubbed electric object along lines directed toward one center region. But magnets always have *two* regions (poles) toward which other magnets are attracted.

Gilbert went beyond summarizing the known facts of electricity and magnets. He suggested new research problems that were pursued by others for many years. For example, he proposed that while the poles of two lodestones might either attract or repel each other, electric bodies could never exert repelling forces. However, in 1646, Sir Thomas Browne published the first observation of electric repulsion. In order to systematize such accounts, scientists introduced a new concept, *electric charge*. In the next section, you will see how this concept can be used to describe the forces between electrically charged bodies.

10.2 ELECTRIC CHARGES AND ELECTRIC FORCES

As Gilbert strongly argued, the behavior of electrified objects must be learned in the laboratory rather than by just reading about it. This section, therefore, is only a brief outline to prepare you for your own experience with the phenomena.

As discussed earlier, amber, when rubbed, acquires in a seemingly mysterious way the property of picking up small bits of grain, cork, paper, hair, etc. To some extent, all materials show this effect when rubbed, including rods made of glass or hard rubber, or strips of plastic. There are two other important basic observations:

1. When two rods of the same material are both rubbed with another material, the rods *repel* each other. Examples that were long ago found to work especially well are two glass rods rubbed with silk cloth, or two hard rubber rods rubbed with fur.
2. When two rods of *different* material are rubbed (e.g., a glass rod rubbed with silk, and a rubber rod rubbed with fur), the two rods may *attract* each other.

Electric Charges

These and thousands of similar experimentally observable facts about electrified objects can be summarized in a systematic way by adopting a very simple model. While describing a *model* for electrical attraction and repulsion, remember that this model is *not* an experimental fact which you can observe separately. It is, rather, a set of invented ideas which help to describe and summarize observations. It is easy to forget this important difference between experimentally observable facts and invented explanations. Both are needed, but they are not the same thing! The model adopted, based upon experimental evidence, consists of the concept of *charge*, along with three rules regarding charges. An object that is rubbed and given the property of attracting small bits of matter is said "to be electrically charged" or "to have an electric charge." All objects showing electrical behavior are found to have either one or the other of the two kinds of charge. The study of their behavior is known as *electrostatics*, since the charges are usually static, that is, not moving. (The study of moving charges is known as *electrodynamics*.)

The three empirical rules regarding electrostatic charges are:

1. There are only two kinds of electric charge.
2. Two objects charged alike (i.e., having the same kind of charge) repel each other.
3. Two objects charged oppositely attract each other.

When two different uncharged materials are rubbed together (e.g., the glass rod and the silk cloth), they acquire opposite kinds of charge. Benjamin Franklin, who did many experiments with electric charges, proposed a mechanical model for such phenomena. In his model, charging an object electrically involved the transfer of an "electric fluid" that was present in all matter. When two objects were rubbed together, some electric fluid from one object passed into the other. One body then had an extra amount of electric fluid and the other had a lack of that fluid. An excess of fluid produced one kind of electric charge, which Franklin called "positive." A lack of the same fluid produced the other kind of electric charge, which he called "negative."

Previously, some theorists had proposed a different, "two-fluid" model involving both a "positive fluid" and a "negative fluid." In that model, normal matter contained equal amounts of these two fluids, so that they canceled out each other's effects. When two different objects were rubbed together, a transfer of fluids occurred. One object received an excess of positive fluid, and the other received an excess of negative fluid.

There was some dispute between advocates of one-fluid and two-fluid models, but both sides agreed to speak of the two kinds of electrical charges as either *positive* (+) or *negative* (−). It was not until the late 1890s that experimental evidence gave convincing support to any model for "electric charge." Franklin thought of the electric fluid as consisting of tiny particles, and that is the present view, too. Consequently, the word "charge" is often used in the plural. For example, we usually say "electric charges transfer from one body to another."

What is amazing in electricity, and indeed in other parts of physics, is that so few concepts are needed to deal with so many different observations. For example, a third or fourth kind of charge is not needed in addition to positive and negative. Even the behavior of an *un*charged body can be understood in terms of + and − charges. Any piece of matter large enough to be visible can be considered to contain a large amount of electric charge, both positive and negative. If the positive charge is equal to the negative charge, the piece of matter will appear to have zero charge, no charge at all. The effects of the positive and negative charges simply cancel each other when they are added together or are acting together. (This is one advantage of calling the two kinds of charge positive and negative rather than, say, x and y.) The electric charge on an object usually means a slight *excess* (or net) of either positive or negative charge that happens to be on that object.

The Electric Force Law

What is the "law of force" between electric charges? In other words, how does the force depend on the *amount* of charge and on the *distance* between the charged objects?

FIGURE 10.3 Benjamin Franklin (1706–1790), American statesman, inventor, scientist, and writer, was greatly interested in the phenomenon of electricity. His famous kite experiment and invention of the lightning rod gained him wide recognition. Franklin is shown here observing the behavior of a bell whose clapper is connected to a lightning rod.

Joseph Priestley (1773–1804), a Unitarian minister and physical scientist, was persecuted in England for his radical political ideas. One of his books was burned, and a mob looted his house because of his sympathy with the French Revolution. He moved to America, the home of Benjamin Franklin, who had stimulated Priestley's interest in science. Primarily known for his identification of oxygen as a separate element that is involved in combustion and respiration, he also experimented with electricity. In addition, Priestley can claim to be the developer of carbonated drinks (soda-pop).

The first evidence of the nature of such a force law was obtained in an indirect way. About 1775, Benjamin Franklin noted that a small cork hanging near an electrically charged metal can was strongly attracted to the can. But when he lowered the cork by a thread into the can, he found that no force was experienced by the cork no matter what its position inside the can. Franklin did not understand why the walls of the can did not attract the cork when it was inside but did when it was outside. He asked his friend Joseph Priestley to repeat the experiment.

Priestley verified Franklin's results and went on to reach a brilliant conclusion from them. He remembered from Newton's *Principia* that gravitational forces behave in a similar way. Inside a hollow planet, the net gravitational force on an object (the sum of all the forces exerted by all parts of the planet) would be exactly zero. This result also follows mathematically from the law that the gravitational force between any two individual pieces of matter is inversely proportional to the square of the distance between them

$$F \propto \frac{1}{R^2}.$$

Priestley therefore proposed that forces exerted by charges vary inversely as the square of the distance, just as do forces exerted by massive bodies. The force exerted between bodies owing to the fact that they are charged is called "electric" force, just as the force between uncharged bodies is called "gravitational" force. (Remember that all forces are known by their mechanical effects, by the push or acceleration they cause on material objects.)

Priestly had based his proposal on reasoning by analogy, that is, reasoning from a parallel, well-demonstrated case. Such reasoning alone could not *prove* that electrical forces are inversely proportional to the square of

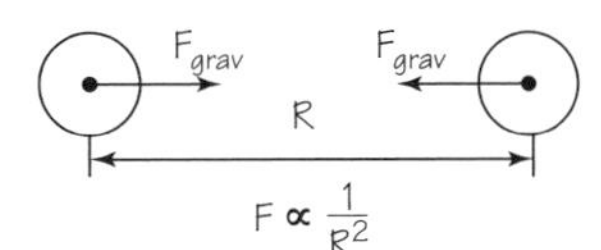

FIGURE 10.4 Two bodies under mutual gravitation.

the distance between charges. But it strongly encouraged other physicists to test Priestley's hypothesis by experiment.

The French physicist Charles Coulomb provided direct experimental evidence for the inverse-square law for electric charges suggested by Priestley. Coulomb used a *torsion balance* which he had invented (see Figure 10.5). A horizontal, balanced insulating rod is suspended by a thin silver wire. The wire twists when a force is exerted on the end of the rod, and the twisting effect can be used as a measure of the force.

Coulomb attached a charged body, a, to one end of the rod and placed another charged body, b, near it. The electrical force F_{el} exerted on a by b caused the wire to twist. By measuring the twisting effect for different separations between the centers of spheres a and b, Coulomb found that the force between spheres varied in proportion to $1/R^2$, just as Priestley had deduced

$$F_{\mathrm{el}} \propto \frac{1}{R^2},$$

where R represents the distance between the centers of the two charges. *The electric force of repulsion for like charges, or attraction for unlike charges, varies inversely as the square of the distance between charges.*

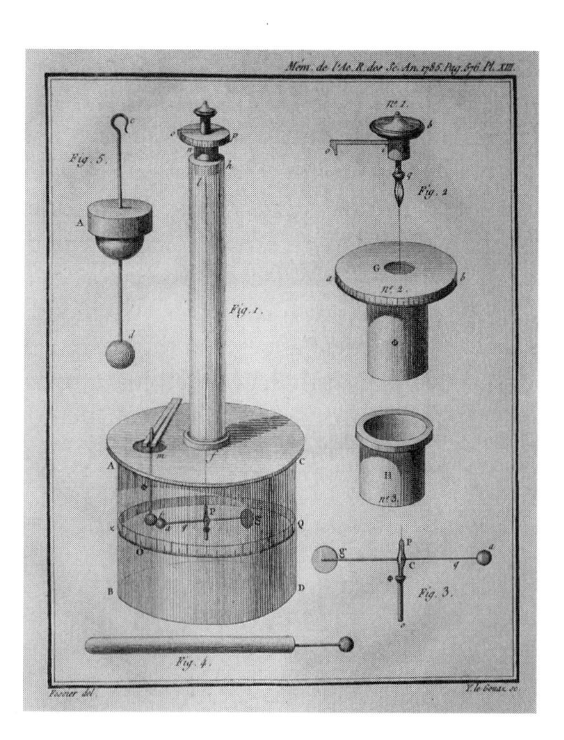

FIGURE 10.5 Coulomb's torsion balance.

Coulomb also demonstrated how the magnitude of the electric force depends on the magnitudes of the charges. There was not yet any accepted method for measuring quantitatively the amount of charge on an object. (In fact, nothing said so far would suggest how to measure the magnitude of the charge on a body.) Yet Coulomb used a clever technique based on symmetry to compare the effects of different amounts of charge. He first showed that if a charged metal sphere touches an uncharged sphere of the same size, the second sphere becomes charged also. You might say that, at the moment of contact between the objects, some of the charge from the first "flows" or is "conducted" to the second. Moreover, after contact has been made, the two spheres are found to share the original charge *equally*. (This is demonstrated by the observable fact that they exert equal forces on some third charged body.) Using this principle, Coulomb started with a given amount of charge on one sphere. He then shared this charge by contact among several other identical but uncharged spheres. Thus, he could produce charges of one-half, one-quarter, one-eighth, etc., of the

FIGURE 10.6 Charles Augustin Coulomb (1738–1806) was born into a family of high social position and grew up in an age of political unrest. He studied science and mathematics and began his career as a military engineer. While studying machines, Coulomb invented his torsion balance, with which he carried out intensive investigations on the mechanical forces caused by electrical charges. These investigations were analogous to Cavendish's work on gravitation.

original amount. In this way, Coulomb varied the charges on the two original test spheres independently and then measured the change in force between them using his torsion balance.

Coulomb found that, for example, when the charges on the two spheres are both reduced by one-half, the force between the spheres is reduced to one-quarter of its previous value. In general, he found that the magnitude of the electric force is proportional to the *product* of the charges. The symbols q_a and q_b can be used for the net charge on bodies a and b. The magnitude F_{el} of the electric force that each charge exerts on the other is proportional to $q_a \times q_b$. This may be written in symbols as

$$F_{el} \propto q_a q_b.$$

Coulomb summarized his two results in a single equation that describes the electric forces two small charged spheres A and B exert on each other

$$F_{el} = k\frac{q_a q_b}{R^2}.$$

In this equation, R represents the distance between the centers of the two charged spheres, and k is a constant whose value depends on the units of charge and of length that are used. This form of the electric force law between two electric charges is now called *Coulomb's law*. Note one striking fact about Coulomb's law: It has exactly the same form as Newton's law of universal gravitation

$$F_{grav} = G\frac{m_1 m_2}{R^2}!$$

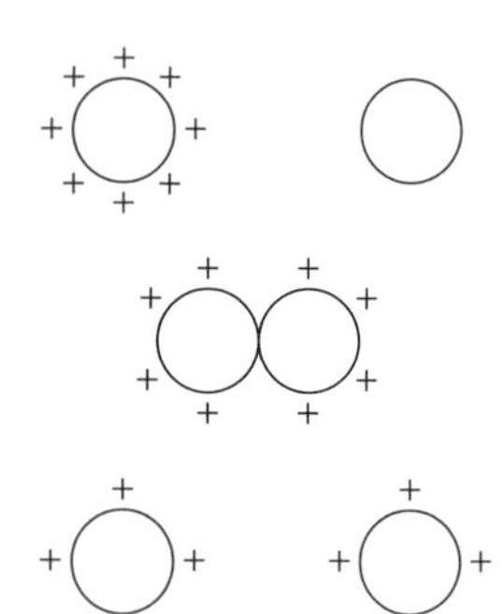

FIGURE 10.7 The equal sharing of charge between a charged and an uncharged sphere.

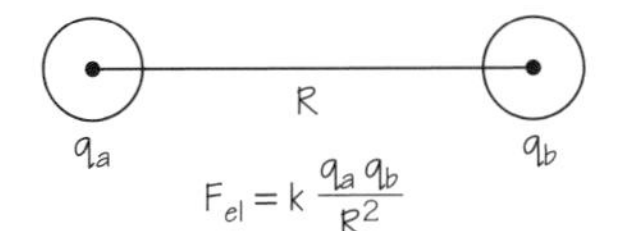

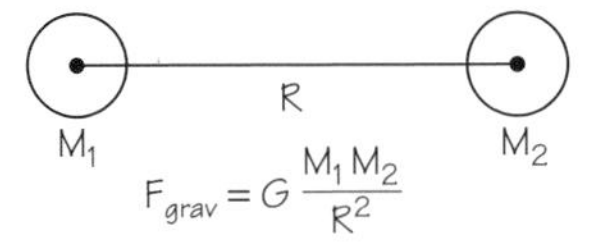

FIGURE 10.8 Magnitudes of electrical and gravitational forces between two spheres.

Here m_1 and m_2 are two masses separated by the distance R between their centers. Yet these two great laws arise from completely different sets of observations and apply to completely different kinds of phenomena. Why they should match so exactly is, to this day, a fascinating puzzle.

The Unit of Charge

Coulomb's law can be used to define a unit of charge, as long as we have defined the units of the other quantities in the equation. For example, we can assign k a value of exactly 1 and then define a unit charge so that two unit charges separated by a unit distance exert a unit force on each other. There actually is a set of units based on this choice. However, another system of electrical units, the "mksa" system, is more convenient to use. In this system, the unit of charge is derived not from electrostatics, but from the unit of current—the *ampere* (A). (See Section 10.5.) The resulting unit of charge is called (appropriately) the *coulomb* (C). It is defined as the amount of charge that flows past a point in a wire in 1 s when the current is equal to 1 A.

The ampere (A), or "amp," is a familiar unit frequently used to describe the current in electrical appliances. The effective amount of current in a common 100-W light bulb in the United States is approximately 1 A. Therefore, the amount of charge that goes through the bulb in 1 s is about 1 C. It might seem that a coulomb is a fairly small amount of charge. However, 1 C of *net* charge collected in one place is unmanageably large! In the light bulb, 1 C of negative charge moves through the filament each second. However, these negative charges are passing through a (more or less) stationary arrangement of *positive* charges in the filament. Thus, the *net* charge on the filament is zero.

Taking the coulomb (1 C) as the unit of charge, you can find the constant k in Coulomb's law experimentally. Simply measure the force between

FIGURE 10.9 A typical bolt of lightning represents about 40,000 amperes (on average) and transfers about 50 coulombs of charge between the cloud and the ground.

known charges separated by a known distance. The value of k turns out to equal about nine billion newton-meters squared per coulomb squared, or in symbols

$$k = 9 \times 10^9 \text{ N m}^2/\text{C}^2.$$

In view of this value of k, two objects, each with a *net* charge of 1 C, separated by a distance of 1 m, would exert forces on each other of nine billion N. (See if you can verify this from Coulomb's law.) This electric force is roughly as large as the gravitational force of one million tons! We can never directly observe such large electric forces in the laboratory because we cannot actually collect so much net charge (just 1 C) in one place. Nor can we exert enough force to bring two such charges so close together. The mutual repulsion of like charges is so strong that it is difficult to keep a charge of more than one-thousandth of a coulomb on an object of ordinary size. If you rub a pocket comb on your sleeve enough to produce a spark when the comb is brought near a conductor (such as a sink faucet), the net charge on the comb will be far less than one-millionth of a coulomb. Lightning discharges usually take place when a cloud has accumulated a net charge of a few hundred coulombs distributed over its very large volume.

Electrostatic Induction

As noted, and as you have probably observed, an electrically charged object can often attract small pieces of paper. But the paper itself has no net charge; it exerts no force on other pieces of paper. At first sight then, its attraction to the charged object might seem to contradict Coulomb's law. After all, the force ought to be zero if either q_a or q_b is zero.

To explain the observed attraction, recall that uncharged objects contain equal amounts of positive and negative electric charges. When a charged body is brought near a neutral object, it may rearrange the positions of some of the charges in the neutral object. The negatively charged comb does this when held near a piece of paper. Some of the negative charges in the paper shift away from the comb, leaving a corresponding amount of positive charge near the comb. The paper still has no *net* electric charge. But some of the positive charges are slightly *closer* to the comb than the corresponding negative charges are. So the attraction to the comb is greater than the repulsion. (Remember that the force gets weaker with the square of the distance, according to Coulomb's law. The force would be only one-fourth as large if the distance were twice as large.) In short, there is a net

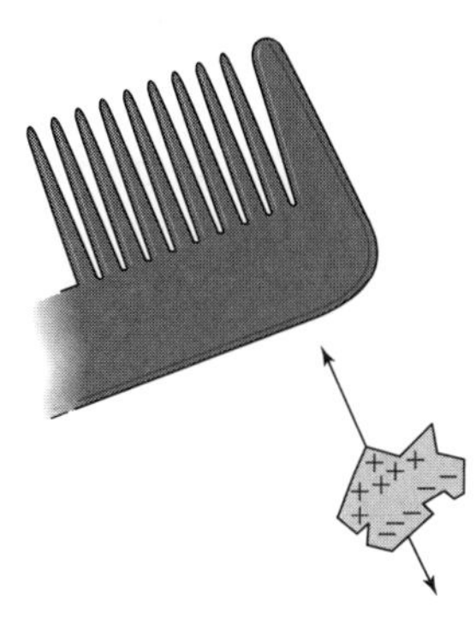

FIGURE 10.10 Electrostatic induction in neutral paper near a charged comb.

attraction of the charged body for the neutral object. This explains the old observation of the effect rubbed amber had on bits of grain and the like.

To put the observation another way: A charged body *induces* a shift of charge in or on the nearby neutral body. Thus, the rearrangement of electric charges inside or on the surface of a neutral body caused by the *influence* of a nearby charged object is called *electrostatic induction*. In Chapter 12, you will see how the theory of electrostatic induction played an important role in the development of the theory of light as an electromagnetic wave.

10.3 FORCES AND FIELDS

Gilbert described the action of the lodestone by saying it had a "sphere of influence" surrounding it. By this he meant that any other magnetic body coming inside this "sphere" would be attracted. In addition, the strength of the attractive force would be greater at places closer to the lodestone. In modern language, we should say that the lodestone is surrounded by a *magnetic field*. We can trace the magnetic field, for instance, of a bar magnet, by placing many small bits of iron fillings in the vicinity of a bar magnet that is on a table or other surface.

The world "field" is used in many different ways. Here, some familiar kinds of fields will be discussed, and then the idea of physical fields as used in science will be gradually developed. This exercise should remind you that most terms in physics are really adaptations of commonly used words, but with important changes. Velocity, acceleration, force, energy, and work are such examples, too.

One ordinary use of the concept of a field is illustrated by the "playing field" in various sports. The football field, for example, is a place where teams compete according to rules that confine the important action to the area of the field. "Field" in this case means a region of interaction.

In international politics, people speak of "spheres" or "fields" of influence. A field of political influence is also a region of interaction. But unlike a playing field, it has no sharp boundary line. A country usually has greater influence on some countries and less influence on others. So in the political sense, "field" refers also to an *amount* of influence, more in some places and less in others. Moreover, the field has a *source*, that is, the country that exerts the influence.

There are similarities here to the concept of field as used in physics. But there is also an important difference. To define a field in physics, it must

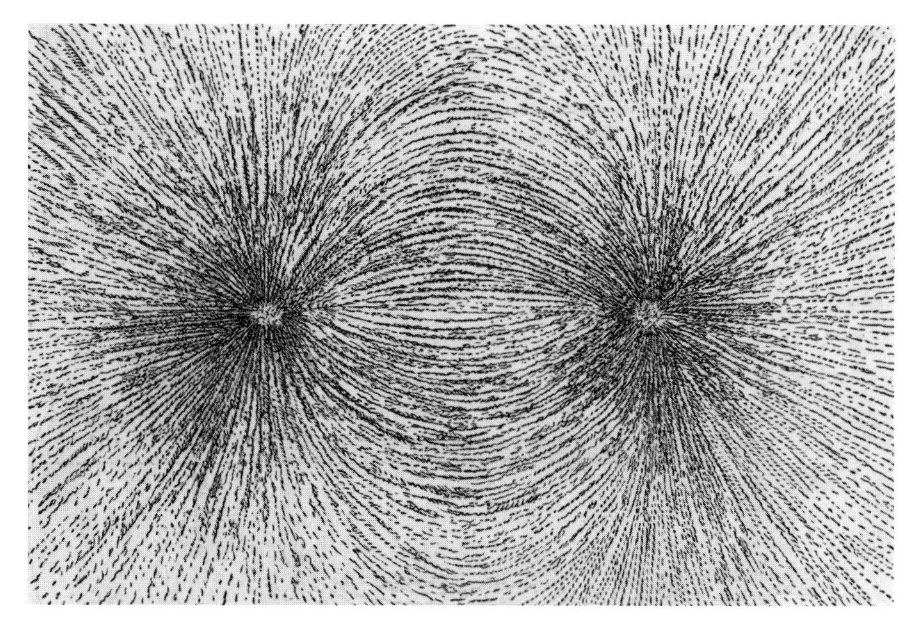

FIGURE 10.11 Iron fillings on a surface above a bar magnet align to show magnetic field lines.

be possible to assign a *numerical value* of field strength to every point in the field. This part of the field idea will become clearer if you consider some situations that are more directly related to the study of physics. First think about these situations in everyday language, then in terms of physics:

(a) You are walking along the sidewalk toward a street lamp; you observe that the brightness of the light is increasing.
(b) You are standing on the sidewalk as an automobile passes by with its horn blaring; you observe that the sound gets louder and then softer.

You can also describe these experiences in terms of fields:

(a) The street lamp is surrounded by a field of illumination. The closer you move to the lamp, the stronger is the field of illumination as registered on your eye or on a light meter (photometer) you might be carrying. For every point in the space around the street lamp, you can assign a number that represents the strength of the field of illumination at that place.
(b) The automobile horn is surrounded by a sound field. You are standing still in your frame of reference (the sidewalk). A pattern of field

values goes past you with the same speed as the car. You can think of the sound field as steady but moving with the horn. At any instance, you could assign a number to each point in the field to represent the intensity of sound. At first the sound is faintly heard as the weakest part of the field reaches you. Then the more intense parts of the field go by, and the sound seems louder. Finally, the loudness diminishes as the sound field and its source (the horn) move away.

Notice that each of the above fields is produced by a single course. In (a) the source is a stationary street lamp; in (b) it is a moving horn. In both cases the field strength gradually increases as your distance from the source decreases. One numerical value is associated with each point in the field.

So far, our examples have been simple *scalar* fields. No direction has been involved in the value of the field at each point. Figure 10.12 shows maps of two fields for the layer of air over North America on two consecutive days. There is a very important difference between the field mapped on the left and that mapped on the right. The *air pressure field* (on the left) is a scalar field; the *wind velocity field* (on the right) is a vector field. For each point in the pressure field, a single number (a scalar quantity) gives the value of the field at that point. But for each point in the wind velocity field, the value of the field is given by both a numerical value (magnitude) and a *direction*, that is, by a *vector*.

> Note that meteorologists have a convention for representing vectors different from the one we have been using. What are the advantages and disadvantages of each?

These field maps can help in more or less accurately predicting what conditions might prevail in the field on the next day. Also, by superimposing the maps for pressure and wind velocity, you can discover how these two kinds of fields are related to each other.

Physicists actually use the term "field" in three different senses:

1. the value of the field *at a point* in space;
2. the set or collection of all values everywhere in the space where that field exists;
3. the region of space in which the field has values other than zero.

In reading the rest of this chapter, you will not find it difficult to decide which meaning applies each time the term is used.

The Gravitational Force Field

Before returning to electricity and magnetism, let us illustrate a bit further the idea of a field. A good example is the gravitational force field of the Earth. The force exerted by the Earth on any object above its surface acts

PRESSURE AND VELOCITY FIELDS

These maps, adapted from those of the U.S. Weather Bureau, depict two fields, air pressure at the Earth's surface and high-altitude wind velocity, for two successive days. Locations at which the pressure is the same are connected by lines. The set of such pressure "contours" represents the overall field pattern. The wind velocity at a location is indicated by a line (showing direction) and (not visible here) feather lines—one for every 10 mi/hr. (The wind velocity over the tip of Florida, for example, is a little to the east of due north and is approximately 30 mi/hr.)

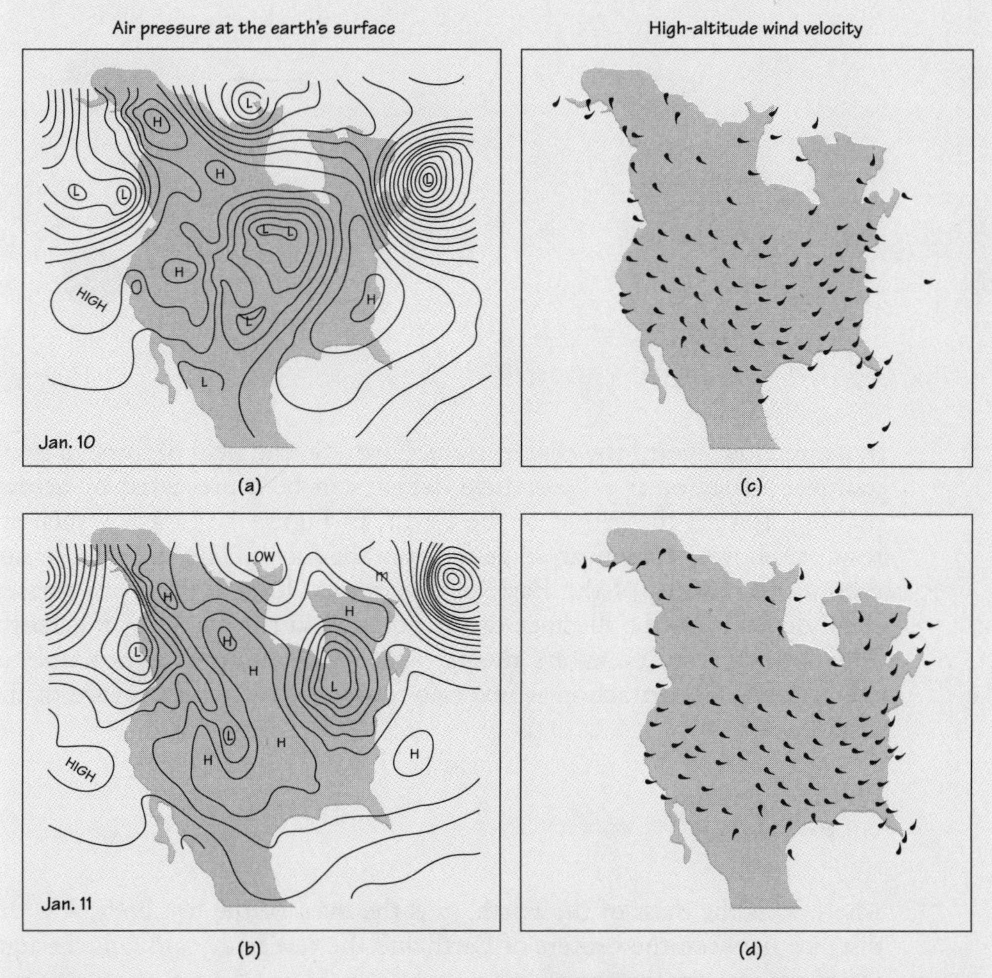

FIGURE 10.12 Weather Bureau maps of air pressure fields and wind velocity fields on two consecutive days.

FIGURE 10.13 Gravitational force field.

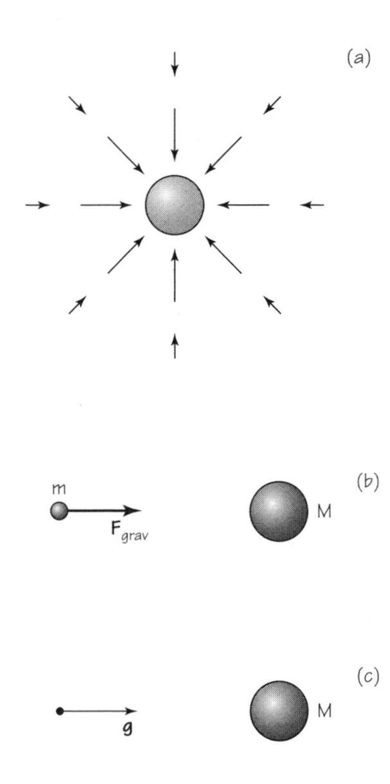

in a direction toward the center of the Earth. So the field of force of gravitational attraction is a *vector* field, which can be represented by arrows pointing toward the center of the Earth. In Figure 10.13 a few such arrows are shown, some near, some far from the Earth. The strength, or numerical magnitude, of the Earth's gravitational force field at any chosen point depends on the distance of the point from the center of the Earth. This follows from Newton's theory, which states that the magnitude of the gravitational attraction is inversely proportional to the square of the distance R:

$$F_{\text{grav}} = G \frac{Mm}{R^2},$$

where M is the mass of the Earth, m is the mass of the test body, R is the distance between the centers of Earth and the test body, and G is the universal gravitational constant.

In this equation, F_{grav} also depends on the mass of the test body. It would be more convenient to define a field that depends only on the properties of the source, whatever the mass of the test body. Then you could think of that field as existing in space and having a definite magnitude and direction at every point. The mass of the test body would not matter. In fact, it would not matter whether there were any test body present at all. As it happens, such a field is easy to define. By slightly rearranging the equation for Newton's law of gravitation, you can write

$$F_{grav} = m\frac{GM}{R^2}.$$

Then define the gravitational field strength $\mathbf{g}$ around a spherical body of mass M as having a magnitude GM/R^2 and a direction the same as the direction of $\mathbf{F}_{grav}$, so that

$$\mathbf{F}_{grav} = m\mathbf{g},$$

where the magnitude of $\mathbf{g}$ is GM/R^2. Thus, note that the magnitude of $\mathbf{g}$ at a point in space is determined by the source mass M and the distance R from the source, and does *not* depend on the mass of any test object.

The total or net gravitational force at a point in space is usually determined by more than one source. For example, the Moon is acted on by the Sun as well as by the Earth and to a smaller extent by the other planets. In order to define the field resulting from any configuration of massive bodies, take $\mathbf{F}_{grav}$ to be the *net* gravitational force due to *all* sources. Then *define* g in such a way that you can still write the simple relationship $\mathbf{F}_{grav} = m\mathbf{g}$; that is, define $\mathbf{g}$ by the equation

$$\mathbf{g} = \frac{\mathbf{F}_{grav}}{m}.$$

Thus, the gravitational field strength at any point is the *ratio* of the net gravitational force $\mathbf{F}_{grav}$ acting on a test body at that point to the mass m of the test body. The direction of the vector $\mathbf{g}$ is the same as that of $\mathbf{F}_{grav}$.

Electric Fields

The strength of any force field can be defined in a similar way. According to Coulomb's law, the electric force exerted by one relatively small charged body on another depends on the product of the *charges* of the two bodies. Consider a charge q placed at any point in the electric field set up by an-

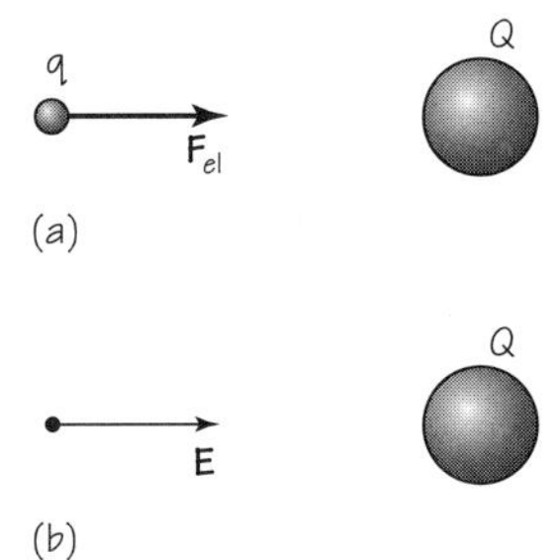

FIGURE 10.14 Electric force of charge Q on charge q (a), and the corresponding field (b).

other charge Q. Coulomb's law, describing the force F_{el} experienced by q, can be written as

$$F_{el} = K\frac{Qq}{R^2},$$

or

$$F_{el} = q\left(\frac{kQ}{R^2}\right).$$

As in the discussion of the gravitational field, the expression for force here is divided into two parts. One part, kQ/R^2, depends only on the charge Q of the source and distance R from it. This part can be called "the electric field strength owing to Q." The second part, q, is a property of the body being acted on. Thus, the vector electric field strength $\mathbf{E}$ owing to charge Q is *defined* as having magnitude kQ/R^2 and the same direction as $\mathbf{F}_{el}$. The electric force is then the product of the test charge and the electric field strength

$$\mathbf{F} = q\mathbf{E}$$

and

$$\mathbf{E} = \frac{\mathbf{F}}{q}.$$

Recall that $\mathbf{F}_{el}$ is *called* an "electric" force because it is caused by the presence of charges. But, as with all forces, we know it exists and can measure it only by its mechanical effects on bodies.

The last equation *defines* $\mathbf{E}$ for an electric force field. Thus, the electric field strength $\mathbf{E}$ at a point in space is the *ratio* of the net electric force $\mathbf{F}_{el}$ acting on a test charge at that point to the magnitude q of the test charge. (Note that $\mathbf{E}$ is quite analogous to $\mathbf{g}$ defined earlier, but for a different field.) This definition applies whether the electric field results from a single point charge or from a complicated distribution of charges. This, true for all fields we shall encounter, is a "superposition principle." Fields set up by many sources of the same sort superpose, forming a single net field. The vector specifying the magnitude of the net field at any point is simply the vector sum of the values of the fields due to each individual source. (Once more, one marvels at the simplicity of nature and the frugality of concepts needed to describe it.)

So far, we have passed over a complication not encountered in dealing with gravitation. There are *two* kinds of electric charge, positive (+) and negative (−). The forces they experience when placed in the same electric field are opposite in direction. By agreement, scientists define the direction of the vector **E** as the direction of the force exerted by the field on a *positive* test charge. Given the direction and magnitude of the field vector at a point, then by definition the force vector $\mathbf{F}_{el}$ acting on a charge q is $\mathbf{F}_{el} = q\mathbf{E}$. A positive charge, say +0.00001 C, placed at this point will experience a force $\mathbf{F}_{el}$ in the same direction as **E** at that point. A negative charge, say −0.00001 C, will experience a force of the same magnitude, but in the *opposite* direction. Changing the sign of q from + to − automatically changes the direction of $\mathbf{F}_{el}$ to the opposite direction.

10.4 ELECTRIC CURRENTS

Touching a charged object to one end of a metal chain will cause the entire chain to become charged. The obvious explanation is that the charges move through and spread over it. Electric charges move easily through some materials, called *conductors*. Metal conductors were most commonly used by the early experimenters, but salt solutions and very hot gases also conduct charge easily. Other materials, such as glass and dry fibers, conduct charge hardly at all. Such materials are called nonconductors or *insulators*. Dry air is a fairly good insulator. (Damp air is not; you may have difficulty keeping charges on objects in electrostatic experiments on a humid day.) If the charge is great enough, however, even dry air will suddenly become a conductor, allowing a large amount of charge to shift through it. The heat and light caused by the sudden rush of charge produces a "spark." Sparks were the first obvious evidence of moving charges. Until late in the eighteenth century, a significant flow of charge, that is, an *electric current*, could be produced only by discharging a highly charged object. In studying electric currents, Benjamin Franklin believed the moving charges to be positive. Because of this, he defined the direction of flow of an electric current to be the direction of flow of positive charges. Today, we know that the moving charges in a current can be positive or negative or both. In most wires, the flowing charges are negative electrons. However, ever since Franklin's early work, *the direction of flow of an electric current is defined as the direction of flow of positive charges, regardless of the actual sign of the moving charges.* This is acceptable because the flow of negative charges in one direction is electrically equivalent to the flow of positive charges in the other direction.

In 1800, Alessandro Volta discovered a much better way of producing electric currents than using short-lived discharge devices. Volta's method involved two different metals, each held with an insulating handle. When put into contact and then separated, one metal took on a positive charge and the other a negative charge. Volta reasoned that a much larger charge could be produced by stacking up several pieces of metal in alternate layers. This idea led him to undertake a series of experiments that produced an amazing result, as reported in a letter to the Royal Society in England in March of 1800:

> Yes! the apparatus of which I speak, and which will doubtless astonish you, is only an assemblage of a number of good conductors of different sorts arranged in a certain way. 30, 40, 60 pieces or more of copper, or better of silver, each in contact with a piece

FIGURE 10.15 Alessandro Volta (1745–1827) was given his title of Baron by Napoleon in honor of his electrical experiments. He was Professor of Physics at the University of Pavia, Italy. Volta showed that the electric effects previously observed by Luigi Galvani, in experiments with frog legs, were due to the metals and not to any special kind of "animal electricity."

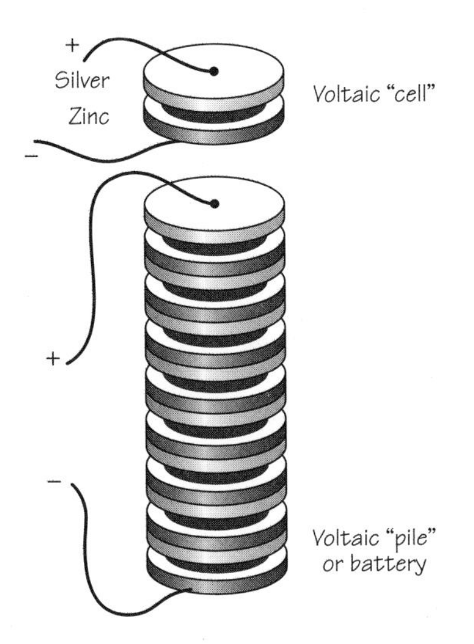

FIGURE 10.16 Voltaic cell and battery.

> of tin, or what is much better, of zinc, and an equal number of layers of water or some other liquid which is a better conductor than pure water, such as salt water or lye and so forth, or pieces of cardboard or of leather, etc., well soaked with these liquids.

Volta piled these metals in pairs, called "cells," in a vertical arrangement known as a "pile." Volta showed that one end, or "terminal," of the pile was charged positive, and the other charged negative. He then attached wires to the first and last disks of his apparatus, which he called a "battery." Through these wires, he obtained electricity with exactly the same effects as the electricity produced by rubbing amber, or by friction in electrostatic machines.

Most important of all, if the ends of the wires from Volta's battery were connected together, or attached to a conducting object, the battery produced a more or less *steady* electric current through the wires for a long period of time. This arrangement is known today as a *circuit*. The current, which flows through the wires of the circuit from the positive side of the battery to the negative side (by the earlier definition) is known as a *direct current*, or DC current. (A current that alternates in direction is known as an *alternating current*, or AC current. Most household circuits around the world provide AC current.) In addition, unlike the older charge devices, Volta's battery did not have to be charged from the outside after each use. Now the properties of electric currents as well as of static electric charges could be studied in a controlled manner. (Far better batteries have been produced as well. But one may say that Volta's invention started the series of inventions of electrical devices that have so greatly changed civilization.)

10.5 ELECTRIC POTENTIAL DIFFERENCE

Sparks and heat are produced when the terminals of an electric battery are connected. These phenomena show that energy from the battery is being transformed into light, sound, and heat energy. The battery itself converts chemical energy to electrical energy. This, in turn, is changed into other forms of energy (such as heat) in the conducting path between the terminals. In order to understand electric currents and how they can be used to transport energy, a new concept, which has the common name *voltage*, is needed.

In Chapter 5 we defined a *change in potential energy* as equal to the work required to move an object without friction from one position to another. For example, a book's gravitational potential energy is greater when the book is on a shelf than when it is on the floor. The increase in potential energy is equal to the work done in raising the book from floor to shelf. This difference in potential energy depends on three factors: the mass m of the book, the magnitude of the gravitational field strength g, and the difference in height d between the floor and the shelf.

Similarly, the *electrical* potential energy changes when work is done in moving an electric charge from one point to another in an electric field. Again, this change of potential energy $\Delta(PE)$ can be directly measured by the work that is done. The magnitude of this change in potential energy, of course, depends on the magnitude of the test charge q. Dividing $\Delta(PE)$ by q gives a quantity that does not depend on how large q is. Rather, it depends only on the intensity of the electric field and on the location of the beginning and end points. The new quantity is called *electric potential difference*. Electric potential difference is defined as *the ratio of the change in electrical potential energy* $\Delta(PE)$ *of a charge* q *to the magnitude of the charge*. In symbols

$$V = \frac{\Delta(PE)}{q}.$$

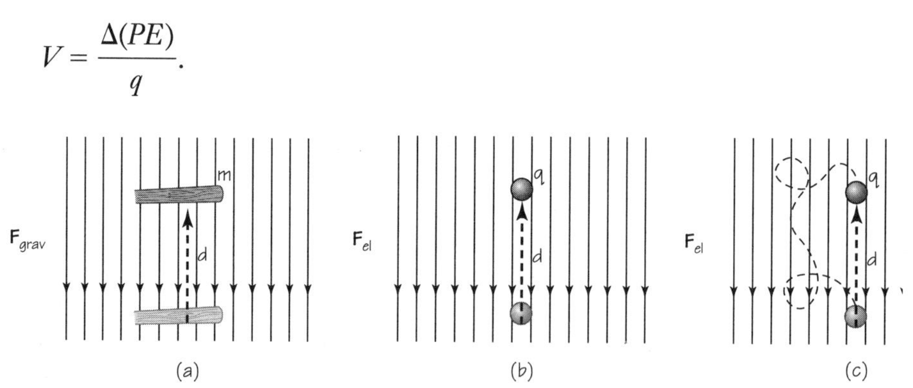

FIGURE 10.17 Gravitational and electrical potentials. The electric potential difference is the same between two points, regardless of the path.

As is true for gravitational potential energy, there is no absolute zero level of electric potential energy. The *difference* in potential energy is the significant quantity. The symbol v is used both for "potential difference" as in the equation above, and as an abbreviation for volt, the unit of potential difference (as in 1 V = 1 J/C).

The units of electric potential difference are those of energy divided by charge, or joules per coulomb. The term used as the abbreviation for joules per coulomb is *volt* (V). The electrical potential difference (or *voltage*) between two points is 1 V if 1 J of work is done in moving 1 C of charge from one point to the other

$$1 \text{ volt} = 1 \text{ joule/coulomb} \equiv 1 \text{ J/C}.$$

The potential difference between two points in a steady electric field depends on the location of the points. It does *not* depend on the *path* followed by the test charge. Whether the path is short or long, direct or roundabout, the same work is done per unit charge. Similarly, a hiker does the same work against the gravitational field per kilogram of mass in the pack he or she is carrying, whether climbing straight up or spiraling up along the slopes. Thus, the electrical potential difference between two points in a field is similar to the difference in gravitational potential energy between two points.

A simple case will help you to see the great importance of this definition of potential difference. Calculate the potential difference between two points in a uniform electric field of magnitude E produced by oppositely charged parallel plates. Work must be done in moving a positive charge q from one point to the other directly against the direction of the electric force. The amount of work required is the product of the force F_{el} exerted on the charge (where $F_{el} = qE$), and the distance d through which the charge is moved in the same direction. Thus,

$$\Delta(PE) = qEd.$$

Substituting this expression for $\Delta(PE)$ in the definition of electric potential difference gives, for the simple case of a uniform field,

$$V = \frac{\Delta(PE)}{q}$$

$$= \frac{qEd}{q}$$

$$= Ed.$$

In practice it is easier to measure electric potential difference V (with a voltmeter) than to measure electric field strength E. The relationship just

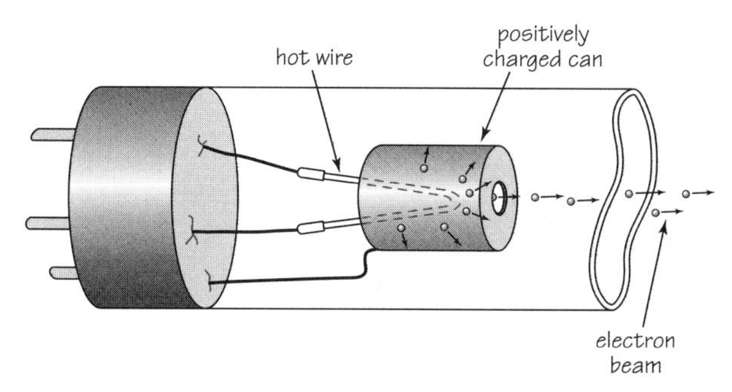

FIGURE 10.18 Electrically charged particles (electrons) are accelerated in an "electron gun" as they cross the potential difference between a hot wire (filament) and a "can" in an evacuated glass tube.

given is often useful in the form $E = V/d$, which can be used to find the intensity of a uniform electric field.

Electric potential energy, like gravitational potential energy, can be converted into kinetic energy. A charged particle placed in an electric field, but free of other forces, will accelerate. In doing so, it will increase its kinetic energy at the expense of electric potential energy. (In other words, the electric force on the charge acts in such a way as to push it toward a region of lower potential energy.) A charge q "falling" through a potential difference V increases its kinetic energy by qV if nothing is lost by friction.

$$\Delta(KE) = qV.$$

The amount of *increase* in kinetic energy is equal to the *decrease* in potential energy. So the sum of the two at any moment remains constant. This is just one particular case of the general principle of energy conservation, even though only electric forces are acting.

The conversion of electric potential energy to kinetic energy is used in *electron accelerators* (a common example is a television picture tube or some computer monitors). An electron accelerator usually begins with an electron "gun." The "gun" has two basic parts: a wire and a metal can in an evacuated glass tube. The wire is heated red-hot, causing electrons to escape from its surface. The nearby can is charged positively, producing an electric field between the hot wire and the can. The electric field accelerates the electrons through the vacuum toward the can. Many electrons stick to the can, but some go shooting through a hole in one end of it. The stream of electrons emerging from the hole can be further accelerated or focused by additional charged cans.

Such a beam of charged particles has a wide range of uses both in technology and in research. For example, a beam of electrons can make a fluorescent screen glow, as in a television picture tube or an electron microscope, or they can produce X rays for medical purposes or research. If a beam of heavier charged particles is used, they can break atoms apart for the study of their composition. When moving through a potential difference of 1 V, an electron with a charge of 1.6×10^{-19} C increases its kinetic energy by 1.6×10^{-19} J, in accord with the equation

$$\Delta(KE) = qV.$$

This amount of energy is called an *electron volt*, which is abbreviated eV. Multiples are 1 keV (= 1000 eV), 1 MeV (= 10^6 eV), and 1 GeV (= 10^9 eV). Energies of particles in accelerators are commonly expressed in such multiples. In a television tube, the electrons in the beam are accelerated across an electric potential difference of about 25,000 V. Thus,

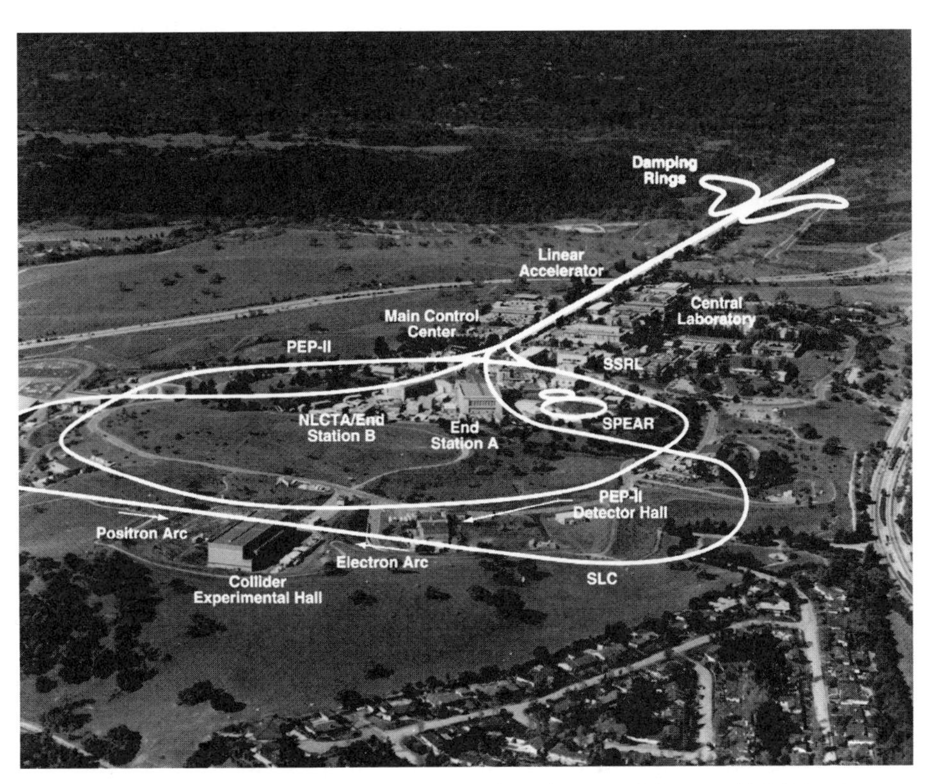

FIGURE 10.19 Stanford Linear Accelerator, with underground features highlighted.

each electron has an energy of about 25 keV. Large accelerators now operating can accelerate charged particles to kinetic energies of about 800 GeV.

10.6 ELECTRIC POTENTIAL DIFFERENCE AND CURRENT

The acceleration of an electron by an electric field in a vacuum is the simplest example of a potential difference affecting a charged particle. A more familiar example is electric current in a metal wire. In this arrangement, the two ends of the wire are attached to the two terminals of a battery. Chemical changes inside a battery produce an electric field that continually drives charges to the terminals, making one charged negatively, leaving the other charged positively. The "voltage" of the battery tells how much energy per unit charge is available when the charges move in any external path from one terminal to the other, for example, along a wire.

> In metallic conductors, the moving charge is the negative electron, with the positive atom fixed. But all effects are the same as if positive charge were moving in the opposite direction. By an old convention, the latter is the direction usually chosen to describe the direction of current.

Electrons in a metal do not move freely as they do in an evacuated tube, but continually interact with the metal atoms. If the electrons were really completely free to move, a constant voltage would make them *accelerate* so that the current would increase with time. This does not happen. A simple relation between current and voltage, first found by Georg Simon Ohm, is at least approximately valid for most metallic conductors: *The total current* I *in a conductor is proportional to the potential difference* V *applied between the two ends of the conductor*. Using the symbol I for the current, V for the potential difference, and $\propto$ for proportionality, we may write

$$V \propto I$$

or

$$V = \text{constant} \times I.$$

This simple relation is called *Ohm's law*. It is usually written in the form

$$V = IR,$$

where R is a constant called the *resistance* of the conducting path. It is measured in units of ohm, symbol Ω (Greek letter omega). Ohm's law may be stated a different way: *The resistance* R *is constant for different values of voltage and current.*

FIGURE 10.20 Georg Simon Ohm (1789–1854).

Resistance depends on the material and dimensions of the path, such as the length and diameter of a wire. For example, a thin tungsten wire has a much larger resistance than the same length of a fat copper wire. But Ohm's law assumes that the resistance R of a given conducting path does not depend on the current or the voltage. In fact, resistance is not strictly constant for any conducting path; it varies with changes in temperature, for example. But Ohm's law applies closely enough for practical technical work, though it does not have the general validity of the law of universal gravitation or Coulomb's law.

10.7 ELECTRIC POTENTIAL DIFFERENCE AND POWER

Suppose a charge could move freely from one terminal to the other in an evacuated tube. The work done on the charge would then simply increase the kinetic energy of the charge. However, a charge moving through some material such as a wire transfers energy to the material by colliding with atoms. Thus, at least some of the work goes into heat energy owing to the increased vibration of the atoms. A good example of this process is a flashlight bulb. A battery forces charges through the filament wire in the bulb. The electric energy carried by the charges is converted to heat energy in the filament. The hot filament in turn radiates energy, a small fraction of which is in the form of visible light. Recall now that "voltage" (electric potential difference) is the amount of *work* done per unit of charge transferred. So the product of voltage and current gives the amount of *work* done per unit *time*:

$$V\text{ (joules/coulomb)} \times I\text{ (coulombs/second)} = VI\text{ (joules/second)}.$$

Work done per unit time is called *power*, symbol P. The unit of electrical power, equal to 1 J/s, is called a *watt* (W). Using the definition of ampere (1 C/s) and volt (1 J/C), the equation for power P provided to an electrical circuit is

$$P\text{ (watts)} = V\text{ (volts)} \times I\text{ (amperes)}.$$

Thus, a 60-W bulb, connected to a 110-V circuit, carries 0.5 A of current.

What energy transformation does this equation imply? During their motion, the charges do work against material by colliding with atoms in the conductor. Thus the electric energy of the charges is converted into heat energy. Using the relation $P = VI$ and substituting IR for V, we obtain the important equation

$$P = IR \times I,$$

$$P = I^2R.$$

Thus, *the heat produced each second (given by* P, *the energy available per second) by a current is proportional to the square of the current.* Joule was the first to find this relationship experimentally. The discovery was part of his series of researches on conversion of different forms of energy (Chapter 6). The fact that heat production is proportional to the *square* of the current is very important in making practical use of electric energy. You will learn more about this in Chapter 11.

10.8 CURRENTS ACT ON MAGNETS

Early in the eighteenth century, reports began to appear that lightning changed the magnetization of compass needles and made magnets of knives and spoons. Some researchers believed that they had magnetized steel needles by discharging a charged object through them. These reports suggested

FIGURE 10.21 Needle-like iron oxide crystals in the magnetic field of a bar magnet. The bar magnet is under a paper on which the iron oxide crystals have been spread.

that electricity and magnetism were closely related in some way. But the casual observations were not followed up with deliberate, planned experiments that might have led to useful concepts and theories.

None of these early reports surprised the nineteenth-century Nature Philosophers in Europe. They were convinced that all phenomena observed in nature were only different effects of a single "force." Their belief in the *unity* of physical forces naturally led them to expect that electrical and magnetic forces were associated or related in some way.

The first concrete evidence of a connection between electricity and magnetism came in 1820, when the scientist Hans Christian Oersted of Copenhagen, Denmark performed an extremely important series of experiments. Oersted placed a long horizontal wire directly beneath a magnetic compass needle. The wire thus lay along the Earth's magnetic north–south line, with the magnetic needle naturally lined up parallel to the wire. When Oersted connected the wire to the terminals of a battery, the compass needle swung toward an east–west orientation, nearly perpendicular to the wire! Charge at rest does not affect a magnet. But here was evidence that charge in motion (a current) does exert an odd kind of force on a magnet.

Oersted's results were the first ever found in which a force did *not* act along a line connecting the sources of the force. (Forces between planets, between electric charges, or between magnetic poles all act along such a line.) The force exerted between the current-carrying wire and each magnetic pole of the compass needle is not along the line from the wire to the pole. In fact, for the needle to twist as it does, the force must be acting *perpendicular* to such a line. The magnetic needle is *not* attracted or repelled by the wire, but is *twisted* sideways by forces on its poles.

This was a totally new kind of effect. No wonder it had taken so long before anyone found the connection between electricity and magnetism. Closer examination revealed more clearly what was happening in this experiment. The long, straight, current-carrying wire sets up a magnetic field around it, given the symbol **B**. This field turns a small magnet so that the north–south line on the magnet is tangent to a circle whose center is at the wire and whose plane lies *perpendicular* to the wire. Thus, the current produces a *circular* magnetic field, not a centrally directed magnetic field as had been expected.

The direction of the magnetic field vector at each point is defined as *the direction of the force on the north-seeking pole of a compass needle placed at that*

HANS CHRISTIAN OERSTED

Hans Christian Oersted (1777–1851), a Danish physicist, studied the writings of the Nature Philosopher Schelling and wrote extensively on philosophical subjects himself. In an essay published in 1813, Oersted predicted that a connection between electricity and magnetism would be found. In 1820, he discovered that a magnetic field surrounds an electric current when he placed a compass over a current-carrying wire. In later years he vigorously denied the suggestion of other scientists that his discovery of electromagnetism had been accidental.

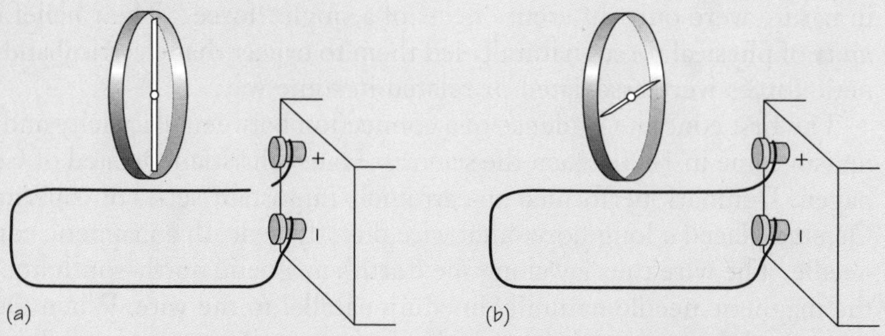

FIGURE 10.22 Oersted's experiment. (a) No current. Compass needle put across a wire points N–S. (b) Current flows. Compass needle is twisted to E–W position by magnetic force produced by current in top part of wire.

FIGURE 10.23

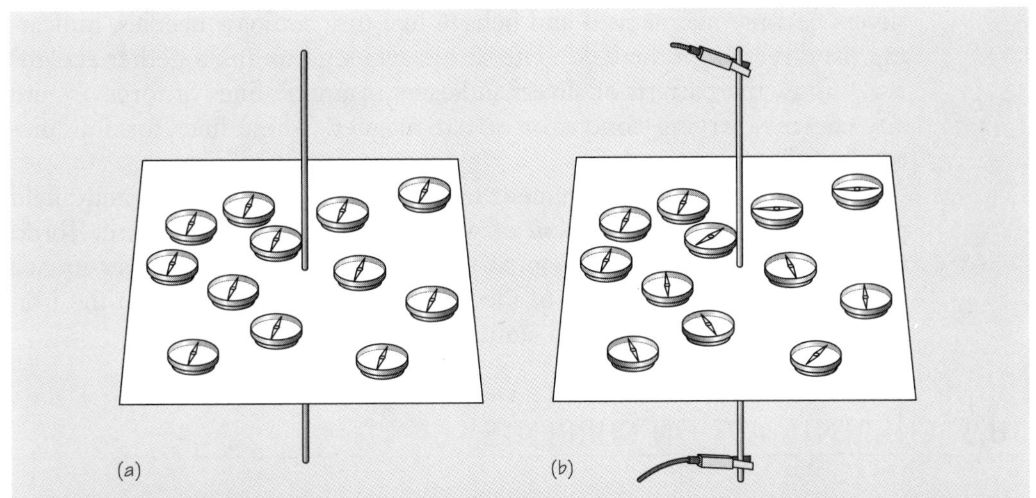

FIGURE 10.24 Representation of an array of tiny compasses on a sheet of cardboard placed perpendicular to a brass rod. When there is a strong current in the rod as in (b), the compass needles are deflected from their normal north-south line by the magnetic field set up by the current. This experiment, too, indicates that the lines of magnetic force owing to the current are circular around the rod.

point. Conversely, the force on the south-seeking pole will be in a direction exactly opposite to the field direction. A compass needle will respond to these forces on each end by turning until it points as closely as possible in the direction of the field. We noted earlier that with a bar magnet you can obtain the "shape" of the magnetic field by sprinkling tiny slivers of iron on a sheet placed over the magnet. In the same way, the magnetic field around a current-carrying wire can be visualized (see Figure 10.26). The

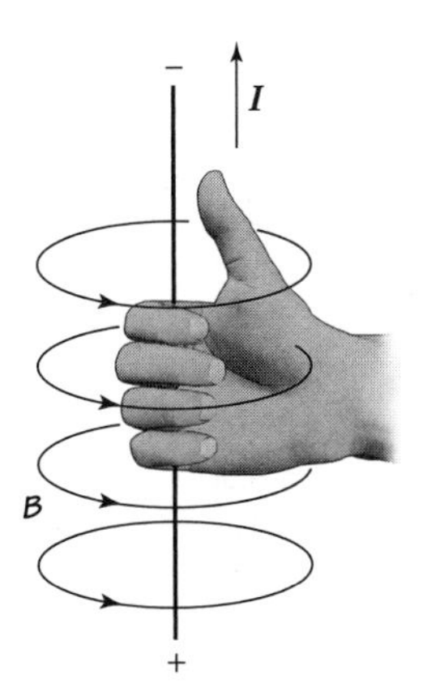

FIGURE 10.25 Remember this useful rule: If the thumb points in the direction of the flow of charge, the fingers curl in the direction of the lines of the magnetic field. The magnitude of the magnetic field is discussed in Sec. 14.13. Use the right hand for positive charge flow, left hand for negative charge flow.

slivers become magnetized and behave like tiny compass needles, indicating the direction of the field. The slivers also tend to link together end-to-end. Thus, the pattern of slivers indicates magnetic lines of force around any current-carrying conductor or bar magnet. These lines form a "picture" of the magnetic field.

You can use a similar argument to find the "shape" of a magnetic field produced by a current in a *coil* of wire, instead of a straight wire. To do this, bend the wire into a loop so that it goes through the paper in two places. The magnetic effects of the different parts of the wire on the iron slivers produce a field pattern similar to that of a bar magnet.

10.9 CURRENTS ACT ON CURRENTS

Oersted's experiment was one of those precious occasions when a discovery suddenly opens up an exciting new subject of research—a whole "ocean of ignorance" waiting to be conquered. In this case, no new equipment was needed. At once, dozens of scientists throughout Europe and America began intensive studies on the magnetic effects of electric currents. The work of André-Marie Ampère (1775–1836) stands out among all the rest. Ampère was called the "Newton of electricity" by James Clerk Maxwell, who decades later constructed a complete theory of electricity and magnetism. Ampère's work is filled with elegant mathematics. Without describing his theory in detail, we can trace some of his ideas and review some of his experiments.

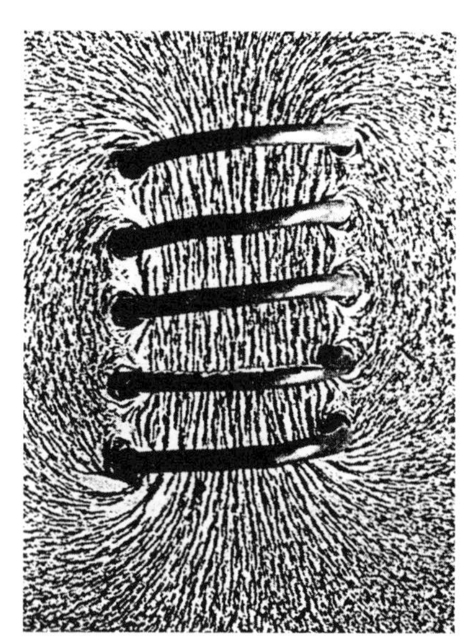

FIGURE 10.26 Iron fillings in the magnetic field produced by current in a coil of wire.

FIGURE 10.27 André-Marie Ampère (1775–1836) was born in a village near Lyons, France. There was no school in the village, and Ampère was self-taught. His father was executed during the French Revolution, and Ampère's personal life was deeply affected by his father's death. Ampère became a professor of mathematics in Paris and made important contributions to physics, mathematics, and the philosophy of science.

Ampère's thoughts raced forward as soon as he heard Oersted's news. He began with a line of thought somewhat as follows: Magnets exert forces on each other, and magnets and currents exert forces on each other. Do currents then exert forces on other currents? The answer is not necessarily yes. Arguing from symmetry is inviting, and often turns out to be right. But the conclusions to which such arguments lead are not logically or physically necessary. Ampère recognized the need to let experiment answer his question. He wrote:

> When Monsieur Oersted discovered the action which a current exercises on a magnet, one might certainly have suspected the existence of a mutual action between two circuits carrying currents; but this was not a necessary consequence; for a bar of soft iron also acts on a magnetized needle, although there is not mutual action between two bars of soft iron.

Ampère put his hunch to the test. On September 30, 1820, within a week after word of Oersted's work reached France, Ampère reported to the

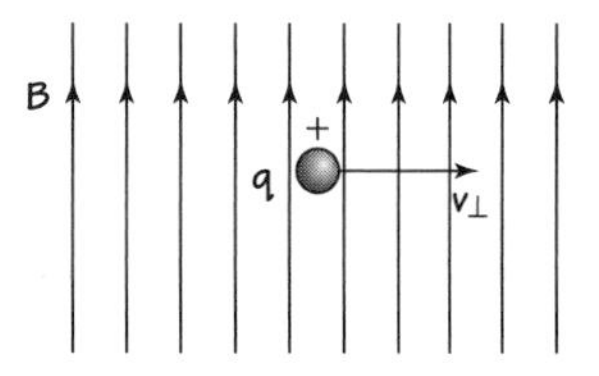

(a) When the charge q moves with velocity **v** perpendicular to **B**,

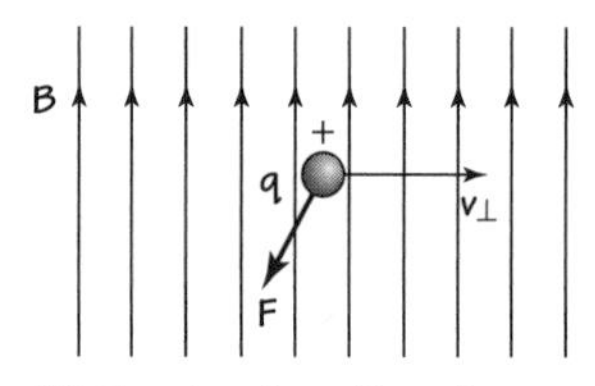

(b) there is a force **F** as shown, directly out of plane of paper, proportional to q, $\mathbf{v}_{\perp}$, and **B**.

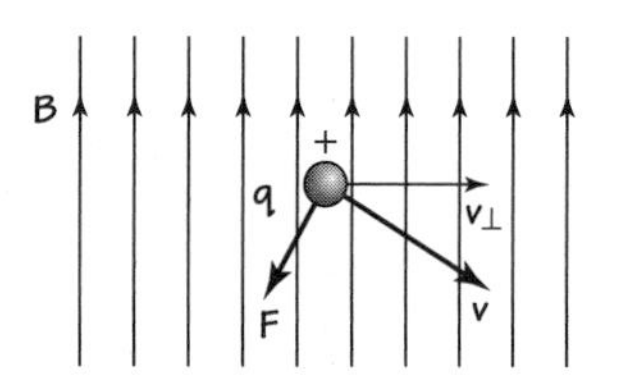

(c) If **v** is not perpendicular (⊥) to **B**, there is a smaller force, proportional to **v** instead of $\mathbf{v}_{\perp}$.

FIGURE 10.28 Magnetic field and force on moving charge.

French Academy of Sciences. He had indeed found that two parallel current-carrying wires exert forces on each other. They did so even though the wires showed no evidence of *net* electric charges.

Ampère made a thorough study of the forces between currents. He investigated how they depend on the distance between the wires, the relative positions of the wires, and the amount of current. In the laboratory, you can repeat these experiments and work out the "force law" between two currents. We need not go into the quantitative details here, except to note that the force between currents is easy to measure, and in turn can be used to measure how much current is flowing. In fact, the magnetic force between currents is now the quantity preferred for *defining* the unit of current. This unit is called the *ampere*, as mentioned in Section 10.2. One ampere (1 A) is defined as the amount of current in each of two long, straight, parallel wires, set 1 m apart, that causes a force of exactly 2×10^{-7} N to act on each meter of each wire.

Question: How many amps of current are flowing through the filament of a 100-W electric light bulb when connected to a 120-V outlet?

Answer: Referring to the relationship between power (in watts), voltage (in volts), and current (in amps) in Section 10.7:

$$P = VI,$$

or

$$I = \frac{P}{V},$$

$$I = \frac{100 \text{ W}}{120 \text{ V}} = 0.83 \text{ W/V} = 0.83 \text{ A}.$$

10.10 MAGNETIC FIELDS AND MOVING CHARGES

In the last two sections, the interactions of currents with magnets and with each other were discussed. The concept of *magnetic field* greatly simplifies the description of these phenomena.

As you saw in studying Coulomb's law, electrically charged bodies exert forces on each other. When the charged bodies are at rest, the forces are "electric" forces, or Coulomb forces. "Electric fields" act as the sources of these forces. But when the charged bodies are moving (as when two parallel wires carry currents), new forces in *addition* to the electric forces are present. These new forces are called "magnetic" and are caused by "magnetic fields" set up by the moving charges.

Magnetic interaction of moving charged bodies is not as simple as electric interaction. Remember the description of Oersted's experiment. The direction of the force exerted by a field set up by a current on a magnet needle is perpendicular both to the direction of the current and to the line between the magnet needle and the current. For the moment, however, it is not necessary to examine the forces on current-carrying conductors. After all, the force on a wire is believed to be caused by forces on the individual electric charges moving in it. How do such individual charges behave when moving freely in an external magnetic field? Once some simple rules have been established for the behavior of free charged particles, current in wires will be discussed further in the next chapter. There you will see how these simple rules are enough to explain the operation of electric generators and electric motors. (You will also see how these inventions have changed civilization.)

The rules summarized in the remainder of this section are best learned in the laboratory. All you need is a magnet and a device for producing a beam of charged particles, for example, the "electron gun" in an oscilloscope tube.

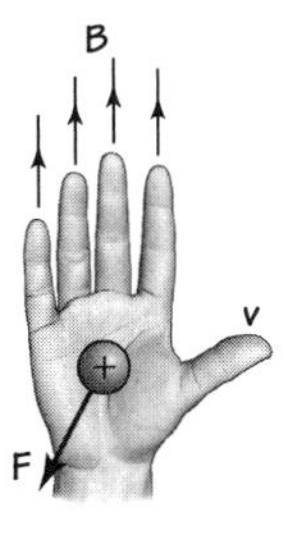

FIGURE 10.29 Remember this useful rule: If your fingers point along **B** and your thumb along v, F will be in the direction your palm would push. For positive charges use the right hand, and for negative use the left hand.

The Force on a Moving Charged Body

Suppose you have a fairly uniform magnetic field, symbol **B**, produced either by a bar magnet or by a current in a coil. How does this external field act on a moving, charged body (say, an electron)? You can find by experiment that the charge experiences a force and that the force depends on three quantities:

(1) the charge q of the body;
(2) the velocity $\mathbf{v}$ of the body; and
(3) the strength of the external field **B** through which the body is moving.

The force depends not only on the *magnitude* of the velocity, but also on its *direction*. If the body is moving in a direction *perpendicular* to the field **B**, the magnitude of the force is proportional to *both* of these quantities; that is,

$$\mathbf{F} \propto qv\mathbf{B}$$

which can also be written as

$$\mathbf{F} = kqv\mathbf{B},$$

where k is a proportionality constant that depends on the units chosen for **F**, q, v, and **B**.

But if the charge is moving in a direction *parallel* to **B**, there is no force on it, since the angle is zero between the field and the velocity vector of the charge. For all other directions of motion, the magnitude of the force is somewhere between the full value and zero. In fact, the force is proportional to the *component* of the velocity that is perpendicular to the field direction. We give this the symbol $v_\perp$. Therefore, a more general expression for the force is

$$\mathbf{F} \propto qv_\perp\mathbf{B}$$

or

$$\mathbf{F} = kqv_\perp\mathbf{B},$$

where k is the same constant as before. *The direction of the force is always perpendicular to the direction of the field. It is also perpendicular to the direction of motion of the charged body.*

The force exerted by an external magnetic field on a moving charged particle can be used to *define* the unit of magnetic field **B**. This is done by taking the proportionality constant k as equal to one. This definition is convenient here, since we are dealing mainly with how magnetic fields act on moving charges (rather than with forces between bar magnets). So in the special case when **B** and v are *at right angles* to each other, the magnitude of the deflecting force becomes simply

$$F = qvB.$$

The Path of a Charged Body in a Magnetic Field

The force on a moving charged body in a magnetic field is always "off to the side"; that is, the force is perpendicular to the body's direction of motion at every moment. Therefore, the magnetic force does not change the *speed* of the charged body. This is analogous to the case of the gravitational force from the Sun acting on a planet in (near enough) circular orbit. Rather, in each case the central force changes the *direction* of the velocity vector, but not its magnitude. If a charged body is moving exactly perpendicular to a uniform magnetic field, there will be a constant sideways push. The body will move along a circular path, in a plane perpendicular to the direction of the magnetic field. If **B** is strong enough, the particle will be trapped in a circular orbit.

What if the charged body's velocity has some component along the direction of the field but not exactly parallel to it? The body will still be deflected into a curved path (see Figure 10.30), but the component of its motion *along* the field will continue undisturbed. So the particle will trace out a coiled (helical) path (Figure 10.30b). If the body is initially moving exactly parallel to the magnetic field (Figure 10.30a), there is no deflecting force at all, since $v_\perp$ is zero.

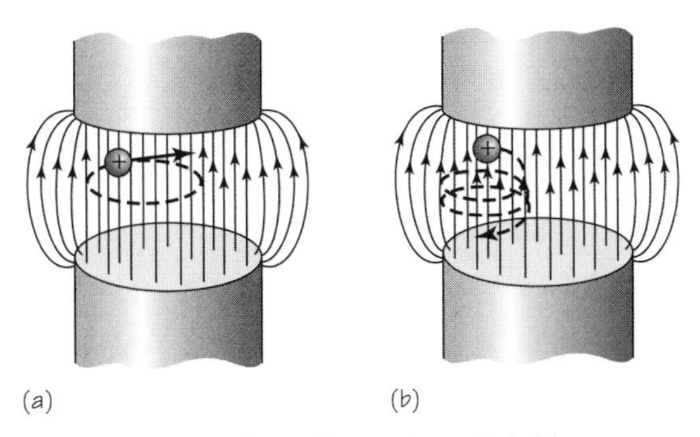

FIGURE 10.30 Charged particle in **B** field.

Some important examples of the deflection of charged particles by magnetic fields include particle accelerators and bubble chambers. One example of "coiled" motion is found in the Van Allen radiation belts, a doughnut-shaped region encircling the Earth and extending from a few hundred kilometers to about fifty thousand kilometers above the Earth's surface. A stream of charged particles, mainly from the Sun, but also from outer space, continually sweeps past the Earth. Many of these particles are deflected into spiral paths by the magnetic field of the Earth and become "trapped" in the Earth's field. Trapped particles from the geomagnetic tail sometimes spiral their way toward the Earth's magnetic poles. When they hit the atmosphere, they excite the atoms of the gases to radiate light. This is the cause of the aurora ("northern lights" and "southern lights").

Since magnets acts on currents and currents act on currents, the forces produced by magnets and currents can be used to perform work by pushing or pulling on the currents on which they act. The harnessing of this work in useful forms had profound economic and social consequences. This behavior of magnets and currents made possible the invention of the electric motor, which turns electrical energy into mechanical work, and the electric generator, which turns mechanical work into electrical energy. Together these devices helped to open the electric age, in which we still live.

FIGURE 10.31 The glow of the aurora borealis is produced when the upper atmosphere is excited by charged particles moving through the Earth's magnetic field.

SOME QUANTITATIVE EXPRESSIONS IN THIS CHAPTER

Magnitude of gravitational force on m_1 or m_2:

$$F_{\text{grav}} = G\frac{m_1 m_2}{R^2}.$$

Magnitude of electric force on q_1 or q_2:

$$F_{\text{el}} = k_{\text{el}}\frac{q_1 q_2}{R^2}.$$

Magnitude of magnetic force on a moving charge, if v and B are perpendicular:

$$F = qvB.$$

Gravitational field strength:

$$\mathbf{g} = \frac{\mathbf{F}_{\text{grav}}}{m}.$$

Electric field strength, **E**, is defined as the electric force per unit of test charge:

$$\mathbf{E} = \frac{\mathbf{F}_{\text{e}}}{q}.$$

Ohm's law:

$$V = IR.$$

Potential difference, V, is defined as the ratio of the change in electrical potential energy of charge to the magnitude of the charge:

$$V = \frac{\Delta PE}{q}.$$

Electric power:

$$P = VI \qquad \text{or} \qquad P = I^2R.$$

SOME IMPORTANT NEW UNITS

One amp of current = one coulomb of charge per second, 1 A = 1 C/s.

1 volt = 1 joule/coulomb, 1 V = 1 J/C.

1 ohm = 1 volt/amp, 1 Ω = 1 V/A.

SOME NEW IDEAS AND CONCEPTS

AC current
ampere
charge
conductors
Coulomb's law
current
DC current
electric field
electric field strength
electrodynamics
electrostatic induction
electrostatics
insulators
lines of force
lodestone
magnetic field
Ohm's law
pole
potential difference
voltage

FURTHER READING

G. Holton and S.G. Brush, *Physics, The Human Adventure* (Piscataway, NJ: Rutgers University Press, 2001), Chapter 24.

B. Franklin, *The Autobiography of Benjamin Franklin* (New York: Dover, 1996), and many other editions.

STUDY GUIDE QUESTIONS*

10.1 Gilbert's *Magnets*

1. What are amber and loadstone?
2. What are some of the strange properties of amber and loadstone?
3. How did Gilbert make it plausible that the Earth behaves like a spherical loadstone?

---

* These questions are intended as an aid to study. Your instructor may ask you to work on these in groups, or individually, or as a class.

4. How does the attraction of objects by amber differ from the attraction by lodestone?

10.2 Electric Charges and Electric Forces

1. What are some important observations about static charges?
2. What are the three general rules about electric charges?
3. Why did people think that electricity involves a fluid?
4. How did Priestley discover that the electric force is proportional to $1/R^2$?
5. How did Coulomb arrive at the electric force law?
6. Compare and contrast the electric, magnetic, and gravitational forces.
7. If the distance between two charged objects is doubled, how is the electric force between them affected? How is the force affected if the charge on one of them is cut to one-quarter its former size.
8. State Coulomb's law in your own words. What is the direction of the force between two charged particles? What is the value of k?

10.3 Forces and Fields

1. What is the difference between a scalar field and a vector field?
2. Describe how you can find by experiment the magnitude and directions of:
   (a) the gravitational field at a certain point in space;
   (b) the electric field at a certain point in space.
3. On a sheet of paper draw a circle to represent a ball of charge. Label it a negative charge, i.e., $-q$. Then choose three points on the paper and draw the direction of the electric field at each point. Finally, draw the direction of the force vector on a negative charge at each point.
4. A negative test charge is placed in an electric field where the electric field vector is pointing downward. What is the direction of the force on the test charge?
5. What is the electric field at a point if a test particle of 3×10^{-5} C experiences a force of 10^{-2} N upward?

10.4 Electric Currents

1. Describe Volta's pile and how it operated.
2. What were some of the advantages of the pile?
3. What is the difference between AC and DC current?

10.5 Electric Potential Difference

1. How is the electric potential difference between two points defined?
2. Does the potential difference between two points depend on:
   (a) the path followed as a charge moves from one point to the other?
   (b) the magnitude of the charge that moves?
3. How is the unit of electron-volt (eV) defined?
4. What are eV units used in some situations instead of Joules? What are some situations in which they are used?

10.6 Electric Potential Difference and Current

1. How does the current in a conductor change if the potential difference between the ends of the conductor is doubled?
2. How would you test whether Ohm's law applies to a given piece of wire?

10.7 Electric Potential Difference and Power

1. What happens to the electrical energy used to move a charge through a conducting material?
2. How does the energy per second converted to heat in a conductor change if the current in the conductor is doubled?
3. Show that the kilowatt-hour is a measure of energy and not of power.

10.8 Currents Act on Magnets

1. Describe, with drawings, Oersted's discovery. Why is it important to this day?
2. Under what conditions can electric charges affect magnets?
3. What was surprising about the force a current exerted on a magnet?
4. What are the shape and direction of the magnetic field produced near a straight wire carrying a steady current?

10.9 Currents Act on Currents

1. What was Ampère's hunch?

10.10 Magnetic Fields and Moving Charges

1. Which of the following affect the magnitude of the deflecting force on a moving charged particle?
   (a) The component of the velocity parallel to the magnetic field?
   (b) The component of the velocity perpendicular to the magnetic field?
   (c) The magnetic find itself?
   (d) The magnitude of the charge?
   (e) the sign of the charge?
2. What are differences between deflecting forces on a charged object due to:
   (a) gravity?
   (b) an electric field?
   (c) a magnetic field?

DISCOVERY QUESTIONS*

1. What is an electromagnet and how does it work?
2. A battery is used to light a flashlight bulb. Trace as many of the energy conversions as you can in this circuit.

* These questions go beyond the text itself.

3. What are the magnetic fields around the wires carrying steady currents in the figures below?

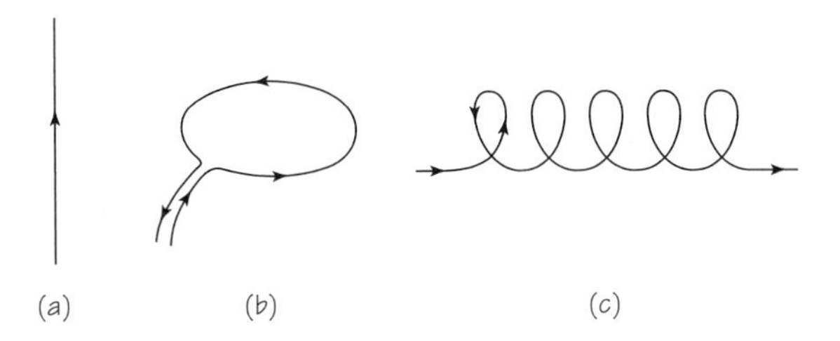

4. Show that for a charge Q moving a distance D under the influence of an electric field E, its kinetic energy will be $KE = QED$.

Quantitative

In the following, take the value of k_e in Coulomb's law to be about nine billion newton-meters squared per coulomb squared, or $k_e = 9 \times 10^9$ N m²/C². The value of the gravitational constant G is 6.67×10^{-11} N m²/kg².

1. If you compare the gravitational force between two 1-kg masses with the electric force between two 1-C charges at the same distance, which force is larger, and by how much?
2. What is the electric force between two positive charges of 1 C placed 1 m from each other?
3. How does the force in Question 2 compare with the gravitational force between two 1-kg masses placed 1 m apart?
4. What is the electric field at a point where a test charge of 3×10^{-5} C experiences a force of 10^{-2} N upward?
5. The electric charge on an electron is -1.6×10^{-19} C. How many electrons would be required to make -1 C of charge?
6. An electron crosses a potential of 10 V. What is the change in its kinetic energy, in eV?
7. An alpha particle (helium nucleus of charge $+2e$) crosses the same potential of 10 V. What is the change in its kinetic energy, in eV? How many joules is this?
8. Find out the electric power ratings of some common household devices that you would normally use. During the course of 1 day, note which devices you use and for approximately how long. Include light bulbs, TV, computer, a hair dryer, electric water heater, washer and dryer, etc. Obtain your total energy consumption for that day. How many hours could your total energy consumption keep a 100-W light bulb burning?
9. A household consumed 5500 kilowatt-hours in 1 month.
   (a) How many joules of energy is this?
   (b) How many hours could this amount of energy keep a 100-W light bulb burning?
10. What is the power used by a circuit in which 3 A flow across a 240-V supply? What is the resistance of this circuit?

CHAPTER 11

The Electric Age

11.1 TRANSPORTING ENERGY FROM ONE PLACE TO ANOTHER

In Chapter 6, we discussed the development of steam engines during the eighteenth and nineteenth centuries. These engines enabled industrialization by making available the vast stores of energy contained in coal, wood, and oil. By burning fuel, chemical energy is converted into heat energy, which in turn can be used to boil water to produce steam. By letting the steam expand against a piston or a turbine blade, heat energy can be converted to mechanical energy. In this way, a steam engine can power machinery.

Steam engines had two major defects, however. First, the mechanical energy was available only at the place where the steam engine was located. Second, practical steam engines were big, hot, and dirty. As the use of machines run by steam engines increased, people were crowded together in factories, and their homes stood in the shadow of the smoke stacks. Even steam-powered locomotives, though useful for transportation, were limited by their size and weight. They also added further air pollution.

Using one central power plant for sending out energy for use at a distance could partially overcome these defects. The energy transmitted by the central power plant could drive machines of any desired size and power at the most practical locations. After Volta's development of the battery, many scientists and inventors speculated that electricity might provide such a means of distributing energy and running machines. But the energy in batteries is quickly used up unless it is delivered at a low rate. A better way of generating electric currents was needed. When such a way was found, it changed the whole shape of life in homes, factories, farms, and offices. It even changed the very appearance of cities and landscapes.

In this chapter you will see another example of how discoveries in basic physics have given rise to new technologies. These technologies have revolutionized and benefited modern civilization. But they have brought some new problems in their turn.

The first clue to the broader use of electricity came from Oersted's discovery that a magnetic needle is deflected by a current supplied by a battery to a conductor. It was quickly realized that since an electric current can exert a force on a magnet, the force might be harnessed to perform useful work. In addition, many physicists naturally speculated that a magnet could somehow produce a current in a wire. (Such reasoning from symmetry is common in physics and is often useful.) Soon after the news of Oersted's discovery reached Paris, the French physicists Biot, Savart, and Ampère began research on the interactions of electricity and magnetism. (Some of their results were mentioned in Chapter 10.) A flood of other experiments and speculations on electromagnetism poured from all over the world into the scientific journals. Yet the one key discovery—how to generate an ample and continuous electric current—still eluded everyone.

11.2 FARADAY'S FIRST ELECTRIC MOTOR

Scientific journals regularly print brief announcements of the technical details of new discoveries. From time to time they also provide valuable in-depth surveys of recent broad advances in science. The need for such a review article is especially great after a burst of activity of the kind that followed Oersted's discovery of electromagnetism in 1820.

In 1821, the editor of the British journal *Annals of Philosophy* asked the English physicist Michael Faraday to review the experiments and theories of electromagnetism that had appeared in the previous year. In preparing for this review, Faraday made a remarkable discovery. Repeating Oersted's experiment (Section 10.9), he put a compass needle at various places around a current-carrying wire. Faraday was particularly struck by one fact: The force exerted by the current on each pole of the magnet tended to carry the pole along a circular line around the wire. As he expressed it later, the wire is surrounded by *circular lines of force:* a circular magnetic field.

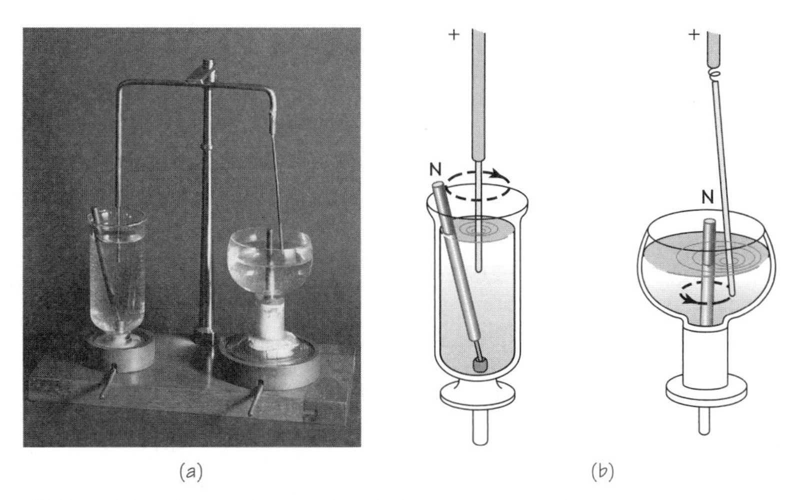

FIGURE 11.1 Two versions of Faraday's electromagnetic rotator. In each, the cup was filled with mercury so that a current could be passed between the base and overhead support. In one version (a), the north end of a bar magnet revolves along the circular magnetic lines of force surrounding the fixed current. In the other version (b), the rod carrying the current revolves around the fixed bar magnet, moving always perpendicular to the magnetic lines of force coming from the pole of the magnet.

Faraday then constructed an "electromagnetic rotator" based on this idea. It worked. Though very primitive, it was the first device for producing continuous motion by the action of a current: the first electric motor.

As in many other cases, Faraday was also guided by the idea that for every effect of electricity on magnetism, there must exist a corresponding effect of magnetism on electricity. Of course, it was not always so obvious what form the corresponding effect would take. But it led him to design an arrangement in which the magnet was fixed and the current-carrying wire rotated around it. (If a current exerts a force on a magnet, the magnet should exert an equal force on the current, according to Newton's third law.) This device, too, was eventually developed into a type of electric motor.

11.3 THE DISCOVERY OF ELECTROMAGNETIC INDUCTION

Armed with his idea involving "lines of force" of electric and magnetic fields, Faraday joined the search for a way of producing currents by magnetism. Scattered through his diary in the years after 1824 are many descriptions of

MICHAEL FARADAY

Michael Faraday (1791–1867) was the son of an English blacksmith. In his own words:

> My education was of the most ordinary description, consisting of little more than the rudiments of reading, writing and arithmetic at a common day-school. My hours out of school were passed at home and in the streets.

At the age of 12 he went to work as an errand boy at a bookseller's store. Later he became a bookbinder's assistant. When Faraday was about 19 he was given a ticket to attend a series of lectures given by Humphry Davy at the Royal Institution in London. The Royal Institution was an important center of research and education in science, and Davy was Superintendent of the Institution. Faraday became strongly interested in science and undertook the study of chemistry by himself. In 1813, he applied to Davy for a job at the Royal Institution and Davy hired him as a research assistant. Faraday soon showed his genius as an experimenter. He made important contributions to chemistry, magnetism, electricity, and light, and eventually succeeded Davy as Superintendent of the Royal Institution.

FIGURE 11.2

Because of his many discoveries, Faraday is generally regarded as one of the greatest experimental scientists. Faraday was also a fine lecturer and had an extraordinary gift for explaining the results of scientific research to nonscientists. His lectures to audiences of young people are still delightful to read. Two of them, "On the Various Forces of Nature" and "The Chemical History of a Candle," have been republished in paperback editions.

Faraday was a modest, gentle, and deeply religious man. Although he received many international scientific honors, he had no wish to be knighted, preferring to remain without title.

such experiments. Each report ended with a note: "exhibited no action" or "no effect."

Finally, in 1831, came the breakthrough. Like many discoveries that follow much research and discussion among scientists, this one was made almost at the same time by two scientists working independently in different countries. Faraday was not quite the first to produce electricity from magnetism. *Electromagnetic induction*—the production of a current by magnetism—was actually discovered first by the American scientist Joseph Henry. At the time Henry was teaching mathematics and philosophy at an academy in Albany, New York. Unfortunately for the reputation of American science, teachers at the Albany Academy were expected to spend all their time on teaching and related duties. There was little time left for research. Henry had hardly any opportunity to follow up his discovery, which he made during a one-month vacation. He was not able to publish his work until a year later. In the meantime, Faraday had made a similar discovery and published his results.

But Faraday is known as the discoverer of electromagnetic induction not simply because he was the first to publish his results. More importantly, he conducted exhaustive investigations into all aspects of the subject. His earlier experiments and his ideas about lines of force had suggested that a current in one wire should somehow induce a current in a nearby wire. Oersted and Ampère had shown that a *steady* electric current produced a *steady* magnetic field around the circuit carrying the current. Perhaps a steady electric current could somehow be generated if a wire were placed near or around a very strong magnet. Or a steady current might be produced in one wire by a large steady current in another wire nearby. Faraday tried all these possibilities, with no success.

The solution Faraday found in 1831 came partly by accident. He was experimenting with two wire coils that had been wound around an iron ring (see Figure 11.3). He noted that a current appeared in one coil—called the *secondary*—while the current in the other coil—called the *primary*—was being switched on or off. When a current was turned on in the primary coil A, a current was induced in secondary coil B, but it lasted only a moment. As soon as there was a steady current in coil A, the current in coil B disappeared. When the current in coil A was turned off, a current again appeared briefly in coil B.

To summarize Faraday's result:

> A current in a stationary wire can induce a current in another stationary wire only while the current is changing. A steady current in one wire cannot induce a current in another wire.

Faraday was not satisfied with merely observing and reporting his accidental arrangement and its important result. Guided by his concept of "lines of force," he tried to find out the basic principles involved in electromagnetic induction.

According to Faraday's theory, the changing current in coil A would change the lines of magnetic force in the whole iron ring. The change in

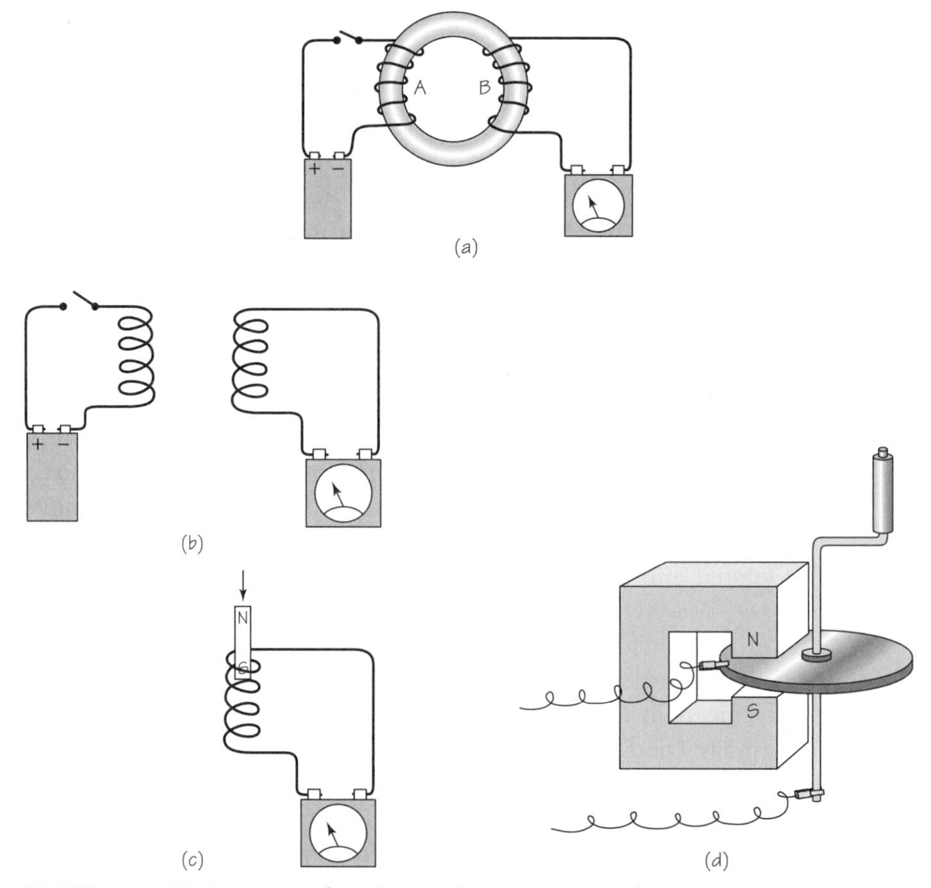

FIGURE 11.3 Various ways of producing electromagnetic induction.

lines of magnetic force in the part of the ring near coil B would then induce a current in B. But if this was really the correct explanation, Faraday asked himself, should it not be possible to produce the same effect in another way? In particular, he asked:

1. Is the iron ring really necessary to produce the induction effect? Or does the presence of iron merely strengthen an effect that would also occur without it?
2. Is coil A really necessary? Or could current be induced simply by changing the magnetic lines of force through coil B in some other way, such as by moving a simple magnet relative to the wire?

Faraday answered these questions almost immediately by performing further experiments. First, he showed that the iron ring was not necessary.

Starting or stopping a current in one coil of wire would induce a momentary current in a nearby coil, with only air (or a vacuum) between the coils. (See the drawing in Figure 11.3b. Note that there is no battery in the circuit at the right in (b), only a meter to measure the induced current.) Second, he studied what happened when a bar magnet was inserted into or removed from a coil of wire. He found that a current was induced at the instant of insertion or removal. (See drawing (c).) In Faraday's words:

> A cylindrical bar magnet . . . had one end just inserted into the end of the helix cylinder; then it was quickly thrust in the whole length and the galvanometer [current-meter] needle moved; when pulled out again the needle moved, but in the opposite direction. The effect was repeated every time the magnet was put in or out. . . .

Note that his arrangement amounted to a primitive *electric generator*; it provided electric current by having some mechanical agent move a magnet.

Having done these and many other experiments, Faraday started his general *principle of electromagnetic induction*. Basically, it is that *changing lines of magnetic force can induce a current in a wire*. The needed "change" in lines of force can be produced either by a magnet moving relative to a wire or by a changing current in another circuit. In the case of the moving magnet, Faraday described the wire in which current was induced as being "cut across" by lines of force from the magnet. In the case of the effect caused by a changing current in another circuit, the lines of force from the latter "cut across" the wire. He later used the word *field* to refer to the arrangement and intensity of lines of force in space. Using this term, one can say a current can be induced in a circuit by changes set up in a magnetic field

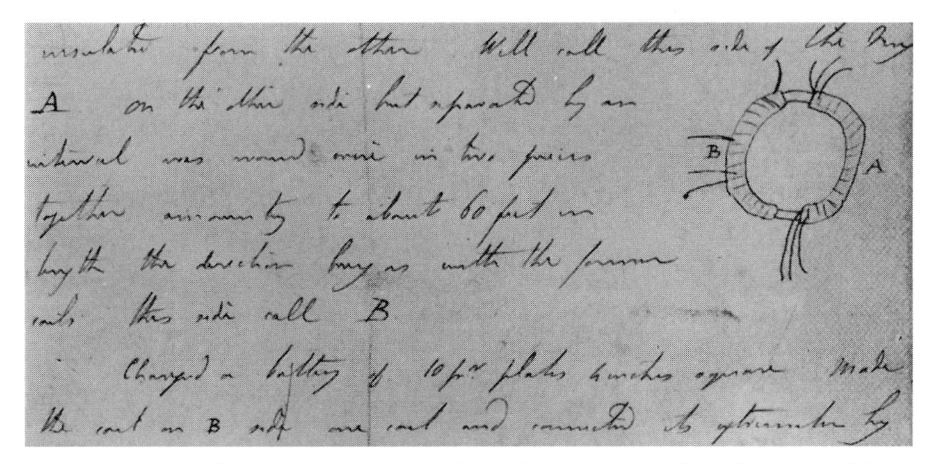
insulated from the other. Will call this side of the ring A on the other side but separated by an interval was wound wire in two pieces together amounting to about 60 feet in length the direction being as with the former coils this side call B

Charged a battery of 10 pr. plates 4 inches square Made the coil on B side one coil and connected its extremities by

FIGURE 11.4 Detail of a page in Faraday's diary where he recorded the first successful experiment in electromagnetic induction.

around the circuit. Such changes may result either from relative motion of wire and field or simply from a change in intensity of the inducing field.

So far, Faraday had produced only momentary surges of current by induction. This was hardly an improvement over batteries as a source of current. Was it possible to produce a continual current by electromagnetic induction? To do this would require a situation in which magnetic lines of force were *continually changing* relative to the conductor. Using a simple magnet, the relative change could be produced either by keeping the magnet or the conductor in motion. This is just what Faraday did. He placed a copper disk between the poles of a magnet, and applied mechanical energy to keep it turning. (See Figure 11.3d.) A steady current was produced in a circuit connected to the disk through brass contacts or "brushes." His device—called the "Faraday disk dynamo"—was the first constant-current electric generator. While this particular arrangement did not turn out to be very practical, it showed at last that continuous generation of electricity was possible.

These first experimental means of producing a continuous current were important aids to understanding the connection between electricity and magnetism. Moreover, they suggested the possibility of eventually generating electricity on a large scale. The production of electrical current involves changing energy from one form to another. When electrical energy appears, the law of conservation of energy requires that it be at the cost of some other form of energy. In the electric battery, chemical energy (the energy of formation of chemical compounds) is converted to electrical energy. Batteries are useful for many portable applications (automobiles, computers, cell phones, for example). But it is not practical to produce large amounts of electrical energy by this means. There is, however, a vast supply of mechanical energy available from many sources. Electrical energy could be produced on a large scale if some reasonably efficient means of converting mechanical energy to electrical energy were available. This mechanical energy might be in the form of wind, or falling water, or continuous mechanical motion produced by a steam engine. The discovery of electromagnetic induction showed that, at least in principle, it is possible to produce electricity by mechanical means. In this sense, Faraday can rightly be regarded as the founder of the modern electrical age.

11.4 GENERATING ELECTRICITY: THE GENERATOR

Faraday had shown that when a conducting wire moves relative to a magnetic field, an electric current is produced. Whether it is the wire or the magnetic field that moves does not matter. What counts is the relative motion of one with respect to the other. Once the principle of electromagnetic induction was known, experimenters tested many combinations of wires and magnets in relative motion. One basic type of *generator* (or "dynamo," as it was often called) was widely used in the nineteenth century. In fact, it remains the basic model for many generators today.

This form of generator is basically a coil of wire made to rotate in a steady magnetic field, thereby "cutting" the lines of force, using Faraday's image. The coil is connected to an external circuit by sliding contacts. In Figure 11.5, the "coil" is shown for simplicity as a single rectangular loop of wire. This loop rotates around an axis xy between the north and south poles of a magnet. Two conducting rings, labeled d and e are permanently attached to the loop and, therefore, also rotate around the axis. Conducting contacts, called "brushes," here labeled g and f, complete a circuit through a meter (h) that indicates the current produced. The complete circuit is abdfhgea. (Note that one part of the wire goes through ring d without touching it and connects to e.)

Initially, the loop is at rest between the magnetic poles and no charge flows through it. Now suppose the loop is rotated counterclockwise. The wire's long sides a and b now have a component of motion perpendicular to the direction of the magnetic lines of force; that is, the wire "cuts across" lines of force. This, according to the principle of electric induction, is the condition for inducing an electric current in the loop. The greater the rate at which the lines are cut, the greater the induced current.

To understand better what is going on in the wire, one should understand its operation in terms of the force on the movable charges in the wire. It is the movement of these charges, which are, by convention, assumed to be positive, that forms the current. The charges in the part of the loop labeled b are being physically moved together with the loop across the magnetic field. Therefore, they experience a magnetic force given by qvB (as described in Section 10.11). This force pushes the charges in the wire "off to the side"—that is, perpendicular to both the velocity vector and the magnetic field vector. In this situation, "off to the side" is *along the wire*.

What about side a? It is also moving through the field and "cutting" lines of force, but in the opposite direction. So the charges inside a experience a push along the wire in the direction opposite to those in b. This is just what is needed; the two effects reinforce each other in generating a current around the whole loop. The "push" on the charges that produces the current can also be regarded as resulting from a potential difference ("voltage") induced in the loop of wire. Thus, a generator produces both "voltage" and current.

The generator just described produces *alternating current* (abbreviations AC). The current is called "alternating" because it regularly reverses

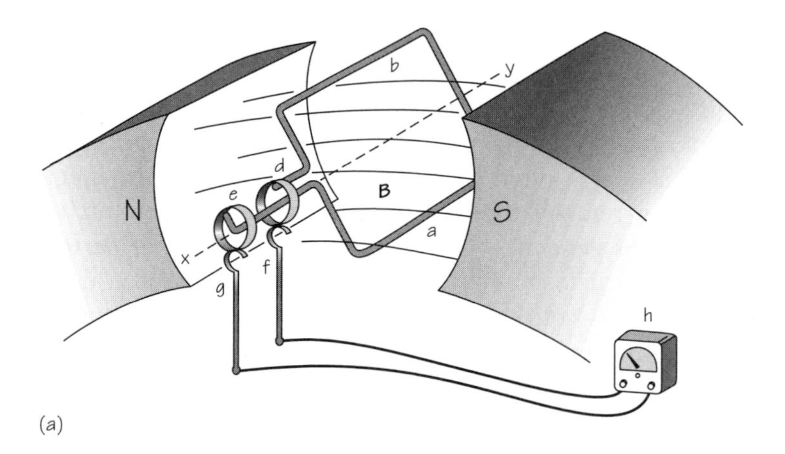

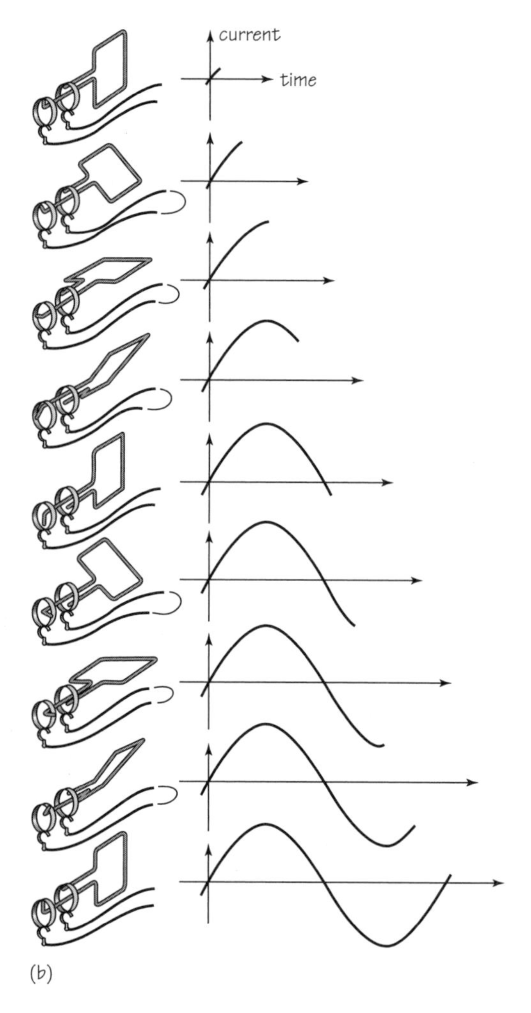

FIGURE 11.5 An alternating current generator. The graph shows the electric current generated as successive positions of the single loop are reached.

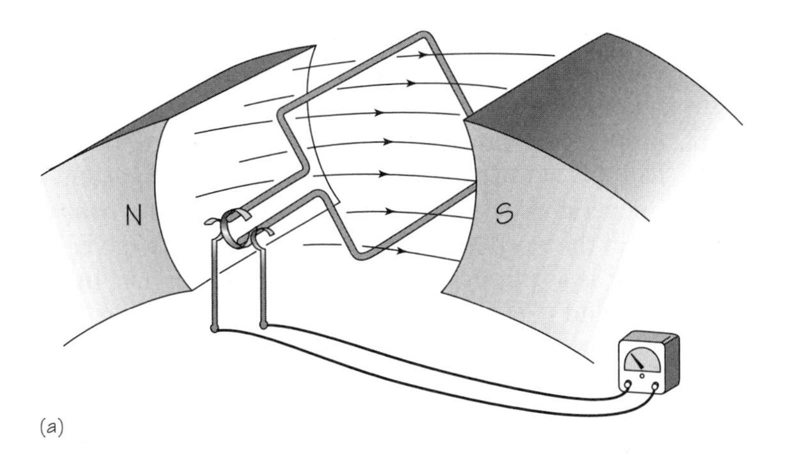

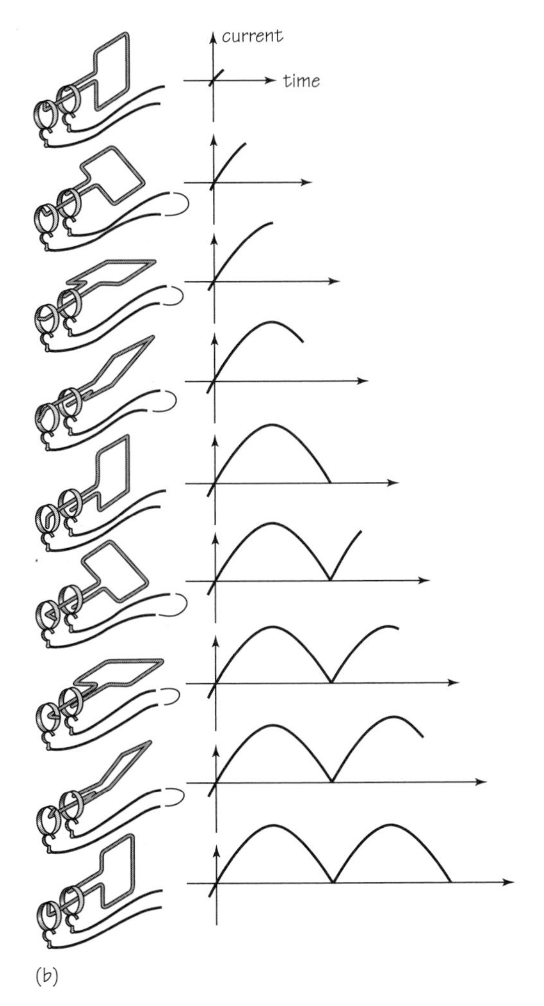

FIGURE 11.6 A direct current generator. The graphs show the current at different positions of the loop.

(alternates) its direction. At the time this kind of generator was first developed, in the 1830s, alternating current could *not* be used to run machines. Instead *direct current* (DC) was needed.

In 1832, Ampère announced that his talented instrument maker, Hippolyte Pixii, had solved the problem of generating direct current. Pixii modified the AC generator by means of a device called the *commutator*. The name comes from the word *commute*, to interchange or to go back and forth. The commutator is a split cylinder inserted in the circuit. In the AC generator, brushes f and g are always connected to the same part of the loop. But with the commutator, the brushes *reverse connections* each time the loop passes through the vertical position. Just as the direction of current induced in the loop is at the point of reversing, the contacts reverse. As a result, the current in the outside circuit is always in the same direction; in short, it is a direct current. Although the current in the outside circuit in that system is always in the same direction, it is not constant. It rises and falls rapidly between zero and its maximum value. In working generators, many sets of loops and commutators are connected together on the same shaft. In this way, their induced currents reach their maximum and zero values at different times. The *total* current from all of them together is then more uniform.

Whether a generator delivers alternating current or direct current, the electric power (energy per unit time) produced at every instant is given by the same equation developed in Section 10.7:

$$P = V \cdot I = I^2R.$$

For example, suppose that a wire (e.g., the filament wire in a light bulb) with resistance R is substituted for the meter at h. If the current generated in the circuit at a given time is I, the electrical energy per unit time delivered to the wire is given by I^2R. For alternating current, the power output varies from instant to instant. But the *average* output power is simply $P = (I^2)_{av}R$. This electrical energy, of course, does not appear by itself, without any source. That would violate the law of conservation of energy. In the generator, the "source" of energy is clearly the mechanical energy which is supplied to keep the coils rotating in the magnetic field. This mechanical energy may be provided by a steam or gasoline engine, or by water power, wind power, or even by human exertion. The electric generator is, in principle, simply a device for converting mechanical energy into electrical energy.

11.5 PUTTING ELECTRICITY TO WORK: THE MOTOR

After the invention of the electric motor in 1873, the greatest obstacle to its practical use was the lack of cheap electric current to run it. The chemical energy in a battery was quickly exhausted. The electric generator,

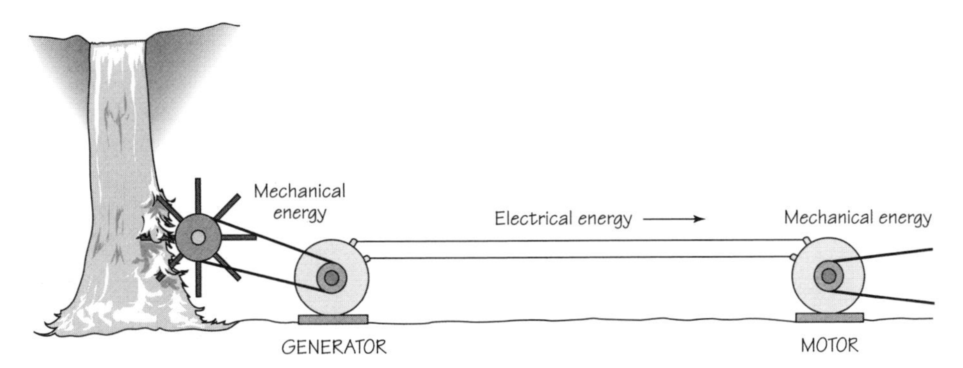

FIGURE 11.7 Generator and motor.

invented almost simultaneously by Faraday and Henry in 1832, was at first not at all economical in producing electrical current when mechanical energy was expended on it. Generators that used mechanical power efficiently to produce electric power were needed. But to design such generators required an understanding of the details of operation, and this understanding took nearly 50 years to acquire.

A chance event marked the effective take-off of the electric power age. This event was an accidental discovery at the Venice Exhibition of 1873. The story goes that an unknown worker at the Exhibition just happened to connect two dynamos (generators) together. The first dynamo, which was mechanically driven, generated current, and this current then passed through the coils of the second dynamo. Amazingly, the second dynamo then ran as an electric motor, driven by the electricity generated by the first dynamo.

The exhibitors at the Exhibition immediately utilized the accidental discovery that a generator run in reverse by the input of current could function as a motor through the output of mechanical work. They used a small artificial waterfall to drive the generator. Its current then drove the motor, which in turn operated a device that did mechanical work. *This, in effect, is the basic operation of a modern electrical transmission system*. A turbine driven, say, by steam or moving water, drives a generator which converts the mechanical energy to electrical energy. Conducting wires transmit the electricity over long distances to motors, toasters, electric lights, computers, etc. These devices in turn convert the electrical energy to mechanical energy, heat, or light.

The steam *turbine*, invented by the English engineer Charles Parsons in 1884, has now largely replaced older kinds of steam engines. At present, steam turbines drive the electric generators in most electric-power stations. These steam-run generators supply most of the power for the machinery of modern civilization. Even in nuclear power stations, the nuclear energy

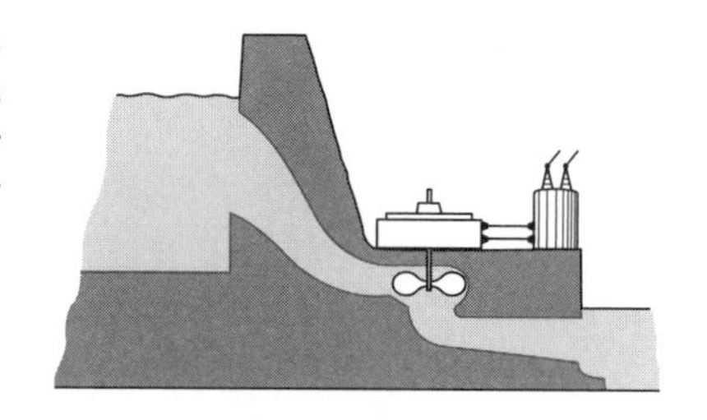

FIGURE 11.8 The general principle of hydroelectric power generation is illustrated in this sketch. Water flowing from a higher to a lower level turns turbine blades attached to a generator shaft. The details of construction vary widely.

is generally used to produce steam, which then drives turbines and electric generators. (Other means may be used to drive turbines such as wind and water. See Section 11.10.)

The basic principle of the Parsons turbine is simpler than that of the Newcomen and Watt engines. A jet of high-pressure steam strikes the blades of a rotor, driving the rotor around at high speed. The steam expands after passing through the rotor, so the next rotor must be larger. This accounts for the characteristic shape of turbines. Large electric-power station turbines using more than 500,000 kg of steam an hour can generate electrical energy at a rate greater than one billion joules per second.

The development of electrical generators shows an interaction of science and technology different from that of the development of steam engines. As was pointed out in Chapter 7, the early steam engines arose from the efforts of practical inventors. These inventors had no knowledge of the current theory of heat (thermodynamics). But their development of the steam engine, and attempts by Sadi Carnot and others to improve its efficiency through theoretical analysis, contributed greatly to the establishment of

FIGURE 11.9 Power station with a set of water-driven electric generators.

FIGURE 11.10 Steam turbine rotors.

thermodynamics. In that case, the advance in technology came before the advance of science. In the case of electromagnetism, the reverse occurred. Ampère, Faraday, Kelvin, Maxwell, and others had built up a large amount of scientific knowledge before any serious practical application succeeded. The scientists, who understood electricity better than anyone else, were not especially interested in commercial applications. And the inventors, who hoped to make huge profits from electricity, knew very little scientific theory. After Faraday announced his discovery of electromagnetic induction, people started making generators to produce electricity immediately. But it was not until decades later that inventors and engineers understood enough to work with such necessary concepts as lines of force and field vectors. With the introduction of the telegraph, telephone, radio, and alternating-current power systems, a much greater mathematical knowledge was needed to work with electricity. Universities and technical schools started to give courses in electrical engineering. Gradually, a group of specialists developed who were familiar with the physics of electricity and who also knew how to apply it.

11.6 THE ELECTRIC LIGHT BULB

The enormous growth of the electrical industry resulted from the great public demand for electrical products. One of the first commercially successful electrical products was the electric light bulb, which brought the benefits of electricity to the general public and made electricity the dominant form of early private energy consumption. The success of the light bulb mirrors in some respects the success of such other technological inventions as the microchip or television. It is an outstanding case study of the relationship between physics, industry, and society.

At the beginning of the nineteenth century, most buildings and homes were lit by candles and oil lamps. It is said that, during his school years, Abraham Lincoln did most of his reading by the fireplace. There was almost no street lighting in cities except for a few lamps hung outside houses at night. The natural gas industry was just starting to change this situation. London got its first street lighting system in 1813, when gas lights were installed on Westminster Bridge. However, the social effects of gas lighting were not all beneficial. For example, gas lighting in factories enabled employers to extend an already long and difficult working day into one still longer.

As the British chemist Humphry Davy and other scientists showed, light can be produced simply by heating a wire to a high temperature by passing a current through it. This method is known as *incandescent* lighting. The major technical drawback here was that the wire filament gradually burned up. The obvious solution was to enclose the filament in a glass container from which all the air had been removed. But this was easier said than done. The vacuum pumps available in the early nineteenth century could not produce a strong enough vacuum for this purpose. It was not until 1865, when Hermann Sprengel in Germany invented an improved vacuum pump, that the modern electric light bulb became possible. (Sprengel's pump also greatly aided Crookes and others in scientific experiments leading to important discoveries in atomic physics. These discoveries will be discussed in Chapter 13.)

Thomas A. Edison was not the first to invent an incandescent light, nor did he discover any essentially new scientific principles. What he did was to develop a practical light bulb for use in homes. Even more important, he worked out a distribution system for electricity. His system not only made the light bulb a practical device, but it opened the way for mass consumption of electrical energy in the United States.

Edison started by making an important business assumption about *how* people would want to use their light bulbs. He decided that each customer

FIGURE 11.11 Edison in his laboratory.

must be able to turn on and off any single bulb without affecting the other bulbs connected to the circuit. This meant that the bulbs must be connected "in parallel," like the rungs of a ladder, rather than "in series." (See Figure 11.13.)

The choice of parallel rather than series circuits had important technical consequences. In a series circuit, the same current goes through each bulb. In a parallel circuit, only part of the total current available from the source goes through any one bulb. To keep the total current needed from being too large, the current in each bulb has to be small.

As noted in Section 10.7, the heating effect of a current depends on both the resistance of the wire and the amount of current. The rate at which heat and light energy are produced is given by $P = I^2R$. According to this relationship, the rate goes up directly as the resistance, but increases as the *square* of the current. Therefore, in order to produce more light, most inventors used high-current, low-resistance bulbs and assumed that parallel circuits would not be practical. Edison realized that, in addition to the practicality of parallel circuits, a small current can have a large lighting effect if the resistance is high enough.

So Edison began a search for a suitable high-resistance, nonmetallic substance for his light-bulb filaments. To make such a filament, he first had to

FIGURE 11.12 Lewis Howard Latimer (1848–1928), the son of an escaped slave, became one of the original associates of Thomas Edison. Latimer was an inventor, patent authority, poet, draftsman, author, and musician.

bake or "carbonize" a thin piece of a substance. Then he sealed it inside an evacuated glass bulb with wires leading out. His assistants, including Lewis Howard Latimer, tried more than 1600 kinds of material:

> paper and cloth, thread, fishline, fiber, celluloid, boxwood, coconut shells, spruce, hickory, hay, maple shavings, rosewood, punk, cork, flax, bamboo, and the hair out of a redheaded Scotchman's beard.

Edison made his first successful high-resistance lamp in October 1879 with carbonized cotton thread in a high-vacuum sealed bulb. It burned continuously for 2 days before it fell apart. The following year, Edison produced lamps with filaments made from bamboo and paper. The Edison Electric Light Company began to install lighting systems in 1882. After only 3 years of operation, the Edison company had sold 200,000 lamps. It had a virtual monopoly of the field and paid big dividends to its stockholders.

The electric light bulb has changed somewhat since Edison's original invention. For example, the carbonized filaments of the older lamps were replaced in newer bulbs by thin tungsten wires. Tungsten had the advantages of greater efficiency and longer life.

The widespread use of light bulbs confirmed the soundness of Edison's assumptions about what people would buy. It also led to the rapid development of systems of power generation and distribution. The need for more power for lighting spurred the invention of better generators, the harnessing of water power, and the invention of the steam turbine. Success in providing large quantities of cheap energy made other uses of electricity practical. Once homes were wired for electric lights, the current could be used to run sewing machines, vacuum cleaners, washing machines, toasters, and (later on) refrigerators, radios, television sets, and computers. Once electric power was available for public transportation, cities could grow rapidly in all dimensions. Electric elevators made high-rise buildings practical,

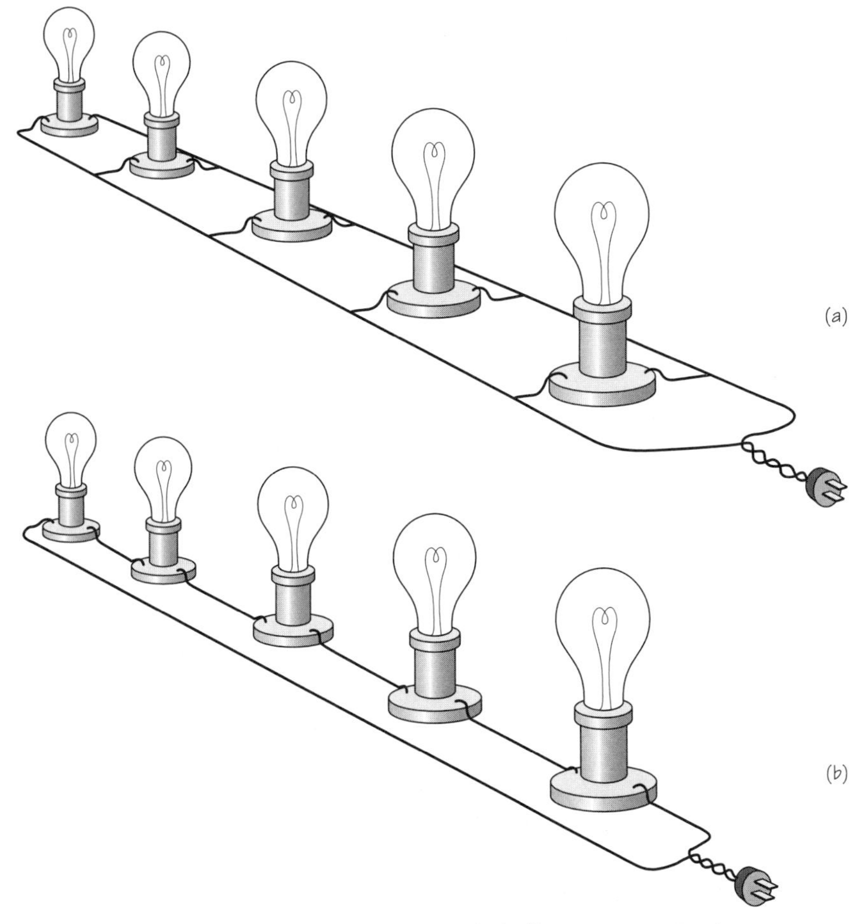

FIGURE 11.13 (a) Bulbs in a parallel circuit; (b) bulbs in a series circuit.

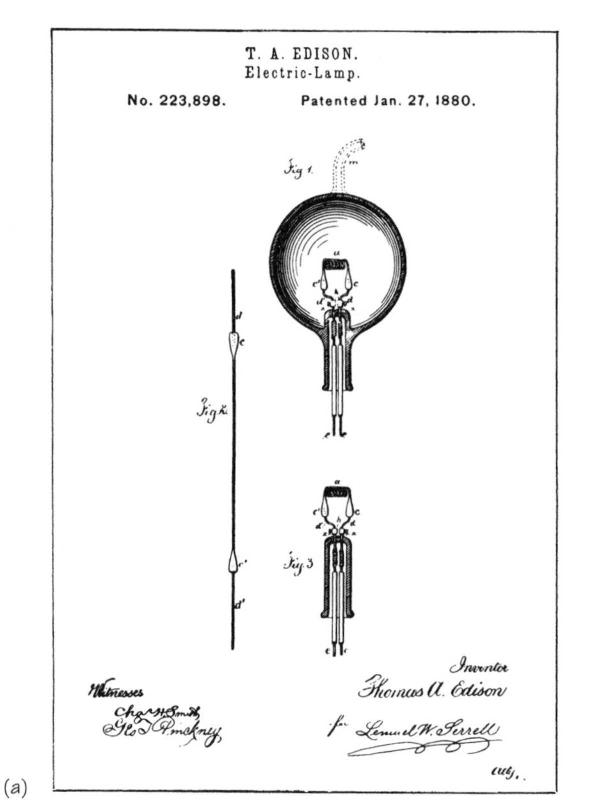

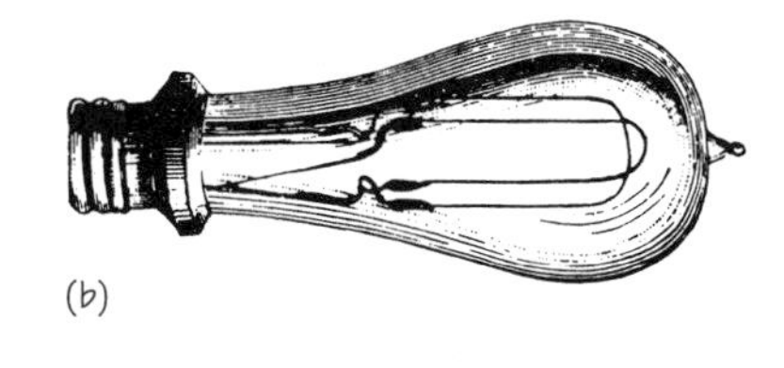

FIGURE 11.14 Two drawings of Edison's invention.

while electric tramways and subways rapidly transported people from their homes to jobs and markets.

We are now so accustomed to more sophisticated applications of electricity that it is hard to realize the impact of something as simple as the light bulb. But most people who lived through the period of electrification, which was as late as the 1930s and 1940s in many rural areas of the United States, agreed that the electrical appliance that made the greatest difference in their lives was the electric light bulb, which made the evenings fit for finishing chores indoors, as well as for leisure activities or reading. The last was one of the reasons given by President Franklin D. Roosevelt for his interest in promoting large-scale rural electrification, as in the Tennessee Valley Authority (TVA). In addition, of course, electric machines have been able to lighten many heavy physical labors and to make possible work that human or animal strength could never have accomplished.

11.7 AC VERSUS DC: THE NIAGARA FALLS POWER PLANT

Section 11.4 stated that the earliest electric generators produced alternating current (AC), which could be changed into direct current (DC) by the use of a commutator. Throughout most of the nineteenth century, most engineers believed that only DC was useful in practical applications of electricity. However, as the demand for electric power increased, some disadvantages of DC became evident. One problem was that the commutator complicated the mechanical design of generators, especially if the ring had to rotate at high speed. This difficulty became even more serious after the introduction of steam turbines in the 1890s, since turbines work most effectively at high speeds. Another disadvantage was there was no convenient way of changing the generated voltage of a DC supply.

Why should it be necessary to change the voltage with which current is driven through a transmission system? One reason involves the amount of power lost in heating the transmission wires. The power output of a generator depends (as indicated in Section 10.8) on the output *voltage* of the generator and the amount of *current*:

$$P_{\text{total}} = VI.$$

The power made available by the generator is transmitted to the line and to the consumer. (For this reason, commercial power is often given in units of volt-amps.) The same amount of power can be delivered at smaller I if V is somehow made larger. As noted in Section 10.7, when there is a current I in a transmission wire of resistance R, the portion of the power lost as heat in transmission is equal to the resistance times the square of the current:

$$P_{\text{heat loss}} = I^2R.$$

The power finally available to consumers is $P_{\text{total}} - P_{\text{heat loss}}$. For transmission lines of a given resistance R, where the value of R is fixed by the wires themselves, the current I should be as small as possible in order to minimize the power loss. Obviously, therefore, electricity should be transmitted at low current and at high voltage.

However, most generators cannot produce electricity at very high voltages. To do so would require excessively high speeds of the moving parts. Some way of "stepping up" the generated electricity to a high voltage for transmission is needed. But some way of "stepping down" voltage again at

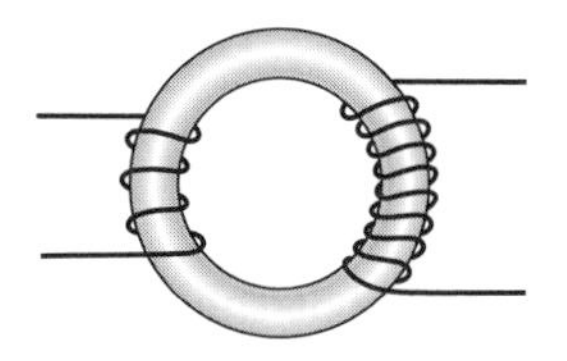

FIGURE 11.15 Transformer coils.

the other end, where the consumer uses the power, is also needed. For most applications of electricity, especially in homes, it is neither convenient nor safe to use high voltages. In short, *transformers* are needed at both ends of the transmission line.

A transformer can easily be made by a simple change in Faraday's induction coil (Section 11.4). Recall that Faraday wound a coil of wire—called the *secondary* coil—around one side of an iron ring. He then induced a current in this secondary coil by changing a current in another coil—the *primary* coil—wound around the other side of the ring. A current is induced in the secondary coil whenever the primary current changes. If the primary current is changing all the time, then a current is continually induced in the secondary. An AC applied to the primary coil (e.g., from a generator without a commutator) induces an AC in the secondary coil.

After the American engineer George Westinghouse saw an AC electrical system in Italy he bought the American patent rights for it. When the

FIGURE 11.16 The commercial distribution of AC electric power requires elaborate transmission facilities. Generator output voltages of about 10^4 volts are stepped up to about 10^5 volts for transmission, stepped down to about 10^4 volts for local distribution, and further stepped down to about 10^2 volts by neighborhood power-pole transformers. Within the home, they may be stepped down further (often to 6 volts for doorbells and electric trains) and stepped up by transformers in radio and TV sets for operating high-voltage tubes.

FIGURE 11.17 Dam on Niagara River.

Westinghouse Electric Company introduced the AC system to the United States with improved transformers in 1886, the Edison Electric Light Company (which later merged into General Electric) already had a monopoly on electric lighting and had already invested heavily in DC generating plants and distribution systems for most of the large cities. A bitter public controversy erupted between the two companies. Edison attempted to show that AC was unsafe because of the high voltage (tension) used for transmission in "high-tension" wires. In the middle of the dispute, the New York State legislature accepted Edison's suggestion of electrocution as a means of capital punishment. This event seems to have added to the popular fear of high voltage. Nevertheless, Westinghouse's AC system continued to grow. There were no spectacular accidents, and the public began to accept AC as reasonably safe.

The final victory of the AC system in the United States was assured in 1893 when businessmen in Buffalo, New York, chose AC for the new hydroelectric plant at Niagara Falls. It was a close decision. AC could be generated and transmitted more efficiently, but the demand for electricity in 1890 was mainly for lighting. This meant that there would be a peak demand in the evening. The system would have to operate at less than full capacity during the day and late at night. Because of this variation, some engineers believed that a DC system would be cheaper to operate. However, European systems began demonstrating that this was not the case, and expert opinion gradually changed in favor of AC over DC. Today, electric power in the United States is delivered to homes and factories in an AC that alternates at a frequency of 60 cycles per second, or 60 Hz. Most of the rest of the world uses AC current at 50 Hz.

It turned out that the critics had been wrong about the variation of demand for electricity throughout the day, as electricity found many uses besides lighting. During periods of lower demands, hydroelectric plants often use their excess capacity to store energy by pumping water up into reservoirs. The potential energy thus created can be used later to produce electricity during periods of peak demand as the potential energy is converted back into mechanical energy to run the generators. (We return to hydroelectric power in Section 11.10.)

11.8 THE ENERGY PICTURE TODAY

The average human being requires in his or her nutrition about eight million joules, or 2000 Calories, of energy for an average work day. (Nutrient labels required on all food packaging are based on an assumed diet of 2000 Calories, which is 2000 kcal.) This is roughly equivalent to the energy output of a 100-W light bulb burning for 24 hr. For millions of years the average daily consumption of energy per person on this planet remained at this amount. Then, about 10,000 years ago, the amount of energy use began to rise with the establishment of empires, the domestication of animals for work in agriculture, and the introduction of machines—all of which consumed additional energy. Energy consumption has risen steadily ever since, until today the amount of energy consumption per day, averaged over the world's population, is about one billion joules per person per day—over one hundred times the subsistence level. This figure is even more astounding when we realize that the number of people on Earth has also increased dramatically, and that most of those people do not live in energy-intensive industrial societies. The world population today is about six billion people, compared with only about one million people 10,000 years ago.

What are these billions of joules of energy per day used for, and where do they come from? Of course, the picture differs from nation to nation and between industrial and developing nations. We focus here on the United States economy, which, according to recent estimates, consumes about 85×10^{18} J of energy per year—or about one hundred billion joules per person per day. The table below shows the distribution of this energy and its usage by each sector of the United States economy in the mid-1990s.

Distribution of Energy Usage in the United States Economy in Units of 10^{18} J

Sector of economy	*Useful work and heating*	*Wasted energy*	*Total energy consumed*
Industrial	12	12	24
Residential and commercial	12	5	17
Transportation	5	19	24
Power generation and transmission	*	20	20
Column total	29	56	85
Percent of total energy	34%	66%	100%

* 10×10^{18} J for this purpose were included above in the useful energy consumed by the *industrial* and *residential/commercial* sectors.

Derived from A. Hobson, *Physics: Concepts and Connections* (Englewood Cliffs, NJ: Prentice-Hall, 1999), p. 448.

The table indicates that nearly two-thirds of the energy consumed in the United States is wasted. This waste arises not only from careless use of the energy provided, but also in large part from the restrictions on the efficiencies of heat engines owing to the second law of thermodynamics. The impact of the second law is greatest in the transportation and power generation sectors of the economy, which rely heavily on heat engines. For instance, the transportation sector, which displays the lowest energy efficiency (20.8%) of the above sectors of the economy, is dominated by the internal combustion engine. The average automobile cruising without acceleration at the most efficient highway speed (55 mi/hr) has an energy efficiency of only about 13%; that is, of the 70 kW of energy obtained from the fuel tank, 61 kW are unused in some fashion and only 9 kW are converted into the useful work needed to overcome air resistance and rolling resistance. Highway speeds above 55 mi/hr yield significantly lower efficiencies for such vehicles. Sport utility vehicles and light trucks, which currently represent a large fraction of new private passenger vehicles, have even lower overall efficiencies than automobiles. (Heat-engine efficiencies are further discussed in the next section.)

Energy Resources

Where does the energy consumed come from? The pie graph and table on next page show the main sources of energy in the United States and their percentage contributions to the economy (as of 2000). (Following the categories used by the Department of Energy, "renewables" include all sources other than fossil fuels and nuclear power.)

Energy Sources for the United States Economy, 2000

	Percent
Fossil Fuels	
Crude oil	35.2
Coal	27.9
Natural gas	19.4
Natural gas from plant liquid	2.52
Nuclear Power	7.77
Renewables	
Hydroelectric	3.25
Wood and biomass	3.53
Geothermal	0.33
Solar	0.076
Wind	0.038

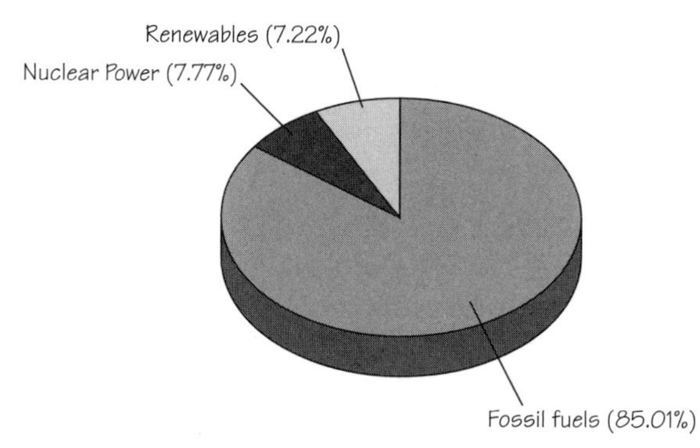

Source: US Dept. of Energy, http://www.eia.doe.gov/emeu/aer/overview.html

As you can see, 85% of the nation's energy needs are provided by three fossil fuels—oil, coal, and natural gas—in which chemical energy from the remains of organisms that lived millions of years ago is stored. The remaining 15% is provided by nuclear power and renewables. All of the nuclear and hydroelectric power and most of the coal is used for the production of electricity through steam-powered generators. Some of the oil and natural gas is also used for this purpose, especially in the Northeast, where it is easier and cheaper to import natural gas and oil, e.g., from Canada, than it is to ship coal from the Midwest. However, all of the wood and most of the oil and natural gas are used for nonelectric purposes—wood and gas for heating, oil for transportation.

While the United States is heavily dependent upon fossil fuels for its energy needs, the outlook is not promising. Some analysts have pointed out that industries in the more developed nations have used up in 200 years most of the reserves of chemical energy accumulated over the last 200 *million* years. According to current estimates, the world's remaining supply of coal will last only about another two centuries at the current rate of usage, while the estimated remaining supply of oil will last only another 50 years or so at the current usage rate. If such estimates are accurate, they indicate that we cannot continue the current energy scenario much longer. For example, the United States is over 50% dependent on imported energy sources, but many of the reserves are in politically unstable regions. Developing nations are demanding an increasing share of the fossil fuels in order to attain the benefits of industrialization for their economies. And the burning of fossil fuels is contributing to pollution and to significant climate change through the greenhouse effect leading to global warming. This is further discussed in Section 12.5.

11.9 CONSERVATION

Nuclear energy was once thought to be the answer to our increasing energy needs, and it may well be in the future—especially considering the urgency of supplying energy for commercial use. However, in the wake of the Three Mile Island and Chernobyl accidents and in view of the possibility of a terrorist attack on nuclear reactors, the American public has turned against nuclear energy, and federal regulators have not renewed the licenses of reactors located near densely populated areas. All active construction of new fission reactors has ceased in the United States, often at great losses to the utility companies (which have shifted the losses to their consumers through higher rates). Although there is only a little air pollution (not radioactive) from nuclear reactors, there is the problem of heat pollution as well as radioactive waste. All of the equipment, spent fuel rods, water, and even the clothing of the workers are radioactive and must be safely stored for thousands of years (until several half-lives of the radioactive elements involved have passed). These problems have both a technological as well as a political component, and until these can be resolved, fission reactors will probably not be the long-term solution to this nation's energy needs. As further discussed in Chapter 18, fusion energy, which would have none of the problems of present-day nuclear reactors, is not yet a practical reality, but it may well be in the years ahead.

> Note that other industrial nations, in particular France and Japan, rely heavily on nuclear energy for their electrical needs and continue to build new plants.

With nuclear energy not a long-term solution for the United States at this time, there are at least two other ways in which the United States, which currently has the largest economy and the greatest total energy consumption on Earth, can provide for its long-term energy needs:

- conservation of energy through reduction of energy losses;
- the opening of renewable and other alternative sources of energy.

As with the steam engine and the early electric age, the problem of energy usage is primarily a scientific and technological one, but the fundamental decisions are ultimately made by the public through their representatives and, especially, through individual consumer choices. (Increasing production of crude oil or natural gas from national reserves cannot be a long-term solution.)

Efficiency of a Power Plant

An electric power plant, whether powered by fossil fuels (coal, oil, or gas) or nuclear fuel, needs both a heat engine and a generator in order to produce electricity. The thermodynamic limit of the efficiency of a heat engine sets very severe constraints on how much of the energy released from burning the fuel is ultimately available as electrical energy. (This limit does not, of course, apply to hydroelectric plants.)

As discussed in Section 6.3, any engine that converts heat into mechanical work must also release heat into the environment. A diagram of this process, which can be applied to Watt's steam engine or to a large steam turbine, is sketched in Figure 11.18. T_1 and T_2 are the temperatures of the hot and cold reservoirs, respectively; Q_1 is the heat fed into the engine; Q_2 is the waste heat released into the environment; and W is the useful mechanical work obtained from the engine.

The second law of thermodynamics states that in the best possible circumstances the efficiency (eff) of the heat engine can be no greater than

$$\text{eff} = 1 - \frac{T_1}{T_2}.$$

This law is discussed in detail in Section 6.3.

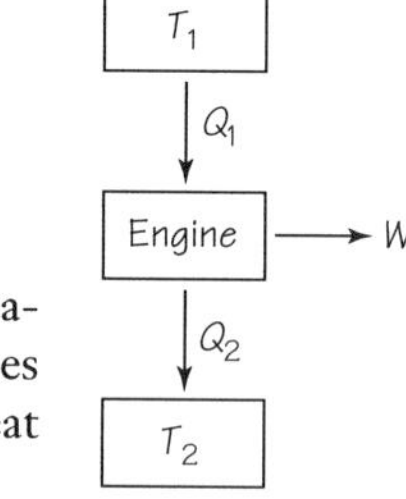

FIGURE 11.18 Block diagram of physical quantities in the operation of a heat engine.

What does this mean for a power plant? Fuel is burned in a combustion chamber, the chemical (or nuclear) energy is converted into thermal energy that keeps the combustion chamber at the temperature T_1. Water, heated by the combustion chamber in the boiler, circulates through the plant as steam during parts of its route and as liquid during others. In most plants, very high-pressure steam is created in the boiler. This steam pushes against the blades of a turbine, doing work on the turbine, and leaves the turbine as steam at a much lower pressure and temperature. The electric generator converts the mechanical work done on the turbine into electric energy; this process is not restricted by the second law because no thermal energy is involved. Finally, the steam must be condensed so that the water can retrace its route through the plant. This is done by allowing heat Q_2 to escape to the environment at temperature T_2. The whole process is shown schematically in Figure 11.18.

For the best efficiency (as close to 1 as possible), T_2 should be as low as possible, and T_1 as high as possible, so that T_2/T_1 is as small as possible. However, T_2 is fixed by the environment, since cooling air or water must be used at whatever temperature is available outside. This is generally about 20–25°C (about 300 K). T_1 is limited by technology and chemistry. Metals weaken and melt when they get too hot. For a modern fossil-fuel plant, T_1 may be as high as 500°C (about 770 K). In a nuclear power plant, caution suggests more conservative limits, and therefore T_1 is typically 400°C (about 670 K). The lower temperature is necessary, in particular, to avoid damaging the fuel rods.

The maximum possible efficiencies of fossil-fuel and nuclear-power plants is thus

$$\text{eff} = 1 - \frac{T_2}{T_1} = 1 - 300/730 \simeq 0.59 \quad \text{(fossil-fuel plant)}$$

$$= 1 - 300/670 \simeq 0.55 \quad \text{(nuclear-power plant)}$$

(The sign $\simeq$ means "approximately equal to.") Therefore, *even if there were no losses of any kind whatsoever*, a power plant could only turn about half of the thermal energy into electrical energy. For each joule of electrical energy produced, 2 J of energy will have to be provided originally by the fuel. The remaining joule will be released to the environment (into a river, the ocean, or the air) as thermal pollution.

The preceding paragraph describes the maximum possible efficiency of a perfect Carnot engine. Real power plants are significantly less efficient. Modern fossil-fuel plants can achieve about 38% or 40% in practice; nuclear plants, because of the lower value of T_1, can manage about 30%. Older fossil plants have efficiencies of 30% or less. These additional losses are

FIGURE 11.19 The Hong Kong skyline at night.

due to the fact that turbines are not ideal Carnot engines (they have friction; some heat simply leaks through them without doing any work at all) and to the fact that there are losses in generators, transformers, and power lines. A useful rule of thumb is that the overall efficiency of a power plant is about 33%.

What this analysis shows is, very roughly, that any time you use 1 J of electrical energy, about 3 J of thermal energy were produced at the power plant, and 2 J were released into the environment, mostly near the plant. For example, if you heat a room with a small electric heater, about three times as much fuel has to be burned to produce the needed energy when the same fuel is burned directly within the room itself (in a gas stove, for example). Through state and federal incentive programs, some progress is being made in convincing consumers to buy, and manufacturers to produce, more energy-efficient appliances. But this cannot significantly reduce the overall waste of energy in electrical production and transmission. This is the trade-off for the fact that electricity is such a convenient source of power. Only new technologies and discoveries can make a major difference in reducing this waste.

Because it is largely a consumer decision, the reduction of waste in the area of transportation may occur more readily. Transportation as a whole accounts for about 70% of the crude oil consumed in the United States

FIGURE 11.20 Average gas mileage of U.S. vehicles over time (at moderate highway speed).

and for about one-third of the total energy lost. The fuel efficiency of automobiles is controlled by technology, as discussed earlier, but advances in technology are strongly driven by consumer choice. On this point the track record in the United States has been spotty at best. Automobile efficiency began to go up following the oil crisis of the mid-1970s, when the public supported improvements in fuel efficiency and conservation. The picture changed beginning in the mid-1980s, when gas prices turned far lower than those in any other industrial nation, and when conservation became less

TWO AMERICAN TECHNOLOGIES

Telephone Technology

The word telephone comes from the Greek roots *tele*, "far," and *phone*, "sound." It was used as far back as the seventeenth century to refer to the communication device using a string with a mouthpiece and hearing devices at either end. However, in its modern use it refers to the invention by the Scottish-born American inventor Alexander Graham Bell.

Bell had a lifelong preoccupation with speech and communication, and while professor of "Vocal Physiology and Elocution" at Boston University, he began work on a machine that would transmit the voice along electric wires. In 1875, drawing on Hans Christian Oersted's and Michael Faraday's work on electromagnetism, Bell placed a metal diaphragm near an electromagnet; and then used the sound of a human voice to vibrate the diaphragm, which in turn changed the magnetic field, thus inducing a varying current.

The following year Bell improved on this model by placing a wire in mercury, which could be used in a circuit to vary resistance and therefore produce an undulating current when vibrated by a voice. In March 1876, in Boston, MA, Watson heard from the device Bell saying in the next room, "Mr. Watson, come here. I want to see you."

Bell received patent no. 174,465 for the development of this device to transmit speech sounds over electric wires, which is said to be one of the most valuable patents in history. Over the next few years, Bell was lionized as the inventor of the telephone. In tribute, all telephone service was suspended for a minute during his burial.

Automobile Technology

By 1910, America had overtaken Europe as the leading producer of automobiles. The most successful of this first generation of American automobile manufacturers was Henry Ford, who opened his first plant in 1903 in Detroit, Michigan. The automobile industry was competitive, and the main reason Ford succeeded where others failed seems to have been his central objective to manufacture quality automobiles at a price which the average American family could afford.

With low cost–high quality in mind, Ford introduced the Model T in 1908, which could be purchased at the time for $825, with the advertising claim "No car under $2000 offers more, and no car over $2000 offers more except the trimmings." In 1910 Ford opened the Highland Park assembly plant, which was a 60-acre site in Detroit. Continuous conveyor belts were introduced at this plant in 1912, and moving assembly lines in 1913. These methods of mass production enabled Ford to reduce the cost of the Model T further by more than half, while increasing the number of units produced. Not only did these methods of mass production help to increase the standard of living of the average American family, but with the introduction of the (then remarkable) 8-hour day into his plants, Ford also increased the demand for his motorcars by increasing factory workers' pay and leisure time.

Further Reading

R.V. Bruce, Alexander Graham Bell and the Conquest of Solitude, In: C.W. Pursell, Jr., ed., *Technology in America* (Cambridge, MA: MIT Press, 1989), pp. 105–116.

J.J. Flink, Henry Ford and the Triumph of the Automobile, In: C.W. Pursell., Jr., ed., *Technology in America* (Cambridge, MA: MIT Press, 1989), pp. 172–173.

popular. At that point the United States began to grow even more dependent on imported oil from the Middle East and elsewhere, imported oil now exceeding 50% of total crude-oil consumption. During the same period, the average fuel usage calculated for all automobiles remained constant at about 25 miles per gallon, while the amount of gasoline used increased owing to more driving.

The picture changed for the worse during the 1990s with the popularity of less efficiency sport utility vehicles (SUVs), the low prices for gas during that decade, and a general public apathy about conservation efforts. Some states have passed laws requiring at least a minimum number of "zero emission vehicles" on the road by the year 2010. At present, manufacturers and the government are devoting increased financial resources to research and development of battery-operated cars, buses, and trucks as well as hybrid (electric and internal combustion) vehicles.

11.10 RENEWABLE AND ALTERNATIVE ENERGY SOURCES

Electric generators are not the only means of producing electricity, nor is electricity the only means of producing heat and light. The burning of wood, coal, and even oil and gas can produce rudimentary heat and light without electricity, as does the use of solar heating. In solar heating, direct sunlight is used to light and warm buildings during the day. It is also used to heat coils of water on buildings. The heated water is then circulated into a building to provide heat as well as hot water. More indirectly, solar power is also used to grow foodstuffs, which of course is also an essential source of energy for us.

Solar power is receiving intense study as a means of producing electricity because of the vast amount of energy potentially available. Just outside the Earth's atmosphere the Sun's radiation provides 1360 W of power to each square meter of surface. By the time the radiation reaches the Earth's surface, much of it is lost, because of atmospheric absorption and clouds. Since the Earth rotates, more or less direct sunlight is available in any particular spot on the Earth's surface for only about 8 hr (on the average) each day. Depending upon the location, then, between 150 and 450 W/m^2 are delivered to ground level when averaged over a 24-hr period. This power can be used to boil water with which to power an ordinary steam turbine generating plant, or to heat water for household use.

Sunlight can also be converted directly into electricity in photocells or "solar" cells by a process called *photovoltaic conversion*. The basic operation

FIGURE 11.21 Major electric transmission lines in the continental United States.

FIGURE 11.22 Solar energy collectors in the Mojave Desert of California.

of the photocell is explained in Chapter 16. Such devices are used today mostly to power individual buildings, pocket calculators, and satellites in outer space. However, even in the best photocell is unable to turn all the energy that strikes it into electrical energy, for much of the light is reflected, transmitted, or turned into wasted heat within the cell.

When solar power could solve many of the problems resulting from fuel shortages and avoid the environmental pollution of other energy sources, the generation of even a significant fraction of our nation's electric power directly from the Sun is dependent on finding ways of building far less expensive collection systems for accepting the Sun's energy, and storage systems for providing the electricity when it is needed (at night, for example, or on cloudy days). Probably the most widely practiced use of solar energy will remain for the foreseeable future at the local level—the heating of houses and the production of hot water for home use.

One of the most attractive alternative energy sources for electricity generation is *wind energy*, the kinetic energy of moving air set into motion by the effects of solar energy on the Earth's atmosphere. The energy in the wind can generate electricity by turning the blades of a propeller attached to an electric generator. Such devices are being used to provide the elec-

FIGURE 11.23 Wind turbines in California for generating electricity.

tricity for cities such as San Jose, California, through the turning of a vast array of propeller-driven generators (wind turbines) in a mountain pass outside the city. Such arrays could also be used successfully in other windy areas of a country, supplying a fraction of the energy needs in those regions. However, winds are never steady and some regions of the nation do not have sufficient winds to generate much electricity. Moreover, some people regard the sight of large arrays of wind turbines across the countryside as a blight upon the landscape. In addition, measures must be taken to keep birds away from the spinning propellers. Thus, while wind energy offers the possibility of large-scale energy production for local use, its prospects at this time are limited as a large-scale commercial source of energy.

Another means of providing the work necessary to turn the coils in an electric generator is to use the potential energy of falling water in a *hydroelectric power* plant. As discussed earlier (Section 11.7), water from a dam or a water fall flows over the blades of a crank shaft connected to a turbine, turning the kinetic energy of the water into the mechanical work necessary to generate electric current. In this way, the Niagara Falls power plant discussed earlier, as well as the Hoover Dam and others, provide

FIGURE 11.24 The out-flow end of water-driven electric generators producing power at Wilson Dam (part of the Tennessee Valley Authority). The plant can generate electric energy at a rate of over 100,000,000 watts.

enormous amounts of hydroelectric power. However, also like wind energy, hydroelectric power is limited only to regions of the country near such a dam or a water fall. In addition, there are often important environmental effects caused by the damming of a river: the impounding of nutrient-rich sediments that are denied to downstream marine life; the disruption of bird or fish migrations, such as the return of salmon to upstream spawning grounds; and the destruction of scenery. Because of the practical and environmental shortcomings, hydroelectric power still accounts for only about 3.25% of this nation's total energy needs.

Strange as it may seem, at present the only other means of providing the mechanical work necessary to turn the crank in a generator is to make use of steam-engine technology—using heat to boil water to produce steam that performs the work necessary to generate an electric current. As noted earlier, the burning of fossil fuels—coal, oil, and natural gas—is still the most widely used means of heating water to produce the steam. But there are at least four other possible heat sources—again, all with significant disadvantages. These alternatives are:

- geothermal energy—tapping the heat inside the Earth arising from geological pressure and radioactive decay;
- biomass energy—the heat released in the chemical breakdown of organic materials, such as wood, compost heaps, and garbage;
- solar-thermal energy—focusing the Sun's rays to heat water to produce the necessary steam (discussed above);
- nuclear energy—using the heat produced in nuclear fission to boil water (Section 11.5, further discussed in Chapter 18).

The first three of these alternatives, as well as wind and hydroelectric power, are classified as *renewable energy sources*. In contrast, fossil fuels and nuclear fission are not renewable, because there is only a limited supply of these resources. When they are used up, they're gone forever. The dilemma is that while the renewable sources involve less pollution and will not be depleted in the near term, they cannot yet supply our total energy need. Moreover, all forms of energy conversion have economic and/or environmental consequences.

In the end, the energy problem will have to be solved through partnership comprising scientific research and technology community and the

FIGURE 11.25 Geothermal power plant in Iceland.

public. The balance among these three can change very rapidly. New breakthroughs, such as the harnessing of fusion energy, may be expected, and new demands by the public expressed through voting patterns and consumer choices can always lead to new choices for our energy needs and our energy future.

FURTHER READING

M. Brower, *Cool Energy: Renewable Solutions to Environmental Problems* (Cambridge, MA: MIT Press, 1992).

R. Rhodes and D. Bollen, The Need for Nuclear Power, *Foreign Affairs*, Jan./Feb. 2000, 30–44.

SOME NEW IDEAS AND CONCEPTS

AC current
commutator
conservation
DC current
electromagnetic induction
fossil fuels
generator
high-tension wire
incandescent lighting
lines of force
motor
parallel circuit
photovoltaic conversion
renewable energy sources
series circuit
transformer
turbine
wind turbine

STUDY GUIDE QUESTIONS

11.1 Transporting Energy from One Place to Another

1. What were some of the deficiencies of the steam engine? In what general way might they be remedied?

11.2 Faraday's First Electric Motor

1. Why does the magnetic pole of Faraday's "electromagnetic rotator" move in a circle around a fixed wire?

11.3 The Discovery of Electromagnetic Induction

1. Why is Faraday considered the discoverer of electromagnetic induction?
2. What is the general definition of electromagnetic induction?

11.4 Generating Electricity: The Generator

1. What is the position of a rotating loop when it generates maximum current? minimum current? Why?
2. What is the purpose of the commutator?
3. Where does the energy delivered by the generator come from?

11.5 Putting Electricity to Work: The Motor

1. How does interaction between science and technology differ in the development of electrical technology from the development of steam-engine technology?
2. How would you make an electric motor out of a generator?
3. What prevented the electric motor from being an immediate economic success?
4. What chance event led to the beginning of the electric power age?

11.6 The Electric Light Bulb

1. What device was essential to the development of the incandescent lamp?
2. Why did Edison require a substance with a high resistance for his light-bulb filaments?
3. What were some of the major effects the introduction of electric power had on everyday life?

11.7 AC versus DC: The Niagara Falls Power Plant

1. What were some of the disadvantages of DC generation?
2. What factors made Edison's recommendation for the use of DC for the Niagara Falls system less attractive?
3. Give one reason why it is more economical to transmit electric power at high voltage and low current than at low voltage and high current.
4. Why will transformers not operate if steady DC is furnished for the primary coil?

11.8 The Energy Picture Today

1. Why is there growing concern about the energy future of the United States and the world?
2. How is the energy used by the United States economy distributed among the different sectors?
3. Where does the energy used by the United States economy come from?
4. Why is there concern about these energy sources, and what are some of their other disadvantages?

11.9 Conservation

1. Why isn't nuclear fission energy at present the answer to our energy needs?
2. What limits the efficiency of electric power plants? How efficient are they at best?

3. What are the current possibilities and prospects for reducing waste energy in personal transportation?
4. Where does the ultimate decision lie regarding conservation of energy resources?

11.10 Renewable and Alternative Energy Sources

1. List all of the various energy options discussed in this section. Describe what each one entails in your own words.
2. Using your list of energy sources in Question 1, indicate the following:
   (a) Which ones are renewable and which are not?
   (b) Which ones involve direct production of electricity?
   (c) Which ones can provide heat and light without the use of electricity?
   (d) Which ones can provide the mechanical work to run a generator?
   (e) Which ones can provide the heat to create steam to generate electricity?
3. Using your list in Question 1, list the advantages and disadvantages of each of these energy sources.
4. Why haven't any of these alternatives replaced fossil fuels as the main energy source for advanced industrial economies?

DISCOVERY QUESTIONS

1. During the course of 1 day, try to take careful note of all of the electrical appliances and devices that you use. Then try to imagine what your life would be like if none of these devices existed.
2. During the course of 1 day, try to take careful note of all of the different ways in which you use energy. Note also the types of energy you use. Then try to imagine what your life would be like if the only energy you could use is what you could obtain only from food and the natural environment.
3. What sources of energy were there for industry before the electrical age? How was the energy transported to where it was needed?
4. Oersted discovered that a magnetic needle was affected by a current. Would you expect a magnetic needle to exert a force on a current? Why? How would you detect this force?
5. In which of these cases will electromagnetic induction occur?
   (a) A battery is connected to a loop of wire that is being held near another closed loop of wire.
   (b) A battery is disconnected from a loop of wire held near another loop of wire.
   (c) A magnet is moved through a loop of wire.
   (d) A loop of wire is held fixed in a steady magnetic field.
   (e) A loop of wire is moved across a magnetic field.
6. It was stated on page 516 that the output of a DC generator can be made smoother by using multiple windings. If each of two loops set at an angle were

connected to commutators as shown, what would the output current of the generator be like?

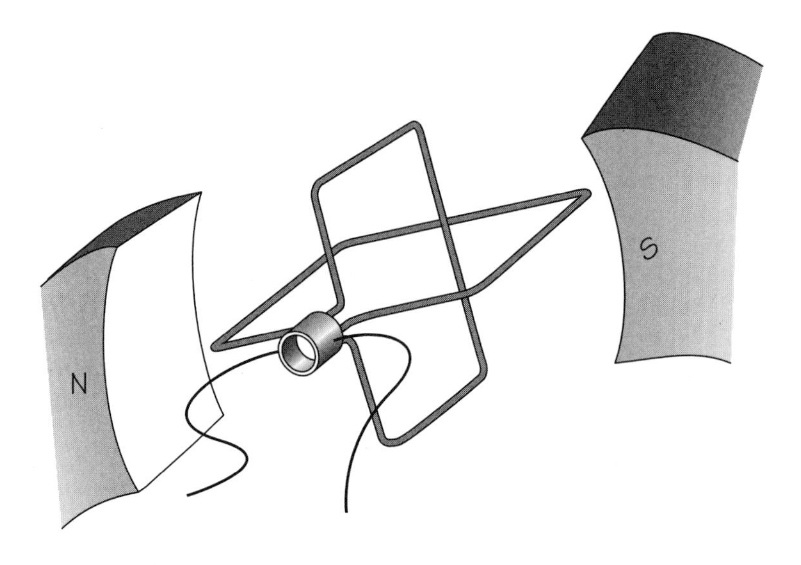

7. Trace the energy conversions in a system of a generator and motor, from heat input to work output. In what ways are the first and second laws of thermodynamics obeyed?
8. Why is a generator coil harder to turn when it is connected to an appliance to which it provides current, such as a lamp, than when it is disconnected from any appliance?
9. Suppose two vertical bar magnets, each held by one end at the same level, are dropped simultaneously. One of them passes through a closed loop of wire on the way down. Which magnet reaches the ground first? Why?
10. Comment on the advisability and possible methods of getting out of a car over which a high-voltage power line has fallen.
11. Using the data given in Section 11.8, compare the amount of energy required for the subsistence of an average human being per day with the amount of energy consumed per person per day in the United States. How would you account for the difference?
12. For each of the sources of energy listed in this chapter list the advantages and disadvantages.
13. On the basis of this chapter, and your own readings, how would you assess the outlook for the world's future energy needs? How would you assess the outlook for the available solutions, namely other energy sources and conservation?
14. Look up the history of a recent new technology, such as the Intel processor or Microsoft's computer operating systems, or a successful Internet company, and compare their sources and strategies for success with Edison's.
15. This chapter has described several instances in which science, technology, and society have interacted with each other. Summarize some of these interactions

and notice any similarities and differences. Can you draw any conclusions from this?

16. This chapter has described several great technological breakthroughs associated with fundamental scientific discoveries. Make a list of some of these discoveries and breakthroughs, then list the benefits to humankind as well as any real or potential detriments. In each case what are the respective responsibilities, if any, of the scientists, the engineers, the industries, the governors, and the general public in ensuring that the detriments are avoided and the benefits are achieved?

CHAPTER 12

Electromagnetic Waves

12.1 FARADAY'S SUGGESTION

On April 11, 1846, the distinguished physicist Charles Wheatstone was scheduled to give a lecture at the Royal Institution in London. Michael Faraday was to introduce Wheatstone to the audience. At the last minute, just as Faraday and Wheatstone were about to enter the lecture hall, Wheatstone got stage fright, turned around, and ran out into the street. Faraday had to improvise and give a lecture himself. Normally, Faraday discussed in public only his actual experiments. But on this occasion he revealed certain speculations which, as he later admitted, he would never have made public had he not suddenly been forced to speak for an hour—although these speculations soon changed physics.

Faraday's speculations dealt with the nature of light. Faraday, like Oersted before him, believed that all the forces of nature are somehow connected. Electricity and magnetism, for example, could not be separate forces that just happen to exist in the same universe. Rather, they must be different forms of one basic phenomenon. This belief paralleled that of Immanuel Kant, Friedrich Wilhelm Joseph von Schelling, and other German nature philosophers at the beginning of the nineteenth century. It had inspired Oersted to search in the laboratory for a connection between electricity and magnetism. Eventually he found such a connection in his discovery

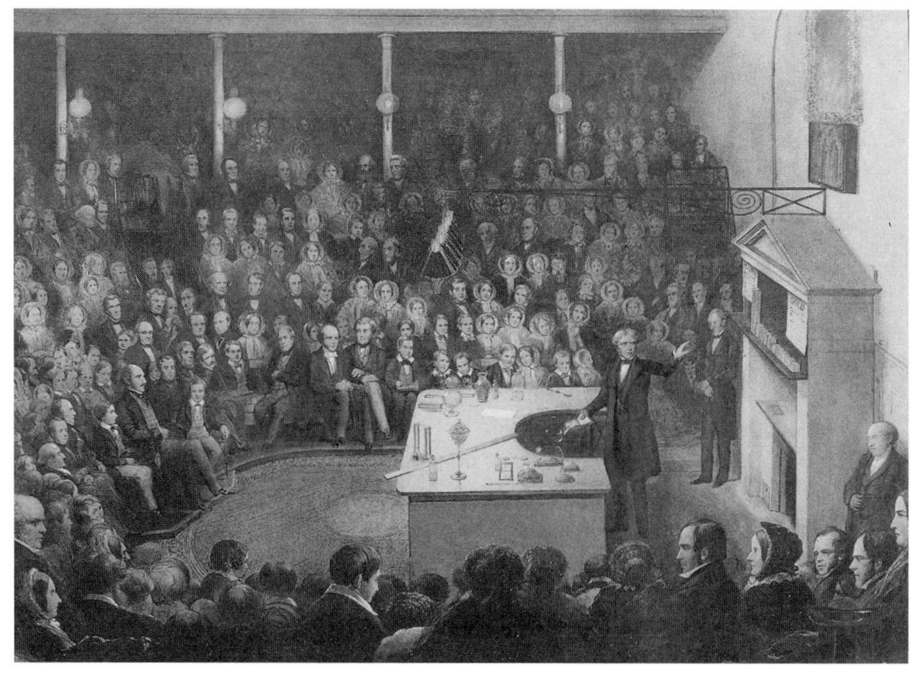

FIGURE 12.1 Faraday's Christmas Lecture at the Royal Institution, London (painting by Blaikley).

that an electric current in a conductor can turn a nearby magnet (see Chapter 10).

Faraday, too, had been guided by a belief in the unity of natural forces. Could *light* be another form of this basic "force"? Or rather, to use more modern terms, is light a form of *energy*? If so, scientists should be able to demonstrate experimentally its connection with other forms of energy such as electricity and magnetism. Faraday did succeed in doing just this. In 1845, he showed that light traveling through heavy glass had its plane of polarization rotated by a magnetic field applied to the glass.

Having shown a definite connection between light and magnetism, Faraday could not resist going one step further in his unrehearsed lecture the following year, revealing thoughts he had held privately. Perhaps, he suggested, light itself is a vibration of magnetic lines of force. Suppose, for example, that two magnetized or charged objects are connected by a magnetic or electric line of force. If one of them moved, Faraday reasoned, a disturbance would be transmitted along the line of force. Furthermore, if light waves were vibrations of lines of force, then a hypothetical elastic substance such as "ether" would not be needed in order to explain the propagation of light. The concept of the ether could be replaced if it could be shown that lines of force themselves have the elastic properties needed for wave transmission.

Faraday could not make his idea more precise. He lacked the mathematical skill needed to prove that waves could propagate along lines of electric or magnetic force. Other physicists in Britain and Europe might have been able to develop a mathematical theory of electromagnetic waves. But at the time these scientists either did not understand Faraday's concept of lines of force or did not consider them a good basis for a mathematical theory. Ten years passed before James Clerk Maxwell, a Scottish mathematical physicist, saw the value of the idea of lines of force and started using mathematics to express Faraday's concepts.

12.2 MAXWELL'S PRINCIPLES OF ELECTROMAGNETISM

The work of Oersted, Ampère, Henry, and Faraday had established two basic principles of electromagnetism:

1. *An electric current in a conductor produces magnetic lines of force that circle the conductor.*
2. *When a conductor moves across externally set-up magnetic lines of force, a current is induced in the conductor.*

In the 1860s, James Clerk Maxwell developed a mathematical theory of electromagnetism. In it, he added to and generalized these principles so that they applied to electric and magnetic fields in conductors, in insulators, and even in space free of matter. In 1855, less than 2 years after completing his undergraduate studies at Cambridge University, Maxwell had already presented to the Cambridge Philosophical Society a long paper titled "On Faraday's Lines of Force." It described how these lines are constructed:

> . . . if we commence at any point and draw a line so that, as we go along it, its direction at any point shall always coincide with that of the resultant force at that point, this curve will indicate the direction of that force for every point through which it passes, and might be called on that account a *line of force*. We might in the same way draw other lines of force, till we had filled all space with curves indicating by their direction that of the force at any assigned point.

JAMES CLERK MAXWELL

James Clerk Maxwell (1831–1879) was born in Edinburgh, Scotland, in the same year Faraday discovered electromagnetic induction. Unlike Faraday, Maxwell came from a well-off family. He was educated at the Edinburgh Academy and the University of Edinburgh. He showed a lively interest in how things happened when he was scarcely three years old. As a child, he constantly asked, "What's the go of that?" He studied mechanisms, from a toy top to a commercial steam engine, until he had satisfied his curiosity about how they worked. His formal studies, begun at the Academy in Edinburgh and continued through his work as an undergraduate at Cambridge, gave Maxwell experience in using mathematics to develop useful parallels among apparently unrelated occurrences. His first publication appeared in the proceedings of the Royal Society of Edinburgh when he was only 14 years old. In the 1870's, Maxwell organized the Cavendish Laboratory at Cambridge University, which soon became a world center for physics research. Maxwell was one of the main contributors to the kinetic theory of gases, to statistical mechanics and thermodynamics, and also the theory of color vision. His greatest achievement was his electromagnetic theory. Maxwell is generally regarded as the most profound and productive physicist between the time of Newton and Einstein.

FIGURE 12.2

Maxwell stated that his paper was designed to

> show how, by a strict application of the ideas and methods of Faraday, the connection of the very different orders of phenomena which he has discovered may be clearly placed before the mathematical mind.

During the next 10 years, Maxwell created his own models of electric and magnetic induction. In developing his theory, he first proposed a mechanical model for the electrical and magnetic quantities observed experimentally by Faraday and others. Maxwell then expressed the operation of the model in a group of equations that gave the relations between the electric and magnetic fields. He soon found these equations to be the most useful way to represent the theory. Their power allowed him eventually to discard the mechanical model altogether. Maxwell's mathematical view is still considered by physicists to be the proper approach to the theory of electromagnetic phenomena. If you go on to take another physics course after this introductory one, you will find the development of Maxwell's mathematical model (Maxwell's equation, using vector calculus) to be one of the high points of the course.

Maxwell's work contained an entirely new idea of far-reaching consequences: *An electric field that is changing with time must be accompanied by a magnetic field.* Not only do steady electric currents passing through conductors (a "conduction current") produce magnetic fields around the conductors, but changing electric fields in insulators such as glass, air, or even empty space also produce magnetic fields.

It is one thing to accept this newly stated connection between electric and magnetic fields. But it is harder, and more interesting, to *understand* the physical necessity for such a connection. The remainder of this section is intended to make it clearer.

An uncharged insulator (such as glass, wood, paper, or rubber) contains equal amounts of negative and positive charges. In the normal state, these charges are distributed evenly. Thus, the *net* charge is zero in every region of the material. But when the insulator is placed in an electric field, these charges are subjected to electrical forces. The positive charges are pushed in one direction, the negative in the opposite direction. Unlike the charges in a conductor, the charges in an insulating material are *not* free to move far through the material. The charges can be displaced only a small distance before restoring forces in the insulator balance the force of the electric field. If the strength of the field is increased, the charges will be displaced further. But the changing displacement of charges that accompanies a changing electric field in an insulator briefly forms a *current*. Maxwell called this current a *displacement current*. He assumed that this momentary displacement current in an insulator surrounds itself with a magnetic field just as a conduction current of the same magnitude does.

In an insulator, the displacement current is defined as *the rate at which the charge displacement changes.* This rate is directly proportional to the rate at which the electric field is changing in time. Thus, the magnetic field that circles the displacement current can be considered a consequence of the time-varying electric field. Maxwell assumed that this model, developed for matter, also applies to space free of matter (though at first glance this may seem absurd). Therefore, under all circumstances, as noted above, an electric field that is changing with time surrounds itself with a magnetic field. Previously, it was thought that the only current that produced a magnetic field was the current in a conductor. Now Maxwell predicted that a magnetic field would also arise from a changing electric field, even in empty space. Unfortunately, this field was very small in comparison to the magnetic field produced by the current in the conductors of the apparatus. So it was not at that time possible to measure it directly. But as you will see, Maxwell predicted consequences that soon could be tested.

According to Maxwell's theory, then, the two basic principles of electromagnetism should be expanded by adding a third principle:

3. *A changing electric field in space produces a magnetic field.*

The induced magnetic field vector **B** is in a plane perpendicular to the changing electric field vector **E**. The magnitude of **B** depends on the rate at which **E** is changing—not on **E** itself, but on $\Delta \mathbf{E}/\Delta t$. Therefore, the higher the rate of alteration of **E**, the greater the field **B** so induced.

For instance, consider a pair of conducting plates connected to a source of current. Charges are moved onto or away from plates through the conductors connecting them to the source. Thus, the strength of the electric field **E** in the space between the plates changes with time. This changing electric field produces a magnetic field **B** as shown. (Of course, only a few of the infinitely many lines for **E** and **B** are shown.)

An additional principle, known before Maxwell, assumed new significance in Maxwell's work because it is so symmetrical to Statement 3 above:

4. *A changing magnetic field in space produces an electric field.*

The induced electric field vector **E** is in a plane perpendicular to the changing magnetic field vector **B**. Similarly to Principle 3, the magnitude of **E** depends on the rate at which **B** is changing—not on **B** itself, but on $\Delta \mathbf{B}/\Delta t$. For instance, consider the changing magnetic field produced by temporarily increasing the current in an electromagnet. This changing magnetic field

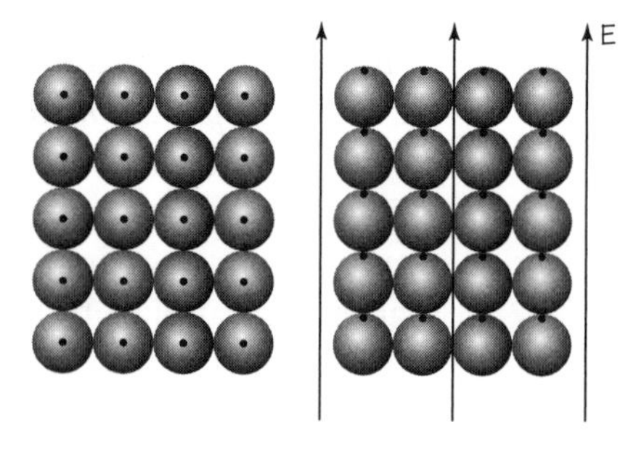

FIGURE 12.3 When an electric field is set up in an insulating material (as in the diagram here), the positive and negative charges, which are bound to one another by attraction, are displaced. This displacement forms a current (the positive charges are represented by dots, and the negative charges by shaded circles).

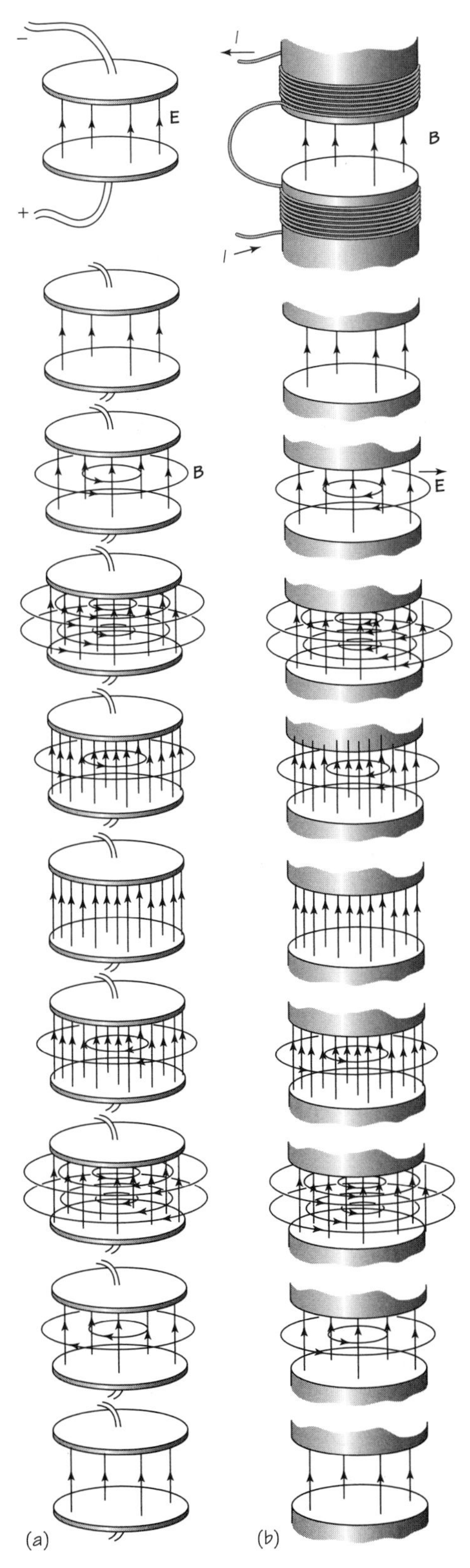

FIGURE 12.4 (a) A changing electric field produces a magnetic field: When the electric field **E** between a pair of charged plates starts to increase in intensity, a magnetic field **B** is induced. The faster **E** changes, the more intense **B** is. When **E** momentarily has reached its maximum value, **B** has decreased to zero momentarily. When **E** diminishes, a **B** field is again induced, in the opposite direction, falling to zero as **E** returns to its original strength. (b) A changing magnetic field produces an electric field. When the magnetic field **B** between the poles of an electromagnet starts to increase, an electric field **E** is induced. The faster **B** changes, the more intense **E** is. When **B** momentarily has reached its maximum value, **E** has decreased to zero momentarily. When **B** diminishes, an **E** field is again induced, in the opposite direction, falling to zero as **B** returns to its original strength.

induces an electric field in the region around the magnet. If a conductor happens to be lined up in the direction of the induced electric field, the free charges in the conductor will move under the field's influence. Thus, a current in the direction of the induced field will arise in the conductor. This electromagnetic induction had been discovered experimentally by Faraday (Section 11.3).

Maxwell's theory, involving the total set of relations between electric and magnetic fields in space, was not at once directly testable. When the test finally came, it concerned his prediction of the existence of waves traveling as interrelating electric and magnetic fields, that is, as *electromagnetic waves.*

12.3 THE PROPAGATION OF ELECTROMAGNETIC WAVES

Suppose in a certain region of space, an electric field is created that changes with time. According to Maxwell's theory, an electric field **E** that varies in time simultaneously induces a magnetic field **B** that also varies with time. (The strength of the magnetic field also varies with the distance from the region where the changing electric field was created.) Similarly, a magnetic field that is changing with time simultaneously induces an electric field that changes with time. (Here, too, the strength of the electric field also changes with distance from the region where the changing magnetic field was created.) The electric and magnetic field changes occur together, much like the "action" and "reaction" of Newton's third law.

As Maxwell realized and correctly predicted, the mutual induction of varying electric and magnetic fields should set up an unending sequence of events. First, a time-varying electric field in one region produces a time-

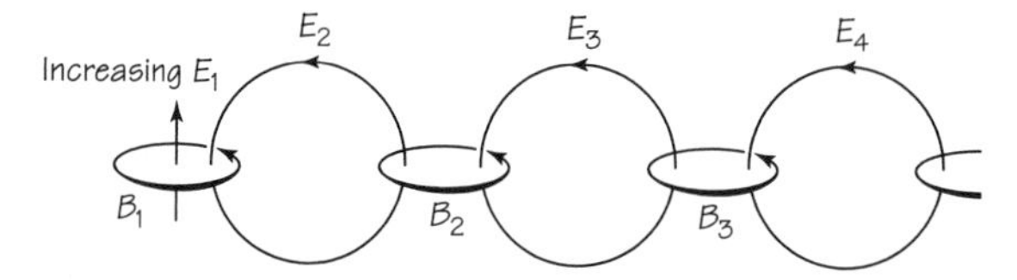

FIGURE 12.5 Electric and magnetic fields linked by induction. An increasing electric field at the left (or current) surrounds itself with a magnetic field. As the current changes, it induces an interlinking electric field. The chain-like process continues with finite velocity. This is only a symbolic picture of the process, which propagates itself in all directions in space.

and space-varying magnetic field at points near this region. But this *magnetic* field produces a time- and space-varying *electric* field in the space surrounding it. And this electric field produces time- and space-varying magnetic fields in its neighborhood, and so on. Thus, suppose that an electromagnetic disturbance is started at one location, say by vibrating charges in a hot gas or in the transmitter wire of a radio or television station. This disturbance can travel to distant points through the mutual generation of the electric and magnetic fields. The fluctuating, interlocked electric, and magnetic fields *propagate* through space as a *wave*. This wave is an *electromagnetic wave*, a disturbance in the electric and magnetic field intensities in space.

Maxwell had shown that in an electromagnetic disturbance **E** and **B** should be perpendicular to each other and to the direction of propagation of the wave. Therefore, in the language of Chapter 8, electromagnetic waves are *transverse*. And as was noted in Chapter 8, it was long known that light waves are transverse.

In Part One, Chapter 8, we showed that waves occur when a disturbance created in one region produces at a later time a disturbance in adjacent regions. Snapping one end of a rope produces, through the action of one part of the rope on the other, a displacement at points farther along the rope and at a later time. Dropping a pebble into a pond produces a disturbance that moves away from the source as one part of the water acts on neighboring parts. Analogously, time-varying electric and magnetic fields produce a disturbance that moves away from the source as the varying fields in one region create varying fields in neighboring regions.

The Speed of Electromagnetic Waves

What determines the speed with which electromagnetic waves travel? For mechanical waves the speed of propagation is determined by the stiffness and density of the medium (see Section 8.3). Speed increases with increasing stiffness, but decreases with increasing density. It is given by the expression

$$\text{speed} \propto \frac{\text{stiffness}}{\text{density}}.$$

This relationship between wave speed, stiffness, and density holds for mechanical wave motions and for many other types of waves. Only the barest outline of how Maxwell proceeded beyond this point is given here. First, he assumed that a similar "stiffness and density" relation would hold for electromagnetic waves. Then he computed what he thought to be the "stiffness" and "density" of electric and magnetic fields propagating through the hypothetical ether. In finding values for these two properties of the elec-

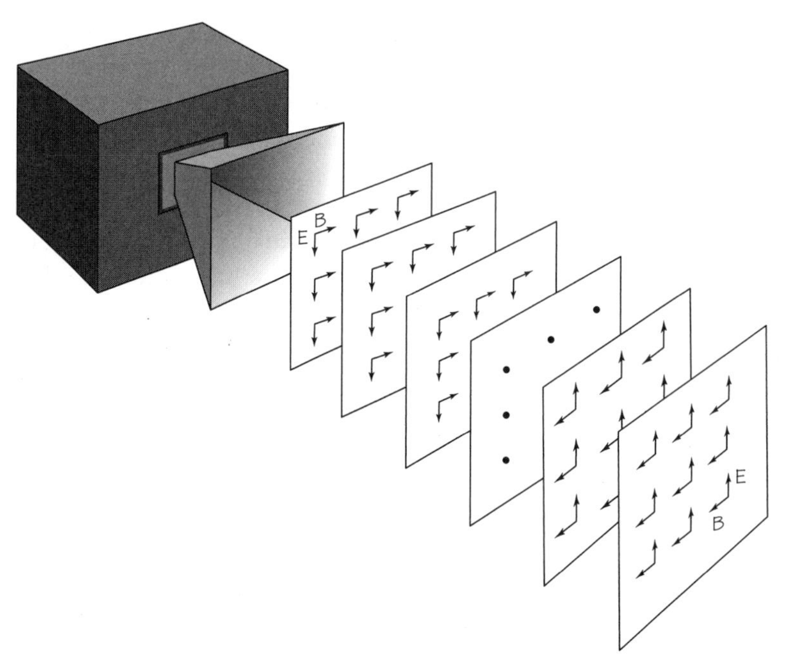

FIGURE 12.6 In a microwave oscillator, which you may see in your laboratory work, electric oscillations in a circuit are led onto a rod in a metal "horn." In the horn, they generate a variation in electric and magnetic fields that radiates away into space. This drawing represents an instantaneous "snapshot" of almost plane wave fronts directly in front of such a horn.

tric and magnetic fields, Maxwell was guided by his mechanical model representing the ether. In this model, stiffness was related to the electric field, and density to the magnetic field. Next, he proved mathematically that the ratio of these two factors, which should determine the wave speed, is the same for all strengths of the fields. Finally, Maxwell demonstrated that the speed of the waves (if they exist!) is a definite quantity that can be deduced from measurements in the laboratory.

The necessary measurements of the factors involved actually had been made 5 years earlier by the German scientists Weber and Kohlrausch. Using their published values, Maxwell calculated that the speed of the supposed electromagnetic waves (in metric units) should be about 311,000,000 m/s. He was immediately struck by the fact that this large number was very close to a measured speed already well known in physics. In 1849, Armand Fizeau had measured the speed of *light* and had obtained a value of about 315,000,000 m/s. (Today, the measured speed of light in vacuum is known to be 299,792,458 m/s.) The close similarity could have been a chance occurrence. But Maxwell believed that there must be a deep underlying reason for these two numbers being so nearly the same. The significance for physics seemed obvious to him. Making an enormous leap of the imagination, he wrote:

> The velocity of the transverse undulations in our hypothetical medium, calculated from the electromagnetic experiments of MM. Kohlrausch and Weber, agrees so exactly with the velocity of light calculated from the optical experiments of M. Fizeau, that we can scarcely avoid the inference that *light consists in the transverse undulations of the same medium which is the cause of electric and magnetic phenomena*.

It was already long known that light waves are transverse. When Maxwell found that in an electromagnetic disturbance **E** and **B** should be perpendicular to each other and to the direction of propagation of the wave, he concluded that electromagnetic waves are also transverse.

Here, then, in Maxwell's statement, was *an explanation of light waves and at the same time a joining of the previously separate sciences of electricity and magnetism with optics—a new synthesis.* Maxwell realized the importance of his discovery. Now he set to work making the theory mathematically sound and freeing it from his admittedly artificial model based on the ether hypothesis.

Classical Physics

Maxwell summarized his synthesis of electromagnetism and optics in his monumental *Treatise on Electricity and Magnetism*, published in 1873. After the experimental confirmation of his work (see Section 12.14), Maxwell's synthesis was seen as a great event in physics. In fact, physics had known no greater time since the 1680s, when Newton was writing his monumental work on mechanics. Of course, Maxwell's electromagnetic theory had arisen in Maxwell's mind in a Newtonian, mechanical framework. But it had grown out of that framework, becoming another great general physical theory, independent of its mechanical origins.

Like Newtonian mechanics, Maxwell's electromagnetic field theory succeeded spectacularly. You will see something of that success in the next few sections. The success occurred on two different levels: the practical and the theoretical. Practically, it led to a host of modern developments, such as radio and television. On the theoretical level, it led to a whole new way of viewing phenomena. The Universe was not only a Newtonian machine of whirling and colliding parts; it included fields and energies that no machine could duplicate. Maxwell's work formed a basis of the special theory of relativity. Other physical theories were nourished by it also. In a sense, the work of Maxwell and Newton, as well as that of Carnot and other founders of thermodynamics, enabled a fairly complete understanding of events in the physical world that surrounds us, from the motions of space satellites, cars, and atoms in gases to the behavior of light and other electromagnetic waves. The physics of the everyday, visible world, based upon the work of Newton and Maxwell, has remained to this day and is often known as *classical physics*. It is based on three steps of verification: Galileo's and Newton's verification of terrestrial and celestial phenomena; Oested's fusion of electric and magnetic phenomena; and Maxwell's addition of light to Oested's verification

Eventually, however, as research pushed into unfamiliar realms of nature at the scale of the very small (inside atoms), the very fast (approaching the speed of light), and the very large (the size of the Universe), results accumulated that could not be explained using classical physics. Something more was needed. Starting about 1925, after a quarter century of discovery, the development of quantum mechanics led to a larger synthesis, which included Maxwell's electromagnetism (see Chapter 15).

12.4 HERTZ'S EXPERIMENTAL CONFIRMATION

Did Maxwell himself establish without doubt that light actually does consist of electromagnetic waves, or even that electromagnetic waves exist at all? No. Most physicists remained skeptical for several years. The fact that the ratio of two quantities determined by electrical experiments came out nearly equal to the speed of light certainly suggested *some* connection between electricity and light. No one would seriously argue that this was only a coincidence. But stronger evidence was needed before the rest of Maxwell's theory, with its notion of the displacement current, could be accepted.

What further evidence was needed to persuade physicists that Maxwell's theory was correct? Maxwell showed that his theory could explain all the known facts about electricity, magnetism, and light. But so could other theories, although with less sweeping connections between their separate parts. To a modern physicist, the other theories proposed in the nineteenth century seem much more complicated and artificial than Maxwell's. But at the time, Maxwell's theory seemed strange to physicists who were not accustomed to thinking in terms of fields. It could be accepted over other theories only if it could be used to predict best some newly discovered property of electromagnetism or light.

Maxwell himself made two such predictions from his theory. He did not live to see them verified experimentally in 1888, for he had died in 1879, at the age of 48. Maxwell's most important prediction was that electromagnetic waves of many different frequencies could exist. All such waves would propagate through space at the speed of light. Visible light itself would correspond to waves of only a small range of high frequencies (from 4×10^{14} Hz to 7×10^{14} Hz), the range of frequencies detectable by the human eye. (Recall that the unit "cycles per second" is called the hertz, symbol Hz, after Heinrich Hertz.)

To test Maxwell's predictions required inventing apparatus that could both produce and detect electromagnetic waves, preferably of frequencies other than light frequencies. This was first done by the German physicist Heinrich Hertz, whose contribution was triggered by a chance observation. In 1886, Hertz noticed a peculiar effect produced during the sparking of an induction coil. As was well known, sparks sometimes jump the air gap between the terminals of an induction coil (see drawing). You will recall (Chapter 11) that an induction coil can be used to produce high voltages if there are many more turns of wire on one side than on the other. Ordinarily, air does not conduct electricity. But when there is a very large potential difference between two wires a short distance apart, a conducting pathway may form briefly as air molecules are ionized. A short burst of electricity then may pass through, attended by a visible spark. Each visible spark produced is actually a series of many small sparks, jumping rapidly back and forth (oscillating) between the terminals. Hertz found that he could control the spark's frequency of oscillation by changing the size and shape of metal plates attached to the spark gap of the induction coil.

Hertz then took a simple piece of wire and bent it so that there was a short gap between its two ends. When it was held near an induction coil, a *spark jumped across the air gap in the wire just when a spark jumped across the terminals of the induction coil.* This was a surprising phenomenon. To explain it, Hertz reasoned that as the spark jumps back and forth across the gap of the induction coil, it must set up rapidly changing electric and magnetic

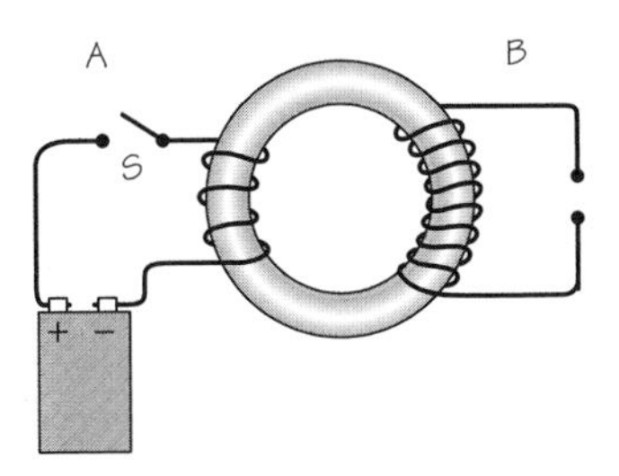

FIGURE 12.7 Operation of the induction coil: Starting and stopping the current in coil A with a vibrating switch S produces a rapidly changing magnetic field in the iron core. This rapidly changing field induces high-voltage peaks in the many-turn coil B and can cause a spark to jump across the air gap. Spark coils for use in car engines operate this way.

FIGURE 12.8 Heinrich Hertz (1857–1894) was born in Hamburg, Germany. During his youth, Hertz was mainly interested in languages and the humanities, but he was attracted to science after his grandfather gave him some experimental apparatus. Hertz did simple experiments in a small laboratory which he had fitted out in his home. After completing secondary school (and a year of military service) he undertook the serious study of mathematics and physics at the University of Berlin in 1878. In 1882, Hertz devoted himself to the study of electromagnetism, including the recent and still generally unappreciated work of Maxwell. Two years later he began his famous series of experiments with electromagnetic waves. During the course of this work, Hertz discovered the photoelectric effect, which has had a profound influence on modern physics. His early death is thought to have been caused by poisoning of vapor from mercury, then much used in laboratories without precautions.

fields. According to Maxwell's theory, these changes propagate through space as electromagnetic waves. (The frequency of the waves is the same as the frequency of oscillations of the sparks.) When the electromagnetic waves pass over the bent wire, which acted as a detector, they set up rapidly changing electric and magnetic fields there, too. A strong electric field around the detector produces a spark in its air gap, just as the transmitter field did between the terminals of the induction coil. Since the field is rapidly changing, sparks can jump back and forth between the two ends of the wire. Hertz's observation of the induced spark in the detector was the first solid clue that electromagnetic waves exist.

Hertz showed that the electromagnetic radiation coming from his induction coil has all the usual properties of light waves. It can be reflected at the surface of solid bodies, including metallic conductors. In addition, the angle of reflection is equal to the angle of incidence. The electromagnetic radiation can be focused by concave metallic mirrors. It shows diffraction when it passes through an opening in a screen. It displays all interference phenomena, including standing waves. Also, electromagnetic waves are refracted by prisms made of glass, wood, plastic, and other nonconducting material. By setting up a standing-wave pattern by using a large metal reflector, Hertz was also able to determine the distance between consecutive nodes and thus measure the wavelength. He determined the frequency of the oscillating electric current through an analysis of his circuits. Thus, he was able to determine the speed of his waves and found it to be the same value that Maxwell had predicted: the speed of light!

Hertz's experiments dramatically confirmed Maxwell's electromagnetic theory, by showing that electromagnetic waves actually exist, that they do travel with the speed of light, and that they have the familiar characteristics of light. Now physicists rapidly accepted Maxwell's theory and applied it with great success to the detailed analysis of a wide range of phenomena.

Thus, at the end of the nineteenth century, Maxwell's electromagnetic theory stood with Newton's laws of mechanics as an established part of the foundations of physics.

12.5 THE ELECTROMAGNETIC SPECTRUM

Hertz's induction coil produced electromagnetic radiation with a wavelength of about 1 m. This is about one million times the wavelength of visible light. Later experiments showed that a very wide and continuous range of electromagnetic wavelengths (and frequencies) is possible. The entire possible range is called the *electromagnetic spectrum*. The electromagnetic spectrum should not be confused with the *visible spectrum*, which includes only the frequencies of visible light. In principle, the electromagnetic spectrum ranges from close to 0 Hz to infinite Hz, but in practice the range of frequencies from about 1 Hz to 10^{26} Hz, corresponding to wavelengths in the range from 10^8 m to 10^{-18} m, has been studied. Many of these frequency regions have been put to practical use.

As shown in the illustration, light, heat, radio waves, and X rays are names given to radiations in certain regions of the electromagnetic spectrum. In each of these regions radiation is produced or observed in a particular way. For example, visible light may of course be perceived directly through its effect on the retina of the eye. But to detect radio waves requires electronic equipment. The named regions overlap. For example, some radiation is called "ultraviolet" or "X ray," depending on where it lies on the total spectrum or how it is produced.

All waves in the electromagnetic spectrum, although produced and detected in various ways, behave as predicted by Maxwell's theory. All electromagnetic waves travel through empty space at the same speed—the speed of light, 3×10^8 m/s. They all carry energy; when they are absorbed, the absorber is heated, as, for example, is food in a microwave oven. Electro-

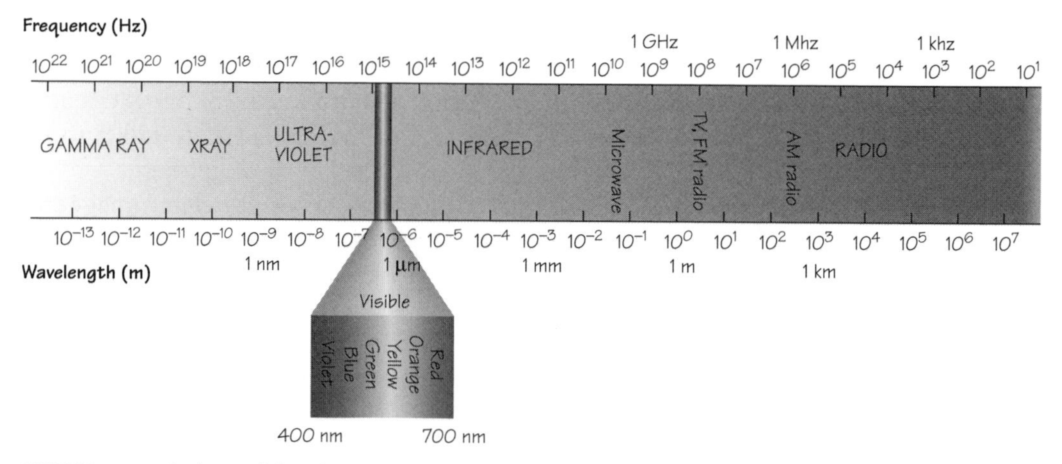

FIGURE 12.9 A chart of the electromagnetic spectrum, with visible light occupying the range between 400 nm and 700 nm in wavelength (1 nanometer = 10^{-9} m).

magnetic radiation, whatever its frequency, can be emitted only if energy is supplied to the source of radiation, which is, ultimately, a charge that is undergoing acceleration. This charge acceleration can be produced in many ways. For example, heating a material will increase the vibrational energy of charged particles. Also, one can vary the motion of charges on an electric conductor—an *antenna*—or cause a charged particle to change its direction. In these and other processes, work is done by the force that is applied to accelerate the electric charge. Some of the energy supplied to the antenna in doing this work is "radiated" away; that is, it propagates away from the source as an electromagnetic wave.

The work of Maxwell and Hertz opened up a new scientific view of nature. It also prepared for a rapid blooming of new technologies, such as radio, TV, radar, etc. We review below some of the indirect, technological consequences of a scientific advance.

Radio waves (λ = 10 m to 10,000 m; f = 10^4 Hz to 10^7 Hz). Electromagnetic waves in this region are reflected quite well by electrically charged layers of ions that exist in the upper atmosphere. This reflection makes it possible to detect radio waves at great distances from the source. Since radio signals have wavelengths from tens to thousands of meters, such waves can easily diffract around relatively small obstacles such as trees or buildings. But large hills and mountains may cast "dark" shadows.

Radio waves that can cross large distances, either directly or by relay, are very useful for carrying information. They are used not only for radio transmissions but also to carry cellphone communications via geographic "cells" centered around a single radio transmitter. A cellphone is technically a cellular radio transceiver, since it receives and sends radio signals.

In December 1901, the Italian inventor Guglielmo Marconi successfully detected radio waves sent from Newfoundland to Ireland. Marconi's work, perfected later on Cape Cod, Massachusetts, showed that long-distance radio communication was possible, because the waves were reflected by the previously unsuspected layers of ionized particles in the upper atmosphere and therefore could be received at great distances despite the curvature of the Earth's surface.

Radio communication is accomplished by changing the signal according to an agreed code that can be deciphered at the receiving end. The first radio communication was achieved by turning the signal on and off in an agreed pattern, such as Morse code (which is recognized today as a digital code). Later, sounds were coded by continuous variations in the amplitude (i.e., the intensity) of the broadcast wave. This is known as amplitude modulation, or AM. Later still, the information was coded as frequency variations in the broadcast wave, known as frequency modulation, or FM. In broadcast radio and television, the "decoding" is done in the receiver serving the loudspeaker or TV monitor, since the output message from the receiver takes the same form that it had at the transmitter. Radio stations regularly announce their frequencies in megahertz (MHz) for the FM band and kilohertz (kHz) for the AM band.

Because signals from different radio stations should not be received at the same spot on the dial, it is necessary to apportion the allowed frequencies of transmission within a region covered by radio signals. The International Telecommunication Union (ITU) controls radio transmission and other means of international communication. Within the United States, the Federal Communications Commission (FCC) regulates radio transmission. In order to reduce the interference of one station's signal with another, the FCC assigns suitable frequencies to radio stations (and other

FIGURE 12.10 Radio waves bouncing off the ionosphere.

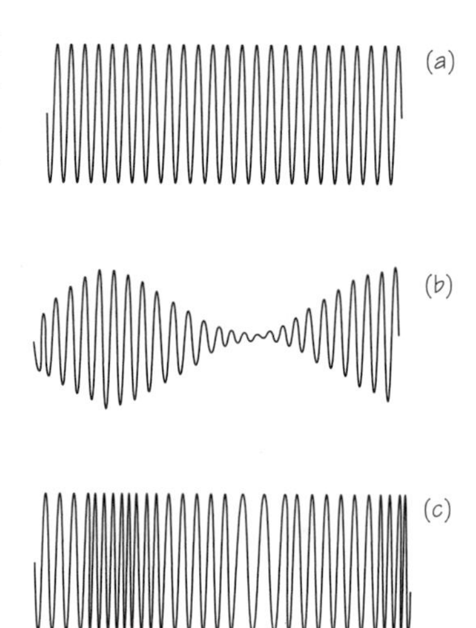

FIGURE 12.11 (a) A "carrier" radio wave; (b) AM (amplitude modulation): information is coded as variations in the amplitude (or intensity) of the carrier; (c) FM (frequency modulation): information is coded as variations in the frequency of the carrier.

transmitters). It also limits their power or the power radiated in particular directions, and may restrict the hours of transmission.

Television, FM, and Radar (λ about 1 m; f about 10^8 Hz). Waves at high frequencies of about 10^8 Hz are not reflected by the layers of electric charge in the upper atmosphere known as the *ionosphere*. Rather, the signals travel in nearly straight lines and pass into space. Thus, they can be used in communication between the Earth and orbiting satellites. But on Earth, TV signals cannot be received directly between points more than about 80 km apart, even if there are no mountains in the way. Instead, communications satellites are used to relay the signals, either directly to the receiver in a home equipped with a satellite dish, or to a cable-company receiver, which then relays the signal to its customers over a large region using cables.

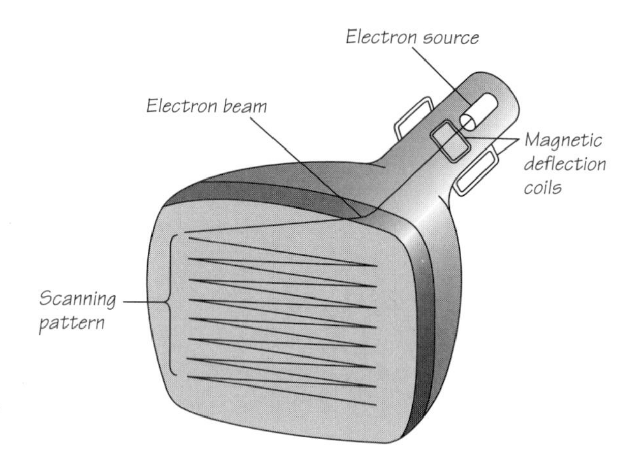

FIGURE 12.12 Schematic diagram of a TV picture tube.

■ TELEVISION

How It Works

Television—nearly every family in the United States owns at least one set, yet most do not know how it works.

The possibility of television emerged on the technological horizon with two developments: the discovery of *phosphors*, substances which glow with visible light when exposed to radiation, for example, when hit by a beam of electrons; and the use of radio waves to send signals from a broadcast station which could then be received in homes and be made to control thin beams of electrons. Combining these two elements allowed researchers, over the course of several decades, to develop one of the world's most widely used household appliances.

All television sets, except the recent plasma screens and digital sets, operate on the same general principles. A composite video signal, received from a broadcasting station or television cable, incorporates three parts: the luminance signal, which dictates the intensity, or brightness, of the electron beam; the chrominance signal, a sine wave that incorporates information about which colors the resulting image should be made up of; and the synchronization signal, which synchronizes the other signals to ensure they scan simultaneously.* Black and white televisions lack the chrominance signal, and in these, the intensity of the beam determines different shades of black and white on the screen.

The received *video* signal feeds into the cathode ray tube (CRT), the central feature of standard television sets. A "cathode" refers to a negatively charged end, and in the CRT, the cathode is a heated filament located inside a glass tube containing a vacuum. Electrons come off the cathode into the vacuum, forming a narrow beam, which is attracted by one or more positively charged "anodes." Inside a television's CRT, one anode focuses the beam while another accelerates the electrons from one end of the tube to the other, where they hit a flat screen at the end. The phosphor-coated screen glows briefly at whichever spot the electron beam hits it.

Engineers developed a way to control where the electron beam hits the screen by placing copper deflection coils inside the tube; these coils create magnetic fields that shift the beam in different directions. The electron beam quickly scans the screen line by line, eventually forming a still picture composed of 525 thin horizontal lines, which our brains interpret as a whole image. This scanning process occurs 60 times every second, although television sets now use a process called interlacing, which means that the beam only "paints" every other line of the screen each scan, but the eye processes images at a speed too slow to notice this effect.

On the black and white set, the flat screen at the end of the CRT is coated with a white phosphor which glows in different shades of white or gray each place the electrons hit it. Color television sets contain some additional features. Inside the CRT, three electron beams scan the screen simultaneously. The screen is coated with red, green, and blue phosphors arranged in tiny dots or stripes. A thin metal screen, the shadow mask, which contains small holes aligned with the phosphors, covers the screen to ensure that each electron beam strikes only phosphor stripes of the right color. The different electron beams directed to phosphors producing different colors scan the screen simultaneously, and light up various combinations of tiny red, green, and blue dots, or pixels (short for "picture elements"), which we see assembled into a fully colored picture. If you use a magnify-

---

* D. Macauley, *The Way Things Work* (Boston, MA: Houghton Mifflin, 1998), p. 246.

ing glass close to a television screen, you will probably be able to see the individual pixels that make up the image.

Through this process, a television set composes rapidly one still image after another on the screen at the rate of at least 15 images, or frames, per second, which your brain perceives as a fluid moving scene.

TELEVISION: THE INVENTION

The research and development of television technology happened along two paths, mechanical and electronic.

Mechanical Television

Television was inspired by human vision, in which thousands upon thousands of electrical circuits are used to relay information in the optic nerve from the retina to the brain. Designs based on so many electrical circuits, however, were far too complicated to ever reach fruition. In 1880 Maurice Leblanc in France and W.E. Sawyer in the United States suggested instead that if each element of a picture could be rapidly scanned, electrical transmission of pictures could be achieved using only one circuit between the transmitter and receiver. Paul Nipkow of Germany adopted this scanning technique in 1884.

The picture to be televised is focused on a rotating disk. When the first aperture in the disk has scanned a line of the picture, the next aperture scans the parallel line directly below. At each rotation of the disk another line is scanned until the whole picture has been examined. In this design, more apertures mean more lines, and hence greater detail of the picture. As it passes through each aperture, the light differs according to the light and shade of the picture, it then passes through a photoelectric cell where it is changed into an electrical image, and then translated into electrical impulses. These impulses are then sent down a circuit to a receiver, where they pro-

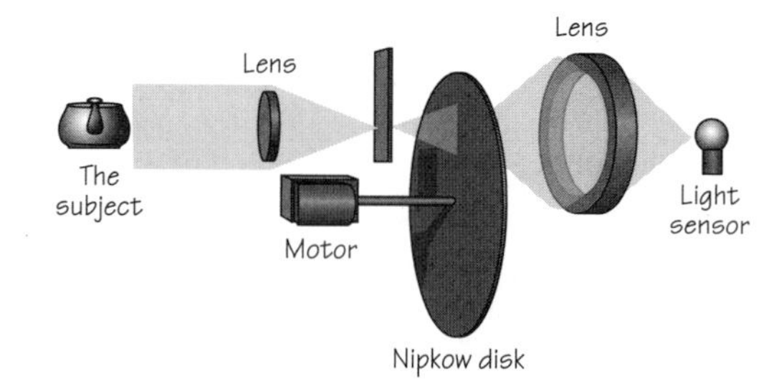

FIGURE 12.13 Diagram of Nipkow disk apparatus.

duce light in something like a gas-discharge lamp. The light from this lamp is then projected onto another disk, similar to the transmitter, and by a reversal of the transmission process, the brightness of each segment is reproduced and the original picture is reconstructed. In order for this system to work, the rotation of the disks must be synchronized, and provided they are rotating at a sufficient speed, persistence of vision allows the observer to see a whole image rather than a series of moving segments.

Until the arrival of electronic scanning, it was this mechanical system, characterized by the Nipkow disk, which dominated the development of television. However, the system was crude, with the small number of lines resulting in poor definition and the small number of rotations causing a flicker. Therefore research turned toward electronic systems.

Electronic Systems

One of the leading players in the development of electronic television was Vladimir Kosma Zworykin. He was convinced that electronic television was the path he should follow, and while working for the Radio Corporation of America (RCA), he developed the Iconoscope which used a CRT for transmitting images.

At the wide end of the Ionoscope is a sheet of mica, on one side of which is a signal plate, and the other a silver mosaic treated with cesium vapor and oxygen, in which each element is encircled by an aluminum oxide insulator. This combination of elements provides a surface from which electrons are easily liberated when exposed to light. When an image is focused onto the mosaic surface, it takes on a positive charge that corresponds to the light distribution. When the electron beam passes over the mosaic, the charge of each section changes in proportion to the amount of light that is falling on it. This change is transferred to the signal plate, which takes on a series of voltages corresponding to the light along the particular line.

Although it marked a major breakthrough in electronic television, the Iconoscope was not sensitive enough to produce consistently clear images, and in 1940 it was replaced by the Orthicon, based on a similar design but with a rigid set of squares. The Orthicon, however, was made obsolete in the same year, by a tube developed by Corning Glass Company, which was three to ten times more sensitive than the Orthicon and thirty to five hundred times more sensitive than the Iconoscope.

FIGURE 12.14 Vladimir Zworykin, one of the inventors of television, is shown here holding a cathode-ray tube.

At the beginning of the 1950s, television technology based on the electrical system had advanced so far that engineers began looking toward developing a color system. There were two possibilities. The first was a system based on a frame-by-frame sequential transmission of signals, each corresponding to the primary colors, but this design was incompatible with the current black and white transmissions. The more complex system proposed at the time was one in which signals representing the three primary colors were transmitted simultaneously. This second system was developed by the National Television Systems Committee (NTSC) in the United States and is now the basis for color systems all over the world. By the early 1960s, color television was becoming a consumer success in the United States.

Recent developments in television technology have focused on the development of High Definition (HDTV), in particular on the use of digital technology. Instead of conventional analog technology, which transmits signals in the form of waves, the new digital HDTV system transmit pictures as digital data, which is then translated by computers within the digital television. As well as clearer pictures and better sound, digital television also has the potential to transform and manipulate pictures as well as to receive them, and could perform the job of both a computer and a television set.

Further Reading

D.E. Fisher and M.J. Fisher, *Tube: The Invention of Television* (New York: Harcourt Brace, 1996).

Television, in which both sound and picture (in three primary colors) are transmitted, uses both frequency and amplitude modulations. The frequency of the wave is changed in a way that is analogous to the sound, while the picture is transmitted via amplitude modulations. This is called an *analogue wave*. However, in a recent development, the analogue TV signals are gradually being replaced by *digital* signals for digital TV. Here the analogue sound and picture waves are approximated by series of 1's and 0's (or, on and off voltages) that are converted into electromagnetic pulses and compressed at the sender, then transmitted to the receiver where they are decompressed and reconverted into continuous sound and light waves.

Signals at wavelengths of only about 1 m are not diffracted much around objects that have dimensions of several meters, such as cars, ships, or aircraft. Thus, the reflected portions of signals of wavelengths from 1 m down to 1 mm can be used to detect such objects. The interference between the direct waves and reflection of these waves by passing airplanes can distort a television picture considerably. The signal also may be radiated in the form of pulses. If so, the time t between the emission of a pulse and the reception of its echo measures the distance l of the reflecting object ($l = 2\ c/t$). This technique is called "*ra*dio *d*etection *a*nd *r*anging," or *radar* (see Section 8.9). By means of the reflection of a beam that is pulsed, both the direction and distance of an object, such as an aircraft, can be measured. This helps enormously in regulating traffic at busy airports. But initially it had an even more important role in alerting fighters in the United Kingdom during World War II of the approach of German aircraft, e.g., during the "Blitz" meant to destroy London.

Microwave radiation ($\lambda = 10^{-1}$ m to 10^{-4} m; $f = 10^9$ Hz to 10^{12} Hz). Electromagnetic waves in this region also do not bounce off the ionosphere, but instead pass easily right through it. These waves can thus be used for communicating with devices far beyond the Earth's atmosphere, such as those sent to explore space.

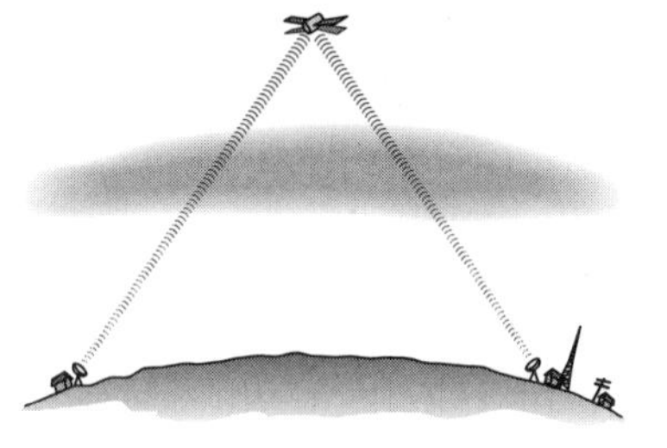

FIGURE 12.15 Satellites are used to relay microwaves all over the world. The microwaves can carry radio or TV information. (Not to scale.)

Microwave radiation also interacts strongly with the charged particles in ordinary matter, and thus has uses other than communication. When irradiated by microwaves, the matter absorbs the energy in the microwaves. This behavior is used in microwave ovens, in which the kinetic energy of the oscillating charges in food appears as heat, warming the food very quickly. Water, for example, readily absorbs radiation with a wavelength on the order of 10 cm. Thus, any moist substance placed in a region of intense microwave radiation of this wavelength (meat, soup, or a cake batter, for example) will become hot very quickly. Because the heat is generated within the substance itself, rather than conducted inward from the outside, foods can be cooked rapidly in a microwave oven. It is, however, important to keep the radiation confined to the oven because when such microwaves are emitted their radiation can damage living tissue.

Infrared radiation and the greenhouse effect (λ about 10^{-4} m to 10^{-6} m; f about 10^{12} Hz to 10^{14} Hz). Radiation in this region of the electromagnetic spectrum, just below the red end of the visible spectrum, is often called "thermal radiation," because it transmits heat. Because of the oscillation of charges within molecules due to heat energy, all warm objects, such as a glowing fireplace or warm-blooded creatures, emit infrared electromagnetic radiation. This is also how heat is transmitted from the Sun to the Earth, and it is one way in which living creatures can be detected at night by nocturnal predators or by humans using special "night vision" apparatus. Thus, warm objects and animals can be detected or photographed "in the dark" using infrared-sensitive equipment or film.

FIGURE 12.16 A photograph made with film sensitive only to infrared radiation.

In an actual greenhouse, the largest contribution to keep the interior heated is the reduction of convection, which is the same principle as used for wine-cooler jackets.

The environmental and technological problem of the phenomenon of *global warming* is associated with infrared radiation. Since the Earth is warmed by the Sun, the surface of the Earth also emits infrared radiation. Much of this radiation is dissipated into outer space, but some of it is naturally trapped by water vapor in the atmosphere. Like a blanket, the water vapor in the air reflects many of the infrared rays back to the Earth's surface, thus keeping the Earth's surface at just the right temperature for life to exist. This is known as the *greenhouse effect*, since greenhouses operate in the same way. The glass walls and roof of the greenhouse allow the Sun's visible rays to pass through, but they prevent the invisible infrared rays from escaping, thus warming the inside of the greenhouse where plants can grow year round. (You encounter the same effect when you leave the windows rolled up in a car parked in the sun.)

It has now been substantiated by scientists, including the U.S. National Academy of Sciences, that since the advent of the Industrial Revolution, much more of the infrared radiation emitted back into space by the Earth's surface is becoming trapped in the atmosphere by the gases produced through the burning of fossil fuels that run electric generators and transportation vehicles. The burning of these fuels produces carbon dioxide, sulfur dioxide, water vapor, and other gases and small particles. These gases allow the Sun's visible and ultraviolet rays to reach the ground, but reflect the infrared radiation emitted by the Earth's surface back to the Earth, keeping the infrared rays trapped in the atmosphere and thus causing global warming. (These gases also cause acid rain.)

This enhanced greenhouse effect accounts for an average rise in the Earth's temperature during the past century of only a degree or two Celsius, but scientists using computer models predict that this increase does have a major impact on the ecology on Earth, causing at different locations increased drought or rainfall, more powerful storms, and crop failures. They point to the fact that the average temperature of Earth during the last Ice Age was only about 5°C cooler than it is now. The unusual melting of glaciers and polar ice, the unusually warm winters and summers in recent decades, and the unusually large rainfalls and flooding around the world are signs of even more trouble ahead unless the burning of fossil fuels is curtailed.

During the 1990s a series of international meetings occurred in order to find a way to reduce the emission of "greenhouse gases." Unfortunately, the nations of the world could not agree because of economic reasons. The developing nations do not want to limit their growth, especially after the developed nations had already achieved their growth without limiting their

ASTRONOMY ACROSS THE SPECTRUM

The electromagnetic spectrum comprises more than the rainbow effect produced by passing white light through a prism. Electromagnetic radiation of different wavelengths provides different kinds of information. You are familiar with the effects of various parts of the spectrum: sunburn (ultraviolet rays) (a), visible light (d), heat (infrared) (b) and x rays (c). Scientists make use of electromagnetic radiations in such fields as astronomy, earth and life sciences, and communications.

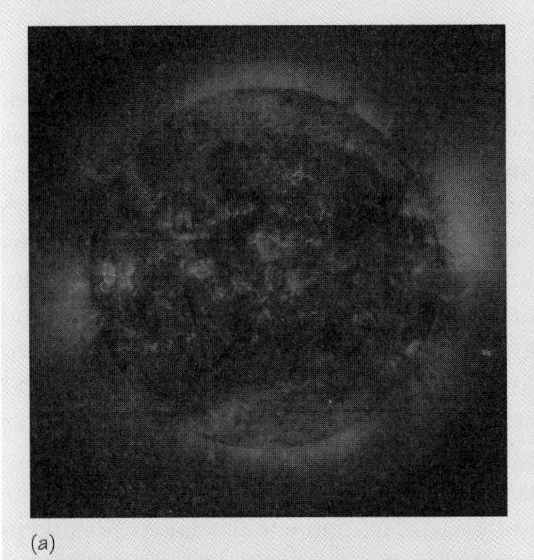

(a)

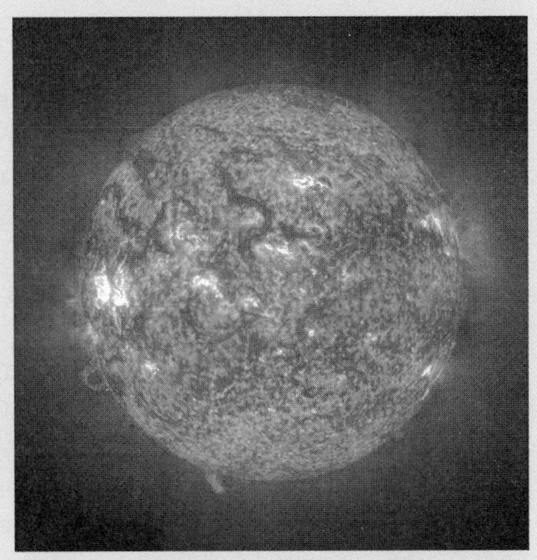

(b)

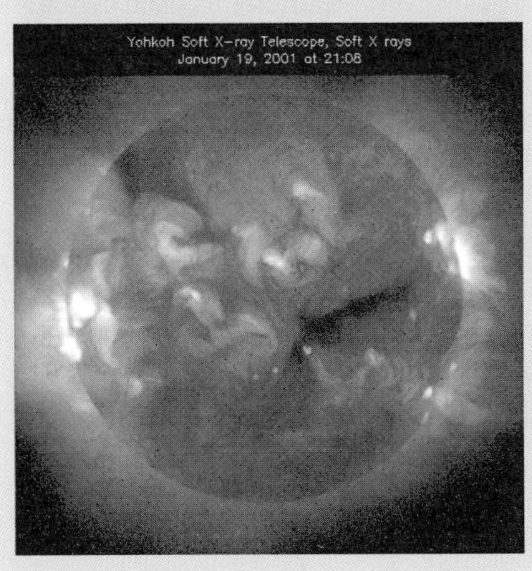

(c)

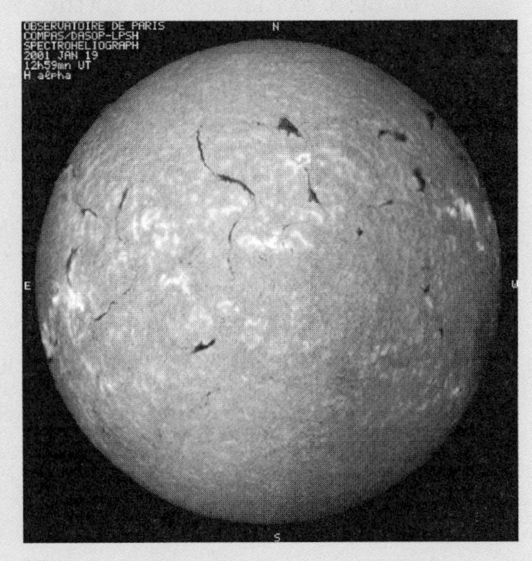

(d)

FIGURE 12.17

emissions. Some politicians and industrialists in the United States, the world's largest economy and thus the largest producer of greenhouse gases, feared the possible economic consequences of agreeing to a suggested 20% reduction in greenhouse-gas emissions by the year 2010. Further natural catastrophes related to possible climate change, and further evidence of global warming, may yet encourage the nations of the world to view the reduction in greenhouse gases as probably the better economic choice. In addition, it has been argued that the investment in devices and engines made to reduce the emission of "greenhouse gases" would stimulate the economy in a major way.

Visible light ($\lambda = 7 \times 10^{-7}$ m to 4×10^{-7} m; $f = 4 \times 10^{14}$ Hz to 8×10^{14} Hz). This small band of frequencies is known as visible light because the visual receptors in the human eye are sensitive only to these frequencies. If light within this band of frequencies is sent through a glass prism or through the raindrops in a cloud, it can be broken down into its constituent frequencies, which are observed by the human eye as the colors of the rainbow. These colors range from red at the low-frequency end (4×10^{14} Hz) to violet at the high-frequency end (8×10^{14} Hz). The main colors, in order, are: red, orange, yellow, green, blue, indigo, and violet (which may be remembered by the acronym ROY G BIV). (See Chapter 8, for more on the behavior of visible light.)

The visible-light frequencies are those at which the Sun copiously radiates energy. In the course of evolution, the human eye has taken advantage of light in that region. In the struggle for survival, humans have evolved as day creatures. However, many nocturnal predators, including cats, dogs, and some marsupials have developed the ability to use small amounts of visible light for night vision, and some are even able to see into the infrared region of light, which is invisible to human eyes.

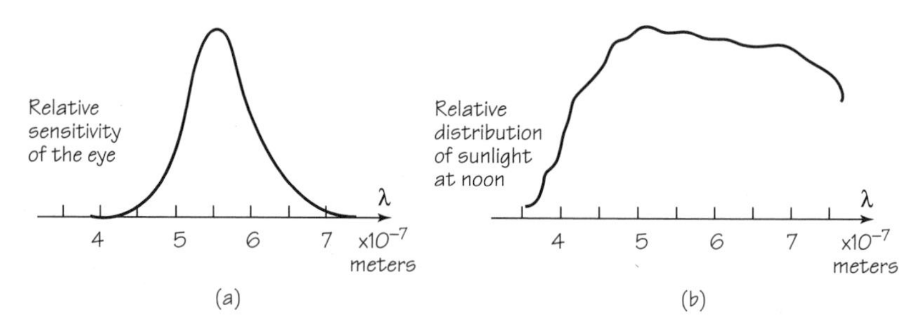

FIGURE 12.18 Sunlight and eye sensitivity versus wavelength.

Ultraviolet waves and ozone depletion (λ about 10^{-7} m to 10^{-8} m; f about 10^{15} Hz to 10^{16} Hz). Electromagnetic waves above the visible range—ultraviolet, X rays, and gamma rays—can damage living tissue, and some can cause cancer and genetic mutations. Although X rays and gamma rays occur naturally in our environment, they are much less abundant than ultraviolet rays, which are contained in the Sun's rays. Fortunately, a layer in the Earth's atmosphere provides some protection against the Sun's damaging ultraviolet rays. This layer contains a molecule called "ozone," a rare form of oxygen.

Normally two oxygen atoms chemically join to form a stable molecule, O_2. However, under certain circumstances, such as those caused by lightning, three oxygen atoms join together loosely to form the molecule, O_3, which is called ozone. At ground level, when ozone combines with exhaust from automobiles or other engines on a hot day, it produces smog that has serious health hazards. However, high in the atmosphere it forms a protection against ultraviolet rays. This molecule vibrates at the frequency of ultraviolet rays and can thus absorb or reflect these rays from the Sun back into space.

Unfortunately, man-made chemicals, widely used in industry, known as CFCs (chlorofluorocarbons), have been destroying the ozone layer since the 1930s, creating an "ozone hole" and allowing more of the damaging ultraviolet rays to reach the Earth's surface. This has caused a noticeable increase in skin cancers and eye cataracts in people, and the endangerment of some sea creatures and crops.

CFCs were developed in the 1930s for use as a coolant in refrigerators and air conditioners, and as a propellant in aerosol sprays. When used for these purposes under normal circumstances CFCs are essentially inert and very stable. Thus they seemed like ideal insulators, coolants, and propellants. However, during the 1980s it was discovered that when CFCs reach the upper atmosphere the chlorine in these molecules destroys ozone molecules in the atmosphere by taking away one of the three oxygen atoms, leaving ordinary oxygen, O_2, behind. Ultraviolet rays then break up the chlorine and oxygen, freeing the chlorine ion to destroy another ozone molecule. One chlorine ion can destroy thousands of ozone molecules. Studies at both the North and South poles of the Earth indicate that the destruction of ozone is worldwide.

The world reaction to the dangerous ozone depletion has been quite different to this problem compared with the reaction to global warming. This is mainly because the chemical industry had found a replacement for CFCs which would not cause any significant economic difficulty. The result was an international treaty that banned the production and sale of CFCs by

2000. Nevertheless, the chlorine already in the atmosphere can continue the destruction of the ozone layer for several decades. The result has been a steady increase in harmful effects of ultraviolet radiation. These effects are predicted to continue for decades to come until most of the chlorine previously emitted into the atmosphere is finally rendered harmless. In the meantime, physicians recommend that in open sunlight everyone wear sunglasses that actually filter out ultraviolet rays (in all forms), and not to stay in the Sun without wearing a hat and applying sunblock lotions to exposed skin.

X rays ($\lambda = 10^{-8}$ m to 10^{-17} m; $f = 10^{16}$ Hz to 10^{25} Hz). Atoms emit X radiation when electrons undergo transitions between the inner shells of the atoms. X rays are also produced by the sudden deflection or stopping of electrons when they strike a metal target. The maximum frequency of the radiation generated is determined by the energy with which the electrons strike the target. In turn, this energy is determined by the voltage through which the electrons are accelerated. So the maximum frequency increases with the accelerating voltage. The higher the frequency of the X rays, the greater is their power to penetrate matter. But the distance of penetration also depends on the nature of the material being penetrated. X rays are readily absorbed by bone, which contains calcium, while they pass much more easily through less dense organic matter such as flesh, which contains mainly the light atoms: hydrogen, carbon, and oxygen.

These properties of X rays, combined with their ability to affect a photographic plate, have led to some of the spectacular medical uses of X-ray

FIGURE 12.19 X-ray photograph of chambered nautilus sea shell.

photography. Because X rays can damage living cells and even cause genetic mutations, they have to be used with great caution and only by trained technicians. Since some kinds of diseased cells are injured more easily by X rays than are healthy cells, a carefully controlled X ray beam is sometimes used to destroy cancerous growths or other harmful cells.

X rays produce interference effects when they fall on a crystal in which atoms and molecules are arranged in a regular pattern. Different portions of the incident beam of X rays are reflected or diffracted from different planes of atoms in the crystal structure. These reflected rays can interfere constructively, and this fact can be used in either of two ways. If the spacing of the atoms in the crystal is known, the wavelength of the X rays can be calculated. Conversely, if the X ray wavelength is known, the distance between crystal planes, and thus the structure of the crystal, can be determined. X rays are now widely used by chemists, physicists, mineralogists, and biologists in studying the structure of crystals and complex molecules. (You will encounter these ideas again in Chapter 13.)

Gamma rays ($\lambda = 10^{-17}$ m and smaller; $f = 10^{25}$ Hz and higher). The gamma-ray region of the electromagnetic spectrum overlaps the X ray region. Gamma radiation is emitted mainly by the unstable nuclei of natural or artificial radioactive materials. They are also a component of so-called cosmic radiation, radiation streaming to the Earth from outer space. Gamma

FIGURE 12.20 Čerenkov radiation in reactor. The glow in the photograph is caused when gamma rays emitted by radioactive cobalt cylinders interact with the surrounding pool of water.

rays are the most energetic radiation known, and, in cosmic radiation, they are produced by the most energy-intensive events in the Universe—the explosions of supernovae and other cataclysmic events. Many of the cosmic events that produce the observed gamma rays are not well understood. A series of gamma-ray sensitive satellites is presently studying these events. (You will learn more about gamma rays in Chapter 17.)

12.6 WHAT ABOUT THE ETHER NOW?

The "luminiferous ether" had been proposed specifically as a medium for the propagation of light waves. Maxwell found that the ether could also be thought of as a medium for transmitting electric and magnetic forces. Later, he realized that he could drop his specific model of the ether entirely if he focused on the mathematical form of the theory. Yet, just before his death in 1879, Maxwell wrote an article in which he still supported the ether (or aether) concept:

> Whatever difficulties we may have in forming a consistent idea of the constitution of the aether, there can be no doubt that the interplanetary and interstellar spaces are not empty, but are occupied by a material substance or body, which is certainly the largest, and probably the most uniform body of which we have any knowledge. . . .

Maxwell was aware of the failures of earlier ether theories. Near the beginning of the same article he said:

> Aethers were invented for the planets to swim in, to constitute electric atmospheres and magnetic effluvia, to convey sensations from one part of our bodies to another, and so on, till all space had been filled three or four times over with aethers. It is only when we remember the extensive and mischievous influence on science which hypotheses about aethers used formerly to exercise, that we can appreciate the horror of aethers which sober-minded men had during the 18th century. . . .

Maxwell had formulated his electromagnetic theory mathematically, independent of any particular model of the ether. Why, then, did he continue to speak of the "great ocean of aether" filling all space? It seemed unthinkable to Maxwell that there could be vibrations without something that vibrates, or waves without a medium. Also, to many nineteenth-century physicists the idea of "action at a distance" seemed absurd. How could one object exert a force on another body far away if something did not transmit the force? One body is said to act *on* another, and the word *on* gives the idea of contact. Thus, according to accepted ways of describing the world in common language, the ether seemed somehow necessary.

Yet 25 years after Maxwell's death the ether concept had lost much of its support. Within another decade, it had vanished from the collection of useful concepts. In part, the success of Maxwell's theory itself helped to undermine the general belief in the existence of an ether, simply because his equations did not depend on details of the ether's structure. In fact, they could be taken to describe the relations between changes of electric and magnetic fields in space without any reference to the ether at all.

FIGURE 12.21 In this chapter, you have read about how mechanical models of light and electromagnetism faded away, leaving a model-less, mathematical (and therefore abstract) field theory. The situation has been likened to Lewis Carroll's Cheshire Cat, which disappears leaving only its grin behind (illustration by John Tenniel).

Another difficulty with belief in the ether was that all attempts to detect the motion of the Earth relative to the ether failed (see Chapter 9). If light is a kind of vibration of an ether that fills all space, then light should travel at a definite speed relative to the ether. But the Earth must also be moving through the ether in its annual orbit around the Sun. Thus, the Earth should be moving like a ship, against an "ether wind" at some times, and with it at other times. Under these conditions, the apparent speed of light should be observed to differ. When the Earth and a beam of light are moving in the same direction through the ether, the observed speed of light should not be the same as when the Earth and the light are moving in opposite directions.

Theorists computed the time required for a pulse of light to make a round trip in a laboratory with and against the ether wind. They compared this interval with the time calculated for a round trip in the absence of an ether wind. The expected time difference was found to be very small; only 10^{-15} s for a round trip of 30 m. This is too short a time difference to measure directly, but it is of the same order as the time for one vibration of visible light. Therefore, the difference might be detected from observations of a properly produced interference pattern. In 1887, the American scientists Albert A. Michelson and Edward Morley used a device sensitive enough to detect an effect only 1% as great as that predicted by the ether theory. Neither this experiment nor the many similar experiments done since then have revealed the existence or expected effects of an ether wind.

Supporters of the ether concept offered various explanations for this unexpected result. For example, they suggested that objects moving at high speeds relative to the ether might change their size in just such a way as to make this relative speed undetectable. But even those who made such attempts to rescue the ether concept felt their proposals to be "ad hoc," forced, and artificial. Finally, a decisive development led scientists to abandon the ether concept. This breakthrough was not a specific experiment, but a brilliant proposal by a 26-year-old man.

The man was Albert Einstein, who, as discussed in Chapter 9, suggested that a new and deep union of mechanics and electromagnetism could be achieved without the ether model, on the basis of the two fundamental postulates of relativity theory: the relativity principle and the constancy of the speed of light. The price of accepting these postulates was, Einstein showed, the necessity of revising some common-sense notions of space and time. Einstein showed that Maxwell's equations are fully consistent with extending the principle of relativity to all physics. This was yet another great synthesis of previously separate ideas, like the syntheses forged by Copernicus, Newton, and Maxwell.

SOME NEW IDEAS AND CONCEPTS

classical physics
displacement current
electromagnetic spectrum
electromagnetic wave
global warming
greenhouse effect
ionosphere
ozone
spectrum
visible spectrum

FURTHER READING

R. Buderi, *The Invention that Changed the World* [Radar]. Sloan Technology Series (New York: Touchstone Books, 1998).

G. Cantor, D. Gooding, F.A. Frank, and J.L. James, *Michael Faraday* (Atlantic Highlands, NJ: Humanity Books, 1996).

D. Park, *The Fire within the Eye: A Historical Essay on the Nature and Meaning of Light* (Princeton: Princeton University Press, 1997).

C. Susskind, *Heinrich Hertz: A Short Life* (San Francisco: San Francisco Press, 1995).

STUDY GUIDE QUESTIONS

12.1 Faraday's Suggestion

1. What did Faraday suggest?
2. What was the motivation for this suggestion?

12.2 Maxwell's Principles of Electromagnetism

1. When there is a changing electric field, what else occurs (according to Maxwell)?
2. What is a displacement current?
3. What are the four principles of electromagnetism?

12.3 The Propagation of Electromagnetic Waves

1. What is an electromagnetic wave?
2. How is an electromagnetic wave formed?
3. What discovery did Maxwell make when he calculated the speed that electromagnetic waves should travel?
4. What did Maxwell infer from this result?
5. What is Maxwell's synthesis?
6. What is classical physics? Is it still valid today? Under what circumstances are revisions required?

12.4 Hertz's Experimental Confirmation

1. What predictions of Maxwell's did Hertz verify?
2. What did Hertz use as a detector of electromagnetic waves?
3. How does visible light fit into Maxwell's theory?
4. How do Faraday, Maxwell, and Hertz represent the stages we often observe in the formation of a new theory in science?

12.5 The Electromagnetic Spectrum

1. What is the electromagnetic spectrum and how does it differ from the visible spectrum?
2. List the types of electromagnetic waves discussed in this section. For each one, indicate where it is on the electromagnetic spectrum, and describe a use, or social impact, or technological application.
3. Why do ordinary radio waves not cast noticeable "shadows" behind such obstacles as trees or small buildings?
4. Why are satellites required for TV transmission over long distances, while some radio waves can be heard at great distances from the source?
5. How is the frequency of X rays related to their penetration of matter?
6. How do the wavelengths used in radar compare with the wavelengths of visible light?
7. What is global warming and why is it occurring, according to most scientists?
8. Why is global ozone depletion a cause for concern?
9. How does ozone depletion occur?
10. How have the nations of the world responded to each of these crises, and what is the reason for any difference?

12.6 What About the Ether Now?

1. Why did Maxwell (and others) adhere for a time to the concept of an ether?
2. Whose argument finally showed that the ether was an unnecessary hypothesis?
3. Did the demise of the ether idea lead to a demise of Maxwell's theory? Explain.

DISCOVERY QUESTIONS

1. A current in a conductor can be caused by a steady electric field. Can a displacement current in an insulator be similarly caused? Explain your answer briefly.
2. What is the "disturbance" that travels in each of the following waves?
   (a) water waves?
   (b) sound waves?
   (c) electromagnetic waves?

3. In Hertz's detector, it is the electric fieldstrength in the neighborhood of the wire that makes the sparks jump. How was Hertz able to show that the waves from the induction-coil-spark gap were polarized?
4. What evidence did Hertz obtain that his induction-coil-generated waves have many properties similar to visible light waves?
5. Give several factors that contributed to the 25-year delay in the general acceptance by scientists of Maxwell's electromagnetic wave theory.
6. What evidence is there for believing that electromagnetic waves carry energy? Does this suggest why the early particle theory of light had some success?
7. You are listening to a radio station while driving in a car when you notice something strange. The radio signal fades out at some points along the road, and later returns to the previous level. What is happening?
8. Examine the wavelengths and frequencies of several different electromagnetic waves. You may notice that as the frequency increases, the wavelength decreases. Why is this so?
9. Why must there be some federal control of the broadcast power, frequency, and direction of signals emitted by radio and TV stations, but no such controls on the distribution of newspapers and magazines?
10. If there are extraterrestrial beings of advanced civilizations, what method for gathering information about Earth-people might they have? Conversely, why is it far more probable that any contact with an extraterrestrial civilization, if it exists, will be made by receiving electromagnetic signals from it rather than actual "visits"?
11. Some scientists have been speculating about setting up an Earth colony on Mars. Mars is not hospitable to human life, but it does have a lot of direct sunlight which is not filtered much by the thin atmosphere there. How could the colonists make use of the sunlight to create a habitable environment? How could they protect themselves from some of the dangers of direct sunlight?
12. Why do you think the human eye is sensitive to the range of light wavelengths to which it is actually sensitive?
13. A sensitive thermometer placed in different parts of the visible light spectrum formed by a quartz prism will show a rise in temperature. This proves that all colors of light produce heat when absorbed. But the thermometer also shows an increase in temperature when it is placed in either of the two dark regions to either side of the end of the visible spectrum. Why is this?
14. During the eighteenth century many scientists were fascinated by what seemed to them a strange phenomenon. In their laboratory was the customary fireplace with a fire in it for warmth. A lens was located on a small table some distance across the room facing the fire. When a sheet of paper was placed behind the lens, the paper burst into flames, to the amazement of all who observed this. What was happening?
15. What was the principal reason for the loss of scientific support for the ether concept?
16. At many points in the history of science, the "natural" or common-sense way

of looking at things has changed greatly. Attitudes toward action-at-a-distance are a case in point. What are some other examples?

17. Can intuition be educated; that is, can feelings about the fundamental aspects of reality be changed? Use attitudes toward action-at-a-distance through the ether as one example, and give others.

Quantitative

1. An electron oscillating on an antenna produces an electromagnetic wave that vibrates at 10,000 Hz.
   (a) What are the wave's frequency, its period, and its wavelength?
   (b) What type of wave is this?
2. Obtain the frequency range of the FM dial on a radio. What is the corresponding range of wavelengths of the FM waves received by the radio?
3. How much time would elapse between the sending of a radar signal from the Earth to the Moon and the detection of its echo on Earth?

CHAPTER 13

Probing the Atom

13.1 THE PERIODIC TABLE

It has been hypothesized for millennia that all matter is made of tiny, indivisible, smallest bits of matter called atoms. Great progress had been made during the nineteenth century in attributing the thermodynamic laws and some of the properties of matter, especially gases, to the kinetic-molecular theory (Chapter 7). In addition, it was known for centuries that there are different types of so-called fundamental "elements" in nature—gold, silver, copper, sodium, etc. There are the smallest units into which substances can be divided by chemical means. Eventually it was found useful to give the elements special symbols, for example, "C" for carbon, "O" for oxygen, "H" for hydrogen, and so on.

For many people, such as the English chemist John Dalton (1766–1844), these different elements indicated that nature is also made up of different types of atoms, one type of atom for each element. Each element was considered a collection of identical, indestructible atoms, and this idea was confirmed in chemical studies during the nineteenth century. When two or more atoms link together, they form a molecule. The molecule may be an element itself if both atoms are the same, such as O_2, or it may be a com-

FIGURE 13.1 John Dalton (1766–1844). Born in Eaglesfield, England, the son of a weaver, Dalton attended a Quaker school until the age of 12. He taught at New College, Manchester. He is best known for his theory that matter is composed of atoms of differing weights that combine in simple, whole-number ratios by weight. He listed the atomic weights of known elements relative to the weight of hydrogen. Dalton's work formed the basis of the periodic table of the elements.

pound if the atoms are different, such as H_2O. Since atoms are not divisible, the idea of joining two atoms of hydrogen to, say, $1\frac{1}{2}$ atoms of oxygen instead of exactly one atom of oxygen is meaningless. Dalton's law of fixed proportions follows quite naturally from the hypothesis that elements are made up of identical, indestructible atoms. When elements combine to form compounds, precisely the same ratio of masses of the constituents is required each time to make a particular compound. For example, 23.0 g of sodium always combine with 35.5 g of chlorine to produce 58.5 g of salt. If you start with 25.0 g of sodium and 35.5 g of chlorine, you still get only 58.5 g of salt, but there will be 2.0 g of sodium left over. This rule, the *law of fixed proportions*, applies to all compounds.

The law of fixed proportions indicated that the atoms of the elements that combine to form compounds such as NaCl do not have the same mass. Since only 23.0 g of sodium combined with 35.5 g of chlorine to form 58.5 g of NaCl, the chlorine atoms must be more massive than the sodium atoms. In fact, if one sodium atom has 23.0 "units" of mass, then one chlorine atom would have 35.5 units of mass. The units of mass could be grams, kilograms, pounds, or whatever. For convenience, they are called *atomic mass units*, for which the standard symbol is u.

We now know that there are just 92 different elements that occur naturally throughout nature, and in fact throughout the entire Universe. These have a variety of atomic mass units (amu, abbreviated to u). Today, the atomic mass units are measured relative to one type, or isotope, of carbon known as carbon-12 (^{12}C), which is *defined* as having atomic mass of 12.0000 u. The atomic masses of all other elements can then be given relative to carbon-12. For instance, relative to carbon-12, natural sodium has an atomic mass of 22.99 u, hydrogen has an atomic mass of 1.008 u, oxygen has an atomic mass of 15.999 u, uranium has a mass of 238.03 u, and so on. Some of these elements have similar properties: some are gases at room temperature; some are solid metals. There are differences in densities, melting and boiling points, electrical conductivity, and so on.

In the 1869, the Russian chemist Dimitri Mendeleev (1834–1907) arranged the 63 elements then known in a table according to their physical properties from the lightest (hydrogen) to the heaviest (uranium). He found that when arranged by chemical properties, the elements tended to line up vertically and to vary horizontally in a periodic fashion, forming what is now called the *periodic table*. Although some elements were missing at the time and were discovered only later, Mendeleev courageously assigned a position on the table to each element known to him, leaving blanks

FIGURE 13.2 Dimitri Mendeleev (1834–1907), founder of the modern periodic table.

where he hypothesized that an element had not yet been discovered. These elements were discovered later, confirming Mendeleev's hypothesis!

Only one element can occupy a particular position on the periodic table. Mendeleev assigned a number to each element at each position. This number, called the *atomic number*, now goes from 1 for hydrogen to 92 for uranium. It is given the symbol Z. The atomic number immediately identifies the element, since it is unique for each element. It is always an integer and is given at the top of the space in most periodic tables. Thus, when someone refers to element $Z = 3$, we know immediately that they are referring to the element lithium; element 53 is iodine, and so on.

When the elements are arranged on the modern periodic table as shown in the color insert, Plate 6, the elements below one another in each column or group share physical properties to a remarkable degree, as Mendeleev had discovered. Therefore, these elements can be considered to belong to the same "family" of elements. For instance, Group I on the left contains the family of *alkali* metals: lithium, sodium, potassium, rubidium, and cesium. This is a group of soft metals with very low densities, low melting points, and similar chemical behavior. Another family of elements, called the *halogens*, is found in Group VII: fluorine, chlorine, bromine, and iodine. These elements combine violently with many metals and form

Таблица II.

Вторая попытка Менделѣева найти естественную систему химическихъ элементовъ. Перепечатана безъ измѣнен. изъ „Журнала Русскаго Химическаго Общества", т. III, стр. 31 (1871 г.).

	Группа I.	Группа II.	Группа III.	Группа IV.	Группа V.	Группа VI.	Группа VII.	Группа VIII, переходъ къ группѣ I.
	H=1							
Типическіе элементы.	Li=7	Be=9,4	B=11	C=12	N=14	O=16	F=19	
1-й періодъ. Рядъ 1-й.	Na=23	Mg=24	Al=27,3	Si=28	P=31	S=32	Cl=35,5	
— 2-й.	K=39	Ca=40	?=44	Ti=50?	V=51	Cr=52	Mn=55	Fe=56, Co=59, Ni=59, Cu=63
2-й періодъ. — 3-й.	(Cu=63)	Zn=65	?=68	?=72	As=75	Se=78	Br=80	
— 4-й.	Rb=85	Sr=87	Yt?=88?	Zr=90	Nb=94	Mo=96	—=100	Ru=104, Rh=104, Pd=104, Ag=108
3-й періодъ. — 5-й.	(Ag=108)	Cd=112	In=113	Sn=118	Sb=122	Te=128?	J=127	
— 6-й.	Cs=133	Ba=137	—=137	Ce=138?	—	—	—	— — — —
4-й періодъ. — 7-й.	—	—	—	—	—	—	—	
— 8-й.	—		—	—	Ta=182	W=184	—	Os=199?, Ir=198?, Pt=197, Au=197
5-й періодъ. — 9-й.	(Au=197)	Hg=200	Tl=204	Pb=207	Bi=208	—	—	
— 10-й.		—	—	Th=232	—	Ur=240	—	
Высшая солевая окись	R_2O	R_2O_2 или RO	R_2O_3	R_2O_4 или RO_2	R_2O_5	R_2O_6 или RO_3	R_2O_7	R_2O_8 или RO_4
Высшее водородное соединеніе . . .			(RH_3)	RH_4	RH_3	RH_2	RH	

FIGURE 13.3 Mendeleev's original periodic table of elements as it appeared in 1872.

white, crystalline salts (*halogen* means "salt former"). These salts have similar formulas, such as NaF, NaCl, NaBr, $MgCl_2$, $MgBr_2$, etc.

Occasionally, for reasons then not known, it was necessary to depart from the overall scheme of ordering the elements. For example, the chemical properties of argon (Ar) and potassium (K) demand that they be placed in the eighteenth and nineteenth positions in order to fall into groups characteristic of their families. On the basis of their atomic masses alone (39.948 u for argon; 39.102 u for potassium) their positions would have been reversed.

The beautiful regularity and symmetry of the periodic table indicate that a model of the atoms that make up the different elements will probably also display a similar beauty and symmetry. Such a model did arise during the early twentieth century, and it did not disappoint our expectations. But we shall see that an enormous amount of ingenuity and detective work was required to comprehend the structure of the atom.

13.2 THE IDEA OF ATOMIC STRUCTURE

Chemistry in the nineteenth century had succeeded remarkably in accounting for combining proportions and in predicting chemical reactions. This success had convinced most scientists that matter is indeed composed of atoms. But there remained a related question: Are atoms really indivisible, as had been assumed, or do they consist of still smaller particles?

You can see how this question arose by thinking a little more about the periodic table. Mendeleev had arranged the elements in the order of increasing atomic mass. But the atomic masses of the elements cannot explain the *periodic* features of Mendeleev's table.

Why, for example, do the 3rd, 11th, 19th, 37th, 55th, and 87th elements, with quite different atomic masses, have similar chemical properties (e.g., they burn when exposed to air)?

Why are these properties somewhat different from those of the 4th, 12th, 20th, 38th, 56th, and 88th elements in the list (which react slowly with air or water), but greatly different from the properties of the 2nd, 10th, 18th, 36th, 54th, and 86th elements (which rarely combine with any other element)?

The periodicity in the properties of the elements led to speculation that atoms might have structure, that they might be made up of smaller pieces. The properties changed gradually from group to group. This fact suggested that some unit of atomic structure might be added from one element to the next, until a certain portion of the structure is completed. The completed condition would occur in the atom of a noble gas (Group VIII in Plate 6). In an atom of the next heavier element, a new portion of the structure would be started, and so on. The methods and techniques of classical chemistry could not supply experimental evidence for such structure. In the nineteenth century, however, discoveries and new techniques in physics opened the way to prove that atoms do actually consist of smaller pieces. Evidence piled up to support the conclusion that the atoms of different elements differ in the number and arrangement of these pieces.

13.3 CATHODE RAYS

In 1855, the German physicist Heinrich Geissler invented a powerful vacuum pump. This pump could remove enough gas from a strong glass tube to reduce the pressure to 0.01% of normal air pressure. It was the first major improvement in vacuum pumps after Guericke's invention of the air pump, two centuries earlier. So we noted, using Geissler's new pump made possible the electric light bulb, the electron tube, and other technologically valuable inventions over the next 50 years. It also opened new fields to pure scientific research.

Geissler's friend Julius Plücker connected one of Geissler's evacuated tubes to a battery. He was surprised to find that, at the very low pressure obtained with Geissler's pump, electricity flowed through the tube. Plücker used apparatus similar to sketch (a) on page 591. He sealed a wire into each end of a strong glass tube. Inside the tube, each wire ended in a metal plate, called an electrode. Outside the tube, each wire ran to a source of high voltage. The negative plate inside the tube is called the *cathode*, and the positive plate is called the *anode*. A meter indicated the current going through the tube.

Plücker and his student Johann Hittorf noticed that when an electric current passed through the low-pressure gas in a tube, the tube itself glowed with a pale green color. Several other scientists observed this effect, but two decades passed before anyone undertook a thorough study of the glowing tubes. By 1875, William Crookes had designed new tubes for studying the glow. When he used a bent tube the most intense green glow appeared on the part of the tube that was directly opposite the cathode (at g in sketch (b) on page 591). This suggested that the green glow is produced by something that comes out of the cathode and travels down the tube until it hits the glass. Another physicist, Eugen

Substances that glow when exposed to light, particularly ultraviolet, are called fluorescent. "Fluorescent lights" are essentially Geissler tubes with an inner coating of fluorescent powder.

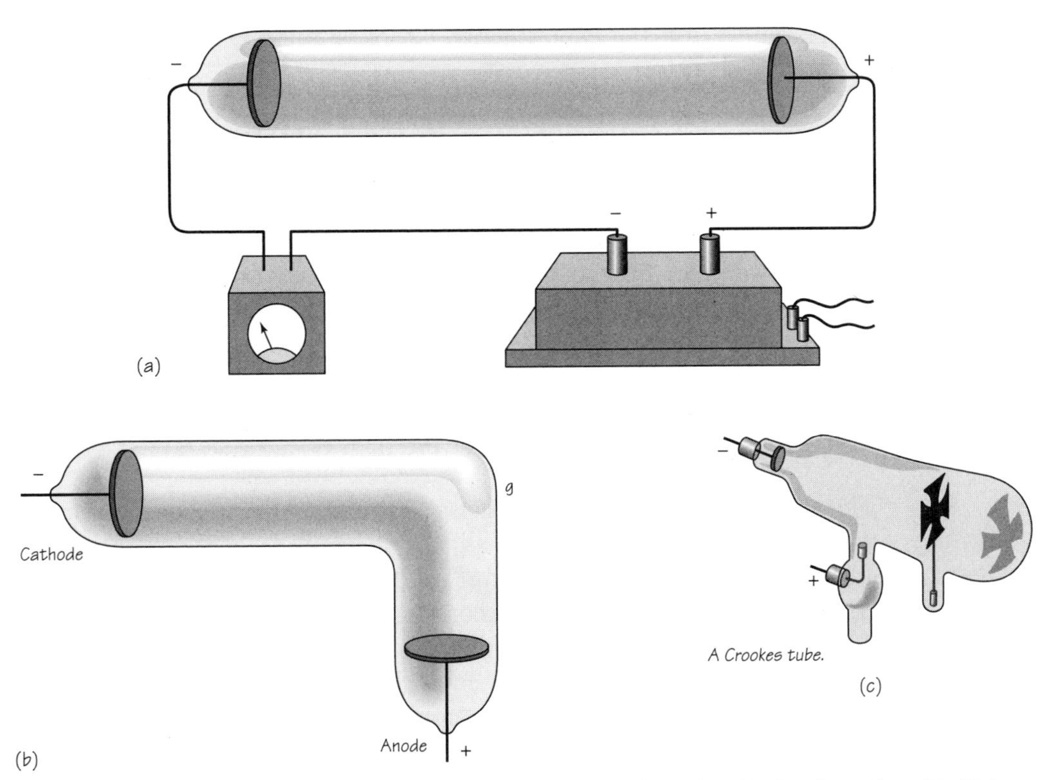

FIGURE 13.4 (a) Geissler-Plücker tube; (b) bent Geissler-Plücker tube; (c) Crookes tube with Maltese cross as barrier.

Goldstein, was also studying the effects of passing an electric current through a gas at low pressure. Goldstein coined a term for whatever it was that appeared to be coming from the cathode, hence the name *cathode rays*. But what could they be?

To study the nature of the rays, Crookes did some clever experiments. He reasoned that if cathode rays could be stopped before they reached the end of the tube, the intense green glow would disappear. He therefore introduced barriers like the Maltese cross (made of metal). A shadow of the barrier appeared in the midst of the green glow at the end of the tube. The cathode seemed to act like a source that radiates a kind of light; the cross acted like a barrier blocking the light. The shadow, cross, and cathode appeared along one straight line. Therefore, Crookes concluded, cathode rays, like light rays, travel in straight lines. Next, Crookes moved a magnet near the tube, and the shadow moved. Thus, he found that magnetic fields deflect cathode rays (which does not happen with light).

In the course of many experiments, Crookes found the following properties of cathode rays:

- No matter what material the cathode is made of, it produces rays with the same properties.
- In the absence of a magnetic field, the rays travel in straight lines perpendicular to the surface that emits them.
- A magnetic field deflects the path of the cathode rays.
- The rays can produce some chemical reactions similar to the reactions produced by light. For example, certain silver salts change color when hit by the rays.
- In addition, Crookes suspected (but did not succeed in showing) that electrically charged objects deflect the path of cathode rays.

The cathode rays fascinated physicists at the time. Some thought that the rays must be a form of light. After all, they have many of the properties of light: they travel in straight lines and produce chemical changes and fluorescent glows just as light does. According to Maxwell's theory, light consists of electromagnetic waves. So the cathode rays might, for example, be electromagnetic waves of frequency much higher than that of visible light.

However, while magnetic fields do not bend light, they do bend the path of cathode rays. Chapter 10 described how magnetic fields exert forces on currents, that is, on moving electric charges. A magnetic field deflects cathode rays in the same way that it deflects negative charges. Therefore, some physicists believed that cathode rays consisted of negatively charged particles.

The debate over whether cathode rays are a form of electromagnetic waves or a stream of charged particles continued for 25 years. Finally, in

FIGURE 13.5 Joseph John Thomson (1856–1940), one of the greatest British physicists, attended Owens College in Manchester, England, and then Cambridge University. He worked on the conduction of electricity through gases, on the relation between electricity and matter, and on atomic models. His greatest single contribution was the discovery of the electron. Thomson was the head of the famous Cavendish Laboratory at Cambridge University, where one of his students was Ernest Rutherford.

Thomson found that

$$\frac{q}{m} = 1.76 \times 10^{11}\ \text{C/kg}.$$

According to Millikan's experiment, the magnitude of q is 1.6×10^{-19} C. Therefore, the mass of an electron is

$$m = \frac{1.6 \times 10^{-19}\ \text{C}}{1.76 \times 10^{11}\ \text{C/kg}}$$
$$= 0.91 \times 10^{-30}\ \text{kg}.$$

(The mass of a hydrogen ion is 1.66×10^{-27} kg. This is approximately the value of one "atomic mass unit".)

1897, J.J. Thomson, head of the famous Cavendish Laboratory at Cambridge University, made a series of experiments that convinced physicists that cathode rays are negatively charged particles.

By then, it was well known that the paths of charged particles are affected by both magnetic and electric fields. By assuming that cathode rays were negatively charged particles, Thomson could predict what should happen when they passed through such fields. For example, an electric field of just the right magnitude and direction should exactly balance the deflection of a beam of cathode rays by a magnetic field. As Thomson discovered, the predictions were verified. *Thomson could therefore conclude that cathode rays were indeed made up of negatively charged particles.* He was then able to calculate, from the experimental data, the ratio of the charge of a particle to its mass. This ratio is represented by q/m, where q is the charge and m is the mass of the particle. Thomson found that the rays coming from cathodes made of different materials all had the same value of q/m: 1.76×10^{11} C/kg.

Thus, it was clear that cathode rays must be made of something all materials have in common. The value of q/m for the cathode-ray particles was about 1800 times larger than the value of q/m for charged hydrogen atoms (ions), which had been known to be 9.6×10^{7} C/kg, as measured in chemical electrolysis experiments. It was therefore obvious that either the charge of the cathode-ray particles is much greater than that of the hydrogen ion, or the mass of the cathode-ray particles is much less than the mass of the hydrogen ion.

To decide between the two possibilities, Thomson also measured the charge q on the negatively charged particles in cathode rays with methods other than deflection by electric and magnetic fields. His experiments were not very accurate, but they were good enough to indicate that the magnitude of the negative charge of a cathode-ray particle was the same or not much different from that of the magnitude of the charge of the positive hydrogen ion in electrolysis. In view of the large value of q/m for the cathode-ray particles, Thomson concluded that the mass m of cathode-ray particles is much less than the mass of hydrogen ions.

In sum, the cathode-ray particles were found to have three important properties:

1. The same types of particles were emitted by a wide variety of different cathode materials.

2. They were much smaller in mass than the hydrogen atom, which has the smallest known mass.
3. They had the same magnitude of charge as the charge on the hydrogen ion.

Thomson therefore concluded that the cathode-ray particles must be a hitherto unobserved constituent of all matter. Since it carries electric charge, the cathode-ray particle was later named the *electron*. It has a negative charge of -1.6×10^{-19} C. The *magnitude* of the electron charge is given the symbol e, where $e = 1.6 \times 10^{-19}$ C.

Thomson's finding meant that the atom was not the ultimate limit to the subdivision of matter. Rather, the electron is part of an atom, and could perhaps even be a basic building block of all atoms. The electron, whose existence Thomson had first proved by quantitative experiment, is one of the fundamental or "elementary" particles of which all matter is made. How do we know this?

13.4 THE SMALLEST CHARGE

In Section 10.2, you read that an electrified comb can pick up a small piece of paper. Obviously, the electric force on the paper must exceed the gravitational force exerted on the paper by the Earth. This observation indicates that electric forces generally are stronger than gravitational forces. Using the same principle, the gravitational force on a microscopically small object (which still contains several billion atoms) can be balanced against the electrical force on the same object when the object has a net electric charge of only a single excess charge. This single charge is the electron. It is one of the basic constituents of all atoms.

The fact that the gravitational force on a small object can be balanced by the electric force is the basis of a method for actually measuring the electron's charge. The method was first employed by the American physicist Robert A. Millikan in 1909.

Suppose a small body of mass m, for example, a tiny drop of oil, has a net negative electric charge of magnitude q. (Millikan used fine droplets of oil from an atomizer. The droplets became charged as they formed a spray. The oil was convenient because of the low rate of evaporation of the droplet.) The negatively charged oil drop is placed in an electric field $\mathbf{E}$ directed downward. A force $\mathbf{F}_{el}$ of magnitude $q\mathbf{E}$ is now exerted on the drop in the *upward* direction. Of course, there is also a downward gravitational force $\mathbf{F}_{grav}$ of magnitude mg on the drop. The oil drop will accelerate

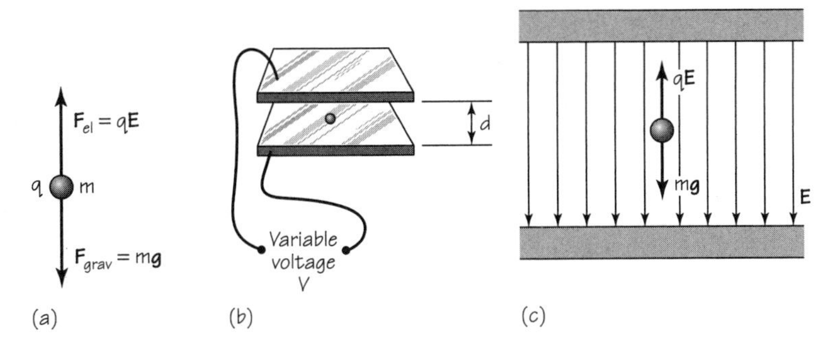

FIGURE 13.6 Schematic diagram of Millikan's oil-drop experiment.

upward or downward, depending on whether the electric force or the gravitational force is greater. By adjusting the magnitude of the electric field strength **E** (i.e., by changing the source that sets up **E**), the two forces, one electric and one gravitational, can be balanced.

What happens when the two forces acting on the drop are balanced? Remember that if a zero *net* force acts on a body, the body can have no acceleration; that is, it would be at rest or continue to move at some constant velocity. In this case, air resistance is also acting as long as the drop moves at all and will soon bring the drop or sphere to rest. The drop will then be in equilibrium. In fact, it will be suspended in mid-air. When this happens, the magnitude of the electric field strength **E** which was applied to produce this condition can be recorded.

Since now the electric force balances the gravitational force, the following must hold:

$$qE = mg.$$

You can calculate the charge q from this equation if you know the quantities E, m, and g, since

$$q = \frac{mg}{E}.$$

Thus, you can find, in the laboratory, what different values of charge q a very small test object can carry.

When you do this, you will discover a remarkable fact: *All possible measurable charges in nature are made up of whole multiples of a smallest charge.* This smallest possible charge is called the *magnitude of the charge on one electron.* By repeating the experiment many times with a variety of small charges,

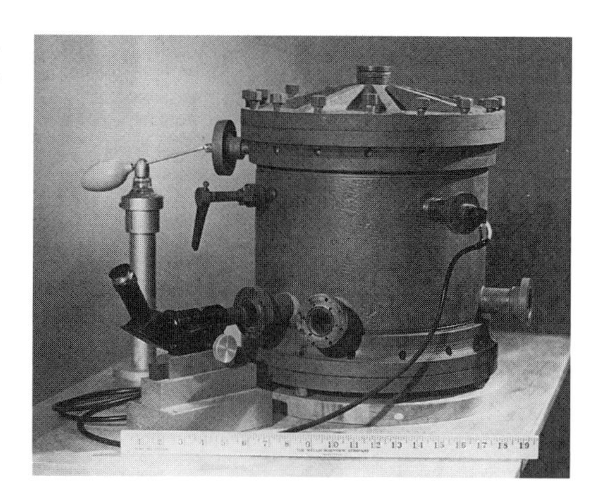

FIGURE 13.7 The original apparatus used by Robert Millikan to conduct his oil-drop experiment.

you can find the value of the charge on one electron. By convention, the charge on an electron is represented by the symbol e.

In effect, this is what Millikan did. He obtained the value of $e = 1.6024 \times 10^{-19}$ C for the electron charge, and he found that the sign of the charge on the electron is negative. Any charge q is therefore given by $q = ne$ where n is the whole number of individual charges, each of magnitude e. Therefore, for example, 1 C is the magnitude of the charge on $1/(1.6 \times 10^{-19})$ electrons. For most purposes you can use the value $e = 1.6 \times 10^{-19}$ C. This value agrees with the results of many other experiments done since then.

No experiment has yet revealed the existence of a smaller unit of charge on ordinary matter, such as an oil drop. However, scientists have found that most elementary particles are in fact composed of smaller entities, known as *quarks*, each of which can have a charge of either $+2/3e$ or $-1/3e$. But quarks cannot exist in isolation, so the smallest charge that can be observed on ordinary mater is still the charge of the electron, e.

13.5 THOMSON'S MODEL OF THE ATOM

What is the atom like? By the beginning of the twentieth century, enough chemical and physical information was available to allow many physicists to devise models of atoms. It was known that negative particles with identical properties (electrons) could be obtained from many different substances and in different ways. This suggested that electrons are parts of all atoms. Electrons are negatively charged. But samples of an element are ordinarily electrically neutral. Therefore, the atoms making up such samples are also presumably neutral. If so, the presence of negative electrons in an atom would seem to require the presence of an equal amount of positive charge.

As mentioned earlier, hydrogen atoms are about 1800 times more massive than electrons, which have a mass of 9.1×10^{-28} g. Thus electrons make up only a very small part of the atomic mass in any atom. Consequently, any model of an atom must take into account the following information:

- an electrically neutral atom contains equal amounts of positive and negative charge;
- the negative charge is associated with only a small part of the mass of the atom, that carried by the electrons.

In addition, any atomic model should answer at least two questions:

1. How many electrons are there in an atom?
2. How are the electrons and the positive charge arranged in an atom?

During the first 10 years of the twentieth century, several atomic models were proposed, but none was satisfactory. Until 1911, the most popular model for the atom was one proposed by J.J. Thomson in 1904. Thomson suggested that an atom consisted of a sphere of positive electricity in which an equal amount of negative charge was distributed in the form of small electrons. From studies of the ionization of atoms in chemistry and through X radiation (discussed in Section 13.8). The amount of positive charge in each atom was surmised to be equal to the atomic number of the atom Z (an integer) times the magnitude of the electron charge e, that is, an atom's positive charge is equal to Ze. In order to render the atom electrically neutral, there were an equal number of negatively charged electrons.

Under these assumptions, Thomson's atom was like a "plum pudding" of positive electricity, with the negative electrons scattered in it like plums or raisins. Thus, hydrogen ($Z = 1$) consisted of one electron, charge $-1e$, embedded in a sphere of positive charge, $1e$. Helium ($Z = 2$) consisted of two electrons embedded in a sphere of positive charge of $2e$, and so on.

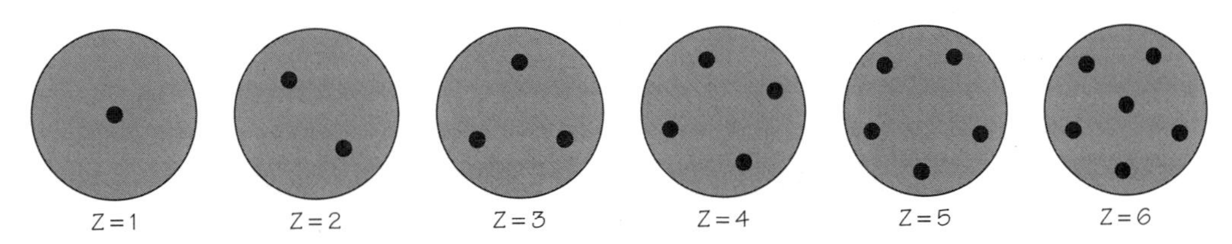

FIGURE 13.8 Thomson's "plum pudding" model of the atom for the first six elements.

The positive "pudding" was assumed to act on the negative electrons, holding them in the atom by electric forces only. Thomson did not know how the positive "pudding" itself was held together, though it should fall apart by electrical repulsion. He took the radius of the atom to be of the order of 10^{-8} cm, based on information from the kinetic theory of gases and other considerations. But his incomplete model was unable to account for either the stability of the atom or for its chemical properties. Much more experimental information and a radical new concept—the concept of the energy quantum—was required to construct a much more satisfactory model of the atom. This concept entered through the use of results from work on an entirely different set of problems—thermal radiation and the puzzle of the photoelectric effect—pursued at that time by Max Planck and Albert Einstein.

13.6 THE PHOTOELECTRIC EFFECT

In 1887, the German physicist Heinrich Hertz was testing Maxwell's theory of electromagnetic waves (see Section 12.4). He noticed that a metallic surface can emit electric charges when light of very short wavelength falls on it. Because light and electricity are both involved, this phenomenon is called the *photoelectric effect*.

When the electric charges so produced passed through electric and magnetic fields, their paths changed in the same ways as the paths of cathode rays. It was therefore deduced that the electric charges emitted by the photoelectric effect consist of negatively charged particles. In 1898, J.J. Thomson measured the value of the ratio q/m for these particles. Using the same method that he had used for cathode-ray particles, Thomson got the same value for the photoelectric particles as he had for the cathode-ray particles. These experiments (and others) demonstrated that photoelectric particles had the same properties as electrons. In fact, physicists now consider that these particles *are* electrons, although they are often referred to as *photoelectrons*, to indicate their origin. Later work showed that all substances (solids, liquids, and gases) exhibit the photoelectric effect under appropriate conditions. However, it is more convenient to study the effect with metallic surfaces.

The photoelectric effect has had an important place in the development of atomic physics because the effect could not be explained in terms of the ideas of classical physics. New ideas had to be introduced to account for the experimental results. In particular, the introduction of the revolutionary concept of *quanta* initiated a new branch of physics called *quantum theory*. This theory and the notion of quanta arose, at least in part, because of the explanation provided for the photoelectric effect.

The basic information for studying the photoelectric effect comes from two kinds of measurements:

1. measurements of the photoelectric current (the number of photoelectrons emitted per unit time); and
2. measurements of the kinetic energies of the photoelectrons after they are emitted.

The photoelectric current can be studied with an apparatus like that sketched in Figure 13.9(a). Two metal plates, C and A, are sealed inside a well-evacuated quartz tube. (Quartz glass is transparent to ultraviolet light as well as to visible light.) The two plates are connected to a source of potential difference (e.g., a high-voltage battery). In the circuit is also an ammeter, which measures the electric current. As long as light strikes plate C, as in sketch (b), electrons with a range of kinetic energies are emitted from

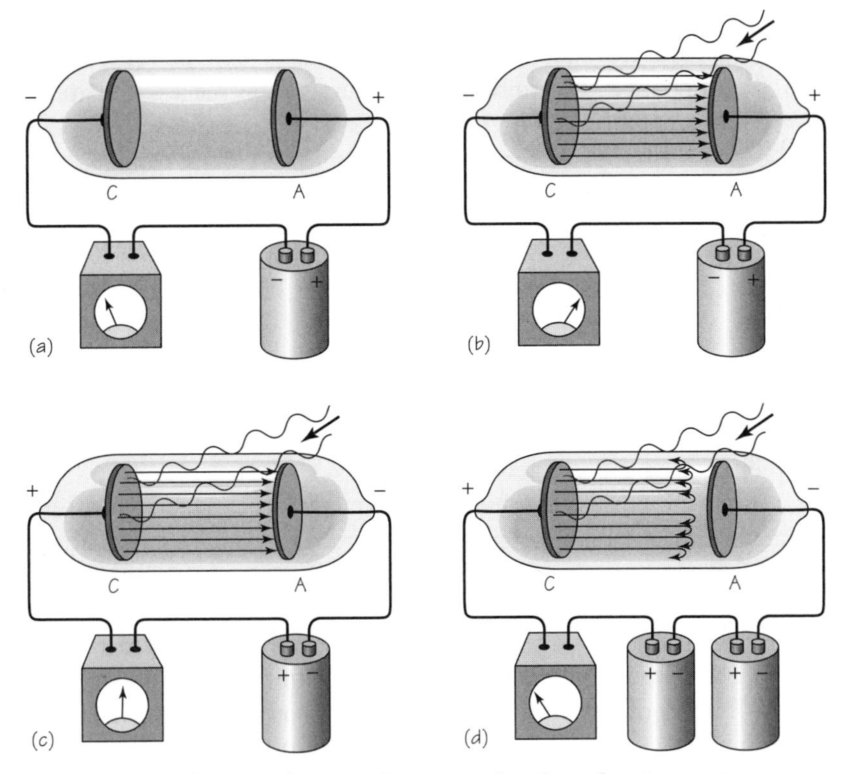

FIGURE 13.9 Schematic diagram of apparatus for photoelectric experiments.

it. If the potential of plate A is positive relative to plate C, these emitted photoelectrons will accelerate to plate A. (Some emitted electrons will reach plate A even if it is not positive relative to C.) The resulting electric current is indicated by the ammeter.

The results of the experiment include the following: the stronger (or more intense) the beam of light of a given color (frequency), the greater the photoelectric current. Any metal used as plate C shows a photoelectric effect, but only if the light has a frequency greater than a certain value. This value of the frequency is called the threshold frequency for that metal. Different metals have different threshold frequencies. But if the incident light has a frequency lower than the threshold frequency, *no* photoelectrons are emitted, no matter how great the intensity of the light or how long the light is left on! This is the first set of surprising discoveries.

The kinetic energies of the electrons can be measured in a slightly modified version of the apparatus, sketched in (c). The battery is reversed so that plate A now tends to repel the photoelectrons. The voltage can be changed from zero to a value just large enough to keep any electrons from reaching plate A, as indicated in (d).

When the voltage across the plates is zero, the meter will indicate a current. This reading shows that the photoelectrons, emerging with kinetic energy from the metallic surface, can reach plate A. As the repelling voltage is increased, the photoelectric current decreases. Eventually a certain voltage is reached at which even the swiftest electrons are repelled and thus the current becomes zero, as indicated in (d). This voltage, which is called the *stopping voltage*, is a measure of the maximum kinetic energy of the emitted photoelectrons (KE_{max}). Using V_{stop} to indicate the stopping voltage, then maximum kinetic energy is given by the relation

$$KE_{max} = eV_{stop},$$

where e is the magnitude of the electron's charge. The measured maximum kinetic energy of the emitted electrons was found to be proportional to the frequency of the incident light.

The above experimental results can be summarized in the following statements. Only the important experimental results are listed here. Their theoretical interpretation will be discussed later.

1. A metal shows a photoelectric effect only if the incident light has a frequency above a certain threshold frequency (symbol f_0).
2. If light of a given frequency does produce a photoelectric effect, the photoelectric current from the surface is proportional to the intensity of the light falling on it.

3. If light of a given frequency releases photoelectrons, the emission of these electrons is *immediate*.
4. The kinetic energies of the emitted electrons display a maximum value, which is proportional to the frequency of the incident light (above the threshold frequency).

The measured time interval between the instant the light strikes the metallic surface and the appearance of electrons is at most 3×10^{-9} s and probably much less. In some experiments, the light intensity used was extremely low. According to the classical wave theory of light, it should take several hundred seconds for an electron to accumulate enough energy from such light to be emitted. But even in these cases, electrons are emitted practically right after the light strikes the surface.

5. The maximum kinetic energy of the photoelectrons increases in direct proportion to the frequency of the light that causes their emission. The maximum *KE* is *not* dependent on the *intensity* of the incident light, as the classical wave theory of light would require.

The way in which the maximum kinetic energy of the electrons varies with the frequency of the incident light is shown in Figure 13.10. The symbols $(f_0)_1$, $(f_0)_2$, $(f_0)_3$ stand for the different threshold frequencies of three different substances. For each substance, the experimental data points fall on a straight line. All the lines have the same slope.

What is most astonishing about the experimental results is that there is a threshold frequency, and that the photoelectrons are emitted if the light frequencies are just above the threshold frequency, no matter how weak the beam of light. But if the light frequencies are just below the threshold

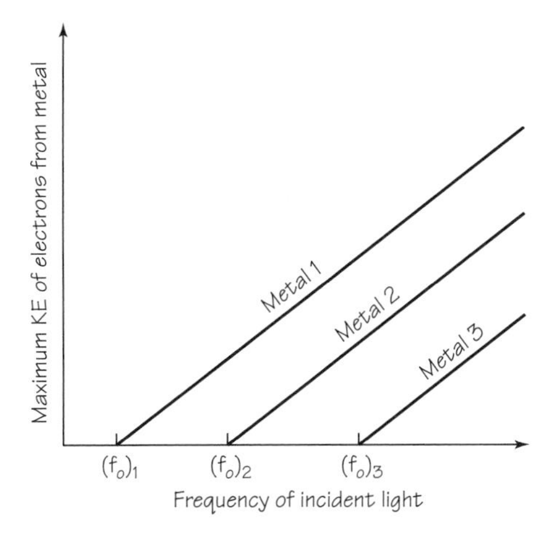

FIGURE 13.10 Photoelectric effect: Maximum kinetic energy of the electrons as a function of the frequency of the incident light. Different metals yield lines that are parallel but have different threshold frequencies.

frequency, no electrons are emitted no matter how great the intensity of the light beam is.

Statements 1, 3, and 4 above could not be explained on the basis of the classical electromagnetic theory of light. How could a low-intensity train of light waves, spread out over a large number of atoms, concentrate, in a very short time interval, enough energy on one electron to knock the electron out of the metal?

Furthermore, the classical wave theory could not account for the existence of a threshold frequency. There seemed to be no reason why a high-intensity beam of low-frequency radiation should not produce photoelectricity if low-intensity radiation of higher frequency could produce it. Neither could classical theory explain why the maximum kinetic energy of the photoelectrons increases directly with the frequency of the light, but is independent of the intensity. Thus, the photoelectric effect posed an important challenge that the classical wave theory of light was not able to meet.

13.7 EINSTEIN'S THEORY OF THE PHOTOELECTRIC EFFECT

The explanation of the photoelectric effect was the major work cited in the award to Albert Einstein of the Nobel Prize in Physics in 1921. Einstein's explanation, proposed in 1905, played a major role in the development of atomic physics. He based his theory on a daring hypothesis, for few of the experimental details were known in 1905. Moreover, the key point of Einstein's explanation contradicted the classical ideas of the time.

Einstein assumed that the energy of light is not distributed evenly over the whole expanding wave front (as the classical theory assumed). Instead, the light energy is concentrated in separate "lumps." In addition, the amount of energy in each of these regions is not just any amount, but a definite amount of energy that is proportional to the frequency f of the light wave. The proportionality factor is a constant (symbol h); it is called *Planck's constant* for reasons which will be discussed later. Thus, in this model, the light energy in a beam of frequency comes in pieces, each of amount

$$E = hf$$

where

$$h = 6.6 \times 10^{-34}\ \mathrm{J/s}.$$

The amount of radiant energy in each piece is called a *light quantum* (*quantum* is Latin for *quantity*), or a quantum of light energy. The quantum of light energy was later called a *photon*.

There is no explanation clearer or more direct than Einstein's. A quote from his first paper (1905) on this subject is given here. Only the notation is changed, in order to agree with modern practice (including the notation used in this text):

> . . . According to the idea that the incident light consists of quanta with energy hf, the ejection of cathode rays [photoelectrons] by light can be understood in the following way. Energy quanta penetrate the surface layer of the body, and their energy is converted, at least in part, into kinetic energy of electrons. The simplest picture is that a light quantum gives up all its energy to a single electron; we shall assume that this happens. . . . An electron provided with kinetic energy inside the body may have lost part of its kinetic energy by the time it reaches the surface. In addition, it is to be assumed that each electron, in leaving the body, has to do an amount of work W (which is characteristic of the body). The electrons ejected directly from the surface and at right angles to it will have the greatest velocities perpendicular to the surface. The maximum kinetic energy of such an electron is
>
> $$KE_{max} = hf - W.$$
>
> If the plate C is charged to a positive potential, V_{stop}, just large enough to keep the body from losing electric charge, we must have
>
> $$KE_{max} = hf - W = eV_{stop},$$
>
> where e is the magnitude of the electronic charge. . . .
>
> If the derived formula is correct, then V_{stop}, when plotted as a function of the frequency of the incident light, should yield a straight line whose slope should be independent of the nature of the substance illuminated.

The first equation in the above quotation is usually called *Einstein's photoelectric equation*. Let us compare Einstein's photoelectric equation with the experimental results, in the order given in the previous section, to test whether or not his theory accounts for those results:

1. According to the photoelectric equation, the kinetic energy of the photoelectrons is greater than zero only when the photon energy hf is greater than the work W, which is the work the electron must do against the forces of attraction of the material in the cathode C as it leaves the metal. The energy required to escape the metal is known as the *work*

function. Therefore, an electron can be emitted only when the frequency of the incident light is greater than a certain lowest value corresponding to the work required to exit from the metal. In symbols, the minimum frequency f_0 is defined by the relation $hf_0 = W$.

2. Next, according to Einstein's photon model of light, it is an individual photon that ejects an electron (if above the threshold frequency of light). Now, the intensity of the light is proportional to the number of the photons in the light beam. In addition, the number of photoelectrons ejected is proportional to the number of photons incident on the surface. Therefore, the number of electrons ejected (and with it the photoelectric current) is proportional to the intensity of the incident light. (However, not every photon in a light beam strikes an electron, emitting it from the metal; only about 1 in 50 photons does so.)
3. In Einstein's model, the light energy is concentrated in a stream of light quanta (photons). So no time is needed for the electron to collect light energy. Instead, the quanta transfer their energy immediately to the photoelectrons, which emerge after the very short time required for them to escape from the surface.
4. Finally, the photoelectric equation predicts that the greater the frequency of the incident light, the greater the maximum kinetic energy of the ejected electrons. According to the photon model, the photon's energy is directly proportional to the light frequency. The minimum energy needed to eject an electron is the energy required for the electron to escape from the metal surface. This explains why light of frequency less than some frequency f_0 cannot eject any electrons. The kinetic energy of the escaping electron is the difference between the energy of the absorbed photon and the energy lost by the electron in escaping the surface.

Testing Einstein's Theory by Experiment

As you can see, Einstein's photoelectric equation agreed qualitatively with the experimental results. But could it hold up under quantitative experimental testing? In particular:

(1) Does the maximum kinetic energy of the electrons vary in direct proportion to the light frequency?
(2) Is the proportionality factor h really the same for all substances?

For 10 years, experimental physicists attempted these quantitative tests. One experimental difficulty was that the value of W for a metal changes greatly if there are impurities (e.g., a layer of oxidized metal) on the sur-

ALBERT EINSTEIN

Albert Einstein (1879–1955) was born in the city of Ulm, in Germany. Like Newton, he showed no particular intellectual promise as a youngster. He received his early education in Germany, but at the age of 15, dissatisfied with the discipline in school and militarism in the nation, he left and went to Switzerland. After graduation from the Polytechnic Institute in Zurich, Einstein (in 1902) found work in the Swiss Patent Office in Berne. This job gave Einstein a salary to live on, to get married, and an opportunity to use his spare time for working in physics on his own. In 1905, he published three papers of immense importance. One dealt with quantum theory and included his theory of the photoelectric effect. Another treated the problem of molecular motions and sizes, and worked out a mathematical analysis of the phenomenon of "Brownian motion." Einstein's analysis and the subsequent experimental work by Jean Perrin, a French physicist, provided a strong argument for the molecular motions assumed in the kinetic theory. Einstein's third 1905 paper provided the theory of special relativity, which revolutionized modern thought about the nature of space and time, and of physical theory itself.

In 1915, Einstein published a paper on the theory of general relativity. In it he provided a new theory of gravitation that included Newton's theory as a special case.

When Hitler and the Nazis came to power in Germany in 1933, Einstein went to the United States and became a member of the Institute for Advanced Study at Princeton. He spent the rest of his working life seeking a unified theory which would include gravitation and electromagnetism. Shortly before the Germans launched World War II, Einstein signed a letter to President Roosevelt, warning of the war potential of an "atomic bomb," for which the Germans had all necessary knowledge and motivation to build. After World War II, Einstein devoted much of his time to organizations advocating world agreements to end the threat of atomic warfare.

face. Finally, in 1916, Robert A. Millikan established that there is indeed a straight-line relationship between the frequency of the absorbed light and the maximum kinetic energy of the photoelectrons, as required by Einstein's equation. If KE_{max} is plotted along the y-axis and f along the x-axis, then Einstein's equation exhibits the familiar form of the equation for a straight line

$$y = mx + b.$$

In a graph of Einstein's equation, the slope should be equal to h, and the KE-intercept should be equal to $-W$. This is exactly what Millikan found.

In order to obtain his data, Millikan designed an apparatus in which the

FIGURE 13.11 Robert Andrews Millikan (1868–1953), an American physicist, attended Oberlin College, where his interest in physics was only mild. After his graduation, he became more interested in physics, taught at Oberlin while taking his master's degree, and then obtained his doctorate degree from Columbia University in 1895. After more study in Germany, Millikan went to the University of Chicago, where he became a professor of physics in 1910. His work on the determination of the electron's charge took place from 1906 to 1913. He was awarded the Nobel Prize in physics in 1923 for this research and for the very careful experiments which resulted in the verification of the Einstein photoelectric equation. In 1921, Millikan moved to what became known as the California Institute of Technology, eventually to become its president.

metal photoelectric surface was cut clean while in a vacuum. A knife inside the evacuated volume was manipulated by an electromagnet outside the vacuum to make the cuts. This rather intricate arrangement was required to achieve a pure metal surface.

The straight-line graphs Millikan obtained for different metals all had the same slope, h, even though the threshold frequencies (related to W) were different. The value of h could be obtained from Millikan's measurements, and it was the same for each metal surface. Also, it agreed very well with a value obtained by means of other, independent methods. Much to Millikan's own surprise (who started out believing in the classical theory of light), his experiment provided a quantitative verification of Einstein's theory of the photoelectric effect. Thus, it can be said that Einstein's equation led to two Nobel Prizes: one to Einstein, who derived it theoretically, and one to Millikan, who verified it experimentally.

The Idea of Energy Quanta

Historically, the first suggestion that the energy in electromagnetic radiation is "quantized" (comes in definite quanta) did not come from the photoelectric effect. Rather, it came from studies of the heat and light radiated by hot solids. Max Planck, a German physicist, introduced the concept of quanta of energy (though in a different context) in 1900, 5 years before Einstein's theory. Thus, the constant h is known as Planck's constant.

Planck was trying to explain how the heat (and light) energy radiated by a hot body is related to the frequency of the radiation. Classical physics (nineteenth-century thermodynamics and electromagnetism) could not account for the experimental facts. Planck found that the facts could be interpreted only by assuming that atoms, on radiating, change their energy not in varying, but in separate, quantized amounts, E. Einstein's theory of the photoelectric effect was actually an extension and application of Planck's quantum theory of thermal radiation. The essential point was that Einstein postulated the change in the atom's energy E being carried off was located in a distinct photon of energy $E = hf$, where f is the frequency of the light emitted by the atom, rather than being spread continuously over the light wave.

The experiments and the theory of radiation are much more difficult to describe than the experiments and theory of the photoelectric effect. By now, many tests have been made of both Planck's and Einstein's conceptions. In all cases, Planck's constant h is regarded as one of the universal constants of nature. It is found to have the same basic position in quantum physics that Newton's universal constant G has in the physics of gravitation. As Planck said, it is an absolute constant in the sense that even if extraterrestrials

FIGURE 13.12 Max Planck (1858–1947), a German physicist, was the originator of the quantum theory, one of the two great revolutionary physical theories of the twentieth century (along with Einstein's relativity theory). Planck won the Nobel Prize in 1918 for his work on quantum theory. He tried for many years to show that this theory could be understood in terms of the classical physics of Newton and Maxwell, but this attempt did not succeed. Quantum physics is fundamentally different because of its postulate that energy in light and matter exists in finite quanta and is not continuously divisible.

exist on other planets, they would find the same value for h (as indeed they would find for G, e, c, and all other physical constants of nature).

The Physicists' Dilemma

The photoelectric effect presented physicists with a real dilemma. According to the classical wave theory, light consists of electromagnetic waves extending continuously throughout space. This theory was highly successful in explaining optical phenomena (reflection, refraction, polarization, interference). Light behaves like a wave experimentally, and Maxwell's theory gives a good account of this wave behavior. But Maxwell's theory could not account for the photoelectric effect. Einstein's theory, which postulated the existence of separate quanta of light energy, accounted for the photoelectric effect. But it could not account for other properties of light, such as interference. The result was that there were two models for light whose basic concepts seemed to contradict each other. According to one, light is a wave phenomenon; according to the other, light has particle-like properties. Each model had its successes and limits. What, if anything, could be done about the contradictions between the two models? You will see later that this problem and its treatment occupy a central position in modern physics.

13.8 X RAYS

In 1895, the German physicist Wilhelm Konrad Röntgen made a discovery that surprised him and all the world. Like the photoelectric effect, it did not fit with accepted ideas about electromagnetic waves, and eventually it too required the introduction of quanta for a complete explanation. The discovery was that of X rays, which were often called "Röntgen" rays, after their discoverer. The consequences for atomic physics, medicine, and technology were dramatic and important.

On November 8, 1895, Röntgen was experimenting with the newly found cathode rays, as were many physicists all over the world. According to a biographer:

> . . . he had covered the all-glass pear-shaped tube [Crookes tube] with pieces of black cardboard, and had darkened the room in order to test the opacity of the black paper cover. Suddenly, about a yard from the tube, he saw a weak light that shimmered on a little bench he knew was nearby. Highly excited, Röntgen lit a match and, to his great surprise, discovered that the source of the mysterious light was a little barium platinocyanide screen lying on the bench.

FIGURE 13.13 Wilhelm Röntgen (1845–1923).

Barium platinocyanide, a mineral, is one of the many chemicals known to *fluoresce* (emit visible light when illuminated with ultraviolet light). But no source of ultraviolet light was present in Röntgen's experiment. Cathode rays had not been observed to travel more than a few centimeters in air. So, neither ultraviolet light nor the cathode rays themselves could have caused the fluorescence. Röntgen therefore deduced that the fluorescence involved rays of a new kind. He named them *X rays*, since the rays were an unknown nature.

In an intensive and thorough series of experiments over the next 7 weeks, he determined the properties of this new radiation. Röntgen reported his results on December 28, 1895, in a paper whose title (translated) is "On a New Kind of Rays."

Röntgen's paper described nearly all of the properties of X rays that are known even now. It described the method of producing the rays and proved that they originated in the glass wall of the tube, where the cathode rays struck it. Röntgen showed that X rays travel in straight lines from their place of origin, and that they darken a photographic plate. He reported in detail the ability of X rays to penetrate various substances, such as paper, wood, aluminum, platinum, and lead. Their penetrating power was greater through light materials (paper, wood, flesh) than through dense materials (platinum, lead, bone). He described and exhib-

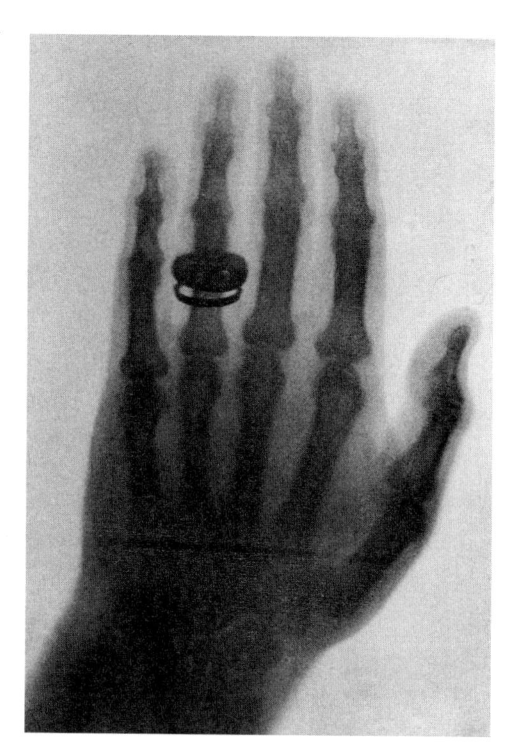

FIGURE 13.14 X-ray image of Frau Röntgen's hand with rings. © Bettmann/CORBIS.

ited photographs showing "the shadows of bones of the hand, of a set of weights inside a small box, and of a piece of metal whose inhomogeneity becomes apparent with X rays." He gave a clear description of the shadows cast by the bones of the hand on the fluorescent screen. Röntgen also reported that the X rays were not deflected by a magnetic field. He also could not show their reflection, refraction, or interference effects in ordinary optical apparatus.

J.J. Thomson discovered one of the most important properties of X rays a month or two after the rays themselves had become known. He found that when the rays pass through a gas, they make it a conductor of electricity. Thomson attributed this effect to "a kind of electrolysis, the molecule being split up, or nearly split up by the Röntgen rays." The X rays, in passing through the gas, knock electrons loose from some of the atoms or molecules of the gas. The atoms or molecules that lose these electrons become positively charged. They are called *ions* because they resemble the positive ions in electrolysis, and the gas is said to be *ionized*. Also, the freed electrons may attach themselves to previously neutral atoms or molecules, giving them negative charges.

Röntgen and Thomson found, independently, that electrified bodies lose their charges when the air around them is ionized by X rays. (It is now easy to see why: the electrified body attracts ions of the opposite charge from the air.) The rate of discharge depends on the intensity of the rays (hence the amount of ionization). This property was therefore used—and still is—as a convenient quantitative means of measuring the intensity of an X ray beam. As a result, careful quantitative measurements of the properties and effects of X rays could be made.

Are X Rays Waves or Particles?

One problem that aroused keen interest following the discovery of X rays concerned the nature of the mysterious rays. Unlike charged particles (electrons, for example) they were not deflected by magnetic or electric fields. Therefore, it seemed that they had to be either neutral particles or electromagnetic waves. It was difficult to choose between these two possibilities. On the other hand, no neutral particles of atomic size (or smaller) that had the penetrating power of X rays were then known. The existence of such particles would be extremely hard to prove, because there was no way of getting at them. On the other hand, if the X rays were electromagnetic waves, they would have to have extremely short wavelengths because only in this case, according to theory, could they have high penetrating power and show no refraction or interference effects with ordinary optical apparatus.

As discussed in Chapter 8, distinctly wave-like properties become apparent only when waves interact with objects, such as slits in a barrier, that are smaller than several wavelengths across. The wavelengths hypothesized for X rays would be on the order of 10^{-10} m (see the diagram of the electromagnetic spectrum in Section 12.5). So a demonstration of their wave behavior would require a diffraction grating with slits spaced about 10^{-10} m apart. Evidence from kinetic theory and from chemistry indicated that atoms were about 10^{-10} m in diameter. It was suggested, therefore, that X rays might be diffracted measurably by crystals in which the atoms form orderly layers about 10^{-10} m apart.

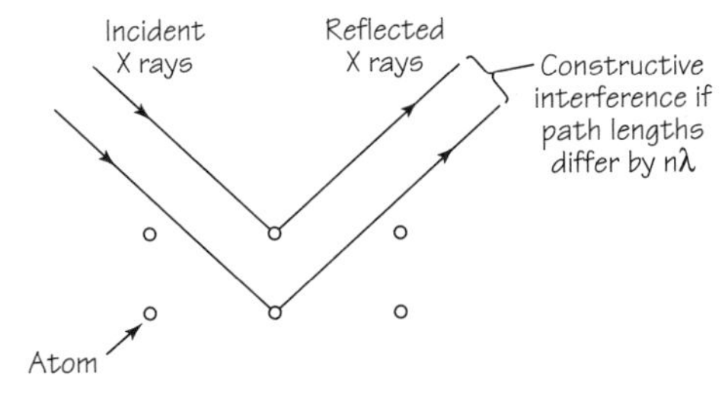

FIGURE 13.15 X-ray diffraction patterns from a metal crystal formed on a photographic film. The black spots are produced by constructive interference of X rays scattered from atoms.

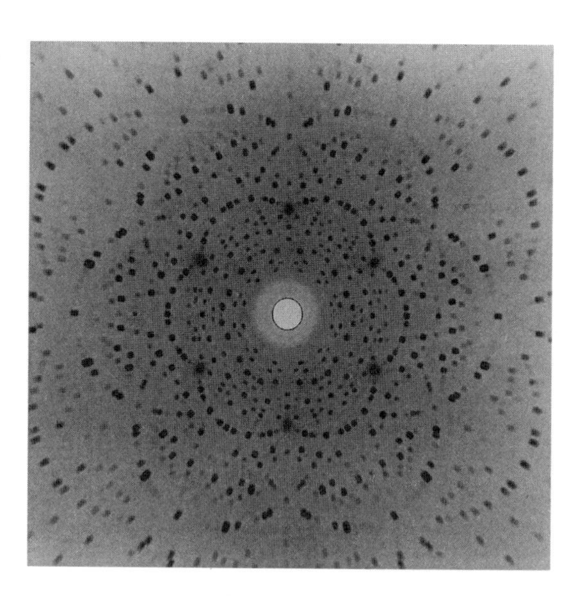

FIGURE 13.16 X-ray diffraction pattern from beryllium crystal.

Such experiments succeeded in 1912. The layers of atoms did act like diffraction gratings, and X rays did, indeed, act like electromagnetic radiations of very short wavelength (like *ultra*-ultraviolet light). These experiments are more complicated to interpret than diffraction of a beam of light by a single, two-dimensional optical grating. The diffraction effect occurs in three dimensions instead of two. Therefore, the diffraction patterns are far more elaborate (see Figure 13.16).

However, in addition to wave properties, X rays were found to have quantum properties, which meant they also exhibited particle-like behavior. For example, they can cause the emission of electrons from metals. These electrons have greater kinetic energies than those produced by ultraviolet light. (The ionization of gases by X rays is also an example of the photoelectric effect. In this case, the electrons are freed from the atoms and molecules of the gas.) Thus, X rays also require quantum theory for the explanation of some of their behavior. So, like light, X rays were shown to have both wave and particle properties.

The Discovery Causes a Sensation

Röntgen's discovery of X rays excited scientists throughout the world. His experiments were immediately repeated and extended in many laboratories in Europe and America. Scientific journals during the year 1896 were filled with letters and articles describing new experiments or confirming the results of earlier experiments. The passage of electricity through gases had been a popular topic for study by physicists; this made widespread experimentation much easier during the years after Röntgen's discovery, because many physics laboratories had cathode-ray tubes and so could produce X rays easily.

The spectacular use of these rays in medicine generated intense interest in X rays. Within 3 months of Röntgen's discovery, X rays were put to practical use in surgical operations in a hospital in Vienna. The use of this new aid to surgery spread rapidly. Since Röntgen's time, X rays have revolutionized some phases of medical practice, especially the diagnosis of some diseases and the treatment of some forms of cancer (because X rays also can destroy malignant tissue). Extremely important uses of X rays occur as well in other fields of applied science, both physical and biological. Among these are the study of the crystal structure of materials; "industrial diagnosis," such as the search for possible defects in materials and engineering structures; the study of what is behind the optically visible surface of old paintings and sculptures; and many others.

The public reaction to the discovery of X rays was sensational also. Many people rushed to have their bodies irradiated by the new rays, thinking they have miraculous properties, while others worried about moral decline if modesty gave way to "X-ray vision." A Berlin newspaper even published a full-body X ray of the Kaiser, giving new meaning in retrospect to the old story of "The Emperor's New Clothes."

From the apparatus used in Röntgen's original discovery, there emerged two paths of development in medical technology. One concentrated on a fluoroscope, the other focused on improving the radiograph; Thomas Edison was crucial in both.

It was Edison's familiarity with the Crookes tube, which was very similar to his 1879 invention of the electric light bulb, that enabled him to make one of the first improvements to X-ray technology. By constructing a tube using thinner glass, Edison found that more X rays could escape. Edison also led the investigation which found that calcium tungstate could produce a clearer image on the fluorescent screen than the previously used barium platinocyanide. Edison took this knowledge and applied it to the manufacture of a "fluoroscope," a device which allowed a person to peer through a box at a screen coated in calcium tungstate, and see a moving image of the inside of their, or another person's, body, as it moved beneath the screen.

A friend of Edison's, Michael Pupin, took this improvement in the fluorescent screen and combined it with a photographic plate, which reduced the patient exposure time for a radiograph from 1 hour to just a few minutes, while also increasing the clarity of the picture. This greatly reduced the danger of damage to tissues.

X-RAY APPLICATIONS

Originally, X rays were produced in Röntgen's laboratory when cathode rays (electrons) struck a target (the glass wall of the tube). Today, X rays usually are produced by directing a beam of high-energy electrons onto a metal target. As the electrons are deflected and stopped, X rays of various energies are produced. The maximum energy a single ray can have is the total kinetic energy the incident electron gives up on being stopped. So the greater the voltage across which the electron beam is accelerated, the more energetic and penetrating are the X rays.

X rays are used in a wide range of fields. Computer-aided tomography (CAT) allows doctors to create three-dimensional images of a patient's body. X rays are also used in art restoration, revealing structural defects in sculptures and enabling the assessment of previous restoration efforts of paintings and frescoes.

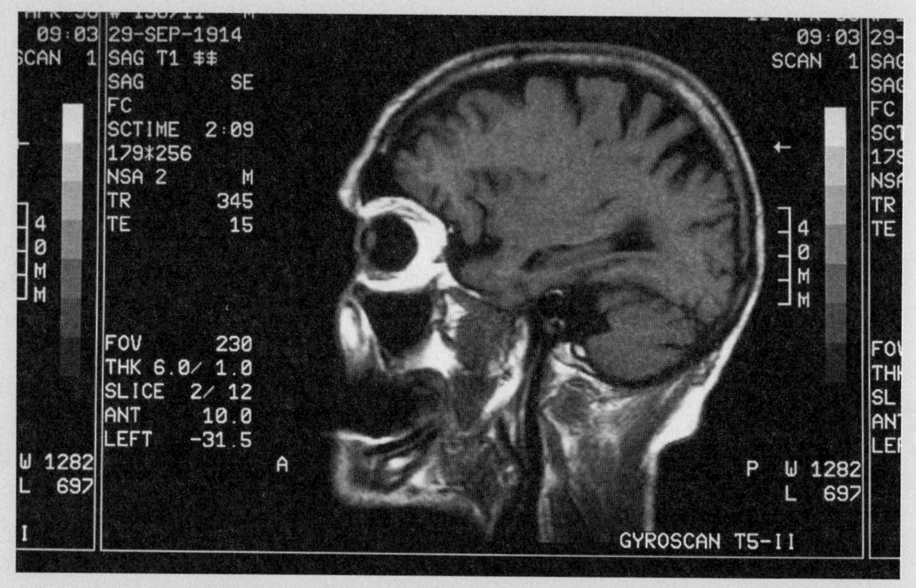

FIGURE 13.17 Computed tomography scan of a person's head in which X rays are used.

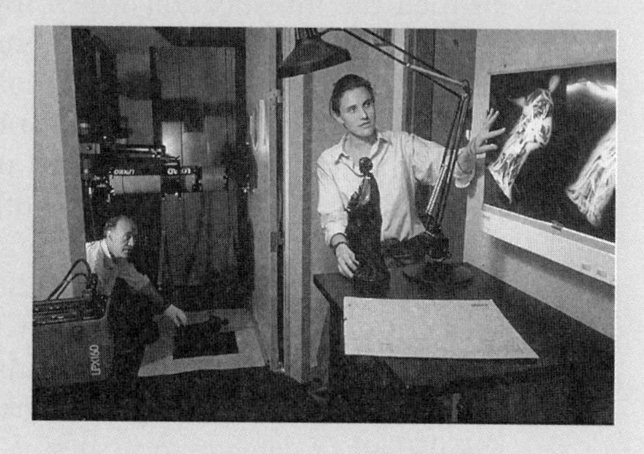

FIGURE 13.18 Another application of X rays. Here, a bronze sculpture by Gian Lorenzo Bernini is being examined by comparison with X-radiograph images (displayed on screen). Such non-invasive procedures have become essential tools in the field of art preservation.

One of the most problematic aspects of X-ray technology for the first two decades was the unreliability of the glass tubes, which often cracked when heated. The high incidence of cracking was eliminated in 1913 when William Coolidge, while working for General Electric, invented the high-vacuum, hot-cathode, tungsten-target X-ray tube. As part of his research into electric-light-bulb filaments, Coolidge found that because tungsten vaporized less than any other metals, it could reduce the buildup of gas residue. Applying this knowledge, Coolidge replaced platinum with tungsten in the cathode-ray tube. When these "Coolidge tubes" came on the market in 1913, they had many benefits over the previous design; they could produce clearer duplications of previous images, be adjusted much more accurately, and because of their increased flexibility, they could arrange to go instantly from high to low penetration.

After World War II the monopoly of X rays as a medical imaging tool began to be challenged by new technologies that combined computers and television to help in obtaining an image of those parts of the human body which X rays had failed to reach. They include CT (computed tomography) scanning, MRI (magnetic resonance imaging), PET (positron emission tomography), and ultrasound.

FIGURE 13.19 Edison examining his hand under a fluoroscope, ca. 1896.

Many of the scientific and technological advances, on which these "daughter technologies" depended, were taking place throughout the first half of this century. However, it took the invention and subsequent improvements in computer technology, before the complex algorithms required for the imaging techniques involved in CT scanning and MRI could be calculated. These advanced medical imaging techniques, whose importance is now established in medical diagnostics, are a world away from the simple technology that allowed Röntgen to see through his skin in 1895, still, despite these advances, X rays remain the most commonly used imaging tools, largely because of the simplicity, and the size, and the financial cost of its "daughter technologies."

SOME NEW IDEAS AND CONCEPTS

atomic mass units
atomic number
cathode
cathode rays
Einstein's photoelectric equation
electron
law of fixed proportions
light quantum
periodic table
photoelectric effect
photoelectron
photon
stopping voltage
Thomson's model of the atom
work function

SOME IMPORTANT EQUATIONS

$$E = hf,$$

$$KE_{max} = hf - W,$$

$$KE_{max} = eV_{stop}.$$

FURTHER READING

G. Holton and S.G. Brush, *Physics, The Human Adventure* (Piscataway, NJ: Rutgers University Press, 2001), Chapter 26.

B.H. Kevles, *Naked to the Bone: Medical Imaging in the Twentieth Century.* Sloan Technology Series (New Brunswick, NJ: Rutgers University Press, 1997).

STUDY GUIDE QUESTIONS

13.1 The Periodic Table

1. How did Mendeleev arrange the elements on the periodic table?
2. Describe some of the common features of different groups of elements.
3. Which element is element 56? What is its atomic number? atomic mass?

13.2 The Idea of Atomic Structure

1. How did the periodic table suggest that atoms might have a structure?
2. What changes with respect to the structure of the atoms would one find as one progressed through the periodic table?

13.3 Cathode Rays

1. What was the most convincing evidence to support the fact that cathode rays were not electromagnetic radiation?
2. What was the reason given for the ratio q/m for electrons being about 1800 times larger than q/m for hydrogen ions?
3. What were two main reasons for Thomson's belief that electrons may be "building blocks" from which all atoms are made?

13.4 The Smallest Charge

1. How can the small oil drops or plastic spheres used in the Millikan experiment experience an electric force upward if the electric field is directed downward?
2. What did the results of the Millikan experiment indicate about the nature of electric charge?
3. In Millikan's experiment, what is necessary to keep a charged droplet suspended in place?

13.5 Thomson's Model of the Atom

1. Describe Thomson's model of the atom. Give the reasoning behind the inclusion of each component and their arrangement.
2. What was one problem with this model?
3. Draw a picture of an atom of oxygen, according to Thomson, and label everything in it.

13.6 The Photoelectric Effect

1. Light falling on a certain metal surface causes electrons to be emitted. What happens to the photoelectric current as the intensity of the light is decreased?
2. What happens as the frequency of the light is decreased?
3. Which observations could not be explained by the classical wave theory of light?
4. Sketch a rough diagram of the equipment and circuit used to demonstrate the main facts of photoelectricity.

13.7 Einstein's Theory of the Photoelectric Effect

1. High-frequency light falls on the metal surface of a cathode in a cathode-ray tube, producing a photocurrent. Use Einstein's theory to describe what happens in each step of this phenomenon.
2. Briefly explain how Einstein's theory accounts for each of the observations listed in Section 13.6.
3. Describe an experiment that tested Einstein's theory. What was the result?
4. Einstein's idea of a quantum of light had a definite relation to the wave model of light. What was it?
5. Why does the photoelectron freed from the metal surface not have as much energy as the quantum of light that causes it to be ejected?
6. What does a stopping voltage of 2.0 V indicate about the photoelectrons emerging from a metal surface?
7. Write down Einstein's photoelectric equation and define the meaning of every symbol in it.
8. Einstein's equation contains the maximum kinetic energy, KE_{max} of the photoelectrons. Under what conditions would the photoelectrons have less than maximum kinetic energy?
9. The success of Einstein's theory posed a dilemma for physicists. What was it?

13.8 X Rays

1. What processes can produce X rays?
2. X rays were the first "ionizing" radiation discovered. What does "ionizing" mean?
3. What are three properties of X rays that led to the conclusion that X rays were electromagnetic waves?
4. What was the experimental and theoretical evidence to support the notion that X rays had very short wavelengths?

DISCOVERY QUESTIONS

1. If classical physics was so successful, why did new theories arise?
2. Evaluate Thomson's "plum-pudding model" as a model of atoms. Was it a good model in terms of what scientists wanted from a model and what they knew at the time about the atom? What were its strengths and weaknesses?
3. A television tube or a computer-monitor tube is often called a CRT, which stands for "cathode-ray tube." Look up the operation of one of these tubes in an encyclopedia or on-line information source. What do these tubes have to do with cathode rays? How do they work?
4. At light frequencies below the threshold frequency no photoelectrons are emitted. What might happen to the light energy?
5. Planck's constant h has the value 6.6×10^{-34} J/s. What does this tell us about the magnitude of the energy carried by a light quantum?

Quantitative

1. The charge on an electron is 1.6×10^{-19} C. To gain an idea of the small size of an electron charge consider the following. A current of 1 A flows for 1 s down a wire. How many electrons enter the wire?
2. How many joules of energy does one photon of ultraviolet light carry? Assume its frequency to be 1.5×10^{15} Hz.
3. An average photon of visible light has a frequency of about 1×10^{14} Hz. To gain an idea of the small size of a quantum of energy, consider the following. A 100-W light bulb is turned on for 1 s. Only about 5% of the electric energy input to the tungsten filament is given off as visible light. How many photons does the bulb emit?
4. For most metals, the work function W in Einstein's photoelectric equation is about 10^{-18} J. Light of what frequency will cause photoelectrons to leave the metal with virtually no kinetic energy? In what region of the spectrum is this frequency?
5. Monochromatic light of wavelength 5×10^{-7} m falls on a metal cathode to produce photoelectrons. The light intensity at the surface of the metal is 10^2 J/m^2 s.
   (a) What is the frequency of the light?
   (b) What is the energy (in joules) of a single photon of the light?
   (c) How many photons fall on 1 m^2 in 1 s?
   (d) If the diameter of an atom is about 10^{-10} m, how many photons fall on one atom in 1 s, on the average?
   (e) How often would one photon fall on one atom, on the average?
   (f) How many photons fall on one atom in 10^{-10} s, on the average?
   (g) Suppose the cathode is a square 0.05 m on a side. How many electrons are released per second, assuming every photon releases a photoelectron? (In fact, only about 1 in 50 photons does so.) How big a current would this be in amperes?

CHAPTER 14

The Quantum Model of the Atom

14.1 SPECTRA OF GASES

One of the first important clues to understanding atomic structure involved the study of the emission and absorption of light by the atoms of different elements. Physicists knew from Maxwell's theory that light is emitted and absorbed only by accelerating charges. This suggested that the atom might contain moving charges. Patterns and regularities in the properties of the light emitted we expected to provide valuable clues about the precise nature of the motions of the moving charges. The results of this study were so important to the unraveling of atomic structure that we review their development here in some detail.

Emission Spectra

It has long been known that light is emitted by gases or vapors when they are excited in any one of several ways:

(1) by heating the gas to a high temperature, as when a volatile substance is put into a flame;
(2) by an electric discharge through gas in the space between the terminals of an electric arc; or
(3) by a continuous electric current in a gas at low pressure, as in the now familiar "neon sign."

The Scottish physicist Thomas Melvill made the pioneering experiments on light emitted by various excited gases in 1752. He put one substance after another in a flame, "having placed a pasteboard with a circular hole in it between my eye and the flame . . . , I examined the constitution of these different lights with a prism." Melvill found that the spectrum of visible light from a hot gas of a single element was different from the well-known rainbow-colored spectrum of a glowing solid or liquid. Melvill's spectrum was not an unbroken stretch of color continuously graded from violet to red. Rather, it consisted of individual patches, each having the color of that part of the spectrum in which it was located. There were dark gaps—missing colors—between the patches. Later, more general use was made of a narrow slit through which to pass the light. Now the spectrum of a gas was seen as a set of bright lines (see Figure 14.1). The bright lines are in fact colored images of the slit. Such spectra show that light from a gas is a mixture of only a few definite colors or narrow wavelength regions of light. These types of spectra are called *emission spectra* or *bright-line spectra*, and their study is known as *spectroscopy*.

Melvill also noted that the colors and locations of the bright lines were different when different substances were put into the flame. For example, with ordinary table salt in the flame, the dominant color was "bright yellow" (now known to be characteristic of the element sodium). In fact, the bright-line spectrum is markedly different for each chemically different gas because each chemical element emits its own characteristic set of wavelengths. In looking at a gaseous source without the aid of a prism or a grating, the eye combines the separate colors. It perceives the mixture as reddish for glowing neon, pale blue for nitrogen, yellow for sodium vapor, and so on.

Some gases have relatively simple emission spectra. Thus, the most prominent part of the visible spectrum of sodium vapor is a pair of bright yellow lines. This is why, for example, the flame in a gas stove turns yellow when soup, or any liquid containing salt, boils over. Sodium-vapor

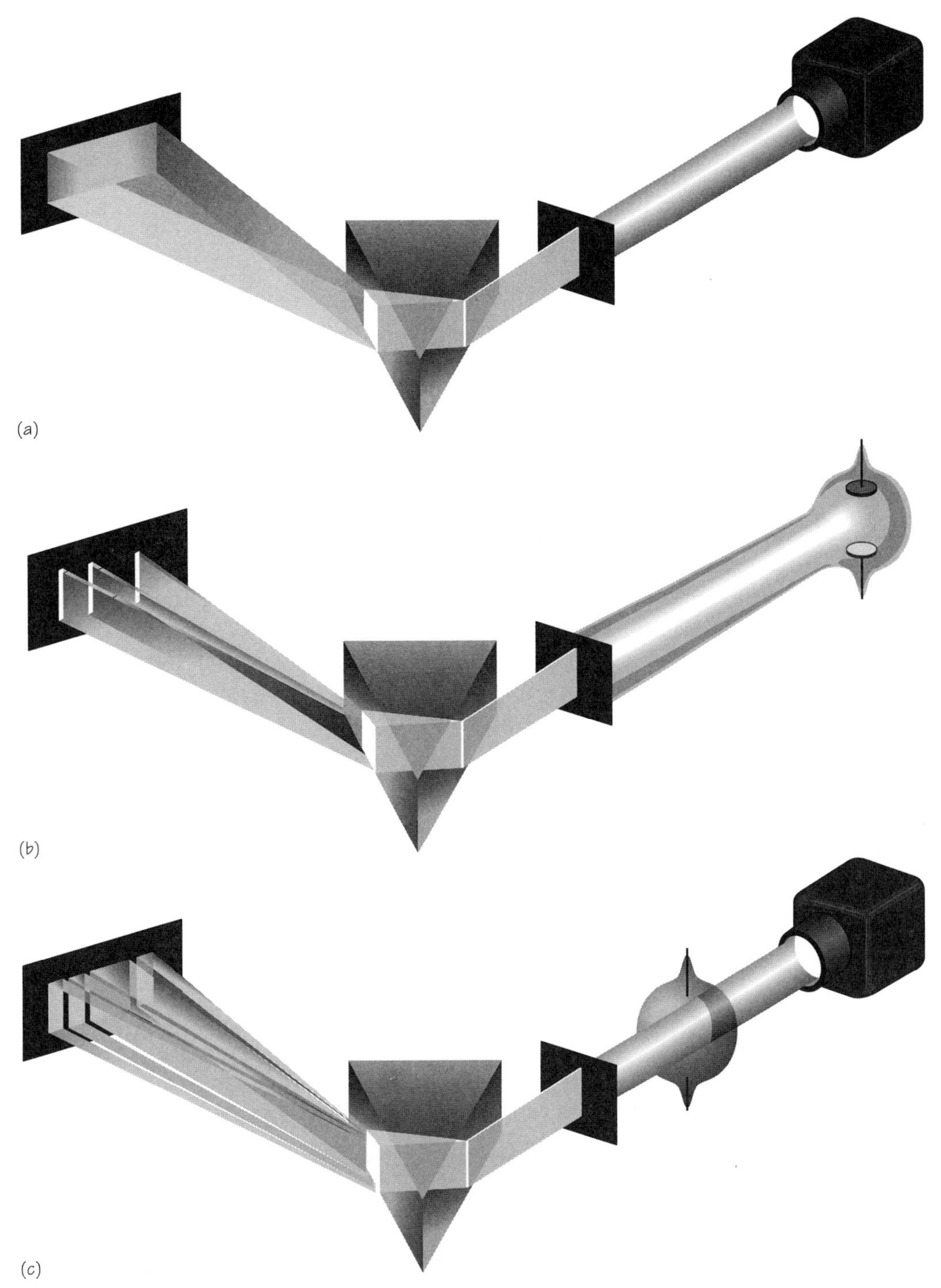

FIGURE 14.1 (a) Hot solids emit all wavelengths of light, producing a continuous spectrum on the screen at left. The shorter-wavelength portions of light are refracted more by the prism than are long wavelengths. (b) Hot gases emit only certain wavelengths of light, producing a bright line spectrum. If the slit had a different shape, so would the bright lines on the screen. (c) Cool gases absorb only certain wavelengths of light, producing a dark line spectrum when "white" light from a hot solid is passed through the cool gas.

lamps are now used in many places as street lights at night. Some gases or vapors have very complex spectra. Iron vapor, for example, has some 6000 bright lines in the visible range alone.

In 1823, the British astronomer John Herschel suggested that each gas could be identified by its unique line spectrum. By the early 1860s, the physicist Gustav R. Kirchhoff and the chemist Robert W. Bunsen, in Germany, had jointly discovered two new elements (rubidium and cesium) by noting previously unreported emission lines in the spectrum of the vapor of a mineral water. This was the first of a series of such discoveries. It started the development of a technique for speedy chemical analysis of small amounts of materials by *spectrum analysis*. The "flame test" you may have performed in a chemistry class is a simple application of this analysis.

Absorption Spectra

In 1802, the English scientist William Wollaston saw in the spectrum of sunlight something that had been overlooked before. Wollaston noticed a set of seven sharp, irregularly spaced *dark* lines, or spaces, across the continuous solar spectrum. He did not understand why they were there and did not investigate further. A dozen years later, the German physicist Joseph von Fraunhofer, using better instruments, detected many hundreds of such dark lines. To the most prominent dark lines von Fraunhofer assigned the letters A, B, C, etc. These dark lines can be easily seen in the Sun's spectrum with even quite simple modern spectroscopes. The letters A, B, C, . . . are still used to identify them. In the spectra of several other bright stars, von Fraunhofer found similar dark lines. Many, but not all, of these lines were in the same positions as those in the solar spectrum. All such spectra are known as *dark-line spectra* or *absorption spectra*.

> Spectrometer or *spectrograph*: A device for measuring the wavelength of the spectrum and for recording the spectra (e.g., on film).

In 1859, Kirchhoff made some key observations that led to better understanding of both the dark-line and bright-line spectra of gases. It was

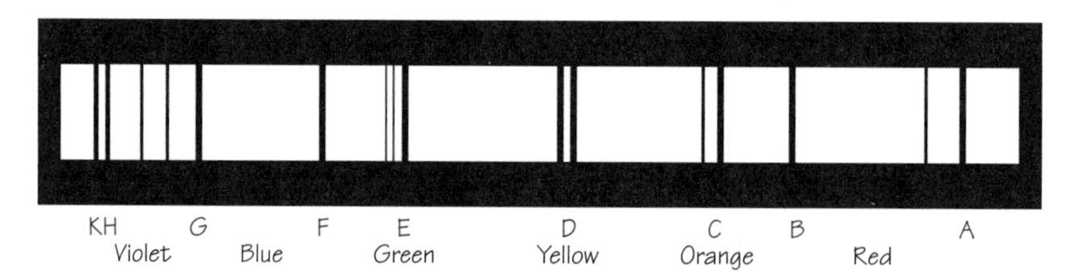

FIGURE 14.2 The Fraunhofer dark lines in the visible part of the solar spectrum. Only a few of the most prominent lines are represented here.

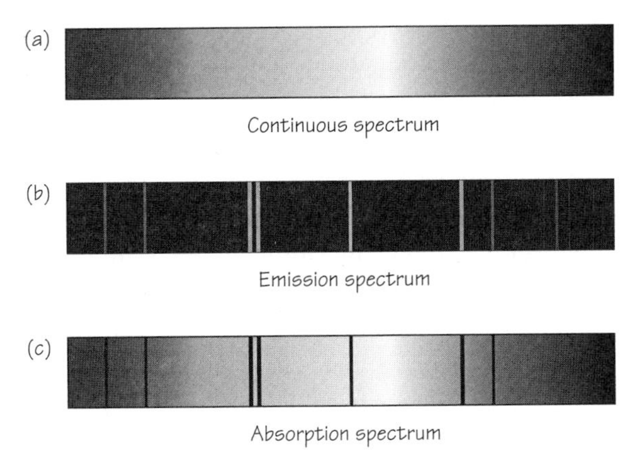

FIGURE 14.3 Emission, absorption, and continuous spectra (see Color Plate 5 for emission spectra of selected elements).

already known that the two prominent yellow lines in the emission spectrum of heated sodium vapor had the same wavelengths as two neighboring prominent dark lines in the solar spectrum. (The solar spectrum lines were the ones to which von Fraunhofer had assigned the letter D.) It was also known that the light emitted by a glowing solid forms a perfectly continuous spectrum that shows no dark lines. Kirchhoff now experimented with light from a glowing solid, as shown in Figure 14.1c. The white light was first passed through cooler sodium vapor and then dispersed by a prism. The spectrum produced showed the expected rainbow pattern, but it had two prominent dark lines at the same place in the spectrum as the D lines of the Sun's spectrum. It was therefore reasonable to conclude that the light from the Sun, too, was passing through a mass of sodium gas. This was the first evidence of the chemical composition of the gas envelope around the Sun.

Kirchhoff's experiment was repeated with various other relatively cool gases, placed between a glowing solid and the prism. Each gas produced its own characteristic set of dark lines. Evidently, each gas in some way absorbs light of certain wavelengths from the passing light. In addition, Kirchhoff showed that the wavelength of each absorption line matches the wavelength of a bright line in the emission spectrum of the same gas. *The conclusion is that a gas can absorb only light of those wavelengths which, when excited, it can emit.* (Note that not every emission line is represented in the absorption spectrum. Soon you will see why.)

Each of the various von Fraunhofer lines across the spectra of the Sun and other stars has now been identified with the action of some gas as tested in the laboratory. In this way, the whole chemical composition of the outer region of the Sun and other stars has been determined. This is really quite breathtaking from several points of view. First, it is surprising that scientists can learn the chemical composition of immensely distant objects—something which earlier thinkers had thought to be, almost by definition, an impossibility. It is even more surprising that chemical materials out there are, as Newton had earlier assumed, the same as those on Earth. (That this is true is clearly shown by the fact that even very complex absorption spectra are reproduced exactly in star spectra.) Finally, this result leads to a striking conclusion: *The physical processes that cause light absorption in the atom must be the same among the distant stars as on Earth.*

In these facts you can see a hint of how *universal* physical laws really are. Even at the farthest edges of the cosmos from which the Earth receives light, the laws of physics appear to be the same as for common materials close at hand in the laboratory! This is just what Galileo and Newton had intuited when they proposed that there is no difference between terrestrial and celestial physics.

14.2 REGULARITIES IN THE HYDROGEN SPECTRUM

Of all the spectra, the emission spectrum of hydrogen is especially interesting for both historical and theoretical reasons. In the visible and near-ultraviolet regions, the emission spectrum consists of an apparently systematic or orderly series of lines (see Figure 14.4). In 1885, Johann Jakob Balmer (1825–1898), a teacher at a girls' school in Switzerland, who was interested in number puzzles and numerology, found a simple formula—an empirical relation—which gave the wavelengths λ of the lines known at the time. The formula is

$$\lambda = b\left(\frac{n^2}{n^2 - 2^2}\right).$$

The quantity b is a constant which Balmer determined empirically and found to be equal to 364.56×10^{-9} m; n is a whole number, different for each line. Specifically, for the equation to yield the observed value for the respective wavelengths, n must be 3 for the first visible (red) line of the hydrogen emission spectrum (named H_α); $n = 4$ for the second (green) line (H_β); $n = 5$ for the third (blue) line (H_γ); and $n = 6$ for the fourth (violet)

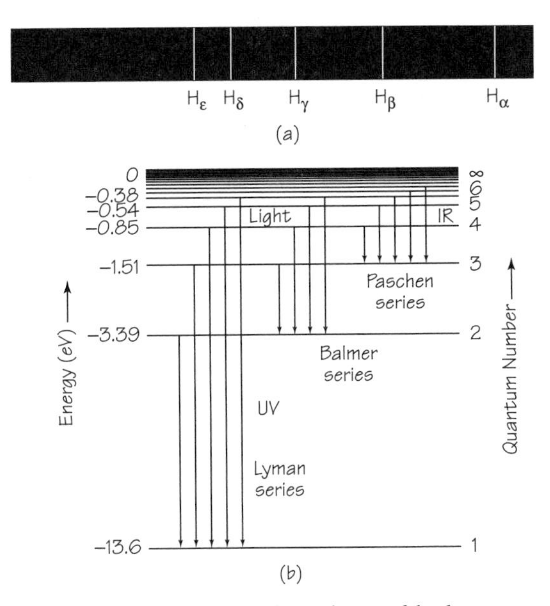

FIGURE 14.4 (a) The Balmer lines of hydrogen as they would appear in a photograph made with film sensitive to some ultraviolet as well as visible light. The lines get more crowded as they approach the series limit in the ultraviolet. (b) In Section 14.8, this scheme will explain the existence of all hydrogen emission lines.

line (H_δ). The table below shows excellent agreement (within 0.02%) between Balmer's calculations from his empirical formula and previously measured values.

Wavelength λ, in nanometers (10^{-9} m), for hydrogen emission spectrum.*

Name of line	n	*From Balmer's formula*	*By Ångström's measurement*	*Difference*
H_α	3	656.208	656.210	+0.002
H_β	4	486.08	486.074	−0.006
H_γ	5	434.00	434.01	+0.01
H_δ	6	410.13	410.12	−0.01

* Data for hydrogen spectrum (Balmer, 1885).

Not until 30 years later did scientists understand why Balmer's empirical formula worked so well, why the hydrogen atom emitted light whose wavelengths made such a simple sequence. But this did not keep Balmer from speculating that there might be other series of unsuspected lines in the hydrogen spectrum. Their wavelengths, he suggested, could be found by replacing the 2^2 in his equation with numbers such as 1^2, 3^2, 4^2, and so on. This suggestion stimulated many scientists to search for such additional spectral series. The search turned out to be fruitful, as you will see shortly.

In order to use modern notation, we rewrite Balmer's formula in a form that will be more useful:

$$\frac{1}{\lambda} = R_H\left(\frac{1}{2^2} - \frac{1}{n^2}\right).$$

In this equation, which can be derived from the previous one, R_H is a constant, equal to $4/b$. It is called the Rydberg constant for hydrogen, in honor of the Swedish spectroscopist J.R. Rydberg. Following Balmer, Rydberg made great progress in the search for various spectral series. The series of lines described by Balmer's formula are called the *Balmer series*. Balmer constructed his formula from the known wavelengths of only four lines in the visible part of the spectrum. The formula could be used to predict that there should be many more lines in the same series (indeed, infinitely many such lines, as n takes on values such as $n = 3, 4, 5, 6, 7, 8, \ldots \infty$). Moreover, every one of the lines is correctly predicted by Balmer's formula with considerable accuracy.

Following Balmer's speculative suggestion of replacing 2^2 by other numbers gives the following possibilities:

$$\frac{1}{\lambda} = R_H\left(\frac{1}{1^2} - \frac{1}{n^2}\right),$$

$$\frac{1}{\lambda} = R_H\left(\frac{1}{3^2} - \frac{1}{n^2}\right),$$

$$\frac{1}{\lambda} = R_H\left(\frac{1}{4^2} - \frac{1}{n^2}\right),$$

and so on. Each of these equations describes a possible series of emission lines. All these hypothetical series of lines can then be summarized by one overall formula

$$\frac{1}{\lambda} = R_H\left(\frac{1}{n_f^2} - \frac{1}{n_i^2}\right),$$

where n_f is a whole number that is fixed for any one series for which wavelengths are to be found. (For example, $n_f = 2$ for all lines in the Balmer series.) The letter n_i stands for integers that take on the values $n_f + 1$, $n_f + 2$, $n_f + 3$, . . . for the successive individual lines in a given series. (Thus, for the first two lines of the Balmer series, n_i is 3 and 4.) The constant R_H should have the same value for all of these hydrogen series.

So far, this discussion has been merely speculative. No series, no single line fitting the general formula, *need* exist, except for the observed Balmer series, where $n_f = 2$. But when physicists began to look for these hypothetical lines with good spectrometers, they found that they do, in fact, exist!

In 1908, F. Paschen in Germany found two hydrogen lines in the infrared. Their wavelengths were correctly given by setting $n_f = 3$ and $n_i = 4$ and 5 in the general formula. Many other lines in this "Paschen series" have since been identified. With improved experimental apparatus and techniques, new regions of the spectrum could be explored. Thus, other series gradually were added to the Balmer and Paschen series. In the table below, the name of each series listed is that of the discoverer.

Series of lines in the hydrogen spectrum.

Name of series	*Date of discovery*	*Region of spectrum*	*Values in Balmer equation*
Lyman	1906–1914	Ultraviolet	$n_f = 1$, $n_i = 2, 3, 4, \ldots$
Balmer	1885	Ultraviolet-visible	$n_f = 2$, $n_i = 3, 4, 5, \ldots$
Paschen	1908	Infrared	$n_f = 3$, $n_i = 4, 5, 6, \ldots$
Brackett	1922	Infrared	$n_f = 4$, $n_i = 5, 6, 7, \ldots$
Pfund	1924	Infrared	$n_f = 5$, $n_i = 6, 7, 8, \ldots$

Balmer hoped that his formula for the hydrogen spectra might be a pattern for finding series relationships in the spectra of other gases. This suggestion also bore fruit. Balmer's formula itself did not work directly in describing spectra of gases other than hydrogen. But it did inspire formulas of similar mathematical form that successfully described order in portions of many complex spectra. The Rydberg constant R_H also reappeared in such empirical formulas.

However, no model based on classical mechanics and electromagnetism could be constructed that would explain the spectra described by these formulas. What you have already learned in Chapter 13 about quantum theory suggests one line of attack. Obviously, the emission and absorption of light from an atom must correspond to a decrease and an increase of the atom's energy. If atoms of an element emit light of only certain frequencies, then the energy of the atoms must be able to change only by certain amounts. These changes of energy must involve rearrangement of the parts of the atom.

14.3 RUTHERFORD'S NUCLEAR MODEL OF THE ATOM

As so often, the next step arose from completely unrelated research. Ernest Rutherford, an outstanding physicist in the Cavendish Laboratory at Cambridge, provided a new basis for atomic models during the period 1909–1911. Rutherford was interested in the rays emitted by radioactive substances, especially α (alpha) rays. As you will see in Chapter 17, α rays consist of positively charged particles. These particles are positively charged helium ions with masses about 7500 times greater than the electron mass. Some radioactive substances emit α particles at very high rates and energies. Such particles are often used as projectiles in bombarding samples of elements. The experiments that Rutherford and his colleagues did with α particles are examples of a highly important kind of experiment in atomic and nuclear physics: the scattering experiment.

In a scattering experiment, a narrow, parallel beam of "projectiles" (e.g., α particles, electrons, X rays) is aimed at a target. The target is usually a thin foil or film of some material. As the beam strikes the target, some of the projectiles are deflected, or scattered, from their original direction. The scattering is the result of the interaction between the particles in the beam and the atoms of the material. A careful study of the projectiles after scattering can yield information about the projectiles, the atoms, and the interaction between them. If you know the mass, energy, and direction of the projectiles and see how they are scattered, you can deduce properties of the atoms that scattered the projectiles.

Rutherford noticed that when a beam of α particles passed through a thin metal foil, the beam spread out. This scattering may be thought of as caused by electrostatic forces between the positively charged α particles and the charges that make up atoms. Atoms contain both positive and negative charges. Therefore, an α particle undergoes both repelling and attracting forces as it passes through matter. The magnitude and direction of these forces depend on how closely the particle approaches the centers of the atoms among which it moves. When a particular atomic model is proposed, the extent of the expected scattering can be calculated and compared with experiment. For example, the Thomson model of the atom predicted al-

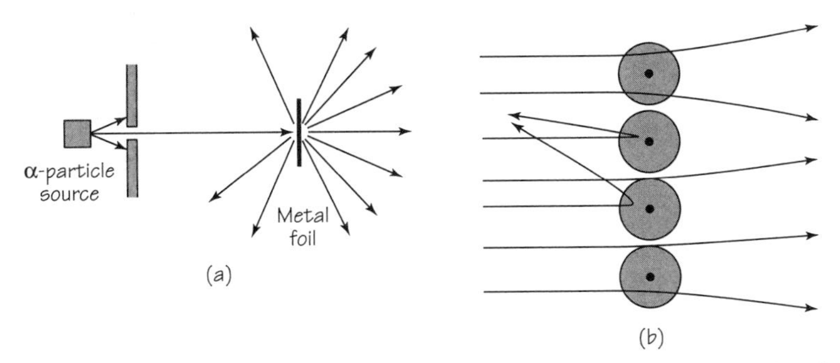

FIGURE 14.5 Alpha particle scattering showing deflection by the nuclei of the metal atoms. In somewhat the same way, you could (in principle) use á scattering experiment to discover the size and shape of an object hidden in a cloud or fog by directing a series of projectiles at the unseen object and tracing their paths back after deflection.

most no chance that an α particle would be deflected by an angle of more than a few degrees.

The breakthrough which led to the modern model of the atom followed a discovery by one of Rutherford's assistants, Hans Geiger. Geiger found that the number of particles scattered through angles of 10° or more was much greater than the number predicted by the Thomson model. In fact, a significant number were scattered through an angle greater than 90°, that is, many α particles virtually bounced right back from the foil. This result was entirely unexpected. According to Thomson's model, the atom should have acted only slightly on the projectile, rather like a cloud in which fine dust is suspended. Some years later, Rutherford wrote:

> . . . I had observed the scattering of α-particles, and Dr. Geiger in my laboratory had examined it in detail. He found, in thin pieces of heavy metal, that the scattering was usually small, of the order of one degree. One day Geiger came to me and said, "Don't you think that young Marsden, whom I am training in radioactive methods, ought to begin a small research?" Now I had thought that, too, so I said, "Why not let him see if any α-particles can be scattered through a large angle?" I may tell you in confidence that I did not believe that they would be, since we knew that the α-particle was a very fast, massive particle, with a great deal of [kinetic] energy, and you could show that if the scattering was due to the accumualted effect of a number of small scatterings, the chance of an α-particle's being scattered backward was very small. Then I remember two or three days later Geiger coming to me in great excitement and saying, "We have been able to get some of the α-particles coming backward . . ."

FIGURE 14.6 Ernest Rutherford (1871–1937) was born, grew up, and received most of his education in New Zealand. At age 24 he went to Cambridge, England, to work at the Cavendish Laboratory under J.J. Thomson. From there he went to McGill University in Canada, then home to be married and back to England again, to Manchester University. At these universities, and later at the Cavendish Laboratory where he succeeded J.J. Thomson as director, Rutherford performed important experiments on radioactivity, the nuclear nature of the atom, and the structure of the nucleus. Rutherford introduced the concepts "alpha," "beta," and "gamma" rays, "protons," and "half-life."

> It was quite the most incredible event that has ever happened to me in my life. It was almost as incredible as if you fired a 15-inch shell at a piece of tissue paper and it came back and hit you. On consideration, I realized that this scattering backward must be the result of a single collision, and when I made calculations I saw that it was impossible to get anything of that order of magnitude unless you took a system in which the greater part of the mass of the atom was concentrated in a minute nucleus. It was then that I had the idea of an atom with a minute massive centre, carrying a charge.

These experiments and Rutherford's interpretation marked the origin of the modern concept of the *nuclear atom.* Look at the experiments and Rutherford's conclusion more closely. Why must the atom have its mass and positive charge concentrated in a tiny nucleus at the center about which the electrons are clustered?

He writes that a possible explanation of the observed scattering is that the foil contains concentrations of mass and charge, that is, positively charged *nuclei.* These nuclei are much more dense than anything in Thomson's atoms. An α particle heading directly toward one of them is stopped and turned back. In the same way, a ball would bounce back from a rock but not from a cloud of dust particles. The drawing in Figure 14.7 is based

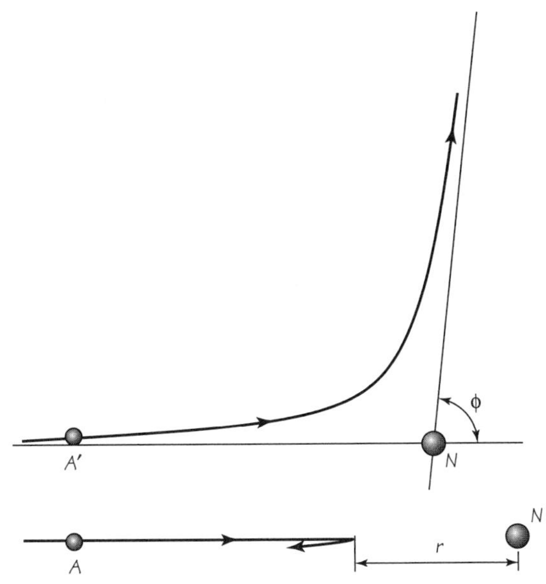

FIGURE 14.7 Paths of two alpha particles A and A′ approaching a nucleus *N*.

on one of Rutherford's diagrams in his paper of 1911, which laid the foundation for the modern theory of atomic structure. It shows two positively charged α particles, A and A′. The α particle A is heading directly toward a massive nucleus N. If the nucleus has a positive electric charge, it will repel the positive α particle. Because of this electrical repulsive force, A will slow to a stop at some distance r from N and then move directly back. A′ is an α particle that is *not* headed directly toward the nucleus N. It is repelled by N along a path which calculation shows must be a hyperbola. The deflection of A′ from its original path is indicated by the angle ϕ.

Rutherford considered the effects on the α particle's path of the important variables: the particle's speed, the foil thickness, and the quantity of charge Q on each nucleus. According to Rutherford's model, most of the α particles should be scattered through small angles, because the chance of approaching a very small nucleus nearly head-on is so small. But a significant number of α particles should be scattered through large angles.

Geiger and Marsden tested these predictions with the apparatus sketched in Figure 14.8. The lead box B contains a radioactive substance (radon) that emits α particles. The particles emerging from the small hole in the box are deflected through various angles ϕ in passing through the thin metal foil F. The number of particles deflected through each angle ϕ is found by letting the particles strike a zinc sulfide screen S. Each α particle that strikes the screen produces a scintillation (a momentary pinpoint of fluorescence). These scintillations can be observed and counted by looking through the

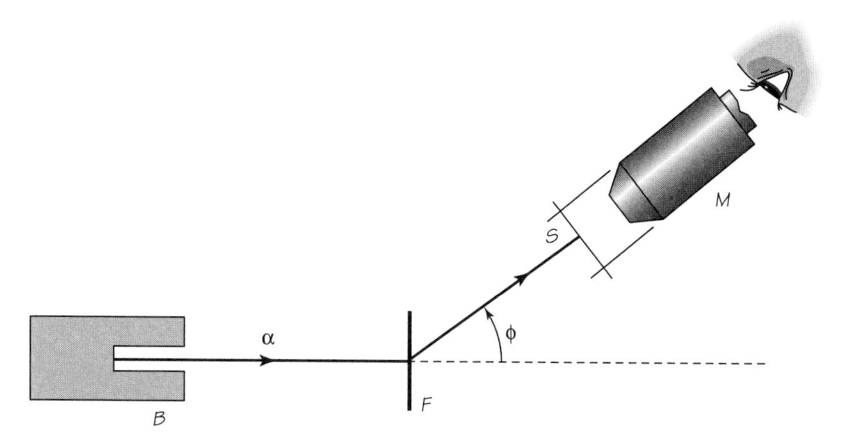

FIGURE 14.8 Rutherford's scintillation apparatus was placed in an evacuated chamber so that the alpha particles would not be slowed down by collisions with air molecules.

microscope M. The microscope and screen can be moved together along the arc of a circle. In later experiments, the number of α particles at any angle ϕ was counted more conveniently by a counter invented by Geiger (see Figure 14.9). The Geiger counter, in its more recent versions, is now a standard laboratory item.

Geiger and Marsden found that the number of α particles counted depended on the scattering angle, the speed of the particles, and the thickness of the foil. These findings agreed with Rutherford's predictions and

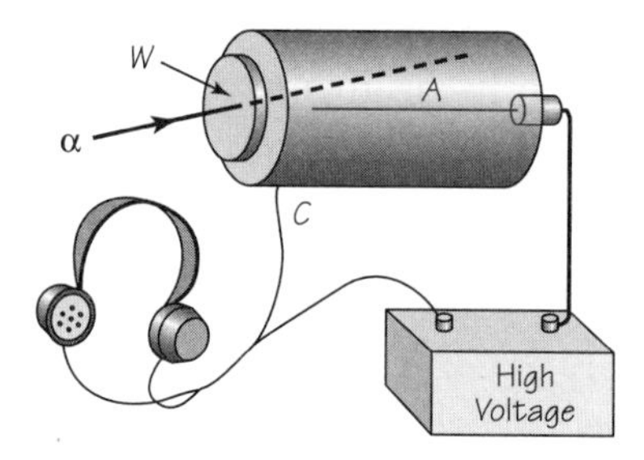

FIGURE 14.9 A Geiger counter that consists of a metal cylinder C containing a gas and a thin wire A that is insulated from the cylinder. A potential difference slightly less than that needed to produce a discharge through the gas is maintained between the wire (anode A) and cylinder (cathode C). When an alpha particle enters through the thin mica window (W), it frees a few electrons from the gas molecules. The electrons are accelerated toward the anode, freeing more electrons along the way by collisions with gas molecules. The avalanche of electrons constitutes a sudden surge of current that can be amplified to produce a click in the headphones or to operate a register

supported a new atomic model, in which most of the mass and all positive charge occupy a very small region at the center of the atom.

14.4 NUCLEAR CHARGE AND SIZE

Despite the success of Rutherford's model in dealing with α-scattering data, a problem remained. There still was no way to measure independently the charge Q on the nucleus. However, the scattering experiments had confirmed Rutherford's predictions about the effect of the speed of the α particle and the thickness of the foil on the angle of scattering. As often happens when part of a theory is confirmed, it is reasonable to proceed temporarily as if the whole theory were justified; that is, pending further proof, one could assume that the value of Q needed to explain the observed scattering data was the correct value of Q for the actual nucleus, as determined by Coulomb's law and the motion of the α particles. On this basis, Rutherford compiled scattering data for several different elements, among them carbon, aluminum, and gold. The following positive nuclear charges yielded the best agreement with experiments: for carbon, $Q = 6e$; for aluminum, $Q = 13e$ or $14e$; and for gold, $Q = 78e$ or $79e$, where e is the magnitude of the charge of one electron ($e = 1.6 \times 10^{-19}$ C). Similarly, values were found for other elements.

The magnitude of the positive charge of the nucleus was an important and welcome piece of information about the atom. The atom as a whole is of course electrically neutral. So if the nucleus has a positive charge of $6e$,

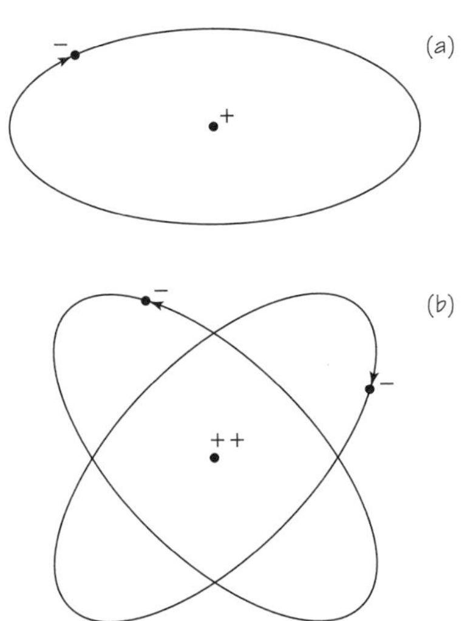

FIGURE 14.10 Sketch of simple atomic structure: (a) hydrogen, (b) helium.

$13e$, $14e$, etc., the number of negatively charged electrons surrounding the nucleus must be 6 for carbon, 13 or 14 for aluminum, etc. Thus, for the first time, scientists had a good idea of just how many electrons an atom may have.

An even more important fact was soon noticed. For each element, the value for the nuclear charge, in multiples of e, was close to the atomic number Z, the place number of that element in the periodic table! The results of scattering experiments with α particles were not yet precise enough to make this conclusion with certainty. But the data indicated that *each nucleus has a positive charge* Q *numerically equal to* Ze.

This suggestion made the picture of the nuclear atom much clearer and simpler. On this basis, the hydrogen atom ($Z = 1$) has one electron outside the nucleus. A helium atom ($Z = 2$) has in its neutral state two electrons outside the nucleus. A uranium atom ($Z = 92$) has 92 electrons. Additional experiments further supported this simple scheme. The experiments showed that it was possible to produce singly ionized hydrogen atoms, H^+, and doubly ionized helium atoms, He^{++}, but neither H^{++} nor He^{+++}. Evidently, a hydrogen atom has only one electron to lose, and a helium atom only two. Unexpectedly, the concept of the nuclear atom thus provided new insight into the periodic table of the elements. The nuclear concept suggested that *the periodic table is really a listing of the elements according to the number of electrons around the nucleus, or equally well according to the number of positive units of charge on the nucleus.*

These results cleared up some of the difficulties in Mendeleev's periodic table. For example, the elements tellurium and iodine had been assigned positions $Z = 52$ and $Z = 53$ on the basis of their chemical properties. This positioning contradicted the order of their atomic weights. But now Z was seen to correspond to a fundamental fact about the nucleus. Thus, the reversed order of atomic weights was understood to be not a basic fault in the scheme.

As an important additional result of these scattering experiments, Rutherford could estimate the size of the nucleus. Suppose an α particle is moving directly toward a nucleus. Its kinetic energy on approach is transformed into electrical potential energy. It slows down and eventually stops, like a ball rolling up a hill. The distance of closest approach can be computed from the known original kinetic energy of the α particle and the charges of α particle and nucleus. The value calculated for the closest approach is approximately 3×10^{-14} m. If the α particle does not penetrate the nucleus, this distance must be at least as great as the sum of the radii of α particles and nucleus; thus, the radius of the nucleus could not be larger than about 10^{-14} m. But 10^{-14} m is only about 1/1000 of the known radius of an atom. Furthermore, the total volume of the atom is proportional to the cube of its radius. So it is clear that *the atom is mostly empty, with the nucleus occupying only one-billionth of the space!* This explains how α particles or electrons can penetrate thousands of layers of atoms in metal foils or in gases, with only an occasional large deflection backward.

Successful as this model of the nuclear atom was in explaining scattering phenomena, it raised many new questions: What is the arrangement of electrons about the nucleus? What keeps the negative electron from falling into a positive nucleus by electrical attraction? Of what is the nucleus composed? What keeps it from exploding on account of the repulsion of its positive charges? Rutherford realized the problems raised by these questions and the failure of his model to answer them. But he rightly said that one should not expect one model, made on the basis of one set of puzzling results which it explains well, also to handle all other puzzles. Additional assumptions were needed to complete the model and answer the additional questions about the details of atomic structure. The remainder of this chapter will deal with the theory proposed by Niels Bohr, a young Danish physicist who joined Rutherford's group just as the nuclear model was being announced.

14.5 BOHR'S THEORY: THE POSTULATES

Assume, as Rutherford did, that an atom consists of a positively charged nucleus surrounded by a number of negatively charged electrons. What, then, keeps the electrons from falling into the nucleus, pulled in by the electric force of attraction? One possible answer is that an atom may be like a planetary system, with the electrons revolving in orbits around the nucleus. As you may know (see Section 3.12), a ball whirling on a string or a planet orbiting the Sun must be subject to an attractive force toward the center. Otherwise, the ball or planet would fly away on a straight line, according to Newton's first Law of Motion. This force toward the center is often called a *centripetal force*. For planets, this force arises from the gravitational attraction of the Sun on the planet. For electrons in atoms, Rutherford suggested that, instead of the gravitational force, the electric attractive force between the nucleus and an electron would supply a centripetal force. This centripetal force would tend to keep the moving electron in orbit.

This idea seems to be a good start toward a theory of atomic structure. But a serious problem arises concerning the stability of a "planetary" atom. According to Maxwell's theory of electromagnetism, a charged particle radiates energy when it is accelerated. An electron moving in an orbit around a nucleus continually changes its direction, hence also its velocity vector. In other words, it is *always being accelerated* by the centripetal electric force.

(a)

(b)

(c)

FIGURE 14.11 Niels Bohr (1885–1962): (a) pictured with his wife, Margrethe, on their wedding day; (b) ca. 1917; (c) in his later years. Bohr was born in Copenhagen, Denmark, and became a professor at the university there. He received the Nobel Prize in physics in 1922 for his work described in this chapter. He helped found the new quantum mechanics, was a leading contributor to theories of nuclear structure and nuclear fission, and helped press for peaceful uses of nuclear energy.

The electron, therefore, should lose energy by emitting radiation and thus being drawn steadily closer to the nucleus. (Somewhat similarly, an artificial satellite loses energy, because of friction in the upper atmosphere, and gradually spirals toward the Earth.) Within a very short time, the energy-radiating electron should actually be pulled into the nucleus. According to classical physics, mechanics, and electromagnetism, a planetary atom would not be stable for more than a very small fraction of a second.

The idea of a planetary atom was nevertheless appealing. Physicists continued to look for a theory that would include a stable planetary structure and predict separate line spectra for the elements. Niels Bohr, then an unknown Danish physicist who had just received his doctorate, succeeded in constructing such a theory in 1912–1913. This theory was called the Bohr model or quantum model of the atom, because it incorporated the quantum idea of Einstein and Planck. It was widely recognized as a major victory. Although it had to be modified later to account for many more phenomena, it showed how to attack atomic problems by using quantum theory. Today, it seems a rather naive way of thinking about the atom, compared with more recent quantum-mechanical theories. But in fact, considering what it was designed to do, Bohr's theory is an impressive example of a successful physical model. Since Bohr incorporated Rutherford's idea of the nucleus, the model that Bohr's theory discusses is often called the Rutherford–Bohr model.

Bohr introduced two bold new postulates specifically to account for the existence of stable electron orbits and for separate emission spectra for each element. These postulates may be stated as follows:

1. Contrary to the predictions of classical physics—which after all had been tested only for relatively large-scale circumstances—there are states for an atomic system in which electromagnetic radiation simply does not occur, despite any acceleration of the charged particles (electrons). These states are called the *stationary states* of the atom.
2. Any emission or absorption of radiation, either as visible light or other electromagnetic radiation, corresponds to a sudden transition of the charge between two such stationary states. The radiation emitted or absorbed has a frequency f determined by the relation $hf = E_i - E_f$. (In this equation, h is Planck's constant, and E_i and E_f are the energies of the atom in the initial and final stationary states, respectively.)

Quantum theory had begun with Planck's idea that atoms emit light only in definite amounts of energy. This concept was extended by Einstein's idea that light travels only as definite parcels, quanta, of energy. Now it was extended further by Bohr's idea that atoms exist in a stable condition only in

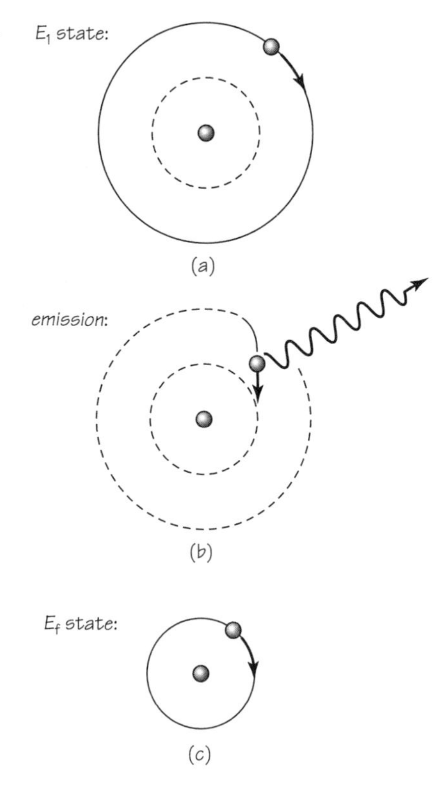

FIGURE 14.12 An electron changing orbital states with the emission of a photon.

definite, "quantized" energy states. But Bohr also used the quantum concept in deciding which of all the conceivable stationary states were actually possible. An example of how Bohr did this is given in the next section.

For simplicity, the hydrogen atom, with a single electron revolving around the nucleus, is used. Following Bohr, we assume that the possible electron orbits are simply circular. Light is emitted by the atom when it changes from one state to another (see Figure 14.12). (The details of some additional assumptions and calculations are worked out in the *Student Guide*.) Bohr's result for the possible stable orbit radii r_n was $r_n = a \cdot n^2$, where a is a constant $(h^2/4\pi^2 mkq_e^2)$ that can be calculated from known physical values, and n stands for any whole number, 1, 2, 3. . . .

14.6 THE SIZE OF THE HYDROGEN ATOM

Bohr's result is remarkable. In hydrogen atoms, the possible orbital radii of the electrons are whole multiples of a constant which can at once be evaluated; that is, n^2 takes on values of 1^2, 2^2, 3^2, . . . , and all factors to the left of n^2 are quantities known previously by independent measurement! Calculating the value $(h^2/4\pi^2mkq_e^2)$ gives 5.3×10^{-11} m. Therefore, according to Bohr's model, the radii of stable electron orbits should be $r_n = 5.3 \times 10^{-11}$ m $\times n^2$, that is, 5.3×10^{-11} m when $n = 1$ (first allowed orbit), $4 \times 5.3 \times 10^{-11}$ m when $n = 2$ (second allowed orbit), $9 \times 5.3 \times 10^{-11}$ m when $n = 3$, etc. *In between these values, there are no allowed radii.* In short, the separate allowed electron orbits are spaced around the nucleus in a regular way, with the allowed radii quantized in a regular manner. Emission and absorption of light should therefore correspond to the transition of the electron from one allowed orbit to another. Emission of light occurs when the electron "drops" from a higher energy state to a lower state; absorption of light occurs when the electron "jumps" from a lower-energy state up to a higher-energy state.

This is just the kind of result hoped for. It tells which radii are possible and where they lie. But so far, it had all been model building. Do the orbits in a real hydrogen atom actually correspond to this model? In his first paper of 1913, Bohr was able to give at least a partial "yes" as an answer. It had long been known that the normal "unexcited" hydrogen atom has a radius of about 5×10^{-11} m (i.e., the size of the atom obtained, for example, by interpreting measured characteristics of gases in terms of the kinetic theory). This known value of about 5×10^{-11} m corresponds excellently to the prediction from the equation for orbital radius r if n has the lowest value, namely 1. Now there was a way to understand the size of the neutral, unexcited hydrogen atom. For every such atom, the size corresponds to the size of the innermost allowed electron orbit.

14.7 OTHER CONSEQUENCES OF THE BOHR MODEL

With his two postulates, Bohr could calculate the radius of each permitted orbit. In addition, he could calculate the total energy of the electron in each orbit, i.e., the energy of the stationary state.

The results that Bohr obtained may be summarized in two simple formulas. As you saw, the radius of an orbit with quantum number n is given by the expression

$$r_n = n^2 r_1,$$

where r_1 is the radius of the first orbit (the orbit for $n = 1$) and has the value 5.3×10^{-9} cm or 5.3×10^{-11} m.

The energy (the sum of kinetic energy and electric potential energy) of the electron in the orbit with quantum number n can also be computed from Bohr's postulates. As pointed out in Chapter 6, it makes no sense to assign an absolute value to potential energy. In this case, only *changes* in energy have physical meaning. Therefore, any convenient zero level can be chosen. For an electron orbiting in an electric field, the mathematics is particularly simple if, as a zero level for energy, the state $n = \infty$ is chosen. At this level, the electron would be infinitely far from the nucleus (and therefore free of it). The energy for any other state E_n is then the *difference* from this free state. The possible energy states for the hydrogen atom are therefore

$$E_n = \frac{1}{n^2}E_1,$$

where E_1 is the total energy of the atom when the electron is in the first orbit ($n = 1$). E_1 is the lowest energy possible for an electron in a hydrogen atom. Its value is -13.6 eV (the negative value means only that the energy is 13.6 eV less than the free state value E_∞). This is called the *ground state*. In that state, the electron is most tightly "bound" to the nucleus. The value of E_2, the first "excited state" above the ground state, is, according to the above equation,

$$E_2 = \frac{1}{2^2} \times (-13.6 \text{ eV}) = -3.4 \text{ eV}.$$

This state is only 3.4 eV less than in the free state.

According to the formula for r_n, the first stationary orbit, defined by $n = 1$, has the smallest radius. Higher values of n correspond to orbits that have larger radii. The higher orbits are spaced further and further apart, and the force field of the nucleus falls off even more rapidly. So the work required to move out to the next larger orbit actually becomes smaller and smaller. Also, the jumps in energy from one level of allowed energy E to the next become smaller and smaller.

14.8 BOHR ACCOUNTS FOR THE SERIES SPECTRA OF HYDROGEN

The most spectacular success of Bohr's model was that it could be used to explain all emission (and absorption) lines in the hydrogen spectrum; that is, Bohr could use his model to derive, and so to explain, the Balmer formula for the series spectra of hydrogen!

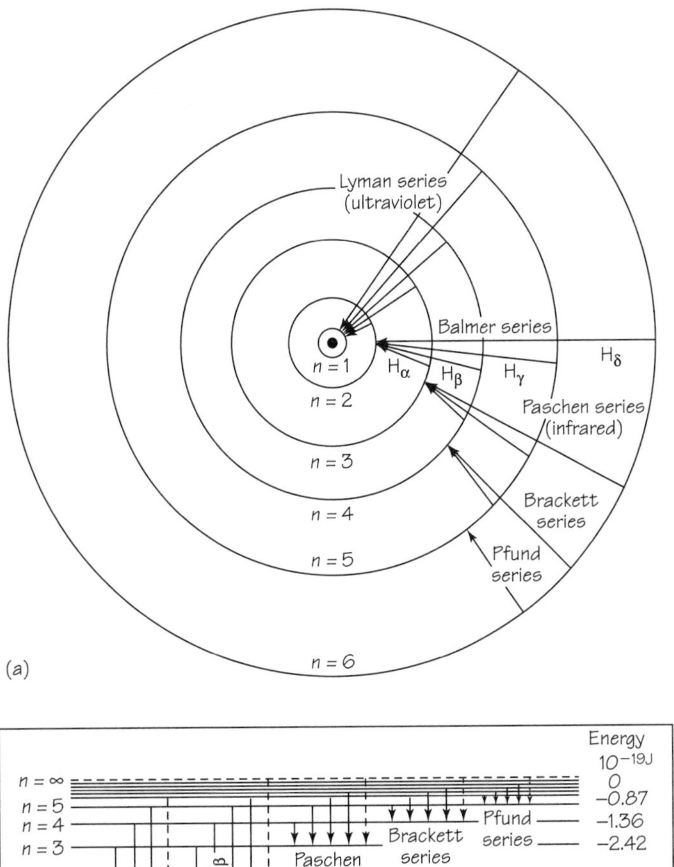

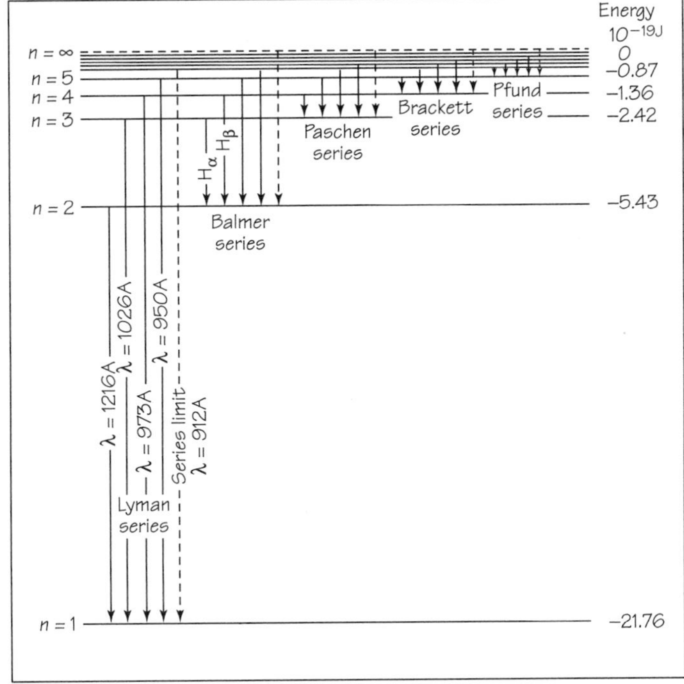

FIGURE 14.13 (a) A schematic diagram of transitions between stationary states of electrons in hydrogen atom, giving rise to five of the series of emission spectra lines. (b) Energy-level diagram for the hydrogen atom. Possible transitions between energy states are shown for the first few levels (from n = 2 to n = 3 to n = 2 or n = 1, etc.). The dotted arrow for each series indicates the series limit, a transition from the state where the electron is completely free from the nucleus.

By Bohr's second postulate, the radiation emitted or absorbed in a transition in an atom should have a frequency determined by

$$hf = E_1 - E_2.$$

If n_f is the quantum number of the final state and n_i is the quantum number of the initial state, then according to the result for E_n:

$$E_f = \frac{1}{n_f^2}E_1 \quad \text{and} \quad E_i = \frac{1}{n_i^2}E_1.$$

The frequency of radiation emitted or absorbed when the atom goes from the initial state to the final state is therefore determined by the equation

$$hf = \frac{E_1}{n_i^2} - \frac{E_1}{n_f^2} \quad \text{or} \quad hf = E_1\left(\frac{1}{n_i^2} - \frac{1}{n_f^2}\right).$$

In order to deal with wavelength λ (as in Balmer's formula) rather than frequency, we use the relationship between frequency and wavelength given in Chapter 8. The frequency of a line in the spectrum is equal to the speed of the light wave divided by its wavelength: $f = c/\lambda$. Substituting c/λ for f in the last equation and then dividing both sides by the constant hc (Planck's constant times the speed of light), gives

$$\frac{1}{\lambda} = \frac{E_1}{hc}\left(\frac{1}{n_i^2} - \frac{1}{n_f^2}\right).$$

According to Bohr's model, then, this equation gives the wavelength λ of the radiation emitted or absorbed when a hydrogen atom changes from one stationary state with quantum number n_i to another with n_f.

How does this prediction from Bohr's model compare with the long-established *empirical* Balmer formula for the Balmer series? This, of course, is the crucial question. The Balmer formula, given in Section 14.2, in modern terms is

$$\frac{1}{\lambda} = R_H\left(\frac{1}{2^2} - \frac{1}{n^2}\right).$$

You can see at once that the equation for the wavelength λ of emitted (or absorbed) light derived from the Bohr model is exactly the same as Balmer's formula, if $n_f = 2$ and $R_H = -E_1/hc$.

The Rydberg constant R_H was long known from spectroscopic measurements to have the value of $1.097 \times 10^7\ \mathrm{m}^{-1}$. Now it could be compared with the value for $-E_1/hc$. (Remember that E_1 is negative, so $-E_1$ is positive.) Remarkably, there was fine agreement. R_H, previously regarded as just an experimentally determined constant, was now shown to be a number that could be calculated from known fundamental constants of nature, namely, the mass and charge of the electron, Planck's constant, and the speed of light.

More important, you can now see the *meaning*, in physical terms, of the old empirical formula for the lines (H_α, H_β, . . .) in the Balmer series. All the lines in the Balmer series simply correspond to transitions from various initial states (various values of n_i larger than 2) to the same final state, for which $n_f = 2$. Thus, photons having the frequency or wavelength of the line H_α are emitted when electrons in a gas of hydrogen atoms "jump" from the state $n = 3$ to the state $n = 2$, as shown in the diagrams in Figure 14.14. The H_β line corresponds to "jumps" from $n = 4$ to $n = 2$, and so forth.

When Bohr proposed his theory in 1913, emission lines in only the Balmer and Paschen series for hydrogen were known definitely. Balmer had suggested, and the Bohr model agreed, that additional series should exist. Further experiments revealed the Lyman series in the ultraviolet portion of the spectrum (1904–1914), the Brackett series (1922), and the Pfund series (1924), both of the latter series being in the infrared region of the spectrum. In each series, the measured frequencies of every one of the lines were found to be those predicted by Bohr's theory, and (equally important) no lines existed that were *not* given by the theory. Similarly, Bohr's model could explain the general formula that Balmer guessed might apply for all spectral lines of hydrogen. As described in empirical terms in Section 14.2, the lines of the Lyman series correspond to transitions from various initial states to the final state $n_f = 1$; the lines of the Paschen series correspond to transitions from various initial states to the final state $n_f = 3$; and so on, as indicated by the equation on page 644 from Bohr's model:

$$\frac{1}{\lambda} = \frac{E_1}{hc}\left(\frac{1}{n_i^2} - \frac{1}{n_f^2}\right) \quad \text{or} \quad \frac{1}{\lambda} = R_H\left(\frac{1}{n_f^2} - \frac{1}{n_i^2}\right).$$

The general scheme of possible transitions among the first six orbits is shown in Figure 14.14a. Thus, the theory not only related known information about the hydrogen spectrum, but also predicted correctly the wavelengths of previously unknown series of lines in the spectrum. Moreover, it provided a reasonable physical model; Balmer's general formula had provided no physical reason for the empirical relationship among the lines of each series.

FIGURE 14.14 Diagrams of an atom of mercury undergoing impacts by electrons of energies of 4.0 eV, 5.0 eV, 6.0 eV.

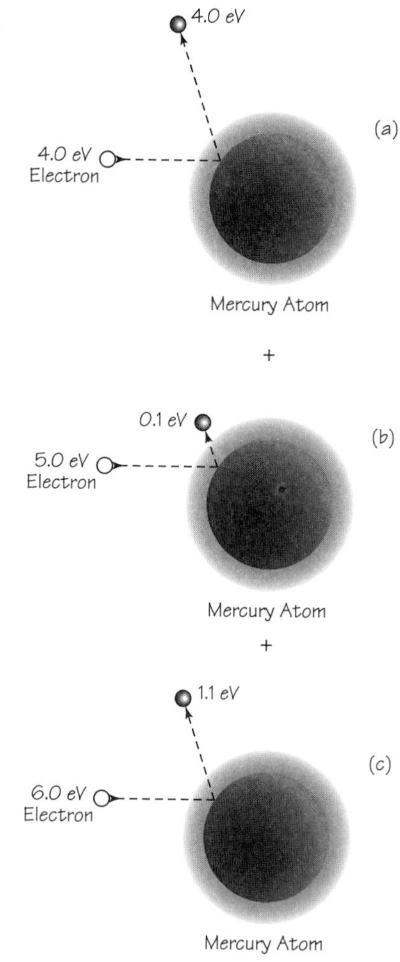

The schematic diagram shown on page 643 is useful as an aid for the imagination. But it has the danger of being too specific. For instance, it may lead one to think of the emission of radiation as actual "jumps" of electrons between orbits. In Chapter 15 you will see why it is impossible to detect an electron moving in such orbits. A second way of presenting the results of Bohr's theory yields the same facts but does not adhere as closely to a picture of orbits. This scheme is shown in Figure 14.13b. It focuses not on orbits but on the corresponding possible energy states. These energy states are all given by the formula $E_n = 1/n^2 \times E_1$. In terms of this *mathematical model*, the atom is normally unexcited, with an energy E_1 about -13.6 eV (or -22×10^{-19} J).

Absorption of energy can place the atoms in an excited state, with a correspondingly higher energy. The excited atom is then ready to emit light, with a consequent reduction in energy. The energy absorbed or emitted always shifts the total energy of the atom to one of the values specified by the formula for E_n. Thus, the hydrogen atom may also be represented, not by orbits, but by means of an energy-level diagram.

14.9 DO STATIONARY STATES REALLY EXIST?

The success of Bohr's theory in accounting for the spectrum of hydrogen left this question: *Could experiments show directly that atoms do have only certain, separate energy states?* In other words, are there really gaps between the energies that an atom can have? A famous experiment in 1914, by the German physicists James Franck and Gustav Hertz (a nephew of Heinrich Hertz), showed that these separate energy states do indeed exist.

Franck and Hertz bombarded atoms with electrons from an "electron gun," a hot wire that emitted electrons which were then accelerated through a hole leading into an evacuated region where they were aimed at a target. (A similar type of electron gun is used today in TV tubes and computer monitors.) Franck and Hertz were able to measure the energy lost by the electrons in collisions with the target atoms. They could also determine the energy gained by the atoms in these collisions.

In their first experiment, Franck and Hertz bombarded mercury vapor contained in a chamber at very low pressure. The procedure was equivalent to measuring the kinetic energy of electrons on leaving the electron gun, and again after they had passed through the mercury vapor. The only way electrons could lose energy was in collisions with the mercury atoms. Franck and Hertz found that when the kinetic energy of the electrons leaving the gun was small (up to several electron volts), the electrons still had almost exactly the same energy after passage through the mercury vapor as they had on leaving the gun. This result could be explained in the following way. A mercury atom is several hundred thousand times more massive than an electron. When it has low kinetic energy, the electron just bounces off a mercury atom, much as a golf ball thrown at a bowling ball would bounce off. A collision of this kind is called an "elastic" collision (discussed in Chapter 6). In an elastic collision, the mercury atom (bowling ball) takes up only a negligible part of the kinetic energy of the electron (golf ball), so that the electron loses practically none of its kinetic energy.

But when the kinetic energy of the electrons was raised to 5 eV, the experimental results changed dramatically. When an electron collided with a mercury atom, the electron lost almost exactly 4.9 eV of energy. When the energy was increased to 6.0 eV, the electron still lost just 4.9 eV of energy

Physicists now know two ways of "exciting" an atom: by absorption and by collision. In absorption, an atom absorbs a photon with just the right energy to cause a transition from the lowest energy level to a higher one. Collision may involve collision with an electron from an electron gun or collisions among agitated atoms (as in a heated enclosure or a discharge tube).

in collision, being left with 1.1 eV of energy. These results indicated that a mercury atom cannot accept less than 4.9 eV of energy. Furthermore, when the mercury atom is offered somewhat more energy, for example, 5 eV or 6 eV, it still accepts only 4.9 eV. The accepted amount of energy cannot go into kinetic energy of the mercury because the atom is so much more massive than the electron. Therefore, Franck and Hertz concluded that the 4.9 eV is added to the internal energy of the mercury atom; that is, the mercury atom enters a stationary state with enemy 4.9 eV greater than that of the lowest energy state, with no allowed energy level in between.

What happens to this extra 4.9 eV of internal energy? According to the Bohr model, this amount of energy should be emitted as electromagnetic radiation when the atom returns to its lowest state. Franck and Hertz looked for this radiation, and they found it! They observed that the mercury vapor, after having been bombarded with electrons, emitted light at a wavelength of 253.5 nm. This wavelength was known to exist in the emission spectrum of hot mercury vapor. The wavelength corresponds to a frequency f for which the photon's energy, hf, is just 4.9 eV (as you can calculate). This result showed that mercury atoms had indeed gained (and then radiated away) 4.9 eV of energy in collisions with electrons.

Later experiments showed that mercury atoms bombarded by electrons could also gain other sharply defined amounts of energy, for example, 6.7 eV and 10.4 eV. In each case, the subsequently emitted radiation corresponded to known lines in the emission spectrum of mercury. In each case, similar results were obtained: the electrons always lost energy, and the atoms always gained energy, both only in sharply defined amounts. Each type of atom studied was found to have separate energy states. The amounts of energy gained by the atoms in collisions with electrons always corresponded to the energy of photons in known spectrum lines. Thus, this direct experiment confirmed the existence of separate stationary states of atoms as predicted by Bohr's theory of atomic spectra. This result provided strong evidence of the validity of the Bohr theory.

14.10 CONSTRUCTING THE PERIODIC TABLE

In the Bohr model, atoms of the different elements differ in the charge and mass of their nuclei and in the number and arrangement of the electrons. Bohr, along with the German physicist Arnold Sommerfeld, came to picture the electronic orbits, not only as circular but also as elliptical orbits, and not as a series of concentric rings in one plane, but as patterns in three dimensions.

How does the Bohr model of atoms help to explain chemical properties? Recall that the elements hydrogen (atomic number $Z = 1$) and lithium ($Z = 3$) are somewhat alike chemically. (Refer to the periodic table on the color plate in this book.) Both have valences of 1. Both enter into compounds of similar types, for example, hydrogen chloride (HCl) and lithium chloride (LiCl). There are also some similarities in their spectra. All this suggests that the lithium atom resembles the hydrogen atom in some important respects. Bohr speculated that two of the three electrons of the lithium atom are relatively close to the nucleus, in orbits resembling those of the helium atom ($Z = 2$), forming, as one may call it, a "shell" around the nucleus. But the third electron is in a circular or elliptical orbit outside the inner system. Since this inner system consists of a nucleus of charge $+3e$ and two electrons each of the charge $-e$, its net charge is $+e$. Thus, the lithium atom may be roughly pictured as having a central core of charge $+e$. Around this core one electron revolves, somewhat as for a hydrogen atom. This similar physical structure, then, is the reason for the similar chemical behavior.

Referring to the periodic table, you will see that helium ($Z = 2$) is a chemically inert noble gas. These properties indicate that the helium atom is highly stable, having both of its electrons closely bound to the nucleus. It seems sensible, then, to regard both electrons as moving in the same *innermost "shell"* group or on orbits around the nucleus when the atom is unexcited. Moreover, because the helium atom is so stable and chemically inert, we may reasonably assume that this shell cannot hold more than two electrons. This shell is called the K-shell. The single electron of hydrogen is also said to be in the K-shell when the atom is unexcited. Lithium has two electrons in the K-shell, filling it to capacity; the third electron starts a new shell, called the L-shell. This single outlying and loosely bound electron is the reason why lithium combines so readily with oxygen, chlorine, and many other elements.

Sodium ($Z = 11$) is the next element in the periodic table that has chemical properties similar to those of hydrogen and lithium. This similarity suggests that the sodium atom is also hydrogen-like in having a central core about which one electron revolves. Moreover, just as lithium follows helium in the periodic table, sodium follows the noble gas neon ($Z = 10$). You may assume that two of neon's 10 electrons are in the first (K) shell, while the remaining eight electrons are in the second (L) shell. Because of the chemical inertness and stability of neon, we may further assume that these eight electrons fill the L-shell to capacity. For sodium, then, the eleventh electron must be in a third shell, called the M-shell.

LASERS

An atom in an excited state gives off energy by emitting a photon, a quantum of electromagnetic radiation, according to Bohr's second postulate. Although Bohr's specific model of the atom has been vastly extended and incorporated into models based on a different approach (see Chapter 15), this postulate is still valid.

As you have seen, atoms can acquire internal energy, that is, be brought to an excited state, in many ways. In the Franck–Hertz experiment, inelastic collisions provided the energy; in a cool gas displaying a dark-line spectrum, it is the absorption of photons; in a spark or discharge tube, it is collisions between electrons and atoms. There are other mechanisms as well.

Once an atom has acquired internal energy, it can also get rid of it in several ways. An atom can give up energy in inelastic collisions, or (as discussed above) it can emit energy as electromagnetic radiation. There are many different kinds of inelastic collisions; which one an atom undergoes depends as much on its surroundings as on the atom itself.

There are also two different ways an atom can emit radiation. Spontaneous radiation is the kind considered elsewhere in this chapter. At some random (unpredictable) moment, the previously excited atom emits a photon (of frequency ν) and changes its state to one of lower energy (by an amount ΔE). If, however, there are other photons of the appropriate frequency ($f = \Delta E/h$) in the vicinity, the atom may be *stimulated* to emit its energy. The radiation emitted is at exactly the same frequency, polarization, and phase as the stimulating radiation. That is, it is exactly in step with the existing radiation. In the wave model of light, you can think of the emission simply increasing the amplitude of the oscillations of the existing electromagnetic field within which the emitting atom finds itself.

Stimulated emission behaves very much like the classical emission of radiation discussed in Chapter 12. A collection of atoms stimulating one another to emit radiation behaves much like an antenna. You can think of the electrons in the different atoms as simply vibrating in step just as they do in an ordinary radio antenna, although much, much faster.

Usually atoms emit their energy spontaneously long before another photon comes along to stimulate them. Most light sources therefore emit incoherent light, that is, light made up of many different contributions, differing slightly in frequency, out of step with each other, and randomly polarized.

Usually, most of the atoms in a group are in the ground state. Light that illuminates the group is more likely to be absorbed than to stimulate any emission, since it is more likely to encounter an atom in the ground state than in the appropriate excited state. But suppose conditions are arranged so that more atoms are in one of the excited states than are in the ground state. (Such a group of atoms is said to be inverted.) In that case, light of the appropriate frequency is more likely to stimulate emission than to be absorbed. Then an interesting phenomenon takes over. Stimulated emission becomes more probable the more light there is around. The stimulated emission from some atoms therefore leads to a chain reaction, as more and more atoms give up some of their internal energy to the energy of the radiation. The incident light pulse has been amplified. Such an arrangement is called a *laser* (*light amplification* by *stimulated emission* of *radiation*).

Physicists and engineers have developed many tricks for producing "inverted" groups of atoms, on which laser operation depends. Exactly what the tricks are is not important for the action of the laser itself, although without them the laser would be impossible. Sometimes it is possible to maintain the inversion even while the laser is working; that is, it is possible to supply enough energy by the mechanisms that excite the atoms (inelastic collisions with other kinds of atoms, for example) to compensate for the energy emitted as radiation. These lasers can therefore operate continuously.

There are two reasons laser light is very desirable for certain applications. First, it can be extreme *intense*; some lasers can emit millions of joules in minute fractions of a second, as all their atoms emit their stored energy at once. Second, it is *coherent*; the light waves are all in step with each other. Incoherent light waves are somewhat like the waves crisscrossing the surface of a pond in a gale. But coherent waves are like those in a ripple tank, or at a beach where tall breakers arrive rhythmically.

The high intensity of some lasers can be used for applications in which a large amount of energy must be focused on a small spot. Such lasers are used in industries for cutting and welding delicate parts. In medicine, they are used, for example, to reattach the retina (essentially by searing a very small spot) in the eye.

The coherence of lasers is used in applications that require a stable light source emitting light of a precisely given frequency and polarization in one precise direction. Surveyors can use lasers to lay out straight lines, since the coherent beam spreads out very little with distance. Telephone companies can use them to carry signals in the same way they now use radio and microwaves.

FIGURE 14.15 NOVA laser at Lawrence Livermore National Laboratory. The five tubes are lasers focused on a single point.

Passing on to potassium ($Z = 19$), the next alkali metal in the periodic table, you may again picture an inner core and a single electron outside it. The core consists of a nucleus with charge $+19e$. There are two, eight, and eight electrons occupying the K-, L-, and M-shells, respectively. The nineteenth electron revolves around the core in a fourth shell, called the N-shell. The atom of the noble gas argon, with $Z = 18$, comes just before potassium in the periodic table. Argon again represents a tight and stable electron pattern, with two in the K-shell, eight in the L-shell, and eight in the M-shell.

These qualitative considerations lead to a consistent picture of electrons distributed in groups, or shells, around the nucleus. The arrangement of electrons in the noble gases may be considered particularly stable. For each new alkali metal in Group IA of the periodic table, a new shell is started. Each alkali metal atom has a single electron around a core that resembles the pattern for the preceding noble gas. You may expect this outlying electron to be easily "loosened" by the action of neighboring atoms, and this agrees with the facts. The elements lithium, sodium, and potassium are alkali metals. In compounds or in solution (as in electrolysis), they may be considered to be in the form of ions such as Li^+, Na^+, and K^+. Each ion lacks one electron and so has one positive net charge $+e$. In the neutral atoms of these elements, the outer electron is relatively free to move about. This property has been used as the basis of a theory of electrical conductivity. According to this theory, a good conductor has many "free" electrons that can form a current under appropriate conditions. A poor conductor has relatively few "free" electrons. The alkali metals are all good conductors. Elements whose electron shells are all filled are very poor conductors; they have no "free" electrons. In Chapter 10, you saw how electrical conduction takes place in metals. It is because metals have many "free" electrons that they are conductors. We will return to this in Chapter 16.

In Group II of the periodic table, you would expect those elements that follow immediately after the alkali metals to have atoms with two outlying electrons. For example, beryllium ($Z = 4$) should have two electrons in the K-shell, thus filling it, and two in the L-shell. If the atoms of all these elements have two outlying electrons, they should be chemically similar, as indeed they are. Thus, calcium and magnesium, which belong to this group, should easily form ions such as Ca^{++} and Mg^{++}, each with a positive net charge of $+2e$. This is also found to be true.

As a final example, consider those elements that immediately precede the noble gases in the periodic table. For example, fluorine atoms ($Z = 9$) should have two electrons filling the K-shell but only seven electrons in the L-shell, one less than enough to fill it. If a fluorine atom captures an additional electron, it should become an F^- ion with one negative net charge. The L-shell would then be filled, as it is for neutral neon ($Z = 10$), and you would expect the F^- ion to be relatively stable. This prediction agrees with observation. Indeed, all the elements immediately preceding the inert gases tend to form stable, singly charged negative ions in solution. In the solid state, you would expect these elements to lack free electrons. In fact, all of them are poor conductors of electricity.

As indicated in Figure 14.16, based on an illustration from Bohr's work in 1922, the seven main shells, K, L, M, . . . , Q, divide naturally into orbits or subshells. The shells fill with electrons so that the total energy of the atom is minimized. The periodicity results from the completion of the subshells.

Bohr's table, still useful, was the result of physical theory and offered a fundamental *physical basis* for understanding chemistry. For example, it showed how the structure of the periodic table follows from the shell structure of atoms. This was another triumph of the Bohr theory.

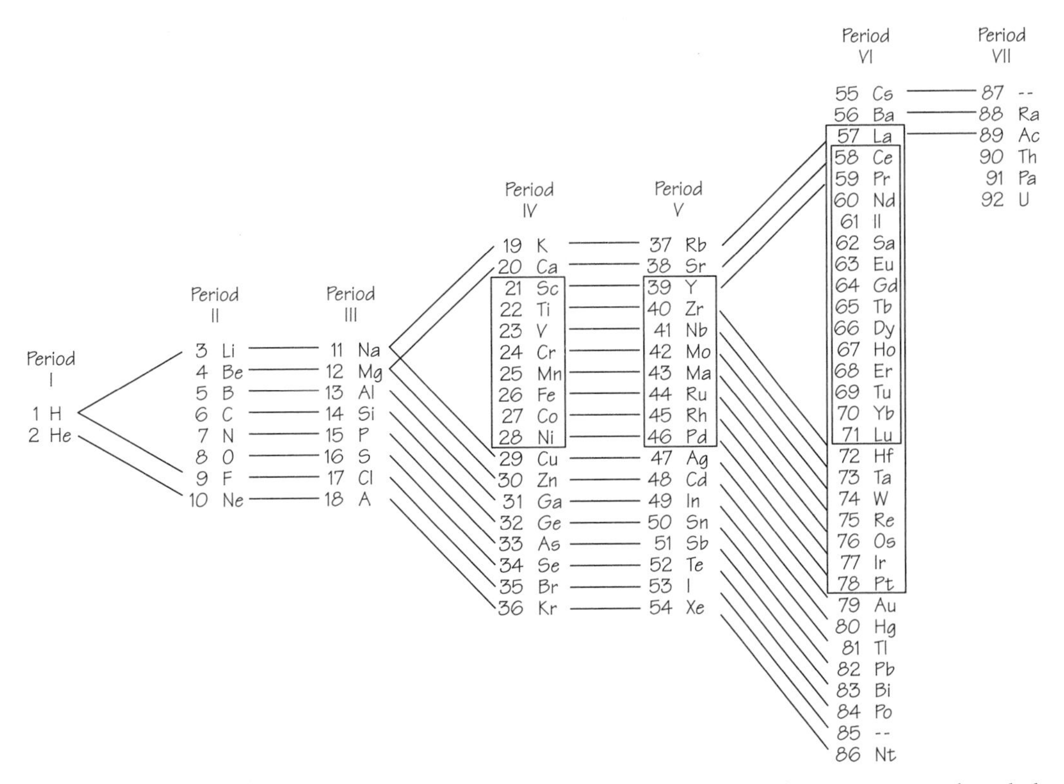

FIGURE 14.16 Bohr's periodic table of the elements (1921). Some of the element names and symbols have since been changed. Masurium (43) had been falsely identified at the time. The place is taken by technetium (43).

14.11 EVALUATING THE BOHR MODEL

In March 1913, Bohr wrote to his mentor Rutherford, enclosing a draft of his first paper on the quantum theory of atomic constitution. Rutherford replied in a letter, the first part of which is quoted here:

> Dear Dr. Bohr:
>
> I have received your paper and read it with great interest, but I want to look it over again carefully when I have more leisure. Your ideas as to the mode of origin of spectra in hydrogen are very ingenious and seem to work out well; but the mixture of Planck's ideas with the old mechanics make it very difficult to form a physical idea of what is the basis of it. There appears to me one grave difficulty in your hypothesis, which I have no doubt you fully realize, namely, how does an electron decide what frequency it is going to vibrate at when it passes from one stationary state to the other? It seems to me that you would have to assume that the electron knows beforehand where it is going to stop. . . .

Every Model and Every Theory Has Its Limits

The Bohr theory achieved great successes in the years between 1913 and 1924. But it also contained unanswered questions and unresolved problems, as Rutherford so keenly observed. As time progressed, further problems arose for which the theory proved inadequate.

Bohr's theory accounted very well for the spectra of atoms with a single electron in the outermost shell. However, serious differences between theory and experiment appeared in the spectra of atoms with two or more electrons in the outermost shell. Experiments also revealed that when a sample of an element is placed in an electric or magnetic field, its emission spectrum shows additional lines. For example, in a magnetic field each line is split into several lines. The Bohr theory could not account, in a quantitative way, for some of the observed splittings. Furthermore, the theory supplied no method for predicting the relative brightness (intensity) of spectral lines. These relative intensities depend on the probabilities with which atoms in a sample undergo transitions among the stationary states—high probabilities resulting in more intense lines. Physicists wanted to be able to calculate the probability of a transition from one stationary state to another. They could not make such calculations with the Bohr theory.

By the early 1920s it was clear that the Bohr theory, despite its remarkable successes, was limited. To form a theory that would solve more problems, Bohr's theory would have to be revised or replaced—incidentally, a reminder that a main purpose of science today is always to prepare the ground for better science tomorrow. But the successes of Bohr's theory did show that a better theory of atomic structure would still have to account for the existence of stationary states, which are separate, distinct atomic energy levels. Therefore, such a theory would still have to be based on quantum concepts—a reminder that new theories tend to evolve by incorporating what was good in old ones, rather than by a revolutionary overthrow of the old theory.

Besides the inability to predict certain properties of atoms at all, the Bohr theory had two additional shortcomings. First, it predicted some results that did not agree with the experiment (such as incorrect spectra for elements with two or three electrons in the outermost electron shells). Second, it predicted results that could not be tested in any known way (such as the details of electron orbits). Although orbits were easy to draw on paper, they could not be observed directly. Nor could they be related to any observable properties of atoms. Planetary theory has very different significance when applied to a planet in an observable orbit than when applied to an electron in an atom. The precise position of a planet is important, especially in experiments such as photographing an eclipse, or a portion of the surface of Mars from a satellite. But the moment-to-moment position of an electron in its orbit has no such meaning because it has no relation to any experiment physicists have been able to devise. It thus became evident that the Bohr theory led to some questions that could not be answered experimentally.

In the early 1920s, physicists, especially Bohr himself, began to work seriously on revising the basic ideas of the theory. One fact that stood out was, as Rutherford had pointed out, that the theory started with a mixture of classical and quantum ideas. An atom was assumed to act according to the laws of classical physics up to the point where these laws did not work. Beyond this point, quantum ideas were introduced. The picture of the atom that emerged was an inconsistent mixture. It combined ideas from classical physics with concepts for which there was no place in classical physics. The orbits of the electrons were determined by the classical, Newtonian laws of motion, much like the orbits of planets around the Sun. But of the many theoretical orbits, only a small portion were regarded as possible. Even these few orbits were selected by rules for which there was no room in classical mechanics. Again, the frequency calculated for the orbital revolution of electrons was quite different from the frequency of light emitted or absorbed when the electron moved from or to this orbit. Also, the decision that the number n could never be zero seemed arbitrary but it was necessary to prevent the model from collapsing by letting the electron fall on the nucleus. It became evident that a better theory of atomic structure would need a more consistent foundation in quantum concepts.

The contribution of Bohr's theory may be summarized as follows. It provided some excellent answers to the questions raised about atomic structure in Chapter 13. Although the theory turned out to be inadequate, it drew attention to how quantum concepts can be used. It indicated the path that a new theory would have to take. A new theory would have to supply the right answers that the Bohr theory gave, but it would also have to supply the right answers for the problems the Bohr theory could not solve. One of the most fascinating aspects of Bohr's work was the proof that physical and chemical properties of matter can be traced back to the fundamental role of integers (quantum numbers such as $n = 1, 2, 3, \ldots$). As Bohr said, "The solution of one of the boldest dreams of natural science is to build up an understanding of the regularities of nature upon the consideration of pure number." You can catch here an echo of the hope of Pythagoras and Plato, of Kepler and Galileo.

Since the 1920s, a successful new theory of atomic structure has been developed and generally accepted by physicists. It is part of *quantum mechanics*, so called because it is a new mechanics built directly on quantum concepts. It goes far beyond understanding atomic structure. In fact, it is the basis of the modern conception of events on a submicroscopic scale. Some aspects of this theory will be discussed in the next chapter. Significantly, Bohr himself was again a leading contributor.

SOME NEW IDEAS AND CONCEPTS

absorption spectra
Balmer series
Bohr model
electron shells
emission spectra
ground state
nuclear model
nucleus
quantum mechanics
spectroscopy
stationary states

SOME IMPORTANT EQUATIONS

$$\frac{1}{\lambda} = R_H\left(\frac{1}{2^2} - \frac{1}{n^2}\right), \qquad n = 3, 4, 5, 6, \ldots$$

$$\frac{1}{\lambda} = R_H\left(\frac{1}{n_f^2} - \frac{1}{n_i^2}\right),$$

$$r_n = a \cdot n^2,$$

$$E_n = \left(\frac{1}{n^2}\right)E_1.$$

FURTHER READING

G. Holton and S.G. Brush, *Physics, The Human Adventure* (Piscataway, NJ: Rutgers University Press, 2001), Chapter 28.

C.H. Townes, *How the Laser Happened: Adventures of a Scientist* (New York: Oxford University Press, 1999).

STUDY GUIDE QUESTIONS

14.1 Spectral of Gases

1. How is the emission spectrum of an element related to its absorption spectrum?
2. What can you conclude about a source if its light gives a bright-line spectrum?
3. What can you conclude about a source if its light gives a dark-line spectrum?
4. What evidence is there that the physics and chemistry of materials at great distances from Earth are the same as those of matter close at hand? What does this fact say about the structure of the Universe?
5. An unknown gas is contained in a glass tube. Give two ways in which it could be identified using spectroscopy.

14.2 Regularities in the Hydrogen Spectrum

1. What is the Balmer series? How is it summarized by Balmer's formula?
2. What evidence did Balmer have that there were other series of lines in the hydrogen spectrum, with terms such as 3^2, 4^2, etc., instead of 2^2?
3. How are the other series summarized by the extension of Balmer's formula?
4. Often discoveries result from grand theories (like Newton's) or from a good intuitive grasp of phenomena (like Faraday's). What led Balmer to his relation for hydrogen spectra?
5. From the Balmer formula in the last form given, is there any upper limit to the wavelengths of light emitted by the hydrogen atom?

14.3 Rutherford's Nuclear Model of the Atom

1. Describe Rutherford's experiment and its "incredible" result. What did Rutherford conclude from this experiment?
2. Why was Rutherford as surprised with this result as he would have been if a 15-in shell fired at a piece of tissue paper "came back and hit you"?
3. Why are α particles scattered by atoms? Why is the angle of scattering mostly small but sometimes large?
4. What was the basic difference between Rutherford's and Thomson's models of the atom?

14.4 Nuclear Charge and Size

1. What does the "atomic number" of an element refer to, according to the Rutherford model of the atom?
2. What is the greatest positive charge that an ion of lithium (the next heaviest element after helium) could have?
3. How did the scattering of α rays help Rutherford estimate the size of the nucleus?
4. How big is the nucleus? How does this compare with the size of the atom? What does this say about the interior of the atom?
5. How does one find by experiment the size of the nucleus?

14.5 Bohr's Theory: The Postulates

1. State Bohr's two quantum postulates in your own words.
2. In what ways do these postulates contradict Newton's mechanics and Maxwell's electromagnetic theory?
3. What was the main evidence to support the fact that an atom could exist only in certain energy states?
4. How did Bohr deal with the idea that as long as an electron is steadily orbiting a nucleus, it does not radiate electromagnetic energy?

14.6 The Size of the Hydrogen Atom

1. According to Bohr, why do all unexcited hydrogen atoms have the same size?
2. Why does the hydrogen atom have just the size it has?

14.7 Other Consequences of the Bohr Model

1. What happens to the electron in the hydrogen atom as n goes to infinity?
2. How is the ground state defined?
3. What happens to the radii of the Bohr orbits as n increases?
4. What happens to the stationary-state energies as n increases?
5. Why do the energies of the stationary states have negative values?

14.8 Bohr Accounts for the Series Spectra of Hydrogen

1. In general terms, how did Bohr account for Balmer's formula?
2. Balmer had predicted accurately that there might be other spectral series of hydrogen 30 years before Bohr did. Why is Bohr's prediction considered more significant?

3. How does Bohr's model account for absorption spectra?
4. In Section 14.1 you saw that an absorption spectrum does not contain all the lines of the corresponding emission spectrum. Based on the Bohr model, why is this so?
5. Why is it correct to say that the hydrogen atom can have an infinity of emission lines?

14.9 Do Stationary States Really Exist?

1. Briefly describe the Franck–Hertz experiment and the conclusion Franck and Hertz obtained in answering the question in the title of this section.
2. How much kinetic energy will an electron have after a collision with a mercury atom if its kinetic energy before collision is:
   (a) 4.0 eV?
   (b) 5.0 eV?
   (c) 7.0 eV?

14.10 Constructing the Periodic Table

1. Describe in your own words what happens to the structure of the atoms of different elements as you progress through the periodic table.
2. Draw a sketch of the atoms in the first two rows of the periodic table and label everything in your picture. Include the nucleus, its charge, and the various electron shells.
3. Why do the next heavier elements after the noble gases easily become positively charged?
4. Why do the elements in the next to last column of the periodic table easily become negatively charged?
5. What is special about the noble-gas elements?
6. Why are there only two elements in Period I, eight in Period II, eight in Period III, etc.?

14.11 Evaluating the Bohr Theory

1. Evaluate Bohr's theory of the atom. Was it a good theory? What were some of its advantages? What were some of its problems?
2. Why did some physicists begin looking for a quantum mechanics?
3. How did they expect this theory to differ from Bohr's theory?

DISCOVERY QUESTIONS

1. (a) Suggest experiments to show which of the Fraunhofer lines in the spectrum of sunlight result from absorption in the Sun's atmosphere rather than from absorption by gases in the Earth's atmosphere.
   (b) How might one decide, from spectroscopic observations, whether the Moon and the planets shine by their own light or by reflected light from the Sun?

2. Theoretically, how many series of lines are there in the emission spectrum of hydrogen? In all these series, how many lines are in the visible region?
3. As indicated in the figure on page 643, the lines in one of hydrogen's spectral series are bunched closely at one end. Does the formula

$$\frac{1}{\lambda} = R_H\left(\frac{1}{n_f^2} - \frac{1}{n_i^2}\right)$$

suggest that such bunching will occur?
4. Physicists generally suppose that the atom and the nucleus are each spherical. They assume that the diameter of the atom is of the order of 10^{-10} m and that the diameter of the nucleus is of the order of 10^{-14} m.
   (a) What are the evidences that these are reasonable suppositions?
   (b) What is the ratio of the diameter of the nucleus to that of the atom?
5. Make an energy-level diagram to represent the results of the Franck–Hertz experiment.
6. Many substances emit visible radiation when illuminated with ultraviolet light. This phenomenon is an example of fluorescence. Stokes, a British physicist of the nineteenth century, found that in fluorescence, the wavelength of the emitted light usually was the same or longer than the illuminating light. How would you account for this phenomenon on the basis of the Bohr theory?
   Sometimes in fluorescence the wavelength of the emitted light is *shorter* than the illuminating light. What may cause this?
7. Use the periodic table to predict the electron structure of element 19. Why does it have chemical properties similar to those of elements 1, 3, and 11?
8. Write an essay on the successes and failures of the Bohr model of atoms. Can it be called a good model? a simple model? a beautiful model?

Quantitative

1. What would be the radius of a hydrogen atom if its electron is orbiting in state: $n = 2$? $n = 5$? $n = 10$? Do you see a pattern in these results?
2. The constant R in the Balmer formula has the value 1.1×10^7 m. What are the wavelengths of the first two Balmer lines? What are the wavelengths of the lines for $n = 10$ and $n = 20$? Do you see a pattern to these results? Where does each of these lines lie in the electromagnetic spectrum?
3. The "Lyman series" for hydrogen involves a "jump" of the electron to the state $n = 1$. What are the wavelengths of the first two lines of the Lyman series? Why is the word "jump" only to be taken as a metaphor?
4. The nucleus of a hydrogen atom has a radius of about 1.5×10^{-15} m. Imagine the atom magnified so that the nucleus has a radius about the size of a grain of dust, or 0.1 mm. What would be the size of the corresponding hydrogen atom in the ground state? Attempt to construct a scale model of the radius of such a "hydrogen atom" on a long sidewalk or in a stadium.

CHAPTER 15

Quantum Mechanics

15.1 THE QUANTUM

The discovery of the quantum of energy in the early years of the twentieth century provided an explanation of the photoelectric effect and it enabled the success of Bohr's quantum model of the atom. This model and the other successes of that time contributed to what is known as "quantum theory."

Nevertheless, the existence of the energy quantum, whether in light or in atoms, posed a serious problem for physics, since it was incompatible with both Newton's mechanics and Maxwell's electromagnetic wave theory. In these theories energy is always continuous and infinitely divisible. But these "classical" theories were constructed on the basis of events occurring in the visible, human-scale world, ranging from planets to microscopic objects. Perhaps, then, we should not be surprised that nature might behave differently when we enter into regions far removed from everyday experience, such as the interior of atoms or the sub-microscopic structure of minute quantities of light. And when we are surprised, it only impresses upon us even more how intricate and exquisite nature really is.

By the mid-1920s scientists realized that the quantum theory was inadequate and that a new theory was needed to encompass the quantum world at the subatomic level, a new *quantum mechanics* in which the quantum is built into the foundations of physics from the beginning. An important clue to the new mechanics came from the further study of particles and waves.

15.2 THE PARTICLE-LIKE BEHAVIOR OF LIGHT

Einstein's hypothesis of light quanta created a dilemma for physicists. While Einstein's work indicated that light behaves like particles in such experiments as the photoelectric effect, light clearly behaved like waves in Young's important double-slit experiment. As you may recall from Chapter 8, when a beam of light shines on two narrow slits near each other, the light emerging from the slits interferes and forms on a screen the alternating bright and dark bands that are characteristic of the interference of waves. Particles cannot form this pattern. Moreover, Maxwell's electromagnetic theory accounted for electromagnetic radiation as a wave phenomenon, and this theory was supported by Young's experiment and many other experiments.

On the other hand, Einstein's account of the photoelectric effects showed that light behaved as if it consisted of particle-like light quanta—later called "photons." Each photon has energy $E = hf$, where h is Planck's constant and f is the frequency of the light. Einstein himself pointed out that, since photons carry energy, this energy, while the photon moves at the speed of light, is equivalent to a certain amount of mass, according to his famous formula

$$E = mc^2.$$

The amount of mass is just

$$m = \frac{E}{c^2}.$$

Here c is the speed of light.

If a photon has energy, and the energy is equivalent to an amount of mass, does it also have momentum? The photoelectric effect did not tell anything about the momentum of a photon.

The magnitude of the momentum p of a body is defined as the product of its mass m and speed v in the same direction; $p = mv$. (See Chapter 5.) Replacing m with its energy equivalent E/c^2 gives

$$p = \frac{Ev}{c^2}.$$

Note that this equation is an expression for momentum p but that it contains no direct reference to mass. Now suppose this same equation is applied to the momentum of a photon of energy E. Since a photon moves at the speed of light, v would be replaced by the speed of light c to give

$$p = \frac{Ec}{c^2} = \frac{E}{c}.$$

Remember, $E = hf$ for a light quantum. If you substitute this expression for E in $p = E/c$, you get the expression

$$p = \frac{hf}{c}.$$

Using the wave relation that the speed equals the frequency times the wavelength, $c = f\lambda$, you can express the momentum of a photon as

$$\boxed{p = \frac{h}{\lambda}.}$$

FIGURE 15.1 Arthur H. Compton (1892–1962).

Does it make sense to define the momentum of a photon in this way? It would, if the definition helps in understanding experimental results. The first successful use of this definition was in the analysis of a now-famous phenomenon discovered by Arthur H. Compton, the *Compton effect*, which we describe below.

Consider a beam of light (or X rays) striking the atoms in a target, such as a thin sheet of metal. According to classical electromagnetic theory, the light will be scattered in various directions, but its frequency will not change. The absorption of light of a certain frequency by an atom may be followed by reemission of light of a different frequency. But if the light wave is simply *scattered*, then according to classical theory the frequency should not change.

According to quantum theory, however, light is made up of photons. According to relativity theory, photons have momentum. Therefore, Compton reasoned, in a collision between a photon and an atom, the law of conservation of momentum should apply. According to this law (Chapter 5), when a body of small mass collides with a massive object at rest, it simply bounces back or glances off. It experiences very little loss in speed and so very little change in energy. But if the masses of the two colliding objects are not very different, a significant amount of energy can be transferred in the collision. Compton calculated how much energy a photon should lose in a collision with an atom, if the photon's momentum is h/λ. He concluded that the change in energy is too small to observe if a photon simply bounces off an entire atom. But if a photon strikes an electron, which has a small mass, the photon should transfer a significant amount of energy to the electron.

In 1923, Compton was able to show that X rays did in fact behave like particles with momentum $p = h/\lambda$ when they collided with electrons. Compton measured the wavelength (or frequency) of the incident and the scattered X rays and thus was able to determine the X ray photon's change in momentum. By measuring separately the momentum of the scattered electron, he was able to verify that $p = h/\lambda$, using the law of conservation of momentum. For this work, Compton received the Nobel Prize in 1927.

Thus, Compton's experiment showed that a photon can be regarded as a particle with a definite momentum as well as energy. It also showed that collisions between photons and electrons obey the laws of conservation of momentum and energy.

As noted in Section 13.5, in the discussion of the photoelectric effect, light has particle-like properties. The momentum expression and the Compton effect gave additional evidence for this fact. To be sure, photons are not like ordinary particles, if only because photons do not exist at speeds other than that of light. (There can be no resting photons and, therefore, no rest mass for photons.) But in other ways, as in their scattering behav-

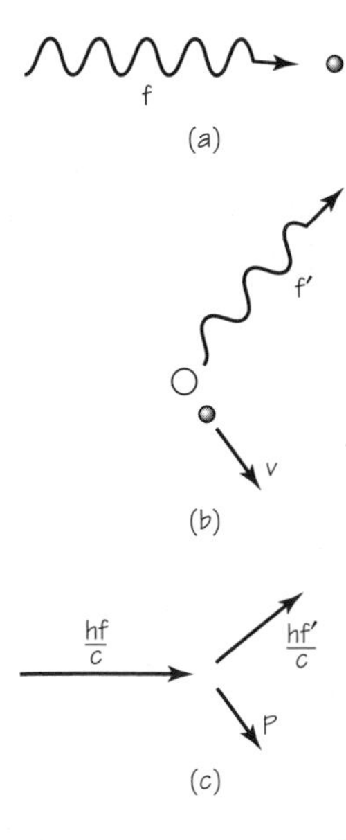

FIGURE 15.2 Compton's experiment: (a) X ray of frequency f approaches an electron; (b) X ray is scattered, leaving at lower frequency f', and electron recoils at velocity $\mathbf{v}$; (c) the momentum before "collision" (hf/c) is equal to the vector sum of the momentum of both the scattered photon and the recoiling electron.

ior, photons do act much like particles of matter. For example, they have momentum as well as energy. Yet they also act like waves, having frequency and wavelength. In other words, electromagnetic radiation in some experiments exhibits behavior similar to what is thought of as particle behavior, and in other experiments, its behavior is similar to what is thought of as wave behavior. This pattern of behavior is often referred to as the *wave–particle dualism of radiation.* Is a photon a wave or a particle? The only answer, Bohr pointed out, is that it can *act* like either, depending on what is being done with it.

15.3 THE WAVE-LIKE BEHAVIOR OF PARTICLES

In 1923, the French physicist Louis de Broglie suggested that the wave–particle dualism that applies to electromagnetic radiation might also apply to electrons and other atomic particles. Perhaps, he said, the wave–particle dualism is a fundamental property of all quantum processes. If so, particles that were always thought of as material particles can, in some circumstances, act like waves. De Broglie sought an expression for the wavelength that might be associated with the wave-like behavior of an electron. He found such an expression by means of a simple argument.

The momentum of a photon of wavelength λ is $p = h/\lambda$. De Broglie thought that this relation might also apply to electrons with the momen-

LOUIS VICTOR DE BROGLIE

Prince Louis Victor de Broglie (1892–1987), whose ancestors served the French kings as far back as the time of Louis XIV, was educated at the Sorbonne in Paris. He proposed the idea of wave properties of electrons in his doctoral thesis.

FIGURE 15.3

tum $p = mv$. He therefore boldly suggested that the wavelength of an electron is

$$\lambda = \frac{h}{mv},$$

where m is the electron's mass and v its speed.

What does it mean to say that an electron has a wavelength equal to Planck's constant divided by its momentum? As before, if this statement is to have any physical meaning, it must be possible to test it by some kind of experiment. Some wave property of the electron must be measured. The first such property to be measured was *diffraction*.

The relationship $\lambda = h/mv$ indicates that the wavelengths associated with electrons will be very short, even for fairly slow electrons. An electron accelerated across a potential difference of only 100 V would have a wavelength of only 10^{-10} m. So small a wavelength would not give measurable diffraction effects on encountering even a microscopically small object (say, 10^{-5} m).

By 1920, it was well known that crystals have a regular lattice structure. The distance between rows of planes of atoms in a crystal is about 10^{-10} m. After de Broglie proposed that electrons have wave properties, several physicists suggested that the existence of electron waves might be shown by using crystals as diffraction gratings. Experiments begun in 1923 by C.J. Davisson and L.H. Germer in the United States yielded diffraction patterns similar to those obtained earlier for X rays (Section 13.7). See Figure 15.4.

The Davisson–Germer experiment showed two things. First, electrons *do* have wave properties, otherwise they could not display the diffraction pattern of waves. One may say that an electron moves along the path taken by the de Broglie wave that is associated with the electron. Second, electron wavelengths are correctly given by de Broglie's relation, $\lambda = h/mv$. These results were confirmed in 1927 when G.P. Thomson, the son of J.J. Thomson, directed an electron beam through thin gold foil. Thomson found a pattern like the one shown here. It resembles diffraction patterns produced by light beams going through thin slices of materials. By 1930, diffraction from crystals had been used to demonstrate the wave-like behavior even of helium atoms and hydrogen molecules. (It can be said that

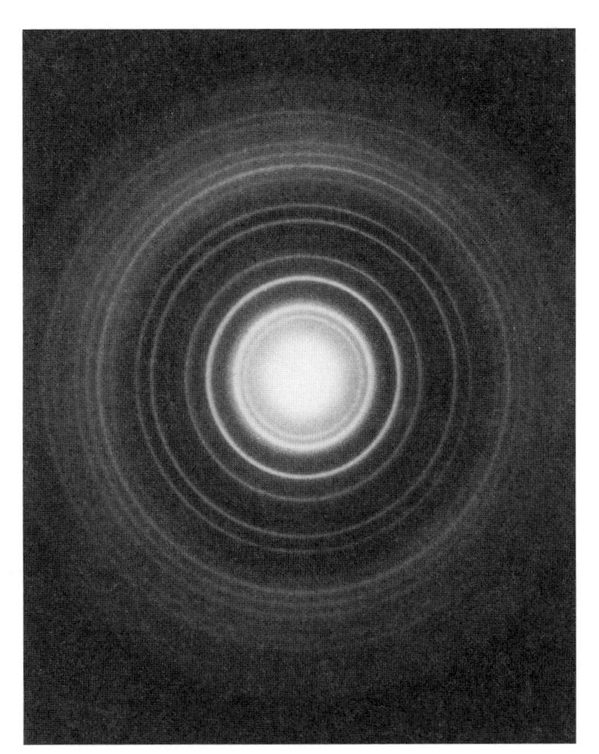

FIGURE 15.4 Diffraction pattern produced by directing a beam of electrons through polycrystalline aluminum (that is, many small crystals of aluminum oriented at random). With a similar pattern using gold foil, G.P. Thomson demonstrated the wave properties of electrons (28 years after their particle properties were first demonstrated by his father, J.J. Thomson).

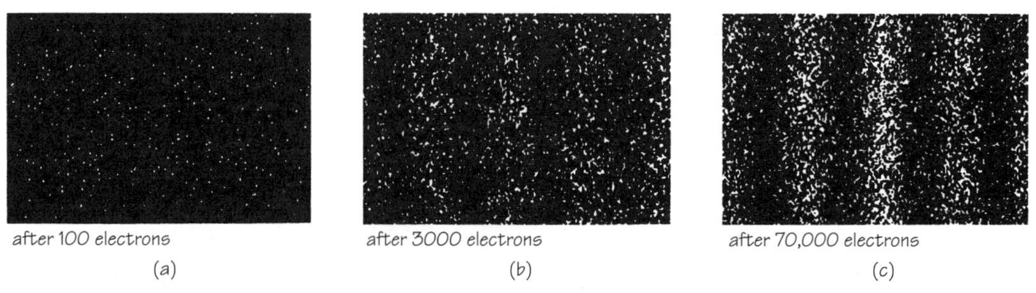

FIGURE 15.5 Various numbers of electrons passed through two slits to reach a screen forming an interference characteristic of waves.

J.J. Thomson discovered the electron *particle*, and his son showed it to be a *wave* as well.)

The experiments confirming de Broglie's hypothesis showed that the wave–particle dualism is a general property not only of radiation but also of matter. Scientists now customarily refer to electrons and photons as "particles" while recognizing that both have properties of waves as well.

Electron Waves and Atomic Structure

Bohr had postulated that the quantity mvr, which is called the "angular momentum" of the orbiting electron in the hydrogen atom, where r is the radius of the electron's orbit, m is the electrons mass, and v is its linear speed, can have only certain, quantized values. These quantized values help to define the stationary states. De Broglie's relation, $\lambda = h/mv$, has an interesting yet simple application that supports this postulate and sheds light on the existence of stationary states. Bohr assumed that mvr can have only the values

$$mvr = \frac{nh}{2\pi},$$

where, as before, $n = 1, 2, 3, \ldots$.

Now, suppose that an electron wave is somehow spread over an orbit of radius r so that, in some sense, it "occupies" the entire orbit. Can *standing waves* be set up as indicated, for example, in Figure 15.6? If so, the circumference ($2\pi r$) of the orbit must be equal in length to a whole number (n) of wavelengths, that is, to $n\lambda$. The mathematical expression for this condition of "fit" is

$$2\pi r = n\lambda.$$

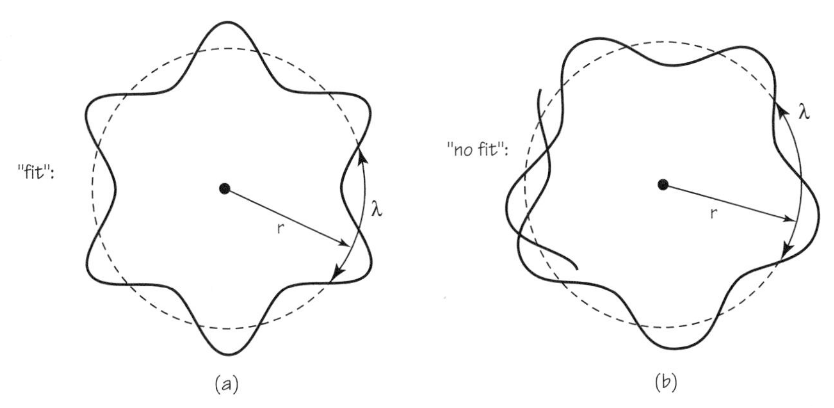

FIGURE 15.6 Only certain wavelengths will "fit" around a circle to form a standing wave.

Replacing λ by h/mv according to de Broglie's relation gives

$$2\pi r = \frac{nh}{mv}$$

or

$$mvr = \frac{nh}{2\pi}.$$

This is exactly equivalent to Bohr's quantization condition! The de Broglie relation for electron waves, and the idea that electrons have orbits that allow standing waves, allows us to *derive* the quantization of the electron orbits that Bohr had to *assume*.

The result obtained so far indicated that one could picture the electron in the hydrogen atom in two ways: as a *particle* moving in an orbit with a certain quantized value of mvr, or as a standing de Broglie-type *wave* occupying a certain region around the nucleus.

15.4 CONSTRUCTING QUANTUM MECHANICS

By the mid-1920s it was clear that "things" (electrons, atoms, molecules) long regarded as particles also show wave properties. This fact is the basis for the currently accepted theory of atomic structure. This theory, *quantum mechanics*, was introduced in 1925. Its foundations were developed very

FIGURE 15.7 Erwin Schrödinger (1887–1961) was born in Austria. He developed wave mechanics in 1926 and then fled from Germany in 1933 when Hitler and the Nazis came to power. From 1940 to 1956, he was professor of physics at the Dublin Institute for Advanced Studies.

rapidly during the next few years, primarily by Born, Heisenberg, Schrödinger, Bohr, Jordan, and Dirac. At first, the theory appeared in two different mathematical forms, proposed independently by Werner Heisenberg and Erwin Schrödinger. Heisenberg provided the basis for quantum mechanics emphasizing the particle aspect of quantum objects, while Schrödinger emphasized the wave aspect. Since Schrödinger's form of the theory is closer to the ideas of de Broglie (discussed in the previous section), it is often referred to as *wave mechanics*. Eventually Schrödinger proved that in fact these two forms of quantum mechanics are equivalent, that is, different ways of expressing the same relationships. Schrödinger's formulation is now predominant, although the symbols used in his equations are now interpreted differently from Schrödinger's original usage, as discussed presently.

Schrödinger sought to express the dual wave–particle nature of matter mathematically. Maxwell had formulated the electromagnetic theory of light in terms of a wave equation. Physicists were familiar with this theory and its applications. Schrödinger reasoned that the de Broglie waves associated with electrons could be described in a way analogous to the classical waves of light. Thus, there should be a wave equation that holds for matter waves, just as there is a wave equation for electromagnetic waves. This mathematical part of wave mechanics cannot be discussed adequately without using advanced mathematics, but the physical ideas involved require only a little mathematics and are essential to understanding modern physics. Therefore, the rest of this chapter will discuss some of the physical ideas of the theory so as to indicate that they are indeed reasonable. Some of the results of the theory and some of the significance of these results are also considered.

Schrödinger successfully derived an equation for the *matter waves* (de Broglie waves) that are associated with moving electrons. This equation, which has been named after him, defines the wave properties of electrons and also predicts particle-like behavior. The Schrödinger equation for an electron bound in an atom has a solution only when a constant in the equation has the whole-number values 1, 2, 3. . . . These numbers correspond to different energies. Thus, the Schrödinger equation predicts that only certain electron energies are possible in an atom. In the hydrogen atom,

FIGURE 15.8 P.A.M. Dirac (1902–1984), an English physicist, was one of the developers of modern quantum mechanics. In 1932, at the age of 30, Dirac was appointed Lucasian Professor of Mathematics at Cambridge University, the post held by Newton. Dirac is pictured here with Werner Heisenberg and Erwin Schrödinger while in Stockholm in 1933 on the occasion of the award of the Nobel Prize.

for example, the single electron can be in only those states for which the energy of the electron has the numerical values

$$E_n = \frac{k^2 2\pi^2 me^2}{n^2 h^2},$$

with n having only whole-number values. But these are just the energy values that are found experimentally, and just the ones given by the earlier Bohr theory! In addition, these states correspond to the picture of standing electron waves in the various stationary, as discussed in the previous section.

In Schrödinger's theory, these results follow directly from the mathematical formulation of the wave and particle nature of the electron. Bohr had to assume the existence of these stationary states at the start and make no assumptions about the allowable orbits. In Schrödinger's theory, however, the stationary states and their energies are *derived* from the theory. The new theory yields all the results of Bohr's theory, with none of the Bohr theory's inconsistent hypotheses. The new theory also accounts for certain experimental information for which Bohr's theory failed to account. For instance, it allows one to compute the intensity of a spectral line, which is understood as the probability of an electron changing from one energy state to another.

After the unification of Schrödinger wave mechanics with Heisenberg's formulation, quantum mechanics still contained the Schrödinger equation, but it not longer provided a physical model or visualizable picture of the atom. The planetary model of the atom has been given up and has not been replaced by another simple picture. There is now a highly successful *mathematical* model for the atom, but no easily visualized *physical* model. The concepts used to build quantum mechanics are more abstract than those of the Bohr theory. But the mathematical theory of quantum mechanics is much more powerful than the Bohr theory in predicting and explaining phenomena. Many problems that were previously unsolvable were rapidly solved with quantum mechanics (see the table on p. 673). Physicists have learned that the world of atoms, electrons, and photons cannot be thought of in the same mechanical terms as the world of everyday experience. Instead, the study of atoms presents some fascinating new concepts, such as those discussed below. What has been lost in easy visualizability is made up for by an increase in fundamental understanding about nature at the most fundamental level.

What does it mean to "visualize" or "picture" something? One answer is that it means relating an abstract idea to something that you are familiar with from everyday life; for example, a particle is like a baseball or a marble. But why should there be anything from everyday life that is exactly like an electron or an atom?

The following table shows the rapid pace of development in accounting for some previously unsolvable problems and inexplicable phenomena following the formulation of quantum mechanics in 1926.

1926	Quantum mechanics of hydrogen atom	W. Pauli
1926	Quantum mechanics of helium atom	W. Heisenberg
1926	Hydrogen molecule	W. Heisenberg
1926	Effect of magnetic fields on spectral lines (Zeeman effect)	W. Heisenberg and P. Jordan
1927	Molecules in general	M. Born and J.R. Oppenheimer
1928	Magnetism (ferromagnetism)	W. Heisenberg
1928	Electron theory of metals	F. Bloch
1928	Relativistic quantum theory of electrons	P.A.M. Dirac

15.5 THE UNCERTAINTY PRINCIPLE

Up to this point, it has been assumed that any physical property of an object can be measured as accurately as necessary. To reach any desired degree of accuracy would require only a sufficiently precise instrument. Wave mechanics showed, however, that even in thought experiments with ideal instruments there are limits to the accuracy of measurements that can be achieved.

For example, think how you would go about measuring the positions and velocity of a car moving slowly along a driveway. You could mark the position of the front end of the car at a given instant by making a scratch on the ground. At the same time, you could start a stopwatch. Then you could run to the end of the driveway, where you have previously placed another mark. At the instant when the front of the car reaches this point, you stop the watch. You then measure the distance between the marks and get the average speed of the car by dividing the distance traveled by the time elapsed. Since you know the direction of the car's motion, you know the average velocity. Thus, you know that at the moment the car reached the second mark it was at a certain distance from its starting point and had traveled at a certain average velocity. By going to smaller and smaller intervals, you could also get the instantaneous velocity at any point along its path.

How did you get the needed information? You located the car by sunlight that was bounced off the front end into your eyes. The light permitted you to see when the car reached a mark on the ground. To get the average speed, you had to locate twice where the front end was.

But suppose that you had decided to use reflected radio waves instead of visible light. At 1000 kHz, a typical value for radio signals, the wavelength is 300 m. This wavelength is very much greater than the dimensions of the car. Thus, it would be impossible to locate the position of the car with any accuracy. The wave would reflect from the car ("scatter" is a better term) in all directions. It would also sweep around any human-sized device you may wish to use to detect the wave direction. The wavelength has to be comparable with or smaller than the dimensions of the object before the object can be located well.

Radar uses wavelengths from about 0.1 cm to about 3 cm, so a radar apparatus could be used instead of sunlight. But even radar would leave uncertainties as large as several centimeters in the two measurements of position. The wavelength of visible light is less than 10^{-6} m. For visible light, then, you could design instruments that would locate the position of the car to an accuracy of a few thousandths of a millimeter.

Now think of an electron moving along an evacuated tube. You will try to measure the position and speed of the electron. But you must change your method of measurement. The electron is so small that you cannot locate its position by using ordinary visible light. (The wavelength of visible light, small as it is, is still at least 10^4 times greater than the diameter of an atom.)

FIGURE 15.9 Werner Heisenberg (1901–1976), a German physicist, was one of the developers of modern quantum mechanics (at the age of 23). He was the first to state the uncertainty principle. After the discovery of the neutron in 1932, he proposed the proton-neutron theory of nuclear structure.

You are attempting to locate the electron within a tiny region, say the size of an atom, about 10^{-10} m across. So you need a light beam whose wavelength is about 10^{-10} m or smaller. But a photon of such a short wavelength λ (and high frequency f) has very great momentum (h/λ) and energy (hf). Recalling Compton's work (Section 15.2), you know that such a photon will give the electron a strong kick when it is scattered by the electron. As a result, the velocity of the electron will be greatly changed, into a new and unknown direction. (This is a new problem, one you did not even think about when measuring the position of the car!) Therefore, when you receive the scattered photon, you can deduce from its direction where the electron *once was*; in this sense you can "locate" the electron. But in the process you have changed the velocity, hence the momentum, of the electron (in both magnitude and direction). In short, the more accurately you locate the electron by using photons of shorter wavelength, the less accurately you can know its momentum. You could try to disturb the electron less by using less energetic photons. But because light exists in quanta of energy hf, a *lower-energy* photon will have a *longer* wavelength. This would create greater uncertainty about the electron's *position*! In other words:

> It is impossible to measure both the position and the momentum of a subatomic particle, in the same instant to unlimited accuracy. The more accurate is the measurement of the momentum, the less accurate is the measurement of the position in that instant, and vice versa.

This conclusion is expressed in the *uncertainty principle*, first stated by Werner Heisenberg in 1927. The uncertainty principle can be expressed quantitatively in two simple mathematical expressions, known as *uncertainty relations*, which, as indicated by the above example, are necessary conclusions drawn from experimental facts about measurements involving quantum objects.

In this example, let Δx represent the uncertainty in the measurement of the position of the object, and let Δp_x be the uncertainty in the measurement of the momentum of the object in the x direction at the same instant. Heisenberg's principle says that the product of these two uncertainties must be equal to, or greater than, Planck's constant divided by 4π:

$$\Delta x \cdot \Delta p_x \geq \frac{h}{4\pi}.$$

There are similar relations for the momenta in the y and z directions and the uncertainties in the measurements of the position along the y and z coordinates.

This uncertainty relation says that if we use a short-wavelength photon in an attempt to measure the position of an electron to very high accuracy, so that Δx is very small, then the uncertainty in the momentum measurement Δp must be at least $h/4\pi\Delta x$. This means that as Δx gets smaller, Δp has to get larger. Again one can see this also from the Compton effect; the electron would bounce away faster the shorter the wavelength (the higher the energy) of the measuring photon. In fact, if we measure the position so accurately that there is no uncertainty at all in the position, then Δx would be zero. But to do this we would have been forced to use a photon whose wavelength λ was zero! Such a photon has infinite energy! In that case, the uncertainty in the momentum of the electron would be infinite, or undefined.

On the other hand, if we allowed the uncertainty in the position measurement to become very large, then the uncertainty in the momentum measurement would become very small, since the photon would have long wavelength (low momentum). If Δx became so large that it was infinite, or undefined, then Δp would become zero. We could measure the momentum in that instant with absolute precision. But we cannot measure both the position and the momentum to absolute precision at the *same* time. Instead, the uncertainty relation forces us into a trade-off. When the precision of one variable goes up, the other must go down, and vice versa.

Similar considerations result in a second uncertainty relation involving the variables time and energy. Let Δt represent the uncertainty in the measurement of time, and ΔE the uncertainty in the measurement of the energy of a quantum object at the same instant. Heisenberg's uncertainty principle expresses the uncertainty relation for these two uncertainties

$$\Delta t \cdot \Delta E \geq \frac{h}{4\pi}.$$

Similarly to the above relationship, this relationship states:

> It is impossible to measure both the time and the energy of a quantum object in the same instant to unlimited accuracy. The more accurate is the measurement of the energy, the less accurate is the measurement of the time in that instant, and vice versa.

15.6 ORIGINS AND A CONSEQUENCE OF THE UNCERTAINTY PRINCIPLE

The uncertainty principle and the two resulting uncertainty relations hold for any object—even our earlier experiment on the car. But the limitations imposed by the uncertainty principle have no practical consequence for such massive objects as cars or baseballs moving at everyday speeds. This is because the amounts of uncertainty involved are too small to be noticed. It is only on the atomic scale that the limitations become evident and important.

Whose Fault Is It?

It is important to understand that the uncertainties mandated by the uncertainty principle are not the fault of the experimenter or of the instruments we use. We can *never* build instruments to get around the reciprocal uncertainties in measurements imposed by the uncertainty relations. This is because Heisenberg's uncertainty relations are a direct consequence of quantum mechanics and the wave–particle duality, and they will remain valid as long as quantum mechanics remains an acceptable theory. In fact, the uncertainty relations may be even more fundamental than quantum mechanics, for these relations seem to be connected with the very existence of the quantum itself. You have seen the role that Planck's constant h has played in defining the light quantum and in accounting for the stationary states in the Bohr atom. In addition, the constant h appears in both of the basic equations for the energy and the momentum of a photon, $E = hf$ and $p = h/\lambda$, and in many other quantum equations as well. It also appears in the two uncertainty relations. If h were 0, the quantum of energy would be zero, so there would be no light quanta, only continuous waves. The momentum of a photon would also be zero, and the uncertainty relations would read

$$\Delta x \cdot \Delta p_x = 0,$$

$$\Delta t \cdot \Delta E = 0.$$

In this case there would be no reciprocal uncertainties in position and momentum, time and energy. We could measure simultaneously the wave and particle features of quantum objects without any problem. But Planck's constant is *not* zero (although it is very small), the quantum does exist, we are faced with the wave–particle duality, quantum mechanics is still an accepted theory, and nature is so arranged as to limit the precision of our measurements of fundamental quantities at the most fundamental level.

The Sizes of Atoms

On the atomic scale one of the main uses made of the uncertainty principle is in general arguments in atomic theory rather than in particular numerical problems. For instance, the uncertainty principle helps to answer a long-standing fundamental question: Why do atoms have the sizes that they do? As you saw previously (Chapter 14), atoms are actually made up mostly of empty space. At the center of each atom is a very tiny nucleus, which carries all of the positive charge and nearly all of the mass of the atom. Surrounding the nucleus are a number of electrons, equal to the positive charge of the nucleus. The electrons are arranged on various quantum orbits. The lowest one is called the "ground state." But even the orbit of the ground state is still far away from the nucleus. In most atoms, the radius of the ground state is about 10^{-8} cm, while the radius of the nucleus is only about 10^{-12} cm. This means that a nucleus occupies only a tiny fraction of the space inside an atom; the rest is empty (except for a few point-sized electrons.)

As you saw with Rutherford's model, the atom should collapse into the nucleus, because the electrons should radiate away their energy and spiral inward. Bohr attempted to account for why this does not happen by postulating the existence of stationary quantum states, while quantum mechanics associated such states with standing waves, the ground state having the smallest standing electron wave in that orbit. But it is the uncertainty principle that explains why we can't have any lower states, and why the negative electrons cannot exist within or on the positive nucleus. A simple application of the uncertainty relation for position and momentum (see Quantitative Discovery Question 3) shows that if an electron is confined to a space of 10^{-8} cm, the size of an average atom in centimeters, then the uncertainty in its speed is less than the speed of light. But if it is confined to a much smaller space, or even down to the size of the nucleus, the uncertainty in its speed would exceed the speed of light, which is 3×10^{10} cm/s. As you know from relativity theory, no material particle may exceed the speed of light. So the space inside the atom between the nucleus and the first quantum state must be left empty.

Why Aren't Atoms Any Larger?

To increase the size of an atom, we would have to bring electrons into much higher quantum states. Aside from requiring the input of a lot of energy, the higher states are not evenly spaced but become farther and farther apart. Thus, the likelihood increases greatly that the electrons in these higher states can escape the atom and become free, so such an atom would not exist for long, which means that in practice most atoms one can study are about 10^{-8} cm in size. You will see in the next chapter that the fact that atoms do have a size of about 10^{-8} cm helps to account for many of the properties of matter that we see around us.

15.7 THE PROBABILITY INTERPRETATION

The wave–particle dualism is also a fundamental aspect of quantum mechanics. In order to explore this dualism further, it is necessary to review some ideas of probability. In some situations, no single event can be predicted with certainty. But it may still be possible to predict the *statistical probabilities* of certain events. For example, on a holiday weekend during which perhaps 25 million cars are on the road in the United States, statisticians, basing themselves on past experience, predict that about 400 people will be killed in car accidents. It is of course not known which cars in which of the 50 states will be involved in the accidents. But the *average* behavior is still quite accurately predictable.

Along somewhat similar lines, physicists think about the behavior of photons and material particles. As you have seen, there are basic limitations on the ability to describe the behavior of an individual particle. But the laws of physics often make it possible to describe the behavior of large collections of particles with good accuracy (as was the case in the kinetic theory of gasses, Chapter 7). Schrödinger's equation for the waves associated with quantum particles gives the *probabilities* for finding the particles at a given place and a given time; it does not give the precise behavior of an individual particle.

In order to see how quantum probability works, consider the situation of a star being photographed through a telescope. As you have already seen, for example, in the case of waves striking a barrier in which there is a single opening about the size of the wavelength (see Chapter 8), the image formed on a screen behind the opening is not a precise point. Rather, it is a *diffraction pattern*, a central spot with a series of progressively fainter circular rings.

The image of a star on the photographic film in a telescope would be a similar pattern. Imagine now that you wish to photograph a very faint star. If the energy in light rays were not quantized, it would spread continuously over ever-expanding wave fronts. Thus, you would expect the image of a very faint star to be exactly the same as that of a much brighter star, except that the intensity of light would be less over the whole pattern. However, the energy of light *is* quantized; it exists in separate quanta, "photons," of a definite energy, which obey Schrödinger's equation. A photon striking a photographic emulsion produces a chemical change in the film at a single location, not all over the image area. That location, however, is not predictable in advance. All that we can predict is the *probability* that the photon might arrive at that location.

If the star is very remote, only a few photons per second may arrive at

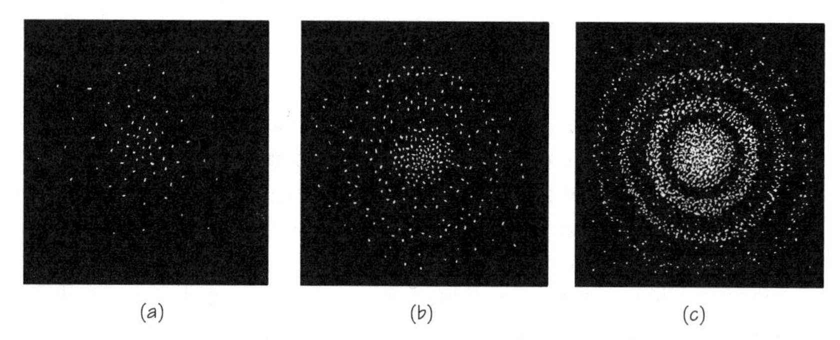

FIGURE 15.10 These sketches represent successive stages of a greatly enlarged image of a distant star on a photographic plate showing the impacts of individual photons.

the film. The effect on the film after a very short period of exposure would be something like the pattern in Figure 15.10(a). As the exposure continues, the effect on the film would begin to look like (b). Finally, after many photons have arrived from the star, a pattern like (c) would be produced, just like the image produced by a much brighter star with a much shorter exposure. (Compare with Figure 15.5.)

For huge numbers of *particle-like* photons, the overall distribution they form is very well described by the distribution expected on the basis of the *wave* intensity of light. For small numbers of quanta, the wave intensity is not very useful for predicting where the photons will go. One might expect them to go mostly to the "high-intensity" parts of the image, but one cannot predict exactly where for an individual photon. These facts fit together beautifully if you consider the wave intensity at a location to indicate the *probability* of a photon going there!

A similar connection can be made for de Broglie waves and particles of matter. For this purpose, rather than considering a diffraction pattern formed by an electron beam, consider an electron wave that is confined to a particular region in space. An example is the de Broglie wave associated with the electron in a hydrogen atom, which is spread out all over the atom. Another example is the de Broglie wave of an electron in a good conductor of electricity. The wave's amplitude at some location represents the probability of the electron being there, if a measurement of the electron's location were to be performed.

According to quantum mechanics, the hydrogen atom does not consist of a localized negative particle (an electron) moving around a nucleus as in the Bohr model. Indeed, the theory does not provide any fixed, easily visualizable picture of the hydrogen atom. A description of the probability distribution is the closest thing to a picture that quantum mechanics provides.

As discussed in Chapter 7 in connection with kinetic theory and disorder, it is easy to predict the average behavior of very large numbers of particles, even though nothing is known about the behavior of any single one of them. Unlike kinetic theory, however, the use of probabilities in quantum mechanics is not for convenience, but seems to be an intrinsic necessity. There is no other way to deal with quantum mechanics. The theory is not really concerned with the position of any individual electron in any individual atom, but rather gives a mathematical representation that can be used to predict interactions with particles, fields, and radiation. For example, it can be used to calculate the probability that hydrogen will emit light of a particular wavelength. The intensity and wavelength of light emitted by a large number of hydrogen atoms can then be compared with these calculations. Comparisons such as these have shown that the theory agrees with experiment.

In most cases, atomic physics deals with the average behavior of many atomic particles. The laws governing this average behavior are those of quantum mechanics. The waves can be considered waves whose amplitudes are a measure of probability. The information (concerning the probability with which a particle will reach some position at a given time) travels through space in probability waves. These waves can interfere with each other in exactly the same way that water waves do. For example, think of a beam of electrons passing through two slits. In such an experiment one can consider the electrons to be waves, and one can compute their interference pattern. The interference pattern is actually a probability pattern that provides the probabilities that individual electrons will arrive at different locations behind the slits. Large constructive interference at a location indicates high probability that electrons will arrive there; large destructive wave interference indicates low or vanishing probability. We cannot say where an individual electron will end up after passing through the slits; all we can know, according to quantum mechanics, is the probability of landing at each location. However, after the passage of many electrons, the statistical buildup of particle-like electrons will provide the familiar interference pattern that we expect to see for waves. As Max Born, the primary founder of the probability interpretation, wrote in 1926:

> The motion of particles conforms to the laws of probability, but the probability itself is propagated in accordance with the law of causality.

15.8 THE COMPLEMENTARITY PRINCIPLE

Quantum mechanics was founded upon the existence of the wave–particle dualism of light and matter, and the enormous success of quantum mechanics, including the probability interpretation, seems to reinforce the importance of this dualism. But how can a particle be thought of as "really" having wave properties? And how can a wave be thought of as "really" having particle properties? One could build a consistent quantum mechanics upon the idea that a light beam or an electron can be described simultaneously by the incompatible wave and particle concepts.

In 1927, Niels Bohr realized that the word "simultaneously" provided the key to a consistent account. He realized that our models, or pictures, of matter and light are based upon their behavior in various experiments in our laboratories. In some experiments, such as the photoelectric effect or the Compton effect, light behaves as if it consists of particles; in other experiments, such as the double-slit experiment, light behaves as if it consists of waves. Similarly, in experiments such as J.J. Thomson's cathode-ray studies, electrons behave as if they are particles; in other experiments, such as his son's diffraction studies, electrons behave as if they are waves. But light and electrons never behave *simultaneously* as if they consist of both particles and waves. In each specific experiment they behave *either* as particles *or* as waves, but never as both.

This suggested to Bohr that the particle and wave descriptions of light and of matter are both necessary even though they are logically incompatible with each other. They must be regarded as being "complementary" to each other—that is, like two different sides of the same coin. This led Bohr to formulate what is called the *Principle of Complementarity*:

> The wave and particle models are both required for a complete description of matter and of electromagnetic radiation. Since these two models are mutually exclusive, they cannot be used simultaneously. Each experiment, or the experimenter who designs the experiment, selects one or the other description as the proper description for that experiment.

Bohr showed that this principle is a fundamental consequence of quantum mechanics. He handled the wave–particle duality, not by resolving it in favor of either waves or particles, but by absorbing it into the foundations of quantum physics. Like the Bohr atom, it was another bold initiative toward the formulation of a new theory, even though this required contradictions with classical physics.

FIGURE 15.11 Max Born (1882–1969) was born in Germany but left for England in 1933 when Hitler and the Nazis came to power. Born was largely responsible for introducing the statistical interpretation of wave mechanics.

It is important to understand what the complementarity principle really means. By accepting the wave–particle duality as a fact of nature, Bohr was saying that light and electrons (or other objects) encompass potentially the properties of both particles and waves—until they are observed, at which point they behave as *if* they are either one or the other, depending upon the experiment and the experimenter's choice. This was a profound statement, for it meant that what we observe in our experiments is not the way nature "really is" when we are not observing it. In fact, nature does not favor any specific model when we are not observing it; rather, it is a mixture of the many possibilities that it could be until we finally do observe it! By setting up an experiment, *we* select the model that nature will exhibit, and *we* decide how photons and electrons and protons and even baseballs (if they move fast enough) are going to behave—either as particles or as waves.

In other words, according to Bohr, *the experimenter becomes part of the experiment!* In so doing, the experimenter interacts with nature, so that we can never observe all aspects of nature "as she really is" by herself. In fact, that phrase, while so appealing, has no operational meaning. Instead, we should say we can know only the part of nature that is revealed by our experiments. (This is no invitation to mysticism. After all, we know even about a good friend only through a patchwork of repeated encounters and discussions, in many different circumstances.) The consequence of this fact, for events at the quantum level, said Bohr, is the uncertainty principle, which places a quantitative limitation upon what we can learn about nature in any given interaction; and the consequence of this limitation is that we must accept the probability interpretation of individual quantum processes. For this reason, the uncertainty principle is often also called the principle of indeterminacy. There is no way of getting around these limitations, according to Bohr, as long as quantum mechanics remains a valid theory.

Such ideas, of course, are totally at odds with how we usually think of nature. We usually think of nature as existing completely independently of us and possessing a definite reality and behavior even when we are not observing it. For instance, you assume that the world outside of the place where you are now sitting still exists pretty much as it was when you last observed it. Certainly this is the way nature always behaves in our everyday experience, and this view is a fundamental assumption of classical physics. It even has a philosophical name; it is called "realism," and for phenomena and objects in the range of ordinary experience is perfectly appropriate. But, as Bohr so often emphasized, we have to be prepared to expect that the quantum world will not be anything like the everyday world in which we live. Max Born, one of the founders of quantum mechanics, has written:

> The ultimate origin of the difficulty lies in the fact (or philosophical principle) that we are compelled to use the words of common language when we wish to describe a phenomenon, not by logical or mathematical analysis, but by a picture appealing to the imagination. Common language has grown by everyday experience and can never surpass these limits. Classical physics has restricted itself to the use of concepts of this kind; by analyzing visible motions it has developed two ways of representing them by elementary processes: moving particles and waves. There is no other way of giving a pictorial description of motions—we have to apply it even in the region of atomic processes, where classical physics breaks down.

Bohr's complementarity principle, Heisenberg's uncertainty principle, and Born's probability interpretation together form a logically consistent interpretation of the meaning of quantum mechanics. Since this interpretation was developed largely in Bohr's institute at the University of Copenhagen, it has been called the *Copenhagen Interpretation* of quantum mechanics. The results of this interpretation have profound scientific and philosophical consequences that have been studied and debated to this day.

15.9 SOME REACTIONS

The idea that the solution of Schrödinger's equation is a wave that represents, not a physical wave, but the probability of finding the associated particle in some specific condition of motion has had great success. In fact, every experiment devised so far to test this interpretation has confirmed these results. Yet many scientists still find it hard to accept the idea that it is impossible to know exactly what any one particle is doing. The most prominent of such disbelievers was Einstein. In a letter to his friend and colleague Max Born, written in 1926, he remarked:

> The quantum mechanics is very imposing. But an inner voice tells me that it is still not the final truth. The theory yields much, but it hardly brings us nearer to the secret of the Old One. In any case, I am convinced that He does not play dice.

FIGURE 15.12 Photograph of participants in the Fifth Solvay Congress, Brussels, 1927, a veritable "Who's Who" of physics in the first half of the twentieth century. Back row, from left: Auguste Piccard, E. Henriot, Paul Ehrenfest, E. Herzen, T. de Donder, Erwin Schrödinger, E. Verschaffelt, Wolfgang Pauli, Werner Heisenberg, Ralph Fowler, Leon Brillouin. Middle row, from left: Peter Debye, Martin Knudsen, William L. Bragg, H.A. Kramers, Paul Dirac, Arthur Holly Compton, Louis de Broglie, Max Born, Niels Bohr. Front row, from left: Irving Langmuir, Max Planck, Marie Curie, H.A. Lorentz, Albert Einstein, Paul Langevin, Charles Guye, Charles Wilson, Owen Richardson.

Thus, Einstein agreed with the usefulness and success of quantum mechanics, but he refused to accept probability-based laws as the final level of explanation in physics. The remark about not believing that God played dice (an expression he used many times later) expressed Einstein's faith that more basic, deterministic laws are yet to be found. By this he meant that if all the conditions of an isolated system are known and the laws describing the interactions are known, then it should be possible to predict precisely, according to "strict causality," what will happen next, without any need for probability.

Some scientists agreed with Einstein, but all scientists do agree that, as a theory, quantum mechanics does work in practice. It gives the right answers to many questions in physics; it unifies ideas and occurrences that were once unconnected; it has produced many new experiments and new concepts; and it has opened the door to many technological advances, from transistors, microprocessors, and superconductors, to lasers and the latest medical-imaging techniques.

On the other hand, there is still vigorous argument about the basic significance of quantum mechanics. It yields probability functions, not precise trajectories. Some scientists see in this aspect of the theory an important indication of the nature of the world. For other scientists, the same fact indicates that quantum mechanics is still incomplete. Some in this second group are trying to develop a more fundamental, nonstatistical theory. For such a theory, the present quantum mechanics is only a special case. As in other fields of physics, the greatest discoveries here may be those yet to be made.

SOME NEW IDEAS AND CONCEPTS

complementarity
Compton effect
Copenhagen interpretation
de Broglie wave
double-slit experiment
photon
probability interpretation
quantum mechanics
Schrödinger equation
uncertainty principle
wave mechanics
wave–particle dualism

SOME IMPORTANT EQUATIONS

$$E = hf,$$

$$m = \frac{E}{c^2},$$

$$p = \frac{h}{\lambda},$$

$$\lambda = \frac{h}{mv},$$

$$\Delta x \cdot \Delta x \geq h/4\pi,$$

$$\Delta t \cdot \Delta E \geq h/4\pi.$$

FURTHER READING

D. Cassidy, *Uncertainty: The Life and Science of Werner Heisenberg* (New York: Freeman, 1992).

R. Crease et al., *The Second Generation: Makers of the Revolution in Twentieth Century Physics* (Piscataway, NJ: Rutgers University Press, 1986).

R.P. Feynman, *Six Easy Pieces* (Reading, MA: Perseus, 1995), Chapter 6.

R.P. Feynman, *QED: The Strange Theory of Light and Matter* (Princeton, NJ: Princeton University Press, 1985).

B. Greene, *The Elegant Universe: Superstrings, Hidden Dimensions, and the Quest for the Ultimate Theory* (New York: Norton, 1999).

J. Gribbon, *In Search of Schrödinger's Cat, Quantum Physics, and Reality* (New York: Bantam, 1984).

R.M. Hazen with M. Singer, *Why Aren't Black Holes Black? The Unanswered Questions at the Frontiers of Science* (New York: Anchor/Doubleday, 1997).

A. Hobson, *Physics: Concepts and Connections*, 2nd ed. (Upper Saddle River: Prentice Hall, 1999), Chapters 13, 14, 18.

G. Holton, The Roots of Complementarity. In: *Thematic Origins of Scientific Thought: Kepler to Einstein* (Cambridge, MA: Harvard University Press, 1988), pp. 99–145.

G. Holton and S.G. Brush, *Physics, The Human Adventure* (Piscataway, NJ: Rutgers University Press, 2001), Chapter 29.

G. Kane, *The Particle Garden: Our Universe as Understood by a Particle Physicist* (Reading, MA: Addison-Wesley, 1995).

H. Kragh, *Quantum Generations: A History of Physics in the Twentieth Century* (Princeton, NJ: Princeton University Press, 1999).

L. Lederman with D. Teresi, *The God Particle* (New York: Dell, 1993).

D. Lindley, *Where Does the Weirdness Go? Why Quantum Mechanics Is Strange, But Not As Strange As You Might Think* (New York: Basic, 1996).

D. Schramm and L. Lederman, *From Quarks to the Cosmos: Tools of Discovery* (New York: Freeman, 1989).

S. Weinberg, *Dreams of a Final Theory* (New York: Pantheon, 1992).

D. Wick, *The Infamous Boundary: Seven Decades of Heresy in Quantum Physics* (New York: Springer-Verlag, 1996).

STUDY GUIDE QUESTIONS

15.1 The Quantum

1. Why did the existence of the quantum of energy pose a major problem for physicists?

15.2 The Particle-Like Behavior of Light

1. Why was momentum conservation not considered in the discussion of the photoelectric effect?
2. How does the momentum of a photon depend on the frequency of the light?
3. How could you figure out how much mass a photon carries?
4. Write down the steps leading the expression for the momentum, $p = h/\lambda$.
5. What did Compton do, and what did the experiment prove?
6. What is the best answer to the question "Is the photon a particle or a wave?"

15.3 The Wave-Like Behavior of Particles

1. How did de Broglie obtain the relation $\lambda = h/mv$ for electrons?
2. Why were crystals used to get diffraction patterns of electrons?
3. How can electron waves be used to explain the origin of the stationary states that Bohr assumed in his theory?
4. What two pictures of the atom were possible as a result of the discovery of electron waves?
5. How would you obtain the wavelength of a moving electron?
6. In what way is nature symmetric in the wave–particle duality?

15.4 Constructing Quantum Mechanics

1. The set of energy states of hydrogen could be derived from Bohr's postulate that $mvr = nh/2\pi$. In what respect was the derivation from Schrödinger's equation better?
2. Quantum (or wave) mechanics has had great success. What is its drawback for those trained to think in terms of physical models?

15.5 The Uncertainty Principle

1. What is the "uncertainty" of a measurement?
2. If photons used in finding the momentum of an electron disturb the electron too much, why cannot the observation be improved by using less energetic photons?

3. If the wavelength of light used to locate a particle is too long, why cannot the location be found more precisely by using light of shorter wavelength?
4. If you measured the position of an electron with very small uncertainty, what consequence would this have for the measurement of its momentum? Explain.
5. What effect does the greater-than-or-equal-to sign, $\geq$, have on the relationship between the two uncertainties?

15.6 Origins and a Consequence of the Uncertainty Principle

1. Whose fault is it that we can't get rid of the uncertainty in our measurements?
2. Why can't atoms have much larger sizes?
3. Why can't electrons exist inside the nucleus?

15.7 The Probability Interpretation

1. In wave terms, the bright lines of a diffraction pattern are regions where there is a high field intensity produced by constructive interference. What does the probability interpretation say about the bright lines of a diffraction pattern?
2. Quantum mechanics can predict only probabilities for the behavior of any one particle. How, then, can it predict many phenomena, for example, half-lives and diffraction patterns, with great certainty?
3. Individual photons are sent onto the double-slits of Young's experiment. They hit the photographic plate in random fashion as quantum mechanics requires. If that is the case, then how is the wave interference pattern formed?
4. Explain in your own words what Born is saying in the quotation.
5. How does the probability interpretation in quantum mechanics differ from the probability interpretation of entropy in the kinetic theory of gases?
6. How can nature be fundamentally random yet at the same time we can predict with great accuracy the interference pattern that is formed in Young's experiment?

15.8 The Complementarity Principle

1. What did Bohr realize when he examined experiments on particles and waves?
2. What does the word "complementary" mean?
3. State the principle of complementarity in your own words.
4. How does the experimenter become part of the experiment?
5. How does this contradict the idea of "realism"?
6. How did Bohr regard the strange behavior of the quantum world?
7. Why can't we know nature as it really is, according to quantum mechanics?

15.9 Some Reactions

1. What was Einstein's objection to the probability interpretation?
2. What did Einstein mean when he said God does not play dice?

DISCOVERY QUESTIONS

1. What would happen to the wave–particle dualism if Planck's constant $h = 0$? What would happen to the uncertainty principle if Planck's constant $h = 0$?
2. Why can't we measure both the position and the time simultaneously with absolutely no uncertainty?
3. How do you think it was possible for Einstein to be a primary founder of quantum theory, yet object to quantum mechanics?
4. Review some of the new ideas and concepts in this chapter. Which ones do you find to be the most startling, and the most disturbing. Evaluate why you find them to be startling and disturbing.
5. Niels Bohr thought the complementarity principle applies also to events in our lives—for example, that one loves one's child and yet has to punish him or her for bad behavior. Can you think of other examples where a "complementarity principle" may apply to ordinary life?
6. What explanation would you offer for the fact that the wave aspect of light was shown to be valid before the particle aspect was demonstrated?
7. Suppose that the only way you could obtain information about the world was by throwing rubber balls at the objects around you and measuring their speeds and directions of rebound. What kind of objects would you be unable to learn about? How does this question relate to this chapter?
8. Some writers have claimed that the uncertainty principle proves that there is free will. Do you think this extrapolation from atomic phenomena to the world of living beings is valid?
9. A physicist has written:

   > It is enough that quantum mechanics predicts the average value of observable quantities correctly. It is not really essential that the mathematical symbols and processes correspond to some intelligible physical picture of the atomic world.

   Do you regard such a statement as acceptable? Give your reasons.
10. Previous chapters have discussed the behavior of large-scale "classical particles" (e.g., tennis balls) and "classical waves" (e.g., sound waves). Such particles and waves in most cases can be described without any use of ideas such as the quantum of energy or the de Broglie matter wave. Does this mean that there is one sort of physics ("classical physics") for the phenomena of the large-scale world and quite a different physics ("quantum physics") for the phenomena of the atomic world? Or does it mean that quantum physics really applies to all phenomena but is no different from classical physics when applied to large-scale particles and waves? What arguments or examples would you use to defend your answer?
11. Some writers have declared that the impossibility of finding classical causality at the level of quantum objects proves "science cannot really know nature." Some also claim that it coincides with aspects of Eastern religions. What would be your responses?

Quantitative

1. Most professional pitchers can throw a baseball at 100 mi/hr (62 km/hr). What would be the de Broglie wavelength of a ball of mass 147 g at that speed? Could this wavelength be detected by crystal diffraction?
2. What is your de Broglie wavelength when you walk at a brisk pace of 4 mi/hr? Why have you not experienced the wave side of yourself?
3. (a) Once it was thought that β rays from the nucleus are electrons that were originally present inside the nucleus. According to the uncertainty principle, what would be the approximate range of speeds (uncertainty of speed) of an electron confined to a nucleus of size 10^{-14} m? The rest mass of an electron is about 9.1×10^{-31} kg; assume it is not subject to relativistic mass increase. How does the result of your calculation compare with the speed of light? What does this tell about the old idea of the presence of electrons in the nucleus?
   (b) What would be the approximate uncertainty in the speed of an electron confined to the first Bohr orbit of a hydrogen atom, where $r = 5.29 \times 10^{-9}$ cm?
   (c) According to the uncertainty principle, what would be the approximate speed of a proton confined to a nucleus of size 10^{-14} m? Neglect relativistic effects. (The rest mass of a proton is about 1.6×10^{-27} kg.) How does this compare with the speed of light?
   (d) On the basis of your results, why can a proton exist in the nucleus if an electron cannot?
4. An electron is fired from an "electron gun" in the CRT of a computer monitor. The electron is aimed at a blue-fluorescent subpixel on the screen of 10^{-6} m in width. What is the minimum uncertainty in the electron's momentum in the horizontal direction along the screen?
5. Calculate the momentum of a photon of wavelength 400×10^{-9} m. How fast would an electron have to move in order to have the same momentum? What would be the wavelength of an electron moving at that speed?

CHAPTER 16

Solids Matter

16.1 THE SUCCESS OF QUANTUM MECHANICS

The triumph of quantum mechanics during the 1920s opened the door to a new understanding of a wide range of phenomena involving the most fundamental properties of matter and radiation, the fundamental constituents of the Universe. The new physics also enabled an amazing variety of technological applications that are continuing to transform our lives—from lasers and microchip transistors to new medical imaging devices, and a host of new materials. Combined with relativity theory, quantum mechanics is the foundation for understanding the interactions of radiation and matter, the behavior of elementary particles at extremely high energies, and many of the stellar phenomena observed in cosmology. Soon after the formulation of quantum mechanics, scientists successfully applied the new theory to a new understanding of the nucleus of the atom that is still used today. This is discussed in Chapters 17 and 18. During the same period, scientists also utilized quantum mechanics to gain a new understanding of the structure and behavior of matter, including, in particular, the thermal and electrical properties of solids. These are the focus of the current chapter.

16.2 FORMING A SOLID

It was well known that matter appears in one of three states, or *phases*—solid, liquid, and gas or vapor. According to the kinetic-molecular theory (Chapter 7), one factor distinguishing each of these states is the amount of energy distributed among the molecules making up the material. In the gaseous state, the molecules have the most energy and in the solid state the least. Transitions from one state to another can be achieved by adding or extracting energy in the form of heat, as is readily observed. But that can't be the whole story; as we shall examine below, at each transition from one state to another, the heat content seems to disappear (if going to states of higher temperature) or to appear without a loss of energy elsewhere (if going to states of lower temperature)! The temperature of the material does not change at all. We need to include an additional concept, the concept of *entropy*, which gives an idea of the level of disorder. Much of the basic understanding of phase transitions and the formation of solids was developed around 1900, but a complete understanding required quantum mechanics over the subsequent decades, and some aspects of these complicated processes are still under investigation.

In order to view the basic process leading to the formation of a solid, we look at the behavior of water, which can also easily be observed in your laboratory. We start with liquid water, which we will cool down by extracting heat. As the water cools, the amount of decline in its temperature, ΔT, is related to the amount of heat extracted, ΔQ, by the expression (Chapter 6):

$$\Delta Q = mc\,\Delta T,$$

where m is the mass of the water and c is the specific heat of liquid water, which in this case is 1 calorie of heat per gram and per degrees centigrade.

As we continue to cool the water, we reach a temperature at which the liquid undergoes a *phase change* from liquid into solid. All liquids eventually freeze into solids, but they do so at very different temperatures. Water at atmospheric temperature will begin to freeze at 0°C (273 K), but liquid helium does not freeze until it reaches 4 K, very near to absolute zero. At this point we observe the effect noted earlier—as we extract heat, the liquid continues to freeze into ice, but the temperature stays the same. The heat extracted seems to have been "hidden" in the liquid and to be associated with the formation of the solid, ice. This "hidden" heat is called the *latent heat of fusion*, named for the linking of the molecules together into a

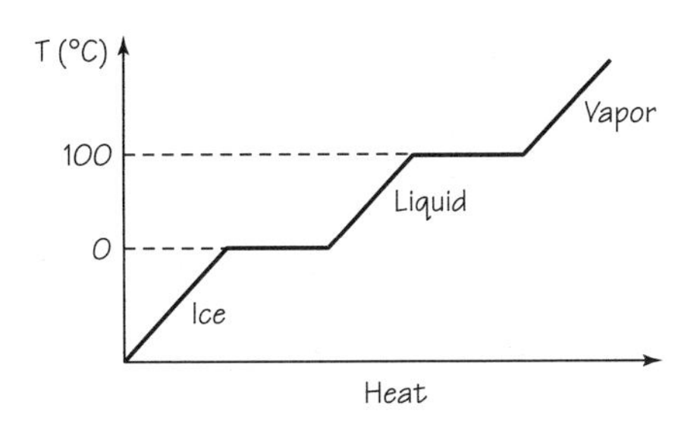

FIGURE 16.1 Three phases of water and its transitions as heat is added.

solid. (This has nothing to do with nuclear fusion.) For water, it is about 80 calories of heat per gram of water.

What is this latent heat and what does it have to do with the freezing of a liquid into a solid? Since the cooling process involves the extraction of energy from the material, the molecules begin to arrange themselves into a state of minimum energy. This is achieved as the molecules group together into an arrangement that is as orderly and as structured as possible. From the earlier discussion about entropy, you know that the state of minimum energy will have the highest order, which means that it has the lowest entropy. The change in the entropy, ΔS, is defined as the change in the heat content per unit of absolute temperature (see Chapter 6):

$$\Delta S = \frac{\Delta Q}{T}.$$

So, the latent heat that is extracted from the water as it freezes, ΔQ, is directly associated with the decreasing entropy of the water molecules as they form the ordered structure of solid ice. Of course, if heat is pumped back into the system, the entropy will start to increase as the molecules begin to vibrate faster and faster. They may eventually break their solid bonds. This is observed as melting. Again the temperature remains constant as the solid is melting under the input of heat energy. During this process, the incoming heat energy becomes latent once again, since the temperature does not change, but the entropy increases and the orderly arrangement within the solid breaks down, forming a pool of water where once we had an orderly ice cube.

As a solid is formed from a liquid, the type of orderly arrangement that it creates depends upon the types of molecules and the nature of the electrical attraction between the molecules. The lowest entropy states would involve the largest number of bonds possible between the molecules. This

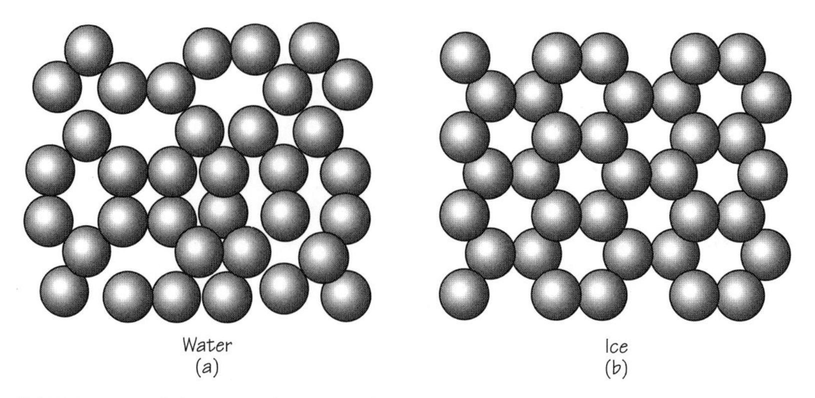

FIGURE 16.2 Schematic drawing of water molecules as liquid and solid.

occurs in the formation of crystals, which involve beautifully ordered arrays of billions and billions of molecules. In these arrays a basic crystal structure involving only a small number of molecules is repeated many times over, like grouted tiles on a floor, or a wallpaper pattern. Water, salt, and many metals form such crystalline structures. Many metal alloys, which involve a mixture of different metals, form what are called "polycrystalline" ("many crystal") solids. Examples include metal door knobs, tableware, and automotive parts.

However, not all solids involve crystalline bonding. There are three other important types of bonding (as well as some exotic types). These include *quasi-crystalline*, *amorphous*, and *composite* bonding. The types of molecules involved and even the way in which the solid was formed determine the type of structure formed. If the cooling from liquid occurs too quickly, the molecules which might normally form a perfect crystal do not have time to arrange themselves into a low-entropy crystal structure. They literally tend to be frozen into place while in the liquid state. In this case there is only local, short-range order and no long-range repetitive array. This structure is called *amorphous*. Some solids are naturally amorphous, even when frozen slowly. These include rubber, glass, and plastics. Carbon in the form of coal is amorphous, but in the form of graphite or diamond it is highly crystalline.

Quasi-crystalline solids were first discovered only in 1983. The molecules are arranged into a very orderly three-dimensional pattern, but the pattern does not repeat itself throughout the crystal as it does in a truly crystalline structure.

Finally, there are the *composite solids*, which involve the bonding together of different materials with different structures. Examples here include wood, cement, and fiberglass, as well as such biological materials as bone, flesh, and muscle. Research in recent years to develop materials that can act as biological "spare parts" has undergone a revolution because of the discovery of composite solids.

16.3 QUANTUM SOLIDS

One of the first recognized successes of the early quantum theory arose, not from the study of radiation, but from the theory of solids. Once again Albert Einstein was at the forefront. After presenting the hypothesis of light quanta in 1905, which explained the photoelectric effect (Section 13.5), Einstein showed in 1906 that Planck's earlier results on the distribution of energy in electromagnetic radiation could be derived using two assumptions. The first, already familiar from his 1905 work, was that light is composed of light quanta; the second was that the light quanta are emitted and absorbed by oscillating charges within matter that are vibrating only with certain quantized energies. This means that light is emitted by oscillating charges within matter, as might be expected from Maxwell's theory, but the energies both of the emitted light and of the oscillators are restricted to certain quantized values, which contradicted classical theory. The oscillators behaved like charges attached to springs, oscillating back and forth with certain frequencies corresponding to certain allowed quantum energies.

In 1904, the German physicist Paul Drude had proposed a model of a crystalline solid in which the atoms are arranged in a lattice such that each atom is bound to its neighboring atoms by bonds that can act like little oscillator springs. Drude showed, using classical physics, that the three-dimensional oscillation of the lattice atoms can account for the observed absorption of infrared radiation by some solids, if the atoms are assumed to be ionized. A similar oscillation of the electrons, attached to the atoms or ions, accounted for the observed absorption of ultraviolet radiation in solids.

In 1907 Einstein applied the quantum hypothesis to Drude's classical oscillators. Even before Drude's model, Einstein had been wondering as early as 1901 whether the internal kinetic energy of solids and liquids might be regarded as the energy of "electric resonators." One outstanding puzzle at that time involved the behavior of the specific heat, C, of solids as the temperature decreased. Normally the specific heat is assumed to be constant in the relationship between the transfer of heat and the change in temperature of a substance

$$\Delta Q = mc\,\Delta T.$$

Applying classical physics to Drude's oscillator model of the atoms in a lattice confirmed that the specific heat C should be a constant for solids at

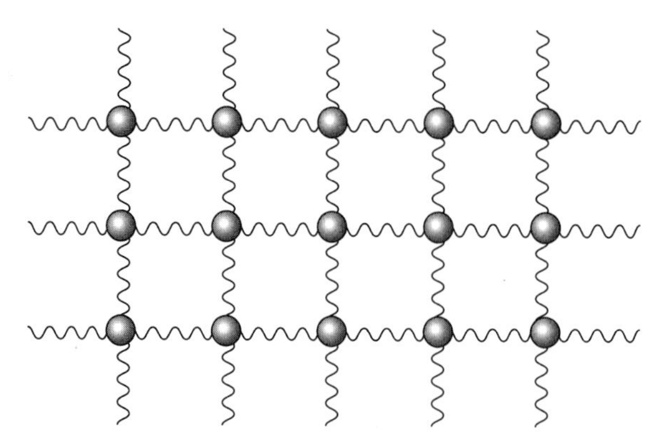

FIGURE 16.3 Drude's model of a crystalline solid.

all temperatures. However, when Einstein applied the quantum hypothesis to the oscillators in Drude's model, he obtained an expression for the specific heat indicating that the value of the specific heat should change with temperature. While remaining constant at room temperature and slightly below, the specific heat would drop precipitously to zero at the temperature declined to zero, as shown in Figure 16.3. Conversely, as the temperature increased from 0 K, Einstein's value for the specific heat of a crystalline substance as a function of its temperature should approach the constant value at room temperature.

Einstein's predictions for the behavior of the specific heat of solids as a function of temperature were based on an application of the quantum hy-

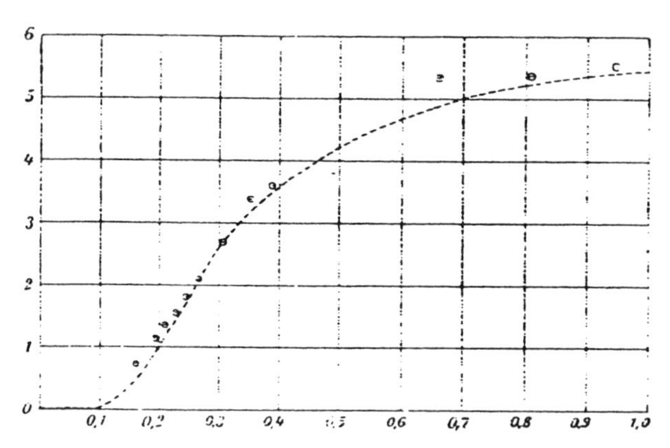

FIGURE 16.4 Einstein's graph of specific heat versus temperature from his 1907 paper applying the quantum hypothesis to Drude's classical oscillators.

pothesis to a model of a crystalline solid in which the lattice atoms behave as three-dimensional oscillators vibrating with quantized energies. The physicists Walther Nernst and Frederick A. Lindemann, working in a low-temperature laboratory in Berlin, astonished their colleagues with a complete confirmation of Einstein's prediction for the way the specific heat of a crystalline solid changed with temperature. It was the first confirmation of the quantum hypothesis outside of the field of radiation, and it clearly indicated that crystalline solids could be treated as an array of oscillating lattice atoms subject to requirements of quantum theory and, later, quantum mechanics. This model has served as the basic model of a crystalline solid ever since. As Nernst and Lindemann wrote, "That the observations in their totality provide a brilliant confirmation of the quantum theory of Planck and Einstein is obvious."

16.4 CONDUCTING ELECTRONS

You have seen in previous chapters how research on electricity and the behavior of electric currents enabled the enormous growth and spread of the electric age in which we live today. Electrical energy is now the dominant form of energy consumption in industrialized countries and is an essential element in the operation of many of the devices we use every day. For decades, one of the most poorly understood yet most practical properties of a piece of solid matter was its ability, or lack of ability, to conduct electricity. Why do some materials conduct electricity, while others act as insulators, and still others (such as semiconductors) act as one or the other depending upon the conditions? How can an understanding of such properties be put to even greater practical use?

You recall from Section 10.7 that the German physicist Georg Ohm discovered an empirical law, known as Ohm's law, relating the current in a material and the voltage applied to the material

$$V = IR.$$

Here V is the potential difference, I is the current, and R is the resistance of the material. The resistance is normally constant for each material, but there is a different resistance for different materials. As the resistance of a material increases, its conductivity goes down, so the amount of current that it conducts must also decrease, according to Ohm's law:

$$I = \frac{V}{R}.$$

Of course, if the resistance of a material is so large that it approaches infinitely, then V/R becomes zero, so $I = 0$. This means that no current flows through the material. Such a material is called an *insulator. An insulator allows virtually no current to get through*. On the other hand, if R ever happened to reach zero, the current would become infinite, or undefined, for even the tiniest voltage. This is what happens in a *superconductor*.

Even without a voltage source, a current in a loop made of a superconductor, once started, will continue in principle undisturbed virtually forever. The first superconductor was discovered in 1911 by Kammerlingh Onnes in Amsterdam while studying the conductivity of mercury at extremely low temperatures. During the late 1980s, one superconductor maintained a current without loss for nearly 3 years. A normal conductor would lose its current within a small fraction of a second without the help of an external voltage. Even with an applied external voltage, you know that a wire of normal resistance carrying a current begins to warm up, indicating that electrical energy is being converted into heat energy in the wire. Wires can become so hot that they glow and emit light; this is put to good use in incandescent light bulbs and in electrical heating elements. This behavior is similar to the conversion of kinetic energy into heat energy due to friction, as when a book slides across a tabletop and comes to rest. It seems that perhaps the current in a non-superconducting wire encounters a type of friction as it moves down the wire, which explains why the wire gets hot. This suggests that in an insulator there is so much internal friction that it does not allow any current to flow, while superconductors have no friction at all, allowing currents to flow indefinitely without any noticeable loss of energy.

Can we utilize the basic oscillator model of a solid to account for Ohm's law and the related electrical observations just described? Soon after Thomson's discovery of the electron, Drude in 1900 and H.A. Lorentz in 1909 developed the first electron model of conductivity in a conducting material. In their model, the solid material is assumed to consist of metal atoms arranged in a crystal structure, which is able to vibrate. However, electrons are the crucial element in understanding conductivity. Conducting metals such as copper and silver have one valence electron that is easily detached. So Drude and Lorentz assumed that these valence electrons provide the conduction current when a voltage is applied. Their model is called the *classical free-electron model*, because it involved only classical physics (no quantum theory) and because the electrons were considered to be little balls of charged matter that are completely free to wander about within the material. They were not subject to repulsive forces from other electrons or to attractions to the metal ions. Only a "boundary force" around the edge of the metal acted on the electrons, preventing them from escaping into space.

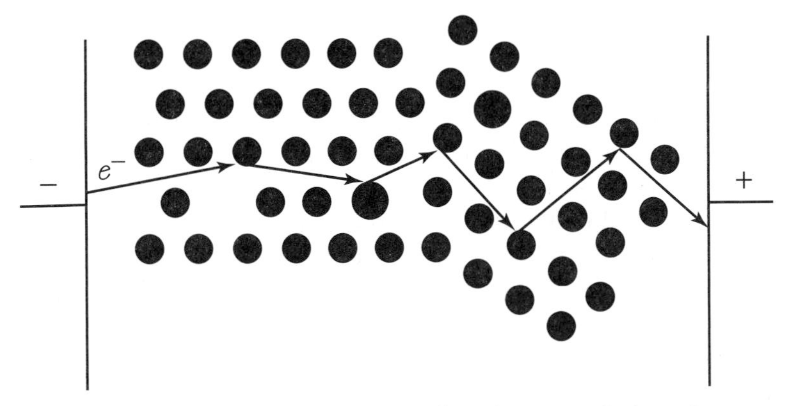

FIGURE 16.5 Schematic representation of an electron path through a conductor (containing vacancies, impurity atoms, and a grain boundary) under the influence of an electric field.

In many respects the electrons in the Drude–Lorentz free-electron model form a type of ideal gas, an "electron gas" inside the metal. Despite the presence of the lattice atoms, the electrons are assumed to move about randomly, like molecules in the kinetic theory of gases, until an electric field is applied. Since the electrons are charged, at the instant the external field is applied they start to move in the direction of the electric force on them, but as they move they do undergo inelastic collisions with the vibrations of the crystal lattice, as well as collisions with impurities in the material and with imperfections in the lattice. All of these work together to slow the electrons' advance and to reduce their kinetic energy. This slowing appears as a type of friction, which we surmised above, and the amount of "friction" is defined as the *resistance*. The lost kinetic energy appears as heat. The more inelastic collisions the electrons experience as they traverse the conductor, the more the resistance they encounter; on the macroscopic scale this is expressed as Ohm's law.

Of course, if the material is made up of atoms that have no valence electrons (which can be shared), such as glass, wood, or plastic, there can be no conduction of electrons, and thus no current when a voltage is applied. The material is an *insulator*.

The Drude–Lorentz free-electron model accounted quite well for Ohm's law, but it could not account for some of the thermal properties of the metals, besides the specific heat, such as the dependence of the resistance upon the temperature. Moreover, with the advent of quantum mechanics it was obvious that a quantum mechanical understanding of conductivity was required, since atoms and electrons obey the laws of quantum mechanics.

16.5 BANDING TOGETHER*

In 1928, just two years after the formulation of quantum mechanics, the German physicist Arnold Sommerfeld modified the classical free-electron model by treating the electrons according to quantum mechanics. But the new theory still contained the unrealistic assumption that the electrons do not interact with the charged lattice icons except to collide with them. As before, Sommerfeld also considered the electrons to be little charged particles of matter.

Beginning in the same year, Felix Bloch, an assistant to Werner Heisenberg in Leipzig, began to make more realistic assumptions in an attempt to formulate a more complete quantum mechanics of electrical conductivity. First, because he wanted to assign a definite momentum and energy to each of the electrons, but not a definite position or a time interval, he chose the wave side of the wave-particle duality. He assumed that the electrons behave, not like particles, but like infinitely extended de Broglie waves. As a result, Bloch did not treat electrons inside metals as a "gas" of particles, but rather as periodic waves extending throughout the periodic crystal lattice. This, it later turned out, helped to explain how electricity can begin to flow in a wire the instant a wire is plugged into a wall socket. If the electrons are viewed as balls of matter, it would take a small amount of time for the current to begin flowing at the rate specified by Ohm's law.

Bloch made a second assumption. He assumed that the positive metal ions, which are arranged in an infinite, periodic array (that is, in a perfect crystal), each exerts an attractive electric force on the negative electrons. This attractive force formed in visual terms a potential energy that looked like a type of "potential well." The wells of neighboring ions then overlapped so that together they formed a periodic arrangement that gave the electron waves a very bumpy ride down the wire.

Bloch then solved the Schrödinger equation for the energies that these types of de Broglie waves (wave functions) could possess while moving in this type of periodic potential. He discovered that the allowed energies of the electrons in the material are joined together into *bands* of quantum states, just as there are certain quantum stationary states within each atom in which the electrons are allowed to exist. Between the bands, as between the quantum states, there is a range of energies in which electrons are forbidden to exist. The bands in the material are actually created by the join-

---

* The remainder of this chapter contains material that is somewhat more advanced than that in the previous sections. Your instructor will inform you how much, if any, of this material is included in the course work or in individual assignments. You may find some of it of interest for independent study.

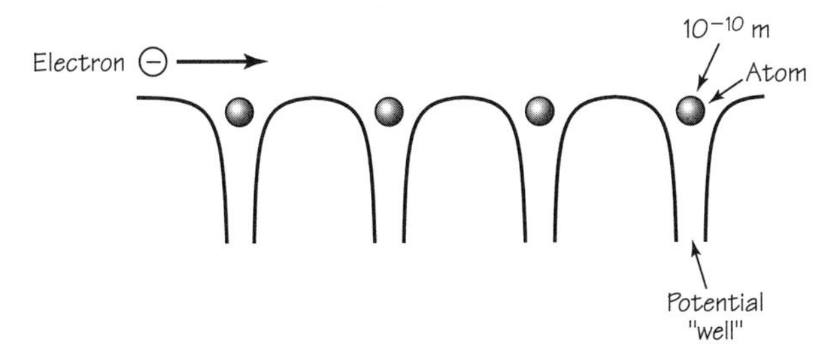

FIGURE 16.6 An electron moving through a region of potential wells.

ing together of the quantum states of the individual atoms. In fact, if there are a total of N identical atoms in the material, then there are N quantum states within each band. According to a rule in quantum mechanics (the Pauli exclusion principle), only two electrons are allowed to occupy any one quantum energy state of a single atom, and this is allowed only because the two electrons spin on their axes in opposite directions.

As extended and refined by Bloch and others during the 1930s, Bloch's theory, known as the *band theory* of metals, accounts very well for the conducting behavior of materials. When atoms are joined together into a crystal, each of the individual quantum states of the atoms joins with the corresponding states in other (identical) atoms in the crystal to form the various energy bands within the material. The electrons in the atoms then fill up the available states within each band. For instance, if there are N atoms in a material, there are N states in a single band and up to $2N$ electrons in each band (since two electrons can occupy one energy state if they spin in opposite directions).

The interesting physics occurs at the top of the filled bands. When an external electric field (voltage) is applied to the material, an electron can respond to the field only if it can move up to a slightly higher quantum state, since it would have slightly more energy in responding to the field.

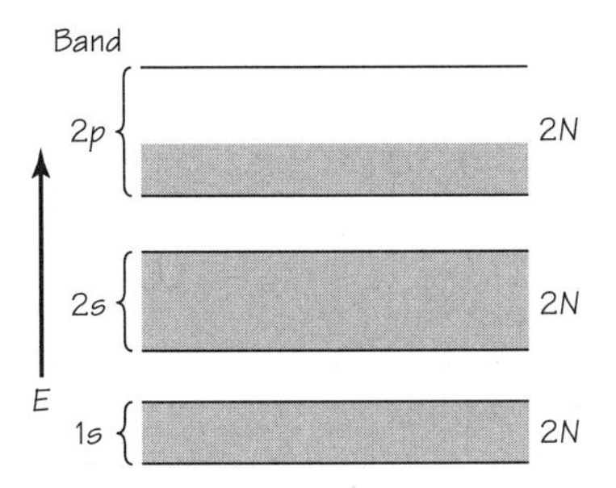

FIGURE 16.7 Expected occupation of the energy bands in a crystal of N atoms with $6N$ electrons. $2N$ electrons go into each band, filling the first two but leaving the third band partly empty.

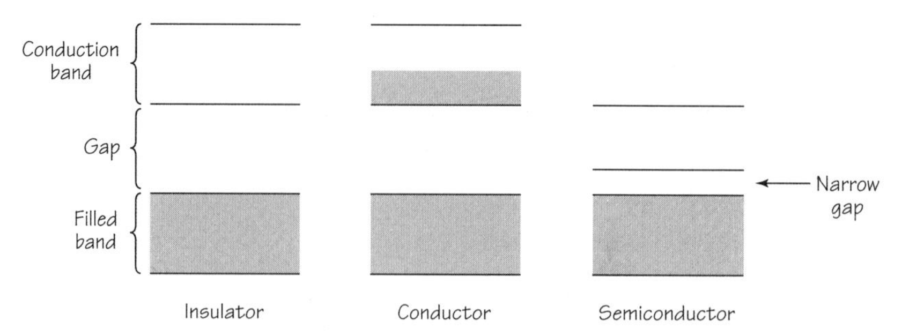

FIGURE 16.8 Bands for insulator, conductor, and semiconductor.

This is possible only if there are some nearby free states to which the electron can jump. This is the case for conducting metals, such as copper and silver, since as their electrons fill in the available states, the highest band is only partially filled. The electrons in this band, which is called the *conduction band*, are then free to be conducted (actually, propagated as waves) down the wire, since there are empty quantum states nearby in their energy band. But the electrons in the filled lower bands are not able to move, since there are no free states nearby.

On the other hand, if the filling of the states within the bands stops exactly at the top of a band, then the electrons cannot respond normally to an electric field by moving down the wire, since there is an energy gap preventing them from reaching a quantum state in which they can move freely. Such a material would then be an insulator. An example is diamond, in which the highest filled quantum state also fills out an energy band. This band is called the *valence band*, since it is occupied by the outer valence electrons of the atoms. The energy gap to the next band, which is empty,

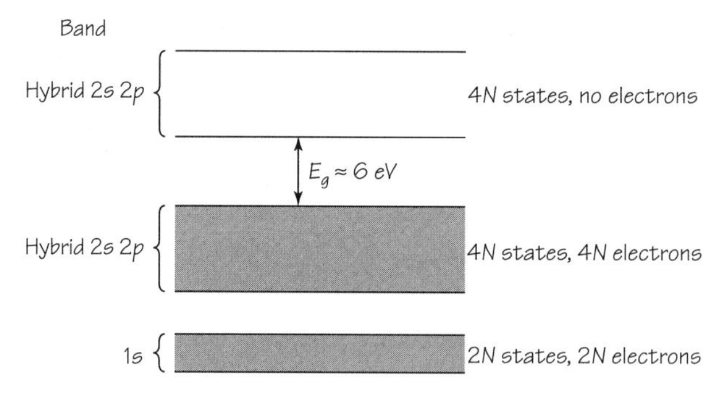

FIGURE 16.9 Actual occupation of the energy bands in a diamond crystal with N atoms ($6N$ electrons). The two lower bands are full and separated by an energy gap Eg from the higher hybrid *2s2p* band, which has no electrons.

is about 6 eV, so the electrons in diamond cannot normally reach the conduction band and diamond is therefore a good insulator.

16.6 SUPERCONDUCTORS

The band theory of metals has been subjected to experimental tests many times and is now the accepted theory of the behavior of conductors and insulators. But it has to be modified when it comes to superconductivity. We noted earlier that electrical resistance is due to collisions of the electrons (whether treated as particles or waves) with impurities, imperfections, and especially the lattice vibrations of the metal crystal. The lattice vibrations of the solid will decrease as the temperature falls, because the entropy, which represents disorder, also decreases. Therefore the resistance should also decrease. This is what is observed, and it is well explained by quantum band theory. But at very low temperatures something strange happens: the resistance of some materials suddenly drops to zero at a certain temperature. As noted before, these materials become *superconductors*, meaning they can conduct currents without resistance or loss of electric energy. The temperature at which superconductivity occurs is called the *critical temperature*. It ranges from 0.015 K for tungsten to as high as 160 K for certain ceramics.

Good conductors usually cannot become superconductors.

You can imagine some of the possible advantages of electrical wires having zero resistance. It is like motion without friction. Among the many actual and potential applications are electric power transmission lines without power loss, faster computers, and super-strong electromagnets. The latter are used today in magnetic resonance imaging (MRI) devices, high-energy particle accelerators, and high-speed levitation trains (often called "mag-lev" for magnetic levitation). Since the boiling temperature of liquid nitrogen, which is plentiful and relatively easy to produce, is 77 K, superconductors with transition temperatures higher than 77 K can be kept suf-

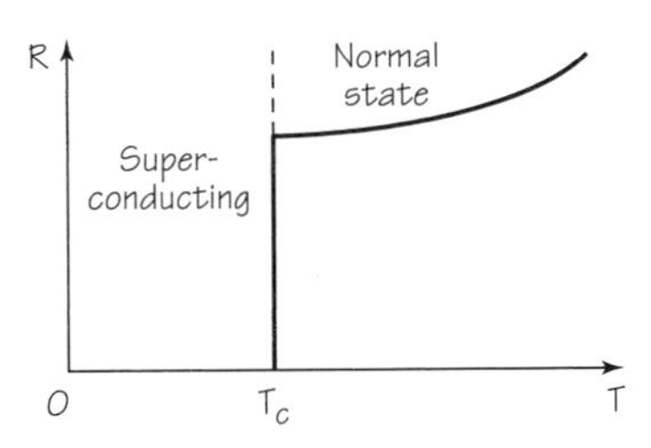

FIGURE 16.10 Transition of resistance (R) to superconductivity at the critical temperature T_c.

FIGURE 16.11 Japanese Mag-Lev train driven by superconducting electromagnets.

ficiently cold by keeping them immersed in liquid nitrogen. However, such materials are brittle, and the superconductivity can be easily destroyed by stray magnetic fields. Researchers today are working to overcome these problems and to create superconducting materials with transition temperatures even closer to room temperature (20°C, or 293 K).

The development of superconducting devices was greatly stimulated after the acceptance of the basic theory of superconductivity proposed in 1957 by John Bardeen, Leon Cooper, and Robert Schrieffer. The authors of the BCS theory, as it is known, received the Nobel Prize for their work in 1972. Their theory is highly technical but the basic idea is that the electron waves in the superconducting state no longer act independently, as in Bloch's model. Instead, they are paired together at the critical temperature so that their wave functions act as one unit as they interact with the crystal lattice. Moreover, all of the electron pairs move together in one collective motion, so that if any single electron is scattered by the lattice it is pulled back into the flow by its partner, and if any pair of electrons is somehow scattered off track, it is pulled back into the collective flow by all of the other pairs. Since there is no scattering or inelastic collisions, there is no resistance, and the material becomes a superconductor.

Although the BCS theory accounted well in general for certain crystalline solids, further refinements are required today for other materials, such as ceramics, and for the more detailed behavior of superconductors.

16.7 SEMICONDUCTORS

The most prevalent use today of semiconductors is forming them into transistors, the basic electronic building blocks of all "solid-state electronics" and computer microchips. Semiconductors are the most important materials in the burgeoning revolution in computers and other electronic devices today. Such applications also arose, beginning in the 1930s, from the quantum mechanics of the band structure of solids.

The most common semiconductors are made of silicon or germanium, elements 14 and 32 on the periodic table. It was known that they form very stable crystal structures that should be insulators but are in fact weak conductors of electricity. For both of these elements, it was found that the numbers of electrons are just enough to fill up to the top of an energy band in each case. This is why they should be insulators, and, in fact, at near absolute zero, 0 K, they are insulators (not superconductors).

At very low temperature, the lattice vibrations in silicon and germanium are minimal, and the electrons at the top of the valence band are not able to obtain enough energy from the lattice vibrations to enable them to jump the energy gap to the next band and become conducting. However, the gap to the next band is very small, only 0.7 eV for germanium and only 1.1 eV for silicon. Because these gaps are so small, at temperatures somewhat above absolute zero the electrons can pick up enough energy from the vibrations of the crystal lattice to jump the gap and land in the empty conduction band. So at room temperature these metals, which by their structure should be insulators, are actually weak conductors.

As failed insulators and poor conductors, silicon and germanium were not much use in electronics until the 1950s when advances were made in

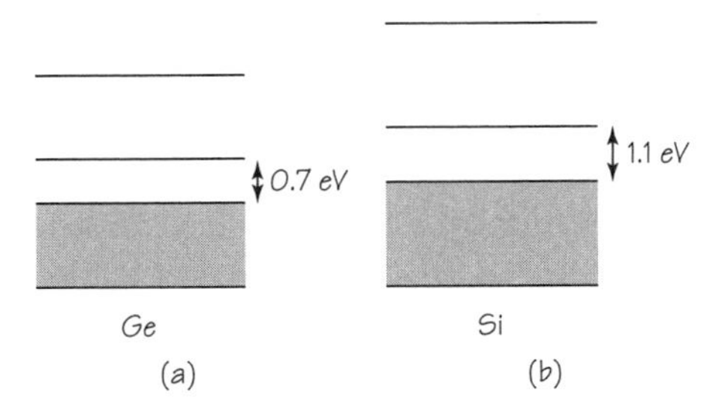

FIGURE 16.12 Energy bands of germanium (Ge) and silicon (Si).

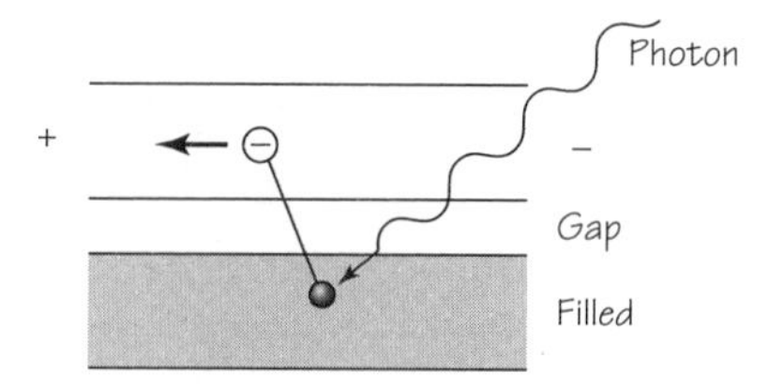

FIGURE 16.13 The photoelectric effect of an electron in a semiconductor jumping into the conduction band with a potential difference across the material.

the controlled introduction of certain impurities into the lattice structure (see Section 16.8 below). But the real use of these two plentiful metals took off during the 1980s with the introduction of mass-production methods for super-thin, microscopically structured layers of silicon and (to a lesser extent) germanium crystals, which, when properly arranged (as also discussed below) can act as transistors. Today, wafer-thin layers of silicon, when made into micro-transistors through the introduction of impurities and broken up into "chips," are the basis of the trillion-dollar-per-year electronics and computer industries, and there is as yet no end in sight to the revolution they have unleashed. And it all depends on the narrow energy gaps in these crystals.

Since germanium and silicon are so sensitive to impurities, their large-scale use as semiconductors did not occur until methods were developed for producing ultra-pure graphite for nuclear reactors and ultra-pure germanium for electronic circuits during World War II. Pure germanium was also used at first in photoelectric cells. A photon from the outside can strike an electron in the valence band of germanium (and later silicon), providing the electron with enough energy to reach the conduction band in a type of internal photoelectric effect. For this to occur, the energy of the photon must be at least 0.7 eV for germanium and 1.1 eV for silicon. From the relationship for the energy of a photon, $E = hf$, these energies correspond to photons possessing frequencies in the infrared range of electromagnetic waves. Any waves (or photons) with frequencies in the infrared or higher range, which includes visible light, will cause electrons to jump into the conduction band and form a current. This is one reason these crystals are known as *semiconductors*, since they are good conductors only when the band gap can be overcome.

This type of light-induced conductivity can be used in a *photoelectric cell*, or *photocell* for short—a cell that produces electricity when light shines on it. You can imagine some of the many possible applications of a photocell. It can be used, for instance, for motion detectors. A light beam shining on a photocell in a circuit will generate a steady current. If somebody walks through the beam, this will interrupt the current, which could either set off an alarm or open an automatic door for the person to exit or enter. Automatic controls on night lights use the same principle. Since photocells are sensitive even to infrared rays, as long as there is sufficient daylight the cell will produce a current. When the sun goes down, the current stops, which signals a circuit to switch on the lights.

16.8 INTRODUCING IMPURITIES

The most extensive use of semiconductors, such as silicon or germanium, including their use as transistors, arises from their behavior when, after being sufficiently purified of atoms other than the basic element (e.g., silicon or germanium), very small amounts of special impurities are carefully introduced. While the methods for first purifying then adding small amounts of impurities to germanium became available after World War II, research on the purification and controlled introduction of impurities for silicon did not succeed until the late 1950s. Since silicon is more abundant than germanium and other semiconductors, it soon replaced germanium as the preferred semiconductor.

Germanium (element 32) and silicon (element 14) each have four valence electrons, which fill up the valence band when these metals form a crystal. Element 33, arsenic, has five valence electrons, as does element 15, phosphorus. If a very small amount of arsenic is added to germanium as the germanium crystal is formed, the arsenic atoms will substitute for a germanium atom in the lattice. The same happens when a small amount of phosphorus is added to silicon. The small amounts are carefully controlled in the production process and are only about 0.0001% of the total atoms. This process is calling *doping*, and the impurities are called *dopants*.

When the corresponding dopant is introduced into the germanium or silicon lattice, four of its valence electrons are bonded to the other atoms in the lattice, leaving the fifth valence electron without a bond. These extra valence electrons from the dopants are so lightly attached to their atoms that they easily absorb vibration energy and jump into the conduction band

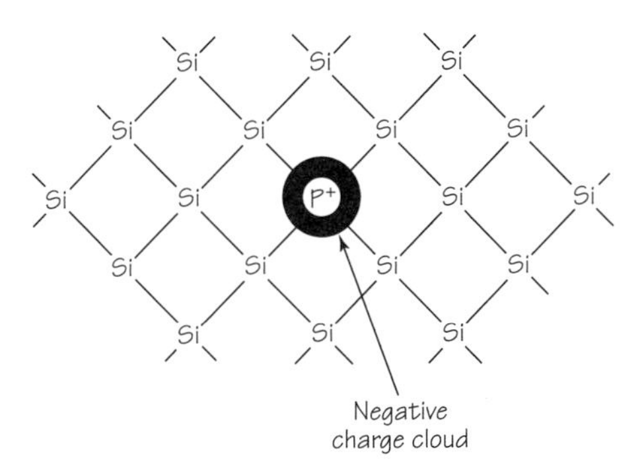

FIGURE 16.14 Two-dimensional representation of a silicon lattice in which a phosphorus atom substitutes a regular lattice atom, and thus introduces a negative charge cloud about the phosphorus atom. Each electron pair between two silicon atoms constitutes a covalent bond.

of the semiconductor. At room temperature they are all in the conduction band, which means that the doped semiconductor now acts like a full conductor. Since the negative electrons donated by the impurities enable the material to conduct, these types of semiconductors are called *n-type semiconductors.*

A similar situation occurs when the semiconductors are doped with impurities one space lower on the periodic table, that is when silicon (14) is doped with element 13, aluminum, and germanium (32) is doped with element 31, gallium. The dopants have only three valence electrons. So, when they are bound into the silicon or germanium lattice, they leave a space in the valence band indicating that they are able to accept an electron from a neighboring silicon or germanium atom. This space is called a *hole*. The hole behaves similarly to an air bubble in a glass of soft drink or beer. When an external electric field is turned on, electrons from the neighboring atoms, which normally cannot move anywhere, can now move over one atom in filling up the hole. But then this leaves a hole in the neighboring atom, which can be filled up by its neighbor, and so on. Similarly, the upward motion of the air bubble in a glass of beer actually involves the downward flow of beer into the space occupied by the bubble, which appears to the eye to be moving upward. In the semiconductor, if the field moves the electrons to the left in the diagram, the hole appears to migrate to the right. By moving to the right, it is behaving like a positive charge, even though it is just a space. Nevertheless, because of the motion of negative charge to the left, the physical effect is the same as if positive charges were flowing to the right in the valence band, below the conduction band, where previously no charges could flow at all. Once again, the very careful introduction of small amounts of impurities has turned the semiconductor into a full conductor. Since the conducting charges appear to be positive, this is known as a *p-type semiconductor.*

These doped semiconductors could be used like any other conductor, but that is hardly worthwhile, since we can produce full conductors much more easily and cheaply, for example, copper wires. Instead, it was recognized during and after WWII that the really useful applications of these types of semiconductors is when they are placed physically next to each other inside electronic devices.

16.9 SEMICONDUCTOR DEVICES

Let's take a p-type semiconductor with a very clean surface and place it right next to the clean surface of a n-type semiconductor. This type of device is called an *n-p junction diode*, or simply a *diode*. What happens? Scien-

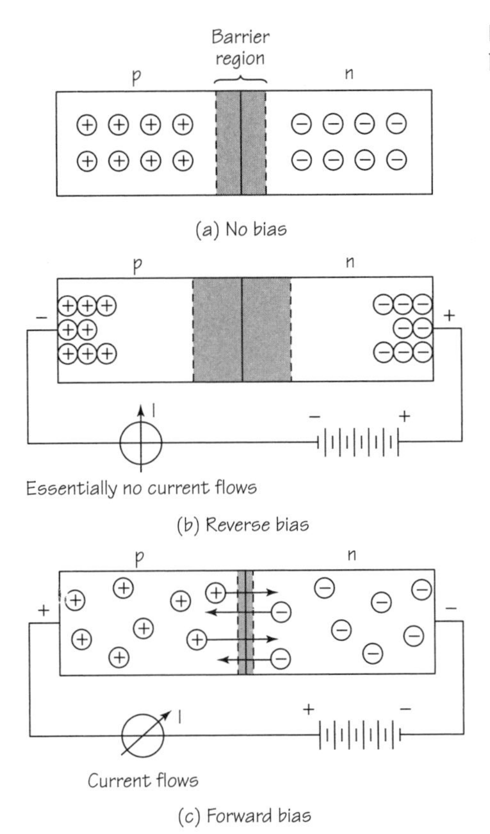

FIGURE 16.15 An *n-p* diode with (a) no bias; (b) reverse bias; and (c) forward bias.

tists realized that the electrons in the conduction band of the n-type semiconductor are able to move across the boundary between the two semiconductors, and fall into the holes in the lower energy valence band of the p-type semiconductor. The electrons and holes begin to disappear. But as the electrons disappear in the n-type, the positive charges of the impurities are no longer balanced by the negative conducting electrons, so the n-type becomes positively charged. The opposite happens in the p-type semiconductor. As the holes disappear, the impurities are left holding an extra electron, which causes the p-side to become negatively charged. The result of all this is that after a very short time a net electric field is set up between the n-type and p-type semiconductors which stops the process of mutual destruction by keeping the electrons on the n-type side separated from the holes on the p-type side, with a "depletion layer" in the middle.

Now, if a potential difference, such as one caused by a small battery, is placed across the device from the p-type side to the n-type side, the equilibrium may or may not be disturbed, depending upon how the potential is placed. If the positive wire is placed on the n-type material and the negative wire on the p-type material, as in Figure 16.15(b) the separation of the electrons and holes is reinforced. So no current will flow. But, if the wires are reversed, so that the positive wire is placed on the p-type material and the negative wire on the n-type side, the negative electrons on the n-type side will be pushed toward the boundary, and the same for the positive holes on the p-type side. If the external potential is greater than the internal potential of these two, a current will flow in the device.

The current will be enhanced if now a photon arrives from the outside and pushes an electron from the n-type material into the conduction band. The n-p diode can then be used as an even more powerful photocell than a simple semiconductor alone. Perhaps the most common use of this type of photocell is in the generation of electricity from solar energy. This can be done for small-scale devices, such as pocket-calculators, or for larger scale energy needs, such as the energy for an entire building. Such photocells are often known as *photovoltaic cells*. The International Space Station and other satellites near the earth's orbit depend for their energy needs upon huge "wings" of photovoltaic cells, unfurled to catch the radiation from the sun. Research continues on how to make even greater use of photoconductivity from the sun's rays to produce electrical energy for our needs on a large scale.

Deep-space satellites use energy from isotope decay, since they are too far from the sun for solar panels to be of much use.

FIGURE 16.16 Artist's conception of International Space Station with its "wings" of photovoltaic cells.

A photocell working in reverse is also useful. If a conduction electron in an n-p diode happens to fall into a hole, it will emit the excess energy as a photon, much as an electron would do in an atom as it jumps to a lower quantum state. If the diode is designed to emit light in the visible range for display purposes, it is called a *light-emitting diode*, or *LED* for short. These LED lights, which are usually red, green, or orange, can be found on most electronic devices, usually to indicate that the device is on or that the circuit is functioning. They can be also used in large-scale devices, such as computer monitors, where they consume less energy and exhibit less distortion than do traditional CRT monitors.

One of the biggest uses of n-p diodes is in circuits where we want the current to flow in one direction only, but not in the opposite direction. This is used, for instance, in logic circuits for computers, in which an answer of "true" or "false," or "yes" or "no," or 0 or 1 can be decided by whether or not the current is allowed to flow through the device. A device that allows a current to flow in one direction only can also be used to convert an alternating current (AC), for instance the current from a wall socket, into direct current (DC) for use in small electronics devices. Commercial electric current in the United States alternates in a wave-like pattern at a rate or frequency of 60 cycles per second (60 Hz). This is shown in Figure 16.17a. If we hook up a closed circuit that includes the AC wall socket, a device for measuring the current (an ammeter), and an n-p semiconductor, the current in the wire will go in only one direction, which occurs only when the voltage is positive on the p-type side and negative on the n-type side. This conversion of AC to DC current is called *rectification*, and devices that rectify are naturally called *rectifiers*. These are useful for devices that accept only DC current.

You can see from Figure 16.17b that the voltage produced by the n-p junction diode is positive only, but it is constantly changing over each hump

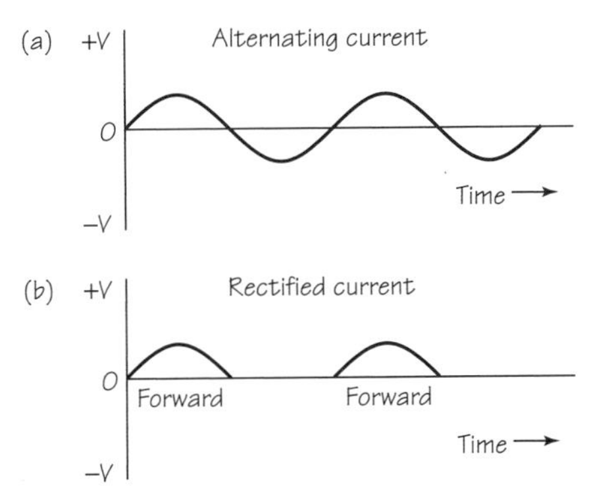

FIGURE 16.17 Graphs of AC and rectified voltages resulting in AC and DC currents.

and hits zero for a long interval. These effects can be reduced by sending the current through additional electronic devices, known as *filters*, that help to smooth out the voltage to a constant value, producing a steady direct current that can then be fed into a DC electronic device—an electronic game, a calculator, a laptop computer, etc.

These properties of semiconductors become even more interesting when we add a third semiconductor to our junction diode. This forms a *transistor*.

16.10 TRANSISTORS

The properties of n-type and p-type semiconductors described above were well known by the end of World War II, during which considerable research was devoted to electronics and the invention of radar. In 1947 three researchers at Bell Laboratories in New Jersey took the idea of the n-p junction diode a step further. William Shockley, Walter Brattain, and John Bardeen (who, alone of the three, later went on to work on the theory of superconductivity), placed two germanium diodes back to back—an n-p diode next to a p-n diode, each sharing the p-type semiconductor. They found that the device could be made to contact electricity or to resist, that is, to block, the flow of electricity, depending upon the charge applied to the middle layer of p-type semiconductor. Such a device is therefore called a *transistor*, since it transmits or resists current depending upon the charge provided to the middle layer. The type of transistor just described is called a *n-p-n bipolar transistor*. A *p-n-p bipolar transistor* is also viable. (There is also a different type of transistor construction, known as the "field effect transistor," which works on the same general principle.)

By the early 1960s, as silicon-based transistors had become smaller, more reliable, and cheaper to make, they began replacing the bulkier and more expensive vacuum tubes previously used for such purposes as amplifying voltages, for instance in stereo sound systems, TVs, and radios, and as switching and logic devices in computers. Shockley, Bardeen, and Brattain received the Nobel Prize in 1956 for an invention that would soon launch the revolution in computer electronics—the transistor. Shockley, based at Stanford University, went on to lay the foundations for the development of Silicon Valley industries in the vicinity of the university. Many of the companies there, including Intel, were founded by people who had originally worked with Shockley.

The revolution in computer electronics was hastened when the need for miniaturized electronic components for space-age missiles helped lead to

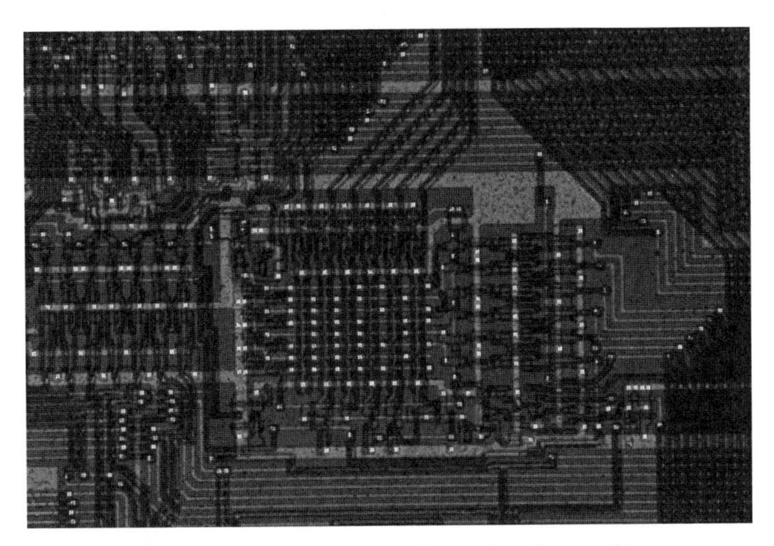

FIGURE 16.18 Close-up of a portion of a semiconductor chip.

the development of the *integrated circuit*. In 1958 Jack Kilby of Texas Instruments developed the idea of integrating transistors and related circuitry on a single silicon chip. A year later, Robert Noyce and Jean Noerni of Fairchild Semiconductor Corporation put the idea into practice by developing the method for integrating the separate elements onto the chip. In 1970 reduction in the size of integrated circuits was introduced with the first microprocessor, and in 1971 the newly formed Intel Corporation introduced the first integrated microprocessor circuit. Continued advances in microprocessor design and manufacturing techniques enabled the mass production of microprocessors for personal computers, cell phones, automobiles, industrial robots, and even toasters and “talking” dolls. By the early 2000s Intel and its competitors could place as many as 42 million microscopic transistors on a single silicon chip, two centimeters square.

Figure 16.19(a) shows a schematic diagram of an n-p-n bipolar transistor with a voltage placed across it. Figure 16.19(b) shows how such a device is actually constructed by “chip makers” such as Intel and Advanced Micro Devices. A thin layer of n-type silicon is placed over a thin “chip” (crystal) of very lightly doped p-type silicon. The p-type silicon, which is sandwiched between the n-type pieces, is called the “base.” Being very thin and only lightly doped, most of its electrons, which will move through it from left to right, will not be captured by the holes, yet they will still provide an internal field to prevent any flow when there is no external field. The entire chip must also be very thin in order to allow the electronic cir-

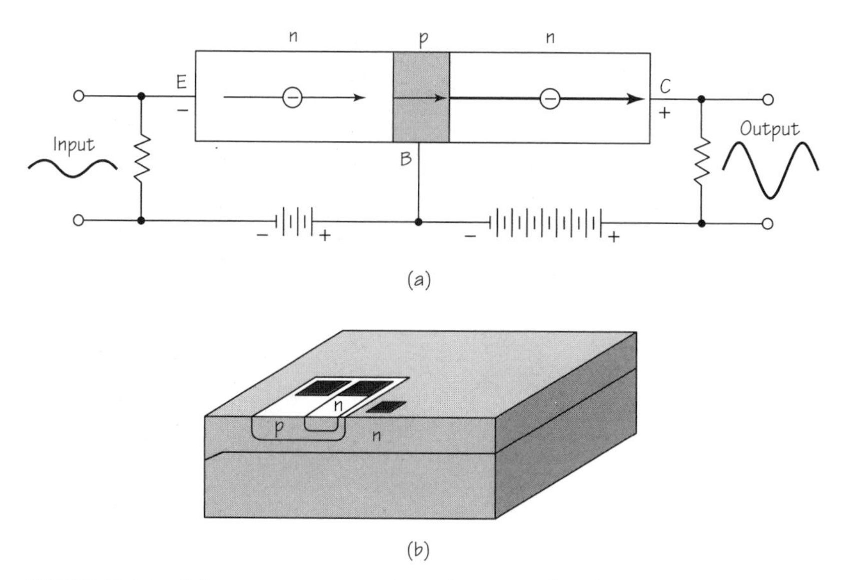

FIGURE 16.19 Schematic representation of an *n-p-n* bipolar transistor (see text). The dark areas in (b) are the contact pads.

cuitry to function as rapidly as possible. Using a photographic emulsion, a microscopic circuit pattern is placed on the surface layer of the n-type silicon such that the n-type surface layer is stripped away at certain places when an acid solution is applied to the surface. This exposes the p-type silicon underneath, so that the electrical connections can be made to form the n-p-n transistor.

What does the transistor actually do, and how does it do it? Think of Figure 16.19(a) as representing two n-p diodes back to back. Remember that if a negative or zero charged wire is placed on the p-type silicon and a positive charged wire is placed on the n-type side, no charge will flow. However, if a positive wire is placed on the base (p-type) and a negative wire on the left n-type silicon, charges will flow from the n-type silicon into the base. If the base is very thin and only lightly doped, the charges will make it through the base into the n-type silicon on the right. Here the situation is reversed. The p-type stays positive, but an even higher voltage positive wire is attached to the end of the right-hand n-type silicon. The purpose of this is to keep the electrons moving from left to right. In addition to the conduction electrons that are already in the right-hand n-type silicon, new electrons are arriving from the n-type and p-type silicon on the left. The amount of flow can be controlled by the positive voltage at the base. The net effect is to amplify (increase) the current from left to right.

16.11 SOME APPLICATIONS OF TRANSISTORS

Amplifying Voltage

If we place a large resistor in a circuit containing a transistor, as in Figure 16.19(a), we can amplify the incoming voltage, since according to Ohm's law voltage is equal to current times resistance. So a current crossing a small resistance corresponds to a small voltage. But if the same current is made to cross a much larger resistance, it will yield a much larger voltage, that is, the small input voltage will be amplified to yield a large output voltage.

For example, the amplified voltage can correspond to a portion of a digital sound wave, in which the analog wave has been approximated by a series of single voltages. This small portion of the wave (one bit) can be represented by a tiny voltage received at the transistor base from a photocell that is stimulated by a laser beam that picked up the signal from a music compact disc (CD). The amplified voltage from the transistor can then be sent to a speaker set, which turns the signal into a vibration that is emitted by the speakers as a pressure wave that your ears detect and your brain interprets as—let's say—the beautiful strains of Beethoven's *Symphonie Pastorale!*

DRAM

As a logic device, a transistor can act like a diode in allowing current to pass or in preventing current flow, depending upon the voltage at the base. By combining p-n-p and n-p-n transistors together, along with diodes, various logic systems and "gates" can be created. One of the most useful

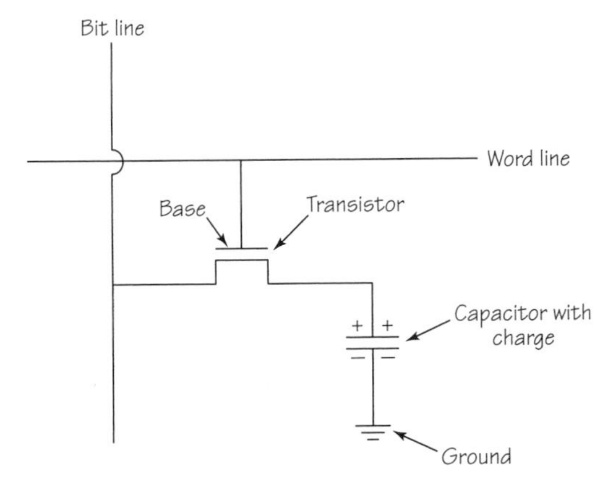

FIGURE 16.20 Schematic representation of a DRAM cell.

applications of this is in the creation of "dynamic random access memory" (DRAM) in computers, devices that store data that can be accessed at will in random order. As is well known, information is stored digitally in computers, that is, it is stored in the form of strings of binary numbers made up of just 1s and 0s. Each binary number (whether it is 1 or 0) is called a "bit." (Eight bits make a "byte.") For instance, the binary number 10101010, which has eight bits (making one byte), is equivalent in the decimal system to the number 170. If you happen to type the number 170 onto a spread sheet and save it, the computer will first convert the decimal number 170 to the binary number 10101010. It will then assign each of the eight bits to a single cell within dynamic random access memory. Each bit will correspond to the charge on a charge device known as a capacitor, which, in principle, is just two plates held next to each other. If the bit is 1, the capacitor will receive a small charge. If the number is 0, it receives no charge. One of these cells is shown in Figure 16.20 with the number 1 on it (plus charge on the capacitor).

Now you ask the computer to retrieve the number 170 from memory. Here's what happens. Your request is sent to the "word line" as a positive voltage, which is then applied to the base of the transistor in each circuit on that line. Since the base is now positive, the charge on the capacitor can flow through the transistor and up through the "bit line" to the processor which detects the current and identifies the number in that cell as a 1. Of course, if there is no charge stored in the capacitor to begin with, there will be no current on the bit line, and that bit will be identified as 0. Figure 16.21 shows what the cells for all eight bits would look like for the decimal number 170. This diagram represents only one byte of memory. Many personal computers today now come with hundreds of "megabytes" (MB), or even gigabytes of memory, miniaturized on microprocessor memory modules. For example, 128 million bytes of memory, or $8 \times 128 \times 10^6$ bits, are 1024×10^6, or about *one billion cells* like Figure 16.21! It is aston-

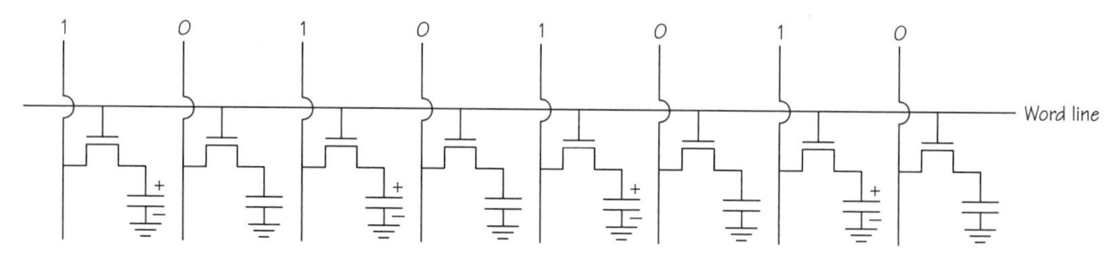

FIGURE 16.21 One byte of DRAM showing the decimal number 170 as the binary number 10101010. The 1 is indicated by the charge on the capacitor. 0 is indicated by no charge on the capacitor.

ishing how far this technology has come in only a few decades since the invention of the transistor in 1947. Where will we be in another 50, or even another 10 years?

SOME NEW IDEAS AND CONCEPTS

amorphous solids
band theory
bit
byte
classical free-electron model
composite solids
conduction band
critical temperature
crystalline solids
diode
dopant
doping
hole
insulator
latent heat of fusion
LED
n-type semiconductor
p-type semiconductor
phase
photocell
photovoltaic cell
quasi-crystalline solids
rectifier
semiconductor
superconductor
transistor
valence band

FURTHER READING

I. Amato, *Stuff: the Materials the World is Made Of* (New York: Avon, 1997).

P. Ball, *Made to Measure: New Materials for the 21st Century* (Princeton, NJ: Princeton University Press, 1997).

M. Campbell-Kelly and W. Aspray, *Computer: A History of the Information Machine.* Sloan Technology Series (New York: Basic Books, 1997).

G. Holton, H. Chang, and E. Jurkowitz, How a Scientific Discovery Is Made: A Case History [High-Temperature Superconductivity], *American Scientist*, **84** (1996), 364–375.

R.E. Hummel, *Understanding Materials Science: History, Properties, Applications* (New York: Springer-Verlag, 1998).

M. Riordan and L. Hoddeson, *Crystal Fire: The Birth of the Information Age* (New York: Norton, 1997).

STUDY GUIDE QUESTIONS

16.1 The Success of Quantum Mechanics

1. Describe some of the successes of quantum mechanics.

16.2 Forming a Solid

1. What happens to the temperature of a material at a phase transition, and why?
2. A container of cold water is placed in a refrigerator. The refrigerator is constantly running but the reading on a thermometer in the water remains unchanged. What do you conclude is happening to the water?
3. What exactly happens when a liquid freezes into a solid?
4. Why would the lowest entropy state for a solid involve the largest number of possible bonds?
5. What is the most prominent characteristic of a crystal structure?
6. Describe the other three main types of solid bonding.

16.3 Quantum Solids

1. What did Einstein first show about the charges in solids?
2. Describe Drude's oscillator model of solids.
3. In what ways did Einstein's model go beyond Drude's?
4. What did Einstein predict in 1907 from his model?
5. Was his prediction confirmed, and why was it so significant?

16.4 Conducting Electrons

1. Using Ohm's law, describe how insulators and conductors differ from each other.
2. Describe the classical free-electron model of a solid.
3. How did the classical model account for conductivity and Ohm's law?
4. Evaluate the classical model as a theory.

16.5 Banding Together

1. How did Sommerfeld modify the classical free-electron model of solids?
2. What assumptions did Bloch make in constructing a quantum model of metals?
3. What did Bloch discover when he solved Schrödinger's equation?
4. Describe the band theory of metals. How are the bands formed?
5. What is the difference between an insulator and an conductor, according to the band theory?

16.6 Superconductors

1. What is superconductivity, and where does it occur?
2. Scientists are now trying to obtain superconducting materials as close as possible to room temperature. What would be some of the advantages to this?
3. Briefly, what is the BCS theory of superconductivity?

16.7 Semiconductors

1. What is a semiconductor?
2. According to the band theory, why are semiconductors "failed insulators and poor conductors"?
3. How might semiconductors be used as photodetectors?

16.8 Introducing Impurities

1. How is an n-type semiconductor formed?
2. What is it that conducts in an n-type semiconductor?
3. How is a p-type semiconductor formed?
4. What is a "hole," and why is considered to be positive if it is really just an empty orbit on an ion?

16.9 Semiconductor Devices

1. What is an n-p junction diode?
2. How does an n-p junction diode reach equilibrium?
3. How does a photovoltaic cell work?
4. How does a LED light work?
5. How might a diode be used as a rectifier?

16.10 Transistors

1. How is a transistor made from semiconductors?
2. What is an integrated circuit?
3. How are millions of microscopic transistors placed on a single silicon chip?
4. What does a transistor actually do and how does it do it?

16.11 Some Applications of Transistors

1. How might a transistor be used as an amplifier?
2. How might a transistor be used to store information in a computer?

DISCOVERY QUESTIONS

1. How do the free-electron theory and the band theory each account for the existence of insulators?
2. Heat is added to a melting ice cube until the cube has completely melted. The heat that goes into the cube is called the "latent heat of fusion," the word *latent* meaning "hidden." Where is the heat hidden?
3. Why is it that the temperature of a melting ice cube and its melted water stay at 0°C until all of the ice has melted?
4. Why can't some materials remain in the liquid state all the way down to absolute zero?
5. Why is it that some elements, such as mercury, are liquid at room temperature?
6. Whenever you plug an appliance into a wall socket the current flows instantly. Should it not take some time for the electrons to begin to flow down the wire?
7. Why did Bloch decide to view the electrons as waves rather than as particles?
8. According to the band theory, how do an insulator, a conductor, and a semiconductor differ from each other?
9. What are some of the applications of semiconductors?

10. Why do impurities have to be added to semiconductors to make them into conductors?

Quantitative

1. Translate the binary number 100 into a decimal number.
2. Translate the decimal number 100 into the binary system.
3. What would a memory circuit look like that is storing the decimal number 100 in binary form?

CHAPTER 17

Probing the Nucleus

17.1 QUESTIONS ABOUT THE NUCLEUS

We saw in Chapter 14 that studies of the atom indicated that the atom consists of a very small, positively charged nucleus surrounded by negatively charged electrons. Experiments on the scattering of α particles revealed that the nucleus has dimensions of the order of 10^{-14} m. Since the diameter of an atom is of the order of 10^{-10} m, the nucleus takes up only a minute fraction of the volume of an atom. The nucleus, however, contains nearly all of the mass of the atom, as was also shown by the scattering experiments.

The existence of the atomic nucleus (named by analogy with the nucleus of a living cell) and its properties raised many questions similar to those raised about the atom. Is the nucleus itself made up of still smaller units? If so, what are these units, and how are they arranged in the nucleus? What methods can be used to get answers to these questions? What experimental evidence can be used as a guide?

The study of the properties and structure of atoms needed new physical methods. The methods that could be used to study the properties of bodies of ordinary size, that is, those with dimensions of the order of centimeters or meters, could not yield information about the structure of atoms. It is reasonable to expect that it is still more difficult to get information about what goes on inside the nucleus, which is such a small part of the atom. New kinds of experimental data must be obtained. New theories must be devised to help correlate and explain the data. In these respects, the study of the nucleus is still another step on the long road from the very large to the very small, along which we have been traveling in this course, following the historical development of physical science.

17.2 BECQUEREL'S DISCOVERY

One of the first and most important clues to an understanding of the nucleus occurred with the discovery of the phenomenon later known as *radioactivity* in early 1896 by the French physicist Henri Becquerel. It was an-

FIGURE 17.1 Henri Becquerel (1852–1908) received the 1903 Nobel Prize in physics together with Pierre and Marie Curie (for the discovery of natural radioactivity).

other of those "accidents" that illustrate how the trained and prepared mind is able to respond to an unexpected observation.

Only 2 months before, in November 1895, Röntgen had discovered X rays. In doing so, he had unwittingly set the stage for the discovery of radioactivity. Röntgen had found that X rays came from the glowing spot on a glass tube where a beam of cathode rays (high-speed electrons) was hitting (see Section 13.7). When the cathode-ray beam was turned off, the spot of light on the face of the glass tube disappeared; the X rays coming from that spot also stopped.

The emission of light by the glass tube when it is excited by the cathode-ray beam is an example of the phenomenon called *fluorescence*, which was well known before Röntgen's work. A considerable amount of research had been done on fluorescence during the latter part of the nineteenth century. A substance is said to be fluorescent if it immediately emits visible light when struck by:

(1) visible light of shorter wavelength;
(2) invisible radiations, such as ultraviolet light; or
(3) the beam of electrons that make up cathode rays.

Fluorescence stops when the exciting light is turned off. (The term *phosphorescence* is generally applied to a related phenomenon, the emission of visible light that continues *after* the exciting light is turned off.)

Röntgen's observation that the X rays came from the spot that also showed fluorescence raised his suspicion that there was a close connection between X rays and fluorescence. Becquerel was fortunate in having the necessary materials and training to study this problem. In addition, he was the son and grandson of physicists who had made important contributions to the field of fluorescence and phosphorescence. In his Paris laboratory, Becquerel had devised an instrument for examining materials in complete darkness a small fraction of a second after they had been exposed to a brilliant light. The question occurred to Becquerel: When bodies are made to fluoresce (or phosphoresce) in the visible region with sufficient intensity, do they also emit X rays in addition to the light rays? He tested a number of substances by exposing them to sunlight; his method of checking whether they also emitted invisible X rays followed Röntgen's idea: Is a well-wrapped photographic plate exposed when near the source of such invisible rays? One of the samples Becquerel used happened to be a salt of the metal uranium, a sample of potassium-uranyl sulfate. In his words:

> I wrapped a . . . photographic plate . . . with two sheets of thick black paper, so thick that the plate did not become clouded by ex-

> posure to the sun for a whole day. I placed on the paper a crust of the phosphorescent substance, and exposed the whole thing to the sun for several hours. When I developed the photographic plate I saw the silhouette of the phosphorescent substance in black on the negative. If I placed between the phosphorescent substance and the paper a coin or a metallic screen pierced with an open-work design, the image of these objects appeared on the negative. The same experiment can be tried with a thin sheet of glass placed between the phosphorescent substance and the paper, which excludes the possibility of a chemical action resulting from vapors which might emanate from the substance when heated by the sun's rays.
>
> We may therefore conclude from these experiments that the phosphorescent substance in question emits radiations which penetrate paper that is opaque to light. . . .

In his published paper, Becquerel was careful to conclude from his experiment only that "penetrating radiations" were emitted from the phosphorescent substance. He did not write that the substance emitted X rays while it phosphoresced, because he had not fully verified that the radiations were X rays—though the radiations were transmitted through the black paper, just as X rays are—or that they were actually related to the phosphorescence (though he strongly suspected that they were). Before he could investigate these possibilities, he made this discovery:

> . . . among the preceding experiments some had been made ready on Wednesday the 26th and Thursday the 27th of February [1896]; and as on those days the sun only showed itself intermittently, I kept my arrangements all prepared and put back the holders in the dark in the drawer of the case, and left in place the crusts of uranium salt. Since the sun did not show itself again for several days, I developed the photographic plates on the 1st of March, expecting to find the images very feeble. On the contrary, the silhouettes appeared with great intensity. I at once thought that the action might be able to go on in the dark. . . .

Further experiments verified this surprising thought. Even when the uranium compound was not being excited by sunlight to phosphoresce, it continually emitted something that could penetrate black paper and other substances opaque to light, such as thin plates of aluminum or copper. Becquerel found that all the compounds of uranium, many of which were not phosphorescent at all, and metallic uranium itself had the same property. The amount of action on the photographic plate did not depend on what the particular compound of uranium was, but only on the amount of uranium present in it!

Becquerel also found that the persistent radiation from a sample of uranium did not appear to change, either in intensity or character, with the passing of time during days, weeks, and months. Nor was a change in the activity observed when the sample of uranium or of one of its compounds was exposed to ultraviolet light, infrared light, or X rays. Moreover, the intensity of the uranium radiation (or "Becquerel rays," as they came to be known) was the same at room temperature (20°C), at 200°C, and at the temperature at which oxygen and nitrogen (air) liquefy, about -190°C. Thus, *these rays seemed unaffected by physical (and chemical) changes of the source.*

Becquerel also showed that the radiations from uranium produced ionization in the surrounding air. They could discharge a positively or negatively charged body such as an electroscope. So the uranium rays resemble X rays in two important respects: their penetrating power and their ionization power. Both kinds of rays were invisible to the unaided eye, but both affected photographic plates. Still, X rays and Becquerel rays differed

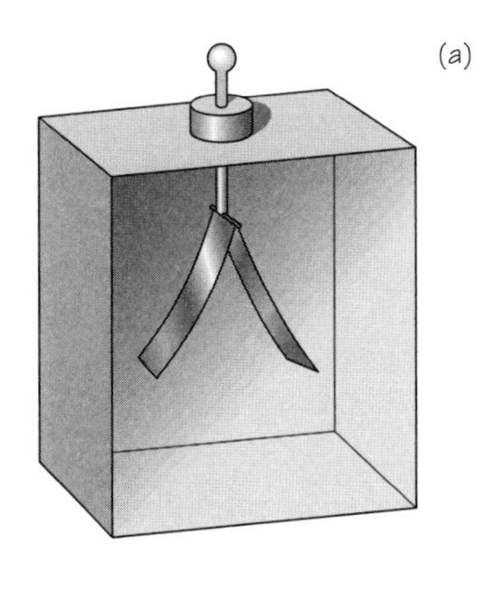

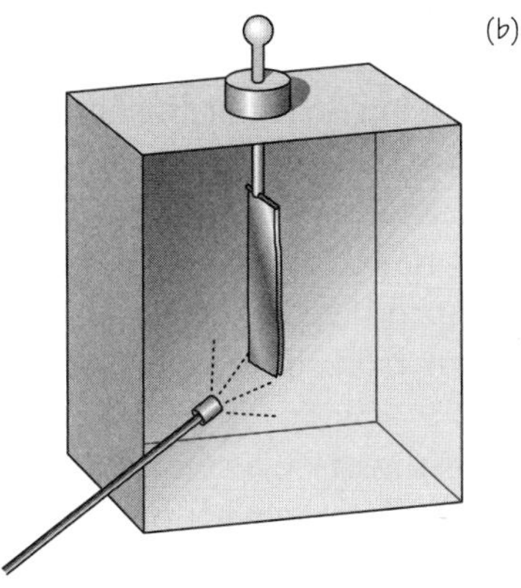

FIGURE 17.2 The ionizing effect of the Becquerel rays could be demonstrated with a charged electroscope. When a sample of uranium is held near the electroscope leaves, the rays cause gas molecules in the air to ionize—that is, to become electrically charged. Ions, with a charge opposite to that on the leaves, drift to the leaves and neutralize their charge. The time taken for the leaves to fall is a measure of the rate of ionization of the gas and, therefore, of the activity of the uranium source.

in at least two important ways: Compared to X rays, these newly discovered rays from uranium needed no cathode-ray tube or even light to start them, and they could not be turned off. Becquerel showed that even after a period of 3 years a given piece of uranium and its compounds continued to emit radiations spontaneously.

The years 1896 and 1897 were years of great excitement in physics, to a large extent because of the interest in the recently discovered X rays and in cathode rays (electrons). It quickly became evident that X rays could be used in medicine, and they were the subject of much research. In comparison, the properties of the Becquerel rays were less spectacular, and little work was done on them in the period from the end of May 1896 until the end of 1897. In any case, it seemed that somehow Becquerel rays were special cases of X-ray emission. Even Becquerel himself turned to other work. But attention began to be attracted by the fact that the invisible rays from the uranium and its compounds appeared spontaneously.

Two questions were asked. First, what was the source of the energy creating the uranium rays and making it possible for them to penetrate opaque substances? Second, did any other of the 70 or more elements known then have properties similar to those of uranium? The first question was not answered for some time, although it was considered seriously. The second question was answered early in 1898 by the Curies, who, by doing so, opened a whole new field of research in physical science.

17.3 THE CURIES DISCOVER OTHER RADIOACTIVE ELEMENTS

One of Becquerel's colleagues in Paris was the physicist Pierre Curie, who had recently married a Polish-born physicist, Marja Sklodowska, who became Marie Curie. Marie Curie undertook a systematic study of the Becquerel rays and looked for other elements and minerals that might emit them. Using a sensitive type of electrometer that her husband had recently invented, she measured the small electric current produced when the rays ionized the air. She assumed this current is proportional to the intensity of the rays (and it actually is). With this new technique, Curie could give a numerical value to the ionizing effect produced by the rays. These values were reproducible within a few percent from one experiment to the next with the same sample.

One of Marie Curie's first results was the discovery that the element thorium (Th) and its compounds emitted radiations with properties similar to those of the uranium rays. (The same finding was made independently in Germany by Gerhardt C. Schmidt, at about the same time.) The fact that thorium emits rays like those of uranium was of great importance; it showed that the mysterious rays were not a property peculiar just to one element. The discovery spurred the search for still other elements that might emit similar rays. The fact that uranium and thorium were the elements with the greatest known atomic masses indicated that the very heavy elements might have special properties different from those of the lighter elements.

The evident importance of the problems raised by the discovery of the uranium and thorium rays led Pierre Curie to lay aside his researches in other fields of physics and to join his wife to work on these new problems. They began a herculean task. First, they found that the intensity of the emission from any thorium compound was directly proportional to the fraction by weight of the metallic element thorium present. (Recall that Becquerel had found a similar result for uranium compounds.) Moreover, the amount of radiation was independent of the physical conditions or the chemical combination of the active elements. These results led the Curies to the conclusion that the emission of the rays depended only on the presence of *atoms* of either of the two elements—uranium or thorium. Atoms of other elements that were present were simply inactive or absorbed some of the radiation.

These ideas were especially important because they helped the Curies interpret their later experiments. For example, in their studies of the radiation from minerals they examined *pitchblende*, an ore containing about 80% uranium oxide (U_3O_8). They found that the emission from pitchblende, as measured by its effect in ionizing air, was about four or five times as great as that to be expected from the amount of uranium in the ore. The other elements known at the time to be associated with uranium in pitchblende, such as bismuth and barium, had been shown to be not active, or as Marie Curie now called it, not "radioactive." If emission of rays is an atomic phenomenon, the unexpected activity of pitchblende could be explained only by the presence of another, hitherto undiscovered, element in pitchblende, an element more active than uranium itself.

To explore this hypothesis, the Curies applied chemical separation processes to a large quantity of pitchblende to try to isolate this hypothetical active substance. After each separation process, the products were tested, the inactive part discarded, and the active part analyzed further. Finally, the Curies obtained a highly active product that presumably consisted mainly of the unknown element. In a note titled "On a New Radioactive Substance Contained in Pitchblende," which they submitted to the French Academy of Sciences in July of 1898, they reported:

> By carrying on these different operations . . . finally we obtained a substance whose activity is about 400 times greater than that of uranium. . . .

MARIE AND PIERRE CURIE

Pierre Curie (1859–1906) studied at the Sorbonne in Paris. In 1878, he became an assistant teacher in the physical laboratory there, and some years later, professor of physics. He was well known for his research on crystals and magnetism. Pierre and Marie Sklodowska were married in 1895 (she was 28 years old). After their marriage, Marie chose radioactivity for her doctoral research. In 1898, Pierre joined his wife in this work. Their collaboration was so successful that in 1903 they were awarded the Nobel Prize in physics, which they shared with Becquerel. Pierre Curie was run over and killed by a horse-drawn vehicle in 1906. Marie Curie was appointed to a professorship at the Sorbonne, the first woman there to have this post.

In 1911, Marie Curie was awarded the Nobel Prize in chemistry for the discovery of the two new elements, radium and polonium. She was the first person to win two Nobel Prizes in science. The rest of her career was spent in the supervision of the Paris Institute of Radium, a center for research on radioactivity and the use of radium in the treatment of cancer. During her visit to the U.S., a group of women

(a)

(b)

FIGURE 17.3 (a) Marie Curie. (b) Marie and Pierre Curie on a bicycling holiday. (*Continued*) presented her with the precious gifts of 1 gram of radium.

Marie Curie died in 1934 of leukemia, a form of cancer of the leukocyte-forming cells of the body, probably caused by overexposure to the radiations from radioactive substances during her research over more than three decades.

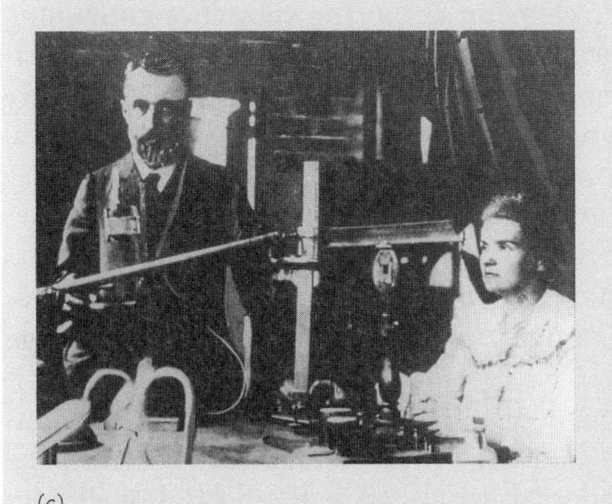

(c)

(d)

FIGURE 17.3 (*Continued*) (c) Marie and Pierre Curie in their laboratory; (d) Marie and Pierre Curie with their daughter, Irène, who later also received a Nobel Prize.

> We believe, therefore, that the substance which we removed from pitchblende contains a metal which has not yet been known, similar to bismuth in its chemical properties. If the existence of this new metal is confirmed, we propose to call it *polonium*, after the name of the native country of one of use.

Six months after the discovery of polonium, the Curies chemically separated another substance from pitchblende. They found the emission from it so intense as to indicate they had found yet another new element, even more radioactive than polonium! This substance had an activity per unit mass nine hundred times that of uranium and was chemically entirely different from uranium, thorium, or polonium. Spectroscopic analysis of this substance revealed spectral lines characteristic of the inactive element barium, but also a line in the ultraviolet region that did not seem to belong to any known element. The Curies reported their belief that the substance, "although for the most part consisting of barium, contains in addition a new element which produced radioactivity and, furthermore, is very near barium in its chemical properties." For this new element, so extraordinarily radioactive, they proposed the name *radium*.

The next step in making the evidence for the newly discovered elements more convincing was to determine their properties, especially their atomic masses. The Curies had made it clear that they had not yet isolated either polonium or radium in pure metallic form, or even yet obtained a pure sample of a compound of either element. From the substance containing the strongly radioactive substance that they called radium, they had separated a part consisting of barium chloride mixed with a presumably very small quantity of radium chloride. Additional separations by chemical means yielded an increasing proportion of radium chloride. The difficulty of this task is indicated by the Curies' remark that radium "is very near barium in its chemical properties," for it is very difficult to separate elements whose chemical properties are similar. Moreover, to obtain their highly radioactive substances in usable amounts, they had to start with a very large amount of pitchblende.

With an initial 100-kg shipment of pitchblende (from which the uranium salt had been removed to be used in the manufacture of glass) the Curies went to work in an abandoned woodshed at the School of Physics where Pierre Curie taught. Having failed to obtain financial support, the Curies made their preparations without technical help in this "laboratory." Marie Curie wrote later:

> I came to treat as many as twenty kilograms of matter at a time, which had the effect of filling the shed with great jars full of precipitates and liquids. It was killing work to carry the receivers, to pour off the liquids and to stir, for hours at a stretch, the boiling material in a smelting basin.

From the mixture of radium chloride and barium chloride they produced, only the average atomic mass of the barium and radium could be computed. At first an average value of 146 u was obtained, as compared to 137 u for the atomic mass of barium. After many additional purifications that increased the proportion of radium chloride, the average value for atomic mass rose to 174 u. (The unit u is the standard unit used today for the atomic mass, the hydrogen atom having an atomic mass of about 1 u.) Continuing the tedious purification process for 4 years, during which she treated several tons of pitchblende residue, Marie Curie was able to report, in July 1902, that she had isolated 0.1 g of radium chloride, so pure that spectroscopic examination showed no evidence of any remaining barium. She calculated the atomic mass of radium

The present yield of radium from 1 ton of high-grade uranium ore is about 0.2 g.

to be 225 u (the present-day value is 226.03 u). The activity of radium is more than a million times that of the same mass of uranium.

17.4 IDENTIFYING THE RAYS

Once the extraordinary properties of radium became known, they excited interest both inside and outside the scientific world, and the number of people studying radioactivity increased rapidly. The main question that attracted attention was: What are the mysterious radiations emitted by radioactive bodies?

In 1899, Ernest Rutherford, whose later development of the theory of the nuclear atom was discussed in Chapter 14, started to seek answers to this question. Rutherford found that a sample of uranium emits at least two distinct kinds of rays: one that is very readily absorbed, which he called for convenience α rays (alpha rays), and the other more penetrating, which he called β rays (beta rays). In 1900, the French physicist P. Villard observed that the emission from radium contained rays much more penetrating than even the β rays; this type of emission was given the name γ (gamma) rays. The penetrating power of the three types of rays, as known at the time, is compared in the table below, first published by Rutherford in 1903.

Thickness of Aluminum Required to Reduce the Radiation Intensity to One-half Its Initial Value

Radiation type	*Thickness of aluminum*
α	0.0005 cm
β	0.05 cm
γ	8 cm

Thus, the "Becquerel rays" were more complex than had been thought even before the nature of α, β, and γ rays was ascertained. Of the three kinds of rays, the α rays are the most strongly ionizing and the γ rays the least. The power of penetration is inversely proportional to the power of ionization. This is to be expected; the penetrating power of the α rays from uranium is low because they expend their energy very rapidly in causing intense ionization. Alpha (α) rays can be stopped—that is, almost all are absorbed—by about 0.0005 cm of aluminum, by a sheet of ordinary writing paper, or by a few centimeters of air. Beta (β) rays are completely stopped only after traveling many meters in air, or 0.05 cm in aluminum. Gamma (γ) rays can pass through many centimeters of aluminum or lead, or through a meter of concrete, before being almost completely absorbed. One consequence of these properties of the rays is that heavy and expensive shielding is sometimes needed in the study or use of radiations in accelerator and nuclear reactors, to protect people from the harmful effects of the rays. The rays ionize and, consequently, break down molecules in living cells, causing radiation "burns," fatal injuries to cells, damage that can lead to the growth of cancer cells, and dangerous mutations in the structure of the DNA molecules. In some cases, these "radiation shields" are as much as 3 m thick, but they are effective, protecting workers at reactor stations and accelerator research facilities.

17.5 THE CHARGE AND MASS OF THE RAYS

Another method used to study the rays emitted in radioactivity was to direct them through a magnetic field to see if they were deflected or deviated from their initial directions by the action of the field. This method came to provide one of the most widely used tools for the study of atomic and nuclear events. It is based on the now familiar fact that a force acts on a charged particle when it moves across a magnetic field. As was discussed in Section 10.11, this force always acts at right angles to the direction of motion of the charged particle. The particle experiences a continual deflection and, if sent into a uniform field at right angles, it moves along the arc of a circle.

This property had been used in the 1890s by J.J. Thomson in his studies of cathode rays. He showed that these rays consist of very small negatively charged particles, or electrons (Chapter 13). Becquerel, the Curies, and others found that the α, β, and γ rays behaved differently from one another in a magnetic field. The behavior of the rays is illustrated in Figure 17.4.

Suppose that some radioactive material, such as a sample of uranium, is placed at the end of a narrow hole in a lead block and that a narrow beam consisting of α, β, and γ rays escapes from the opening. If the beam enters a strong, uniform magnetic field (as in the last two drawings), the three types of rays will go along paths separated from one another. The γ rays continue in a straight line without any deviation. The β rays will be deflected to one side, moving in circular arcs of differing radii. The α rays will be deflected slightly to the other side, moving in a circular arc of large radius; they are rapidly absorbed in the air.

The direction of the deflection of the β rays in such a magnetic field is the same as that observed earlier in Thomson's studies of the properties of cathode rays. It was concluded, therefore, that the β rays, like cathode rays,

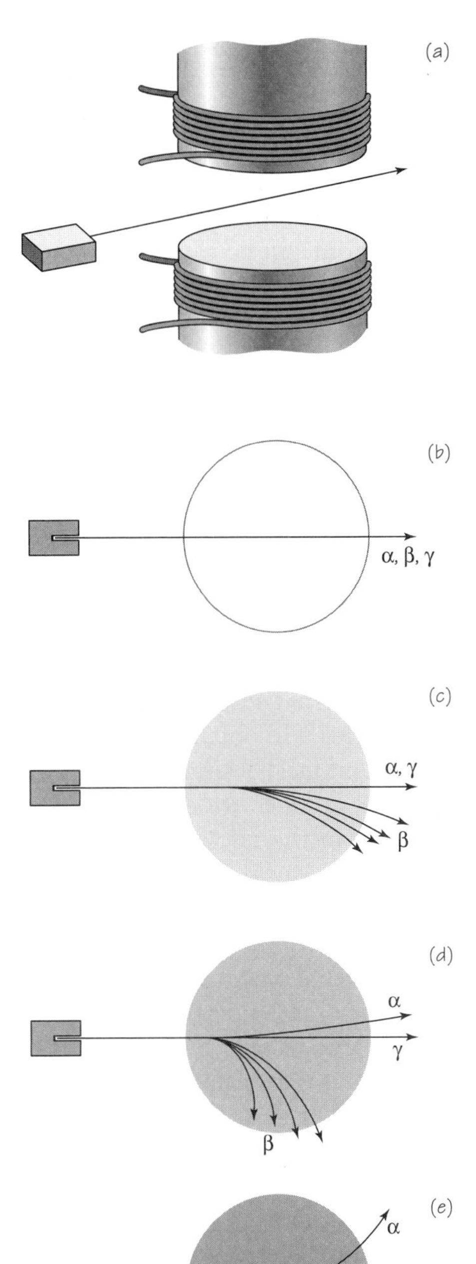

FIGURE 17.4 (a) Alpha, beta, and gamma rays are separated from a sample of radioactive material by their passage through a magnetic field; (b) no magnetic field present; (c) weak magnetic field; (d) stronger magnetic field; (e) very strong magnetic field.

consist of *negatively charged particles*. (The Curies confirmed the negative charge on the β particles in 1900; they caused the beam of the particles to enter an electroscope that became negatively charged.) Since the direction of the deflection of the α rays was opposite that of the β rays, it was concluded that the α rays consist of *positively charged particles*. Since the γ rays were not deflected at all, it was concluded that they are neutral, that is, they have no electric charge. Electromagnetic radiation is neutral, as are particles that carry equal amounts of positive and negative charge. No conclusion could be drawn from this type of experiment as to whether the γ rays are, or are not, particles.

The deflection of a charged particle in electric and magnetic fields depends on both its charge and its mass. Therefore, the ratio of charge (q) to mass (m_q) for β particles can be calculated from measured deflections in fields of known intensity. Becquerel, investigating β particles in 1900, used a procedure that was essentially the same as that used by J.J. Thomson in 1897 to obtain a reliable value for the ratio of charge to mass for the particles in cathode rays. At that time, the fact that a consistent single value of e/m_e had been found established quantitatively the *existence* of the electron (see Section 13.3). (Here e is the electron's charge, and m_e is its mass.) By sending β rays through electric and magnetic fields, Becquerel was able to calculate the speed of the β particles. He obtained a value of q/m_q for β particles which was close to that found by Thomson for the electron. This permitted the deduction that the *β particles are electrons*. (However, they should not be thought of as existing in the nucleus until emitted. Rather, they are generated during the emission.)

The nature of the α radiation was more difficult to establish. It was necessary to use a very strong magnetic field to produce measurable deflections of α rays. The value of q/m_q found for α particles (4.8×10^7 C/kg) was about 4000 times smaller than q/m_q for β particles. The reason for the small q/m_q value could be a small value of q or a large value of m. Other evidence available at the time indicated that magnitude of the charge q for an α particle was not likely to be smaller than that for a β particle. It was therefore concluded that mass m would have to be much larger for the α particle than for the β particle.

The value of q/m_q given above for α particles is just one-half that of q/m_q found earlier for a hydrogen ion. The value would be explained in a reasonable way if the α particle were like a hydrogen molecule minus one electron (H_2^+), or if it were a helium atom (whose mass was known to be about four times that of a hydrogen atom) without its two electrons (He^{++}). Other possibilities might have been entertained. In fact, however, the right identification turned out to be that of α particles with He^{++}. *The α particle was* *found to be the same as a helium nucleus* and, therefore, has a mass of about four atomic mass units. The clever experiment described in the following section provided the final proof.

17.6 RUTHERFORD'S "MOUSETRAP"

The gas helium was first discovered to exist on the Sun through spectroscopic analysis of the Sun's radiation (Section 14.1). Later, it was discovered that helium gas can be found on the Earth, imprisoned in radioactive minerals. In addition, William Ramsey and Frederick Soddy, working in the Cavendish Laboratory in Cambridge, had discovered, in 1903, that helium is given off from a radioactive compound, radium bromide. This led Rutherford, also in Cambridge, to advance the hypothesis that the α particle is a double ionized helium atom (He^{++}), that is, a helium atom minus its two electrons, or, as we would now say, the nucleus of a helium atom. In a series of experiments conducted from 1906 to 1909, Rutherford succeeded in proving the correctness of his hypothesis in several different ways. The last and most convincing of these experiments was made in 1909, with T.D. Royds, by constructing what James Jeans later called "a sort of mousetrap for α particles."

The experiment used the radioactive element radon (Rn). Pierre Curie and André Debierne had discovered radon in 1901. They had found that a radioactive gas is given off from radium. A small amount of the gas collected in this way was found to be a strong α emitter. They showed the gas to be a new element, which they called "radium emanation," later called "radon." Ramsey and Soddy then found that when radon is stored in a closed vessel, helium gas always appears in the vessel also. Thus, helium is given off not only by radium but also by radon.

Rutherford and Royds put a small amount of radon in a fine glass tube with a wall only 0.01 mm thick. This wall was thin enough so that α particles could pass through it, but radon itself could not. The tube was sealed into a thick-walled, outer glass tube that had an electric discharge section at the top (Figure 17.5). They pumped the air out of the outer tube and allowed the apparatus to stand for about a week. During this time, while α particles from the radon passed through the thin walls of the inner tube, a gas was found gradually to collect in the previously evacuated space. Then they pumped mercury in at the bottom of the apparatus to compress the very small quantity of gas and confine it in the discharge tube. When a potential difference was applied to the electrodes of the discharge tube, an

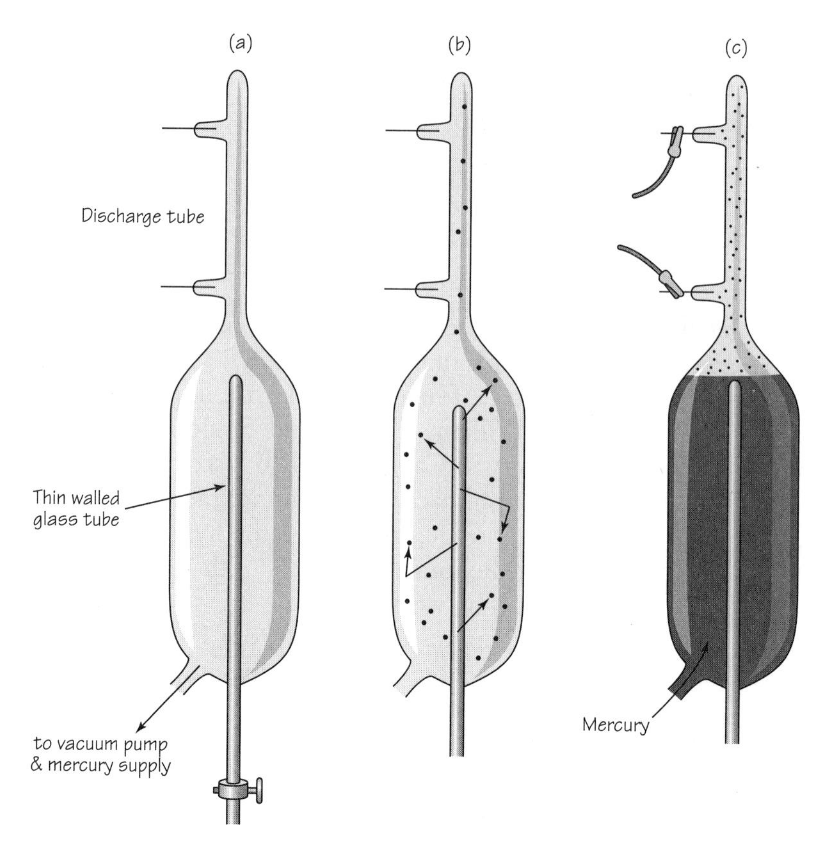

FIGURE 17.5 Rutherford's "mousetrap" for identifying particles.

electric discharge was produced in the gas. They examined the resulting light with a spectroscope, and they found that the spectral lines were characteristic of the element helium. (In a separate control experiment, helium gas itself was put in the inner, thin-walled tube and did not leak through the wall of the inner tube.)

Now it was clear to Rutherford how to interpret his results. He could safely conclude that the helium gas that collected in the outer glass tube was formed from α particles that had passed into the outer tube and picked up some electrons to form helium atoms. Rutherford's result implied conclusions more important than just the identity of α particles. Apparently, an atom of an element (radon) can spontaneously emit a fragment (an α particle) that is the nucleus of *another* element (helium)! This was a startling idea, but only the beginning of more startling things to come.

17.7 RADIOACTIVE TRANSFORMATIONS

The emission of α and β particles raised difficult questions with respect to existing ideas of matter and its structure. The rapid development of chemistry in the nineteenth century had made the atomic–molecular theory of matter highly convincing. According to this theory, a pure element consists of identical atoms, which are indestructible and unchangeable. But if a radioactive atom emits as substantial a fragment as an α particle, which was shown to be an ionized helium atom, can the radioactive atom remain unchanged? That did not seem plausible. Rather, it seemed that there must be a transformation in which the radioactive atom is changed to an atom of a different chemical element.

If an atom emits an α particle, a significant part of its mass—four atomic mass units—will be carried away by the α particle. What about the atoms that emit β particles? The β particle, which was shown to be an electron, is far less massive than the α particle. However, its mass is not zero; so a radioactive atom must also undergo some change when it emits a β particle. It was again difficult to escape the conclusion that radioactive atoms are, in fact, subject to division into two parts of markedly unequal mass, a conclusion contrary to the old, basic concept that the atom is indivisible.

Another fundamental question arose in connection with the energy carried by the rays emitted by radioactive substances. As early as 1903, Rutherford and Soddy, and Pierre Curie and a young coworker, A. Laborde, noted that a sample of radium kept itself at a higher temperature than its surroundings merely by absorbing some of the energy of the α particles emitted by atoms inside the sample. Curie and Laborde found that 1 g of radium can produce about 0.1 kcal of heat per hour. A sample of radium thus can continue to release energy year after year, and evidently for a very long time.

The continuing release of such a quantity of heat could not be explained by treating radioactivity as an ordinary chemical reaction. It was clear that radioactivity did not involve chemical changes in the usual sense. Energy was emitted by samples of pure elements; energy emission by compounds containing radioactive elements did not depend on the type of molecule in which the radioactive element was present. The origin of the production of heat had to be sought in some deep changes *within* the atoms of radioactive elements, rather than in chemical reactions among atoms.

Rutherford and Soddy proposed a bold theory of *radioactive transformation* to explain the nature of these changes. They suggested that when a radioactive atom emits an α or a β particle, it really breaks into two parts: the α or β particle that is emitted, and a heavy, leftover part that is physically and chemically different from the "parent" atom. There was a good deal of evidence for the last part of the assumption. For example, the formation of radon gas from radium was known, as mentioned earlier. When the atomic mass of radon was determined, it turned out to be smaller than that of radium by just four atomic mass units, the mass of an α particle.

The idea of radioactive transformation can be represented by an "equation" similar to the kind used to represent chemical reactions. For example, using the symbols Ra and Rn to represent atoms of radium and radon, and He to represent an α particle after it has picked up two electrons to form a helium atom, the transformation of radium into radon can be expressed as

$$\mathrm{Ra} \rightarrow \mathrm{Rn} + \mathrm{He}.$$

The process can be described as the transformation—also called a "disintegration," "decay," or "transmutation"—of radium into radon, with the emission of an α particle.

Many decay processes, in addition to the example just cited, had been found and studied by the Curies, by Rutherford and his coworkers, and by others, and these processes fitted easily into the kind of scheme proposed by Rutherford and Soddy (who received Nobel Prizes in chemistry for their work). For example, radon is radioactive also, emitting another α particle and thereby decaying into an atom of an element that was called "radium A" at the time. Radium A was later shown to be polonium (Po):

$$\mathrm{Rn} \rightarrow \mathrm{Po} + \mathrm{He}.$$

Polonium is also a radioactive solid. In fact, the original "parent" radium atoms undergo a series or chain of transformations into generation after generation of new, radioactive, so-called *daughter elements*, ending finally with a daughter element that is nonradioactive or, in other words, stable.

17.8 RADIOACTIVE DECAY SERIES

The decay of radium and its daughters was found eventually to lead to a stable end product that was identified by its chemical behavior as *lead*. The chain beginning with radium has 10 members, some emitting α particles and others emitting β particles. Some γ rays are emitted during the decay series, but γ rays do not appear alone; the are emitted only together with an α particle or a β particle.

Rutherford and Soddy suggested that, since radium is always found in uranium ores, such as pitchblende, which the Curies had analyzed, radium itself may be a member of a series starting with uranium as the ancestor of all the members. Research showed that this is indeed the case. Each uranium atom may in time give rise to successive daughter atoms, radium being the sixth generation and stable lead being the fifteenth.

The table shows all the members of the so-called *uranium–radium series.* The meaning of some of the symbols will be discussed in later sections. The number following the name of an element, as in uranium-238, indicates the atomic mass in atomic mass units. Notice that there are heavier and lighter varieties of the element, such as uranium-238 and uranium-235, polonium-218, 214, and 210. (More will be said about these varieties in Section 17.10.) Two other naturally occurring radioactive series have been found; one starts with thorium-232 and the other with uranium-235.

Uranium–Radium Decay Series

Substance	*Decay mode*	*Half-life*
Uranium-238	α, γ	4.51×10^9 yr
Thorium-234	β, γ	24.1 days
Protactinium-234	β, γ	1.18 min
Uranium-234	α, γ	2.48×10^5 yr
Thorium-230	α, γ	8.0×10^4 yr
Radium-226	α, γ	1620 yr
Radon-222	α, γ	3.82 days
Polonium-218	α	3.05 min
Lead-214	β, γ	26.8 min
Bismuth-214	α, β, γ	19.7 min
Polonium-214	α	1.64×10^{-4} s
Lead-210	β, γ	21.4 yr
Bismuth-210	β, γ	5.0 days
Polonium-210	α, γ	138.4 days
Lead-206	stable	stable

Each member of the series differs physically and chemically from its immediate parent above or daughter below; it should therefore be possible to separate the different members in any radioactive sample. This is by no means impossible to do, but the separation problem is made difficult by the fact that the different radioactive species decay at different rates, some very slowly, some rapidly, others at intermediate rates. These rates and their meaning will be discussed in the next section, but the fact that the rates differ gives rise to important effects that can now be discussed.

Two other naturally occurring radioactive series have been found; one starts with thorium-232 and the other with uranium-235.

Growth of Daughter Elements

An interesting effect is provided by that portion of the uranium–radium series that starts with the substance called polonium-218. A pure sample of polonium-218 may be collected by exposing a piece of ordinary material, such as a thin foil of aluminum, to the gas radon. Some of the radon atoms decay into polonium-218 atoms, which then stick to the surface of the foil. The graph in Figure 17.6 shows what becomes of the polonium-218. Polonium-218 (^{218}Po) decays into lead-214 (^{214}Pb), which decays into bismuth-214 (^{214}Bi), which decays into polonium-214 (^{214}Po), then lead-210 (^{210}Pb), then bismuth-210 (^{210}Bi), etc. If the original sample contains 1,000,000 atoms of ^{218}Po when it is formed, after 20 min it will contain about 10,000 ^{218}Po atoms, about 660,0000 ^{214}Pb atoms, about 240,000 ^{214}Bi atoms, and about 90,000 ^{210}Pb atoms. The number of ^{214}Po atoms is negligibly small because most of the ^{214}Po changes into ^{210}Pb in a small fraction of a second.

A sample of pure, freshly separated radium (^{226}Ra) would also change in composition in a complicated way, but much more slowly. Eventually it would consist of a mixture of some remaining radium-226, plus radon-222, polonium-218, lead-214, and all the rest of the members of the chain down to, and including, stable "radium G" (lead-206).

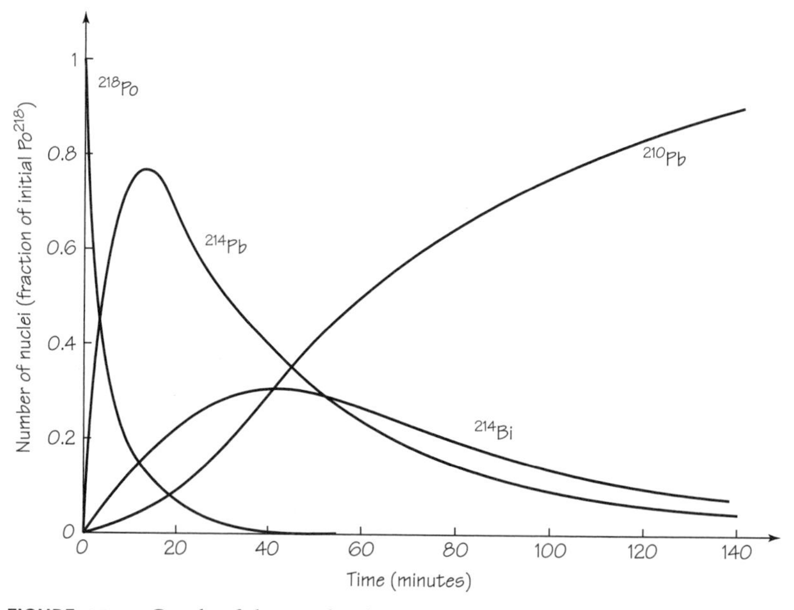

FIGURE 17.6 Graph of decay of polonium-218 with build-up of its daughter elements over time.

Similarly, a sample of pure uranium may contain, after a time, 14 other elements of which 13 (all but the last, stable portion) contribute to the radioactive emission, each in its own way. In all such cases, a complicated mixture of elements results. After starting as a pure α emitter, a sample eventually emits many α particles, β particles, and γ rays, apparently continuously and simultaneously.

It is evident that the separation of the different members of a radioactive chain from one another would be difficult, especially if some members of the chain decay rapidly. The determination of the chemical nature and the radioactive properties of each member required great experimental ingenuity. One successful method depended on the skillful chemical purification of a particular radioactive substance, as the Curies had done with radium and polonium. For example, suppose that a sample has been obtained from which all the radioactive atoms except those of radium-226 have been removed. The sample immediately starts to give off radon gas. The latter can be drawn off and its properties examined before it becomes seriously contaminated by the distintegration of many of its atoms into polonium-218. If this is done, it is found that radon decays (through several transformations) into lead much more quickly than radium decays into radon.

17.9 DECAY RATE AND HALF-LIFE

In the last section, we noted that of 1,000,000 ^{218}Po atoms present in a freshly prepared sample of that radioactive substance, only about 10,000 would remain after 20 min, the rest having decayed into atoms of ^{214}Pb and its daughter products. It would take only 3 min following the preparation of the pure sample of ^{218}Po for 50% of the atoms originally present in the sample to have decayed. In the case of radium (^{226}Ra), it would take 1620 years for half of the radium atoms in a freshly prepared sample of radium to be transformed into radon atoms.

These two examples illustrate the experimental fact that samples of radioactive elements show great difference in their rates of decay. These different rates are the result of *averages* of many individual, different decay events going on at random in a sample. Looking at *one* atom of any radioactive element, one never can tell when it will decay; some may decay as soon as they are produced, while others may never decay. Still, it has been found experimentally that for a large group of atoms of one kind, *the fraction of these atoms that decay per second is unchangeable* and always the same for any large group of atoms of that kind. This fraction is almost completely independent of all physical and chemical conditions, such as temperature, pressure, and form of chemical combination. These remarkable properties of radioactivity deserve special attention, and the meaning of the italicized statement above now will be discussed in detail because it is basic to an understanding of radioactivity.

Say, for example, that 1/1000 of the atoms in a freshly prepared pure sample decay during the first second. Then you would expect that 1/1000 of the remaining atoms will decay during the next second. But also, 1/1000 of the atoms remaining after 10 s will decay during the eleventh second, and so on. In fact, during any subsequent second of time, 1/1000 of the atoms remaining at the beginning of that second will decay, at least until the number of remaining atoms becomes so small that predictions become very uncertain.

Since the *fraction* of the atoms that decay per unit time is a constant for each element, the *number* of atoms that decay per unit time will decrease in proportion to the diminishing number of atoms that have not yet changed. Consequently, if the percentage of surviving, unchanged atoms is plotted as a function of time, a curve like the one in the sketch is obtained. The number of atoms in a sample that decays per unit time is termed the *activity* of the sample. Thus, the graph also represents the way in which the measured activity of a sample would decrease with time.

The curve that shows the number of atoms that have not decayed as a function of time approaches the time axis asymptotically; that is, the number of survivors becomes small, but only approaches zero. This is another way of saying that a definite "lifetime" in which all of the original atoms for a sample will have decayed cannot be assigned.

However, it is possible to specify the time required for any particular *fraction* of a sample to decay, one-half, one-third, or 37%, for instance. For convenience in making comparisons, the fraction one-half has been chosen. Rutherford called the time required for the decay of one-half of the original atoms of a pure sample the *half-life* (symbol $T_{1/2}$). Each kind of radioactive atom has a unique half-life, and thus the half-life of an element can be used to identify a radioactive element. As the table on p. 741 shows, a wide variety of half-lives have been found.

For ^{238}U, the parent of the uranium series, the half-life is 4.5 billion years. This means that after 4.5×10^9 yr, half of the ^{238}U atoms will have decayed. For ^{214}Po, the half-life is of the order of 10^{-4} s; that is, in only 1/10,000 of a second, half of an original sample of ^{214}Po atoms will have decayed. If pure samples of each, containing the same number of atoms, were available, the initial activity (atoms decaying per second) of ^{214}Po would be very strong and that of ^{238}U very feeble. If left for even 1 min, though, the polonium would have decayed so thoroughly and, therefore, the number of its surviving atoms would be so small, that at this point the

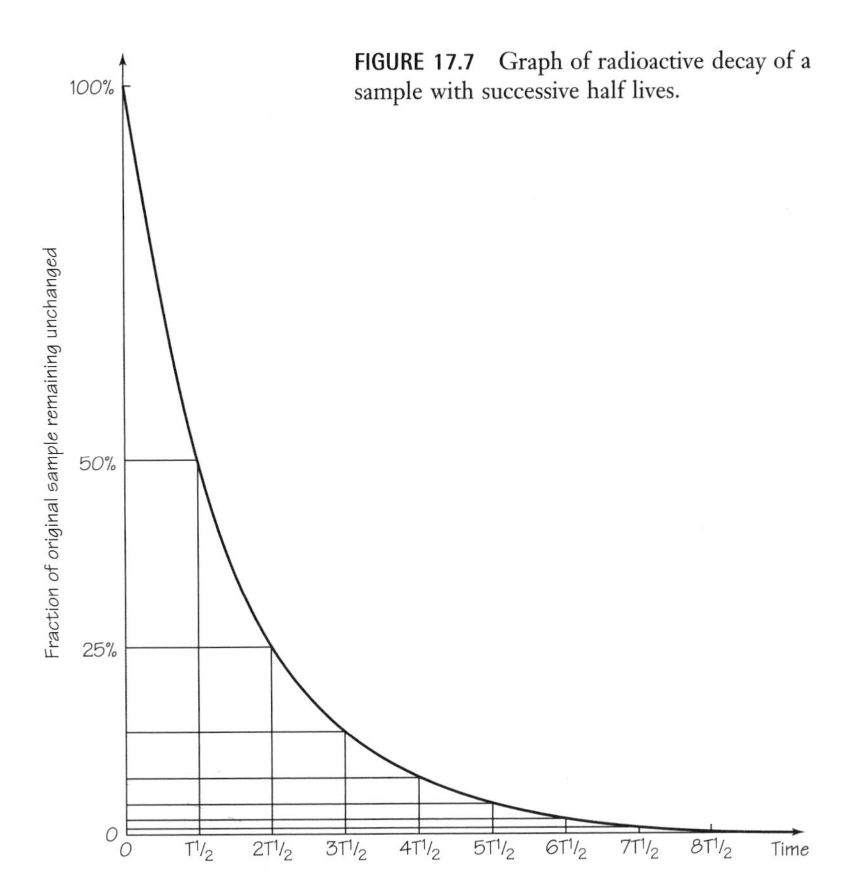

FIGURE 17.7 Graph of radioactive decay of a sample with successive half lives.

activity due to polonium would now be less than the activity of the uranium atoms. Perhaps some radioactive elements, present in great quantities long ago, decayed so rapidly that no measurable traces are now left. On the other hand, many radioactive elements decay so slowly that during any ordinary experimentation time the counting rates that indicate decay seem to remain constant. (That is why Becquerel did not notice any change in the activity of his uranium salt samples.)

The principal advantage of the concept of a half-life lies in the experimental result implied in the graph: For any element of half-life $T_{1/2}$, no matter how old a sample is, on average half of the atoms will still have survived after an additional time interval $T_{1/2}$. Thus, the half-life is not to be thought of as an abbreviation for "half a life." If one-half the original atoms remain unchanged after a time $T_{1/2}$, one-fourth, $\frac{1}{2} \times \frac{1}{2}$, will remain after two consecutive half-life intervals $2T_{1/2}$, one-eighth after $3T_{1/2}$, and so on. Note how different the situation is for a population of, say, human beings instead of radioactive atoms. In a group of N_0 babies, half the number may survive to their 70th birthday; of these $N_0/2$ senior citizens, none is likely to celebrate a 140th birthday. But of N_0 radioactive atoms with a half-life of 70 years, on average $N_0/4$ will have remained intact after 140 yr, $N_0/8$ after 210 yr, etc. To put it differently, *the statistical probability of survival for atoms is unchanged by the age they have already reached.* In humans, of course, the probability of survival (say, for another year) depends strongly on age, and so the concept "half-life" is not usable in this case.

We have been considering the behavior not of individual atoms, but of a very large number of them. As discussed in connection with the behavior of gases in Chapter 7, this method allows us to use laws of statistics to describe the average behavior of the group. If a hundred thousand people were to flip coins simultaneously just once, you could predict with good accuracy that about one-half of them would get heads. But you could not accurately predict that one particular person in this crowd would obtain heads on a single flip. If the total number of coins tossed is small (10), the observed count is likely to differ considerably from the prediction of 50% heads. From experiments in radioactivity, you can predict that a certain fraction of a relatively large number of atoms in a sample will survive in any given time interval (e.g., one-half will survive to reach the age $T_{1/2}$), but not whether a particular atom will be among the survivors. As the sample of survivors decreases in size owing to disintegrations, predictions become less precise. Eventually, when only a few unchanged atoms are left, you could no longer make useful predictions at all. In short, the disintegration law is a *statistical* law and is thus applicable only to large populations of the radioactive atoms. Moreover, it makes no assumptions as to *why* the atoms disintegrate.

In the discussion of the kinetic theory of matter, you saw that it is a hopeless and meaningless task to try to describe the motions of each individual molecule, but you could calculate the *average* pressure of a gas containing a very large number of molecules. Similarly, in dealing with radioactivity, the inability to specify when each of the tremendous number of atoms in a normal sample will disintegrate makes a statistical treatment necessary and useful.

17.10 THE CONCEPT OF ISOTOPES

The discovery that there are three radioactive series, each containing apparently new substances, created a serious problem. In 1910, there were still some empty spaces in the periodic table of the elements, but not enough spaces for the many new substances. The periodic table represents an arrangement of the elements according to their chemical properties, and, if it could not include the radioactive elements, it would have to be revised, perhaps in some drastic and fundamental way.

The clue to the solution of the puzzle lay in the observation that some of the newly found materials that were members of a radioactive series have *chemical* properties identical to those of well-known elements, although some of their *physical* properties are different. For example, what was then called Uranium II, the "great-granddaughter" of Uranium I, was found to have the same chemical properties as Uranium I itself. When both were mixed together, the two could not be separated by chemical means. No chemist has detected, by chemical analysis, any difference between these two substances. But the two substances, now known as uranium-238 and uranium-234, do differ from each other in certain physical properties. As the table on p. 741 shows, uranium-238 and -234 have quite different radioactive half-lives: 4.5×10^9 yr and 2.5×10^5 yr, respectively. The mass of a uranium-234 atom must be smaller than that of a uranium-238 atom by the mass of one α particle and two β particles.

Another pair of radioactive substances, called then radium B and radium G, were found to have the same chemical properties as lead; when mixed with lead they could not be separated from it by chemical means. These substances are now known as lead-214 and lead-206, respectively. Lead-214 is radioactive, and lead-206 is stable. The decay series on p. 741 indicates that the atoms must differ from each other in mass by the mass of two α particles and four β particles. There are many other examples of such physical differences among two or more radioactive substances with the same chemical behavior.

Soddy suggested a solution that threw a flood of light on the nature of matter and on the relationship of the elements in the periodic table. He proposed that a chemical element could be regarded as a pure substance only in the sense that all of its atoms have the same chemical properties; that is, a chemical element may in fact be a *mixture of atoms* having different radioactive behavior and different atomic masses, but all having the same chemical properties. This idea meant that one of the basic postulates of Dalton's atomic theory would have to be changed, namely, the postulate that the atoms of a pure element are alike in *all* respects. According to Soddy, it is only in chemical properties that the atoms of a given element are identical. The several physically different species of atoms making up a particular element occupy the same place in the periodic table, that is, have the same atomic number Z. Because of this, Soddy called them *isotopes* of the element, from the Greek words *isos* and *topos*, meaning *same* and *place*—the same place in the periodic table. Thus, uranium-238 (^{238}U) and uranium-234 (^{234}U) are isotopes of uranium ($_{92}U$); lead-214 (^{214}Pb) and

FIGURE 17.8 Frederick Soddy (1877–1956), an English chemist, studied at Oxford and went to Canada in 1899 to work under Rutherford at McGill University in Montreal. There he worked out his explanation of radioactive decay. Soddy returned to England in 1902 to work with William Ramsay, the discoverer of the rare gases argon, neon, krypton, and xenon. Ramsay and Soddy showed, in 1903, that helium was continuously produced by naturally radioactive substances. In 1921, Soddy was awarded the Nobel Prize in chemistry for his discovery of isotopes. He was a professor of chemistry at Oxford from 1919 to 1936.

lead-206 (^{206}Pb) are isotopes of lead ($_{82}Pb$). They are *chemically the same*; they occupy the same place on the periodic table and have the same atomic number Z, given by the subscript. But they are *physically different*, because they have different atomic masses A, given by the superscript in atomic mass units.

With this idea in mind, the many species of radioactive atoms in the three radioactive series were soon shown by chemical analysis to be isotopes of one or another of the last 11 naturally occurring elements in the periodic table, from lead ($Z = 82$) to uranium ($Z = 92$). For example, the 2nd and 5th members of the uranium series (see the table on p. 741) were shown to be isotopes of thorium, with $Z = 90$; the 8th, 11th, and 14th members turned out to be isotopes of polonium ($Z = 84$). The old names and symbols given to the members of radioactive series upon their discovery were therefore rewritten to represent both the chemical similarity and physical difference among isotopes. The present names for uranium X_1 and ionium, for example, are thorium-234 and thorium-230 (as shown in the table, p. 741).

A modern "shorthand" form for symbolizing any species of atom, called a *nuclide*, is also given in the same table—for example, $^{234}_{90}Th$ and $^{230}_{90}Th$ for two of the isotopes of thorium. The subscript (90 in both cases for thorium) is the atomic number Z, the place number in the periodic table; the superscript (234 or 230) is the mass number A, the approximate atomic mass in atomic mass units.

Note that, when writing the symbol for a nucleus, the atomic mass is always given as a whole number (such as U-238); but the atomic mass of an element as given in the periodic table (see color plate 6) is in most cases a decimal fraction (such as Uranium, 238.03). This is because the atomic mass given in the periodic table refers to the relative mass of the element in its natural state, which is a mixture of the various naturally occurring isotopes of the element. The atomic mass of the natural element is then an average of the atomic masses of the individual isotopes, weighted according to their abundance in relation to the other isotopes. For example, natural hydrogen has 1.0080 atomic mass units. It is a mixture of the abundant isotope $^{1}_{1}H$ and the much less abundant isotopes $^{2}_{1}H$ and $^{3}_{1}H$. (However, the masses of these atoms are *not exactly* 2.0000 u and 3.0000 u, respectively, because of the mass of the electron orbiting the nucleus in the neutral atom.)

17.11 TRANSFORMATION RULES

Two questions then arose: How do changes in chemical nature come about as an atom undergoes radioactive decay? More specifically, what determines whether the atomic number Z increases or decreases in a given radioactive transformation?

In 1913, Soddy in England and A. Fajans in Germany answered these questions independently. They each proposed two rules that systematized all the relevant observations for natural radioactivity. They are called the *transformation rules of radioactivity*. By 1913 Rutherford's nuclear model of the atom was generally accepted. Using this model, one could consider a radioactive atom to have an unstable nucleus that emits an α or a β particle (sometimes with emission of a γ ray). Every nucleus has a positive charge given by Ze, where Z is the atomic number and e is the magnitude of the charge of an electron. The nucleus is surrounded by Z electrons that make the atom as a whole electrically neutral and determine the chemical behavior of the atom. As discussed above, an α particle has an atomic mass of about four units and a positive charge of two units, $+2e$. A β particle has a negative charge of one unit, e, and very little mass compared to an α particle.

With this information in mind, the transformation rules may now be stated as follows:

1. When a nucleus emits an α particle, the mass of the atom decreases by four atomic mass units, and the atomic number Z of the nucleus decreases by two units; the resulting atom belongs to an element two spaces back in the periodic table (see color plates).
2. When a nucleus emits a β particle, the mass of the atom is changed very little, but the atomic number Z increases by one unit; the resulting atom belongs to an element one place forward in the periodic table. (We will discuss, in Section 18.10, how this behavior can be used to produce nonnaturally occurring elements beyond uranium, that is, so-called transuranium elements.)
3. When only a γ ray is emitted, there is no change in the number corresponding to the atomic mass, and none in the atomic number.

The table of the uranium–radium series (p. 741) shows how these rules apply to that series, at least as far as the atomic number is concerned.

These rules, applied to the Rutherford–Bohr model of the atom, help to explain why a change in chemical nature occurs as a result of α or β emission. Emission of an α particle takes two positive charges from the nucleus and four atomic mass units from the atom. An example is the following:

$$^{218}_{84}\text{Po} \rightarrow {}^{214}_{82}\text{Pb} + \alpha.$$

The resulting new atom (^{82}Pb) with its less positive nucleus can hold in its outer shells two fewer electrons than before, so two excess electrons drift away. The chemical behavior of atoms is controlled by the number of electrons; therefore, the new atom acts chemically like an atom of an element with an atomic number two units *less* than that of the parent atom.

On the other hand, in the case of β emission, the nucleus, and with it the whole atom, becomes *more* positively charged, by one unit. An example is the following:

$$^{234}_{90}\text{Th} \rightarrow {}^{234}_{91}\text{Pa} + \beta.$$

The number of electrons that the atom can hold around the nucleus has increased by one. After it has picked up an extra electron to become neutral again, the atom acts chemically as an atom with an atomic number one unit greater than that of the atom before the radioactive change occurred.

By using the transformation rules, Soddy and Fajans were able to determine the place in the periodic table for every one of the substances (or nuclides) in the radioactive series; no revision of the existing periodic table was needed. Many of the nuclides between $Z = 82$ (lead) and $Z = 92$ (uranium) are now known to contain several isotopes each. These results were expected from the hypothesis of the existence of isotopes, but direct, independent evidence was also sought and obtained in 1914.

17.12 SUMMARY OF NOTATION FOR NUCLIDES AND NUCLEAR REACTIONS

We summarize and recapitulate some of the above ideas and notations. (See the next section for some applications.) Because of the existence of isotopes, it is no longer possible to designate an atomic species only by means of the *atomic number* Z. To distinguish among the isotopes of an element some new symbols were introduced. One is the *mass number*, A, defined as the whole number closest to the measured atomic mass. For example, the lighter and heavier isotopes of neon are characterized by the pairs of values: $Z = 10$, $A = 20$, and $Z = 10$, $A = 22$. (An element that consists of a single isotope can, of course, also be characterized by its Z and A values.)

These values of Z and A are determined by the properties of the atomic nucleus. According to the Rutherford–Bohr model of the atom, the atomic number Z is the magnitude of the positive charge of the nucleus in elementary charge units. The mass number A is very nearly equal to (but a bit less than) the atomic mass of the atom, because the total mass of the electrons around the nucleus is very small compared to the mass of the nucleus.

The term *nuclide* is used to denote an atomic species characterized by particular values of Z and A. An *isotope* is then *one* of a group of two or more nuclides, each having the same atomic number Z but different mass numbers A. A radioactive atomic species is called a *radioactive nuclide*, or *radionuclide* for short. A nuclide is usually denoted by the chemical symbol with a subscript at the lower left giving the atomic number and a superscript at the upper left giving the mass number. Thus, in the symbol ${}^{A}_{Z}\mathrm{X}$ for a certain nuclide, Z stands for the atomic number, A stands for the mass number, and X stands for the chemical symbol. For example, ${}^{9}_{4}\mathrm{Be}$ is the nuclide beryllium with atomic number 4 and mass number 9; the symbols ${}^{20}_{10}\mathrm{Ne}$ and ${}^{22}_{10}\mathrm{Ne}$ represent the neon isotopes discussed above. Since the Z value is the same for all the isotopes of a given element, it is often omitted, except when needed for balancing equations (as you will shortly see). Thus, you can write ${}^{20}\mathrm{Ne}$ for ${}^{20}_{10}\mathrm{Ne}$, or ${}^{238}\mathrm{U}$ for ${}^{238}_{92}\mathrm{U}$.

The introduction of the mass number and the symbol for a nuclide makes it possible to represent radioactive nuclides in an easy and consistent way.

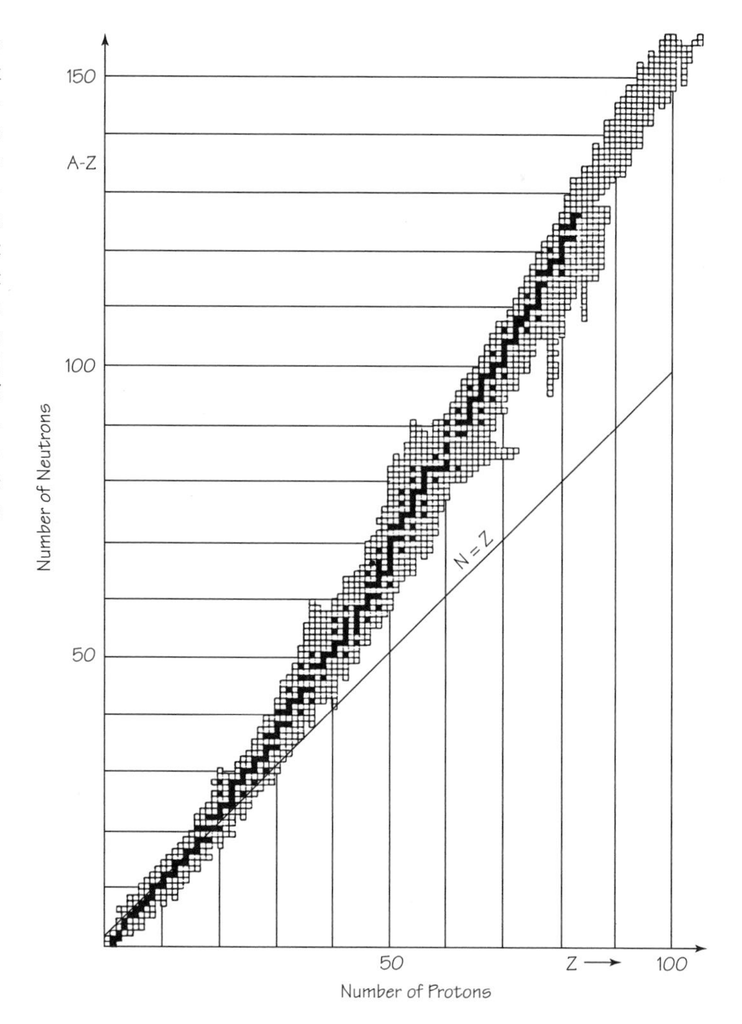

FIGURE 17.10 A chart of the known nuclides. Each black square represents a stable natural nuclide. Each open square represents a known, unstable nuclide, with only a small number of these found naturally (the rest being artificial). Note that all isotopes of a given element are found in a vertical column centered on the element's atomic number, Z. As we will see in the next chapter, the Z number is the number of protons in the nucleus, and the difference between the atomic mass and the atomic number is the number of neutrons.

In addition, radioactive decay can be expressed by a simple "equation" representing the changes that occur in the decay process. For example, the first step in the uranium–radium series, namely, the decay of uranium-238 into thorium-234, may be written

$$^{238}_{92}\mathrm{U} \rightarrow\, ^{234}_{90}\mathrm{Th} + \,^{4}_{2}\mathrm{He}.$$

The symbol $^{4}_{2}\mathrm{He}$ stands for the helium nucleus (α particle); the other two symbols represent the initial and final atomic nuclei, each with the appropriate charge and mass number. The arrow stands for "decays into." The "equation" represents a nuclear *reaction* and is analogous to an equation for a chemical reaction. The atomic numbers on the two sides of the equation must balance because the electric charge of the nucleus must be conserved: In the example above, 92 = 90 + 2. Also, the mass numbers must balance because of conservation of mass: 238 = 234 + 4.

For another example, in the table of the uranium–radium series, $^{234}_{90}\text{Th}$ (thorium-234) decays by β emission, becoming $^{234}_{91}\text{Pa}$ (protactinium-234). Since a β particle (electron) has charge $-e$ and has an extremely small mass, the symbol $^{0}_{-1}e$ is used for it. This β-decay process may then be represented by the equation

$$^{234}_{90}\text{Th} \rightarrow {}^{234}_{91}\text{Pa} + {}^{0}_{-1}e + {}^{-0}_{0}\bar{\nu}.$$

(The $^{-0}_{0}\bar{\nu}$ 0 is another particle, a so-called antineutrino, given off in β decay. It will be discussed in Chapter 18.)

17.13 SOME USEFUL APPLICATIONS OF RADIOACTIVITY

The dangers of exposure to radioactivity were noted briefly in Section 17.4. The "fall-out" of radioactive dust and elements from testing of nuclear weapons on or above ground during the 1950s was so harmful to the public that an international treaty to stop such testing was made. For example, strontium-90 is a radioactive isotope that is produced in fission reactions that can be transported into the upper atmosphere by above-ground nuclear explosions. The element strontium is just below calcium on the periodic table. When strontium-90 falls to the ground as fall-out from nuclear testing, it is ingested by cows eating grass and can replace calcium in the formation of milk, thus getting into the food chain where it can damage the internal organs of children and adults who drink the milk.

These processes of radiation damage to biological organisms are the subject of considerable research today, and they have important applications in agriculture, medicine, and other areas. One important area of research involves the study to find out, for example, how radiation produces genetic changes. Since it has been discovered that many of the key chemical processes in cells are organized by single chains of molecules, including DNA, it is clear that a single particle of radiation can, by breaking a chemical bond in such a chain, cause a permanent and perhaps disastrous change in the cell.

The metabolism of plants and animals is being studied with the aid of extremely small amounts of radioactive nuclides called *isotopic tracers*. A radioactive isotope, for example, ^{14}C, acts chemically (and therefore physiologically) like a stable isotope (^{12}C). Thus, by following a radioactive tracer with detectors, the behavior of a chemical material can be followed as it goes through various metabolic processes. The role of micronutrients (elements that are essential, in extremely small amounts, for the well-being of plants and animals) can be studied in this way. Agricultural experiments with fertilizers containing radioactive isotopes have shown at what point in the growth of a plant the fertilizer is essential. In chemistry, radioactive isotopes help in the determination of the details of chemical reactions and of the structure of complex molecules, such as proteins, vitamins, and enzymes.

Perhaps the most rewarding uses of radioisotopes have been in medical research, diagnosis, and therapy. For example, tracers can help to determine the rate of flow of blood through the heart and to the limbs, thus aiding in the diagnosis of abnormal conditions. Intense doses of radiation can do serious damage to all living cells, but diseased cells are often more easily damaged than normal cells. Radiation can therefore be used to treat some diseases, for example, destroying cancerous tumors. Some parts of the body take up particular elements preferentially. For example, the thyroid gland absorbs iodine easily. Specially prepared radioisotopes of such elements can be administered to the victims of certain diseases, thus supplying desired radiation right at the site of the disease. This method has been used in the treatment of cancer of the thyroid gland, blood diseases, and brain tumors and in the diagnosis of thyroid, liver, and kidney ailments. To destroy a malignancy in a prostate gland, "seeds" containing radioactive materials may be inserted into it.

Some Typical Isotope Applications

Isotope	*Half-life*	*Important uses*
${}^{3}_{1}H$	11 yr	Used as a tag in organic substances.
${}^{14}_{6}C$	5730 yr	Used as a tag in studying the synthesis of many organic substances. When ${}^{14}_{6}C$ is incorporated in food material, its presence can be traced in the metabolic products.
${}^{24}_{11}Na$	15 hr	Useful in a wide variety of biochemical investigations because of its solubility and chemical properties.
${}^{32}_{15}P$	14 days	For the study of bone metabolism, the treatment of blood diseases and the diagnosis of tumors.
${}^{35}_{16}S$	87 days	Has numerous chemical and industrial applications.
${}^{60}_{27}Co$	5.3 yr	Because of its intense γ emission, may be used as a low-cost substitute for radium in radiography and therapy.
${}^{131}_{53}I$	8 days	For the study of thyroid metabolism and the treatment of thyroid diseases.

One of the most well-known applications of radioactivity is in the determination of the age of organic and other materials. The rare isotope of carbon known as carbon-14 is often used for this purpose. This isotope has a half-life of 5730 years. The more abundant isotope of carbon is carbon-12, which is stable. All living things contain the element carbon, and they all absorb carbon from their environment while they are alive. Most of the carbon is in stable form, but a known fraction of it is in the form of carbon-14. When the living thing, such as a muscle or tree, dies, it ceases to absorb any new carbon, while the carbon-14 in its cells undergoes radioactive decay. As time advances, the amount of carbon-14 in the dead tree, or a piece of wood from the tree, decreases owing to its radioactive decay. By comparing the amount of carbon-14 remaining to the amount that is normally found in living trees of that species, scientists can determine its approximate age by referring to the half-life. For instance, if only one-half the original amount of carbon-14 remains, then the tree is about 5730 yr old.

This was the method used to determine the age of the so-called "Ice Man" in Europe, a prehistoric man who was found after having been frozen for centuries in the ice of a glacier in the Alps. The glacier eventually melted, revealing the dead man's body. Scientists determined that the amount of carbon-14 remaining in his body and in the objects found with him was a little more than half of what it would be if he were alive. Thus they placed the time of his death at about 5000 yr ago; it is the oldest preserved human body ever found.

Radioactive carbon-14 has been used with great success to date once-living materials, but it is limited to time spans in the thousands of years. This is because the more half-lives that pass by, the fewer the number of carbon-14 atoms that remain. So the statistical error increases greatly. For much longer time scales, on the order of millions or even billions of years, scientists have turned to the uranium–radium decay series itself. You will notice that the ultimate "ancestor" of this series, uranium-238, has a half-life of about 4.5 billion years. This is also the approximate age of the Earth, so there is still about one-half of the uranium left on Earth compared to the amount there was at Earth's formation some 4.5 billion years ago. (Much of the original uranium may have contributed to the warming of the Earth in the early years, and in fact the total radioactive amount in the Earth has kept it from being much less cold than it otherwise would have been after its formation.) Enough time has passed for the entire uranium–radium series to have become active and so build up a substantial amount of the end product, the stable lead isotope, lead-206. However, lead-208 is the more common isotope of lead, while lead-206 arises only from the uranium–radium decay series. So the amounts of uranium-238 and lead-206 present in ancient rocks in which prehistoric creatures, such as dinosaurs, are found can be used to determine the approximate age of the rocks, hence the approximate age of the dinosaur fossils found in them.

SOME NEW IDEAS AND CONCEPTS

accelerator
activity
alpha rays
beta rays
carbon dating
daughter element
fluorescence
gamma rays
half-life
isotope
isotopic tracers
nuclide
phosphorescence
radioactive transformation
radioactivity
radionuclide
uranium–radium series

SOME FURTHER READINGS

E. Curie, *Madame Curie: A Biography*, V. Sheean, transl. (New York: Da Capo Press, 1986).

G. Holton and S.G. Brush, *Physics, The Human Adventure* (Piscataway, NJ: Rutgers University Press, 2001), Chapter 27.

S. Quinn, *Marie Curie: A Life* (New York: Perseus Press, 1996).

T. Trenn, *The Self-Splitting Atom: The History of the Rutherford–Soddy Collaboration* (London: Taylor and Francis, 1977).

Web Site

Marie Curie: http://www.aip.org/history/curie/contents.htm.

STUDY GUIDE QUESTIONS

17.1 Questions about the Nucleus

1. Why was it difficult to learn about the interior structure of the nucleus?
2. What tools were needed to understand the nucleus?

17.2 Becquerel's Discovery

1. Why was Becquerel experimenting with a uranium compound? Describe his experiment.

2. How did uranium compounds have to be treated in order to emit the "Becquerel rays"?
3. What were the properties of the "Becquerel rays"?
4. In what ways were their properties similar to those of X rays?
5. Why couldn't the "Becquerel rays" be X rays?

17.3 The Curies Discover Other Radioactive Elements

1. How is the radioactive emission of an element affected by being combined into different chemical compounds?
2. Why did the Curies suspect the existence of another radioactive material in uranium ore, in addition to uranium itself?
3. What was the main difficulty in producing a pure sample of the element radium?
4. Why did the Curies conclude that radioactivity is related to the presence of atoms of uranium or thorium in a sample?
5. Make a list of at least a dozen of the elements mentioned in this section, then find them on the periodic table.

17.4 Identifying the Rays

1. List α, β, and γ rays in order of penetrating ability.
2. Why is penetrating power inversely related to ionizing power?
3. Why is shielding required when doing research on radioactive materials?
4. What damage can radioactivity cause to living tissue?

17.5 The Charge and Mass of the Rays

1. What was the evidence to support the theory that β particles are electrons?
2. What observation led to the suggestion that α particles are much more massive than β particles?
3. What evidence led to the tentative conclusion that α particles are helium nuclei?
4. What evidence was available to identify the γ rays? Why was it inconclusive?

17.6 Rutherford's "Mousetrap"

1. What hypothesis were Rutherford and Royds investigating?
2. Briefly describe their experiment.
3. How did Rutherford determine that the gas that appeared in the tube was helium?

17.7 Radioactive Transformations

1. Why was radioactive decay believed not to be an ordinary chemical reaction?
2. What was Rutherford's and Soddy's bold theory of radioactive transformation? Why was it considered bold at the time?

3. Give an example of a radioactive transformation. Why is it contrary to the ideas of nineteenth-century chemistry?

17.8 Radioactive Decay Series

1. Give at least three reasons for the difficulty of separating decay products.
2. If you start with a sample made entirely of pure uranium-238 atoms, what emission is observed at the start? How will the emission change as time goes on?
3. Look back at the discoveries the Curies made in their study of pitchblende (Section 17.3). How would you account for what they found?
4. What else might the Curies have discovered in pitchblende?
5. Why is it difficult to separate the decay products of uranium?

17.9 Decay Rate and Half-Life

1. Why can one not specify the "lifetime" of a sample of radioactive atoms? of a single radioactive atom?
2. A laboratory starts with a certain amount of a radioactive substance. What fraction of the substance will be left unchanged after a period equal to four times its half-life?
3. Another group in the laboratory has a different amount of the same radioactive substance. What fraction of their original amount will be left unchanged after a period equal to four times its half-life?
4. If, after many half-lives, only two atoms of a radioactive substance remain, what will happen during an additional period equal to one half-life of the substance?
5. How does the concept of statistical probability of survival for atoms differ from that for the survival of humans?

17.10 The Concept of Isotopes

1. Why was it not necessary, after all, to expand the periodic table to fit in the newly discovered radioactive substances?
2. The symbol for the carbon-12 nuclide is $^{12}_{6}C$. What is the approximate atomic mass of carbon-12 in atomic mass units? What is its position in the list of elements?
3. On the periodic table the atomic mass of the element carbon is given as 12.011. Why is this mass different from the atomic mass of the carbon-12 nuclide? What does this number say about the other nuclides of carbon?

17.11 Transformation Rules

1. By how many units does the mass of an atom change during α decay? during β decay?
2. By how many units does the charge of a nucleus change during α decay? during β decay?

3. The following α, β, and γ decays occur. Using the transformation rules, determine the daughter nuclide in each case:
   $^{234}_{92}U \rightarrow ? + \alpha + \gamma,$
   $^{210}_{82}Pb \rightarrow ? + \beta + \gamma.$

17.12 Summary of Notation for Nuclides and Nuclear Reactions

1. Write the complete symbol for the nuclide with atomic mass 194 and atomic number 78. Of which element is it an isotope?
2. Complete the following equation for α decay by replacing the question mark (?). Tell what law or rule you relied on. (X and Y are two different unspecified elements.)
   $^{A}_{Z}X \rightarrow {}^{4}_{2}He + {}^{?}_{Z-2}Y$
3. In the same way, complete the following equation for β decay by replacing the question mark (?):
   $^{A}_{Z}X \rightarrow {}^{0}_{-1}e + {}^{A}_{?}Y + {}^{-0}_{0}\bar{\nu}.$

17.13 Some Useful Applications of Radioactivity

1. Scientists have discovered the remains of an extinct mammal. They measured the amount of carbon-14 remaining in its bone structure and found that it is only one-fourth the amount in the bones of similar living creatures. How old did they estimate the mammal's remains to be?
2. Why can't the carbon-dating method be used to date the ages of dinosaur fossils?

DISCOVERY QUESTIONS

1. Radon gas is known to build up in the basements of some homes. Since radon is a strong α-emitter, it can cause damage to living tissue. It may be one cause of lung cancer. What types of locations would be likely to have such a problem? From the table of half-lives in the uranium–radium decay series, what would you recommend might be done about this?
2. Suppose that after the Curies had isolated polonium and radium in uranium ore (pitchblende), they had found another radioactive substance. What steps might they take to identify whether it is a new element or a previously known element?
3. In researching the properties of an element, a scientist observes emissions from the element. How could he/she determine if these emissions are from the nucleus or from the electron region of the atom?
4. After performing the research in answer to Question 3 the scientist determines that the emissions are coming from the nucleus. How might he/she identify these emissions?

5. Why does the uranium–radium decay series start with the element with the highest atomic number and end at the element with the lowest atomic number of the series?
6. Soddy's proposal of the existence of isotopes meant that for some elements not all the atoms are identical. Explain why this proposal does not require that the atoms of a given element show differences in chemical behavior.
7. A scientist is researching a series of radioactive decays and obtains an unknown element. Explain how he/she could determine whether the element is a new element and should be given a new place in the periodic table, or is simply an isotope of an already known element.
8. Refer to the table showing the uranium–radium decay series. Look at the parent and daughter nuclei of each of the three types of decays—alpha, beta, and gamma. Write down in your own words the general rules about what happens to the atomic mass and atomic number in each type of decay.
9. A scientist has a lump of pure lead-214 in a container. He determines that the lump contains 6.02×10^{23} atoms. The half-life is 26.8 min. After 26.8 min, approximately how many atoms of lead-214 are left? Why is this only approximate?
10. At the National Bureau of Standards, in 1932, 3.8 l of liquid hydrogen was evaporated slowly until only about 1 g remained. This residue allowed the first experimental check on the existence of the "heavy" hydrogen isotope 2H.
    (a) With the help of the kinetic theory of matter, explain why the evaporation should leave in the residue an increased concentration of the isotope of greater atomic mass.
    (b) Why should the evaporation method be especially effective with hydrogen?
11. Supply the missing data indicated by these transformation equations:
    (a) ${}^{212}_{?}\mathrm{Pb} \rightarrow {}^{212}\mathrm{Bi} + ?$.
    (b) ${}^{212}_{?}\mathrm{Bi} \rightarrow ? + {}^{0}_{-1}e$.
    (c) $? \rightarrow {}^{208}\mathrm{Pb} + {}^{4}_{2}\mathrm{He}$.
12. For each part below, select the most appropriate radiation(s): α, β, or γ:
    (a) most penetrating radiation;
    (b) most easily absorbed by aluminum foil;
    (c) most strongly ionizing radiation;
    (d) may require thick "radiation shields" for protection;
    (e) cannot be deflected by a magnet;
    (f) largest radius of curvature when traveling across a magnetic field;
    (g) highest q/m value;
    (h) important in the Rutherford-Royd "mousetrap" experiment;
    (i) involved in the transmutation of radium to radon;
    (j) involved in the transmutation of bismuth-210 to polonium-210.
13. A Geiger counter shows that the rate of emission of β particles from an initially pure sample of a radioactive element decreases to one-half the initial rate in 25 hr.
    (a) What fraction of the original number of radioactive atoms is still unchanged at that time?

(b) What fraction of the original number will have disintegrated in a total of 50 hr?
(c) What assumptions have you made in giving these answers? How might you check them?

Quantitative

1. A scientist has obtained 100 g of bismuth-210, which has a half-life of 5.0 days. Bismuth-210 decays into polonium-210, which has a half-life of 138.4 days. Bismuth-210 decays into lead-206, which is stable. The scientist leaves the bismuth in its container for 140 days. After that time:
(a) approximately how much bismuth-210 is left?
(b) how much polonium-210 is present?
(c) how much lead-206 is present?

CHAPTER 18

The Nucleus and Its Applications

18.1 THE PROBLEM OF NUCLEAR STRUCTURE

The discoveries of radioactivity and isotopes were extraordinary advances. And as usual, they also raised new questions about the structure of atoms, questions that involved the atomic nucleus. We saw in Chapter 17 that the transformation rules of radioactivity could be understood in terms of the Rutherford–Bohr model of the atom. But that model said nothing about the nucleus other than that it is small, has charge and mass, and may emit an α or a β particle. This implies that the nucleus has a structure that changes when a radioactive process occurs. The question arose: Can a theory or model of the atomic nucleus be developed that will explain the facts of radioactivity and the existence of isotopes?

The answer to this question makes up much of *nuclear physics*. The problem of nuclear structure can be broken down into two questions:

(1) What are the building blocks of which the nucleus is made?
(2) How are the nuclear building blocks put together?

The attempt to solve the problem of nuclear structure, although still a frontier activity in physics today, has already led to many basic discoveries and to large-scale practical applications. It has also had important social and political consequences, stretching far beyond physics into the life of society in general, as this text has frequently noted in its earlier chapters.

18.2 THE PROTON–ELECTRON HYPOTHESIS

The emission of α and β particles by radioactive nuclei suggested that a model of the nucleus might be constructed by starting with α and β particles as building blocks. Such a model would make it easy to see, for example, how a number of α particles could be emitted, in succession, in a radioactive series. But not all nuclei are radioactive, nor do all nuclei have masses that are multiples of the α-particle mass. For example, the nucleus of an atom of the lightest element, hydrogen, with an atomic mass of one unit (two units in the case of the heavy isotope), is too light to contain an α particle; so is the light isotope of helium, ${}^{3}_{2}\mathrm{He}$.

A positively charged particle with mass of one unit would seem to be more satisfactory as a nuclear building block. Such a particle does indeed exist: the nucleus of the common isotope of hydrogen, ${}^{1}_{1}\mathrm{H}$. This particle has been named the *proton*, from the Greek word *protos* for "first." Following the Rutherford–Bohr theory of atomic structure, the hydrogen atom thus consists of a proton with a single electron revolving around it.

In the preceding chapter we discussed the experimental result that the atomic masses of the nuclides are very close to whole numbers; hence, the nuclides are written in symbols with whole-number values for A. This result, together with the properties of the proton (e.g., its single positive charge) made it appear possible that all atomic nuclei are made up of protons. Could a nucleus of mass number A consist of A protons? If this were the case, the charge of the nucleus would be A units, but, except for hydrogen, the nuclear charge Z is found to be always less than A, usually less than $\frac{1}{2}A$. To get around this difficulty, it was assumed early that in addition to the protons, atomic nuclei contain just enough electrons to cancel the positive charge of the extra protons; that is, they were supposed to contain $A - Z$ electrons. After all, nuclei emitted electrons in β decay, so, it appeared, electrons must exist within the nucleus. These electrons would contribute only a small amount to the mass of the nucleus, but together with the protons they would make the net charge equal to $+Z$ units, as required.

It seemed plausible to consider the atom as consisting of a nucleus made up of A protons and $A - Z$ electrons, with Z additional electrons outside the nucleus to make the entire atom electrically neutral. For example, an atom of $^{16}_{8}O$ would have a nucleus with 16 protons and 8 electrons, with 8 additional electrons outside the nucleus. This model of the nucleus is known as the *proton–electron hypothesis* of nuclear composition.

The proton–electron hypothesis seemed to be consistent with the emission of α and β particles by atoms of radioactive substances. Since it was assumed that the nucleus contained electrons, explanation of β decay was no problem. When the nucleus is in an appropriate state, it may simply eject one of its electrons. It also seemed reasonable that an α particle could be formed, in the nucleus, by the combination of four protons and two electrons. (An α particle might exist, already formed in the nucleus, or it might be formed at the instant of emission.)

The proton–electron hypothesis is similar to an earlier idea suggested by the English physician William Prout in 1815. On the basis of the small number of atomic masses then known, Prout proposed that all atomic masses are multiples of the atomic mass of hydrogen and that therefore all the elements might be built up of hydrogen. Prout's hypothesis was discarded when, later in the nineteenth century, the atomic masses of some elements were found to be fractional, in particular, those of chlorine (35.46 units) and copper (63.54 units). With the discovery of isotopes, however, it was realized that the fractional atomic masses of chlorine and copper, like that of neon, arise because these elements are *mixtures* of isotopes, with each separate isotope having an atomic mass close to a whole number.

Although the proton–electron hypothesis was satisfactory in some respects, it led to serious difficulties and had to be given up. One of the most serious difficulties arose from Heisenberg's uncertainty principle in quantum mechanics. As we noted (Section 15.6), the confinement of an electron to a space as small as the nucleus would result in the circumstance that at times the electron's speed would be greater than the speed of light, which is not possible according to special relativity theory.

How could scientists account for the circumstance that electrons cannot be confined within the nucleus, yet they emerge from the nucleus in β decay. As he recalled later, Heisenberg and his assistants were contemplating this problem one day while sitting in a café across from a building housing a swimming pool. Heisenberg suggested a possible approach to the problem. "You see people going into the building fully dressed," he said. "And you see them coming out fully dressed. But does that mean that they also swim fully dressed?" In short, you see electrons coming out of the nucleus, and occasionally being captured by the nucleus, but that does not mean that they remain electrons while in the nucleus. Perhaps the electrons are created in the process of emission from the nucleus.

18.3 THE DISCOVERY OF ARTIFICIAL TRANSMUTATION

A path that led to a better understanding of nuclear composition was opened, almost by accident, in 1919. In that year, Rutherford found that when nitrogen gas was bombarded with α particles from bismuth-214, swift particles were produced that could travel farther in the gas than did the α particles themselves. When these particles struck a scintillation screen, they produced flashes of light fainter than those produced by α particles, about the intensity that would be expected for positive hydrogen ions (protons). Measurements of the effect of a magnetic field on the paths of the particles suggested that they were indeed protons. With the skepticism characterizing all good scientific research, Rutherford ruled out, by means of careful experiments, the possibility that the protons came from hydrogen present as an impurity in the nitrogen.

Since the nitrogen atoms in the gas were the only possible source of protons, Rutherford concluded that an α particle, in colliding with a nitrogen nucleus, can occasionally knock a small particle (a proton) out of the nitrogen nucleus. In other words, Rutherford deduced that an α particle can cause the *artificial disintegration* of a nitrogen nucleus, with one of the products of the disintegration being a proton. But this process does not happen easily. The experimental results showed that only one proton was produced for about one million α particles passing through the gas.

Between 1921 and 1924, Rutherford and his coworker James Chadwick extended the work on nitrogen to other elements and found evidence for the artificial disintegration of all the light elements, from boron to potassium, with the exception of carbon and oxygen. (These elements were later shown also to undergo artificial disintegration.)

The next step was to determine the nature of the nuclear process leading to the emission of the proton. Two hypotheses were suggested for this process:

(a) The nucleus of the bombarded atom loses a proton, "chipped off" as the result of a collision with a swift α particle.

(b) The α particle is *captured* by the nucleus of the atom it hits, forming a new nucleus that, a moment later, emits a proton.

It was possible to distinguish experimentally between these two possible cases by using a device called a "cloud chamber," which reveals the path or track of an individual charged particle. The cloud chamber was invented by C.T.R. Wilson and perfected by him over a period of years. In 1911, it became an important scientific instrument for studying the behavior of subatomic particles (see Figure 18.1). If hypothesis (a) holds, the chipped-off proton should create four tracks in a photograph of a disintegration event: the track of an α particle before the collision, the track of the same α particle after collision, and the tracks of both the proton and the recoiling nucleus after collision.

In case (b), on the other hand, the α particle should disappear in the collision, and only three tracks would be seen: that of the α particle before collision and those of the proton and recoil nucleus after the collision.

The choice between the two possibilities was settled in 1925 when P.M.S. Blackett studied the tracks produced when particles passed through nitrogen gas in a cloud chamber. He found, as shown in the photograph in Figure 18.2, that the only tracks in which artificial disintegration could be seen were those of the incident α particle, a proton, and the recoil nucleus. The absence of a track corresponding to the presence of an α particle after the collision proved that the α particle disappeared completely and that case (b) is the correct interpretation of artificial disintegration: *The α particle is captured by the nucleus of the atom it hits, forming a new nucleus which thereupon emits a proton.*

The process in which an α particle is absorbed by a nitrogen nucleus and a proton is emitted may be represented by an "equation" that is analogous to the representation used in Chapter 17 to describe radioactive

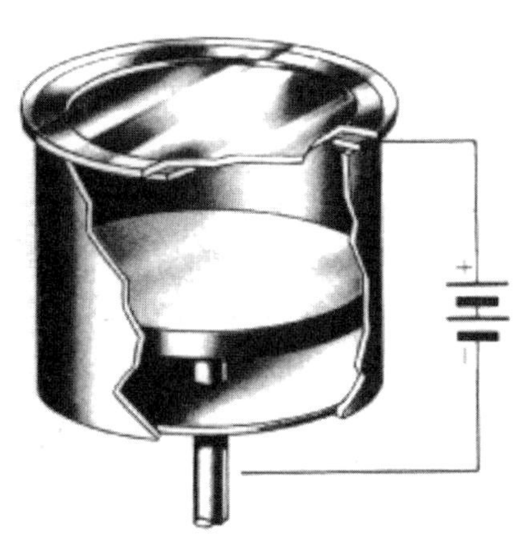

FIGURE 18.1 Cutaway drawing of the Wilson cloud chamber. When the piston is moved down rapidly, the gas in the cylinder cools and becomes supersaturated with water vapor. The water vapor will condense on the ions created along the path of a high-energy charged particle, thereby making the track. For his invention of the cloud chamber, C.T.R. Wilson (1869–1959) of Scotland shared the 1927 Nobel Prize in physics with Arthur H. Compton.

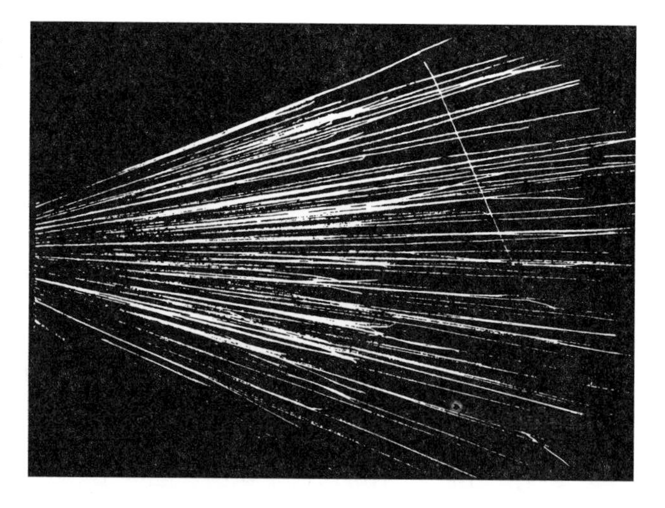

FIGURE 18.2 Alpha-particle tracks from a source at left, in a cloud chamber filled with nitrogen gas. At the right, one alpha particle has hit a nitrogen nucleus; a proton is deflected upward towards the left, and the resulting oxygen nucleus recoils downward to the right.

decay. The equation expresses the fact that the total mass number is the same before and after the collision (i.e., there is conservation of mass number) and the fact that the total charge is the same before and after the collision (there is conservation of charge). The atomic number, the mass number, and the nuclear charge are known for the target nucleus $^{14}_{7}N$, for the incident α particle $^{4}_{2}He$, and for the proton $^{1}_{1}H$. The product nucleus will therefore have the atomic number $7 + 2 - 1 = 8$, which is the atomic number for oxygen, and will have the mass number $14 + 4 - 1 = 17$. Therefore, the product nucleus must be $^{17}_{8}O$, an isotope of oxygen. The disintegration process may therefore be represented by the nuclear reaction

$$^{4}_{2}He + ^{14}_{7}N \rightarrow ^{17}_{8}O + ^{1}_{1}H.$$

This reaction shows that a transmutation of an atom of one chemical element into an atom of another chemical element has taken place. The transmutation did not occur spontaneously, as it does in the case of natural radioactivity; it was produced by exposing target atoms (nuclei) to projectiles emitted from a radioactive nuclide. It was an *artificial transmutation*. In the paper in which he reported this first artificially produced nuclear reaction, Rutherford said:

> The results as a whole suggest that, if α particles—or similar projectiles—of still greater energy were available for experiment, we might expect to break down the nuclear structure of many of the lighter atoms.

(This call for greater energies of "projectiles" was soon answered by the construction of accelerators, see Section 18.7.)

The further study of reactions involving light nuclei led (as you will see in the next section) to the discovery of a new particle, and to a better theory of the constitution of the nucleus. Many types of reactions have been observed with nuclei of all masses, from the lightest to the heaviest, and the possibilities indicated by Rutherford have been realized to an extent far beyond what he could have imagined in 1919.

18.4 THE DISCOVERY OF THE NEUTRON

In 1920, Rutherford suggested that a proton inside the nucleus might have an electron tied to it so closely as to form a neutral particle. Rutherford even suggested the name *neutron* for this hypothetical particle (since it would be neutral in charge). Physicists looked for neutrons, but the search presented at least two difficulties:

(1) They could find no naturally occurring neutron-emitting materials.
(2) The methods used for detecting atomic particles all depended on effects of the electric charge of the particles and so could not be applied directly to neutral particles. Until 1932, the search for neutrons was unsuccessful.

The proof of the existence of neutrons came in 1932 as the climax of a series of experiments on nuclear reactions made by physicists in different countries. The discovery of the neutron is a good example of how physicists operate, how they think about problems, and arrive at solutions. It is an excellent "case history" in experimental science. Working in Germany in 1930, W.G. Bothe and H. Becker found that when samples of boron or of beryllium were bombarded with α particles, they emitted radiations that appeared to be of the same kind as γ rays, at least insofar as the rays had no electric charge. Beryllium gave a particularly marked effect of this kind.

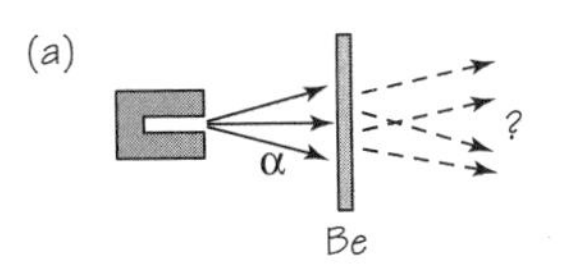

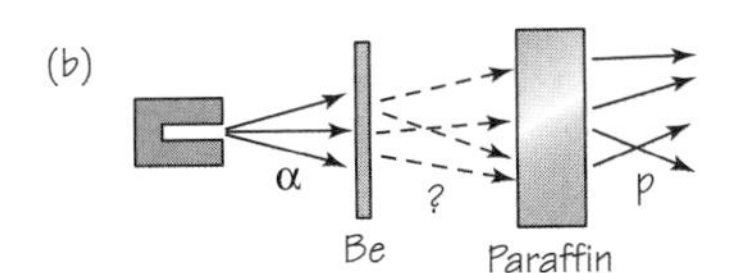

FIGURE 18.3 (a) Alpha particles hitting beryllium with the emission of unknown neutral rays. (b) When paraffin is placed behind the beryllium, protons are ejected.

Observations by physicists in Germany, France, and Great Britain showed that the induced radiation from the beryllium penetrated farther (through lead, for example) than any γ radiation found up to that time. Its interactions with matter showed that it carried energies of about 10 MeV, "MeV" standing for "million electron-volts." (This electron-volt as a unit of energy is discussed in Section 10.6.) The radiation was thus much more energetic than the γ rays (i.e., high-energy photons) previously observed and, as a result, aroused much interest.

Among those who investigated this radiation were the French physicists Frédéric Joliot and his wife Irène Curie, a daughter of the discoverers of radium. They studied the absorption of the radiation in paraffin, a material rich in hydrogen. In the course of their experiments, Joliot and Curie found that the radiation from beryllium, when it fell on paraffin, ejected large numbers of hydrogen nuclei (protons) from the paraffin. The energies of these protons were found to be about 5 MeV. Using the principles of conservation of momentum and energy, they calculated the energy a γ ray would need if it were to transfer 5 MeV to a proton in a collision. The result was about 50 MeV, a value much greater than the 10 MeV that had been measured for the radiation. In addition, the number of protons

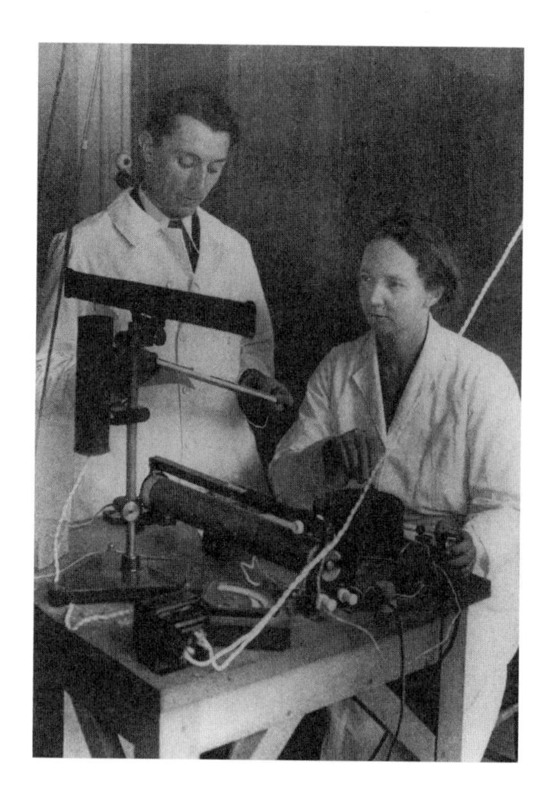

FIGURE 18.4 Irène Curie and Frédéric Joliot in their laboratory. Curie and Joliot were married in 1926 and shared the Nobel Prize for chemistry in 1935.

produced was found to be much greater than that predicted on the assumption that the radiation consisted of γ rays.

These discrepancies (between the results of two sets of experiments and between theory and experiment) left physicists in a dilemma. Either they could conclude that the conservation principles of momentum and of energy did not apply to the collisions between the radiation and the protons in the paraffin, or they could seek another hypothesis about the nature of the radiation. Now, if there is any one thing physicists do not want to do it is to give up the principles of conservation of momentum and of energy. These principles are so basic to scientific thought and have proven so useful for so long and in a vast range of different cases that physicists tried very hard to find an alternative to giving them up.

The English physicist James Chadwick found similarly perplexing results for recoiling nuclei from several other light elements, including helium, lithium, carbon, nitrogen, and argon. In 1932, Chadwick proposed a successful alternative hypothesis about the nature of the radiation. Chadwick's first published report of his hypothesis is reproduced in the *Student Guide*. In a later, more complex paper, "The Existence of a Neutron," he wrote:

> If we suppose that the radiation is not a quantum radiation [γ ray], but consists of particles of mass very nearly equal to that of the proton, all the difficulties connected with the collisions disappear, both with regard to their frequency and to the energy transfers to different masses. In order to explain the great penetrating power of the radiation, we must further assume that the particle has no net charge. We must suppose it to consist of a proton and electron in close combination, the "neutron" discussed by Rutherford [as a speculation] in his Bakerian Lecture of 1920.

Thus, according to Chadwick's hypothesis, when an element such as beryllium is bombarded with α particles, a nuclear reaction can take place that produces neutrons

$${}^{4}_{2}\mathrm{He} + {}^{9}_{4}\mathrm{Be} \rightarrow {}^{12}_{6}\mathrm{C} + {}^{1}_{0}n.$$

Here, the symbol ${}^{1}_{0}n$ represents the neutron postulated by Chadwick, with zero charge and mass number equal to 1. Such neutrons, because they have no electric charge, could penetrate bricks of a material as dense as lead without giving up their energy. When neutrons go through paraffin, there would occasionally be head-on collisions with hydrogen nuclei (protons). The recoiling protons could then be observed because of the ionization they produce. Thus, Chadwick's chargeless particle hypothesis could

FIGURE 18.5 James Chadwick (1891–1974) received the Nobel Prize in physics in 1935 for his discovery of the neutron.

account in a qualitative way for the observed effects of the mysteriously penetrating radiation.

Chadwick's estimate that the particle's mass must be nearly equal to the mass of a proton was made by applying the laws of conservation of momentum and energy to the case of perfectly elastic collisions, that is, simply applying the laws that worked well for the case of interacting billiard balls and other objects treated in "classical" physics. In a perfectly elastic head-on collision between two bodies, as you saw in Chapter 5, almost all of the kinetic energy of the initially moving body will be transferred to the initially stationary body only if the bodies have approximately equal masses. In collisions that are more glancing, i.e., not precisely head-on, less kinetic energy will be transferred. Therefore, on *average*, a kinetic energy of about 5 MeV for the recoiling protons would be about right for collisions produced by neutrons with energies about 10 MeV, if the neutron and proton masses were approximately equal.

Chadwick was able to make a more precise calculation of the neutron's mass by applying the conservation laws to data on collisions with nuclei of different masses; the details of the derivation are shown in the *Student Guide*.

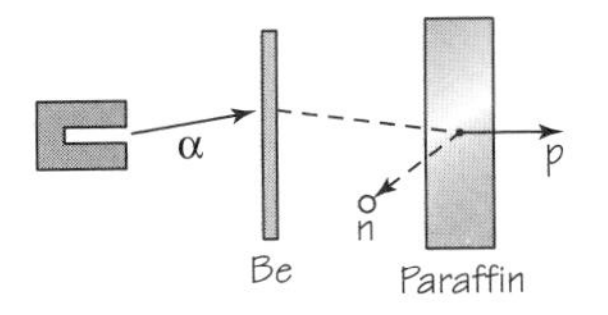

FIGURE 18.6 Experimental setup for alpha particle/beryllium collision producing neutrons that collide with protons in paraffin (compare with Figure 18.3).

Chadwick found the mass of the neutron to be 1.16 u. (The best methods now available for determining the neutron mass give 1.008665 u, based on a scale where ^{12}C is defined to have a mass of 12 u exactly). The difficulties of measuring the kinetic energies of the recoiling nuclei made this only an approximate value, but it was good enough to show that the neutron has a mass very close to that of the proton; thus, Chadwick's hypothesis did indeed offer a satisfactory solution to the problem of the "radiation" emitted when beryllium or boron was bombarded with particles.

Much research has been done since on the properties of neutrons and on the interactions between neutrons and atoms. An entire branch of study called *neutron physics* has arisen. Neutron physics deals with the production of neutrons, their detection, and their interaction with atomic nuclei and with matter in bulk. This research has led, among other things, to the discovery of nuclear fission, to be discussed below.

18.5 THE PROTON–NEUTRON MODEL

The discovery of the neutron, with an atomic mass close to one unit and with no electric charge, confirmed Rutherford's suggestion that the atomic nucleus is made up of protons and neutrons. This hypothesis was soon used as the basis of a detailed theory of the nucleus by Heisenberg in 1932. His work represented another triumph of quantum mechanics.

According to the *proton–neutron model* that arose from the new theory, the nucleus of an atom having atomic number Z and mass number A consists of Z protons and $A–Z$ neutrons. The nuclei of the isotopes of a given element differ only in the number of neutrons they contain. Thus, the nucleus of the hydrogen isotope of mass number 1 contains one proton; the nucleus of the hydrogen isotope of mass number 2 contains one proton and one neutron. (That nucleus is called a deuteron.) The nucleus of the neon isotope ^{20}Ne contains 10 protons and 10 neutrons, while that of ^{22}Ne contains 10 protons and 12 neutrons. The atomic number Z identified with the charge on the nucleus, is the number of protons in the nucleus. The mass number A is the total number of protons and neutrons. The term

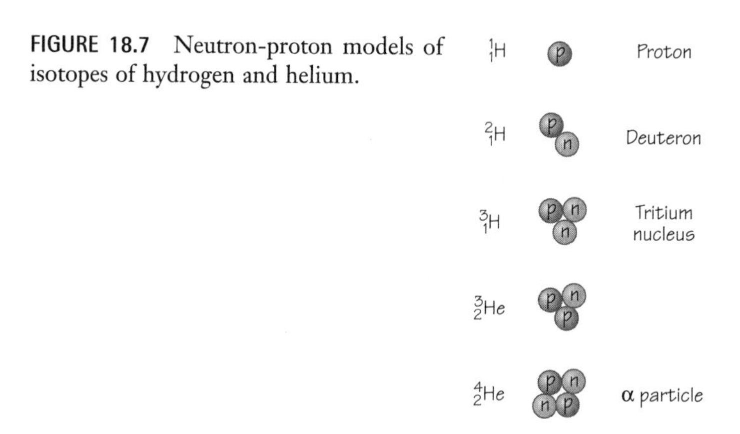

FIGURE 18.7 Neutron-proton models of isotopes of hydrogen and helium.

nucleons refers to both kinds of nuclear particles. *So atomic mass number A turns out to be simply the number of nucleons in the nucleus!*

According to the proton–neutron model, one proton alone forms the common isotope of hydrogen, $^1_1$H. One proton and one neutron yield $^2_1$H, called a deuteron, and the resulting atom is called deuterium. When two deuterium atoms combine with oxygen, they form "heavy water." The atom formed from the rare isotope $^3_1$H is called tritium, a radioactive substance.

Is the proton–neutron hypothesis for the structure of nuclei fully consistent with the facts of radioactivity, such as α and β emission and the transformation rules? If two protons and two neutrons could combine, the resulting particle would have $Z = 2$ and $A = 4$, just the properties of the α particle. The emission of two protons and two neutrons (in the combined form of an α particle) would be consistent with the first transformation rule of radioactivity. (The α particle might exist as such in the nucleus, or it might be formed at the instant of emission; the latter possibility is now considered more likely.)

The neutron–proton hypothesis raised a new question: if the nucleus consists of protons and neutrons, where could a β particle come from in β decay? This question is more difficult to answer than that of the origin of an α particle. The second transformation rule of radioactivity provides a clue: When a nucleus emits a β particle, its charge Z increases by one unit while its mass number A remains unchanged. This would happen if a neutron were to change into a proton and a β particle.

This idea was not a return to the proton–electron hypothesis discussed earlier. Physicists had already come to the conclusion that electrons are not present in the nucleus, so β decay was not considered to be a simple separation of a proton and electron; it would have to be a *transformation* of a neutron that *created* a proton and electron. However, there were additional experimental data that raised difficulties for such a simple transformation idea.

18.6 THE NEUTRINO

The description of β decay in terms of the transformation of a neutron in the nucleus is part of one of the most fascinating stories in modern physics: the prediction and eventual discovery of the particles called the *neutrino* and the *antineutrino*.

Quantitative studies of the energy relations in β decay during the 1920s and 1930s raised a difficult and serious question. Methods were devised for determining the energy change in a nucleus during β decay. According to the principle of conservation of energy, the energy lost by the nucleus should be equal to the energy carried off by the β particle; but the measured kinetic energies of the β particles had a whole range of measured values, all smaller than the amount of energy lost by the nucleus. Some of the energy lost by the nucleus seemed to have disappeared. Measurements made on a large number of β emitters indicated that on the average about two-thirds of the energy lost by the β-decaying nuclei seemed to disappear. Attempts to find the missing energy failed. For example, some physicists thought that the missing energy might be carried off by γ rays; but no such γ rays could be detected experimentally. The principle of conservation of energy seemed to be violated in β decay. Similar discrepancies were found in measurements of the momentum of the emitted electron and the recoiling nucleus.

As in the case of the experiments that led to the discovery of the neutron, physicists tried very hard to find an alternative to accepting a failure of the principles of conservation of energy and momentum. These and related considerations led the Austrian physicist Wolfgang Pauli to suggest that another, hitherto unnoticed, particle is emitted in β decay along with the electron, and that this particle carries off the missing energy and momentum. This hypothetical particle could have no electric charge, because the positive charge of the proton and the negative charge of the β particle together are equal to the zero charge of the original neutron. The mass–energy balance in the decay of the neutron indicated that the rest mass of the hypothetical particle should be very small, much smaller than the mass of an electron and possibly even zero. The combination of zero electric charge and zero or nearly zero mass would make the particle extremely hard to detect.

FIGURE 18.8 Neutrinos were first detected in this tank. Reactions provoked by neutrinos from a nuclear reactor cause flashes of light in the liquid with which the tank is filled. The flashes are detected by the photoelectric tubes that stud the tank wall. This work was done by two American physicists, Clyde Cowan and Frederick Reines (pictured here at a nuclear power plant in South Carolina).

The Italian physicist Enrico Fermi called the suggested particle the *neutrino* ("little neutral one" in Italian). Fermi constructed a theory of β decay based on Pauli's suggestion, in which a neutron decays into a proton, an electron, and a neutrino, here represented by the Greek letter nu (ν):

$$ {}^{1}_{0}n + {}^{1}_{1}p \rightarrow {}^{0}_{1}e + \nu. $$

This theory has been successful in describing the known facts of β decay. From 1934 on, while the difficult hunt for its experimental verification was still in progress, the neutrino was accepted as a "real" particle for two reasons, both theoretical: It saved the principle of conservation of energy in β decay, and it could be used successfully both to describe the result of experiments in β decay and to predict the results of new experiments.

It is now known that a *free* neutron, that is, a neutron separated from an atom, sooner or later decays into a proton, an electron, and a neutrino. (The half-life of a beam of free neutrons has been measured to be 12 min.)

Many unsuccessful attempts were made to detect neutrinos over a period of 25 years. Finally, in 1956, neutrinos were detected in an experiment using the

There is one more complication. It is now known that there are several kinds of neutrinos. The one involved in β decay (as discussed so far) is now referred to as an *antineutrino* and is denoted by the symbol $\bar{\nu}$. The transformation of a neutron during β emission is now written

$$ {}^{1}_{0}n \rightarrow {}^{1}_{1}p + {}^{0}_{1}e + \bar{\nu}. $$

extremely large flow of neutrinos that comes out of a nuclear reactor. The detection of neutrinos is an indirect process that involves detecting the products of a reaction *provoked* by a neutrino. The reaction used was a reverse β decay, the production of a proton from a neutron. Because the proper meeting of a proton, an electron, and a neutrino at the same place and same time is an exceedingly unlikely event—neutrinos can go right through the entire Earth without change—and the resulting neutron difficult to detect, "catching" the neutrinos required a very elaborate and sensitive trap. Again, the faith of physicists in the principle of conservation of energy was justified.

18.7 THE NEED FOR PARTICLE ACCELERATORS

Up to 1932, the study of nuclear reactions was limited by the kind of projectile that could be used to bombard nuclei. Only α particles from the naturally radioactive nuclides could bring about reactions. Progress was limited because α particles could be obtained only in beams of low intensity and with fairly low kinetic energies. These relatively low-energy particles could produce transmutations only in light elements. When heavier elements are bombarded with α particles, the repulsive electric force exerted by the greater charge of the heavy nucleus on an α particle makes it difficult for the α particle to reach the nucleus. The probability of a nuclear reaction taking place becomes very small or zero. Because the interest in nuclear reactions was great, physicists in many countries sought methods of increasing the energy of charged particles to be used as projectiles.

There were advantages to be gained in working with particles like the proton or the deuteron (the nucleus of the deuterium or heavy hydrogen atom) that have only one positive charge. Having only a single charge, these particles would experience smaller repulsive electric forces than would α particles in the neighborhood of a nucleus, and thus would be more successful in getting close enough to produce transmutations, even of heavy (and therefore high-charge) target nuclei. Protons or deuterons could be obtained from positive-ray tubes, but their energies were rather low. Some device was needed to accelerate these particles to higher energies, as Rutherford was among the first to say. Such devices might also offer other advantages. The speed (and energy) of the bombarding particles could be controlled by the experimenter, and very intense projectile beams might be obtained. It would then be possible to find how nuclear reactions depend on the energy of the bombarding particles.

Since 1930 scientists and engineers have invented and developed many devices for accelerating charged particles. In each case, the particles used (electrons, protons, deuterons, α particles, or heavy ions) are accelerated by an electric field. In some cases, a magnetic field is used to control the path of particles, that is, to steer them. The simplest type has a single high-voltage step of about ten million volts, thus increasing electron or proton energies to 10 MeV.

Another type of accelerator has a long series of low-voltage steps applied as the particle travels in a straight line. Some of these machines produce electron energies up to 20 GeV (1 GeV = 10^9 eV, GeV standing for "giga electron-volts"). A third general type uses magnetic fields to hold the particles in a circular path, returning them over and over to the same low-voltage accelerating fields. The first machine of this type was the cyclotron (see Figure 18.9). Some of these accelerators produce 7 GeV electrons or

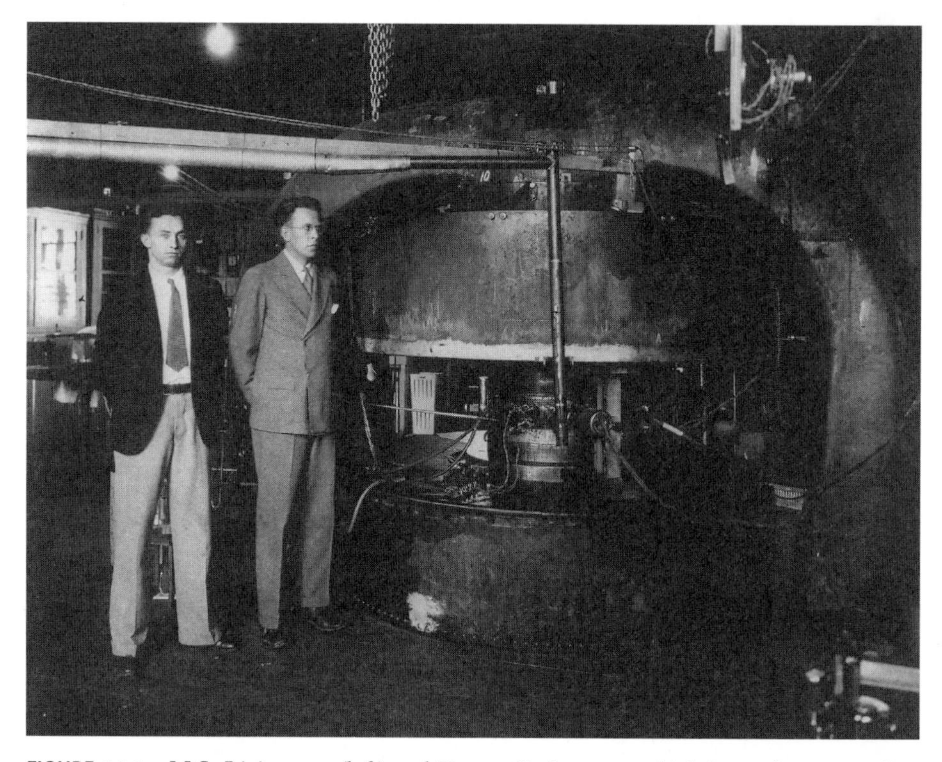

FIGURE 18.9 M.S. Livingston (left) and Ernest O. Lawrence (right) are shown standing beside the magnet for one of the earliest cyclotrons. Lawrence and Livingston invented the cyclotron in 1931, thereby initiating the development of high-energy physics in the United States.

ACCELERATORS

Research into the nature of matter has disclosed the structure of the atom and the atomic nucleus. Much current research is focused on the particles that make up the nucleus. Matter responds to four different types of force: (1) the strong force, (2) the electromagnetic force, (3) the weak force, and (4) the gravitational force. By observing how particles react when influenced by some of these forces, scientists have discovered the existence of many new and seemingly bizarre particles, using particle accelerators of increasingly higher energy. Probing the nature of matter is an international endeavor. For example, at Fermilab (Illinois) during 2001, there were over 2500 users of the accelerators, including 1368 foreign nationals from 25 countries.

(a)

(b)

(c)

FIGURE 18.10 (a) The tunnel of the main accelerator at Fermilab; (b) participants in one of the many teams working at Fermilab; (c) aerial photograph of the Fermilab facility in Illinois.

500 GeV protons. Accelerators producing in excess of 2000 GeV (2 TeV) are being planned at CERN, the European accelerator near Geneva, Switzerland. Accelerators have become basic tools for research in nuclear and high-energy physics. Accelerators are also used in the production of radioactive isotopes and as radiation sources, both for medical and for industrial purposes.

One of the most powerful accelerators currently in use is a 1000 TeV particle accelerator now in operation at the National Accelerator Laboratory (Fermilab) in Batavia, Illinois. Such "machines" are among the most complex and grandiose structures ever built. Indeed, they are monuments to human imagination and ingenuity, the ability to reason and to collaborate in groups—some as many as 500 persons—on peaceful projects that further the understanding of nature. Basically, the "machines" are tools to help physicists find out as much as they can about the structure of nuclear particles and the forces holding them together.

With the discovery of the neutron in 1932, it was then believed that three "elementary" particles act as the building blocks of matter: the proton, the neutron, and the electron. The existence of new particles found later, such as neutrinos and antineutrinos, has been mentioned. As high-energy accelerators became available, additional "elementary" particles were discovered, one after another. These particles are grouped into "families" according to their properties. Most of these particles exist only briefly; typical lifetimes are of the order of 10^{-8} s or less. A whole new field, high-energy physics, has evolved, and the aim of the high-energy physicist of today is to discern the order and structure behind the large number of "elementary" particles that have been discovered.

How do physicists detect these particles? A number of methods by which physicists can observe and measure radioactive emissions have already been mentioned. They include the electroscope and the electrometer employed since the early days of radioactivity, the Geiger counter, and the Wilson cloud chamber. In addition, various types of ionization chambers, scintillation counters, photographic emulsions, semiconductor devices, spark chambers, and bubble chambers are also in use.

18.8 THE ENERGY OF NUCLEAR BINDING

The concepts of atomic and nuclear structure—than an atom consists of a nucleus surrounded by electrons and that the nucleus is made up of protons and neutrons—led to a fundamental question: *Is the mass of a neutral atom equal to the sum of the masses of the protons, neutrons, and electrons that make up the neutral atom?*

This question can be answered precisely because the masses of the proton, the neutron, and the electron are known, as are the masses of nearly all the atomic species. A survey of the known atomic masses has shown that, for each kind of atom, the atomic mass is always *less* than the sum of the masses of the constituent particles when measured in their free states. The simplest atom containing at least one proton, one neutron, and one electron is deuterium, $^2{}_1H$. In this case, the masses (in atomic mass units, or u) of the constituents of a deuterium nucleus, called a deuteron, are

$$\begin{aligned} \text{rest mass of one proton} &= 1.007276 \text{ u}, \\ \text{rest mass of one neutron} &= \underline{1.008665} \text{ u}, \\ \text{total rest mass of particles in free state} &\quad 2.01594 \text{ u}, \\ \text{rest mass of deuteron} &= 2.01355 \text{ u}, \\ \text{difference } (\Delta m) &= 0.00239 \text{ u}. \end{aligned}$$

Although the difference in rest mass, Δm, may appear small, it corresponds to a significant energy difference, because of the factor c^2 in the relation $E = mc^2$, where c is the speed of light (about 3×10^8 m/s). The difference, Δm, in mass, which is called the *mass defect*, corresponds to a difference in the amount of energy ΔE before and after the formation of the nucleus according to the relationship from relativity theory: $\Delta E = \Delta mc^2$. A convenient conversion factor from atomic mass (expressed in atomic mass units) to energy (expressed in million electron volts) is 1 u = 931 MeV. If therefore we consider the formation of a deuterium nucleus from the combination of a proton and a neutron, then an amount of mass 0.00239 u will be "lost" in the process. This mass defect means that an amount of energy equal to (0.00239 u) $\times$ (931 MeV/u) = 2.23 MeV has to be radiated away from this system of combining particles before they settle down as a deuterium nucleus. (In addition, a tiny bit more of energy must also be lost, as a photon, when an electron is bound to an orbital path around this nucleus in forming a deuterium atom.)

The energy equivalent of 1 atomic mass unit:

$$1 \text{ u} = 1.66 \times 10^{-27} \text{ kg},$$

$$\begin{aligned} \Delta E &= \Delta mc^2 \\ &= (1.66 \times 10^{-27} \text{ kg}) \\ &\quad \times (3 \times 10^8 \text{ m/s}) \\ &= 14.9 \times 10^{-11} \text{ J}. \end{aligned}$$

But 1 MeV = 1.60×10^{-12} J:

$$\begin{aligned} \Delta E &= \frac{14.9 \times 10^{-11} \text{ J}}{1.6 \times 10^{-13} \text{ J/MeV}} \\ &= 931 \text{ MeV}. \end{aligned}$$

The expected energy loss calculated from the difference in rest mass can be compared with the result of a direct experiment. When hydrogen is bombarded with neutrons, a neutron can be captured in the reaction

$$^1_0n + {}^1_1H \rightarrow {}^2_1H + \gamma.$$

This reaction produces no particle fragments having large kinetic energy, so the mass of 0.00239 u by which $^2_1$H is lighter than ^1_0n + $^1_1$H must be carried away by the γ ray. The energy of the γ ray has been determined experimentally and found to be 2.23 MeV, just as predicted! This confirms that on forming a nucleus, the constituents give up energy, generally as a gamma ray, corresponding to the amount of mass difference.

The inverse reaction, in which a deuteron is bombarded with γ rays, has also been studied

$$\gamma + {}^2_1\text{H} \rightarrow {}^1_1\text{H} + {}^1_0n.$$

When the energy of the γ rays is less than 2.23 MeV, this reaction cannot occur. But if γ rays of energy 2.23 MeV or greater are used, the reaction can occur; some photons are absorbed, and separate protons and neutrons can be detected.

To summarize: Following the "capture" of a neutron by the nucleus $^1_1$H, energy is liberated in the form of a γ ray. This energy (2.23 MeV) is called the *binding energy* of the deuteron. It can be thought of as the energy released when a proton and neutron bind together to form a nucleus. To get the inverse reaction (when $^2_1$H is bombarded with γ rays), energy must be absorbed. So you can think of the binding energy as also the amount of energy *needed* to break the nucleus up into its constituent nuclear particles.

The concept and observation of binding energy apply, of course, not only to the example just given but to all situations in which simple parts are bound together by some force to form a complex system. For example, the Earth is held in orbit around the Sun and would need to be given a certain additional amount of kinetic energy to escape from the Sun, to which it is now bound by their mutual gravitational attraction. In a hydrogen atom, the electron needs 13 eV before it can escape from the nucleus that

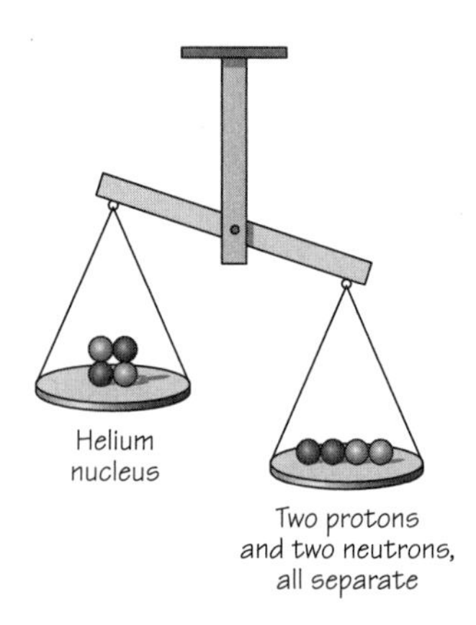

FIGURE 18.11 A case where the whole seems not to be equal to the sum of its parts. Two protons and two neutrons, measured separately, are distinctly more massive than a helium nucleus, which consists of the same particles that are bound together. The particles lose some energy (mass) in binding together to form a nucleus.

binds it by an electric attraction. Conversely, when a bare ^1_1H nucleus captures an electron and becomes a stable, ordinary neutral atom of hydrogen, the system must give up an amount of energy equal to 13.6 eV by radiation, exactly the observed energy of the photon emitted in this process of electron capture. However, only the nuclear binding energies are relatively large enough to represent measurable mass differences.

18.9 NUCLEAR BINDING ENERGY AND STABILITY

The calculation of the nuclear binding energy made for the deuteron can be extended to all other nuclear species, and such calculations have been performed. Figure 18.12 shows in graphic form how the total nuclear bind-

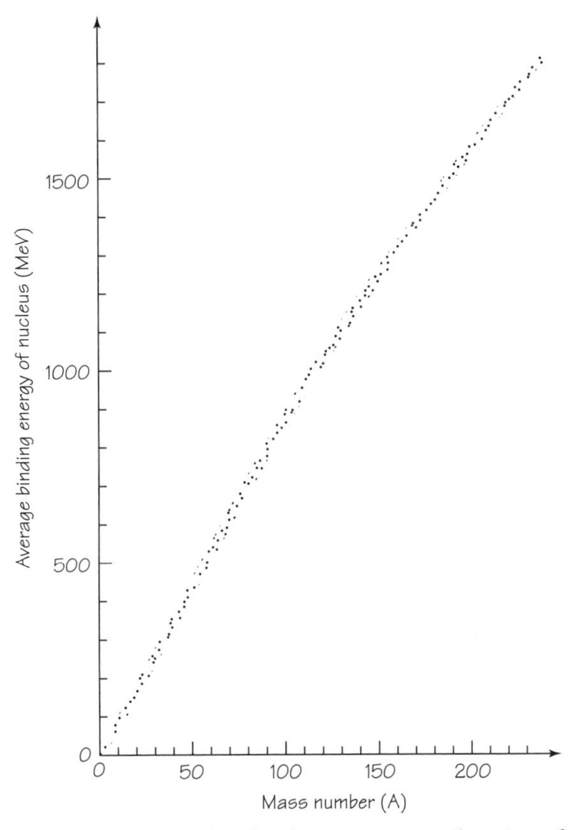

FIGURE 18.12 Nuclear binding energy as a function of the mass number—i.e., the number of particles in the nucleus.

ing energy for stable nuclides increases with increasing atomic mass, as more particles are added to form the nucleus. The term *nucleons* refers to both protons and neutrons; therefore, the binding energy of the nucleus increases with the number of nucleons. But, as you see, the result is not a straight line. Such experimental data have important implications.

The implications can be seen more clearly if the *average binding energy per nucleon* is calculated. In the case of the carbon-12 example, the total binding energy is 92.1 MeV. Since there are 12 nucleons inside the nucleus (six protons and six neutrons), the average binding energy per nucleon is 92.1 MeV/12, or 7.68 MeV. In the graph in Figure 18.13, the experimentally obtained values of the average binding energy per nucleon (in MeV) are plotted against the number of nucleons in the nucleus (mass number, A). Notice the unusually high position (above the curve) of the data point near 7.1 MeV, compared to its neighbors in the periodic table. The point is for ^{4}He. The relatively high value of the binding energy of this nucleus indicates its unusually great stability.

The significance of the graph lies in its striking shape. The binding energy per nucleon starts with a low value for the deuterium nucleus (the first point) and then increases rapidly. Some nuclei in the early part of the curve, for example, ^{4}He, ^{12}C, and ^{16}O, have exceptionally high values as compared with their neighbors. This indicates that more energy would have to be supplied to remove a nucleon from one of these nuclei than from one of

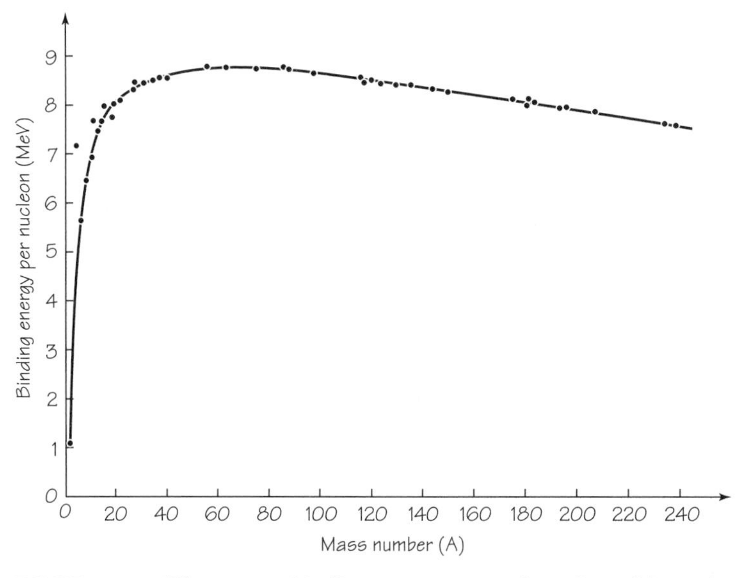

FIGURE 18.13 The average binding energy per nucleon for stable nuclei as a function of the number of particles in the nucleus.

their neighbors. (*Remember*: High binding energy per nucleon means a great deal of energy is needed to take the nucleus apart into its constituent nucleons. In a sense "binding energy" might have been better called "unbinding energy.")

The high binding energy per nucleon of ^{4}He compared with deuterium would mean that if two deuterium nuclei were joined together to form a ^{4}He nucleus, there would be a large amount of excess energy available, which would be emitted to the environment. This excess energy is the source of the enormous energies made available in *fusion*, or *thermonuclear*, reactions, discussed below.

Since they do have such high binding energies, you would expect ^{4}He, ^{12}C, and ^{16}O to be exceptionally stable. There is evidence in favor of this conclusion, for example, the fact that the four particles making up the ^{4}He nucleus are emitted as a single unit, the α particle, in radioactivity.

The experimentally obtained curve of binding energy per nucleon has a broad maximum, extending from approximately $A = 50$ to $A = 90$. Then it drops off for the heavy elements. Thus, $^{63}_{29}$Cu near the maximum is found to have a binding energy per nucleon of about 8.75 MeV, while $^{235}_{92}$U, near the high-A end of the curve, has a value of 7.61 MeV. This indicates that as more nucleons are added to the heavier nuclei, the binding energy per nucleon decreases. It follows that the nuclei in the neighborhood of the maximum of the curve, like those of copper, should be more difficult to break up than heavier nuclei, such as radium and uranium. It also follows that when uranium and other high-A nuclei somehow are made to break up, their fragments are smaller nuclei which possess higher binding energy per nucleon. In such a case there is again excess energy due to the difference in energy between the starting nucleus and its fragments, which is emitted to the environment in the form of kinetic energy of the fragments and gamma radiation. This historically significant process, which involves the splitting of the heaviest nuclei into lighter nuclei, is known as *nuclear fission*. The excess energies available during fission are the source of the enormous energies released in nuclear fission reactions.

The shape of the average binding energy curve, which drops off at both ends, indicates, therefore, that there are two general reaction processes by which one can hope to release energy from nuclei:

(1) combining light nuclei into a more massive nucleus, known as nuclear fusion; or
(2) splitting up heavy nuclei into nuclei of medium mass, which is called nuclear fission.

In either process, the resulting products would have greater average binding energy per nucleon, so energy would be released in the process. Both fusion and fission have been shown to occur, and the technology of fission has been simplified and exploited in many countries. Fission reactions can be made to take place slowly (as in a nuclear power plant) or very rapidly (as in a nuclear explosion).

The idea of binding energy should now make it clear why atomic masses, when precisely measured, are not exactly whole-number multiples of the mass of a hydrogen atom, even though nuclei are just collections of identical protons and neutrons. When those particles combined to make a nucleus, their total rest mass was reduced by an amount corresponding to the binding energy, and the average binding energy varies from nuclide to nuclide, as shown in Figure 18.13.

We now take a closer look at fission and fusion.

18.10 NUCLEAR FISSION: DISCOVERY

The discovery of nuclear fission is an example of an unexpected result with great practical and social implications, yet originally it was obtained during the course of research carried on for reasons having nothing to do with the possible uses society would make of the discovery. It is also an excellent example of the combined use of physical and chemical methods in nuclear research, and of the effectiveness of teamwork.

When Joliot and Curie showed that some products of neutron-induced nuclear reactions are radioactive, Fermi and his colleagues in Rome, Italy, undertook a systematic study of nuclear reactions induced by neutrons. One of the purposes of this research was to produce new nuclides. As a result, many new radioactive nuclides were made and their half-lives determined. One nuclear reaction used successfully in this study was the capture of a neutron followed at once by the emission of a γ ray. For example, when aluminum is bombarded with neutrons, the following reaction occurs:

$$^{1}_{0}n + ^{27}_{13}\mathrm{Al} \rightarrow ^{28}_{13}\mathrm{Al} + \gamma.$$

Aluminum-28 is radioactive, with a half-life of 2.3 min, decaying by β emission into silicon

$$^{28}_{13}\mathrm{Al} \rightarrow ^{28}_{14}\mathrm{Si} + ^{0}_{-1}e + \bar{\nu}.$$

As a result of these two reactions, a nuclide ($^{28}_{14}\mathrm{Si}$) is produced with values of Z and A each greater by one unit than those of the initial nucleus. Fermi thought that if neutrons bombarded uranium, the atomic species having the largest value of Z then known, an entirely *new* element might be formed by the β decay of the heavier uranium isotope

$$ {}^{1}_{0}n + {}^{238}_{92}\text{U} \rightarrow {}^{239}_{92}\text{U} + \gamma. $$

$$ {}^{239}_{92}\text{U} \rightarrow {}^{239}_{93}(?) + {}^{0}_{-1}e + \bar{\nu}, $$

He also speculated that the new nuclide denoted by ${}^{239}_{93}(?)$ in turn might also undergo β decay, producing a second element beyond uranium

$$ {}^{239}_{93}(?) \rightarrow {}^{239}_{94}(??) + \bar{\nu}. $$

In this way, two new elements might be produced, one with $Z = 93$, one with $Z = 94$. If these reactions could really be made to occur, the result would be the artificial production of an element, or elements, not previously known to exist: *transuranium elements.*

Fermi found in 1934 that the bombardment of uranium with neutrons actually produced new radioactive elements in the target, as shown by the emission of rays and a decay activity that revealed new, relatively short half-lives. The new elements were at first assumed to be the hypothesized transuranium elements.

Fermi's results aroused much interest, and in the next 5 years a number of workers experimented with the neutron bombardment of uranium. Many

FIGURE 18.14 Enrico Fermi (1901–1955). Born in Rome, Italy, Fermi received the Nobel Prize for Physics in 1938 for his work on bombarding nuclei with the neutrons. Fermi fled Italy in 1938 and moved to the United States, where he continued work on nuclear structure and participated in the Manhattan Project. The equation Fermi wrote is incorrect. It is reported that after Fermi wrote the equation he turned to the audience to acknowledge the error when this picture was taken. He then erased it.

different radioactive half-lives were found for the radiation from the target, but attempts to identify these half-lives with particular elements led to great confusion. The methods used were similar to those used in the study of the natural radioactive elements (Section 17.7). But the difficulty of identification was even greater because a radioactive nuclide formed in a nuclear reaction is usually present in the target area only in an extremely small amount, possibly as little as 10^{-12} g; special techniques to separate these small quantities had to be developed.

The reason for the confusion was found late in 1938 when Otto Hahn and Fritz Strassmann, two German chemists, showed definitely that one of the supposed transuranium elements had the chemical properties of an isotope of *barium* ($^{139}_{56}Ba$), with a half-life of 86 min. Another nuclide resulting from the neutron bombardment of uranium was identified as lanthanum ($^{140}_{57}La$), with a half-life of 40 hr.

The production of the nuclides $^{139}_{56}Ba$ and $^{140}_{57}La$ from uranium, a nuclide with the atomic number 92 and an atomic mass of nearly 240, required an unknown kind of nuclear reaction, one in which the heavy nucleus is split almost in half. Nothing like it had been known to exist before. However, these two nuclides could not be the two halves, since the sum of their atomic numbers and masses exceeded those of uranium. Perhaps barium and lan-

FIGURE 18.15 Lise Meitner and Otto Hahn. Meitner, born in Austria, joined Hahn in 1908 in a research collaboration that lasted 30 years. In 1938, Meitner was forced to leave Germany by the Nazi regime. She was in Sweden when she published (along with her nephew, Otto Frisch) the first report recognizing and describing the existence of nuclear fission.

thanum were each only one of the two products of two different splittings of uranium. If such splitting processes really occurred, it should also be possible to find "the other half" of each splitting, that is, to find two other nuclides with masses between 90 and 100 and atomic numbers of about 35. Indeed, Hahn and Strassmann were able to find in the target material a radioactive isotope of strontium ($Z = 38$) and one of yttrium ($Z = 39$) which fulfilled these conditions, as well as isotopes of krypton ($Z = 36$) and xenon ($Z = 54$). It was clear from the chemical evidence that the uranium nucleus, when bombarded with neutrons, can indeed split into two nuclei of intermediate atomic mass.

Although Hahn and Strassmann showed that isotopes of intermediate mass did appear, they hesitated to state the conclusion that the uranium nucleus could indeed be split, since such an idea was so startingly new. In their historic report, dated January 9, 1939, they said:

> On the basis of these briefly presented experiments, we must, as chemists, really rename the previously offered scheme and set the symbols Ba, La, Ce in place of Ra, Ac, Th. As nuclear chemists with close ties to physics, we cannot decide to make a step so contrary to all existing experience of nuclear physics. After all, a series of strange coincidences may, perhaps, have led to these results.

The step which Hahn and Strassmann, as chemists, could not bring themselves to take was understood to be necessary by two Austrian physicists, Lise Meitner and her nephew, Otto R. Frisch, on January 16, 1939, both then in Sweden as forced exiles from Germany. They suggested that the incident neutron provoked a disintegration of the uranium nucleus into "two nuclei of roughly equal size," a process they called *nuclear fission* by analogy to the biological division, or fission, of a living cell into two parts.

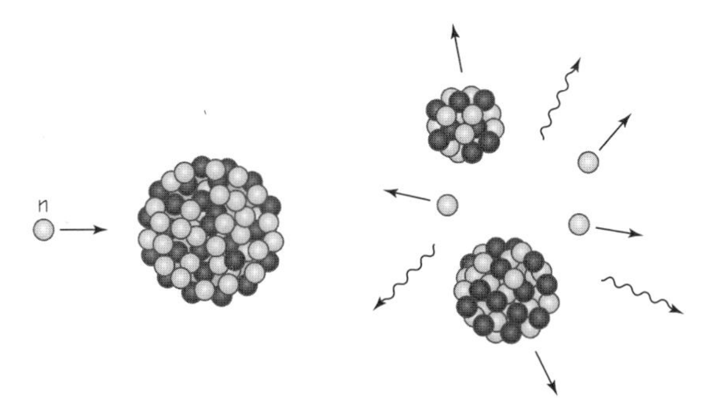

FIGURE 18.16 Schematic diagram representing uranium fission.

FIGURE 18.17 Otto R. Frisch (1904–1979).

On the basis of comparing the low average binding energy per nucleon of uranium with the higher average binding energy per nucleon of the products, they predicted that the fragments would have high kinetic energy resulting from the excess energy emitted in the fission process. This was soon verified experimentally.

Shortly afterward, it was found that transuranium elements may, after all, *also* be formed when uranium is bombarded with neutrons. In other words, the capture of a neutron by uranium sometimes leads to fission and sometimes leads to β decay. The β decay results in the formation of isotopes of elements of atomic number 93 and 94, later named *neptunium* and *plutonium* (after the two planets in the solar system beyond Uranus). The presence of both types of reaction, fission and neutron capture followed by β decay, had been responsible for the earlier difficulty and confusion in the analysis of the effects of neutrons on the uranium target. Now, the interpretation of the experiments opened two new fields of scientific endeavor: the physics and chemistry of the transuranium elements, and the study of nuclear fission.

The discovery of nuclear fission caused research on it all over the world, and much new information was obtained within a short time. It was found that the uranium nucleus, after capturing a neutron, can split in fact into any one of more than 40 different pairs of fragments. Radiochemical analysis showed that nuclides resulting from fission have atomic numbers between 30 and 63 and mass numbers between 72 and 158.

Yet nuclides of medium mass are not the only fission products. In a finding that turned out to have extraordinary importance, neutrons also were discovered to result from fission; the average number of neutrons emitted is usually between two and three per fissioned nucleus. The following reaction indicates only one of the many ways in which a uranium nucleus can split.

$${}^{1}_{0}n + {}^{235}_{92}\mathrm{U} \rightarrow {}^{141}_{56}\mathrm{Ba} + {}^{92}_{36}\mathrm{Kr} + 3\,{}^{1}_{0}n.$$

$^{141}_{56}Ba$ and $^{92}_{36}Kr$ are not "natural" nuclides and are not stable; they are radioactive and decay by β emission. For example, $^{141}_{56}Ba$ can decay into $^{141}_{59}Pr$ by successive emission of three β particles, as shown by the following scheme (the numbers in parentheses in are the half-lives):

$$_{56}Ba^{141} \xrightarrow[(18\text{ min})]{} {}_{57}La^{141} + {}_{-1}e^0 \quad (a)$$

$$\xrightarrow[(3.6\text{ hr})]{} {}_{58}Ce^{141} + {}_{-1}e^0 \quad (b)$$

$$\xrightarrow[(32\text{ days})]{} {}_{59}Pr^{141} + {}_{-1}e^0 \quad (c)$$

Similarly, $^{92}_{36}Kr$ is transformed into $^{92}_{40}Zr$ by four successive β decays.

It has been found that only certain nuclides can undergo fission. For those that can, the probability that a nucleus will split when bombarded depends on the energy of the neutrons used in the bombardment. The nuclides $^{235}_{92}U$ and $^{239}_{94}Pu$ can undergo fission when bombarded with neutrons of *any* energy, even 0.01 eV or less. On the other hand, ^{238}U and ^{232}Th undergo fission only when bombarded with neutrons having kinetic energies of 1 MeV or more. (As noted previously, ^{239}Pu, which is highly fissionable, is produced by the capture of a neutron by ^{238}U and the subsequent emission of two β particles.)

The energy released in the fission of a heavy nucleus is about 200 MeV. This value can be calculated either by comparing atomic rest masses of reactants and products or from the average binding energy curve of the graph in the previous section. The energy release in fission per atom is more than a million times larger than in chemical reactions, and more than 20 times larger than in the more common nuclear reactions, where it is usually less than 10 MeV.

There was one more important result that became obvious to researchers everywhere: Under appropriate conditions the neutrons released in fission can, in turn, cause fission in neighboring uranium atoms, and thereby a process known as a *chain reaction* can develop in a sample of uranium. The combination of the large energy release in fission and the possibility of a chain reaction is the basis of the large-scale use of nuclear energy.

18.11 CONTROLLING CHAIN REACTIONS

For a chain reaction in a sample of uranium to continue at an even rate, there must be a favorable balance between the net *production* of neutrons by fissions and the *loss* of neutrons due to the following three processes:

1. capture of neutrons by uranium without fission resulting;
2. capture of neutrons by other materials in the sample (such as rods of boron or cadmium) or in the structure containing the sample;
3. escape of neutrons from the sample without being captured.

If too *many* neutrons escape from or are absorbed in the structure or assembly (called a *reactor*), there will not be enough to sustain the chain reaction. If too *few* neutrons escape or are absorbed, the reaction will continue to build up more and more. The design of nuclear reactors as energy sources involves finding proper sizes, shapes, and materials to maintain or control a balance between neutron production and neutron loss.

Since the nucleus occupies only a tiny fraction of an atom's volume, the chance of a neutron colliding with a uranium nucleus is small, and a neutron can go past the nuclei of billions of uranium (or other) atoms while moving a few centimeters. If the reactor assembly is small, a significant percentage of the fission neutrons can escape from the assembly without caus-

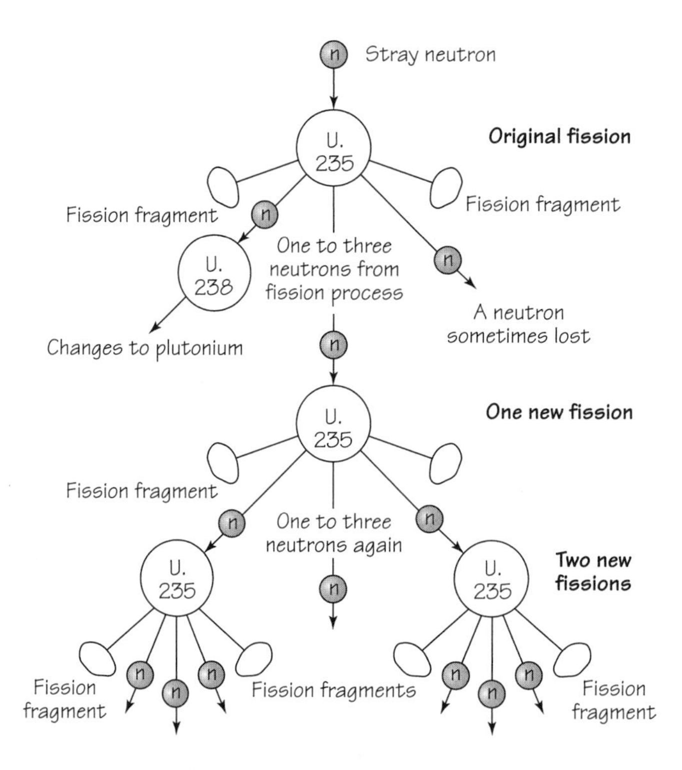

FIGURE 18.18 This diagram indicates what happens in a chain reaction resulting from the fission of uranium-235 atoms (not shown are other emissions, such as alpha, beta, and gamma rays).

ing further fissions. The "leakage" of neutrons can be so large that a chain reaction cannot be sustained. The number of neutrons produced is proportional to the *volume*, but the number of neutrons that escape is proportional to the *surface area*. As the linear size L of the assembly is increased, the volume and area increase in proportion to L^3 and L^2, respectively, so that neutron production increases with size more rapidly than neutron escape does.

For a given combination of materials (uranium and other structural materials that may be needed), there is a size of the reactor, called the *critical* size, for which the net production of neutrons by fission is just equal to the loss of neutrons by nonfission capture and escape. If the size of the reactor assembly is smaller than this critical size, a chain reaction cannot be sustained. The design of a reactor of reasonable dimensions, with given materials, which will correspond to critical size, is an important part of research in the field of *nuclear engineering*.

Another important consideration in the design of nuclear reactors is the fact that the fission is much more probable when ^{235}U is bombarded with *slow* neutrons than when it is bombarded with fast neutrons. The neutrons released in fission generally come out at very high speeds, having kinetic energies from about 0.01 MeV to nearly 20 MeV, with an average kinetic energy of about 2 MeV. The fast neutrons can be slowed down in the reactor by the addition of material (called "moderator") to which the neutrons can lose energy in collisions. The material should be relatively low in atomic mass so that the neutrons will transfer a significant fraction of their energy in collisions; but the material should not also capture and absorb many neutrons, thus taking them out of the reaction. Pure carbon in the form of graphite and also water and beryllium meet these requirements. As moderators, they slow down, or moderate, the newly produced neutrons to lower speeds at which the probability of causing additional fission is high. Although nuclear reactors can be built in which the fissions are induced by fast neutrons, it has been easier to build reactors with materials in which the fissions are induced by slow neutrons.

Although nuclear reactors can be built in which the fissions are induced by fast neutrons, it has been easier to build reactors with materials in which the fissions are induced by slow neutrons.

Hydrogen atoms in water are very effective in slowing down neutrons because the mass of a hydrogen nucleus (a single proton) is nearly the same as that of a neutron and because the number of hydrogen atoms per unit volume is high. A neutron can lose a large fraction of its energy in a collision with a hydrogen nucleus. Only about 20 collisions are needed, on average, to slow down the fast neutron to energies under 1 eV. However, neutrons can also be captured by the hydrogen nucleus in the reaction

$$^1_0n + ^1_1H \rightarrow ^2_1H + \gamma.$$

FIGURE 18.19 Lise Meitner (1878–1968).

The probability of this reaction occurring instead of an elastic collision is high enough so that it has been found impossible to achieve a chain reaction with natural uranium and ordinary water. But the absorption of a neutron by a deuterium nucleus ($^2_1$H), such as the nucleus of the heavy isotope of hydrogen, found in so-called *heavy water*, has an extremely small probability. Neutrons do not lose as much energy per collision with ^{2}H nuclei, but this disadvantage is compensated for by the much lower absorption rate. Therefore, a chain reaction can be achieved easily with natural uranium and heavy water. Reactors, with natural uranium as the fuel and heavy water as the moderator, have been built in the United States, Canada, France, Sweden, Norway, and other countries, and were attempted to be built by German scientists during World War II.

The contrast between the nuclear properties of hydrogen $^1_1$H and deuterium ($^2_1$H or $^2_1$D) has important implications for the development of nuclear reactors. Heavy water is expensive to produce, but when it is used with natural uranium (mostly ^{238}U), a chain reaction can be achieved efficiently. Although the uranium isotope ^{238}U normally absorbs neutrons rather than fissioning, the heavy water slows the neutrons below the energy at which they will be captured by the plentiful ^{238}U nuclei. A slow neutron will simply bounce off the ^{238}U nuclei it encounters until it is eventually absorbed by a rare ^{235}U nucleus, causing the nucleus to fission.

Ordinary water can be used as moderator in a uranium reactor *if* uranium enriched in the isotope ^{235}U is used instead of natural uranium. Many reactors "fueled" with enriched uranium and moderated with ordinary water have been built in the United States. Such reactors are called *light-water reactors*. In fact, this general reactor type is the preferred design for the commercial production of energy (electricity), since it is less expensive

FIGURE 18.20 The Chicago pile No. 1 used by Enrico Fermi and his associates when they first achieved a self-sustaining nuclear reaction on December 2, 1942. Alternate layers of graphite, containing uranium metal and/or uranium oxide, were separated by layers of pure solid graphite blocks. Graphite was used as a moderator to slow down neutrons in order to increase the likelihood of fission. Courtesy of Argonne National Laboratory.

to build and less likely to yield as by-products fissionable materials that could be used for nuclear weapons.

Carbon in the form of ultra-pure graphite has been used as a moderator in many reactors, including the earliest ones. Its atoms being more massive, it is not as good at slowing down fast neutrons as are light water and heavy water; about 120 collisions with carbon atoms are needed to slow down a fast neutron with an initial energy of 2 MeV to the desired energy of about 0.025 eV; in heavy water only about 25 collisions are needed. But although carbon in the form of pure graphite is not the best moderator and absorbs some neutrons, it does permit a chain reaction to occur when lumps of natural uranium (cylindrical rods containing uranium pellets, for example) are arranged in a large mass of graphite. The determination of just how this could be done was one of the main problems that had to be solved before the world's first chain reaction was achieved in December 1942 by a team working under Enrico Fermi at the University of Chicago. (It was a crucial experiment because until its success it was by no means certain that a chain reaction was really possible in practice.) Many graphite-moderated reactors are now in operation throughout the world. Their chief purpose will be discussed in the next section.

The control of a nuclear reactor is relatively simple. Lest fission is occurring too frequently, a few *control rods* are inserted into the reactor. The rods consist of a material (such as cadmium or boron) that absorbs slow neutrons, thereby reducing the number of neutrons in the reactor. Removal of the control rods will allow the rate of fission in the reactor to increase. The sketch (Figure 18.21) illustrates the basic reactions that occur in a nuclear reactor in which uranium is the fissionable material.

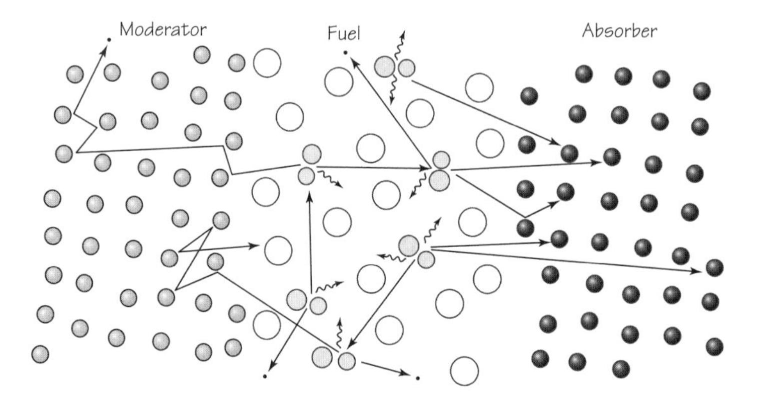

FIGURE 18.21 Schematic diagram of three types of functions fulfilled by parts of a nuclear reactor.

18.12 NUCLEAR POWER PLANTS

Nuclear reactors are useful for energy production because of the large amount of kinetic energy that the moderator in a reactor obtains from the fission neutrons in inelastic collisions. The fission neutrons are slowed by the moderator and return to the uranium to find more ^{235}U nuclei in which to induce fission. The kinetic energy lost by the fission neutrons as they are slowed by the moderator is gained by the molecules of the moderator and appears as heat. The heat generated can—and must—be pumped away from the reactor core by cool water which is thereby made to boil. The resulting steam can then be used to turn a turbine connected to the coils of an electric generator, thus producing electricity.

The main difference between a nuclear power plant and a fossil-fuel power plant is that heat produced by nuclear fission replaces heat produced by chemical reactions in the burning of fossil fuel. In both instances the heat is used to generate steam, as in steam-engine technology, to perform the useful work needed. In both instances the work is used, not to drive an engine directly, but to generate electricity.

The advantages of nuclear-powered production of electricity in a well-shielded and well-run reactor are obvious. The reaction produces no greenhouse gas emissions or other polluting gases. A nuclear reactor does not require the burning of fossil fuels, which are approaching depletion, and

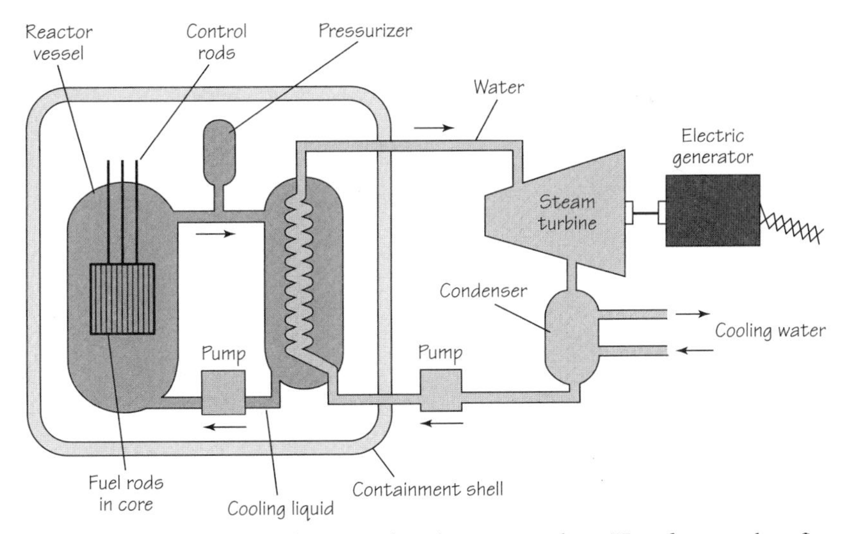

FIGURE 18.22 Schematic diagram of nuclear power plant. Heat from nuclear fission is used to boil water. The steam is used to turn a turbine which rotates coils in a magnetic field to generate electricity.

the dependence on the importation of expensive foreign fuel reserves is diminished. Nevertheless, the world supply of uranium is not unlimited, and reactors do require the disposal of long-lived radioactive waste, as well as the safe disposal of the heated water and all equipment and clothing exposed to radiation.

The ever-increasing use of electrical energy is an important aspect of modern life. As discussed in Chapter 11, every possible source of energy that might be used to meet the increasing demand for electricity is at present problematic.

The development of nuclear power in the United States was slower to develop than was expected at the end of World War II. But during the 1960s nuclear electric power became economically competitive with hydroelectricity and electricity from fossil fuels. By the beginning of 1978, 65 nuclear reactors were operating with over 47 million kilowatts capacity, about 9% of the nation's total electric power production.

However, the picture changed dramatically during the 1980s as the result of the public's increasing concern for safety, especially in the wake of the accidents at Three Mile Island and Chernobyl. More recently, the pos-

FIGURE 18.23 Reactor Number One at Great Britain's first nuclear power station at Calder Hall (opened in 1956). The large towers are cooling towers. The reactor is in the large building.

sibility that terrorists might crash an airplane into a nuclear reactor, with potentially devastating consequences, has underscored public concern about reactor safety. Plants under construction near densely populated areas where rapid evacuation is impossible, have been discontinued. At the same time, the Nuclear Regulatory Commission imposed stricter safety provisions during the 1990s on the operation of the plants and the disposal of radioactive waste. With the drop in the price of imported oil at the time, the operation of a nuclear power plant was no longer economically competitive with a power plant using imported fossil fuel. As a result, no new nuclear power plants have been built in the United States since the 1980s, and those still in operation may well be phased out in the years ahead.

Fusion energy, which has been the subject of intense, sophisticated research, is nevertheless still not a technical reality, but probably remains the main hope for an eventual large-scale solution to our rapidly increasing energy needs.

18.13 NUCLEAR WEAPONS

The large-scale use of nuclear energy in chain reactions was accomplished in the United States between 1939 and 1945. The work was done under the pressure of World War II, as a result of the cooperative efforts in which government agencies brought together large numbers of scientists and engineers. The workers in the United States included Americans, Britons, and European refugees from fascist-controlled countries. They hoped to obtain a nuclear weapon—if one could be made—before the Germans, who were known to be working on one, and in fact had started earlier. Many of the scientists hoped that the very existence of such a weapon would make future wars unlikely. A number of others petitioned that the Government would not use such a weapon on civilian targets, but only as a demonstration on uninhabited areas.

In a so-called *atomic bomb* (more properly a nuclear fission bomb), an extensive chain reaction occurs throughout the material in a few millionths of a second, thereby resulting in the explosive release of an enormous amount of energy. This reaction differs from the controlled nuclear reactor, in which the operating conditions are so arranged that the energy from fission is released at a much slower and essentially constant rate. In the controlled reactor, the fissionable material is mixed with other materials in such a way that, on average, only *one* of the neutrons emitted in fission causes the fission of another nucleus. In this way, the chain reaction just sustains itself. In a nuclear bomb, on the other hand, the fissionable material is not

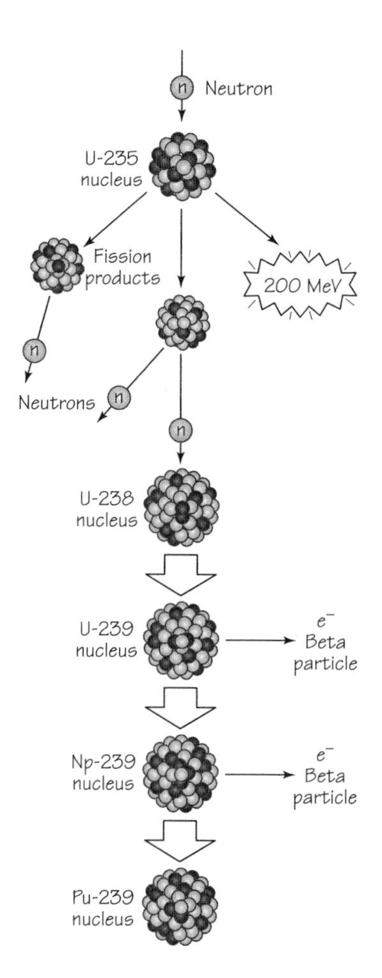

FIGURE 18.24 The "pile reactions" to produce plutonium-239.

mixed with a moderator, and the device is designed so that nearly all of the neutrons emitted in each fission can cause fissions in other nuclei.

Nuclear reactors were first used during World War II in the United States to produce raw materials for one kind of nuclear bomb, namely to manufacture highly fissionable plutonium, $^{239}_{94}$Pu, from the uranium isotope ^{238}U through β decay. Such reactors are called *breeder reactors*. These reactors are designed in such a way that some of the neutrons from the fission of ^{235}U are slowed down sufficiently *not* to cause fission in ^{238}U atoms. (In natural uranium, only about 0.75% of the atoms are ^{235}U.) Instead, the neutrons are absorbed by ^{238}U nuclei to form ^{239}Pu through the sequence of β-decay reactions described in Section 18.10. Some "nonnuclear" nations using nuclear reactors for generating electricity may have obtained weapons-grade plutonium from their reactors in this way. The United

States and other nations have been negotiating with these nations to provide more up-to-date light-water reactors containing less ^{238}U, along with other economic aid, in exchange for dismantling their old heavy-water reactors that can produce plutonium.

^{239}Pu behaves in many ways like ^{235}U. Both materials can sustain a rapid, uncontrolled chain reaction. Both isotopes were used to power the first nuclear weapons used in August 1945 in order to end World War II at President Truman's decision—a war, unleashed against the Allies, and that had already cost many millions of lives and devastated much of Europe and Japan. A single nuclear bomb, using pure ^{235}U, was dropped to destroy the city of Hiroshima, Japan, on August 6, 1945. Another bomb, using ^{239}Pu, destroyed the city of Nagasaki 3 days later. The war ended officially on September 2, 1945.

Since the end of World War II in 1945, other countries besides the United States have made nuclear weapons, including the United Kingdom, Russia (the former Soviet Union), France, India, Pakistan, and China. The enormous death-dealing capability of these weapons, and the ever-larger numbers of nuclear bombs of many varieties that have been accumulating

FIGURE 18.25 Robert Oppenheimer (1904–1967).

FIGURE 18.26 Maj. Gen. Leslie R. Groves (r), Chief of the Manhattan Engineering District where the first nuclear bomb was developed, and J. Robert Oppenheimer, Director of Los Alamos Atomic Bomb Project and Physicist at California Technological Institute, view the base of the steel tower on which the first atomic bomb hung when tested near Alamogordo, New Mexico in July, 1945. The intense heat of the bomb melted the tower and seared the surrounding sands into jade green, glass-like cinders.

all over the globe, have made more dangerous the tensions existing throughout the world, and have emphasized critically the need for the peaceful settlement of international disputes.

Tensions between Western nations and the Soviet Union and its allies reached a frightening level during the depths of the so-called Cold War, especially during the 1950s and early 1960s. As nuclear weapons became ever more powerful, the potential for destruction became ever more immediate with the development of intercontinental ballistic missiles (ICBMs). By the late 1950s, an ICBM launched by one nation and carrying a nuclear bomb could in principle reach any point on the globe in less than an hour. There was no defense against such an attack. Even defending missiles sent to intercept an attacking missile that was moving at thousands of kilometers per hour could be overwhelmed by "multiple reentry warheads" and by decoys, emitted by the single attacking missile.

Without the possibility of defense, the United States prepared the population for possible nuclear war, and both sides instituted a policy known appropriately by the acronym MAD, for "mutually assured destruction." Any nuclear attack by one side in the Cold War would result in "massive retaliation" by the other side—the launching of every available nuclear weapon against the attacking nation, resulting in the likely total destruction of both sides, perhaps even most of the population of the entire world. The prospects for such a scenario became frighteningly more likely during the Missile Crisis in 1962, when the Soviet Union placed long-ranged nuclear-armed missiles in Cuba. When the United States blockaded ("quarantined") Cuba from Soviet ships, traditionally an act of war, the world held its breath as Soviet ships carrying more nuclear weapons approached Cuba. Finally, the ships turned back, averting a possible nuclear war, and the world breathed a sigh of relief.

Since then the world's nuclear powers became more realistic in their search for a way to control nuclear weapons. But the fears of the public remained, as reflected in such classic films as *Dr. Strangelove* and *Fail-Safe*. However, efforts to attempt a defense against incoming nuclear missiles have continued. One such attempt, originating during the 1980s, has involved the design of high-powered lasers, controlled by fast computers, which would target and destroy an incoming enemy missile. This technology, officially called the Strategic Defense Initiative (SDI) but popularly known as "star wars" (after the well-known films of the same name) and its more recent successors, have so far proved unsuccessful. Moreover, weapons of mass destruction could well be deployed by ships or trucks, therefore not vulnerable to a defense relying only upon missiles.

From the very beginning, scientists have been prominently involved in activities to alert their government and fellow citizens to the moral and practical problems raised by the nuclear arms race. One of their earliest successes involved international limits on nuclear testing. In order to develop ever more powerful and efficient weapons, test explosions are often

FIGURE 18.27 Energy released from nuclear fission: the first underwater test of an atomic bomb at Bikini Atoll in the Pacific Ocean in July 1946.

required, and most of these were performed above ground in remote areas during the 1950s. As noted previously, in the explosion of a nuclear bomb, large amounts of radioactive fission products are scattered. These materials can be blown by winds from one part of the world to another and carried down from the atmosphere by rain or snow. This is known as *fallout*.

Partly as the result of public petitions and protests organized by scientists—spearheaded by the chemist Linus Pauling, who received the Nobel Peace Prize for his efforts—the United States, Soviet Union, and most other nations agreed in 1963 to a moratorium on further above-ground testing. This greatly reduced the amount of radioactive pollution in the atmosphere. Testing continued, however, below ground. Further international treaties have placed further curbs on nuclear testing, but they have not yet eliminated all testing, despite the end of the Cold War. The United States and other nations rely increasingly on computer simulations to maintain their arsenals, but some nations have insisted on the right to continue underground testing.

Since the end of the Cold War and the breakup of the Soviet Union, tensions have diminished, but the dangers of the use of nuclear weapons remain high. There are fears that some of the weapons in the large arsenal of nuclear weapons stored in the former Soviet Union may find their way into the hands of bellicose nations or terrorist organizations. Developing nations are slowly obtaining the ability to produce nuclear weapons, and nations such as India and Pakistan, which have been long-time enemies, have both tested nuclear weapons and long-range missiles capable of delivering them.

Finally, the Cold-War weapons production has left the enormous problems and high costs of radioactive waste cleanup at weapons production facilities and the disposal of huge amounts of weapons-grade plutonium. Plutonium is extremely poisonous, and it is relatively long-lived, having a half-life of 24,000 years. Although most advanced nuclear weapons are now powered by fusion reactions involving hydrogen (see below), they are triggered by plutonium-based fission reactions. Disposing of the many tons of highly fissionable, long-lived plutonium, while keeping it out of the hands of terrorists and nonnuclear nations, is a major challenge for both sides of the former Cold War.

As in the past, the decisions by politicians and industrial leaders that will be necessary in the future development and uses of controlled and uncontrolled nuclear energy cannot be made on the basis of physics alone. Scientists can help to illuminate alternatives on which essentially political decisions can be based, and there are several dozen organizations founded by scientists in the U.S.A. alone that have been and are continuing to educate and advise. But science should not be used by itself to choose among the alternatives. Responsible scientific opinion should be supplemented by political insight and a broad humanistic view of society.

18.14 NUCLEAR FUSION

A fusion reaction involves the joining together of two light nuclei into a heavier nucleus. The reaction results in higher binding energies per nucleon when light nuclei are combined. As a consequence, a large amount of energy is released.

Fusion reactions have been produced in the laboratory by bombarding appropriate light target materials with, for example, high-energy deuterons from a particle accelerator. In these reactions, nuclei result that are heavier than the nuclei of either the "projectiles" or the targets; there are usually also additional particles released, as well as energy. Some typical examples of fusion reactions, together with the energy liberated in each reaction, are

$$^2_1\text{H} + {}^2_1\text{H} \rightarrow {}^3_1\text{H} + {}^1_1\text{H} + 4\ \text{MeV},$$

$$^2_1\text{H} + {}^2_1\text{H} \rightarrow {}^3_2\text{He} + {}^1_0n + 3.2\ \text{MeV},$$

$$^2_1\text{H} + {}^3_1\text{H} \rightarrow {}^4_2\text{He} + {}^1_0n + 17.6\ \text{MeV},$$

$$^2_1\text{H} + {}^3_2\text{He} \rightarrow {}^4_2\text{He} + {}^1_1\text{H} + 18.3\ \text{MeV}.$$

In the first of the above equations, the heavier product nucleus is an isotope of hydrogen, called *tritium*, with mass number $A = 3$. It has been found in small traces in nature, is radioactive with a half-life of about 12 years, and decays by β emission into $^3_2$He, an isotope of helium. When a target containing tritium is bombarded with deuterons, $^4_2$He can be formed, as in the third equation above, liberating 17.6 MeV of energy. Of this energy, 14.1 MeV appears as kinetic energy of the neutron and 3.5 MeV as kinetic energy of the product nucleus.

Although the energy liberated in a single fusion is less than that in a single fission, the energy released *per unit mass* is much greater. The mass of about 50 helium atoms is approximately equal to the mass of one uranium atom; 50×17.6 MeV $= 1040$ MeV, compared to 200 MeV for a typical fission.

The fusion of tritium and deuterium offers the possibility of providing large sources of energy, for example, in electric power plants. Deuterium occurs in water with an abundance of about one part in seven thousand hydrogen atoms and can be separated from the lighter isotope. Four liters of water contain about 0.13 g of deuterium, which can now be separated at a cost of about 8 cents. If this small amount of deuterium could be made to react under appropriate conditions with tritium (perhaps produced by the reaction discussed above), the energy output would be equivalent to that from about 1140 l of gasoline. The total amount of deuterium in the oceans is estimated to be about 10^{17} kg, and its energy content would be about 10^{20} kW-yr. If deuterium and tritium could be used to produce energy, they would provide an enormous source of energy.

There are, of course, some difficult problems to be solved before fusion reactions are likely to be useful as steady sources of energy. The nuclei which react in the fusion processes are positively charged and repel one an-

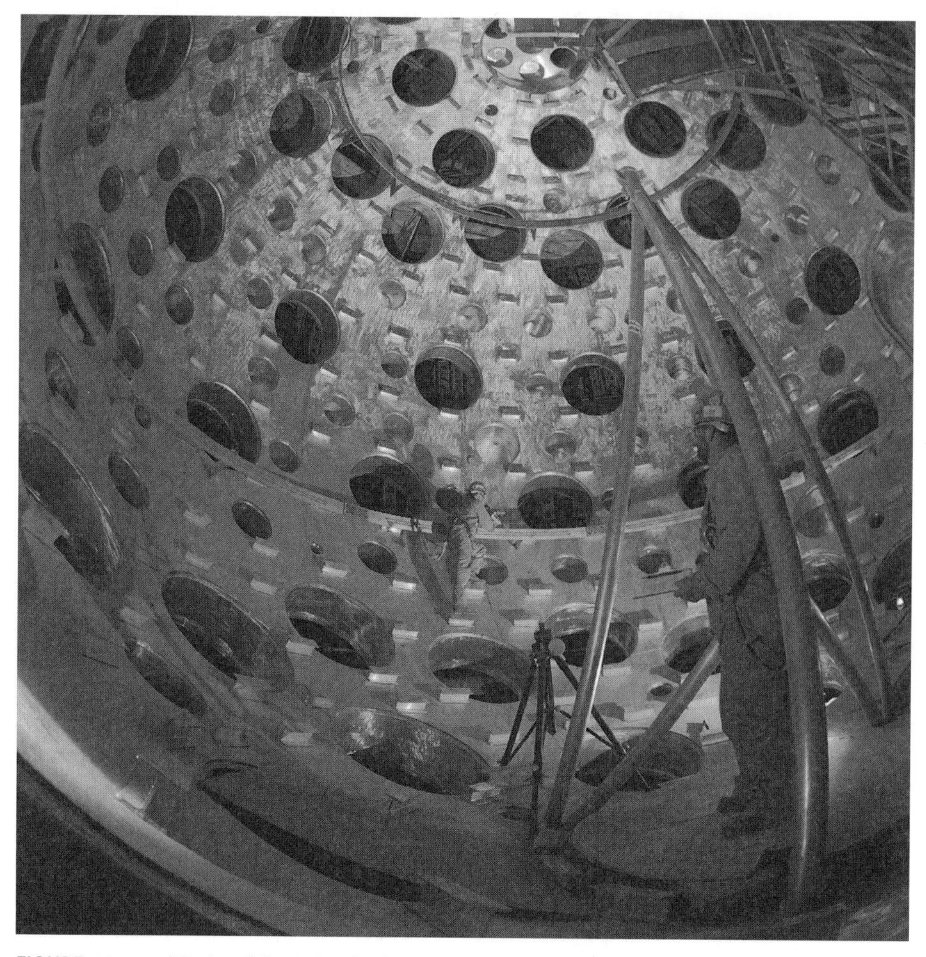

FIGURE 18.28 National Ignition facility. Lawrence Livermore National Laboratory, California.

FIGURE 18.29 Maintenance workers inside the vacuum chamber of the Tokamak Fusion Test Reactor, Princeton.

other because of the repulsive electric force. The nuclei must, therefore, be made to collide with a high relative speed to overcome the repulsive force tending to keep them apart. The nuclear force, which holds neutrons and protons together in the nucleus, is much stronger than the electric force—and it is often called the *strong force*—but it has a very short range. Its effect extends only about 10^{-14} m, about the size of a nucleus. The fusing nuclei must therefore approach within this range in order for the attractive nuclear force to overcome electric repulsion. The nuclei must also be confined in a region where they can undergo many collisions without escaping, or being absorbed by the walls bounding the region, or losing energy by collisions with too many "cooler" (less energetic) molecules. There must be enough collisions per unit time so that fusion can occur at a rate that will yield more energy than that needed to cause the collisions. The combination of these requirements means that the nuclei must be contained at a temperature of the order of 100 million degrees.

At the temperature required for fusion, the atoms have been stripped of their electrons, and the resulting nuclei and separated electrons are said to form a *plasma*. A *plasma* is an ionized gas in which positively and negatively charged particles move about freely. No wall made of ordinary material can

FIGURE 18.30 Subramanyan Chandrasekhar (1910–1995) made major contributions to fields ranging from magnetohydrodynamics to relativity to black hole theory. For his theoretical work on the physical processes of importance to the structure and evolution of stars he was awarded the Nobel Prize for physics in 1983.

contain a hot plasma at 10^8K (the wall would be vaporized instantly!). But the charged particles of a plasma can, in theory, be contained in an appropriately designed magnetic field. The first problem to be solved, therefore, is to contain the plasma of deuterium and tritium nuclei in a magnetic field, while accelerating the nuclei by means of an electric field to the required kinetic energy (or temperature). The behavior of the charged particles in a plasma is complicated; there are many kinds of instabilities that make the plasma difficult to contain properly and long enough. These problems of the release of energy to form a *controlled* and sustained fusion reaction have not yet been solved on a practical scale, but research on them is being carried on in many countries. Significant advances have been made during the last few years in containment of the plasma and in reaching high temperatures using intense, focused laser beams. There are still difficult technological problems to be overcome, and it may be a generation before electric power will be produced by fusion at costs that will compete with electricity from fossil fuels. Although the effort and expenses are great, the possible payoff in terms of future power resources is enormous. Fusion-supplied energy, without the dangerous by-products of fission, and in principle inexhaustible and cheap, can herald a vast change in human civilization, worldwide.

Fusion Reactions in Stars

Fusion reactions are actually quite common in nature, although not on Earth. They are the source of power generated by the Sun and all the many billions of stars throughout the Universe. In a sense, one can say that fusion energy is nature's primary energy source. On the scale of stars, confinement of the plasma is accomplished by gravitational attraction.

One of the most fascinating aspects of nuclear physics is the study of fusion reactions in different types of stars. The Sun is an example. In the Sun, the fusion process results in the production of a helium nucleus from four protons. The net results of the reactions can be written as

$$4\,{}^{1}_{1}\mathrm{H} \rightarrow {}^{4}_{2}\mathrm{He} + 2\,{}^{0}_{+1}e + 26\ \mathrm{MeV},$$

where ${}^{0}_{+1}e$ is an "anti-electron," also known as a positron. The reaction does not take place in a single step but can proceed through different sets of reactions whose net results are summarized in the above equation. In each case, the overall amount of energy released is 26 MeV.

The fusion of four protons into a helium nucleus is the main source of the energy of the Sun. Chemical reactions cannot provide energy at large enough rates (or for long enough duration!) to account for energy pro-

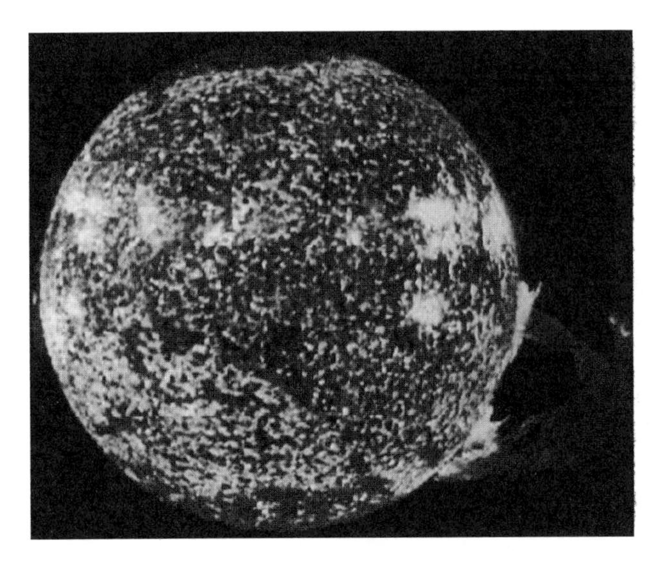

FIGURE 18.31 An X-ray photograph of the Sun. Nuclear fusion is the source of energy in our Sun and powers billions of stars throughout the Universe.

FIGURE 18.32 Cecilia Payne-Gaposchkin (1900–1979), the first person to receive a PhD in astronomy from Harvard University, discovered that stars are primarily made of hydrogen and have varying temperatures.

duction in the Sun, but nuclear fusion reactions can. Hydrogen and helium together make up about 99% of the Sun's mass, with approximately twice as much H as He. Fortunately, there is enough hydrogen to supply the Sun's energy for several billion years to come.

By which of the several possible sets of reactions does the transformation of hydrogen into helium take place? The direct process of four protons colliding to form a helium nucleus has been ruled out because the probability for such a reaction under solar conditions is too low. It may happen, but not often enough for the amount of energy released. A more likely set of reactions is as follows: When the temperature is about 10^7 K, the kinetic energies are large enough to overcome the electric repulsion between protons, and fusion of two protons (^1_1H) takes place. The nuclear reaction results in a deuteron (^2_1H), a positron ($_{+1}^{\ 0}e$), and a neutrino. As soon as the deuteron is formed, it reacts with another proton, resulting in helium-3 (^3_2He) and a γ ray. The helium-3 nuclei fuse with each other, forming α particles and two protons. In each of these reactions, energy is released, resulting in 26 MeV for the complete cycle of four protons forming a helium nucleus.

The rates of the reaction depend on the number of nuclei per unit volume and on the temperature. The higher the temperature, the faster the

FIGURE 18.33 *View of Les Saintes-Maries-de-la-Mer* by Vincent Van Gogh. Courtesy of Oskar Reinhart Collection "Am Römerholz," Winterthur, Switzerland.

thermal motion of the particles and the more frequent and energetic the collisions. At the temperature of the Sun's interior, which has been estimated to be 10 million to 20 million degrees, the kinetic energies resulting from the thermal motion are in the neighborhood of 1 keV.

Fusion reactions are nature's primary source of energy for the Universe as a whole. It is reasonable to hope that in the future they will be ours as well.

FURTHER READING

L. Badash, *Scientists and the Development of Nuclear Weapons* (New York: Prometheus, 1995).

B. Greene, *The Elegant Universe* (New York: Vintage, 2000), (especially recommended as guide to current ideas, from quarks to superstring theory).

G. Holton, Scientists Organizing to Fulfill Their Civic Responsibility, *Physics and Society*, **28** (1999), 11–13.

G. Holton and S.G. Brush, *Physics, The Human Adventure*, "Construction of the Elements and the Universe" (Piscataway, NJ: Rutgers University Press, 2001), Chapter 32.

Physics Today **50**, no. 6 (1997). Issue devoted to the problem of nuclear waste.
R. Rhodes, *The Making of the Atomic Bomb* (New York: Touchstone Books, 1995).
R. Rhodes, *Dark Sun: The Making of the Hydrogen Bomb* (New York: Touchstone Books, 1996).
J. Rotblat, *Scientists and the Quest for Peace: A History of the Pugwash Conferences* (Cambridge, MA: MIT Press, 1972).
E. Segrè, *Enrico Fermi, Physicist* (Chicago: University of Chicago Press, 1995).
R. Sime, *Lise Meitner: A Life in Physics* (Chicago: University of Chicago Press, 1996).
S. Weart, *Nuclear Fear: A History of Images* (Cambridge, MA: Harvard University Press, 1989).
H.G. Wells, *The World Set Free* (New York: Harcourt,1914). A remarkably prescient science-fiction book.

SOME NEW IDEAS AND CONCEPTS

artificial transmutation
binding energy
breeder reactor
chain reaction
control rods
critical mass
fission
fusion
heavy water
hydrogen bomb
light-water reactor
moderator
neutrino
neutron
nuclear bomb
nuclear proliferation
nucleon
nuclide
positron
proton
proton–electron hypothesis
proton–neutron model
reactor
strong force
transuranium elements

STUDY GUIDE QUESTIONS

18.1 The Problem of Nuclear Structure

1. What was one of the main questions raised about the nucleus?
2. What are the two main areas in which research has been pursued?

18.2 The Proton–Electron Hypothesis

1. Alpha and gamma rays are emitted by some radioactive nuclei. Why couldn't they be used as building blocks of the nucleus?
2. Why was the idea of hydrogen atoms being a basic building block of all atoms given up in the nineteenth century?
3. How could protons and electrons be used to build up the nuclei of atoms?

4. On the basis of the proton–electron hypothesis, what would a nucleus of $^{12}_{6}C$ contain?
5. Does the proton–electron hypothesis work out for, say, $^{4}_{2}He$?
6. Why did this model ultimately fail?

18.3 The Discovery of Artificial Transmutation

1. What evidence showed that the bombarding α particle was temporarily absorbed by the nitrogen nucleus rather than simply broken up and bounced off?
2. Why was the reaction called "artificial transmutation"?

18.4 The Discovery of the Neutron

1. What was Rutherford's hypothesis about protons and electrons in the nucleus?
2. Why could the neutral penetrating radiation from bombarded beryllium not be considered γ rays?
3. Why did the mass of a neutron have to be found by measurements on protons ejected by the neutrons in collision?
4. How could the principles of conservation be used to find the mass of the neutron?

18.5 The Proton–Neutron Model

1. Briefly describe the proton–neutron model of the nucleus.
2. According to the proton–neutron model, what is contained in the nucleus of $^{14}_{7}N$?
3. How does this model account for the existence of isotopes?
4. Describe an ordinary helium atom in terms of the three elementary particles: protons, neutrons, and electrons.
5. If nuclei do not contain β particles, how can β emission be explained?
6. What happens inside the nucleus in β emission? As a result, what happens to every isotope that emits a β ray?

18.6 The Neutrino

1. Why was an almost undetectable particle invented to patch up the theory of β decay?
2. What is the almost undetectable particle? Has it been detected?
3. $^{214}_{82}Pb$ undergoes β decay with a half-life of 26.8 min. From the information given on β decay, what would be the daughter nucleus? Write the nuclear equation for the β decay of $^{214}_{82}Pb$.

18.7 The Need for Particle Accelerators

1. Why can low-energy α particles cause transmutations only in nuclei of relatively small atomic number?
2. Why are protons more effective projectiles for producing nuclear reactions than are α particles or heavy ions?

3. What are some of the devices for producing high-energy particles to be used as projectiles?
4. What are some devices for detecting nuclear reactions induced by such projectiles?

18.8 The Energy of Nuclear Binding

1. When energy is "liberated" during a nuclear reaction, what becomes of it?
2. What is the definition of binding energy for the case of the deuteron nucleus?
3. Which would have more mass:
   (a) a deuteron, ${}_{1}^{2}H$?
   (b) a proton and a neutron moving freely and independently of each other?
4. Explain the difference in mass between 3(a) and 3(b).

18.9 Nuclear Binding Energy and Stability

1. Which would be more stable, a nuclide with a high *total binding energy* or a nuclide with a high *average binding energy per nucleon*?
2. Where on the periodic table are elements for which (a) fission and (b) fusion processes can take place?
3. Using the graph of binding energy per nucleon, explain why energy is emitted in fission reactions and in fusion reactions.

18.10 Nuclear Fission: Discovery

1. What happens in nuclear fission?
2. Why was Fermi bombarding uranium with neutrons?
3. How did he make use of β decay in his research?
4. What two successive reactions can result in the appearance of a transuranium element?
5. Describe in your own words the sequence of events that leads to the element plutonium.
6. Why couldn't the observed lanthanum and barium be the products of the fissioning of a single uranium nucleus?
7. How did the physicists Meitner and Frisch explain the appearance of lanthanum and barium in the samples obtained by the chemists Hahn and Strassmann?
8. What product of the fission process makes a chain reaction possible?

18.11 Controlling Chain Reactions

1. A low-speed neutron is fired at a group of uranium isotopes. Describe what can happen if:
   (a) the neutron hits a ^{238}U isotope;
   (b) the neutron hits a ^{235}U isotope.
2. What is a moderator? Why is it needed?
3. What is an advantage and a disadvantage of using regular water as a moderator in nuclear reactors?
4. How can the rate of reaction be controlled in a reactor?
5. What is the difference between a light-water reactor and a heavy-water reactor?

6. Why are light-water reactors usually chosen for delivery to other nations?
7. Describe in your own words how a nuclear reactor works.

18.12 Nuclear Power Plants

1. How is a nuclear reactor used to produce electricity?
2. How does the operation of a nuclear power plant compare with the operation of a fossil-fuel electric-power plant?
3. What are the advantages and disadvantages of nuclear and fossil-fuel power plants?
4. Why did scientists in the U.S. during World War II agree to develop atomic weapons?

18.13 Nuclear Weapons

1. What are breeder reactors? What do they breed and why?
2. With the Cold War over, is the world now safe from the use of nuclear weapons? Explain.
3. Where do the decisions ultimately lie regarding the uses of nuclear energy?

18.14 Nuclear Fusion

1. Why are very high temperatures required to cause fusion reactions?
2. How could extremely hot gases be kept from contacting the wall of a container?
3. In what way has fusion energy been used by humankind?
4. How does the Sun make use of fusion energy?
5. Is the ratio of the amount of hydrogen to the amount of helium in the Sun increasing or decreasing?

DISCOVERY QUESTIONS

(Consult the periodic table as needed.)

1. When ordinary chemical reactions take place, such as the fusion of hydrogen and oxygen to form water, why do we not observe a loss of mass similar to the loss of mass when neutrons and protons fuse together to form a nucleus?
2. Complete the following nuclear equations:
   (a) ${}^{6}_{3}\mathrm{Li} + {}^{1}_{1}\mathrm{H} \rightarrow {}^{4}_{2}\mathrm{He} + (\ \ )$;
   (b) ${}^{9}_{4}\mathrm{Be} + {}^{1}_{1}\mathrm{H} \rightarrow {}^{4}_{2}\mathrm{He} + (\ \ )$;
   (c) ${}^{9}_{4}\mathrm{Be} + {}^{1}_{1}\mathrm{H} \rightarrow (\ \ ) + {}^{2}_{1}\mathrm{H}$;
   (d) ${}^{11}_{5}\mathrm{B} + {}^{4}_{2}\mathrm{He} \rightarrow {}^{14}_{7}\mathrm{N} + (\ \ )$.
3. Complete the following nuclear equations, then describe in words what is happening in each case:
   (a) ${}^{4}_{2}\mathrm{He}^{4} + {}^{10}_{5}\mathrm{B} + \rightarrow (\ \ ) + {}^{1}_{1}\mathrm{H}$;
   (b) ${}^{1}_{1}\mathrm{H} + {}^{9}_{4}\mathrm{Be} \rightarrow (\ \ ) + {}^{2}_{1}\mathrm{H}$;
   (c) ${}^{4}_{2}\mathrm{He} + (\ \ ) \rightarrow {}^{35}_{17}\mathrm{Cl} + {}^{1}_{1}\mathrm{H}$;
   (d) ${}^{2}_{1}\mathrm{H} + {}^{27}\mathrm{Al} \rightarrow (\ \ ) + {}^{4}_{2}\mathrm{He}$;
   (e) ${}^{1}_{0}n + {}^{27}\mathrm{Al} \rightarrow {}^{28}\mathrm{Al} + (\ \ )$.

4. How many electrons are there in a neutral atom of:
   (a) platinum-196;
   (b) gold-198;
   (c) mercury-198;
   (d) mercury-199.
5. Why would it be difficult to explain the nucleus of $^{235}_{92}U$ as a mixture of α particles and electrons?
6. Describe the following nuclear reactions in words:
   $^{1}_{0}n + ^{27}_{13}Al \rightarrow ^{27}_{12}Mg + ^{1}_{1}H$,
   $^{27}_{12}Mg^{7} \rightarrow ^{27}_{13}Al + ^{0}_{-1}e + \nu' + \gamma \quad (T_{1/2} = 9.5 \text{ min})$.
7. How may the discovery of artificially radioactive nuclides have helped the development of theories of nuclear structure?
8. Complete the following table:

	A	Z	*Number of protons*	*Number of neutrons*	*Number of electrons in atom*
^{1}H					
^{2}H					
^{3}H					
^{4}He					
^{7}Li					
^{13}C					
^{238}U					
^{234}Th					
^{230}Th					
^{214}Pb					
^{206}Pb					

9. Write a set of equations that describe the decay of the fission product $^{92}_{36}Kr$ into $^{92}_{40}Z$.
10. Why are the high temperatures produced by the explosion of a fission bomb necessary to initiate fusion in a thermonuclear device?
11. It is generally agreed that stars are formed when vast clouds of hydrogen gas collapse under the mutual gravitational attraction of their particles. How might this process lead to fusion reactions beginning in such stars? (*Hint:* The cloud has gravitational potential energy.)
12. A team of scientists announces that it has discovered a possibly new source of cheap, nonpolluting, renewable energy that will solve all of our energy problems. They caution that further research will be required to determine if it is indeed feasible, and much work will be needed to render it commercially viable. However, there is one problem: the possibility exists that this new source of energy might also be turned into a new military weapon of enormous destructive power. The scientists declare that they are very eager to solve the world's energy problems, but they are worried that if it does prove feasible,

this source might also lead to a new weapon of mass destruction. They have turned to the public for advice.
   (a) As an informed member of the public, what do you recommend?
   (b) Assume you are a member of the scientific team. What are your thoughts on the issue?
   (c) Set up a debate in your class or group on these issues.
13. Write an essay on one of the following topics:
   (a) The various ways an informed citizen can help assure that technological innovations will be made and used in a manner benefitting society as a whole.
   (b) The differences between technology and basic science.
   (c) The responsibilities of scientists to society.
   (d) The responsibilities of society to further science.
   (e) The fields of physics or related sciences in which you may want to do further study.
14. In studying this Part Two of the text, you have followed some of the immense transformation of humankind's culture, from the pre-scientific period to current research questions. After thoughtful reflection on this experience, write a page or two summarizing the stages in this adventure of the creative mind.

Quantitative

1. Compare the mass of a helium nucleus with the sum of the masses of two hydrogen nuclei, two neutrons. What conclusions do you draw from your result?
2. Suppose that a nucleus of $^{13}_{6}C$ is formed by adding a neutron to a $^{12}_{6}C$ atom. Neglecting any kinetic energy the neutron may have, calculate the energy that becomes available to the nucleus because of the absorption of that neutron to make $^{13}_{6}C$. The atomic masses of ^{12}C and ^{13}C (in an unexcited state) are, respectively, 12.000000 u (by definition, an international convention), and 13.003354 u.
3. The atomic mass of ^{4}He is 4.00260 u; what is the average binding energy per nucleon?
4. Use the graph on page 784 to find the binding energies for ^{235}U, ^{141}Ba, and ^{92}Kr. Use these values to show that the energy released in the fission of ^{235}U is approximately 200 MeV.
5. Fusion reactions in the Sun convert a vast amount of hydrogen into radiant energy each second. Knowing that the energy output of the Sun is 3.90×10^{26} J/s, calculate the rate at which the Sun is losing mass.

Illustration Credits

FRONT COVER	*The Ancient of Days* by William Blake, © The British Museum. Bubble Chamber © Kevin Fleming/CORBIS.
BACK COVER	Portrait of Cecilia Payne-Gaposchkin by Patricia Watwood, courtesy of Harvard University.

PROLOGUE TO PART 1

Image of Albert Einstein™ licensed by the Hebrew University of Jerusalem, represented by The Roger Richman Agency, Inc., www.albert-einstein.com

CHAPTER 1

Figure 1.1	Courtesy of History of Science Collections, University of Oklahoma.
Figure 1.3	Courtesy of History of Science Collections, University of Oklahoma.
Figure 1.5	Courtesy of Reuters/ARC/Timepix.
Figure 1.17	Courtesy of Zoology Section of "La Specola" Museum of Natural History of the University of Florence.
Figure 1.19	Courtesy of Harvard College Observatory.
page 53	Courtesy of NERAIL Photo Archive.
page 54	Courtesy of Jacky Naegelan/Reuters/TimePix.
page 54	© Cory Sorensen/CORBIS.

CHAPTER 2

Figure 2.1	Courtesy of Davison Art Center, Wesleyan University.
Figure 2.2	Courtesy of AIP Emilio Segrè Visual Archives
Figure 2.3	Courtesy of Jagiellonian Library, Krakow.

Figure 2.8	Courtesy of History of Science Collections, University of Oklahoma.
Figure 2.9	Courtesy of NASA.
Figure 2.14	Courtesy of Mount Wilson Palomar Observatory.
Figure 2.16	Courtesy of British Museum.
Figure 2.17	© Roger Ressmeyer/CORBIS
Figure 2.18	© Bettmann/CORBIS
Figure 2.22	Courtesy of Jagiellonian Library, Krakow.
Figure 2.26	Courtesy of History of Science Collections, University of Oklahoma.
Figure 2.27	Courtesy of Mansell/Timepix.
Figure 2.28	Courtesy of Royal Observatory, Edinburgh/SPL/Photo Researchers, Inc.
Figure 2.29	Courtesy of NASA.
Figure 2.30	Courtesy of Harvard College Observatory.
Figure 2.37	Courtesy of Sidney Harris.
Figure 2.38	Courtesy of Instituto e Museo di Storia della Scienza, Firenze, Italy.
Figure 2.39	Courtesy of History of Science Collections, University of Oklahoma.
Figure 2.40	Courtesy of NASA.
Figure 2.42	Courtesy of History of Science Collections, University of Oklahoma.
Figure 2.43	Courtesy of Mount Wilson Palomar Observatory.
Figure 2.45	Courtesy of Lowell Observatory.
Figure 2.47	Courtesy of AIP Emilio Segrè Visual Archives

CHAPTER 3

Figure 3.2	© Jonathan Blair/CORBIS.
Figure 3.5	Courtesy of Harvard College Observatory.
Figure 3.6	Courtesy of David Couzens.
Figure 3.7	© Tecmap Corporation/CORBIS
Figure 3.19	Courtesy of NASA.
Figure 3.21	© Mike King/CORBIS.
Figure 3.23	© The Harold & Esther Edgerton Family Trust, courtesy of Palm Press, Inc.
Figure 3.31	Courtesy of NASA.
Figure 3.32	Courtesy of NASA.
Figure 3.34	Courtesy of NASA.

Figure 3.35	Courtesy of NASA.
Figure 3.36	© Wally McNamee/CORBIS.
Figure 3.41	Courtesy of NASA.

CHAPTER 4

Figure 4.1	Courtesy of AIP Emilio Segrè Visual Archives.
Figure 4.2	Courtesy of Harvard College Observatory.
Figure 4.3	Courtesy of Harvard College Observatory.
Figure 4.9	Courtesy of Harvard College Observatory.
Figure 4.10	© Bettmann/CORBIS.

CHAPTER 5

Figure 5.2	© The British Museum.
Figure 5.5	Courtesy of Metropolitan Museum of Art, Purchase, Mr. and Mrs. Charles Wrightsman Gift, in honor of Everett Fahy, 1977.
Figure 5.6	Courtesy of Cornell University Library, Division of Rare and Manuscript Collections.
Figure 5.14	Courtesy of Royal Netherlands Academy of Sciences and Letters.
Figure 5.15	Courtesy of Burndy Library, Dibner Institute for the History of Science and Technology, Cambridge, Massachusetts.
Figure 5.16	© Michael Nicholson/CORBIS.
Figure 5.17	© Duomo/CORBIS.
Figure 5.18	Courtesy of Nat Farbman/Timepix.
Figure 5.21	© Owen Franken/CORBIS.
Figure 5.22	Courtesy of Albert B. Gregory.
Figure 5.24	© The Harold & Esther Edgerton Family Trust, courtesy of Palm Press, Inc.
Figure 5.25	Courtesy of David Couzens.

CHAPTER 6

Figure 6.1	© Bettmann/CORBIS.
Figure 6.2	© CORBIS.
Figure 6.4	Courtesy of Houghton Library, Harvard University.
Figure 6.5a	© Stephen Frink/CORBIS.
Figure 6.6	Courtesy of Stan Sherer.
Figure 6.10	© Bettmann/CORBIS.

Figure 6.11	Courtesy of Science Museum/Science & Society.
Figure 6.14	Courtesy of American Society of Agricultural Engineers.
Figure 6.15	Courtesy of AIP Emilio Segrè Visual Archives.
Figure 6.16	Courtesy of Harvard College Observatory.
Figure 6.17	Courtesy of AIP Emilio Segrè Visual Archives.

CHAPTER 7

Figure 7.2	© Bettmann/CORBIS.
Figure 7.4	Courtesy of AIP Emilio Segrè Visual Archives, Zeleny Collection.
Figure 7.6	Courtesy of Historisches Museum, Basel.
Figure 7.15	Courtesy of David Couzens.
Figure 7.16	© The Harold & Esther Edgerton Family Trust, courtesy of Palm Press, Inc.
Figure 7.17	University of Vienna, courtesy of AIP Emilio Segrè Visual Archives.
Figure 7.19	Courtesy of David Couzens.
Figure 7.22	© Peter Turnley/CORBIS.
Figure 7.23	Courtesy of David Couzens.

Excerpt from "West-Running Brook" from The Poetry of Robert Frost, edited by Edward Connery Lathem. Copyright 1928, © 1969 by Henry Holt and Company, © 1956 by Robert Frost. Reprinted by permission of Henry Holt and Company, LLC.

CHAPTER 8

Figure 8.1	Courtesy of David Couzens.
Figure 8.2	Courtesy of David Couzens.
Figure 8.6	© CORBIS.
Figure 8.24	Courtesy of United States Navy.
Figure 8.36	Courtesy of United States Navy.
Figure 8.38	© Kevin Fleming/CORBIS.
Figure 8.39	Courtesy of David Couzens.
Figure 8.41	Courtesy of Bausch & Lomb.
Figure 8.44	Courtesy of AIP Emilio Segrè Visual Archives.
Figure 8.47	Smithsonian Institution, courtesy of AIP Emilio Segrè Visual Archives.
Figure 8.49	General Electric Research Laboratory, courtesy of AIP Emilio Segrè Visual Archives.
Figure 8.50	Courtesy of Houghton Library, Harvard University.

CHAPTER 9

Figure 9.1a Image of Albert Einstein™ © CORBIS, licensed by the Hebrew University of Jerusalem, represented by The Roger Richman Agency, Inc., www.albert-einstein.com

Figure 9.2b/c Image of Albert Einstein™ licensed by the Hebrew University of Jerusalem, represented by The Roger Richman Agency, Inc., www.albert-einstein.com

Figure 9.16 Courtesy of Fermilab.

Figure 9.17 © Kevin Fleming/CORBIS.

PROLOGUE, PART 2

Image of Albert Einstein™ © CORBIS, licensed by the Hebrew University of Jerusalem, represented by The Roger Richman Agency, Inc., www.albert-einstein.com

Portrait of Isaac Newton property of Massachusetts Institute of Technology, Burndy Library; courtesy of AIP Emilio Segrè Visual Archives.

CHAPTER 10

Figure 10.1 Courtesy of Burndy Library, Dibner Institute for the History of Science and Technology, Cambridge, Massachusetts.

Figure 10.3 Courtesy of Burndy Library, Dibner Institute for the History of Science and Technology, Cambridge, Massachusetts.

Figure 10.5 Courtesy of Burndy Library, Dibner Institute for the History of Science and Technology, Cambridge, Massachusetts.

Figure 10.6 Courtesy of AIP Emilio Segrè Visual Archives, E. Scott Barr Collection.

Figure 10.9 © W. Perry Conway/CORBIS.

Figure 10.11 © Bettmann/CORBIS.

Figure 10.15 Courtesy of AIP Emilio Segrè Visual Archives, Landè Collection.

Figure 10.19 Courtesy of Stanford Linear Accelerator Center.

Figure 10.20 Courtesy of AIP Emilio Segrè Visual Archives, E. Scott Barr Collection.

Figure 10.23 Courtesy of Nationalhistoriske Museum, Frederiksborg, Hillerod.

Figure 10.27 Courtesy of AIP Emilio Segrè Visual Archives.

Figure 10.31 Courtesy of Jan Curtis.

CHAPTER 11

Figure 11.1a	Courtesy of Deutsches Museum, Munich.
Figure 11.2	Courtesy of AIP Emilio Segrè Visual Archives.
Figure 11.4	The Royal Institution, London, UK/Bridgeman Art Library International, Ltd.
Figure 11.9	© CORBIS.
Figure 11.10	Courtesy of Ted Russell/Timepix.
Figure 11.11	Courtesy of Thomas Alva Edison Foundation.
Figure 11.12	Courtesy of Queens Borough Public Library, Long Island Division, and Latimer Family Collection.
Figure 11.17	Courtesy of Manfred Krutein/Photovault.com.
Figure 11.19	© Kevin R. Morris/CORBIS.
Figure 11.21a/b	Courtesy of National Renewable Energy Laboratory.
Figure 11.22	Courtesy of National Renewable Energy Laboratory.
Figure 11.23	Courtesy of National Renewable Energy Laboratory.
Figure 11.24	Courtesy of Tennessee Valley Authority.
Figure 11.25	© Roger Ressmeyer/CORBIS.

CHAPTER 12

Figure 12.1	Courtesy of Royal Institution/Bridgeman Art Library International, Ltd.
Figure 12.2	Courtesy of AIP Emilio Segrè Visual Archives.
Figure 12.8	Property of Deutsches Museum, Munich; courtesy of AIP Emilio Segrè Visual Archives, Physics Today Collection.
Figure 12.14	© Bettmann/CORBIS.
Figure 12.16	Courtesy of SE-IR Corporation.
Figure 12.17	Courtesy of NASA.
Figure 12.19	Courtesy of General Electric Co.
Figure 12.20	Courtesy of Brookhaven National Laboratory.

CHAPTER 13

Figure 13.1	Courtesy of Science Museum/Science and Society.
Figure 13.2	Courtesy of AIP Emilio Segrè Visual Archives, W.F. Meggers Collection.
Figure 13.3	Courtesy of Othmer Library of Chemical History, Chemical Heritage Foundation, Philadelphia, Pennsylvania.
Figure 13.5	Courtesy of Cavendish Library, Cambridge.

Figure 13.6	Courtesy of California Institute of Technology Archives.
Figure 13.11	Courtesy of AIP Emilio Segrè Visual Archives.
Figure 13.12	Courtesy of AIP Emilio Segrè Visual Archives, gift of Jost Lemmerich.
Figure 13.13	Photograph by Aufnahme u. Verlag v. A. Baumann Nacht, München, courtesy of AIP Emilio Segrè Visual Archives, Landè Collection.
Figure 13.14	© Bettmann/CORBIS.
Figure 13.16	Courtesy of Eastman Kodak Company.
Figure 13.17	© Steve Chenn/CORBIS.
Figure 13.18	Courtesy of Straus Center for Conservation, Harvard University Art Museums.
Figure 13.19	Courtesy of U.S. Department of the Interior National Park Service, Edison National Historic Site.

CHAPTER 14

Figure 14.6	Courtesy of AIP Emilio Segrè Visual Archives.
Figure 14.11a	Courtesy of AIP Emilio Segrè Visual Archives, Margrethe Bohr Collection.
Figure 14.11b	Courtesy of AIP Emilio Segrè Visual Archives, W.F. Meggers Gallery of Nobel Laureates.
Figure 14.11c	Courtesy of AIP Emilio Segrè Visual Archives, Margrethe Bohr Collection.
Figure 14.15	Courtesy of University of California, Lawrence Livermore National Laboratory, and the Department of Energy, under whose auspices the work was performed.

CHAPTER 15

Figure 15.1	Courtesy of AIP Emilio Segrè Visual Archives, W.F. Meggers Collection.
Figure 15.3	Courtesy of AIP Emilio Segrè Visual Archives.
Figure 15.4	Courtesy of Prof. Harry Meiners, Rensselaer Polytechnic Institute.
Figure 15.7	Photo by Francis Simon, courtesy of AIP Emilio Segrè Visual Archives.
Figure 15.8	Max Planck Institute, courtesy of AIP Emilio Segrè Visual Archives.
Figure 15.9	Courtesy of AIP Emilio Segrè Visual Archives.
Figure 15.11	Courtesy of AIP Emilio Segrè Visual Archives.
Figure 15.12	Institut International de Physique Solvay, courtesy of AIP Emilio Segrè Visual Archives.

CHAPTER 16

Figure 16.4	From *Collected Papers of Albert Einstein*, Volume 2, edited by John Stachel and David Cassidy (Princeton University Press, 1990).
Figure 16.11	© Michael S. Yamashita/CORBIS.
Figure 16.16	Courtesy of NASA.
Figure 16.18	© Charles O'Rear/CORBIS.

CHAPTER 17

Figure 17.1	Courtesy of Burndy Library, Dibner Institute for the History of Science and Technology, Cambridge, Massachusetts; courtesy of AIP Emilio Segrè Visual Archives.
Figure 17.3a	Courtesy of AIP Emilio Segrè Visual Archives, W.F. Meggers Collection.
Figure 17.3b	Courtesy of AIP Emilio Segrè Visual Archives, E. Scott Barr Collection.
Figure 17.3c	Courtesy of AIP Emilio Segrè Visual Archives.
Figure 17.3d	Courtesy of AIP Emilio Segrè Visual Archives.
Figure 17.8	Courtesy of University of Pennsylvania Library, Edgar Fahs Smith Collection.

CHAPTER 18

Figure 18.4	Berkeley National Laboratory, University of California, Berkeley; courtesy of AIP Emilio Segrè Visual Archives.
Figure 18.5	Photo by Börtzells Esselte, Nobel Foundation; courtesy of AIP Emilio Segrè Visual Archives, Weber and Fermi Film Collections.
Figure 18.8	Courtesy of Los Alamos National Laboratory.
Figure 18.9	Berkeley National Laboratory, University of California, Berkeley; courtesy of AIP Emilio Segrè Visual Archives.
Figure 18.10a/b/c	Courtesy of Fermilab.
Figure 18.14	Courtesy of Argonne National Laboratory.
Figure 18.15	Courtesy of AIP Emilio Segrè Visual Archives.
Figure 18.17	Courtesy of AIP Emilio Segrè Visual Archives.
Figure 18.19	Courtesy of AIP Emilio Segrè Visual Archives, Herzfeld Collection.
Figure 18.20	Courtesy of Argonne National Laboratory.
Figure 18.23	© Hulton-Deutsch Collection/CORBIS.
Figure 18.25	Courtesy of AIP Emilio Segrè Visual Archives.

Figure 18.26	© Bettmann/CORBIS.
Figure 18.27	© Bettmann/CORBIS.
Figure 18.29	© Roger Ressmeyer/CORBIS.
Figure 18.30	Photo by K.G. Somsekhar, AIP Emilio Segrè Visual Archives.
Figure 18.31	Courtesy of NASA.
Figure 18.32	Painting by Patricia Watwood, courtesy of Harvard University.
Figure 18.33	Courtesy of Oskar Reinhart Collection "Am Römerholz," Winterthur, Switzerland.

COLOR PLATES

Plate 1	© Roger Antrobus/CORBIS.
Plate 4	The Museum of Modern Art, New York. Lillie P. Bliss Collection. Photograph © 2001, The Museum of Modern Art, New York.
Plate 5	Courtesy of J. Talbot, University of Ottawa.

Index